SHORT PROTOCOLS IN NEUROSCIENCE

SYSTEMS AND BEHAVIORAL METHODS

A Compendium of Methods from Current Protocols in Neuroscience

EDITORIAL BOARD

Jacqueline N. Crawley

Charles R. Gerfen

Michael A. Rogawski

David R. Sibley

Phil Skolnick

Susan Wray

Published by John Wiley & Sons, Inc.

Cover illustration: © Corbis

Copyright © 2007 by John Wiley & Sons, Inc. All rights reserved.

Published by John Wiley & Sons, Inc., Hoboken, New Jersey.
Published simultaneously in Canada.

Reproduction or translation of any part of this work beyond that permitted by Section 107 or 108 of the 1976 United States Copyright Act without the permission of the copyright owner is unlawful. Requests for permission or further information should be addressed to the Permissions Department, John Wiley & Sons, Inc., 111 River Street, Hoboken, NJ 07030, tel: (201) 748-6011, fax: (201) 748-6008, e-mail: permreq@wiley.com.

While the authors, editors, and publisher believe that the specification and usage of reagents, equipment, and devices, as set forth in this book, are in accord with current recommendations and practice at the time of publication, they accept no legal responsibility for any errors or omissions, and make no warranty, express or implied, with respect to material contained herein. In view of ongoing research, equipment modifications, changes in governmental regulations, and the constant flow of information relating to the use of experimental reagents, equipment, and devices, the reader is urged to review and evaluate the information provided in the package insert or instructions for each chemical, piece of equipment, reagent, or device for, among other things, any changes in the instructions or indication of usage and for added warnings and precautions. This is particularly important in regard to new or infrequently employed chemicals or experimental reagents.

NOTE: All protocols using live animals must be reviewed and approved by an Institutional Animal Care and Use Committee (IACUC) and must follow officially approved procedures for the care and use of laboratory animals.

This book was edited by Drs. Crawley, Gerfen, Rogawski, Sibley, and Wray in their private capacities. No official support or endorsement by the National Institutes of Health is intended or should be inferred.

Library of Congress Cataloging-in-Publication Data

Short protocols in neuroscience : systems and behavioral methods / Jacqueline N. Crawley ... [et al.].

   p. ; cm.

Includes bibliographical references and index.

  ISBN-13: 978-0-471-78397-8 (cloth)
  ISBN-10: 0-471-78397-8 (cloth)

  1. Neurosciences—Laboratory manuals.   I. Crawley, Jacqueline N.

   [DNLM: 1. Neurosciences—Laboratory Manuals. 2. Laboratory Techniques and Procedures—Laboratory Manuals.   WL 25 S5595 2006]

  QP357.S56 2006

  573.8'519—dc22

                                                       2006019138
                                                              CIP

Printed in the United States of America

10 9 8 7 6 5 4 3 2 1

CONTENTS

Preface	xiii
Contributors	xvii

1 Neurophysiology

1.1	**Fabrication of Patch Pipets**		**1-3**
		Basic Protocol 1: Pulling Single-Channel and Whole-Cell Electrodes with an Automated Puller	1-3
		Support Protocol 1: Obtaining Optimal Tip Geometry	1-4
		Support Protocol 2: Calibrating the Puller Filament	1-5
		Support Protocol 3: Cleaning the Glass	1-7
		Support Protocol 4: Noise Considerations for Single-Channel Patch Pipets	1-7
		Support Protocol 5: Considerations for Whole-Cell Pipets	1-9
		Basic Protocol 2: Preparing Pipet Tips with Elastomer Coating	1-10
		Support Protocol 6: Considerations for Pipet Coating	1-12
		Basic Protocol 3: Fire Polishing the Pipet	1-13
		Support Protocol 7: Constructing Fire-Polishing Apparatus Components	1-14
		Basic Protocol 4: Pipet Filling	1-17
		Basic Protocol 5: Mounting and Testing the Pipet Setup	1-18
1.2	**Whole-Cell Voltage Clamp Recording**		**1-18**
		Strategic Planning	1-19
		Basic Protocol 1: Patch-Clamp Technique Setup	1-22
		Basic Protocol 2: Data Acquisition and Pulse Sequences	1-25
		Basic Protocol 3: Analysis of Patch Clamp Data	1-28
1.3	**Preparation of Hippocampal Brain Slices**		**1-29**
		Basic Protocol: Preparation of Acute Mammalian Hippocampal Slices	1-29
1.4	**Synaptic Plasticity in the Hippocampal Slice Preparation**		**1-33**
		Types of Synaptic Plasticity	1-33
		The Use of Field Potential Recording to Study LTP	1-36
		The Use of Field Potential Recording to Study LTD	1-41
		Metaplasticity	1-42
		Basic Protocol: Extracellular LTP in the Hippocampal Slice Preparation	1-42
		Studying LTP and LTD Using Intracellular Recording	1-45
		Special Considerations When Working with Transgenic Mice	1-49
		Data Acquisition and Analysis	1-49
1.5	**Recording Synaptic Currents and Synaptic Potentials**		**1-50**
		Basic Protocol: Recording Synaptic Events in Brain Slices	1-51
1.6	**Voltage Clamp Recordings from *Xenopus* Oocytes**		**1-52**
		Basic Protocol 1: Two-Electrode Voltage Clamp (TEVC) Recording	1-53
		Support Protocol: Fabrication of Agarose Cushion Microelectrodes (ACEs) and Agarose/KCl Bridges for TEVC	1-56
		Basic Protocol 2: Patch Clamp Recordings from Oocytes	1-58
1.7	**Preparation and Maintenance of Organotypic Slice Cultures of CNS Tissue**		**1-60**
		Basic Protocol: Roller-Tube Cultures	1-61
		Alternate Protocol: Interface Cultures	1-62
1.8	**Patch-Pipet Recording in Brain Slices**		**1-64**
		Basic Protocol 1: The Blind Technique	1-64
		Basic Protocol 2: The DIC Technique	1-65
1.9	**Single-Channel Recording**		**1-68**
		Strategic Planning	1-68
		Basic Protocol 1: Patch Formation for Single-Channel Recording	1-71
		Basic Protocol 2: Data Acquisition and Pulse Sequences	1-73

1.10	**Acute Isolation of Neurons from the Mature Mammalian Central Nervous System**	**1-74**
	Basic Protocol: Acute Isolation of Neurons from the Mature Mammalian Central Nervous System	1-74
1.11	**Patch-Clamp Recording from Neuronal Dendrites**	**1-76**
	Basic Protocol 1: Dendritic Recording	1-76
	Basic Protocol 2: Simultaneous Dendritic and Somatic Recording	1-79
	Support Protocol: Preparation of Brain Slices	1-80

2 Neurochemistry/Neuropharmacology

2.1	**Scintillation Proximity Assay**	**2-2**
	Basic Protocol 1: Saturation Analysis of [^{3}H]5-HT Binding to the 5-HT$_{1E}$ Receptor	2-2
	Support Protocol 1: Optimization of Membrane Protein Concentration	2-5
	Support Protocol 2: Optimization of Amount of WGA SPA Beads	2-6
	Support Protocol 3: Optimization of Incubation Time	2-7
	Basic Protocol 2: Pharmacological Profile for the 5-HT$_{1E}$ Receptor Using SPA	2-7
	Basic Protocol 3: Rapid Measurement of cAMP Accumulation in CHO Cells Stably Expressing the Muscarinic M$_1$ Receptor	2-10
	Basic Protocol 4: Time-Course Analysis of Rhinovirus 3C Protease Using SPA	2-13
	Basic Protocol 5: Uptake of [^{14}C]$_P$-Hydroxy-Loracarbef Using Cytostar-T Plates	2-14
	Basic Protocol 6: Muscarinic M$_1$ Receptor-Mediated [γ-^{35}S]GTP Binding: An SPA Approach Using a Specific Anti-G Protein Antibody	2-16
	Basic Protocol 7: Miniaturization of Receptor-Radioligand Interactions Using 384-Well FlashPlates	2-20
2.2	**Methods for Sample Preparation for Direct Immunoassay Measurement of Analytes in Tissue Homogenates: ELISA Assay of Amyloid β-Peptides**	**2-25**
	Basic Protocol 1: Isolation and Homogenization of Tissues for Analysis of Amyloid β-Peptides	2-25
	Basic Protocol 2: Example of ELISA Technique for the Analysis of Amyloid Peptides Aβ(1-x) and Aβ(1-42) from Brain Homogenates	2-26
2.3	**Assessment of Cell Viability in Primary Neuronal Cultures**	**2-28**
	Basic Protocol 1: Assay of Cell Viability by Measurement of Mitochondrial Activity (MTT Reduction)	2-30
	Basic Protocol 2: Assay of Cell Viability by Measurement of Lactate Dehydrogenase Release into the Bathing Medium	2-30
	Alternate Protocol 1: Simplified LDH Assay for Cells Grown in 96-Well Plates	2-31
	Basic Protocol 3: Assay of Cell Viability by Colloidal Dye (Trypan Blue) Exclusion	2-32
	Alternate Protocol 2: Rapid Trypan Blue Staining	2-32
	Basic Protocol 4: Assay of Cell Viability Using Propidium Iodide	2-32
	Alternate Protocol 3: Live/Dead Staining with Standard Epifluorescence Microscopy	2-33
	Support Protocol: Preservation of Stained Cells	2-33
2.4	**Measurement of NO and NO Synthase**	**2-36**
	Basic Protocol 1: Chemiluminescent Detection of NO	2-36
	Support Protocol 1: Preparation of a Reducing Agent for Conversion of Nitrite to NO	2-38
	Support Protocol 2: Preparation of Nitrate Reducing Agent	2-39
	Support Protocol 3: Deproteination of Samples for Nitrate/Nitrite Assay	2-39
	Basic Protocol 2: Nitrite and Nitrite/Nitrate Assay by the Griess Method	2-40
	Basic Protocol 3: L-Citrulline Assay for Nitric Oxide Synthase (NOS) Activity	2-41
	Support Protocol 4: Preparation and Regeneration of Cation-Exchange Columns	2-43

	Support Protocol 5: Purification of Radiolabeled L-Arginine	2-44
	Support Protocol 6: Nitric Oxide Synthase (NOS) Preparation	2-45
2.5	**Measurement of Oxygen Radicals and Lipid Peroxidation in Neural Tissues**	**2-46**
	Basic Protocol 1: Salicylate Trapping and HPLC Assay of Hydroxyl Radical	2-46
	Basic Protocol 2: Spectrophotometric Assay of Lipid-Conjugated Dienes	2-48
	Basic Protocol 3: HPLC Assay of Vitamin E	2-49
	Basic Protocol 4: HPLC Assay of Glutathione	2-51
	Basic Protocol 5: HPLC-Chemiluminescence Assay of Lipid Hydroperoxides	2-53
	Support Protocol: Xylenol Orange Determination of Hydroperoxide Content in Standards	2-55
	Basic Protocol 6: Thiobarbituric Acid Assay of Malondialdehyde	2-56
	Basic Protocol 7: HPLC Assay of Malondialdehyde Using UV Detection	2-56
	Alternate Protocol: HPLC Assay of TBA-Malondialdehyde Adduct Using Fluorescence Detection	2-58
	Basic Protocol 8: Immunoassay of 8-Isoprostaglandin $F_{2\alpha}$	2-59
2.6	**Overview of Microdialysis**	**2-60**
	Description of the Microdialysis Process	2-60
	Empirical Methods for Determining in Vivo Relative Recovery	2-62
	Factors Affecting in Vivo Recovery	2-64
	Perfusate Composition	2-67
	Strategy for Determining Experimental Design	2-68
2.7	**Microdialysis in Rodents**	**2-70**
	Basic Protocol 1: Preparation of Dialysis Probe/Guide Cannula Assembly: Construction of a Concentric Probe (Design 1)	2-70
	Alternate Protocol 1: Construction of a Side-by-Side Probe (Design 2)	2-71
	Alternate Protocol 2: Construction of a Horizontal Probe	2-74
	Basic Protocol 2: Implantation and Tethering of Dialysis Probe/Guide Cannula Assembly in the Rat: Implantation of Concentric or Loop Probe/Guide Assemblies	2-74
	Alternate Protocol 3: Implantation of a Horizontal Probe	2-77
	Basic Protocol 3: Surgical Implantation of Dialysis Probe/Guide Cannula Assembly in the Mouse	2-79
	Basic Protocol 4: Microdialysis in Vitro	2-80
	Basic Protocol 5: In Vivo Microdialysis	2-81
	Support Protocol 1: In Vivo Determination of Extracellular Concentration	2-83
	Support Protocol 2: Determining Extracellular Concentration (C_{Ext})	2-83
2.8	**Microdialysis in Nonhuman Primates**	**2-84**
	Basic Protocol 1: Preparation of Dialysis Probe	2-84
	Basic Protocol 2: Attachment of Guide Assembly to the Monkey Skull	2-88
	Support Protocol 1: Determination of Skull Coordinates by MRI	2-90
	Support Protocol 2: Postoperative Verification of Guide Assembly Position by MRI	2-91
	Basic Protocol 3: In Vivo Microdialysis	2-91
	Support Protocol 3: Anatomical Verification of the Probes	2-94
	Alternate Protocol: Modifications for Using In Vivo Microdialysis in the Awake Monkey	2-94
2.9	**Detection and Quantification of Neurotransmitters in Dialysates**	**2-95**
	Basic Protocol 1: Detection and Quantification of Dopamine and Serotonin by HPLC/EC	2-96
	Basic Protocol 2: Detection and Quantification of Acetylcholine by HPLC/EC	2-97
	Alternate Protocol: HPLC/EC Detection of Acetylcholine Using Acetylcholine/Choline Assay Kit	2-99
	Support Protocol: Preparation of Immobilized Enzyme Reactor (IMER) for Detection of Acetylcholine	2-99
	Basic Protocol 3: Detection of Amino Acids in Microdialysates by HPLC and Fluorometric Labeling with o-Phthaldialdehyde	2-100

2.10	**Saturation Assays of Radioligand Binding to Receptors and Their Allosteric Modulatory Sites**	**2-103**
	Basic Protocol 1: Saturation Analysis of Ligand Binding to the Benzodiazepine Site	2-103
	Basic Protocol 2: Saturation Analysis of Ligand Binding to the $GABA_A$ Receptor	2-106
2.11	**Uptake and Release of Neurotransmitters**	**2-109**
	Basic Protocol 1: Study of Uptake and Release of Dopamine in Intact Attached Cells Expressing the Recombinant Dopamine Receptor	2-109
	Alternate Protocol 1: Study of Dopamine Uptake in Detached Cells	2-111
	Basic Protocol 2: Study of Dopamine Uptake in Synaptosomes	2-111
	Basic Protocol 3: Study of Dopamine Release from Synaptosomes	2-112
	Alternate Protocol 2: Detection of Uptake or Release of Dopamine by HPLC with Electrochemical Detection (HPLC-EC)	2-113
	Alternate Protocol 3: Using a Superfusion Apparatus for Time Sampling	2-114
	Basic Protocol 4: Examination of the Dopamine Transporter with Radioligands	2-116
	Support Protocol: Establishing Initial Binding-Assay Parameters	2-117
2.12	**Measurement of Chloride Movement in Neuronal Preparations**	**2-119**
	Basic Protocol 1: Uptake/Efflux of Radioisotopic Cl^- in Synaptoneurosomes	2-119
	Basic Protocol 2: Uptake of Radioisotopic Cl^- in Primary Cultures	2-122
	Basic Protocol 3: Efflux of Cl^- in Synaptoneurosomes Measured with Fluorescent Dyes and Photometry	2-123
	Support Protocol 1: Preparation of Synaptoneurosomes	2-124
	Basic Protocol 4: Efflux of Cl^- in Primary Cultures Measured with Fluorescent Dyes and Photometry	2-125
	Basic Protocol 5: Uptake of Cl^- in the Acute Brain Slice Measured with Fluorescent Dye and Epifluorescence Imaging	2-126
	Alternate Protocol: Uptake of Cl^- in the Acute Brain Slice Measured with Fluorescent Dye and Confocal Imaging	2-127
	Support Protocol 2: Preparation of MEQ-Loaded Acute Brain Slices	2-128
	Support Protocol 3: Calibration of Fluorescent Dye in Synaptoneurosomes and Cultured Cells	2-129
	Support Protocol 4: Calibration of Fluorescent Dye in Brain Slices	2-130
2.13	**Measurement of Cation Movement in Primary Cultures Using Fluorescent Dyes**	**2-131**
	Strategic Planning	2-131
	Basic Protocol: Measurement of $[Ion^+]_i$ Using Ratiometric Dyes	2-132
	Calibration for Ratiometric Dyes	2-133
	Support Protocol 1: Cell-Free Calibration of Ca^{2+} Dyes	2-134
	Support Protocol 2: In Situ Calibration of Ca^{2+} Dyes	2-135
	Support Protocol 3: In Situ Calibration of Na^+ Dyes	2-136
	Support Protocol 4: Cell-Free Calibration of Mg^{2+} Dyes	2-137
	Alternate Protocol: Calibration and Measurement Using Nonratiometric $[Ca^{2+}]_i$ Dyes	2-137
2.14	**Measurement of Second Messengers in Signal Transduction: cAMP and Inositol Phosphates**	**2-138**
	Basic Protocol 1: Determination of Cyclic Amp Concentration in Cultured Cells	2-138
	Basic Protocol 2: Measurement of [^{3}H]Cyclic Amp Accumulation in Synaptoneurosomes After Prelabeling with [^{3}H]Adenine	2-140
	Basic Protocol 3: Measurement of Phosphoinositide Turnover in Synaptoneurosomes	2-141
2.15	**In Vivo Measurement of Blood-Brain Barrier Permeability**	**2-143**
	Basic Protocol 1: Blood-Brain Barrier Influx Measurement: In Situ Brain Perfusion Procedure for Rats	2-143
	Alternate Protocol: Blood-Brain Barrier Influx Measurement: Simplified In Situ Brain Perfusion for Rats	2-149
	Basic Protocol 2: Blood-Brain Barrier Influx Measurement: In Situ Brain Perfusion Procedure for Mice	2-149

	Support Protocol 1: Construction of PE50-Stainless Steel Connectors and Cut 22-G Hypodermic Needles	2-151
	Support Protocol 2: Construction of In Situ Perfusion Cannulas	2-152
	Support Protocol 3: Experimental Determination of Intravascular Capillary Volume Using Radiolabeled Inulin	2-152
	Support Protocol 4: Experimental Determination of Cerebral Perfusion Fluid Flow Using Radiolabeled Diazepam	2-153
	Support Protocol 5: Use of the In Situ Perfusion Technique for Mechanistic Assessment of Carrier-Mediated or Saturable Transport	2-154
	Basic Protocol 3: Blood-Brain Barrier Influx Measurement: Intravenous Administration/Multiple Time Point Procedure for Rats	2-154
	Support Protocol 6: Construction of Jugular Vein Catheters	2-160
	Support Protocol 7: Capillary Depletion Method	2-160
	Basic Protocol 4: Blood-Brain Barrier Efflux Measurement: The Brain Efflux Index Method	2-161
	Support Protocol 8: Construction of Cerebrospinal Fluid Collection Units	2-165

3 Behavioral Neuroscience

3.1	**Assessment of Developmental Milestones in Rodents**	**3-2**
	Strategic Planning	3-2
	Basic Protocol 1: Physical Landmarks of Rodent Development	3-2
	Basic Protocol 2: Developmental Reflexes in Rodents	3-5
	Basic Protocol 3: Development of Locomotor Behavior in Rodents	3-7
3.2	**Locomotor Behavior**	**3-8**
	Basic Protocol: Locomotor Assessment by Direct Observation Using Interval and Ordinal Scales	3-9
	Alternate Protocol: Locomotor Quantification by Photocell-Based Systems	3-11
	Support Protocol: Data Analysis	3-12
3.3	**Motor Coordination and Balance in Rodents**	**3-12**
	Basic Protocol 1: Assessing Balance Using a Rotarod	3-13
	Basic Protocol 2: Beam Walking	3-14
	Basic Protocol 3: Footprint Analysis	3-16
	Support Protocol: Data Analysis	3-17
3.4	**Basic Measures of Food Intake**	**3-19**
	Basic Protocol	3-19
3.5	**Sexual and Reproductive Behaviors**	**3-22**
	Basic Protocol 1: Assessment of Mounts, Intromissions, and Ejaculations in Male Rats	3-23
	Basic Protocol 2: Assessment of Lordosis in Female Rats	3-25
	Support Protocol 1: Castration	3-26
	Support Protocol 2: Ovariectomy	3-28
	Support Protocol 3: Testosterone Treatment Using Subcutaneous Implant	3-30
	Support Protocol 4: Testosterone Treatment by Subcutaneous Injection	3-31
	Support Protocol 5: Estradiol and Progesterone Treatment	3-31
3.6	**Parental Behaviors in Rats and Mice**	**3-32**
	Description of Parental Behaviors	3-32
	Assessment of Parental Behaviors	3-33
	Basic Protocol 1: Assessment of Parental Behaviors Using Direct Periodic Spot-Check Observation	3-33
	Basic Protocol 2: Assessment of Parental Behavior Using Continuous Observation	3-34
	Basic Protocol 3: Assessment of Nest Construction	3-35
	Alternate Protocol: Assessment of Nest Construction—Mice	3-36
	Basic Protocol 4: Assessment of Retrieval	3-36

	Basic Protocol 5: Induction and Assessment of Parental Behavior in Nonlactating Rats and Mice (Sensitization)	3-37
	Basic Protocol 6: Hormonal Induction of Maternal Behavior in Virgin Rats	3-37
	Assessment of Maternal Preferences and Motivation in Rats and Mice	3-38
	Basic Protocol 7: Test of the Unconditioned Preference for Pups or Pup-Related Cues	3-38
	Basic Protocol 8: Conditioned Place Preference Tests	3-41
	Basic Protocol 9: Tests of Operant Responding	3-42
	Basic Protocol 10: Using a T-Maze Extension of the Home Cage to Assess Maternal Motivation	3-45
3.7	**Application of Experimental Stressors in Laboratory Rodents**	**3-46**
	Strategic Planning	3-46
	Basic Protocol 1: Restraint Stressor	3-47
	Basic Protocol 2: Electric Footshock Stressor	3-48
	Basic Protocol 3: Swim Stressor	3-48
	Basic Protocol 4: Social Isolation Stressor	3-49
	Basic Protocol 5: Resident/Intruder Stressor	3-50
	Basic Protocol 6: Maternal Separation Stressor	3-50
	Basic Protocol 7: Sleep Deprivation Stressor	3-51
	Support Protocol 1: Colony Maintenance	3-52
	Support Protocol 2: Rat Handling	3-52
	Support Protocol 3: ACTH/Corticosterone Determinations	3-53
3.8	**Rodent Models of Depression: Forced Swimming and Tail Suspension Behavioral Despair Tests in Rats and Mice**	**3-54**
	Basic Protocol: Behavioral Despair Test in the Rat	3-54
	Alternate Protocol 1: Forced Swimming Test in the Mouse	3-55
	Alternate Protocol 2: Tail Suspension Test in Mice	3-56
3.9	**Rodent Models of Depression: Learned Helplessness Induced in Mice**	**3-58**
	Basic Protocol: Induction of a Learned Helplessness Effect in Mice	3-58
	Alternate Protocol: Administration of Uncontrollable Shock	3-60
	Support Protocol: Description and Fundamental Characteristics of Shuttle Boxes for Escape Testing	3-61
3.10	**Animal Tests of Anxiety**	**3-62**
	Basic Protocol 1: Geller-Seifter Conflict	3-62
	Basic Protocol 2: Social Interaction	3-63
	Basic Protocol 3: Light/Dark Exploration	3-64
	Basic Protocol 4: Elevated Plus-Maze Test	3-64
	Basic Protocol 5: Defensive Burying	3-67
	Basic Protocol 6: Thirsty Rat Conflict	3-69
3.11	**Assessment of Spatial Memory Using the Radial Arm Maze and Morris Water Maze**	**3-70**
	Basic Protocol 1: Use of Radial Arm Maze Task to Test Basic Working Memory	3-70
	Alternate Protocol 1: Use of Radial Arm Maze Task to Test Working Versus Reference Memory	3-72
	Basic Protocol 2: Use of Morris Water Maze Task to Test Spatial Memory	3-73
	Alternate Protocol 2: Use of Water Maze Task for Spatial Probe Trial	3-74
	Alternate Protocol 3: Use of Water Maze Task to Test Working Memory	3-75
3.12	**Cued and Contextual Fear Conditioning in Mice**	**3-76**
	Basic Protocol: One-Trial Cued and Contextual Fear Conditioning in Mice	3-76
	Alternate Protocol 1: Pre-Exposure Experiments	3-80
	Alternate Protocol 2: Discrimination Protocol	3-82
	Alternate Protocol 3: Contextual Learning without the Auditory Cue	3-82
	Alternate Protocol 4: Immediate Shock Control Protocol	3-82
3.13	**Conditioned Flavor Aversions: Assessment of Drug-Induced Suppression of Food Intake**	**3-83**
	Basic Protocol	3-83

3.14	**Measurement of Startle Response, Prepulse Inhibition, and Habituation**	**3-84**
	Strategic Planning	3-84
	Basic Protocol 1: Basic Test of Acoustic Startle Reactivity	3-86
	Alternate Protocol 1: Between-Subjects Tests of Startle Reactivity	3-87
	Alternate Protocol 2: Between-Subjects Tests of Startle Habituation	3-88
	Basic Protocol 2: Testing Prepulse Inhibition of Startle	3-88
	Support Protocol: Rat Handling	3-89
3.15	**Latent Inhibition**	**3-91**
	Basic Protocol 1: Latent Inhibition (LI) in the Conditioned Emotional Response (CER) Procedure	3-91
	Alternate Protocol 1: Adding Drugs, Lesions, or Other Manipulations: Disruption and Potentiation of LI	3-96
	Basic Protocol 2: LI in a Two-Way Active Avoidance Procedure	3-98
	Alternate Protocol 2: LI in Two-Way Avoidance with Context Shift	3-99
	Alternate Protocol 3: Adding Treatments in the Two-Way Avoidance Procedure: Disruption and Potentiation of LI	3-100

4 Preclinical Models of Neurologic and Psychiatric Disorders

4.1	**Preclinical Models of Parkinson's Disease**	**4-3**
	Strategic Planning	4-3
	Basic Protocol 1: Combined ICA and Intravenous Administration of MPTP: The Overlesioned (Bilateral Asymmetric) Primate Model	4-4
	Alternate Protocol 1: Systemic MPTP Lesion in Primates	4-7
	Basic Protocol 2: Unilateral 6-OHDA Lesion in Primates	4-9
	Support Protocol 1: Evaluation of Changes in Motor Behavior in Response to L-Dopa	4-11
	Support Protocol 2: Monitoring Activity to Assess MPTP-Treated Monkeys	4-12
	Support Protocol 3: Rotational Behavior as a Measure of Unilateral Nigrostriatal Lesions	4-13
	Basic Protocol 3: 6-OHDA Lesions in Rats	4-13
	Basic Protocol 4: MPTP Lesion in Mice	4-15
	Support Protocol 4: Monitoring Activity in MPTP-Treated Mice	4-16
4.2	**Rodent Models of Global Cerebral Ischemia**	**4-16**
	Basic Protocol 1: Gerbil Bilateral Carotid Artery Occlusion (BCAO) Model to Test Systemically Active Neuroprotective Agents	4-17
	Alternate Protocol 1: Gerbil BCAO Model to Test Neuroprotective Agents that Do Not Penetrate the Brain	4-21
	Alternate Protocol 2: Gerbil BCAO to Induce Ischemic Tolerance	4-22
	Basic Protocol 2: Use of 4-Vessel Occlusion (4-VO) Model to Study Neuronal Degeneration and Test the Effects of Neuroprotective Agents Against Global Cerebral Ischemia	4-22
	Support Protocol 1: Hematoxylin and Eosin Staining of Brain Tissue	4-24
	Support Protocol 2: Measurement of Locomotor Activity After Ischemia in Gerbils	4-25
	Support Protocol 3: Fabrication of Atraumatic Clasps for Rat 4-VO	4-25
4.3	**Rodent Models of Focal Cerebral Ischemia**	**4-28**
	Basic Protocol 1: The Intraluminal Suture Model of Middle Cerebral Artery Occlusion (MCAO) to Test Neuroprotective Agents	4-28
	Support Protocol 1: Fabrication of Intraluminal Monofilament Occluders	4-32
	Basic Protocol 2: Middle Cerebral Artery Occlusion (MCAO) Using Stereotaxic Infusion of Endothelin 1	4-33
	Support Protocol 2: Use of Horizontal and Inclined Balance Beam to Assess Sensorimotor Performance After Et-1 MCAO	4-35
	Support Protocol 3: Use of the Staircase Test to Measure Skilled Paw Use After Et-1 MCAO	4-36

	Basic Protocol 3: The Tamura Model of Middle Cerebral Artery Occlusion (MCAO) to Test Neuroprotective Agents	4-37
	Basic Protocol 4: The Spontaneously Hypertensive Rat Model of Middle Cerebral Artery Occlusion (MCAO) to Test Neuroprotective Agents	4-39
	Support Protocol 4: Cresyl Violet Staining of Brain Tissue	4-41
4.4	**Inducing Photochemical Cortical Lesions in Rat Brain**	**4-44**
	Basic Protocol	4-44
4.5	**Traumatic Brain Injury in the Rat Using the Fluid-Percussion Model**	**4-52**
	Basic Protocol	4-52
4.6	**Models of Neuropathic Pain in the Rat**	**4-57**
	Basic Protocol 1: The Chronic Constriction Injury (CCI) Model of Neuropathic Pain	4-58
	Basic Protocol 2: The Partial Sciatic Ligation (PSL) Model of Neuropathic Pain	4-60
	Basic Protocol 3: The Spinal Nerve Ligation (SNL) Model of Neuropathic Pain	4-60
	Support Protocol 1: Behavioral Assays: Heat-Hyperalgesia	4-61
	Support Protocol 2: Behavior Assays: Mechano-Hyperalgesia	4-63
	Support Protocol 3: Behavioral Assays: Mechano-Allodynia	4-63
	Support Protocol 4: Behavioral Assays: Cold Allodynia	4-64
4.7	**Dural Inflammation Model of Migraine Pain**	**4-65**
	Basic Protocol: Assessment of Dural Extravasation in Rats as a Model of Migraine Pain	4-65
	Alternate Protocol: Assessment of Dural Extravasation in Guinea Pigs	4-69
4.8	**Animal Models of Painful Diabetic Neuropathy: The STZ Rat Model**	**4-71**
	Basic Protocol: Induction of Diabetes with STZ	4-71
4.9	**Models of Nociception: Hot-Plate, Tail-Flick, and Formalin Tests in Rodents**	**4-72**
	Basic Protocol 1: Measurement of Acute Pain Using the Hot-Plate Test	4-72
	Basic Protocol 2: Measurement of Acute Pain Using the Tail-Flick Test	4-73
	Basic Protocol 3: Measurement of Persistent Pain Using the Formalin Test	4-75
4.10	**Place Preference Test in Rodents**	**4-78**
	Basic Protocol 1: Place Preference Test in Rats	4-78
	Basic Protocol 2: Place Preference Test in Mice	4-79
4.11	**Intravenous Self-Administration of Ethanol in Mice**	**4-83**
	Basic Protocol	4-83
4.12	**Preclinical Models to Evaluate Potential Pharmacotherapeutic Agents in Treating Alcoholism and Studying the Neuropharmacological Bases of Ethanol-Seeking Behaviors in Rats**	**4-87**
	Basic Protocol 1: Train Rats to Initiate Ethanol-Maintained Responding on an FR-4 Schedule	4-87
	Support Protocol 1: Blood Alcohol Content (BAC)	4-90
	Basic Protocol 2: Train Rats to Initiate Saccharin-Maintained Responding on an FR-4 Schedule	4-90
	Basic Protocol 3: Train Rats to Lever Press Concurrently for Ethanol and Saccharin Under an FR-4 Schedule	4-91
	Basic Protocol 4: Train Rats to Lever Press Concurrently for Alcohol and an Isocaloric Alternative Solution Under an FR-4 Schedule	4-91
4.13	**Experimental Autoimmune Encephalomyelitis (EAE)**	**4-93**
	Strategic Planning	4-93
	Basic Protocol: Active Induction of EAE in Mice	4-95
	Alternate Protocol: Adoptive Transfer of EAE in Mice	4-96
4.14	**Models of Amyotrophic Lateral Sclerosis**	**4-98**
	Basic Protocol 1: Spinal Cord Organotypic Cultures	4-99
	Basic Protocol 2: Assessment of the Clinical Efficacy of Therapeutic Agents in a Mutant SOD1 Transgenic Mouse Model of Amyotrophic Lateral Sclerosis	4-103

4.15		**Measurement of Panic-Like Responses Following Intravenous Infusion of Sodium Lactate in Panic-Prone Rats**	**4-107**
		Basic Protocol 1: Priming of Anxiety in Rats by Administration of Urocortin in the Basolateral Nucleus of the Amygdala	4-107
		Alternate Protocol: Lactate Challenge in Bicuculline Methiodide-Primed Rats	4-112
		Basic Protocol 2: Developing Panic-Prone State in Rats by Administration of L-Allylglycine in the Dorsomedial Hypothalamus	4-112
		Support Protocol: Behavioral Assessment of Rats in the Social Interaction Test	4-115
4.16		**Chemoconvulsant Model of Chronic Spontaneous Seizures**	**4-116**
		Basic Protocol: Kainate Induction of Chronic Spontaneous Seizures in Rats	4-116

Appendices

1		**Reagents and Solutions**	**A1-1**
2		**Animal Techniques**	**A2-1**
	2A	High Precision Stereotaxic Surgery in Mice	A2-1
		Basic Protocol: Site-Specific Central Microinjection in the Anesthetized Mouse	A2-1
		Alternate Protocol: Cannula Implantation and Central Microinjection in the Behaving Mouse	A2-3
		Support Protocol: Cannula, Wire Plug, and Injection Needle Manufacture	A2-4
	2B	Mouse and Rat Anesthesia and Analgesia	A2-6
		Basic Protocol 1: Injectable Anesthesia for Mouse and Rat	A2-6
		Basic Protocol 2: Inhalant Anesthesia Using Isoflurane for Mouse and Rat	A2-8
		Basic Protocol 3: Analgesia for Mice and Rats	A2-10
	2C	Animal Health Assurance	A2-16
		Quarantine and Stabilization	A2-16
		Health Monitoring	A2-16
		Testing of Biological Material	A2-17
	2D	Handling and Restraint	A2-19
		Basic Protocol 1: Mouse Handling and Manual Restraint	A2-19
		Basic Protocol 2: Rat Handling and Manual Restraint	A2-19
		Alternate Protocol 1: Rodent Restrainers	A2-20
	2E	Animal Identification	A2-21
		Basic Protocol 1: Ear Notch or Punch for Mouse, Rat, and Hamster	A2-21
		Basic Protocol 2: Ear Tag for Mouse, Rat, and Hamster	A2-21
		Basic Protocol 3: Tattoo for Mouse and Rat	A2-22
		Basic Protocol 4: Subcutaneous Transponder for Mouse and Rat	A2-22
3		**Selected Suppliers of Reagents and Equipment**	**A3-1**
		References	

Index

Preface

Neuroscience is one of the most dynamic disciplines in biology. The inherent interest of the nervous system is not alone sufficient to explain the explosion of work in this field. The last decade has seen considerable conceptual progress, but technical advances leading to the introduction of new methods in laboratories across the world have also played a major part in driving the field forward. The rapid transfer of technical information and experience is the partner of profound conceptual change.

The techniques for investigating the nervous system that have been developed require familiarity with electrophysiology, mouse genetics, optics, brain anatomy, and many other methods. More than most areas of biology, neuroscience is a multidisciplinary activity. Although it is hard to be comprehensive in this diverse field, our goal in creating this manual is to provide a source for core techniques in the systems and behavioral aspects of neuroscience.

The methods in *Short Protocols in Neuroscience: Systems and Behavioral Methods* are shortened versions of methods published in *Current Protocols in Neuroscience* (CPNS). This compendium includes step-by-step descriptions for a broad range of systems and behavioral neuroscience methods. The methods are complete and easy to follow, and the book is easier to use at the bench than the parent volumes. *Short Protocols* is ideal for graduate students and postdoctoral fellows who are familiar with the background and theory found in CPNS, and sufficient detail is provided to allow experienced investigators to use it as a stand-alone bench guide.

Although mastery of the techniques presented in this manual, and its companion, *Short Protocols in Neuroscience: Cellular and Molecular Methods*, will allow the reader to design and complete experiments in neuroscience, this manual is not intended to be a substitute for an advanced training program in neuroscience, nor for any number of comprehensive textbooks in the field. In addition to becoming thoroughly knowledgeable of the terms and concepts of neuroscience, readers should gain firsthand experience in basic techniques and safety procedures by working in a supervised neuroscience laboratory.

HOW TO USE THIS MANUAL

Format and Organization

Subjects in this manual are organized by chapters, and protocols are contained in units. Protocol units, which constitute the bulk of the book, generally describe a method and include one or more protocols with listings of materials and steps. Recipes for unique reagents and solutions are included in *APPENDIX 1*, and literature references are included in the *References* section at the end of the book. The book also includes several overview units which contain theoretical discussions that lay the foundation for subsequent protocol units.

Protocols

Many units in the manual contain groups of protocols, each presented with a series of steps. One or more *basic* protocols are presented first in each unit and generally cover the recommended or most universally applicable approaches. *Alternate* protocols are provided where different equipment or reagents can be employed to achieve similar ends, where the starting material requires a variation in approach, or where requirements for the end product differ from those in the basic protocol. *Support* protocols describe additional steps that are required to perform the basic or alternate protocols; these steps are separated from the core protocol because they might be applicable to other uses in the manual, or because they are performed in a time frame separate from the basic protocol steps.

Reagents and Solutions

Reagents required for a protocol are itemized in the materials list before the procedure begins. Many are common stock solutions, others are commonly used buffers or media, while others are solutions unique to a particular protocol. All recipes are presented in *APPENDIX 1*. It is important to note that the names of some of these special solutions might be similar from unit to unit (e.g., SDS sample buffer) while the recipes differ. To minimize confusion, the appropriate unit for a given recipe, except for those for commonly used solutions, is given parenthetically in the appendix.

NOTE: Deionized, distilled water should be used in the preparation of all reagents and solutions and for all protocol steps. When sterile solutions are required, that will be indicated in the protocol.

Equipment

When special equipment is required for an experiment, that equipment is listed in the Materials list for the protocol. The Materials list does not include every item used; rather it includes items that might not be readily available in the laboratory, items that have particular specifications, and items that require special preparation or temperatures. Standard pieces of equipment used in the modern neuroscience laboratory are listed in the accompanying box. These items are used frequently in the protocols in this manual and are not necessarily listed in the Material list.

Commercial Suppliers

Throughout the manual we have recommended commercial suppliers of chemicals, biological materials, and equipment. In some cases, the noted brand has been found to be of superior quality, or it is the only suitable product available in the marketplace. In other cases, the experience of the author

---

Standard Equipment

Applicators, cotton-tipped and wooden
Autoclave
Balances, analytical and preparative
Beakers
Bench protectors, plastic-backed (including "blue pads")
Biohazard disposal containers and bags
Bottles, glass and plastic
Bunsen burners
Cell harvester, for determining radioactivity uptake in 96-well microtiter plates
Centrifuges, low-speed (6,000 rpm) and high-speed (20,000 rpm) refrigerated centrifuges and an ultracentrifuge (20,000 to 80,000 rpm) are required for many procedures. At least one microcentrifuge that holds standard 0.5- and 1.5-ml microcentrifuge tubes is essential. It is also useful to have a tabletop swinging-bucket centrifuge with adapters for spinning 96-well microtiter plates.
Clamps
Cold room or cold box, 4°C
Computer (IBM-compatible or Macintosh) and printer
Containers, assortment of plastic and glass dishes for gel and membrane washes
Coplin jars or staining dishes, glass, for 75 × 25–mm slides
Cryovials, sterile—e.g., Nunc
Cuvettes, plastic disposable, glass, and quartz
Darkroom and developing tank, or X-Omat automatic X-ray film developer (Kodak)
Desiccators (including vacuum desiccators) and desiccant
Dry ice
Filtration apparatus, for collecting acid precipitates on nitrocellulose filters or membranes
Flasks, glass (e.g., Erlenmeyer, beveled shaker)

Forceps
Fraction collector
Freezers, −20° and −80°C
Geiger counter
Gel dryer
Gel electrophoresis equipment, at least one full-size horizontal apparatus and one horizontal minigel apparatus, one vertical full-size and minigel apparatus for polyacrylamide protein gels, and specialized equipment for two-dimensional protein gels
Gloves, plastic and latex, disposable and asbestos
Graduated cylinders
Heating blocks, variable temperature up to 100°C; these thermostat-controlled metal heating blocks that hold test tubes and/or microcentrifuge tubes are very convenient for carrying out enzymatic reactions
Heat-sealable plastic bags and sealing apparatus
Hemacytometer
Hoods, chemical (fume), microbiological safety, and laminar flow (tissue culture)
Hot plates, with or without magnetic stirrer
Ice buckets
Ice maker
Incubators, 37°C for bacteria and humidified 37°C, 5% CO_2 for tissue culture
Kimwipes, or equivalent lint-free tissues
Lab coats
Light box, for viewing gels and autoradiograms
Liquid nitrogen and Dewar flask
Lyophilizer
Magnetic stirrers, (with heater is useful)
Markers, including indelible markers and china-marking pencils
Microcentrifuge, Eppendorf-type, maximum speed 12,000 to 14,000 rpm
Microcentrifuge tubes, 1.5-ml and 0.5-ml

continued

of that protocol is limited to that brand. In the latter situation, recommendations are offered as an aid to the novice neuroscientist in obtaining the tools of the trade. Experienced investigators are therefore encouraged to experiment with substituting their own favorite brands.

Addresses, phone numbers, facsimile numbers, and URLs of all suppliers mentioned in this manual are provided in APPENDIX 3.

References

Short Protocols gives only a limited number of the most fundamental references as background. These are listed at the end of each unit. Full bibliographic information for these and cited references are given in the *References* at the end of this book. Readers who would like a more extensive entry into the literature for background and application of methods are referred to the appropriate unit in *Current Protocols in Neuroscience*.

Safety Considerations

Anyone carrying out these protocols may encounter the following hazardous or potentially hazardous materials: (1) radioactive substances, (2) toxic chemicals and carcinogenic or teratogenic

---

Standard Equipment, *continued*

Microscope, standard optical model (optionally with epifluorescence or phase-contrast illumination) and inverted microscope for tissue culture
Microscope slides and coverslips
Mortar and pestle
Ovens, drying, hybridization, vacuum, and microwave
Paper cutter, large size, for 46 × 57–cm Whatman paper sheets
Paper towels
Parafilm
pH meter
pH paper
Pipet bulbs, or battery-operated pipetting devices—e.g., Pipet-Aid (Drummond Scientific)
Pipets, Pasteur and graduated, glass and plastic, serological (1-ml to 25-ml)
Pipettors, adjustable delivery, volume ranges 0.5 to 10 µl, 10 to 200 µl, and 200 to 1000 µl. It is best to have one set of these three sizes for each full-time researcher and sets dedicated for radioactive and PCR experiments.
Plastic wrap, UV-transparent (e.g., Saran Wrap)
Pliers, needle nose
Polaroid camera
Power supplies, 300-V power supplies are sufficient for polyacrylamide gels; 2000- to 3000-V is needed for some applications
Racks, for test tubes and microcentrifuge tubes
Radiation shield, Lucite or Plexiglas
Radioactive waste containers, for liquid and solid waste
Razor blades
Refrigerator, 4°C
Ring stands and rings
Rotator, end-over-end
Rubber bands
Rubber policemen

Rubber stoppers
Safety glasses
Scalpels and blades
Scintillation counter
Scissors
Sectioning equipment, cryostat microtome, sliding microtome (with stage and knife), Vibratome
Shakers, orbital and platform, room temperature or 37°C. An enclosed shaker (e.g., New Brunswick Controlled Environment Incubator Shaker) that can spin 4-liter flasks is essential for growing 1-liter *E. coli* cultures; a rotary shaking water bath (New Brunswick R76) is useful for growing smaller cultures in flasks
Spectrophotometer, UV and visible
Speedvac evaporator (Savant)
Stir-bars, assorted sizes
Surgical equipment, scale for weighing animals, electric razor, syringes, hypodermic needles, dissection instruments (scissors, scalpels, forceps, hemostats, retractor, bone drill, sutures), operating microscope (optional), heating pad, sterile gauze
Tape, masking and electrician's
Thermometers
Timer
UV cross-linker (e.g., Stratalinker from Stratagene)
UV light sources, long- and short-wave, stationary or hand-held
UV transilluminator
Vacuum aspirator
Vacuum line
Vortex mixers
Wash bottles, plastic and glass
Water baths, variable temperature up to 80°C
Water purification equipment, e.g., Milli-Q system (Millipore) or equivalent
X-ray film cassettes and intensifying screens

reagents, (3) pathogenic and infectious biological agents, and (4) recombinant DNA constructs. Check the guidelines of your particular institution with regard to use and disposal of these hazardous materials. Only limited cautionary statements are included in this manual. Users are responsible for understanding the dangers of working with hazardous materials and for strictly following the safety guidelines established by manufacturers as well as local and national regulatory agencies. Radioactive substances must be used only under the supervision of licensed users, following the guidelines of the National Regulatory Commission (NRC).

Animal Handling

Many protocols call for use of live animals (usually rats or mice) for experiments. Prior to conducting any laboratory procedures with live subjects, the experimental approach must be submitted in writing to the appropriate Institutional Animal Care and Use Committee (IACUC). Written approval from this committee is absolutely required prior to undertaking any live-animal studies. Some specific animal care and handling guidelines are provided in the protocols where live subjects are used, but check with your IACUC guidelines to obtain more extensive guidelines.

ACKNOWLEDGMENTS

This manual is the product of dedicated efforts by many of our scientific colleagues who are acknowledged in each unit and by the hard work by the Current Protocols editorial staff at John Wiley & Sons. We are extremely grateful for the critical contributions by Virginia Chanda and Gwen Crooks (Series Editors) who kept the editors and the contributors on track and played a key role in bringing the entire project to completion. Other skilled members of the Current Protocols staff who contributed to the project include Kathy Morgan, Sheila Kaminsky, and Joseph White. The extensive copyediting required to produce an accurate protocols manual was ably handled by Marianne Huntley, and electronic illustrations were prepared by Gae Xavier Studios.

RECOMMENDED BACKGROUND READING

Becker, J.B., Breedlove, M.S., and Crews, D. (eds.) 1993. Behavioral Endocrinology. MIT Press, Cambridge, Mass.

Jacqueline N. Crawley, Charles R. Gerfen, Michael A. Rogawski, David R. Sibley, Phil Skolnick, and Susan Wray

Contributors

Karl E. O. Åkerman
Uppsala University
Uppsala, Sweden

William W. Anderson
MRC Centre for Synaptic Plasticity
Bristol, United Kingdom

Paula K. Andrus
Pharmacia & Upjohn, Inc.
Kalamazoo, Michigan

Hymie Anisman
Carleton University
Ottawa, Canada

Jaime Athos
University of Washington
Seattle, Washington

Krys S. Bankiewicz
University of California
San Francisco, California

Anthony W. Bannon
Amgen, Inc.
Thousand Oaks, California

Anthony Basile
National Institute of Diabetes and
 Digestive & Kidney Diseases,
 NIH
Bethesda, Maryland

Bernard Beer
DOV Pharmaceutical, Inc.
Hackensack, New Jersey

Mark H. Bender
Eli Lilly and Company
Indianapolis, Indiana

Gary J. Bennett
McGill University
Montreal, Quebec, Canada

Marcelle Bergeron
Lilly Research Laboratories
Indianapolis, Indiana

Kathleen M. K. Boje
University at Buffalo
School of Pharmacy
Buffalo, New York

Zuner A. Bortolotto
MRC Centre for Synaptic Plasticity
Bristol, United Kingdom

Patricia Brown
Office of Animal Care and Use,
 NIH
Bethesda, Maryland

Geneviéve Brossard
Porsolt and Partners Pharmacology
Boulogne-Billancort, France

Rebecca J. Carter
University of Cambridge
Cambridge, United Kingdom

Dennis W. Choi
Washington University School of
 Medicine
St. Louis, Missouri

Jin Mo Chung
Marine Biomedical Institute
Galveston, Texas

James A. Clemens
Lilly Corporate Center
Indianapolis, Indiana

Graham L. Collingridge
MRC Centre for Synaptic Plasticity
Bristol, United Kingdom

Alex Cummins
National Institute of Mental
 Health, NIH
Bethesda, Maryland

Miles G. Cunningham
National Institute of Neurological
Disorders and Stroke, NIH
Bethesda, Maryland

Nathan Dascal
Tel Aviv University
Ramat Aviv, Israel

Judith A. Davis
National Institute of Neurological
Disorders and Stroke, NIH
Bethesda, Maryland

John Donovan
Wyeth-Ayerst Research
Collegeville, Pennsylvania

Henryk Dudek
Ontogeny, Inc.
Cambridge, Massachusetts

Stephen B. Dunnett
Cardiff University
Cardiff, United Kingdom

Jamie Eberling
University of California at Berkeley
Berkeley, California

Eleanore B. Edson
Stanford University School of
 Medicine
Stanford, California

A. Christine Engblom
Royal Danish School of Pharmacy
Copenhagen, Denmark

Amy J. Eshleman
Oregon Health Sciences University
 and Veterans Affairs Medical
 Center
Portland, Oregon

Michael Graham Espey
National Institute of Diabetes and
 Digestive & Kidney Diseases,
 NIH
Bethesda, Maryland

Christian C. Felder
Lilly Research Laboratories
Indianapolis, Indiana

Sandra E. File
King's College London
London, United Kingdom

Stephanie D. Fitz
Indiana University Medical Center
Indianapolis, Indiana

Alison S. Fleming
University of Toronto at
 Mississauga
Mississauga, Canada

Kevin B. Freeman
The American University
Washington, D.C.

Christelle Froger
Porsolt and Partners Pharmacology
Boulogne-Billancourt, France

Beat H. Gähwiler
University of Zurich
Zurich, Switzerland

Raquelli Ganel
Johns Hopkins University
Baltimore, Maryland

Mark A. Geyer
University of California, San Diego
La Jolla, California

Frank J. Gottron
Washington University School of
    Medicine
St. Louis, Missouri

Nicholas J. Grahame
Indiana University/Purdue
    University at Indianapolis
Indianapolis, Indiana

Steven S. Gross
Cornell University Medical College
New York, New York

John F. Guzowski
University of Arizona
Tucson, Arizona

Edward D. Hall
Parke-Davis Pharmaceutical
    Research
Ann Arbor, Michigan

Carine Hautbois
Porsolt and Partners Pharmacology
Boulogne-Billancourt, France

Christian Heidbreder
Swiss Federal Institute of
    Technology
Zurich, Switzerland

Stephen C. Heinrichs
Neurocrine Biosciences, Inc.
San Diego, California

Jennifer L. Hellier
University of Colorado Health
    Sciences Center
Denver, Colorado

Charles J. Heyser
Franklin & Marshall College
Lancaster, Pennsylvania

Marie Honore
Abbott Laboratories
Abbott Park, Illinois

Paul A. Hyslop
Eli Lilly and Company
Indianapolis, Indiana

Jon R. Inglefield
U.S. Environmental Protection
    Agency
Research Triangle Park, North
    Carolina

John T. R. Isaac
MRC Centre for Synaptic Plasticity
Bristol, United Kingdom

Mandy Jackson
Johns Hopkins University
Baltimore, Maryland

Meyer B. Jackson
University of Wisconsin Medical
    School
Madison, Wisconsin

Aaron Janowsky
Oregon Health Sciences University
    and Veterans Affairs Medical
    Center
Portland, Oregon

Kirk W. Johnson
Eli Lilly and Company
Indianapolis, Indiana

Terrance D. Jones
University of Washington
Seattle, Washington

Harry L. June
Indiana University–Purdue
    University
Indianapolis, Indiana

Steven D. Kahl
Lilly Research Laboratories
Indianapolis, Indiana

Audrey N. Kalehua
Uniformed Services University of
    the Health Sciences
Bethesda, Maryland

Peter Kalivas
Washington State University
Pullman, Washington

Alan R. Kay
University of Iowa
Iowa City, Iowa

Stanley R. Keim
Indiana University Medical Center
Indianapolis, Indiana

Malgorzata Kohutnicka
National Institute of Neurological
    Disorders and Stroke, NIH
Bethesda, Maryland

Bhaskar S. Kolachana
NIMH/Clinical Brain Disorders
    Branch
Bethesda, Maryland

George F. Koob
The Scripps Research Institute
La Jolla, California

David J. Krupa
University of Iowa
Iowa City, Iowa

Eleanor Y. Lee
Uniformed Services University of
    the Health Sciences
Bethesda, Maryland

R. A. Levis
Rush Medical College
Chicago, Illinois

Geoffrey S. F. Ling
Uniformed Services University of
    the Health Sciences
Bethesda, Maryland

Arnold S. Lippa
DOV Pharmaceutical, Inc.
Hackensack, New Jersey

Morgen T. Lippa
DOV Pharmaceutical, Inc.
Hackensack, New Jersey

Joseph S. Lonstein
University of Massachusetts
Amherst, Massachusetts

Daniel V. Madison
Stanford University School of
    Medicine
Stanford, California

Rafael Maldonado
Laboratori de Neurofarmacologia
University Pompeu Fabra
Barcelona, Spain

Annika B. Malmberg
University of California
San Francisco, California

Lance R. McMahon
Texas A&M University
College Station, Texas

Leslie R. Meek
University of Minnesota at Morris
Morris, Minnesota

Zul Merali
University of Ottawa
Ottawa, Canada

Thomas J. Morrow
VA Medical Center
University of Michigan
Ann Arbor, Michigan

A. Jennifer Morton
University of Cambridge
Cambridge, United Kingdom

Dominique Muller
University of Geneva
Geneva, Switzerland

Kim A. Neve
Oregon Health Sciences University and Veterans Affairs Medical Center
Portland, Oregon

Yoshitsugu Oiwa
National Institute of Neurological Disorders and Stroke, NIH
Bethesda, Maryland

Michael J. O'Neill
Eli Lilly and Company
Windlesham, United Kingdom

A. K. Pani
NIH/NIDA Intramural Research Program
Bethesda, Maryland

Lee A. Phebus
Eli Lilly and Company
Indianapolis, Indiana

R. Christopher Pierce
Washington State University
Pullman, Washington

Nicholas Poolos
University of Washington
Seattle Washington

Roger D. Porsolt
Porsolt and Partners Pharmacology
Boulogne-Billancourt, France

Michael K. Racke
University of Texas Southwestern Medical Center at Dallas
Dallas, Texas

Richard A. Radcliffe
University of Colorado
Denver, Colorado

James L. Rae
Mayo Foundation
Rochester, Minnesota

Ian J. Reynolds
University of Pittsburgh School of Medicine
Pittsburgh, Pennsylvania

Anthony L. Riley
The American University
Washington D.C.

Jeffrey D. Rothstein
Johns Hopkins University
Baltimore, Maryland

Sylvain Roux
Porsolt and Partners Pharmacology
Boulogne-Billancourt, France

Tammy J. Sajdyk
Indiana University Medical Center
Indianapolis, Indiana

Rosario Sanchez-Pernaute
National Institute of Neurological Disorders and Stroke, NIH
Bethesda, Maryland

Richard C. Saunders
NIMH/Clinical Brain Disorders Branch
Bethesda, Maryland

Rochelle D. Schwartz-Bloom
Duke University Medical Center
Durham, North Carolina

Ze'ev Seltzer
University of Toronto
Toronto, Ontario, Canada

Anantha Shekhar
Indiana University Medical Center
Indianapolis, Indiana

Toni S. Shippenberg
NIH/NIDA Intramural Research Program
Baltimore, Maryland

Cheryl L. Sisk
Michigan State University
East Lansing, Michigan

Daniel R. Storm
University of Washington
Seattle, Washington

Greg Stuart
John Curtin School of Medical Research
Australian National University
Canberra, Australia

Neal R. Swerdlow
University of California, San Diego
La Jolla, California

Shelley R. Thielen
Indiana University Medical Center
Indianapolis, Indiana

Alexis C. Thompson
NIH/NIDA Intramural Research Program
Baltimore, Maryland

Scott M. Thompson
University of Maryland School of Medicine
Baltimore, Maryland

Laurence Trussell
Oregon Health Sciences University
Portland, Oregon

Olga Valverde
Laboratori de Neurofarmacologia
University Pompeu Fabra
Barcelona, Spain

Jeanne M. Wehner
University of Colorado
Boulder, Colorado

Daniel R. Weinberger
NIMH/Clinical Brain Disorders Branch
Bethesda, Maryland

Ina Weiner
Tel Aviv University
Tel Aviv, Israel

Ben Avi Weissman
Israel Institute for Biological Research
Ness Ziona, Israel

Paul J. Wellman
Texas A&M University
College Station, Texas

Gary L. Wenk
University of Arizona
Tucson, Arizona

Howard S. Ying
Washington University School of
 Medicine
St. Louis, Missouri

CHAPTER 1
Neurophysiology

This manual includes units describing how to make a variety of preparations of the vertebrate central and peripheral nervous systems that are suitable for microelectrode recording. Two widely used preparations are described in UNIT 1.3 (preparation of brain slices) and UNIT 1.10 (isolation of neurons from the mature mammalian CNS). The brain slice allows stable intracellular and patch-clamp recordings to be carried out under conditions where synaptic connectivity is maintained. Slices from some brain regions, notably the hippocampus, permit electrical stimulation of specific synaptic pathways. In recent years, studies in brain slices have added greatly to the understanding of activity-dependent modifications in synaptic strength, which are believed to underlie learning and memory (UNIT 1.4). However, the brain slice is often not optimal for voltage-clamp recording because it may be difficult to control the membrane potential in the widely ramifying processes of a typical neuron. In addition, rapid changes of the extracellular solution, for example to apply drugs or neurotransmitters, are generally not feasible in slice recordings. These limitations can be overcome with isolated cell preparations. Use of neurons in monolayer culture makes it possible to accomplish fast solution changes on a time scale that mimics synaptic activity, and the acutely dissociated neurons described in UNIT 1.10 are often sufficiently electrically compact to provide excellent voltage-clamp results.

UNIT 1.2 describes the steps that go into making whole-cell voltage-clamp recordings with patch pipets (see UNIT 1.1 for fabrication of patch pipets); the practical and theoretical considerations are relevant to a wide variety of preparations, including cells in tissue culture as well as acutely isolated neurons (UNIT 1.10). UNIT 1.8 considers the specific details of carrying out patch-pipet recordings from neurons or glial cells in brain slices (UNIT 1.3); familiarity with the material in UNIT 1.2 is assumed. Two different approaches to recording in brain slices are described. In the blind technique, the recording chamber and optics are similar to those used for conventional intracellular slice recording; as in intracellular recording, success is a hit-or-miss proposition. Using the DIC technique, which employs more sophisticated optics (that tend to restrict access to the preparation), the cell to be recorded from can be selected based upon its visual appearance, and recordings can be readily made not only from the soma, but also from fine processes. UNIT 1.11 describes enhancements to the general techniques for patch clamp recording in the slice that make it possible to obtain recordings from neuronal dendrites not much larger than the tip of the patch pipet (1 to 2 μm diameter). Maintenance of the health of the slice is critical to the success of dendritic recording. This is achieved by perfusing the brain, prior to slicing, with an ice-cold solution that is low in permeant cations. Dendritic recordings are carried out under DIC visualization and can be obtained in conjunction with simultaneous somatic recordings from the same neuron, which enable studies on the way in which dendrites influence the overall excitability of the neuron. Although the methods in this unit are generally applicable to neurons with extensive dendritic arbors, the unit gives an example of how to cut hippocampal slices so that the apical dendrites of CA1 pyramidal neurons are optimally oriented for recording.

UNIT 1.9 provides a thorough discussion of the practical techniques for obtaining single-channel recordings in membrane patches. The current records obtained depict the quantal behavior of channel opening and closing. While single-channel data are often readily acquired, their analysis can be challenging. Indeed, methods for the interpretation of single-channel data continue to evolve. Methods for analyzing the function of synapses are described in UNIT 1.5. A general protocol is provided for the acquisition of stimulus-evoked and spontaneous synaptic events (excitatory and inhibitory postsynaptic currents and potentials) in recordings from brain slices. The methods presented are easily adapted to other preparations, including neuronal cell cultures. While measurements of synaptic responses have been carried out for nearly 50 years, the technique remains the most powerful approach available for characterizing the actions of physiologically released transmitters, interactions among synaptic inputs onto a neuron, and effects of externally or internally perfused agents on synaptic transmission. The technique is also widely used for defining synaptic strength in studies of synaptic plasticity.

An additional in vitro preparation for microelectrode recording that is also finding numerous other applications in neuroscience is the organotypic slice culture, discussed in UNIT 1.7. This preparation differs from conventional primary (dissociated) cell culture in that many aspects of the cytoarchitectural organization of the intact tissue are preserved. The explanted slice can be maintained ex vivo for up to several weeks, during which time cellular physiology, growth, differentiation, connectivity, or pathology can be readily investigated. Slice cultures are suitable for conventional or patch-pipet recording, and exogenous agents such as chemicals, hormones, growth factors, or even gene transfer vectors can easily be applied at any time during the lifetime of the preparation. In addition, organotypic slice cultures may be co-cultured with other organotypic slices or with dissociated cells, allowing for studies of cell-specific interactions. Two different preparative methods are described in UNIT 1.7. In the roller-tube technique, the slices flatten to a near monolayer that is ideal for examination by optical methods. In the interface technique, by contrast, three-dimensional architecture is better preserved, so the preparation is superior for applications requiring more normal tissue organization, such as field-potential recording.

Expression cloning of many ion channels and transporters was originally accomplished using the *Xenopus laevis* oocyte, and the frog oocyte continues to be a favored expression system for functional studies of electrogenic membrane proteins. Today, the oocyte expression system is commonly used by investigators seeking to deduce the functional roles of specific structural features of ion channels or characterize the consequences of a naturally occurring mutation. Because of their large size, *Xenopus* oocytes can be readily microinjected with native mRNA (even from human tissues), or, as is more commonly the case, with mRNA produced by in vitro transcription of recombinant DNA. The recipient oocyte generates large quantities of the encoded proteins which are faithfully assembled and incorporated into its plasma membrane. UNIT 1.6 describes techniques for two-electrode voltage clamp and patch recording from oocytes, which provide sensitive assays of ion channel or transporter behavior.

A key problem in modern neuroscience is the characterization of mechanisms by which the behavior of nervous systems is modified in response to experience. Changes in the magnitude of synaptic transmission (synaptic plasticity) are believed to underlie many such alterations. UNIT 1.4 considers the most intensively studied forms of synaptic plasticity in the vertebrate nervous system, those occurring at glutamate synapses in the hippocampus. The unit describes methods for investigating short-term (paired-pulse or post-tetanic facilitation and depression) and enduring (long-term potentiation and depression) synaptic plasticity in hippocampal brain slices (UNIT 1.3). In such experiments, the strength of synaptic transmission is assessed using field potential or intracellular recording. The unit discusses the specialized protocols required to elicit each form of synaptic plasticity and describes methods for the acquisition and analysis of the data acquired.

Conventional microelectrode recording does not lend itself to simultaneously accessing more than one or a few neurons in a restricted region of the nervous system. This constrains the investigator's ability to discern the richness of activity patterns over the spatial extent of complex neuronal networks. Biocompatible multi-electrode arrays (MEAs), produced through the application of integrated-circuit fabrication technologies, make it possible to carry out long-term electrical recordings from multiple points in nervous systems in two (planar) and even three dimensions. These devices permit stimulation and recording of field potentials—and in some cases single unit activity—over a spatial extent that typically extends for 1 to 2 millimeters, thus allowing the activity of tens or hundreds of neurons to be captured simultaneously. Commonly used MEAs are fabricated with titanium nitride electrodes on glass substrates using photolithography, reactive ion etching, and physical and chemical vapor deposition. Such MEAs have been used to record from neuronal tissue cultures and brain slices, and more recently, organotypic cultures, which retain circuit connectivity similar to the intact tissue and can be maintained in vitro for extended periods (see UNIT 1.7).

Contributor: Michael A. Rogawski

UNIT 1.1
Fabrication of Patch Pipets

BASIC PROTOCOL 1

PULLING SINGLE-CHANNEL AND WHOLE-CELL ELECTRODES WITH AN AUTOMATED PULLER

This protocol describes an approach to optimizing the pipet puller program (see Table 1.1.1) that works for a widely used puller, the Sutter P-59, but it should represent the general approach used with other pullers.

Materials

Pipet glass (Garner Glass)
Pipet puller (Sutter Instrument)
Micropipet storage jar (World Precision Instruments)

1. Run ramp test for the electrode puller to determine the temperature at which the electrode glass softens (see manufacturer's instructions) and use the heat determined for the pulling program.

Table 1.1.1 Sutter Pipet Puller Programs

Line	Heat	Filament	Velocity	Delay	Pull
Program 1 P-2000 Quartz					
1	960	3	40	130	50
2	875	4	40	130	40
Program 2 P-2000					
1	960	3	40	130	50
2	875	4	40	130	100
Program 3 P-97 Schott 8250, Corning 7052, Kimble EN-1					
1	519	5	10	25	
2	519	5	10	25	
3	519	5	10	25	
4	519	10	50	100	
Program 4 P-97 Schott 8330, Kimax, Pyrex					
1	566	5	10	25	
2	566	5	10	25	
3	566	5	10	25	
4	566	10	50	100	

2. Set up a one-line program:

   heat = ramp value
   pull = 5
   velocity = 10
   time = 25.

   Set pressure to 500 (middle of range).

3. With pipet glass in place, start program and see if glass pulls in four loops. If necessary, decrease or increase velocity by one to pull the pipet in four loops.

   The general shape of the pipet is largely determined by the first three loops. The fourth loop can then be modified as needed to affect primarily the region right near the tip.

4. Make an identical three-line program with:

   heat = ramp value
   pull = 5
   velocity = value determined above
   time = 25.

5. Add a fourth line to the program where:

   heat = ramp value
   pull = 10
   velocity = 50
   time = 100.

6. Pull an electrode and check tip under a high-magnification microscope (600 to 1500×). Adjust the pressure by 100, and repeat this step until the desired tip size is achieved. Place electrode into the micropipet storage jar.

SUPPORT PROTOCOL 1

OBTAINING OPTIMAL TIP GEOMETRY

Single-Channel Recording Pipets

For single-channel recordings, background current noise is minimized by using thick-walled glass with favorable electrical properties and by obtaining the highest possible seal resistance. Pull to quite fine tips. Figure 1.1.1 shows several forms of single-channel pipet tips before fire polishing. A single-stage pull program that works well for these electrodes on the Sutter P-97 puller is as follows:

   pressure = 999
   heat = ramp value
   pull = 50 to 125 (for electrodes of 25 to 75 MΩ)
   velocity = 255
   time = 255.

Whole-Cell Recording Pipets

To optimize bandwidth, reduce whole-cell noise, and minimize series resistance errors, blunt, low-resistance pipet tips are required. Use thin-walled glass tubing and pull the pipets to be as blunt as possible and to result in the lowest possible resistance and gigaohm seals. Figure 1.1.2 shows examples of optimal whole-cell recording pipet tips before fire polishing.

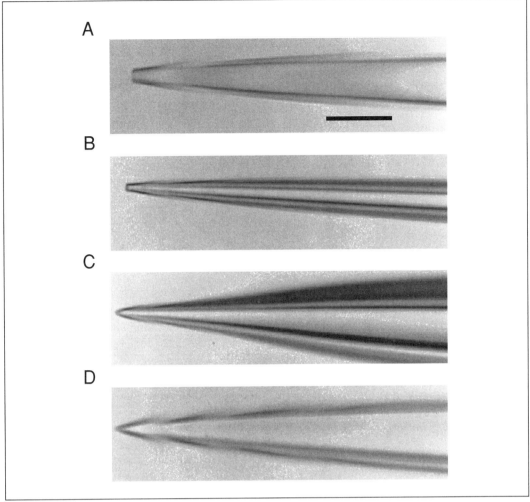

Figure 1.1.1 Single-channel tips before fire polishing. (**A**) Schott 8330 o.d. = 1.65 mm, i.d. = 1.15 mm, Sutter Program 4, pressure = 300. (**B**) Schott 8330 o.d. = 1.5 mm, i.d. = 0.375 mm, Sutter Program 4, pressure = 300. (**C**) Quartz o.d. = 1.5 mm, i.d. = 0.375 mm, Sutter Program 2. (**D**) Quartz o.d. = 1.5 mm, i.d. = 0.75 mm, Sutter Program 2. Tips shaped like those in (C) and (D) would require no fire polishing no matter what glass they were made from. Calibration bar = 10 μm.

SUPPORT PROTOCOL 2

CALIBRATING THE PULLER FILAMENT

Materials

   Fine forceps
   Box filament
   Pipet glass
   Pipet puller

1. Construct a filament former as shown in Figure 1.1.3 by turning on a lathe in a machine shop or by gluing together two pieces of glass of the appropriate lengths and diameters. Clamp the filament former into position as if a pipet were about to be pulled.

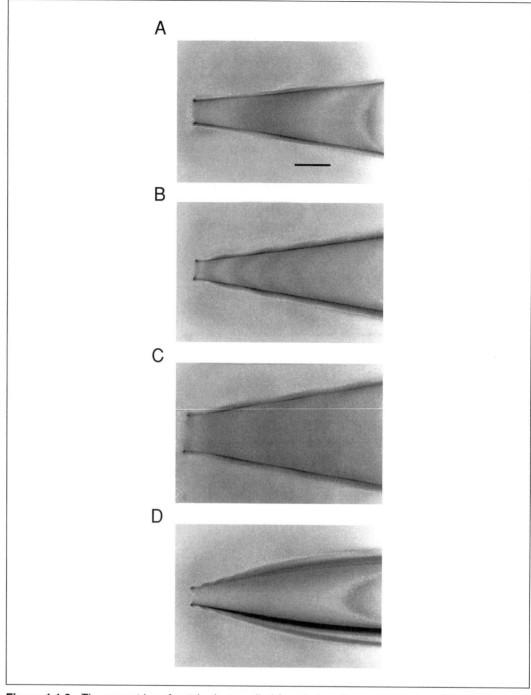

Figure 1.1.2 Tip geometries of patch pipets pulled for whole-cell recording before fire polishing. (**A**) Schott 8330 o.d. = 1.65 mm, i.d. = 1.15 mm, Sutter Program 4, pressure = 600. (**B**) Schott 8250 thin wall o.d. = 1.65 mm, i.d. = 1.30 mm, Sutter Program 3, pressure = 600. (**C**) Schott 8250 thin wall o.d. = 1.65 mm, i.d. = 1.3 mm, Sutter Program 3, pressure = 800. (**D**) Quartz o.d. = 1.50 mm, i.d = 0.75 mm, Sutter Program 1. Quartz geometry is not ideal but is about as good as it is possible to obtain and still hope to obtain gigaohm seals without fire polishing. Calibration bar = 10 μm.

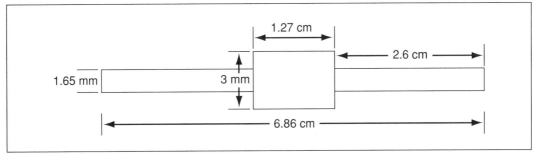

Figure 1.1.3 Drawing of a filament former.

2. Using fine forceps, pull the box filament (that totally surrounds the glass) on both ends until it roughly conforms to the outside dimensions of the filament former. Remove the filament former.

3. Clamp a piece of pipet glass in the puller, centering the glass in the filament with respect to length. Run the ramp test (see manufacturer's instruction manual) to find the heat setting at which the glass softens, and use this temperature in subsequent pulling programs.

SUPPORT PROTOCOL 3

CLEANING THE GLASS

Materials

Pipet glass
Ethanol or methanol
Ultrasonic bath cleaner
100°C oven

1. Immerse pipet glass in a beaker filled with either ethanol or methanol. Sonicate for ~5 min in an ultrasonic bath cleaner, ensuring that there is a steady stream of bubbles traversing the lumen of each piece of pipet glass from bottom to top.

2. After sonication pour off the ethanol or methanol and resonicate the glass in distilled water.

3. Pour off the water and bake the beaker and electrode glass in a 100°C oven for 30 min to 1 hr. Allow sufficient time for the glass to cool before using it in an experiment. If necessary, bake the glass or pulled but unfilled pipets in a 70°C oven for 1 hr immediately prior to use to reduce noise in high humidity environments.

SUPPORT PROTOCOL 4

NOISE CONSIDERATIONS FOR SINGLE-CHANNEL PATCH PIPETS

Distributed *RC* Noise

Distributed *RC* noise, arising from the capacitance of the pipet wall and the resistance of its lumen distributed along the portion of the pipet immersed in the bath, is minimized by using thick-walled pipets and/or heavy elastomer coatings extending as close as possible to the tip, and by shallow depths of immersion of the pipet into the bathing solution. An alternative that can be used with low melting temperature glasses is heat polishing to produce thick walls near the tip, in conjunction with heavy elastomer.

R_e-C_p Noise

In single-channel recording, the capacitance (C_p, 1 fF to 0.25 pF) is in series with the entire resistance of the patch pipet (R_e, almost always >1 MΩ), which has a high thermal voltage noise, producing current noise that is called R_e-C_p noise. Up to frequencies of $\sim 1/(2\pi R_e C_p)$, the power spectral density (in amp^2/Hz) of this noise is given by:

$$S_{ep}^2 = 4\pi^2 e_e^2 C_p^2 f^2$$

where $e_e^2 = 4kTR_e$. The root-mean-square (rms) current noise, i_{ep} (in amp rms), arising from R_e in series with C_p is then given by:

$$i_{ep} = (1.33\pi^2 c_3 e_e^2 C_p^2 B^3)^{1/2}$$

where B is the −3 dB bandwidth in hertz and c_3 is a coefficient that depends on the filter type ($c_3 \cong 1.9$ for an 8-pole Bessel filter). Since tip diameter is a major determinant of both R_e and patch size (and thus C_p), it is not surprising that this noise is determined by the size and geometry of the tip of the pipet. Because higher patch capacitances are associated with lower-resistance pipets, and the equation for i_{ep} depends linearly on C_p and on $R_e^{1/2}$, this source of noise is expected to be minimized by small area patches even if the associated pipet resistance is high.

Seal Noise

With zero applied voltage, the power spectral density of the current noise arising from the membrane-glass seal is expected to be given by $4kTR_e\{Y_{sh}\}$, where $R_e\{Y_{sh}\}$ is the real part of the seal admittance. The minimum value of $R_e\{Y_{sh}\}$ is $1/R_{sh}$, where R_{sh} is the DC seal resistance. Excess low-frequency noise is also possible when current is crossing the seal. It is clear that the amount of noise attributable to the seal will be minimized when the seal resistance is as high as possible. Higher resistance pipets with smaller tip diameters tend to produce the highest resistance seals. However, seal resistances in the range of 100 to 200 GΩ with pipets with resistances of ∼5 MΩ have been observed.

In general, factors that determine seal resistance on a patch-to-patch basis are not well known although there appears to be a clear average trend for small-tipped pipets to produce the highest resistance seals. It is not known at present if achieving teraohm seals depends on the type of cells (or glass) used as well as on small tip diameter.

It should also be noted that very small patches will also tend to minimize the probability of the tiny patch membrane containing other charge-translocating processes such as pumps or exchangers with major benefit likely to be at relative low frequencies (narrow bandwidths). Of course, very small patches also will reduce the likelihood of the patch containing the type of channel the investigator is interested in recording from. This restricts the situations in which such very small-tipped pipets can be used.

SUPPORT PROTOCOL 5

CONSIDERATIONS FOR WHOLE-CELL PIPETS

Dynamic Considerations

In the case of whole-cell voltage clamping, the entire pipet resistance, R_e, is in series with the cell membrane capacitance, C_m. Indeed, following seal formation and disruption of the patch, this resistance usually exceeds the resistance of the pipet as measured prior to attachment to the cell. Thus in this section, R_e represents the measured resistance after the whole-cell recording configuration has been achieved. This is normally strongly dominated by the pipet, but it will also include any intrinsic resistance in series with the cell membrane.

This total series resistance (R_e) has a variety of important effects. When current (I_m) crosses the cell membrane R_e causes voltage errors given by $R_e I_m$. Series resistance compensation circuitry provided in most commercial patch clamp amplifiers can reduce these errors. In addition, any uncompensated series resistance will, in conjunction with C_m, have a filtering effect on the measured current. In the absence of compensation for series resistance, this filtering effect limits the actual bandwidth of measured current to $1/(2\pi R_e C_m)$. The filter is analogous to a simple one-pole RC low-pass filter. As an example, consider that with $R_e = 5$ MΩ and $C_m = 50$ pF, $1/(2\pi R_e C_m) = 640$ Hz. The available bandwidth can be increased by the use of series resistance compensation. If α is defined as the fraction of the series resistance that is compensated ($0 \leq \alpha \leq 1$) and β is defined by $\beta = 1 - \alpha$ (so that β is the fraction of the series resistance that remains uncompensated), then series resistance compensation extends the uppermost usable bandwidth to $1/(2\pi R_e C_m)$. This is an important aspect of whole-cell voltage clamping to always bear in mind. Establishing an external filter bandwidth in excess of the limitations just defined will provide essentially no additional high-frequency information, although it will provide additional noise.

Obviously this problem is minimized by minimizing the product $R_e C_m$, and by using series resistance compensation. In terms of pipet fabrication, this simply means that tips should be as large as practical and pipets should have blunt tips (i.e., the pipet should increase in diameter quickly moving back from the tip). The limitation in terms of tip diameter depends on how easily seals can be formed and a whole-cell recording situation achieved in any particular cell size or type. The largest tipped (and hence lowest resistance) pipets that can produce reliable results will clearly minimize the dynamic problems considered here. This is true in both standard and perforated patch whole-cell recording configurations.

Noise Considerations

Whole-cell pipets produce all of the types of noise considered above with the exception of R_e-C_p noise. However, in the whole-cell situation, these sources of noise are generally unimportant in comparison with the far larger amount of noise caused by the thermal voltage noise of R_e in series with C_m. This noise source is analogous to R_e-C_p noise, but the capacitance C_m is many times larger than C_p. Because of this, the noise produced in a given bandwidth is also much larger. Also, because the time constant $R_e C_m$ is many times larger than R_e-C_p, the bandwidth limitations just described are also significant considerations in terms of noise (unlike the normal situation for R_e-C_p noise). The power spectral density of the noise arising from R_e in series with C_m is given by:

$$S_{em}^2 = \frac{4\pi^2 e_e^2 C_m^2 f^2}{1 + 4\pi^2 \beta^2 R_e^2 C_m^2 f^2}$$

where β is once again the uncompensated fraction of the series resistance as defined above (see Dynamic Considerations) and $e_e^2 = 4kTR_e$ as defined previously (see R_e-C_p Noise). Note that for 100% series resistance compensation ($\alpha = 1$, $\beta = 0$) this equation reduces to $S_{em}^2 = 4\pi^2 e_e^2 C_m^2 f^2$. This is the same form as the expression for the power spectral density of R_e-C_p noise (with C_m substituted for C_p), with noise power increasing as frequency increases in proportion to f^2. When series resistance is not completely compensated ($\beta > 0$), the power spectral density flattens out (to a value of $4kT/\beta^2 R_e$) at a frequency of $1/(2\pi R_e C_m)$; the level of the power spectral density will then be maintained until the signal and noise are rolled off by an external filter, or until the bandwidth limit of the electronics is reached.

Under most circumstances, this source of noise dominates whole-cell voltage clamp whenever the bandwidth is more than a few hundred hertz. As an example, consider a situation with $R_e = 5$ MΩ and $C_m = 50$ pF. Further consider that series resistance compensation has been set to 80%. This provides an actual bandwidth limitation of 3.2 kHz (as opposed to only 640 Hz without series resistance compensation). In this case, the noise from R_e in series with C_m will be ~2.2 pA rms for a bandwidth of 1 kHz (−3 dB, 8-pole Bessel filter), and will increase to nearly 12 pA rms for a bandwidth of 3 kHz. Both of these values are far higher than the noise of any commercially available patch/whole-cell amplifier at the same bandwidths and are also much higher than the noise attributable to the patch pipet per se as described. Only for very small cells (low C_m) with relatively low values of R_e (which are hard to achieve with small cells) will other sources of noise remain significant (provided, of course, that reasonable precautions are taken). However, even with the lowest values of the product $R_e C_m$ that the authors have had experiences which have resulted from small cells ($C_m \cong 6$ pF) with values of R_e as low as 3 to 4 MΩ; this source of noise reaches ~2.5 pA rms in a 5-kHz bandwidth and still dominates total noise at all bandwidths above about a kilohertz or so.

Clearly, minimizing this noise requires that R_e, C_m, or both be minimized. In the case of C_m, this means selecting small cells, and this may not always be practical or possible. Assuming a constant value of R_e and that the channels of interest occur in the same density per unit area in cells of different sizes, then signal-to-noise ratio will be independent of cell size. The most practical way to attempt to minimize this noise source is to use the lowest-resistance pipets that will form seals with the cells and allow the whole-cell recording situation to be reliably achieved. As a practical matter, also be aware of the effective bandwidth limitations described (due to C_m and uncompensated series resistance) and avoid setting the external filter bandwidth higher than this limit.

BASIC PROTOCOL 2

PREPARING PIPET TIPS WITH ELASTOMER COATING

To render the external surface of the pipet hydrophobic for all glasses and to improve the electrical quality of all glasses other than quartz, it is necessary to coat the pipet near the tip with an elastomer. For single-channel pipets, noise and capacitance reduction are the main reasons for coating, and a thick coating to as close to the tip as possible is warranted. With whole-cell pipets coating is required to reduce pipet capacitance. This usually can be achieved by painting the pipet with a thin elastomer coating only in the region where the final pipet taper begins (~2 to 3 mm back from the tip) so as to limit the outside fluid film to just the pipet tip.

Materials

1- to 2-mm o.d. × 5- to 7.6-cm glass tubing or rod (World Precision Instruments)
Pulled pipet (see Basic Protocol 1)

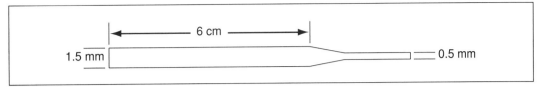

Figure 1.1.4 Effective geometry for wands used to paint elastomer onto patch pipets.

Elastomers RTV615 (General Electric) and Sylgard 184 (Dow Corning): mix thoroughly with a curing agent and store at −20°C in 1.5-ml microcentrifuge tubes

Elastomer R-6101 (Dow Corning): allow frozen stock to reach room temperature before opening; half fill 1-oz. polypropylene jars (Small Parts); tightly seal and place in an oven at 90° to 95°C for 48 hr or until optimal viscosity is reached; store tightly sealed at room temperature for up to 1 year

Dissecting microscope, preferably modified for dark-field illumination
Heat gun (e.g., Master Model 10008, Newark Electronics)

1. Using a Bunsen burner and glass tubing or rod of 1- to 2-mm o.d. × 5 to 7.5 cm in length, pull the glass into two pieces to form a wand (see Fig. 1.1.4) for painting the elastomer near the tip of the pipet.

   If the tip is too long and wispy, simply hold the tip perpendicular to a hard surface and then push the tip against the surface until it breaks off to the desired length and tip diameter. Alternatively, a pulled pipet can also be used as a painting wand.

2. Bring elastomer RTV615 or Sylgard 184 to room temperature before opening to ensure that water does not condense in the elastomer and mix well.

3. Remove pulled pipet from micropipet storage jar and examine the tip up to where the taper first begins using a dissecting microscope. Use a magnification that allows simultaneous observation of the tip of the pipet and the region up to where the taper first begins.

4. From a reservoir (e.g., 1.5-ml microcentrifuge tube) of elastomer, dip the glass painting wand into the elastomer and then put a generous blob of elastomer about halfway up the pipet taper.

5. With the tip of the wand, extend the elastomer to a region above where the pipet taper begins and then to a region as close to the pipet tip as possible. Rotate the pipet about its long axis and be sure to cover all of the pipet glass surface. Hold the pipet tip higher than the rest of the pipet so there is no chance that the elastomer can run (by gravity) into the tip.

   Elastomers RTV615 and Sylgard 184 cure at room temperature. This property can be useful if exceptionally thick coatings are desired, because thick coatings are more easily achieved if the elastomer is highly viscous at the time of painting.

6a. *For elastomers RTV615 and Sylgard 184:* Hold the coated tip extending essentially straight up into the blowing air of a hot heat gun. Continually twirl the pipet so as to uniformly heat the elastomer for ∼5 to 10 sec. Do not overheat!

6b. *For elastomer R-6101:* Hold the coated tip extending down into the blowing air of a hot heat gun, twirling continuously, to result in a much thicker, teardrop shaped coating over the majority of the shank of the pipet.

7. To obtain a thicker coating, repeat steps 5 and 6 as many times as desired.

SUPPORT PROTOCOL 6

CONSIDERATIONS FOR PIPET COATING

Capacity Transients

Coating of pipets fabricated from most types of glass can reduce the amplitude of both the fast and particularly the slow component of the pipet capacity transient. This is because for any given depth of immersion, heavy elastomer coating will reduce the total capacitance of the pipet, particularly the slow component of this transient. Quartz pipets show extremely little slow component in their capacity transients and this is not much changed by the thickness of the elastomer coating; of course; as with other glasses, the total amplitude of the transient (for a given depth of immersion) is reduced in proportion to the reduction of total pipet capacitance.

Noise Considerations

The lowest overall pipet noise for single-channel recording will be obtained from relatively small-tipped pipets made from thick-walled tubing of glass with the lowest possible dissipation factor (with quartz easily being the best) and coated with an elastomer such as R-6101; shallow depths of immersion are extremely useful in reducing pipet noise. The use of an elastomer coating remains a vital part of this low-noise strategy.

Elastomer coating is important to the noise performance of single-channel patch pipets, most importantly to prevent the formation of thin films of solution, which will otherwise form on the outer surface of an uncoated pipet as it emerges from the bath. Such films have a high distributed resistance which is in series with the distributed capacitance of the pipet wall (similar to distributed *RC* noise; see Support Protocol 4). However, the resistance of the thin film is higher than that of the solution in the lumen of the pipet. In an uncoated pipet, it is expected that the power spectral density of this noise will rise at low-to-moderate frequencies and eventually level out at higher frequencies (in the range of kilohertz to several tens of kilohertz), typically in the neighborhood of 100 to 300 fA rms in a 5-kHz bandwidth. This is sufficiently large that in many cases it would dominate overall noise. Coating with an elastomer can essentially eliminate this source of noise.

Elastomer coating can also be quite important to minimizing distributed *RC* noise (see Support Protocol 4) in single-channel patch pipets, and it should be recalled that it can be minimized by using thick-walled pipets. Such a thick wall can be achieved by either starting with thick-walled glass capillaries or applying a heavy layer of elastomer, or both. This reduces the capacitance of the immersed portion of the pipet wall for any given depth of immersion, and thereby reduces distributed *RC* noise. Sylgard 184, for example, has a dielectric constant of ~2.8, which is less than that of any available glass, and very heavy coats of elastomer can readily be applied (see Basic Protocol 2). However, for the lowest noise recordings, using thick-walled glass is most effective in minimizing distributed *RC* noise and preserving o.d./i.d. ratio near the pipet tip during pulling (sharper-tipped pipets seem to more closely preserve the initial o.d./i.d. ratio during pulling). If this ratio is preserved the capacitance of the immersed portion of the pipet is proportional to $1/ln$ (o.d./i.d.). Thus, increasing the o.d./i.d. ratio from 1.4 to 4 reduces this capacitance by about a factor of 4. Even if the initial o.d./i.d. ratio is not preserved during pulling, the relative improvement obtained by increasing the wall thickness of the tubing is roughly the same.

For all glasses except quartz, coating with R-6101 (Dow Corning), Sylgard 184 (Dow Corning) or RTV615 (General Electric) is clearly expected to decrease dielectric noise as well in single-channel patch pipets. The reason for this is that the dissipation factor of these elastomers (e.g., for Sylgard 184 the dissipation factor is reportedly in the range of 0.0009 to 0.002,) is comparable to or less than that of the best glasses described here other than quartz. For pipets fabricated from glasses other than quartz, coating with Sylgard 184 reduces dielectric noise at any particular depth of immersion, and the heavier the coat of elastomer the greater the reduction in noise. For quartz pipets (with quartz having a dissipation factor 20 or more times less than that of Sylgard), coating with Sylgard 184 actually increases the predicted dielectric noise above that expected for an uncoated pipet but somewhat decreases the total noise of the coated quartz pipet from distributed RC noise and particularly thin-film noise. Coating with a suitable elastomer remains absolutely essential even for quartz pipets, and even with a coating of Sylgard, a quartz pipet (of equivalent glass wall thickness) will produce significantly less dielectric noise than any other type of pipet. Finally, if an elastomer can be found with a smaller dissipation factor than that of Sylgard, it could reduce the amount of dielectric noise of quartz pipets further even if the dissipation factor of this elastomer was still greater than that of quartz. Once again it should be recalled that shallow depths of immersion will always minimize dielectric noise.

The importance of pipet noise per se is significantly reduced in the whole-cell pipets, primarily because in most situations the noise of R_e in series with C_m will dominate total noise. Nevertheless, it is normally advisable to use at least a light coating of R-6101, Sylgard 184, or other suitable elastomer on whole-cell pipets to prevent the formation of thin films and their associated noise. In addition, coating reduces the size and complexity of the pipet capacity transient, thereby allowing the pipet capacity compensation circuitry in most patch clamps to more effectively negate this transient. However, the coating of elastomer need not approach the tip particularly closely, and it need not be heavy. With relatively large cells even these moderate precautions may seem unnecessary.

BASIC PROTOCOL 3

FIRE POLISHING THE PIPET

Materials

    100×, long-working-distance metallurgical objective with 210-mm tube length or infinity corrected (e.g., Nikon, Olympus)
    Fire-polishing wire: 0.003-mm platinum-iridium wire (AM Systems)

1. Using a 5 to 10× objective, approximately center the tip of the pipet and the fire-polishing wire in the optical field. Switch the 100× objective in place and reposition the pipet tip and the polishing wire.

2a. *For single-channel pipets:* Position the pipet tip and the fire-polishing wire at least 30 μm apart. Adjust the voltage on the heater to the proper level (determined by trial and error) and turn on the current flow. Under direct observation, round and smooth the tip until a small channel can be observed in the region of the tip where the inner glass walls appear about parallel. See Figure 1.1.5.

2b. *For whole-cell pipets:* Move the pipet tip and fire-polishing wire to ∼150 μm apart under the high power objective. With the tip and wire now much farther apart, increase the heat until obvious changes in the tip geometry slowly begin to occur (∼30 sec). See Figures 1.1.6 and 1.1.7.

The inner walls of the pipet should not become parallel. Fire polish the minimum amount necessary so as to avoid undesirable increases in tip resistance.

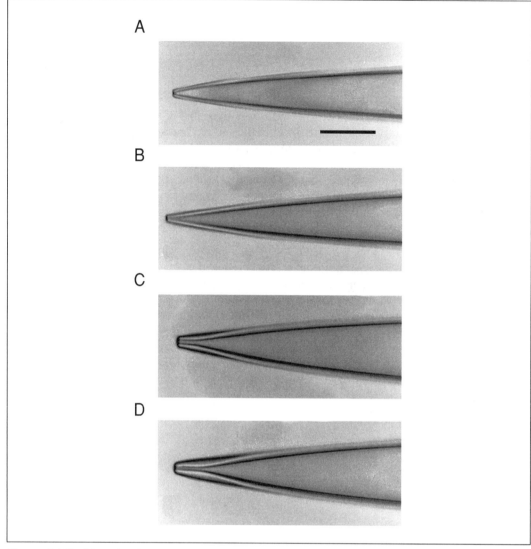

Figure 1.1.5 Pipet tips for single-channel recording following fire polishing. (**A**) Schott 8330 o.d. = 1.65 mm, i.d. = 1.15 mm before fire polishing, Sutter Program 4, pressure = 300. (**B**) Same pipet as (A) after light fire polish. (**C**) Schott 8250 o.d. = 1.65 mm, i.d. = 1.15 mm, Sutter Program 3, pressure = 300 with light fire polish. (**D**) Same pipet as in (C) with additional fire polishing. Note thicker wall near tip and long region of near parallel inner walls. Calibration bar = 10 μm.

SUPPORT PROTOCOL 7

CONSTRUCTING FIRE-POLISHING APPARATUS COMPONENTS

Microforges would do the entire fire polishing job, but they are expensive, so most investigators have chosen to build at least part of the fire-polishing apparatus.

Constructing a Current Source

A current source for fire polishing is easily made from a Variac (Newark Electronics) output fed through a 20/1 step-down filament transformer (Newark Electronics), resulting in a voltage

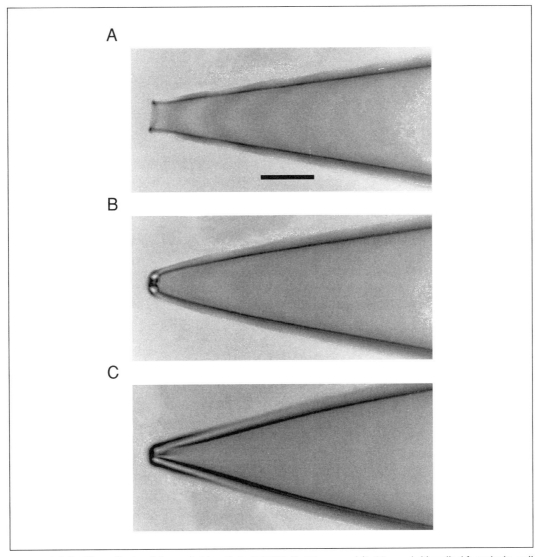

Figure 1.1.6 The effects of fire polishing Schott 8330 (1.65 mm o.d./1.15 mm i.d.) pulled for whole-cell pipets either with the tip near (≈30 microns) or far (≈150 microns) from the fire polishing wire. The unpolished pipet tip, (**A**), narrows and its walls thicken when the tip is closer to the wire, (**B**). When the tip is far away from the wire, its tip simply rounds up (**C**). Calibration bar = 10 μm.

range of 0 to 6 V. Properly constructed fire-polishing heater wires have a resistance of only ∼1 Ω. A voltage range of 0.5 to 1.5 V is the usable range for providing the heat required for fire-polishing.

Constructing a Heater Wire

Two coated solid copper wires of ∼22-G from the filament transformer output are attached with electrical tape to opposite sides of a cylindrical bar mounted on a micromanipulator. Uncoated ends of the wires extend ∼1 in. (2.54 cm) past the end of the bar and are bent twice at right angles to produce an end structure as shown in Figure 1.1.8. The portion of the wire that is visible under the microscope is a 76-μm diameter platinum-iridium wire (A.M. Systems) that is attached to the end structure of the copper wires. Cut a ∼2-in. (5-cm) piece of

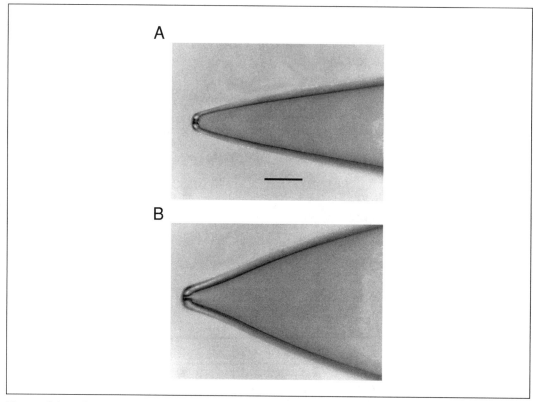

Figure 1.1.7 Pipet tips for whole-cell recording after fire polishing. (**A**) Schott 8330 o.d. = 1.65 mm, i.d. = 1.15 mm, Sutter Program 4, pressure = 600. (**B**) Schott 8250 thin wall o.d. = 1.65 mm, i.d. = 1.30 mm, Sutter Program 3. Pipet in (B) pulled with higher gas pressure (800) than tip in Figure 6.3.2B (600), therefore the blunter taper. Calibration bar = 10 μm.

the platinum-iridium wire and wrap the ends over and over again around the individual pieces of copper wire, leaving ∼1/2 in. (1.3 cm). Bend the remaining wire into a fine hairpin loop (Fig. 1.1.8). Solder the copper and platinum-iridium wires together where the platinum-iridium wire is wound around the copper wire. A proper solder connection will produce <1 Ω of resistance.

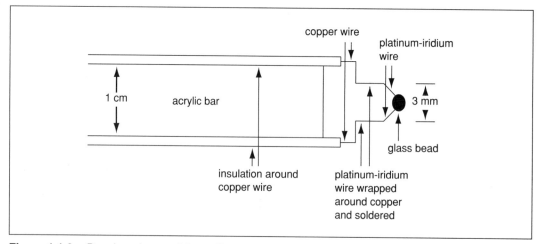

Figure 1.1.8 Drawing of a possible configuration for the fire-polishing wire.

Coat the hairpin loop with glass to keep the platinum from sputtering onto the end of the pipet when it is being fire polished. Use the same glass from which pipets are constructed and pull three to four pipets (see Basic Protocol 1). Tape them, one at a time, to a microscope slide on the stage of the microscope. Using a low-power objective and direct observation, move the pipet tip into the vicinity of the hairpin loop. Turn on the heat at near max (~1 V from the current source) and simply jam the pipet tip against the hot wire to melt the tip which will stay on the hairpin loop. Move the pipet away and turn off the heat. Repeat 3 to 4 times until there is enough glass to make a discrete glass bead at the center of the hairpin. Turn on the heat one more time to allow any projections left by the withdrawal of the pipet to melt into the mass of the bead. The heating wire is then ready for use.

BASIC PROTOCOL 4

PIPET FILLING

Pipets must be filled with salt solutions before use. The particular solution used is dictated by the experiment to be performed, so no attempt is made here to describe the composition of filling solutions.

Materials

    1-ml tuberculin and 10-ml syringes (Becton Dickinson)
    Fire-polished pipet (see Basic Protocol 3)
    Suction apparatus (Fig. 1.1.9)
    2.0-mm holder (World Precision Instruments)
    Needle to fit into the bore of the pipet: e.g., 28-G Microfil (World Precision Instruments), 1.5-in. 22-G Monoject needle, or 1.25-in. 27-G Monoject needle

1. Pull the plunger of a 10-ml syringe to 1 ml, and mount the fire-polished pipet in a suction apparatus constructed from a pipet holder with the suction line plugged and adapted (e.g., with male Luer fitting connected to the syringe by Tygon tubing) to fit a standard 10-ml syringe (see Fig. 1.1.9).

2. After tightening the pipet in place, immerse the tip into a small beaker filled with the proper filling solution. Pull the syringe plunger back to the ~6- to 7-ml mark. Hold for 5 to 30 sec depending on the tip diameter.

3. Remove the pipet. Use a 1-ml tuberculin syringe fitted with the proper gauge and length needle (or a plastic syringe needle) to fit into the bore of the pipet to eject a little fluid to clear any solution that might have been in contact with the metal of the needle for a while.

4. Place the needle into the bore of the pipet from the back until the needle tip bottoms out near the tapered part of the pipet. Inject fluid until the pipet is about half filled and then remove the needle from the pipet.

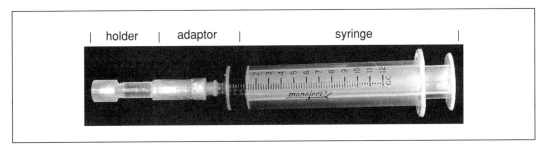

Figure 1.1.9 A syringe and modified pipet holder for drawing suction for pipet filling.

5. Holding the pipet with the thumb, index, and middle fingers of the left hand, gently flick the right index finger against the pipet where it rests against the index and middle fingers of your left hand to knock out any residual bubbles.

6. Check the pipet tip under the dissecting microscope to verify that all bubbles are gone. If bubbles remain, repeat step 5 until the tip is bubble free.

BASIC PROTOCOL 5

MOUNTING AND TESTING THE PIPET SETUP

Materials

Bathing solution appropriate for experiment
Pipet
Suction line connected to a syringe needle
Silver/silver chloride reference electrode
Patch clamp apparatus (see UNIT 1.2)

1. Dry the outer wall of the pipet with a Kimwipe. Using a suction line connected to a syringe needle of the correct gauge and length to fit into the back of the pipet, suck out the excess fluid leaving the level in the pipet just sufficient to immerse the tip of the internal silver/silver chloride reference electrode when it is inserted.

2. Insert the pipet into the holder and tighten in place. Insert the holder into the patch clamp headstage connector. Position the pipet tip just over the chamber being careful not to actually touch the solution in the chamber.

3. Check the noise on the patch clamp noise meter to determine acceptability (e.g., for a resistive headstage with 0.13 pA rms noise in a 5-kHz bandwidth, a 10% to 20% noise increment; for a cooled, integrating headstage, a 50% increment; or other as empirically determined).

4. If the noise increment is too great, clean the holder by disassembling, sonicating in ethanol for 3 to 5 min, and drying at 70°C for 1 hr or longer (see Support Protocol 3). Be sure the holder has cooled to room temperature before using.

5. If the noise is satisfactory, immerse the pipet tip in the bathing solution and measure its resistance using circuitry inherent to the patch clamp. Check the resistance, and if it is not acceptable, get a new pipet and start over again.

Reference: Levis and Rae, 1995

Contributors: J.L. Rae and R.A. Levis

UNIT 1.2

Whole-Cell Voltage Clamp Recording

The voltage clamp is one of the most powerful techniques available for studying functional aspects of voltage-gated channels. It measures current through a cell membrane while controlling the voltage. The whole-cell voltage clamp performs better on relatively small cells and uses a single electrode for both controlling voltage and measuring current (Fig. 1.2.1). It can be used to activate different populations of channels selectively. This protocol will focus on using the whole-cell voltage clamp to study voltage-gated channels, but it should be borne in mind that this technique is readily adapted to the study of ligand-gated channels, synaptic

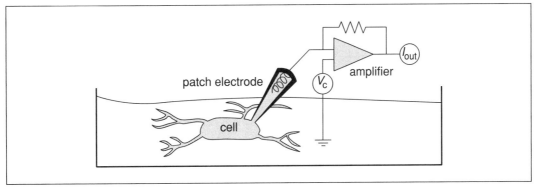

Figure 1.2.1 A sketch of a cell under whole-cell voltage clamp. A single cell is shown with a patch electrode and amplifier. The bathing solution superfuses the preparation while the pipet filling solution perfuses the cell interior. Because the patch-clamp amplifier circuitry maintains the same voltage at the two inputs, the command potential, V_c, will be applied to the cell. The amplifier output is a voltage proportional to the membrane current, and is designated I_{out}.

potentials, and exocytosis. The method described here may also be useful in implementing other techniques, including capacitance measurement, patch clamp recording in brain slices, and analysis of synaptic potentials.

STRATEGIC PLANNING

It is very important to optimize the patch clamp setup prior to experiments. The electronics should be tested on a model cell, the software installed and tested, the microscope optics adjusted, and the vibration isolation table tested for good isolation. Noise from the electronics should be minimized and 60-Hz line interference eliminated. The micromanipulator should be arranged to allow for easy electrode installation and positioning of the electrode tip within the full range of the field of view under the microscope. Cables leading to the amplifier head stage and micromanipulators should be well secured to prevent vibrations.

Electrophysiology Setup

The setup must be equipped with a microscope, oscilloscope, vibration isolation table, micromanipulator, and additional items as mentioned below. The microscope must be of sufficient quality to permit the user to make out as much detail as possible on the cell surface, in order to locate regions where optimal contact can be realized between the electrode tip and the cell membrane. The micromanipulator must have sufficient resolution (<1 μm) to position the electrode tip accurately, and must have little drift (<1 μm), so that position does not change during recordings.

Amplifier

Excellent patch clamp amplifiers are commercially available from Axon Instruments, Instrutech, and ALA Scientific Instruments. For whole-cell voltage-clamp experiments, an amplifier should have a low-noise current-to-voltage converter and a capability for summation of holding potential and stimulus input. It should also have transient cancellation circuitry to provide a means of nulling the capacitative transient that accompanies any voltage change. Series resistance (R_s) compensation is a desirable feature when large currents are anticipated or when rapid settling times are needed.

Computer, Interface, and Software

The computer uses an interface to acquire data through analog-to-digital converters and apply voltage command signals through digital-to-analog converters. Only a single channel of input (membrane current) and a single channel of output (command potential) are essential. One commercially available patch-clamp amplifier, the EPC-9 (Instrutech or ALA Scientific Instruments), is under complete digital control. It has a built-in interface and cannot be used without a computer. Although the computer system performs many of the functions previously left to an oscilloscope, it is still advisable to display signals on an oscilloscope for routine checking of the performance of the system.

A number of excellent software packages have been developed to run voltage-clamp experiments. These include the pClamp package (Axon Instruments; developed for the PC) and Pulse (Instrutech; runs on Macintosh and PC). Axon Instruments and Instrutech also sell interfaces that allow the patch-clamp amplifier and other devices to be controlled by this software. An excellent program for voltage-clamp experiments with Macintosh computers has been developed as a PulseControl extension (down-load free of charge from *http://chroma.med.miami.edu/cap*) of the relatively inexpensive computer program IGOR (Wavemetrics).

Electrode Holder

An electrode holder fitted with a silver chloride-coated silver wire connects the patch electrode to the amplifier head stage. This electrode holder must have a fitting to connect to a flexible plastic tube for the application of suction and pressure during certain steps in the experiment. Such electrode holders are available from the manufacturers of the patch clamp amplifiers mentioned above, as well as from E. W. Wright and World Precision Instruments. It is helpful but not necessary to have a pressure meter of some form to monitor and gauge the pressure. A particularly convenient system can be built with a digital voltage display and a pressure transducer from Sensym.

Patch Electrodes

Patch electrodes must be fabricated, coated, and fire-polished (see UNIT 1.1). Because dirt accumulates on the glass surface of the electrode and prevents sealing, the electrodes should be used the same day. The resistance of a filled electrode provides a useful indicator of the electrode tip size (UNIT 1.1). Electrodes with resistances <1 MΩ are rarely used for whole-cell patch clamping because their tips are too large to obtain gigaseals reliably. With very small cells (<10 μm) it may be necessary to use small-tipped patch electrodes with resistances of $\geq$10 MΩ. Patch electrodes should be fabricated from hard glass.

Reducing electrode capacitance by applying a coating of low-dielectric material (see UNIT 1.1) is very important to allow separation of cell capacitance and R_s and lower the noise level. The most popular coating for patch electrodes is Sylgard (184 silicone elastomer, Dow Corning). When this material is used, the fire-polishing step must be performed after coating to burn off residue at the electrode tip that can interfere with seal formation. Other materials with a low dielectric constant such as Q-dope (GC Electronics), dental wax, or boat varnish have also been used. For some of these coatings time must be allowed for drying.

The coating material should be transparent so one can see whether air bubbles are lodged near the electrode tip after filling. The coating should be applied with the aid of a small glass or plastic tool under a dissecting microscope and cured immediately with an electric heating wire. Incompletely cured coating can create difficulties in forming a gigaseal. For whole-cell voltage clamping it is not critical to cover the glass near the tip because this region has a small area; coating to within 100 to 200 μm is adequate. The coating must be applied high enough up the barrel to prevent any water contact with the glass walls of the electrode, which can turn the entire electrode wall into a large capacitative surface. The layer of coating is usually about as thick as the glass.

Preparation

In making a selection of a biological preparation of cells or slices, one must consider the accessibility of the cell surface and the size and morphology of the cells (see UNITS 1.3 & 1.10). To obtain a gigaseal, cells must have naked surfaces to provide free access for the patch electrode tip. Special conditions, treatments, and manipulations may be necessary to expose the cell surface and facilitate gigaseal formation. Cell size is a consideration because the high capacitance of a large cell makes the settling time longer. The ideal cell size for whole-cell voltage clamping is therefore small, on the order of 10 to 20 µm in diameter. Even small cells can have currents that are large enough to cause significant errors in voltage control, but these problems can often be dealt with effectively by using low-resistance patch electrodes and R_s compensation.

Cell morphology is a consideration because long processes and elaborate morphologies cause space-clamp errors. There are few remedies for these errors, so where possible, morphological simplicity is a major advantage. There is often an enormous trade-off between technical ease of measurement and physiological relevance.

Physiological Bathing Solution

A bathing solution is needed for constant superfusion over the preparation (Fig. 1.2.1). This solution may be used while experiments are in progress or in the initial stages of the experiment prior to switching to a solution designed to enhance a particular component of ionic current (see Experimental Solutions, below). Physiological bathing solutions appropriate for various preparations can be found in the relevant journal articles.

Experimental Solutions

In formulating solutions, one must consider the ease of seal formation, the health of the preparation, and the specific objectives of an experiment. Because small amounts of dirt and debris can prevent gigaseal formation, the patch-electrode filling solution should be filtered using a 0.22-µm filter. This solution perfuses the inside of the cell (Fig. 1.2.1), providing a major technical advantage in isolating specific ionic components of current. For each type of current to be studied it is necessary to provide the appropriate permeant ion, and it may also be necessary to block currents through other channels. It is not possible to provide recipes of solutions for each type of current because the requirements vary so widely between different types of cells and channels. It is therefore important to consult references dealing with a particular current in the relevant preparation (e.g., see Swandulla and Chow, 1992).

One important general objective in the design of a patch pipet solution is to mimic the intracellular milieu. A key consideration in this regard is Ca^{2+} content. EGTA is the most widely used Ca^{2+} buffer. It is common to add 10 mM EGTA and 1 mM $CaCl_2$, which give a calculated free Ca^{2+} concentration of ~25 nM at pH 7.2, based on the relevant stability constants. However, free Ca^{2+} depends on other components of the solution and is quite sensitive to the purity of the EGTA. Lowering EGTA to <1 mM is common, but recordings with low EGTA are generally more difficult to obtain and less stable. Another chelator, BAPTA, binds Ca^{2+} much more rapidly than EGTA and is often used when especially tight control of intracellular Ca^{2+} is desired. Some researchers suspect that BAPTA is unstable and that it causes an increase in R_s. For these reasons, BAPTA-containing solutions should be prepared with a fresh supply of BAPTA less than ~2 weeks prior to use. It is generally not possible to obtain whole-cell patch-clamp recordings with solutions containing added Ca^{2+} and no chelator. Patch pipet filling solutions are generally stored frozen in 1-ml aliquots.

As K^+ is the major monovalent cation of cytoplasm, K^+ salts should be used in the absence of a good reason to do otherwise. When K^+ is replaced, the choice is often Cs^+ or N-methylglucamine, although N-methylglucamine has been reported to alter the behavior of Ca^{2+} channels.

Cytoplasmic Cl^- concentration varies among cell types, and can range from 5 to 40 mM. Although Cl^- is often used as the major anion of patch-pipet filling solutions, it is considered more physiological to replace most of the Cl^- with an organic anion, such as gluconate, aspartate, MOPS, acetate, citrate, or methanesulfonate. F^- is sometimes used to replace Cl^- when the goal is to poison all metabolic processes within a cell and free channels from modulatory influences.

In most experiments, the patch pipet solution is buffered at a pH between 7.1 and 7.3 (usually with 10 mM HEPES). However, the pH can be varied to address questions regarding intracellular pH and membrane function. Titration of the buffer during solution preparation should be done with an acid or base of one of the major ions of the solution (e.g., a CsCl solution should be titrated with CsOH).

The activity of many channels depends on the addition of 1 to 10 mM ATP (Mg^{2+} salt). Even with ATP, rundown of Ca^{2+} current is difficult to avoid, and a number of reagents have been added to patch-pipet filling solutions to prevent this. An ATP-regenerating system consisting of creatine phosphate and creatine phosphokinase has been reported to reduce rundown of Ca^{2+} currents (Forscher and Oxford, 1985). To maintain membrane currents dependent on G-proteins, 100 to 300 μM GTP is commonly added. More exotic additives may be used to test a hypothesis for regulation of channel function by a protein or signaling molecule. In such experiments, the time required for diffusional exchange between the pipet solution and cell interior should be considered. Exchange is roughly exponential in time, with a time constant τ (in seconds) that can be approximated by the following expression (Pusch and Neher, 1988):

$$\tau = 0.042 \, R_S M^{1/3} C_c$$

where M is the molecular weight in Daltons (Da) of the substance in question, R_s is in megaohms, and C_c is the cell capacitance in picofarads (pF). For $R_s = 5$ MΩ and $C_c = 20$ pF, this expression suggests that a 1000-Da molecule will exchange with the cell interior in 42 sec. Such exchange times are typical for whole-cell voltage-clamp recordings.

BASIC PROTOCOL 1

PATCH-CLAMP TECHNIQUE SETUP

Materials

Cells or tissue slice (see UNITS 1.3 & 1.10)
Patch electrode buffer (dependent upon experimental design)
Bathing solution (dependent upon experimental design)
Coated patch electrodes (see Strategic Planning and UNIT 1.1)
Electrophysiology setup (see Strategic Planning)

1. Fire polish and fill a patch electrode with solution by dipping the tip in the solution for a few seconds and then filling from the back by inserting a thin tube. Mount the electrode in the electrode holder, which has been plugged into the amplifier head stage.

2. Apply gentle positive pressure to the patch electrode interior through a tube connected to the electrode holder. With the pressure on (to avoid accumulation of dirt at the air-water interface on the electrode tip, which prevents gigaseal formation), immerse the electrode tip by passing it through the surface of the bathing solution.

3. Calculate the patch electrode resistance by applying a test pulse (typically 5 to 10 msec at 10 mV), using a low gain setting of the amplifier (typically 1 mV/pA), reading the current (e.g., 5 nA), and using Ohm's law to calculate the resistance (2 MΩ ; see Fig. 1.2.2A).

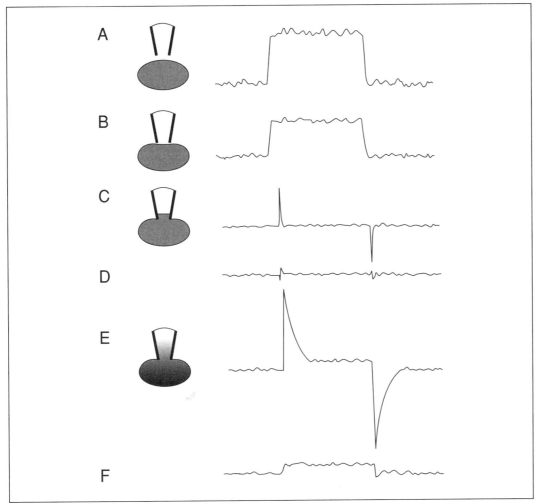

Figure 1.2.2 Test pulses produce different current responses as one proceeds through the establishment of a whole-cell voltage clamp recording. The physical relationship between the patch electrode and the cell is illustrated schematically on the left. The size of the current change produced by the test pulse goes down as the resistance across the patch electrode tip goes up. Thus, a reduction in test-pulse current indicates closer contact between the electrode tip and the cell. (**A**) The electrode is just above the cell, not in direct contact, so the resistance is low (1 to 10 MΩ), and the test pulse current is large. (**B**) The electrode touches the cell surface, the resistance goes up slightly, and the test pulse current gets smaller. (**C**) A gigaseal has formed as the result of gentle suction, which pulls a small patch of membrane up into the electrode tip. The resistance is high (>1 GΩ), so except for the transients, the test pulse current is virtually flat. (**D**) The electrode capacitance transient is nulled. (**E**) Break-in is achieved by strong suction that removes the patch of membrane in the electrode tip, but leaves the seal and cell intact. The resistance goes down and large capacitance transients are seen. Perfusion of the cell interior begins. (**F**) The whole-cell capacitance transient is nulled. Since steps D and F are purely electrical adjustments, the diagram of the cell and patch pipet is the same as in C and E, respectively.

Monitor the resistance of the patch electrode regularly during electrode positioning for increases in resistance, indicating that the electrode tip has clogged and will be less likely to form a gigaseal.

4. Null the junction potential at the tip of the electrode with the appropriate control on the patch clamp amplifier (labeled "Vp-offset" on the EPC-7 and "manual junction null" on the Axopatch-1C).

   The junction potential between the bathing solution and the patch electrode solution should be measured in a separate experiment and used to correct voltages during data analysis according to Neher (1992).

5. With the course control of the micromanipulator, bring the electrode tip into the field of view containing a cell targeted for study (typical time: 1 min). Use the fine control of the micromanipulator to gently touch the electrode tip against the cell surface, check that the resistance has increased, and release the positive pressure (Fig. 1.2.2B).

6. Using resistance and appearance as cues for contact between the cell surface and patch electrode tip, apply gentle suction to the patch pipet interior. The resistance climbs to >1 GΩ (>5 GΩ is more likely to lead to a successful whole-cell recording) to produce a gigaseal (typical time: several seconds to several minutes). Use a high gain setting for the amplifier (50 to 100 mV/pA) to detect a small DC current response ($\sim$1 pA) when the resistance rises.

   Adjusting the holding potential to a negative voltage (e.g., -80 mV) can make sealing faster and improve the success rate of gigaseal formation.

7. After achieving a gigaseal, verify that the test-pulse current is nearly flat except for transient spikes in opposite directions at the beginning and the end of the test pulse (reflecting charging of the electrode capacitance, along with the much smaller capacitance of the membrane patch; Fig. 1.2.2C).

8. Null the transient by adjusting the appropriate patch-clamp amplifier controls (on an EPC-7, C-fast and τ-fast) to give a test-pulse response resembling that shown in Figure 1.2.2D. With the Axopatch 200 null two electrode-charging transients with amplitude (MAG) and time constant (τ) controls (typical time: 15 to 30 sec).

   If these controls cannot be adjusted to null the electrode transients, then either the electrode has not been well coated with Sylgard or the solution level is high enough that the electrode glass above the coating has made contact with the bathing solution. It is also important to maintain a constant depth of bathing solution during perfusion of the recording chamber, because changes in height will change the electrode capacitance (even with good Sylgard coating).

9. Adjust the holding potential to the anticipated resting potential of the cell to avoid a large change in membrane potential upon break-in. Rupture the membrane patch under the electrode top by applying suction while watching the test pulse current. Increase the suction gradually, and stop immediately upon observing a sudden appearance of large spikes at the beginning and the end of the test pulse (Fig. 1.2.2E; typical time: 15 sec to several minutes). Stop the suction as soon as break-in is successful.

   Sometimes break-in occurs spontaneously, so steps 7 and 8 should be performed quickly, and the transient should be checked for changes prior to the application of strong suction.

10. With the holding potential adjusted to a value of -80 to -100 mV (voltage-dependent channels are less active in this range), null the whole-cell charging transient by adjusting the appropriate amplifier controls (on the EPC-7, C-slow and G-series; on the Axopatch 200, Whole Sell Capacitance and Series Resistance) to give the test-pulse response resembling Figure 1.2.2F (typical time: 1 min).

In recordings from cells with complex geometries only the fastest component of the whole-cell charging transient can be nulled. No effort should be made to remove the portion of the slower components associated with process charging.

When the slow whole-cell adjustments fail to remove the transient completely it may be necessary to go back and forth between the fast and slow adjustments until the transient has been eliminated.

11. Write down the cell capacitance (C_c, indicates size of cell or cell body) and R_s off the dials of the controls used to cancel the whole-cell transient. Convert the G-series control on the EPC-7 from conductance to resistance ($R = 1/G$; 1 µS gives 1 MΩ). Calculate the clamp settling time, which is equal to $R_s C_c$. Calculate the cell resistance (R_c) from the amplitude of the test-pulse current using Ohm's law (Fig. 1.2.2F; typical time: 30 sec). Periodically check C_c, R_s, and R_c for changes during the course of a recording to determine how long the recording provides useful data.

12. Apply R_s compensation (optional). Slowly adjust the R_s compensation knob on the patch clamp amplifier to increase the percent compensation while watching the test-pulse (5/sec) response. Record the R_s compensation setting for later correction of the R_s value noted in step 11 (typical time: 15 to 30 sec). Reduce the R_s compensation if oscillations start and allow a 10% to 20% safety margin between the level of R_s compensation and the instability threshold.

13. Choose a minimum acceptable value for R_s, discard recordings with higher values (<5 MΩ when large fast currents are being measured, or as high as 15 MΩ when the demands are less stringent), and initiate a new attempt with another cell.

14. Proceed to data acquisition and measurement of pulse sequences (see Basic Protocol 2).

BASIC PROTOCOL 2

DATA ACQUISITION AND PULSE SEQUENCES

In the study of voltage-gated channels, experiments usually involve application of some form of voltage step, generally in a carefully planned sequence. These pulses are almost always applied by a computer through an analog output connected to the stimulus input of the amplifier. The software packages presently available for voltage-clamp applications provide convenient user interfaces for designing pulse protocols. The software is also used to specify the digitization frequency for data acquisition and the bandwidth of a digital filter. The inverse of the filter corner frequency should be shorter than the time for the fastest channel gating transitions of interest, and the sampling frequency should be two or more times faster than the filter frequency. There is no point in filtering or sampling much faster than the time constant determined by $R_s C_c$, because the clamp is incapable of following more rapid processes.

Pulse Sequences

The pulse sequence is generally the heart of a study of voltage-gated channels, and it is sound practice to attend carefully to its design before starting an experiment. In most experiments, the holding potential and voltage of the test pulse are selected to change the channels from a closed state to an open one. Examples of standard pulse sequences are shown in Figure 1.2.3. To study the voltage dependence of channel activation, a series of steps are applied from a negative holding potential to a series of progressively more positive voltages producing progressively greater activation. Steps should be of sufficient duration to allow current to reach a desired endpoint, and are typically separated by intervals of 10 mV (Fig. 1.2.3A). To study the voltage dependence of inactivation, the holding potential prior to a depolarizing pulse is varied (often referred to as an h-infinity sequence; Fig. 1.2.3B). If a channel is inactivated

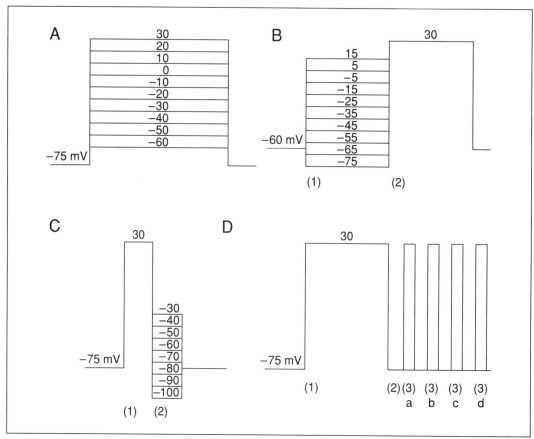

Figure 1.2.3 Pulse sequences for voltage-clamp experiments. Sequences such as these can be designed with software commonly used for whole-cell voltage clamping. (**A**) One of the simplest pulse sequences is a series of steps from the holding potential (−75 mV in this case) to −60, −50,..., and 30 mV. More positive voltage steps generally open more channels, but the voltage range where this occurs will vary depending on the type of channel under study. (**B**) A sequence of prepulses (1) varies the potential prior to a step to 30 mV. The step to 30 mV (2) activates channels that were not inactivated during the preceding pulse. Thus, the current evoked by the 30-mV pulse provides an assay for the extent of inactivation during the prepulse. (**C**) The voltage is varied after a 30-mV pulse to study the voltage dependence of tail currents. The pulse to 30 mV (1) activates the channels. The closing process can then by followed during the subsequent step to voltages ranging from −100 to −30 mV (2). For a K$^+$ channel with $E_k = -80$ mV, this sequence would move the voltage through E_k and reverse the tail currents. (**D**) Pulses presented in pairs separated by different time intervals are designed to study the kinetics of recovery from inactivation. Four pairs of pulses are indicated. The first pulse of the pair to 30 mV (1) is the same each time and inactivates most of the channels. The voltage is then returned to −75 mV (2) for variable amounts of time to allow variable amounts of recovery. Subsequent pulses to 30 mV (3a to 3d) elicit currents that reflect the amount of recovery that has occurred in the intervening interval.

at positive voltages, then each step to an increasingly positive potential will be followed by less current activation during the subsequent test pulse to 30 mV (Fig. 1.2.3B).

Figure 1.2.3C shows a pulse sequence designed to investigate how tail currents vary with voltage. After a positive test pulse to activate current, returning to a negative potential will close the channels. The decaying current after such a pulse is called a tail current. Reversal potentials of tail currents can be determined with a sequence like that in Figure 1.2.3C, and the reversal potential will be that of the channels that are open at the end of the immediately preceding voltage step. Thus, tail-current reversal potentials can be used to characterize the ion selectivity of the channel under study. Tail currents are also very useful in studying the kinetics of channel closure or deactivation.

In some voltage-clamp experiments the voltage is varied linearly rather than stepwise, to produce a voltage ramp. For noninactivating or slowly inactivating channels this method has the advantage of rapidly providing an indication of the voltage range in which a channel activates.

To study the kinetics of recovery from inactivation, the interval between the first inactivating pulse and the second test pulse can be varied (see Fig. 1.2.2D). To study the voltage dependence of recovery from inactivation one would use a similar sequence but with different voltages applied during the recovery interval.

Leak Subtraction

Currents measured during a voltage-clamp experiment are a sum of channel current and leak current. Every membrane passes some current by leakage through voltage- and time-independent pathways. This current may reflect ion permeation through lipid bilayers or through poorly defined pathways involving unspecified membrane proteins. Leak current can be determined with voltage pulses applied in a range where the voltage-dependent channels are closed. The leak pulse sequence should be applied shortly before the experimental pulse sequence to avoid the problems from a slow change in membrane properties.

Leak subtraction is based on the assumption that leak current is linear so that it can be scaled to match the voltage change used to gate channels, and then subtracted from the current activated by a test pulse. This results in what may be considered a more pure channel current. Quite often the leak pulses are applied at a more negative voltage, say -120 to -140 mV. The design of a leak subtraction protocol is best made after examining the voltage dependence of membrane conductances in the cell under study to locate a window of voltage where nothing happens, i.e., where the voltage-gated channels remain closed. This will ensure that the leak pulses do in fact measure leak current and not voltage-gated channel current. The amplitudes of the leak pulses are a small fraction, say 20%, of the amplitudes of pulses used to gate channels. Because leak current is often small, it is common to average several leak pulses. One of the most common leak subtraction protocols is called the P/4 procedure, where current evoked by four successive pulses at one-fourth the amplitude of a test pulse are added together and then subtracted from the test-pulse response. P/N more generally refers to a procedure where the test pulses are divided by an integer N and applied N times to determine the leak current. Because leak subtraction is a common correction in voltage clamping, some commercially available voltage clamp software (e.g., Pulse from Instrutech or ALA Scientific Instruments and pClamp from Axon Instruments) includes an easily implemented leak subtraction feature, which can be performed on-line.

In addition to providing a more accurate measure of channel current, leak subtraction also produces a cosmetic improvement in the appearance of current traces by removing unsightly capacitance artifacts that remain despite adjustment of the transient cancellation circuitry of the patch-clamp amplifier (Basic Protocol 1, step 8). However, the smooth appearance of the current onset after leak subtraction must be viewed with caution. In such cases the current onset may simply reflect the clamp settling time, and therefore will provide little information about channel activation kinetics.

BASIC PROTOCOL 3

ANALYSIS OF PATCH CLAMP DATA

Analysis is an essential part of a voltage clamp study, and the computer programs used for controlling a voltage-clamp experiment have considerable analysis capability, or come with a separate analysis program.

Standard Analysis

The most common first step of analysis is plotting current versus voltage (perhaps from data obtained with a pulse sequence of the form in Fig. 1.2.3A). These current-voltage (I-V) plots generally reveal the voltage range that activates the channel. Further, when the I-V plot exhibits a reversal potential, the permeant ion can also be identified. Thus, such a plot provides an indication of the success of a current separation strategy. Although I-V plots come in a bewildering variety of forms and shapes, the confusion is reduced considerably when one realizes that these plots are usually a product of two very simple functions, single-channel current and channel open probability. Interpretation of an I-V plot in terms of these basic membrane properties is an essential first step in establishing the basic properties of a channel under study.

Once the permeant ion is known, the current can be divided by the driving force (the voltage minus the reversal potential) to give the conductance, which is directly related to the fraction of channels that are open. A plot of conductance versus voltage provides a direct view of the voltage-induced gating of the channel. These plots are generally interpreted in terms of specific models. The Boltzmann equation derives from a simple model for voltage gating based on the assumption that a channel undergoes a voltage-dependent transition between two states (see Hille, 1992). Current can also be fitted to the Boltzmann equation in the analysis of steady-state inactivation using a pulse sequence such as that shown in Figure 1.2.3B. Here the driving force is the same for each pulse, so changes in current will reflect changes in conductance.

In kinetic studies of membrane current, voltage-clamp data is analyzed by fitting a plot of current versus time to one or more exponentials. These fits generally yield time constants for the underlying kinetic processes. Quite often current through different types of channels can be resolved as distinct exponential components of current in a complex multiexponential process. It was found by Hodgkin and Huxley (1952) that currents activate with sigmoidal kinetics. Nearly all voltage-gated channels exhibit similar behavior, with a time course that is best represented by an exponential raised to an integral power. The exponents that characterize this sigmoidicity are also regarded as important parameters reflecting the molecular properties of the channel under study. For further discussion of analysis, see Heinemann (1995).

Cable Analysis (optional)

In cells with processes, cable analysis can be useful either in the characterization of cell morphology or in the evaluation of space-clamp efficacy. The procedure for cable analysis described here is based on the assumption that the processes of the cell behave as a single cylinder with a sealed distal end and an electrotonic length equal to L. A current response is recorded to a 5 to 10 mV test pulse of 20 msec in duration. Because the slow components of the transient (which contain the most information about the process) are small in amplitude, it is generally advisable to average responses to as many as 50 pulses to reduce noise. The pulses can be applied at short intervals so that the entire recording can be completed in <10 sec. The software should make it possible to perform the averaging online. The test pulses should be applied in a voltage range where voltage-dependent currents are not active (typically more negative than −80 mV). To check that the response is purely passive, the decays at the onset and the end of the pulse should be compared. The averaged current is then fitted to a sum of as many exponentials as necessary (usually three or four, not including the rapid cell body-charging component balanced out in step 5). L and τ_m (the membrane time constant) can then be calculated from the slowest time constant, τ_1, and second slowest time constant, τ_2, with the following formulae.

$$L = \frac{\pi}{2}\sqrt{\frac{9 - \frac{\tau_1}{\tau_2}}{\frac{\tau_1}{\tau_2} - 1}}$$

$$\tau_m = \tau_1\left(1 + \left[\frac{3\pi}{2L}\right]^2\right)$$

Contributor: Meyer B. Jackson

UNIT 1.3

Preparation of Hippocampal Brain Slices

BASIC PROTOCOL

PREPARATION OF ACUTE MAMMALIAN HIPPOCAMPAL SLICES

Materials (see APPENDIX 1 for items with ✓)

✓ Dissection buffer, chilled to 4°C and carbogenated with 95% O_2/5% CO_2 gas mixture
  95% O_2/5% CO_2 gas mixture (carbogen)
  Animal
  Anesthetic (e.g., halothane)

19 × 16 × 10–cm Nalgene utility box
Tygon tubing (of an appropriate diameter to fit the connector on the air stone)
Soldering iron
Silicon bathtub sealant (Home Depot)
Aquarium air stone
Large rubber stoppers
Rectangle of 1/8-in.-thick Lucite (see step 1b)
35-mm plastic petri dishes
Whatman no. 1 or no. 2 filter paper cut to 37 × 37–mm square
Time tape
Rongeurs
Tissue sectioner (Stoelting) and double-edged razor blade
Small-animal decapitator (Harvard Apparatus)
2 small Teflon-coated pointed weighing spatulas
Small sharp-nosed dissecting scissors
No. 2 soft artist's paintbrush (e.g., white sable)
Wide-bore pipet (plastic transfer pipet cut off at ∼2/3 of length from the tip, or Pasteur pipet with taper cut off and with cut end fire polished)

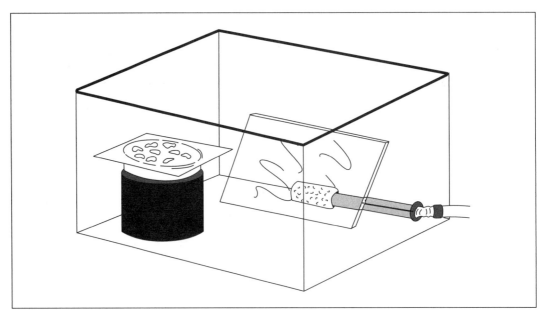

Figure 1.3.1 Diagram of the slice incubation box. Preparation of the box is described in step 1.

1. Prepare the slice incubation box (see Fig. 1.3.1):
   a. Make a hole in the side of a 19 × 16 × 10–cm Nalgene utility box, near the bottom, by melting with a soldering iron. Run a Tygon tube through the hole and seal with silicon bathtub sealant. Inside the box, connect an aquarium air stone to the tube. Outside the box, connect the tube to a 95% O_2/5% CO_2 regulator.
   b. Place one or more large rubber stoppers inside the box to serve as a pedestal for the slices. Fill the box with water to a depth that covers the stone, but not the pedestal. Lean a rectangle of 1/8-in.-thick Lucite against the inside of the box (to act as a splash guard), partially covering the stone.

2. Prepare the slice incubation surface: Place the deeper half of a 35-mm plastic petri dish inside the slice incubation box, on top of the pedestal. Fill the dish to overflowing with dissection buffer (chilled, if desired, and precarbogenated) and set a 37 × 37–mm square of Whatman no. 1 or no. 2 filter paper on top of the liquid-filled dish, making sure that it is centered. Put the lid on the incubation box and make sure carbogen is flowing into the box.

3. Prepare the tissue sectioner: Tape a rectangle of filter paper (sized such that it does not overhang the edges of the cutting surface) onto the cutting surface. Put the razor blade on the chopper and push the blade gently into the filter paper to straighten it; tighten the holding nuts. Make sure the stop beneath the blade arm will not prevent the blade from cutting all the way through the paper.

   CAUTION: *For reasons of safety install the blade the last step before beginning and remove it the first step after finishing the dissection. Make sure that the blade guard is on right side up, or the other cutting edge of the double-edged razor blade will be exposed above the cutting arm.*

4. Anesthetize the animal using a volatile anesthetic such as halothane, and decapitate. Remove the brain from the animal: Slice the scalp down the midline with a scalpel, and then carefully cut the skull along the midline with a fine pair of scissors, taking care not to cut the underlying brain tissue. Remove the skull by pulling it to the sides with rongeurs. Using a fine weighing spatula, scoop the brain out of the brain case.

5. Put the isolated brain into ice-cold carbogenated dissection buffer for a few seconds, then place the brain on a wetted flat absorbent dissection surface (e.g., filter paper). Remove the cerebellum and bisect the remaining brain along the midline with a scalpel (see Fig. 1.3.2A). Put one half back into chilled dissection buffer.

6. Turn the hemibrain so that the medial surface (i.e., the cut surface) is facing up (see Fig. 61.3.2B). Gently hold down the hemibrain with the flat end of one Teflon-coated spatula while using the other spatula to peel the neocortex off of the brain, exposing the hippocampus (see Fig. 1.3.2C).

7. Once the hippocampus is fully exposed, cut the fornix and then gently work one spatula underneath the fimbria, being careful not to cut into the hippocampus. Simply rotate the

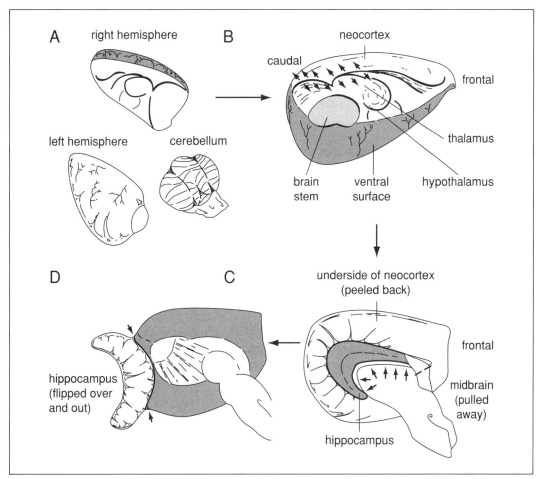

Figure 1.3.2 Dissection of the hippocampus. (**A**) The cerebellum/brain stem is cut off and discarded. The cerebrum is bisected along the midline, separating the two hemispheres. (**B**) The left hemisphere with the medial surface facing up. The neocortex is peeled off toward the caudal surface, and the midbrain is pulled ventrally by pulling in opposite directions at the location marked by the arrows. (**C**) The dentate surface of the hippocampus is revealed. The fornix is cut by pushing the point of the spatula into the brain at the point indicated by the dotted line. The spatula is inserted gently under the fimbria and further under the hippocampus starting at the caudal (temporal) end of the hippocampus and worked toward the septal end (indicated by the arrows). (**D**) The hippocampus is flipped out of the brain by lifting and pushing on the spatula, and then rotating the spatula tip around the long axis of the hippocampus. It is trimmed at the line indicated by the two arrows, and the rest of the brain is pulled away.

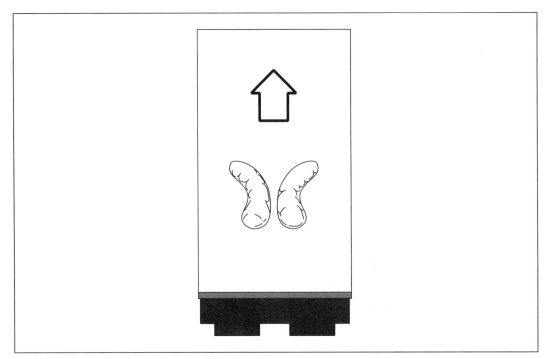

Figure 1.3.3 Placement of the hippocampi on the stage of the tissue sectioner. The left and right hippocampi are shown on the left and right sides of the stage. The direction of stage movement is indicated by the arrow. The plane of the cutting blade is perpendicular to the direction of the arrow.

spatula around the longitudinal axis of the hippocampus to flip it over and out of the brain. Cut along the edges of the hippocampus to separate it from the brain and place the hippocampus into chilled dissection buffer (see Fig.1.3.2D). Repeat with the other hemibrain.

8. Wet the filter paper covering the cutting surface of the tissue sectioner. To produce the right wetness, flood the paper with dissection buffer and make sure it is lying flat on the cutting surface without bubbles underneath. Using a no. 2 artist's paintbrush, brush excess buffer off of the paper until the texture of the paper is just visible when viewed at a shallow angle.

9. Immediately place the hippocampi on the cutting surface, smooth side up, with the septal end closer to the blade. Place them so that the long axis is parallel to the direction of stage movement (i.e., perpendicular to the blade), but let the septal end curve away from the parallel along the natural curve of the hippocampus (see Fig. 1.3.3).

10. Advance the stage of the sectioner 0.5 mm (or desired thickness of slice) and drop the blade from a height of ∼1 to 2 cm to cut the tissue.

11. Gently lift the blade. Using a very wet artist's paint brush, brush the slice off the blade by putting the brush flat against the blade surface, above the slice, and brushing straight down. Place the slice in a dish of chilled carbogenated dissection buffer. Repeat to obtain the desired number of slices or until the entire hippocampus has been sliced.

12. Use a wide-bore pipet to suck the slices out of the dish and deposit them onto the wetted filter paper in the incubation box, simultaneously flooding the paper with excess buffer. While the slices are free to move around in the buffer, push them gently with the paint brush, by their edges only, to distribute them evenly on the filter paper. Place the lid back

on the box, and let the slices sit for ≥1 hr before transferring one to the recording chamber for data collection (*UNIT 1.2*).

References: Alger et al., 1984; Madison, 1992; Yamamoto and McIlwain, 1966a, b

Contributors: Daniel V. Madison and Eleanore B. Edson

UNIT 1.4

Synaptic Plasticity in the Hippocampal Slice Preparation

The principal means by which the nervous system adapts to an organism's external environment and individuals learn from experience is by alterations in the efficiency of communication at their synapses in the brain. This synaptic plasticity is usually studied using electrophysiological techniques (*UNIT 1.5*). The hippocampal slice is the principal preparation for studying synaptic plasticity. This unit highlights the special considerations required to effectively study synaptic plasticity in the hippocampal slice. These considerations also apply to other preparations, such as slice preparations from neocortex and cerebellum.

TYPES OF SYNAPTIC PLASTICITY

Paired-Pulse Plasticity

The simplest forms of plasticity are paired-pulse facilitation (PPF) and its counterpart paired-pulse depression (PPD). Most excitatory synapses in the hippocampus exhibit PPF, which is defined as an increase in the size of the synaptic response to a second pulse delivered within a short interval of time following the first pulse. PPF is maximal at short (e.g., 50 msec) interstimulus intervals and declines exponentially over a period of ∼500 msec. It is well established that PPF is a purely presynaptic phenomenon and is often used to control or monitor for presynaptic changes, for example during long-term potentiation (LTP).

PPF should ideally be studied under voltage clamp conditions (*UNIT 1.2*) with synaptic inhibition blocked pharmacologically (e.g., Isaac et al., 1998). Under these conditions, PPF is a very powerful technique. For example, it can be used in conjunction with minimal stimulation to test for the activation of single fibers, as described (see Studying LTP and LTD Using Intracellular Recording). If membrane potential is allowed to change, then nonlinear summation of synaptic potentials needs to be taken into account. (Nonlinear summation means that the second response appears smaller than it should because it occurs at a membrane potential that is closer to its reversal potential.) If paired pulses are delivered when synaptic inhibition is not blocked, then a complex situation arises where the second pulse is strongly affected by synaptic inhibition. Because several parameters change under these conditions, this approach is not recommended.

In contrast to the other major inputs in the hippocampus, the medial perforant pathway (MPP; Fig. 1.4.2A) projection normally exhibits PPD. This is probably because the initial probability of release (Pr; the likelihood that an invading action potential results in the release of neurotransmitter) is high.

Inhibitory synapses that release γ-aminobutyric acid (GABA) also exhibit PPD. This is an active process controlled by $GABA_B$ autoreceptors. GABA released by the first stimulus feeds back onto presynaptic GABA terminals where it activates $GABA_B$ autoreceptors to suppress subsequent release of GABA. The effect has a characteristic time course determined by the time course of the G protein-coupled receptor. It has an onset latency of ~20 msec, is maximal at ~200 msec, and lasts for ~5 sec. This timing is critical for certain forms of LTP, such as that induced by priming or theta burst patterns of stimuli (see Induction of NMDA receptor-dependent LTP).

Post-Tetanic Potentiation

Following a brief period of high-frequency stimulation (i.e., a tetanus), there is a rapidly decaying potentiation, often referred to as post-tetanic potentiation (PTP). PTP generally lasts for ~1 min following a tetanus and is also thought to be due to presynaptic changes, similar to those that underlie PPF.

Long-Term Potentiation

The most extensively studied form of synaptic plasticity in the brain is LTP, a collective term that refers to a variety of different processes. Generic LTP is induced rapidly, within a minute or so of a conditioning stimulus, and is characterized by a persistent increase in the size of the synaptic response that often lasts for as long as the slice preparation remains viable (i.e., many hours; Fig. 1.4.1A).

Most LTP experiments involve extracellular or intracellular measurement of synaptic transmission mediated by *RS*-2-amino-3-(3-hydroxy-5-methyl-4-isoxazolyl) proprionic acid (AMPA) receptors. Synaptic potentiation refers to an increase in the size of the excitatory postsynaptic current (EPSC) when recorded in voltage clamp mode. Alternatively, there can be an increase in the corresponding excitatory postsynaptic potential (EPSP) when recorded intracellularly in an unclamped cell, or in the field EPSP (fEPSP) when recorded extracellularly as a population event (Fig. 1.4.2).

E-S (EPSP-spike) potentiation refers to the increase in action potential firing for a given size of EPSP or the field potential equivalent, which is an increase in population spike amplitude for a given size of fEPSP. Synaptic and E-S potentiation utilize different mechanisms; for this reason it is not advisable to use population spikes as a measure of LTP unless E-S potentiation is specifically under investigation. Practically all LTP experiments concern synaptic potentiation, and so E-S potentiation will not be considered further here.

LTP is often stable from its induction until recordings are terminated many hours later, so it appears that a single process has been induced. However, there are good reasons to believe that there are several phases to LTP. In some experiments, LTP decays back to baseline within ~30 min. This early decremental form of LTP is sometimes referred to as short-term potentiation (STP). LTP is also often subdivided into early and late LTP (E-LTP and L-LTP) based on sensitivity to protein synthesis inhibitors. E-LTP, which is resistant to protein synthesis inhibition, typically lasts for ~3 hr.

It is also possible to study LTP of synaptic responses mediated by *N*-methyl-D-aspartate (NMDA) receptors. Because NMDA receptors contribute very little to the response evoked by test stimuli under normal experimental conditions, it is necessary to make pharmacological manipulations to study this form of LTP. The best way to study this response is to block AMPA receptor-mediated excitatory synaptic transmission with AMPA receptor antagonists, and block synaptic inhibition with GABA receptor antagonists (Bashir et al., 1991). This allows one to work with synaptic responses (ideally EPSCs) that are mediated purely by NMDA receptors.

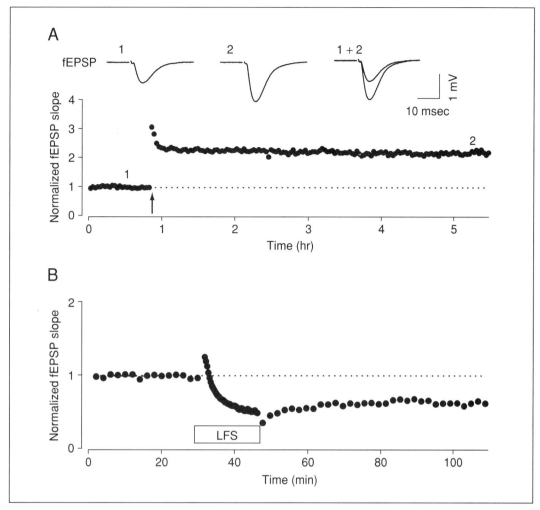

Figure 1.4.1 LTP and LTD studied using field potential recording. (**A**) A typical LTP experiment in a slice obtained from an adult rat. The traces at the top of the graph show the field excitatory postsynaptic potential (fEPSP) responses extracted from the plot at the times indicated on the graph (1, 2). These responses are superimposed in the third trace. A single tetanus (100 Hz, 1 sec, test intensity) was delivered at the time indicated by the arrow. Note that this induced LTP showed no sign of abatement after >4 hr. (**B**) A typical de novo LTD experiment in a slice obtained from a 14-day-old rat induced by low-frequency stimulation (LFS; 900 shocks at 1 Hz, test intensity).

Long-Term Depression

Just as LTP is a collection of related phenomena, so is long-term depression (LTD). It is the converse of LTP, namely a persistent decrease in AMPA (and, when measured, NMDA) receptor-mediated synaptic transmission (measured as field responses, intracellular potentials, or currents; Fig. 1.4.1B). Under certain circumstances, LTD cannot be induced unless LTP is first induced, in which case the process is generally referred to as depotentiation. When LTD is induced in naive tissue, it is sometimes referred to as de novo LTD. These distinctions can be helpful because the mechanisms underlying depotentiation and de novo LTD may be different.

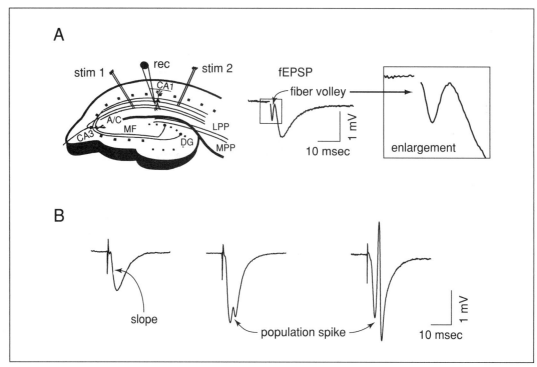

Figure 1.4.2 Field potential recordings in hippocampal slices. (**A**) Electrode placement for two-input LTP or LTD experiments in the CA1 region. The trace is a fEPSP recording from stratum radiatum where there is a pronounced presynaptic fiber volley. The enlargement of this region shows a magnification of the fiber volley. (**B**) fEPSP traces showing (from left to right) a typical fEPSP, the fEPSP evoked at an intensity where a population spike is detectable, and an fEPSP with a large population spike affecting the fEPSP amplitude. Note that large fEPSPs relative to fiber volleys, such as those illustrated in B, are indicative of healthy slices with good connectivity. The trace in A is not representative of a healthy response but is shown to illustrate the fiber volley more clearly. Note also in B the smooth decay of the fEPSP and the single population spike (rather than multiple population spikes) at high-intensity stimulation, indicative of a healthy slice displaying good synaptic inhibition. Abbreviations: A/C, associational-commissural pathway; DG, dentate gyrus; LPP, lateral perforant pathway; MF, mossy-fiber pathway; MPP, medial perforant pathway; rec, recording electrode; stim 1 and stim 2, first and second stimulating electrodes.

Homosynaptic versus Heterosynaptic Plasticity

The forms of plasticity referred to so far are homosynaptic. That is, if two separate inputs converging on the same neuron or set of neurons are stimulated and LTP or LTD is induced in one of them, then there is no change in the other (nonconditioned) input. This input specificity is a hallmark feature of LTP and LTD. However, under certain circumstances, the induction of LTP in one input is associated with LTD in a separate input. Heterosynaptic depression may play an important role in the nervous system but is rarely the subject of study in hippocampal slices and so is not discussed further here.

THE USE OF FIELD POTENTIAL RECORDING TO STUDY LTP

Because of the organization of the hippocampus, field potentials are very large and easy to measure above the noise and, because their origins are well established, relatively easy to interpret.

Recording Configurations

The usual recording arrangement is to place a microelectrode (3 to 6 MΩ), usually filled with a concentrated (e.g., 4 M) NaCl solution, into a dendritic region of the slice and activate the afferent excitatory pathway using a stimulating electrode, such as a twisted insulated bipolar metal electrode (e.g., NiCr) placed approximately the same distance from the cell body region of the slice (Fig. 1.4.2A). The stimulating electrode activates a parallel band of fibers that evoke, first, a presynaptic fiber volley (caused by the inward current of the action potentials in the presynaptic fibers; Fig.1.4.2A) and, second, an fEPSP. LTP is seen as an increase in the size of the fEPSP. When the fEPSP reaches a certain size, a population spike (reflecting the synchronized firing of action potentials) is evoked. The appearance of the population spike complicates the measurement of fEPSP amplitude because it goes in the opposite direction and limits the amplitude of the fEPSP (Fig. 1.4.2B). Therefore, fEPSP slope is the preferred measure of response size as it is generally unaffected by the population spike.

If population spikes are to be measured, it is advisable that the electrode be placed in the cell body region of the slice, at the site of origin of the response. Population spike amplitude is a measure of the number of action potentials evoked and the synchrony of these spikes. It is possible to measure the somatic and dendritic field potentials simultaneously with two independent recording electrodes. However, most investigators simply record the dendritic response, which provides the most pertinent information for LTP studies.

An occasionally used variant of extracellular recording is the grease-gap technique, originally used to record evoked potentials in the spinal cord, but is equally applicable to brain. The potential difference across a grease barrier acts like a large extracellular electrode. The disadvantages of the technique are that the location of the grease barrier is fixed, whereas electrodes can be readily repositioned, and that the precise origin of the potentials is more difficult to establish. Nevertheless, grease-gap recording has the advantage of providing an extremely stable, low-noise configuration and allows agonist-induced and synaptically evoked potentials to be readily recorded at the same time.

Optimizing Synaptic Responses

A critical determinant of the value of LTP or LTD experiments is the quality of the synaptic responses. Care should therefore be taken to optimize the health of the slice, the level of synaptic connectivity, and the positions of the stimulating and recording electrodes. The stimulus frequency should be kept low (0.067 to 0.033 Hz) while optimizing the responses, because increasing the stimulus rate while searching for good responses can lead to metaplasticity (see Metaplasticity). A good synaptic response should comprise a large fEPSP, relative to the fiber volley, with a smooth decaying phase, indicative of good synaptic inhibition (Fig. 1.4.2B).

Baseline Stability

It is imperative that the baseline be absolutely stable if LTP is to be interpreted meaningfully. The duration of a stable baseline that is required before conditioning is somewhat dependent upon how long LTP is to be studied. Consider an LTP experiment lasting 3 hr postconditioning. If a baseline of 30 min is obtained and it drifts upwards by 10% during this time, then by simple linear extrapolation of the baseline, an increase of 60% would be expected 3 hr later. Drifting baselines are probably the biggest source of error in the LTP literature. Obtaining a stable baseline requires patience above all else. It is aided by software that enables on-line measurement of fEPSP slope.

Two-Input Experiments

A useful aid for baseline stability is to record the response to two independent inputs. Changes in the recording electrode position or health of the slice are reflected by changes in the response to both inputs. However, the method is not foolproof because one input can change independently of the other if, for example, a stimulating electrode moves or one set of fibers deteriorates selectively. Two-input experiments provide considerable versatility and are used for many purposes (for example, studying heterosynaptic effects) in addition to checking for baseline stability.

Stimulus Intensity

It is important to decide what stimulus intensity should be used to evoke the test responses and what intensity should be used to provide the conditioning. Wherever possible, it is preferable to use the same stimulus intensity for both test and conditioning stimuli. Altering the stimulus intensity to deliver the conditioning stimulus can lead to errors, particularly if the stimulator uses an analog intensity scale.

How does one determine the optimum stimulus intensity to obtain a baseline response The magnitude of LTP obtained is affected by the size of the test responses. If the stimulus intensity is set to evoke small responses, then there is a large range over which response size can increase and therefore, in percentage terms, LTP can be very large. If the stimulus intensity is set to evoke large responses, then the range for further increases is limited and the maximum obtainable LTP is correspondingly small. Note that the descriptors *small* and *large*, as used here, refer to the size of the response in relationship to the maximum response, because the absolute size of an fEPSP can vary enormously from preparation to preparation (typical maximum fEPSP slopes range from -0.7 to -1.5 V/sec).

Determining the maximum response is not straightforward because the population spike sets the ceiling level on the fEPSP amplitude, and the fEPSP slope increases nonlinearly with large responses. In addition, delivery of strong test responses can affect baseline stability. In particular, high-intensity single-shock stimulation can induce short-term plastic changes in the synaptic response as indicated by an increase in the size of the response. Rather than increasing the stimulus intensity to yield the maximum response, the stimulus intensity can be increased to the point where a population spike is just detectable in the fEPSP record. The test responses can then be set at a fixed proportion (e.g., 50%) of this stimulus intensity. At the 50% stimulus intensity, the responses are on a steep region of the input-output curve and are therefore highly sensitive to changes in both directions (LTP and LTD). This intensity is also very efficient at inducing LTP in healthy slices, obviating the need to alter the stimulus intensity during conditioning.

The type of stimulating electrode and stimulus isolation unit are also important considerations. Insulated metal bipolar electrodes provide a stable arrangement and generate relatively small stimulus artifacts. It is important to deliver a constant stimulus current, but as long as electrode resistance does not change, this can be achieved using a standard stimulation unit in which stimulus voltage is varied. Constant-voltage stimulus isolation units generally perform better than constant-current ones.

Input-Output Curves

Information about synaptic properties can be obtained by constructing input-output curves before and after conditioning. Useful information can be obtained from plots of fiber volley amplitude versus stimulus intensity, fEPSP slope versus stimulus intensity (or better still, versus fiber volley amplitude), and population spike amplitude versus fEPSP slope. However, using high stimulus intensities to construct input-output curves can affect baseline stability, so one should use high intensity stimulation with caution.

Induction Protocols

It is possible to induce LTP using a wide variety of protocols. However, the precise mechanisms of induction and, in certain circumstances, expression differ according to the induction protocol used. It is imperative that these differences be understood, particularly when making comparisons of results between studies. The induction of most forms of LTP in the hippocampus involves the activation of NMDA receptors, as discussed below.

Induction of NMDA Receptor-Dependent LTP

Probably the most common induction protocol is to deliver a single high-frequency train, commonly referred to as a tetanus. A widely used protocol is to deliver 100 shocks at 100 Hz, at test intensity. This protocol can readily induce LTP that lasts many hours (i.e., including L-LTP). The finding that the induction of this form of LTP is blocked by bath application of NMDA receptor antagonists such as 50 μM D-2-amino-5-phosphonopentanoate (D-AP5) demonstrates that it requires the activation of NMDA receptors.

In some studies, multiple tetani of this sort have been delivered over a short period of time. This approach has been used by some investigators in cases where a single tetanus was unable to induce L-LTP, possibly due to compromised slice viability. It is, however, possible that this approach induces a component of L-LTP that is not induced by a single tetanus. This component will be masked if L-LTP is readily evoked by a single tetanus.

LTP can be induced by far fewer than 100 shocks. Two related protocols are theta burst and priming. The former consists of delivering a few brief bursts of high-frequency trains with an interburst interval at a frequency within the theta range. A typical protocol is ten bursts of four shocks (each delivered at 100 Hz) with an interburst interval of 200 msec (a 5 Hz frequency). The advantage of this protocol over a single tetanus is that it mimics more closely the physiological activation patterns of hippocampal neurons during theta activity. In priming, only two bursts are used and the first is replaced by a single stimulus. Thus, a typical protocol is a single shock followed 200 msec (5 Hz) later by a single burst of four shocks delivered at 100 Hz. Indeed, as few as three appropriately timed stimuli (one shock followed 200 msec later by two shocks at 100 Hz) can induce a small LTP.

The cellular basis for the effectiveness of priming and theta burst LTP is that the initial response (priming stimulus or priming burst) depresses synaptic inhibition and the subsequent burst activates the NMDA receptor system. With multiple bursts, the first burst provides priming and the subsequent bursts activate the NMDA receptor system and prime subsequent bursts. Priming is very effective because GABA released from inhibitory interneurons feeds back onto $GABA_B$ autoreceptors to depress the release of GABA. Therefore, priming via activation of $GABA_B$ autoreceptors transiently reduces GABAergic inhibition. Accordingly, $GABA_B$ antagonists, applied in sufficient concentrations to prevent activation of $GABA_B$ autoreceptors, inhibit the induction of priming-induced LTP. However, with a tetanus (e.g., 100 Hz, 1 sec), additional changes occur (i.e., compensation of autoinhibition of GABA release, depolarizing shift in reversal potential of $GABA_A$ responses) that negate the autoreceptor mechanism. Therefore, the effectiveness of $GABA_B$ antagonists is dependent on the induction protocol used. This emphasizes the need to fully understand how the induction protocols work.

In many experiments, LTP is induced in the presence of a $GABA_A$ receptor antagonist, which can have a major effect on the outcome of LTP studies. For example, $GABA_B$ antagonists do not block the induction of LTP when the experiment is performed in the presence of a $GABA_A$ receptor antagonist. Therefore, the use of a $GABA_A$ receptor antagonist has the advantage of simplifying the process under investigation. On the other hand, a wide spectrum of modulatory influences that affect LTP induction via regulation of synaptic inhibition is not evident when LTP is induced in this manner. Most investigators use picrotoxin to block $GABA_A$ receptor-mediated inhibition. Although picrotoxin is not the most selective $GABA_A$ receptor antagonist available, it is stable in solution (a problem with other $GABA_A$ antagonists, for example, bicuculline methochloride) and has been used extensively without obvious nonspecific effects on LTP. Because the blockade of $GABA_A$ receptor-mediated inhibition facilitates the synaptic activation of NMDA receptors, fewer stimuli are required to induce LTP. Picrotoxin essentially acts as the priming stimulus. A potential complication with the use of a $GABA_A$ receptor antagonist is the predisposition to epileptiform activity, which can affect the induction of LTP. When studying LTP in the CA1 region of the slice, this problem is minimized by surgically severing connections with area CA3 (Fig. 1.4.2A), which is particularly prone to spontaneous epileptiform activity that otherwise propagates to area CA1. The simplest approach is to cut off area CA3 with a scalpel shortly after preparing the slice. Another procedure to stabilize the preparation is to raise the divalent cation concentration to, for example, 4 mM Ca^{2+} plus 4 mM Mg^{2+}. This can affect LTP induction in other ways (e.g., elevated Mg^{2+} provides greater inhibition of the NMDA receptor system at resting membrane potentials).

Induction of NMDA Receptor-Independent LTP

Within the hippocampus, all pathways exhibit NMDA receptor-dependent LTP, with the exception of the mossy fiber (MF) pathway. Indeed, a major problem when studying mossy fiber LTP is to ensure that there is no contamination of the LTP under investigation by NMDA receptor-dependent LTP. This is a serious concern because of the extensive innervation of CA3 neurons by projections from other CA3 neurons via associational-commissural fibers. These projections exhibit NMDA receptor-dependent LTP and could be activated directly by an inappropriately positioned stimulating electrode or polysynaptically via mossy fibers driving CA3 neurons. Although it is possible to study pure mossy fiber LTP in the absence of NMDA receptor blockade, the standard procedure is to perform experiments in the presence of an NMDA receptor antagonist (e.g., 50 μM D-AP5) and to ensure its effectiveness by studying associational-commissural LTP in parallel. Mossy fiber LTP is usually induced by one or more tetani. The standard protocol for inducing NMDA receptor-dependent LTP (e.g., 100 Hz, 1 sec, baseline stimulation intensity) is also very effective at inducing mossy fiber LTP (Fig. 1.4.3).

It is also possible to induce LTP without activating NMDA receptors in pathways where NMDA receptor activation is the normal means of inducing LTP (e.g., in area CA1). This NMDA receptor-independent LTP may be subdivided further into two forms. In the first, the LTP process induced is the same as that induced by NMDA receptor activation except that it is obtained by a mechanism that bypasses the need for NMDA receptor activation. For example, activating metabotropic glutamate (mGlu) receptors with the broad spectrum agonist (1S,3R)-1-aminocyclopentane-1,3-dicarboxylic acid (ACPD) can induce LTP, provided connections between CA3 and CA1 neurons are left intact. This form of LTP, which is not blocked by NMDA receptor antagonists, fully cross-saturates with NMDA receptor-dependent LTP induced by a tetanus. The second form of NMDA receptor-independent LTP is a different form of LTP entirely. For example, very-high-frequency stimulation (e.g., 200 Hz) in the presence of NMDA receptor blockers can induce a form of LTP that involves Ca^{2+} entry through voltage-gated Ca^{2+} channels and seems to involve different downstream transduction processes than standard tetanus-induced, NMDA receptor-dependent LTP.

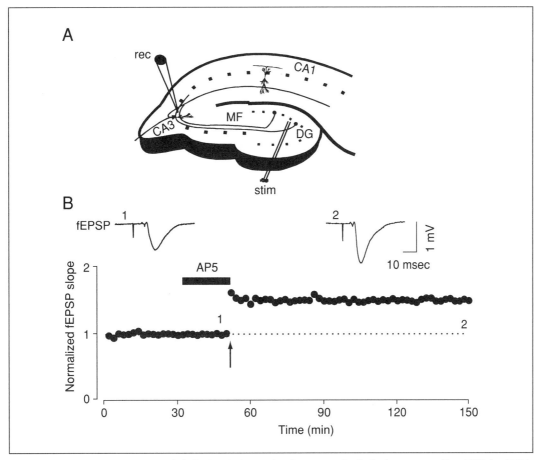

Figure 1.4.3 Mossy fiber LTP. (**A**) Optimal electrode placement for inducing mossy fiber LTP (or LTD). (**B**) A typical mossy fiber LTP experiment. Note that the tetanus (100 Hz, 1 sec, test intensity) was delivered in the presence of the NMDA receptor antagonist D-2-amino-5-phosphonopentanoate (D-AP5; 50 μM). The two traces in B show fEPSPs obtained at the times indicated on the graph (1, 2). Several procedures have been adopted to isolate mossy fiber projections but the best approach is to perform experiments in the presence of a sufficient concentration of an NMDA receptor antagonist to ensure full blockade of NMDA receptor-dependent LTP. By convention, the NMDA receptor-independent LTP then equates with mossy fiber LTP. Use of low stimulus strengths to evoke subthreshold synaptic responses (e.g., traces in B), coupled with appropriate electrode placement, greatly reduces the risk of inadvertently activating non-mossy fiber inputs. This requires very healthy slices with good connectivity. Abbreviations: DG, dentate gyrus; MF, mossy fiber; rec, recording electrode; stim, stimulating electrode.

THE USE OF FIELD POTENTIAL RECORDING TO STUDY LTD

The same considerations (e.g., the need for very steady baselines; advantages of two-input experiments) that apply to the study of LTP also apply to the study of LTD. Generic LTD (i.e., both de novo and depotentiation) can also be subdivided on the basis of sensitivity to NMDA receptor antagonists. NMDA receptor-independent LTD is often sensitive to mGlu receptor antagonists, such as (*S*)-α-methyl-4-carboxyphenylglycine (MCPG).

LTD is induced by more prolonged and lower-frequency stimulation than is used to induce LTP. A typical induction protocol is to deliver 900 stimuli at 1 Hz (i.e., for 15 min). LTD is commonly induced at frequencies between 1 and 5 Hz. A common observation is that de novo LTD is readily induced in slices obtained from young animals (<3 weeks old), but not in tissue obtained from adult animals. Depotentiation is readily induced in young animals, although interpretation of the experiment may be confounded by the simultaneous induction of de novo LTD. Depotentiation can also be obtained fairly readily in slices obtained from adult animals. In addition, de novo LTD can be obtained in slices obtained from adult animals if paired-pulse stimulation, rather than single-shock stimulation, is delivered during the conditioning phase.

LTD can be chemically induced more readily than LTP. For example, perfusion of NMDA can induce a small LTD that occludes NMDA receptor-dependent de novo LTD in slices obtained from young animals. The group I mGlu receptor agonist 3,5-dihydroxyphenylglycine (DHPG) readily induces a different form of LTD in slices obtained from both young and adult animals. DHPG-induced LTD is enhanced in magnitude by treatments that increase cellular excitability, such as the application of picrotoxin or the omission of Mg^{2+} from the perfusate.

METAPLASTICITY

A key feature of synaptic plasticity is that the process itself is plastic. The plasticity of synaptic plasticity has been termed metaplasticity (i.e., higher-order synaptic plasticity) and involves many different processes that are of major significance for LTP and LTD experiments. There are several examples of metaplasticity. First, activation of the NMDA receptor system can block the subsequent induction of NMDA receptor-dependent LTP. Second, activation of group I mGlu receptors can facilitate the subsequent induction of LTP. Third, more covertly, synaptic activation of mGlu receptors renders the subsequent induction of NMDA receptor-dependent LTP insensitive to the mGlu receptor antagonist MCPG (the molecular switch hypothesis; (see *http://www.bris.ac.uk/synaptic*; Fig. 1.4.4). In contrast, in naive tissue MCPG reversibly blocks the induction of LTP. These examples of metaplasticity show that the conditioning effects on LTP (and LTD) can be induced by treatments that have no overt effects on the preparation. It is therefore strongly advised that, unless metaplasticity is the subject of the investigation, all LTP and LTD experiments be conducted on experimentally naive slices.

BASIC PROTOCOL

EXTRACELLULAR LTP IN THE HIPPOCAMPAL SLICE PREPARATION

Materials (see APPENDIX 1 *for items with* ✓)

    Laboratory animal (according to the question under investigation; e.g., rat, mouse, guinea pig, hamster; most commonly Wistar and Sprague-Dawley rats and Sprague-Dawley mice)
    Anesthetic (e.g., a mixture of oxygen and halothane)
✓ Artificial cerebrospinal fluid (aCSF), 4° and 30°C
    Super glue
    Extracellular recording pipets (3 to 6 MΩ; UNIT 1.1) filled with 4 M NaCl or aCSF

0.05 to 0.08 mm diameter insulated NiCr wire (e.g., Advent Research Materials Ltd.)
Borosilicate glass capillary: e.g., 1.2-mm o.d./0.69-mm i.d. (e.g., Harvard Apparatus)
Glass microelectrode puller (e.g., model PP-830, Narishige Scientific Instruments)
Micromanipulators (e.g., Narishige Scientific Instruments)
Stimulus isolation unit for extracellular stimulation (e.g., DS2A Mk.II, Digitimer Ltd.)
Amplifier (either a voltage amplifier or patch clamp amplifier)

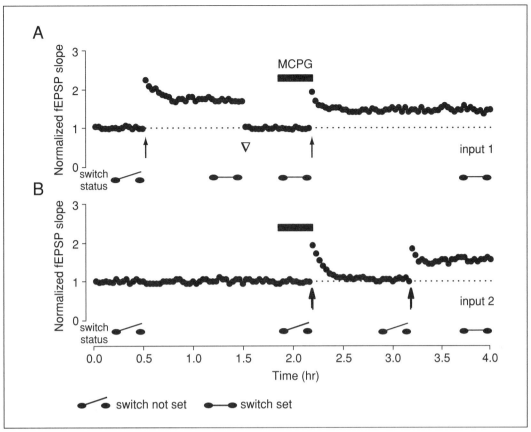

Figure 1.4.4 An example of metaplasticity. A two-input LTP experiment to demonstrate the existence of a molecular switch. For a two-input configuration, see Figure 1.4.2. (**A**) In input 1, the metabotropic glutamate receptor antagonist (S)-α-methyl-4-carboxyphenylglycine (MCPG; 200 μM) failed to block the induction of LTP. This input was conditioned by the first tetanus, which induced LTP prior to the application of MCPG. (**B**) In input 2, MCPG blocked the induction of LTP in a reversible manner. This input was naive (prior to the application of MCPG the pathway had received no experimental manipulation other than baseline stimulation). The data are interpreted as follows: for LTP to occur, a molecular switch needs to be set; this switch is sensitive to MCPG and so if the switch is not already set, MCPG prevents the induction of LTP. Once set, further LTP in the same input is resistant to MCPG. The status of the switch is shown diagrammatically. Arrows show times that tetanus (100 Hz, 1 sec, baseline stimulation intensity) was delivered. The open triangle shows the time when the stimulus intensity was reduced to obtain a new baseline of similar size responses as those obtained prior to the induction of LTP.

Surgical instruments for decapitation or cervical dislocation and dissection
Filter paper
Tissue slicer (e.g., Campden Instruments Ltd. or Leica)
Recording chamber (e.g., Fine Science Tools or Warner Instrument Corp.)
Perfusion system consisting of a container of aCSF placed in a temperature-controlled water bath (30°C) and polyethylene tubing that circulates the pre-warmed aCSF through a peristaltic pump (e.g., Harvard Apparatus) and the recording chamber
Computer with software for on-line acquisition and online and offline analysis of synaptic potentials (e.g., "LTP" Program freely available from the MRC Centre for Synaptic Plasticity, see *http://www.ltp-program.com*; other appropriate software from Axon Instruments or Burleigh Instruments; also see Data Acquisition and Analysis)
Dissecting microscope (e.g., Leica or Nikon)

1. Construct a stimulating electrode by twisting together insulated nickel/chromium wire (wire diameter 0.05 to 0.08 mm). Once twisted, solder the stimulating electrode to a thicker wire, affix it to a micromanipulator, and connect it to the stimulus isolation unit.

2. Construct a glass pipet recording electrode by pulling an appropriate size Borosilicate glass capillary with an electrode puller (*UNIT 1.1*). Fill the recording electrode with 4 M NaCl (for a higher quality signal) or aCSF (for better response in case of electrode leakage), attach it to the electrode holder, and then secure the holder to the head stage which is affixed to one micromanipulator. Connect the electrode to the amplifier.

3. Sacrifice a laboratory rat either by decapitation under anesthetic or by cervical dislocation without anesthetic. Immediately remove the brain using appropriate surgical instruments and place it in chilled (4°C) and oxygenated aCSF. Place the brain in a petri plate (preferably chilled) on filter paper and dissect away the cerebellum and the frontal cortex. Divide the remaining part of the brain at the central sulcus.

4. Using super glue, fix the two hemispheres laterally on the slicer's tissue supporter with the middle region facing upwards. Immediately place the mounted hemispheres inside the slicer's chamber and fill it with the chilled aCSF so the hemispheres are submerged.

5. Slice the hemispheres every 400 μm with the blade moving slowly from the dorsal surface towards the ventral surface of the brain to cut transverse hippocampal slices that keep the lateral perforant pathway (LPP), MPP, MF pathway, and associational-commissural (A/C) pathway (i.e., the trisynaptic pathways) intact (Fig. 1.4.2A). Use one blade for each brain, and slice >200 μm thick for young subjects and 200 to 500 μm thick for adults.

6. Dissect a hippocampal slice from the rest of the brain and leave it to recover at room temperature (25° ± 2°C) for ≥1 hr. Transfer the slice to a recording chamber attached to a perfusion system. Perfuse the slice with prewarmed (30°C) aCSF at a rate of 2 ml/min. Apply and wash out all test compounds using the perfusion system. Store remaining slices up to 10 hr in a petri dish or beaker containing cold aCSF that is continuously oxygenated with 95% O_2/5% CO_2.

   In a submerged recording chamber, the slice is submerged in aCSF ~3 mm below the surface, allowing fast diffusion of drug into the slice and a stable recording environment.

   In an interface recording chamber, the slice lies on a net and the aCSF runs underneath it, allowing easier access for placing electrodes on the slice and yielding a larger signal and less electrical noise with less stimulus intensity required to obtain a reasonable synaptic response. A humid atmosphere is created around the slice by the oxygenation (bubbling) of the solution surrounding the recording chamber.

   In either style chamber under ideal conditions, the slice can be kept in a healthy state for >10 hr.

7. Place the extracellular recording pipet(s) and stimulating electrode(s) in the stratum radiatum (between the thin, very shiny pyramidal cell body layer and the stratum lacunosum) of the hippocampal slice ~500 μm from each other. Carefully place the tip of the recording electrode(s) deep into the slice and position the stimulating electrode(s) so it presses against the surface, using micromanipulators and a microscope for visualization.

   Using two independent stimulating inputs allows one input to be used as an internal control while the other input is used for investigation, the investigation of the effect of drugs on synaptic plasticity or transmission both prior to and after perfusion, and a comparison of the effects of different stimulating protocols in the same slice at the same time.

8. Deliver test pulses (100 μsec and 5 to 25 V; 10 to 50 V for CA1 synapses) to evoke an fEPSP, starting with minimal test pulses and trying one or two tests for each location in the slice and for stimulus intensity. If no response is evoked, relocate the electrodes (often

only the stimulating electrode). Maintain a distance of 500 μm between the electrodes. Start again with a new slice if repeated attempts to evoke synaptic activity fail.

9. When a response is evoked, gradually increase the stimulus intensity until a population spike is generated (1.4.2B, second trace). Decrease the stimulus intensity until it evokes 50% of the maximum response and use this intensity for the whole experiment.

10. Begin stimulating once every 30 sec to obtain a baseline response. Be patient and establish a very steady baseline (usually ≥30 min) before starting any other procedure (Figs. 1.4.1, 1.4.3, and 1.4.4).

11. To induce LTP, deliver a high-frequency tetanus (100 stimuli at 100 Hz) at baseline stimulation intensity, causing field excitatory postsynaptic potential (fEPSP) to increase and the slope to increase and remain increased for the rest of the experiment (Fig. 1.4.1A).

12. To induce LTD, deliver a period of low-frequency stimulation (LFS; 900 stimuli at 1 or 2 Hz) at baseline stimulation intensity, causing the fEPSP to decrease and the slope to decrease and remain reduced for the rest of the experiment (Fig. 1.4.1B).

STUDYING LTP AND LTD USING INTRACELLULAR RECORDING

Methods for Induction of NMDA Receptor-Dependent Homosynaptic LTP

To induce LTP, the voltage-dependent blockade of the NMDA receptor channel by Mg^{2+} needs to be relieved at the same time glutamate is released from the presynaptic terminal. Using intracellular recordings, this can most reliably be achieved by using a pairing protocol in which the cell is artificially depolarized by the recording electrode during low-frequency electrical stimulation of axons. This method sidesteps many of the issues that have to be considered for induction protocols for extracellular recordings. The main issue is that an experimental manipulation that blocks LTP may actually be due to a general change in presynaptic function rather than a specific direct effect on LTP. This would make a high-frequency LTP induction protocol of the sort used for extracellular recordings less effective at inducing LTP. By avoiding the use of high-frequency stimulation with a pairing protocol, this problem is greatly reduced. In voltage clamp, a pairing protocol typically consists of 100 afferent stimuli at 1 Hz with a postsynaptic holding potential of 0 mV. However, because the recording electrode sets the level of postsynaptic depolarization, the frequency of afferent stimulation is not critically important. Pairing is equally efficient with other frequencies (e.g., 0.2, 0.5, 2 Hz). Similar pairing protocols can also be used in current clamp mode in which a depolarizing current injection, used to bring the membrane potential close to 0 mV, is combined with 1 Hz stimulation. For voltage clamp mode, it is not necessary to stimulate at very high frequencies (e.g., 100 Hz), and lower frequencies have the advantage that they are less likely to cause changes in axon excitability. However, in current clamp mode, a tetanic stimulation protocol of 100 Hz for 1 sec in the absence of postsynaptically applied depolarization can be used. This latter protocol has been used in the past because it is identical to the induction protocol for extracellular recordings; however, it has no advantages over a pairing protocol and suffers from the same potential problems as the high-frequency induction protocols used for extracellular recordings as described above.

Under whole-cell recording conditions (UNIT 1.8), the ability to induce LTP is lost soon after the start of the recording due to the dialysis of the cell with the intracellular solution (so-called washout). It is not known precisely which intracellular component(s) that is critical for LTP induction is washed out or inactivated by the whole-cell solution. The rate at which washout occurs depends upon the access resistance (the electrical resistance of the recording electrode in the whole-cell configuration) of the recording (for lower access resistance recordings, LTP washes out faster) and the distance of the synapse from the recording electrode. For a CA1 pyramidal cell (in slices from 2- to 3-week-old rats), LTP washes out within 5 to 10 min from the start of the recording for an access resistance of $\sim$15 MΩ. Washout can be a major problem for studying LTP using whole-cell recordings and limits the length of time one can collect a baseline before inducing LTP. One way to avoid this problem is to use perforated patch clamp recordings or sharp microelectrode recordings. Of particular relevance to washout are studies on whether the introduction of a reagent via the patch electrode into a neuron blocks LTP. For these sorts of studies, it is very important to ensure that LTP induction is blocked by the reagent and is not simply absent due to washout.

Methods for Induction of NMDA Receptor-Dependent Homosynaptic LTD

Induction of LTD requires moderate Ca^{2+} influx through NMDA receptors, although less than for LTP. Similar to LTP, this can most efficiently be achieved for intracellular recordings by a pairing protocol in which the membrane potential is artificially set by the recording electrode during afferent stimulation. Compared with LTP induction, a less-depolarized potential is required to limit the amount of Ca^{2+} entering the cell during the pairing. In voltage clamp, a pairing protocol of 100 to 200 afferent stimuli at 1 or 0.5 Hz with a holding potential of -40 mV is very effective at inducing LTD of naive pathways in hippocampal slices from young rats (up to 18 days old) along with depotentiation of previously established LTP. In current clamp, 600 or 900 stimuli at 1 Hz (similar to the protocol used for induction of LTD using extracellular recordings) induce LTD and depotentiation in slices from young animals. Conveniently, the ability to induce LTD does not appear to wash out even after extended dialysis. This means that baseline data can be collected for longer periods of time and the effects of intracellular infusion of reagents on the ability to induce and express LTD can be easily studied.

Use of Minimal Stimulation and Failures Analysis for Studying the Expression Mechanisms of LTP and LTD

Minimal stimulation means that a very low stimulation intensity—just sufficient to evoke the smallest detectable synaptic response—is used to activate one or a few synapses. Using this approach the stochastic nature of transmitter release is revealed. That is, when an action potential arrives at the presynaptic terminal there is a finite probability that it will cause transmitter release (i.e., probability of release, Pr). This technique also allows one to estimate the number of activated synapses (n) and the size of the synaptic response when one quantum of transmitter (q) is released. Therefore this technique can be used to investigate the mechanism involved in a change in synaptic transmission following experimental manipulation, and thus provides information on the molecular processes underlying the phenomenon. For example, minimal stimulation has been used extensively to study the mechanism of expression of LTP and LTD. For excitatory synapses in CA1, Pr is generally low (<0.5), so when using minimal stimulation, no synaptic response is evoked following a proportion of the stimuli. These are termed failures. During LTP or LTD, changes in failure rate or in the mean amplitude of the successfully evoked synaptic responses (EPSC amplitude excluding failures, a parameter termed potency) can provide information on the underlying expression mechanisms. For LTP and LTD experiments, it is optimal to have a failure rate of $\sim$50% during the baseline period to ensure that equally large changes in failure rate can occur in either direction after induction of plasticity.

In order to use failures analysis to investigate the expression mechanisms of LTP and LTD, it is important to be able to reliably resolve failures from the smallest synaptic events. Whole-cell voltage clamp recordings are used for this purpose; however, voltage control can be poor at synapses remote from the recording electrode (i.e., there is significant voltage escape at the synapses and a significant slowing of the current recorded). This means that for large, highly branched neurons like CA1 pyramidal cells, voltage control must be optimized for the synapses under investigation by reducing the distance from the recording electrode to the synapses. This can be accomplished by either stimulating synapses located on the proximal dendrites close to the soma (for somatic whole-cell recordings) or making whole-cell recordings from the dendrites and activating synapses close to the recording site. To reliably activate synapses close to either a somatic or dendritic recording electrode, it is necessary to place the stimulating electrode close to the recording electrode under visual control using a microscope with high-power water-immersion optics. The best recording configuration for optimizing the voltage control of synapses is a whole-cell dendritic recording from the proximal apical dendrite of a CA1 pyramidal cell together with local minimal stimulation of synapses on that region of the dendrite. Under these conditions, EPSCs resulting from the opening of very few AMPA receptor channels can be reliably resolved from the noise.

The amount of voltage escape at poorly clamped synapses is also dependent upon the size of the synaptic current (the larger the current, the worse the voltage clamp). Therefore, after initially obtaining a response, stimulation intensity should be gradually reduced until an intensity is found that is just above threshold for reliably evoking the smallest EPSCs (e.g., for CA1 cells ∼5 to 15 pA).

Low access resistance is also very important for improving voltage control and should be <20 MΩ for somatic recordings (with recording electrodes optimized, an access resistance of 5 to 10 MΩ should be achievable) and 30 to 40 MΩ for whole-cell dendritic recordings. During an experiment, accurate on-line access resistance measurement is very important because even small changes will affect EPSC amplitude (access resistance often gradually increases, therefore this is vital to monitor during LTD experiments). To estimate access resistance on-line, the peak amplitude of the fast whole-cell capacitance transient should be measured in response to a small (e.g., 2 mV) step. To obtain an accurate measurement of this transient, electrode capacitance must be carefully compensated immediately before breaking in by using the patch clamp fast- and slow-capacitance compensations and using a filter cutoff setting on the patch clamp of ≥10 KHz. Because the fast whole-cell transient is a very-short-duration event, an accurate estimate of the peak amplitude will only be obtained on-line with a software acquisition rate of ≥20 KHz and a 10-KHz filter setting on the patch clamp. Alternatively, a double exponential fit to the decay of the transient can be used to estimate the peak of a transient acquired at 10 KHz and filtered at 5 KHz. However, this latter exponential fitting method should be checked against the value obtained using the former high-frequency acquisition method by measuring the size of the transient on the oscilloscope with a filter setting of 10 KHz.

The voltage control of the synapses under investigation can be objectively estimated by calculating the rise time and decay time constant of the mean EPSC waveform constructed by averaging together a number of individual EPSCs. As the kinetics become faster, the voltage control improves. Furthermore, it is important to check that there are no changes in average EPSC kinetics during an experiment, which would indicate changes in voltage control, most likely due to a change in access resistance.

In order to accurately resolve failures from the smallest EPSCs, it is important to have low background noise. Low-frequency noise (0.1 to 1 KHz) is particularly disruptive because EPSCs are composed of similar frequencies (e.g., typical rise time for EPSCs in CA1 cells is 1 to 3 msec; 0.33 to 1 KHz). This low-frequency noise is often caused by an unstable seal between the electrode and cell (the cell may be unhealthy or the electrode shape not optimized) or noise from the power source (50 or 60 Hz) that can be corrected by proper grounding. Higher-frequency noise is not a problem and can be removed by digital filtering (on-line or off-line) without significantly attenuating the response.

Because the synaptic response to minimal stimulation is highly variable (due to the stochastic nature of transmitter release and quantal variance), it is necessary to collect data from as many trials as possible during the baseline period. Because LTP washes out, a relatively high stimulation frequency (0.33 to 1 Hz) has to be used. The high frequency of stimulation has its own problems in that the synaptic response may gradually run down or axons may stop responding to the stimulation (fiber drop out), causing a sudden step decrease in the response. If either of these phenomena occurs, the experiment should be terminated. When inducing LTP or LTD using minimal stimulation, the frequency of stimulation during the pairing protocol should not be altered from the baseline frequency. This avoids any change in the reliability of axon activation due to the change in frequency that may persist after the end of the pairing protocol.

Fiber failures are a potential problem when interpreting LTP and LTD experiments. These are failures to evoke a postsynaptic response due to a failure of the stimulating electrode to generate an action potential in an axon (rather than a true failure due to low Pr). If a significant proportion of failures are fiber failures, then a change in Pr associated with LTP or LTD will be masked. Fiber failures are detected in two ways. First, they are indicated by experiments in which the size of the response suddenly steps up and down, or when the smallest EPSC evoked is very much larger than normal (e.g., smallest EPSC is 50 to 100 pA for a CA1 pyramidal cell, typically the smallest EPSCs are 5 to 15 pA). Second, the use of paired-pulse stimulation provides a more objective test and allows the fraction of fiber failures to be estimated. If there are no fiber failures, every stimulus will cause an action potential to reach the presynaptic terminal and generate a Ca^{2+} influx. This will happen regardless of whether there is transmitter release (producing a postsynaptic response). If a paired-pulse protocol is applied (50 msec interval) there will be paired-pulse facilitation due to residual Ca^{2+} left over from the first stimulus. Therefore, the average of all the responses to the second stimulus that followed a first stimulus failure should be identical to the average of the responses to the second stimulus that followed a first stimulus nonfailure. Fiber failures are indicated if the mean response following failures is smaller than that following nonfailures. The relative difference in size provides a direct estimate of the number of fiber failures. Fiber failure contamination of <10% is normally acceptable.

Peak-Scaled Nonstationary Fluctuation Analysis

This is a method of analysis for studying the single-channel properties of channels contributing to a synaptic current. This method is used to extract the single-channel conductance of synaptically activated AMPA receptors in minimal stimulation experiments using whole-cell dendritic recordings from the proximal apical dendrites of CA1 pyramidal cells in the hippocampus (see *http://www.bris.ac.uk/synaptic*). Under these recording conditions, the voltage control of locally stimulated CA1 synapses is optimized, and only under these conditions can nonstationary fluctuation analysis (non-SFA) be performed. Each single EPSC waveform is subtracted from an averaged EPSC waveform, and then fluctuation analysis is performed on this difference waveform to obtain single-channel conductance information. To perform this analysis, the EPSCs must have relatively fast rise times (∼1 msec) so that they can be precisely aligned by their rising phase to the rise of the averaged EPSC waveform. If the rise of the individual EPSCs is too slow there will be too much background noise during the rising phase to accurately align the EPSCs. Precise alignment is important because peak-scaled non-SFA requires that there is no difference between the individual EPSCs and the mean EPSC waveform at the peak of the current. To collect enough EPSCs for LTP experiments during a baseline period, failure rate is normally set to ∼30%. This analysis requires absolute stability of access resistance during the entire experiment because the single-channel conductance estimate is very sensitive to changes in this parameter. Non-SFA is sensitive to high-frequency as well as low-frequency noise. Therefore, before the analysis can be performed, it is often necessary to filter the data once they have been collected on the computer. This will remove the high-frequency noise without affecting the EPSC.

SPECIAL CONSIDERATIONS WHEN WORKING WITH TRANSGENIC MICE

LTP and LTD experiments are increasingly being conducted using transgenic mice. Such experiments introduce additional factors that need to be taken into account. First, the viability of hippocampal slices obtained from mice is dependent on the strain of mouse. In particularly sensitive strains (e.g., C57BL/6J, CBA/J, DBA/2J, 129/Sv), it is sometimes helpful to include a broad-spectrum ionotropic glutamate receptor antagonist such as kynurenic acid (1 mM) in the aCSF during the slice preparation phase to limit excitotoxic cell death. An alternative strategy is to add sucrose (~250 mM) to the aCSF to limit cell swelling and death. Second, the level of intrinsic synaptic plasticity may also be strain dependent.

In many experiments, comparisons are made, for example, of the level of LTP between slices. It is essential that transgenic and knockout mice be presented to the experimenter in a randomized and blind manner. Data should be analyzed and decisions made whether to exclude the experiment (e.g., because of a drifting baseline) before the code is broken. Because the experimental variable of interest is genotype, the number of samples should refer to the number of animals (not slices) studied.

DATA ACQUISITION AND ANALYSIS

Appropriately processed data should ideally be captured and displayed on a computer that enables responses to be measured on-line to enable assessment of baseline stability. Therefore, a key issue is the software employed.

Software

A computer program for LTP and LTD experiments should have the ability to capture and analyze synaptic responses recorded extracellularly or intracellularly using sharp electrode or patch clamp recording. An LTP program should be able to measure the peak amplitude of intracellular synaptic potentials, the slope of fEPSPs, and the population spike amplitude of extracellular synaptic potentials. Furthermore, the program should be able to do this analysis on-line, as it is important to assess the progress of the experiment and to be able to manually change amplitude and slope baselines during the experiment. It should also have the capability to slowly (e.g., every 15 to 30 sec) stimulate a single pathway extracellularly or to alternately stimulate two different pathways extracellularly. LTP is normally induced by either a single or multiple train (theta burst) stimulation; LTD is normally induced by low-frequency stimulation at 1 to 2 Hz and with 900 pulses. It is helpful if the software program can produce this stimulation; alternatively, the stimulation can be generated by an external hardware stimulator.

Commercial Software

A number of commercial software packages are available that have at least some of the required capabilities of a good LTP program. These software packages can be divided into those that require programming and those that do not, and those for Intel/Microsoft computers versus Apple Macintosh computers. The software packages that do not require programming include pClamp 8 (Axon Instruments) and Experimenter's Workbench (DataWave Technologies) for Intel/Microsoft computers, and AxoGraph 4 (Axon Instruments) and A/DVANCE (*http://www.kagi.com/douglas/advance*) for Macintosh. Alternatively, custom LTP programs, such as Signal and Spike2 (Cambridge Electronic Design), Acquis1 (Bio-Logic), LabView (National Instruments), and Igor (WaveMetrics), have been written for Intel/Microsoft computers. LabView and Igor software packages can also be used on Macintosh computers.

A Free Program for LTP and LTD Experiments

At the University of Bristol, the freely available "LTP" Program (*http://www.ltp-program.com*) has been developed to handle stimulation, data acquisition, and on-line analysis for studies of LTP, LTD, and related phenomena such as kindling. "LTP" Program is a 32-bit DOS program that can directly access up to 64 MB of memory. It can be used on computers running Windows 95/98 or DOS 6.x/Windows 3.x operating systems, using the following data acquisition boards: Digidata 1200 (Axon Instrument), Labmaster (Scientific Solutions), and the low cost ADC-42 board (Pico Technology).

The "LTP" Program records synaptic activity in extracellular, intracellular, or patch clamp modes for one or two AD channels. The program produces repetitive sweeps with simultaneous data acquisition and stimulation. It can acquire up to 1,000,000 samples at 10 KHz sampling on a 64 MB machine. The stimulation consists of stimulating one or two extracellular pathways (S0 and/or S1) with epoch-like intracellular analog stimulation. The basic protocols are either slow single-pathway (S0) stimulation or slow alternating dual pathway (S0 then S1) stimulation. The program can induce LTP by single trains, repetitive trains (theta burst stimulation), and primed stimulation (a single pulse followed by a short single train). LTD stimulation can be produced by delivering fast, repetitive single-pulse sweeps (up to 2 Hz).

For LTP and LTD experiments, on-line waveform analysis of each S0- and S1-evoked EPSP on both AD channels includes: slope, peak amplitude and latency, population spike amplitude and latency, average amplitude, rise time, decay time, duration, area, and coastline. Cell resistance and patch electrode series resistance (measured using peak or exponential curve fitting) can be monitored continuously throughout the experiment. The sweep data can also be signal averaged and digitally filtered on-line and off-line.

The "LTP" Program can also measure up to 1000 S0- and S1-evoked stimulus responses in a sweep. Not only does this allow on-line measurement of PPF and PPD, it also allows entire trains of synaptic responses to be analyzed. Stimulus artifacts can also be automatically removed to permit accurate determination of synaptic areas and peaks during a train. Enhancements of the "LTP" Program include the use of PCI bus data acquisition boards and a Windows 2000 version of the program.

References: Bear et al., 1996; Fazeli and Collingridge, 1996

Contributors: Zuner A. Bortolotto, William W. Anderson, John T.R. Isaac, and Graham L. Collingridge

UNIT 1.5

Recording Synaptic Currents and Synaptic Potentials

Intracellular recording of postsynaptic currents (PSCs) under voltage-clamp conditions provides the most accurate and direct means for measuring the effects of neurotransmitters and for obtaining information about the type of transmitter used at a synapse, the dynamics of transmitter-receptor interactions, the types and numbers of receptors activated, the effects of

Table 1.5.1 Characteristics of Preparations for Intracellular Recordings of Synaptic Activity

Property	In vivo recordings	Brain slices	Cell cultures
Integrity of axons, dendrites, and synapses	Normal	Axons and dendrites may be severed; synapses intact	Processes intact, but specificity of contacts randomized; apposition of neurons and glia disturbed
Expression of receptors and channels	Normal	Presumed normal	Expression and functionality of receptors and channels may be perturbed
Identification of cell types	Difficult	Feasible	Depends on region
Stability of recordings	Poor (better with patch pipets)	Good	Good
Feasibility of voltage clamp	Poor	Quality of clamp varies	Quality of clamp varies
Ease of solution exchange	Very limited	Limited	Excellent
Preparation of tissue	Challenging	Difficulty varies widely with anatomical region	Very convenient

drugs on transmission, the behavior of functional neural circuitry, and the mechanisms of synaptic plasticity. In addition, recording postsynaptic potentials (PSPs) under current-clamp conditions can indicate the effectiveness of the synapse in a given neural pathway and is necessary to determine how different synaptic events interact electrically—i.e., through the summation of excitatory and inhibitory conductances. In this unit both stimulus-evoked and spontaneous synaptic events are examined. Often, conclusions about synaptic and drug mechanisms are strongest when based upon recording of both types of activity. Individual synaptic events can be measured in a variety of experimental preparations. Table 1.5.1 lists three broad preparation categories and their relative advantages and disadvantages.

BASIC PROTOCOL

RECORDING SYNAPTIC EVENTS IN BRAIN SLICES

Materials

Tissue slice maintained in chamber with standard Ringers solution (see UNITS 1.3 & 1.10)
External and patch-pipet solutions (e.g., see Trussell and Jackson, 1987, Gyenes et al., 1988, and MacDonald et al., 1989)
Test compound (e.g., Research Biochemicals, Tocris Cookson)
Recording solution with selective ion-channel blockers (see literature for experiment-specific information)

Electrophysiology setup for patch-clamp (UNIT 1.2)
Patch pipets (see UNITS 1.1 & 1.8)
Stimulus pipets (e.g., patch pipet)
Stimulus isolators (AMPI; distributed in the U.S. by Pacer Scientific or Grass Instruments)
Acquisition/analysis hardware and software: pCLAMP, AxoBasic, AxoGraph (Axon Instruments) or CDR (Strathclyde Software)
Plotting/curve-fitting software: Origin 5.0 (Microcal Software) or Igor (WaveMetrics)
Tape recorders (VCR with PCM adapter, DAT drive, or audio recorder; optional)

1. Position the stimulus pipet on the tissue slice near, but not necessarily on, the region to be activated before the patch pipet forms a seal. Establish whole-cell recording (see UNITS *1.2 & 1.8*), quickly compensate series resistance, and decide if it is low enough and stable enough for the recording.

2. Identify presynaptic fibers (e.g., see Role and Fischbach, 1987 or Bekkers and Stevens, 1991) and place the stimulus pipet on a presynaptic cell or axon by trial-and-error, being careful not to physically dislodge the patch pipet during the process.

   Stimulus intensities, set to ∼100 to 200 msec and up to 20 V, can destroy the postsynaptic cell if the stimulus pipet is placed too nearby.

3. Record with low-frequency stimulation (0.1 to 1 Hz) for several minutes to establish that the synaptic response is relatively stable. Abandon recording (or record long enough to determine the rate of run-down before applying a drug) if response declines by >10% to 15% during the recording period.

4. Change from standard bath to recording solution (often a standard Ringers solution). After establishing recording, switch the solution to one containing the pharmacological cocktail needed to block unwanted ion channels.

5. Begin recording responses, monitoring the cell continually. Check series resistance and recompensate it every few minutes, noting the values in a notebook. Check for run-down. If testing the effects of a drug, make certain to demonstrate reversibility after washing out the drug (i.e., response returns to within 90% of the control value).

6. *Optional:* Tape record the entire experiment, using a VCR with PCM adapter or with DAT.

Contributors: Laurence Trussell

UNIT 1.6

Voltage Clamp Recordings from *Xenopus* Oocytes

Xenopus oocytes serve as a standard heterologous expression system for the study of cloned (and some endogenous "native") ion channels. The oocyte expression system makes it possible to express membrane proteins in isolation or in combination with other proteins, and to study the properties of mutated and chimeric membrane proteins as well as almost all aspects of ion channel physiology, pharmacology, and biophysics. The central advantage of the *Xenopus* oocyte expression system is the large size of the oocyte (1.0 to 1.2 mm in diameter), facilitating RNA injection, culture, electrophysiological measurements, and any associated biochemical procedures. The large cell size allows penetration with two recording electrodes to perform two-electrode voltage clamp (TEVC) recordings, and, if necessary, the oocyte can also accommodate an additional, third micropipet for intracellular injection of substances during the recording. Healthy cells survive electrode insertion and allow recordings lasting for tens of minutes without appreciable deterioration. Both whole-cell and single-channel current recordings can be performed in *Xenopus* oocytes, as well as an "intermediate" methodology of recording macroscopic currents from large membrane patches. Troubleshooting for patch pipet recording is the same as in other cells (see UNIT *1.8*). Table 1.6.1 at the end of this unit deals only with problems encountered in TEVC procedures.

BASIC PROTOCOL 1

TWO-ELECTRODE VOLTAGE CLAMP (TEVC) RECORDING

This method is conceptually similar to whole-cell patch clamp (UNIT 1.2), allowing the measurement of whole-cell ion-channel currents while controlling the membrane potential. Since very large currents (often reaching many microamperes) may have to be recorded, a high-compliance TEVC amplifier is needed. Suitability for work with *Xenopus* oocytes is usually described in detail in the manufacturer's technical information sheets. For best temporal resolution and minimization of errors, low-resistance current and voltage electrodes and KCl/agarose bridges (which connect the extracellular solution in the experimental chamber with the Ag/AgCl ground electrodes) should be used. A virtual-ground recording configuration, usually recommended and described in detail by TEVC device manufacturers, with separate extracellular ground electrodes for voltage measurement and current injection, is recommended. All these considerations are very important for the study of voltage-dependent currents, where time resolution may be crucial, especially when recording of large currents (>3 µA) is planned. If possible, the levels of channel expression in the oocyte should be designed in such a way that the whole-cell currents do not exceed a few microamperes; for currents of >10 µA, the accuracy of measurement is seriously compromised even with optimal recording arrangements. It is usually impossible to accurately resolve currents flowing during the initial 0.5 to 1.5 msec after a voltage jump; this impedes the study of rapidly activating voltage-dependent ion channels, such as Na^+ and some K^+ channels.

Materials *(see APPENDIX 1 for items with ✓)*

   3 M KCl
   Extracellular medium:
 ✓ ND96 standard extracellular solution
 ✓ High-Ba^{2+} solution
 ✓ High-K^+ solution
   Oocytes injected with RNA, placed in petri dishes with incubation solution

   TEVC apparatus (Axon Instruments, Dagan, NPI Electronic, Warner Instruments)
   Two ACE microelectrodes (see Support Protocol)
   Experimental chamber connected via two agarose/KCl bridges and Ag/AgCl wires to the electronic equipment as shown in Fig. 1.6.1 (also see Support Protocol) and perfusion system (see Fig. 1.6.2; for study of ligand-activated ion channels, a fast-exchange system consisting of a small-volume experimental chamber)
   Electrophysiological setup: computer with an analog signal processing capability and a high-frequency filter (if the TEVC device does not have one built in), oscilloscope and graphic recorder for signal display and archiving, vibration-isolation table, Faraday cage, dissecting binocular microscope with illuminator (40× maximal), and two micromanipulators with three movement axes and at least 10 mm travel and the ability to hold the electrodes at a ~45° angle between the electrode and the plane of the table
   Pasteur pipet with an inner diameter slightly greater than that of the oocyte, thoroughly fire-polished to removed any sharp edges from the tip
   Spiral shield made of silver or another metal with an inner diameter slightly greater than the electrode's (for study of voltage-operated channels)

1. Add 3 M KCl solution to the electrodes through their blunt ends, so that the level of KCl will be 1.5 to 2 cm above the tip. Place the microelectrodes into electrode holders (with the Ag/AgCl wire of the holder immersed 3 to 5 mm into the KCl solution in the electrode) and fix them with the holder's stopper.

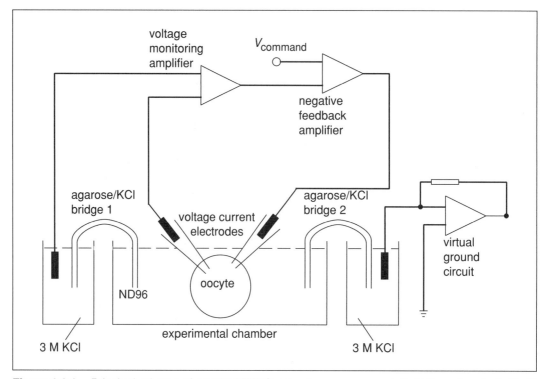

Figure 1.6.1 Principal scheme of a typical TEVC recording arrangement. Bold lines denote Ag/AgCl electrodes. The oocyte is placed in the middle of an experimental chamber filled with a recording solution (ND96). The level of solution is shown by dashed line. The experimental chamber is connected, via KCl/agarose bridges 1 and 2, to two compartments filled with 3 M KCl. The Ag/AgCl electrode of the left compartment is connected via bridge 1 to one of the inputs of the voltage-measuring amplifier built into the TEVC device. The second input of this amplifier (usually denoted as "voltage electrode" input on commercial TEVC devices) is connected via an Ag/AgCl electrode to the voltage recording glass microelectrode. The voltage measured by this device is fed into one of the inputs of the negative feedback amplifier, where it is compared with the $V_{command}$ fed into the second input of this amplifier. $V_{command}$ can be set manually by using a knob provided for this purpose on the front panel of the TEVC device, or input from an external source, typically driven by the computer. The output of the negative feedback amplifier (usually denoted as "current electrode" output on commercial TEVC devices) is connected to the current microelectrode via an Ag/AgCl electrode. The bath solution is also connected, via bridge 2, to the right 3 M KCl compartment, and from there, via an Ag/AgCl electrode, to the virtual ground circuit. The latter is supplied within a separate small box, or is built into the main body of the TEVC device. The outputs of the voltage-monitoring amplifier (measuring the membrane voltage, V_m) and of the virtual ground circuit (measuring the current flowing through the bath to the oocyte; the membrane current, I_m) are connected to a computer and (optionally) additional equipment such as VCR, oscilloscope, or chart recorder. The connections to the computer and additional equipment are not shown here. Commercially available TEVC devices always have special outputs for monitoring V_m and I_m that are used to make these connections.

2. *Optional:* Shield the electrode:
   a. Wrap one of the electrodes with long thin strips of Parafilm (stretch Parafilm before wrapping). Holding the holder in one hand, carefully slip the electrode into the shield.
   b. Alternatively, shield the electrode by wrapping aluminum foil above the Parafilm sheath. The shield should not touch the solution in the bath.
3. Insert the electrode holders into the relevant outlets of the voltage and current headstages.

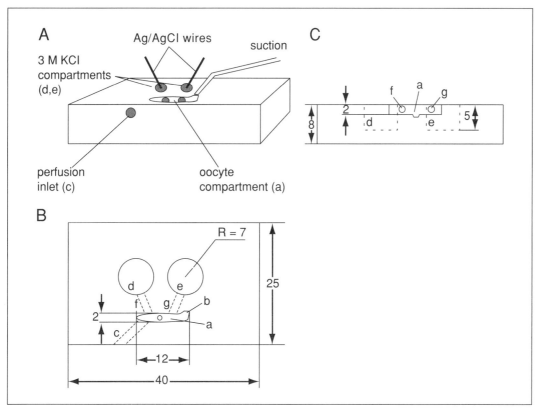

Figure 1.6.2 The recommended oocyte perfusion experimental chamber. (**A**) general view; (**B**) view from above; (**C**) front view on a longitudinal section made along the median line of the oocyte compartment. Dimensions are given in mm. The chamber is made of Perspex. The oocyte compartment (a) is 2 mm deep, 2 mm wide, and ~12 mm long and has a cavity in the middle of the compartment's floor, ~0.2 mm deep and 0.3 to 0.4 mm in diameter. The diameters of all tubes drilled within the body of the chamber (c, f, g) is 1.2 or 1.3 mm. The KCl compartments (d, e) are 7 mm in diameter and 5 mm deep. These and other dimensions are indicated in the drawing.

4. Fill the experimental chamber with ND96 (always recommended until the wounds produced by the electrodes during their insertion recover, typically 1 to 2 min). With the Pasteur pipet, pick up an oocyte from the petri dish together with 20 to 30 µl of solution and (holding the Pasteur pipet vertically until the oocyte sinks to the very end) transfer it with minimal incubation solution to the chamber so the oocyte rests in the cavity of the chamber's floor.

5. Move the electrodes toward the chamber until they are in the solution. Compensate the junction potential by adjusting the DC offset, and measure the resistance (*UNIT 1.8*), typically between 0.2 and 1.5 MΩ for the voltage electrode and between 0.1 MΩ (for >3 µA currents) and 0.8 MΩ (for currents <2 µA) for the current electrode. Focus the microscope on the tips of the electrodes and check whether KCl is leaking (seen as a stream of denser solution coming out of the electrode). If such leak is detected, discard the electrode.

6. Advance the voltage electrode until it almost touches the surface of the cell. Adjust the placement of the electrode until its imaginary continuation points to the center of the oocyte, i.e., at an angle of ~90° between the electrode axis and the surface of the oocyte. Place the current electrode opposite the voltage electrode (on the other side of the oocyte), almost touching the cell's surface.

7. Push the voltage electrode forward (supporting the oocyte with the current electrode), making a depression, until the electrode pierces the membrane and its tip is inside the cell (verified by a negative voltage reading, usually between −20 and −50 mV). Withdraw the electrode 10 to 20 μm back, if necessary, to remove the depression.

8. Start perfusing the chamber with ND96 at a slow rate, ∼1 chamber volume/min, and observe the membrane potential for more negative readings within the next 10 to 20 sec, indicating recovery from the wound caused by the electrode insertion. Push the current electrode forward until it penetrates the plasma membrane; withdraw slightly if necessary.

9. Start the voltage clamp by setting the holding potential (routinely between −60 and −80 mV, close to the resting potential of the nerve or muscle cells originating the ion channels) at the desired level. Wait an additional 20 to 60 sec for the holding current to stabilize. Estimate the input resistance of the oocyte (R_m) from the values of the recorded resting potential (V_r), holding potential (V_h), and holding current (I_h) by the equation $R_m = (V_h - V_r)/I_h$. For instance, if $V_r = -30$ mV, $V_h = -80$ mV, and $I_h = -100$ nA, then $R_m = 0.5$ MΩ.

10. Set the optimal gain of the TEVC amplifier by shifting the voltage from V_h by 50 to 60 mV, e.g., to −140 mV, in 10-msec pulses, and observing both voltage and current (the voltage step is accompanied by a capacitative current which decays with a monotonous multiexponential time course). Increase the gain slowly, watching the capacity transient become shorter and the rise time of the voltage pulse become steeper. Stop increasing the gain when the capacity transient acquires an "overshooting" component, i.e., when the monotonous decay is broken, and the current crosses the steady-state level before stabilizing (this is the first warning that the system is becoming unstable and is on the verge of oscillating). Reduce the gain until the overshoot disappears.

    For recording ion-channel activity at a constant potential, such as with neurotransmitter-evoked currents, a higher gain can be used. Increase the gain as long as no high-frequency oscillations of V_m are apparent.

11. If necessary, switch the perfusion system to a high-Ba^{2+} (for the study of Ca^{2+} channels) or a high-K^+ (for the study inwardly rectifying K^+ channels) solution and begin experimental protocols.

SUPPORT PROTOCOL

FABRICATION OF AGAROSE CUSHION MICROELECTRODES (ACEs) AND AGAROSE/KCl BRIDGES FOR TEVC

Low-resistance microelectrodes (0.1 to 1 mΩ) are needed to inject large currents during TEVC in the oocytes, and it is often sufficient to break the extreme end of the tip of a regular microelectrode to reduce its resistance recordings. To prevent leakage of KCl into the cell and permit stable, long-lasting recordings when using these electrodes, the tip is filled with a 1% agarose/3 M KCl gel, and the rest of the electrode is filled with a regular 3 M KCl solution (Schreibmayer et al., 1994). ACEs do not become clogged by the cytoplasm or organelles, as often happens with regular electrodes and can often be used repeatedly in many oocytes, sometimes for many days of measurement. Due to their outstanding stability, it is also advisable to use ACEs for voltage monitoring. Note that the excessive heat produced by very large currents (>10 μA) may eventually burn the agarose cushion dry, leading to a significant increase in electrode resistance and requiring electrode replacement.

Standard agarose/KCl (or agar/KCl) bridges can be prepared at the same time as the electrodes, since they are filled with the same gel. Since KCl diffuses into the bath from the bridges, thereby raising the K$^+$ concentration, bridges with 3 M NaCl instead of KCl may be used for the measurement of K$^+$ channel currents.

Materials

Agarose (molecular biology grade)
3 M KCl and/or 3 M NaCl solutions (in H$_2$O)

Standard glass tubes for preparation of microelectrodes (UNIT 1.1), soft or borosilicate glass, 1.6 to 2 mm o.d., 1.2 to 1.6 mm i.d.
Microelectrode puller (preferably vertical)
Microelectrode storage jar and a dry microelectrode storage stand (WPI)
10-ml syringe with a 19- to 21-G needle
10-ml syringe with a 17- to 19-G needle and a rubber or silicon tubing attached to the end of the needle, ~1 cm long, with i.d. slightly smaller than the o.d. of the glass tubes

To prepare the ACEs

1a. Set the microelectrode puller at parameters approximately suitable for the first pull, as described for patch pipets (UNIT 1.1). Place the glass tube into the puller. Do not use any spacers.

2a. Turn on the heat of the filament and pull the glass to produce two almost identical electrodes 6 to 10 mm in length from the beginning of the narrowing to the tip. Regulate the length of the narrow part of the electrode by adjusting the heat of the filament. Prepare as many electrodes as are needed for several experiments and place them in a dry holder or stand. Do not add KCl at this stage.

3a. Break the tips of the electrodes to give an outer diameter of the broken tip between 2 and 5 µm by lightly tapping on a piece of thin paper held vertically in the air. Place the electrodes (wide-tipped electrodes for current injection and narrow-tipped ones for monitoring voltage) back in the dry holder.

To prepare the U-shaped tubes for agarose/KCl bridges

1b. Cut the standard glass tubes into ~10 cm pieces. Prepare several tubes at once.

2b. Holding with forceps at the ends of the tube, heat the center portion of the tubing in the flame of a gas or alcohol burner and bend it until the tube becomes U-shaped. Let the glass cool before putting it down on any surface.

3b. Cut the protruding ends of the tube by first scoring the surface with a cutting stone or file and then breaking by bending, leaving ~8 to 10 mm on each side (see Fig. 1.6.1).

4. Weigh 0.2 g agarose into a 50-ml beaker and add 20 ml 3 M KCl solution. Place the beaker on a hot plate and heat the mixture while stirring until the agarose is dissolved and the solution becomes transparent. Do not boil. Remove the beaker from the hot plate just before the solution starts boiling, but do not let it cool down. Keep it hot by placing the intermittently on the hot plate, continuously on a 90°C hot plate, or in 90°C water.

5. Insert the blunt end of a previously prepared broken-tip ACE (step 3a) into the rubber tubing connected to the 10-ml syringe, and immerse the tip of the ACE into the hot agarose/KCl solution. Pull the plunger of the syringe to lower the pressure in the ACE so the agarose/KCl solution slowly crawls into the narrow end of the electrode. Bring the solution up to 3 to 6 mm from the tip, release the plunger, and remove from the solution. Allow the solution in the electrode to cool and polymerize. Remove any agarose stuck to the out side of the ACE with a piece of facial tissue or with fine forceps under a dissecting microscope.

6. Place the electrode into the microelectrode storage jar, with the tip immersed in KCl solution. Discard electrodes in which the level of the gel in the electrode goes up within several hours.

7. After preparing the ACEs, take up a few milliliters of hot agarose/KCl solution into a 10-ml syringe fitted with a 19- to 21-G needle and fill several U-shaped glass tubes by injecting the hot solution into them. Place the bridges on a dry surface for several minutes to cool and allow the gel to polymerize. Store the bridges up to 2 months at room temperature or (preferably) 4°C in a tightly closed vessel filled with 3 M KCl.

8. Several hours after preparation, or just before the experiment, add the necessary amount of 3 M KCl solution to the ACEs through their blunt ends, so that the level of KCl in the ACE will be 1.5 to 2 cm from the tip. If there is an air bubble between the agarose/KCl cushion and the KCl solution added from above, tap lightly on the electrode several times with a finger, while holding the blunt end with the other hand, until the bubble is removed.

BASIC PROTOCOL 2

PATCH CLAMP RECORDINGS FROM OOCYTES

Patch clamp recording in oocytes is no different from that in any other cells (UNITS 1.1–1.3 & 1.8–1.10), except that whole-cell recording is practically impossible. Although special methods for fabrication of patch pipets with very large openings have been developed and can be used with oocytes, low-resistance soft glass or thin-walled borosilicate glass pipets made according to standard protocols are normally sufficient. To make such electrodes, the heating during the second pull should be lowered to a minimum when preparing the electrodes.

The only difference between patch clamping small cells and oocytes is that oocytes require additional microsurgery just before the experiment to remove the outer, noncellular vitelline layer to expose the plasma membrane. The devitellinized oocyte should be handled with extreme care. Exposure of any part of the oocyte to air, or an unintentional touch with forceps, leads to an immediate rupture of the plasma membrane.

Materials (see APPENDIX 1 for items with ✓)

✓ Hypertonic solution 1 or 2
   Two 35-mm petri dishes filled with hypertonic solution
   Pasteur pipet with an inner diameter slightly greater than that of the oocyte, thoroughly fire-polished to remove any sharp edges from the tip
   Dissecting binocular microscope with amplification up to 40×
   Two pairs of fine (no. 55) forceps (it may be necessary to further sharpen the forceps on a fine sharpening stone under a dissecting microscope)
   Experimental bath (constructed in-house; see Fig. 1.6.2) filled with bath solution (experiment-specific)
   Standard patch clamp setup with an inverted microscope or (less appropriately) a dissecting binocular microscope with amplification of 50×

1. Using a Pasteur pipet, pick up an oocyte and, holding the Pasteur pipet vertically until the oocyte sinks to the very end, place it in a petri dish containing hypertonic solution for 2 to 5 min.

2. Under the dissecting microscope, with amplification set at 30× to 40×, turn the oocyte with its light (vegetal) hemisphere upwards by lightly pushing with forceps. Grasp a portion of the vitelline membrane with one pair of forceps, then a portion nearby with the other pair. Tear apart as far as possible. Repeat several times if necessary.

If about half of the vitelline membrane is already removed, the other half can sometimes be removed by catching a patch in the remaining piece of vitelline membrane, raising the oocyte slightly, and waving it cautiously within the solution a few times.

3. Gently aspirate the oocyte into the narrow end of the Pasteur pipet. Make sure that it is surrounded by 2 to 4 mm of solution on both sides, to avoid contact with air. Transfer the oocyte into the first washing petri dish, filled with bath solution, carefully pull it into the Pasteur pipet, and then push it out. Repeat this washing procedure one or two more times. Transfer in the same way to a second washing petri dish filled with bath solution, and repeat the washing procedure.

4. Transfer the oocyte to the experimental bath and place it into the recording setup. Immediately turn the oocyte around or move it, if necessary, before it sticks to the glass and plastic (within a few tens of seconds). Use within 1 hr. See UNITS *1.1–1.3 & 1.8–1.10* for patch-clamp recording procedures.

References: Schreibmayer et al., 1994; Stuhmer, W. 1998

Contributor: Nathan Dascal

Table 1.6.1 Troubleshooting Guide for TEVC Recording

Problem	Possible cause	Solution
The resistance of the agarose cushion electrodes (ACEs) is higher than expected (e.g., >1.5 MΩ) even though the tips have been broken as directed	An air bubble between the agarose cushion and the KCl solution above	Continuously tap on the electrode until the bubble floats up, or replace the electrode
	Electrode tip is clogged	Replace
	AgCl coating of the Ag wire is too coarse	Replace the Ag/AgCl wire
The potential recorded by the voltage electrode in the bath solution drifts	AgCl coating of the Ag wire is insufficient	Replace the Ag/AgCl wire
The resting potential of the oocytes is adequate initially (negative to −20 mV) and then deteriorates	One of the electrodes is too coarse, or KCl is leaking too strongly	Replace first the current electrode (usually the coarser one), and if this does not help, then replace the voltage electrode
	Solution is contaminated or has a wrong pH	Make new solution
	Oocytes are fragile (rare)	Discard this batch of oocytes and discontinue the experiment
After turning on the voltage clamp and establishing a holding potential, the membrane current is very large (several hundred nanoamperes or more)	Cell is bad; R_m is too low	Replace the cell
	One of the electrodes is too coarse so that the cell has been damaged	Replace the cell and the electrode

continued

Table 1.6.1 Troubleshooting Guide for TEVC Recording (*continued*)

Problem	Possible cause	Solution
During TEVC, upon performing a voltage jump, the membrane resistance suddenly drops and the cell starts to deteriorate	The negative feedback loop became unstable, currents oscillate	Reduce the gain of the amplifier to a minimum. Raise again slowly, watching voltage and current records, to maximum possible level before oscillations start.
During TEVC, upon stepping the voltage, the rise time of voltage record is too low (apparent time constant of several msec)	Voltage clamp amplifier gain is too low	Increase gain as much as possible
	Current electrode clogged	Replace electrode
	Currents too large	Decrease electrochemical driving force by reducing voltage or ion concentrations, or use oocytes with lower channel-expression levels
When recording voltage- or ligand-evoked currents, the membrane voltage does not remain at the desired level, but deviates by 1 mV or more	Amplifier is unable to inject the current necessary to sustain the desired voltage ("bad clamp")	Increase the gain of the amplifier to maximum possible
		Use current electrode with a lower resistance
		Decrease electrochemical driving force by reducing ion concentrations or voltage, or use oocytes with lower channel expression levels
The amplitude of the recorded K^+ currents is reduced within several minutes of experiment without a clear reason	KCl is leaking from the KCl/agarose bridge, increasing extracellular K^+ concentration and reducing driving force	Constantly perfuse the bath at a slow rate, 3 to 4 bath volumes/min, or exchange KCl/agarose bridges for NaCl/agarose bridges
The amplitude of recorded currents (e.g., via voltage-dependent Ca^{2+} channels) increases or decreases when the perfusion rate is changed	Some channels are sensitive to mechanical stimuli and their activity is altered by membrane stress caused by changes in perfusion rate	Do not change the perfusion rate during the experiment

UNIT 1.7

Preparation and Maintenance of Organotypic Slice Cultures of CNS Tissue

Organotypic slice cultures are the method of choice for maintaining sections of the nervous system under conditions that allow long-term survival and a high degree of cellular differentiation and organization resembling that of the original tissue.

NOTE: Maintain strictly sterile conditions during all phases of handling whether or not antibiotics are added to the culture medium. Perform all tissue manipulations inside a laminar flow hood. Use ethanol followed by flaming to sterilize all instruments and open bottles.

BASIC PROTOCOL

ROLLER-TUBE CULTURES

Materials (*see* APPENDIX 1 *for items with* ✓)

    70% and 95% ethanol
    25 µg/ml poly-D-lysine (30,000 to 70,000 mol. wt.; Sigma-Aldrich), sterile
    Chicken plasma (Cocalico), centrifuged 30 min at 2500 × g, 4°C, before use
✓ ~150 U/ml thrombin solution
    Gey's balanced salt solution (BSS) containing 33.3 mM glucose
    Experimental rats (fetal to 7-day-old)
✓ Serum-free medium
✓ Roller-tube culture medium
    Antimitotic drugs: 0.1 to 1.0 µM each of cytosine arabinoside, uridine, and
        5-fluoro-2-deoxyuridine (Sigma-Aldrich) in roller-tube culture medium

Surgical instruments including:
    Large scissors
    Fine scissors
    Razor blades (Martin or Fine Sciences Tools)
    2 curved fine forceps (Dumont-type, no. 7)
    2 straight fine forceps (Dumont-type, no. 5)
    2 holders for razor blade knives, curved (#BA 290, Aesculap)
    4 metal spatulas with polished edges (Merck)
Teflon holder (60 × 40 × 20–mm, with grooves for holding 20 coverslips), custom made, for sterilizing coverslips
12 × 24–mm glass coverslips, 1-mm thickness (Kindler)
Bottle-top filters: e.g., Nalgene Steritop-GP
40-, 58-, and 92-mm petri dishes
Stereo dissecting microscope
Plastic foil (Plastoscreen 250 from Muhlebach or Aclar from Ted Pella)
McIlwain-type tissue chopper (Brinkmann)
Screw-top, flat-sided 16 × 110–mm tissue culture tubes (Nunc, no. 156758)
36°C incubator without CO_2 or humidification
Roller-drum (New Brunswick Scientific, Lab-Line Instruments, or Schutt Labortechnik)

1. Sterilize the surgical instruments, preferably in an oven or by dipping in ethanol and flaming.

2. Clean glass coverslips, using a custom-made rectangular Teflon holder with grooves to keep the coverslips separated, by washing them in 95% ethanol, rinsing in distilled water, and then washing a second time in 95% ethanol. Dry them (e.g., with a hair dryer) and sterilize in glass petri dishes in an autoclave at 120°C.

3. Mount coverslips in the custom-made Teflon holder and dip the holder with the sterile coverslips into sterile 25 µg/ml poly-D-lysine solution. Rinse twice in water, and then dry at <50°C.

4. Clean bench surfaces with detergent and then with 70% ethanol. Thaw an aliquot of 150 U/ml thrombin. Prepare a number of 92-mm petri dishes and add several drops of Gey's BSS enriched with glucose.

5. Sacrifice the rats by decapitation, either with or without anesthesia according to institutional regulations, using a quick cut with a large scissors at the level of the foramen magnum. Open the skull with fine scissors and carefully remove the brain using small spatulas.

6. Place the brain under a dissecting microscope in a laminar flow hood and carefully dissect the tissue to be cultured, using razor blades and metal spatulas with polished edges. Free the tissue of blood vessels and meninges and place it on plastic foil (Plastoscreen 250 or Aclar). Cut into 250- to 400-μm thick slices using a McIlwain-type tissue chopper. (See Gähwiler et al., 1998 for a detailed description of dissection procedures.)

7. Transfer the tissue to a 58-mm petri dish containing Gey's BSS with glucose and place it under a dissecting microscope so that slices can be gently separated under visual control with spatulas.

8. *Optional:* To reduce proliferation of nonneuronal cells, irradiate the slices at the time of explantation with X rays or γ rays (800 to 1200 R delivered continuously from a therapeutic X-ray machine or ^{137}Cs source, immediately before embedding in the plasma clot in step 9).

9. Using a spatula, spread a 20-μl drop of chicken plasma over the surface of a poly-D-lysine-coated glass coverslip. Transfer individual slices to the glass coverslips and embed them in the plasma. Coagulate the plasma by thoroughly stirring in (using a spatula) 30 μl of ∼100 U/ml thrombin in Gey's BSS (diluted from 150 U/ml). Place tissue sections from multiple brain regions side by side at a distance of 0.01 to 1 mm together on the coverslip if coculturing.

10. When a firm clot has been formed (in ∼20 sec to 1 min), place the coverslips bearing the slices in screw-top, flat-side culture tubes containing 750 μl roller-tube culture medium and close the tubes. Place the tubes in a roller-drum housed in an incubator at 35° to 36°C. Rotate at 10 revolutions/hr to ensure proper aeration and feeding. Ensure that the roller-drum is tilted 5° so that slices are immersed in the medium only during half of each rotation (see Fig. 1.7.1A).

   Since the tubes are closed, O_2/CO_2 and humidity need not be regulated.

11. Feed the cultures either once weekly (cultures) or twice weekly (cocultures) by pouring off the old medium and replacing with fresh medium. After 4 to 5 days in vitro, expose the cultures to antimitotic drugs for 16 to 20 hr to reduce proliferation of nonneuronal cells, return to fresh roller-tube culture medium, and maintain under these conditions (e.g., for several weeks in the case of hippocampal slice cultures).

ALTERNATE PROTOCOL

INTERFACE CULTURES

Materials (see APPENDIX 1 for items with ✓)

    70% and 100% ethanol
    Experimental rats or mice (neonatal animals between 0 and 14 days old recommended)
✓ Dissection medium
✓ Interface culture medium

    Surgical instruments:
        Large scissors
        Fine scissors
        Razor blades
        Curved fine forceps (Dumont-type, no. 7)
        Straight fine forceps (Dumont-type, no. 5)
        Plastic spatula with polished edges (Merck)
    Biopore-CM membrane (10 × 10–cm sheets: Millipore)
    8-mm hole punch

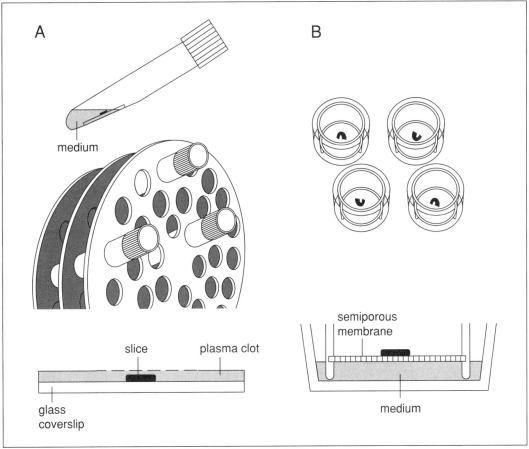

Figure 1.7.1 Techniques used for preparation of organotypic cultures. (**A**) In roller-tube cultures, slices are embedded in a plasma clot on glass coverslips. The clot is lysed during the first week in culture due to the release of proteolytic substances from the cultured slice (lower part of panel). The glass coverslips that bear the cultures are placed in flat-sided plastic culture tubes that contain a small amount of culture medium. The slow rotation results in a continuous alternation of feeding and aeration. (**B**) In interface cultures, the slice is placed directly on a semiporous membrane, and medium is added to the bottom of the culture dish. In these stationary cultures, the slices are immersed in the medium on one side and accessible to oxygen from the other side (modified from Gähwiler et al., 1997).

35- and 100-mm petri dishes (Falcon)
McIlwain-type tissue chopper (Brinkmann)
Cut, fire-polished Pasteur pipets
6-well, 35-mm multiwell culture plates (Falcon)
Millicell-CM inserts (Millipore)
Stereo dissecting microscope
33° and 36°C humidified 5% CO_2 incubators

1. Sterilize all instruments, preferably in an oven or by dipping in ethanol and flaming. Prepare 8-mm diameter confetti of Biopore-CM membrane using a paper punch. Dip them in 70% and then 100% alcohol and allow to dry in a 100-mm petri dish sealed with Parafilm. Sterilize by exposing to UV light for 20 min.

2. Sacrifice the experimental animals by decapitation using a quick cut with large scissors at the level of the foramen magnum. Open the skull with fine scissors and carefully remove the brain using small spatulas.

3. Place the brain under a dissecting microscope in a laminar flow hood and carefully dissect the tissue to be cultured using razor blades and spatulas with polished edges. Free the tissue of blood vessels and meninges and then place it on a Teflon plate adapted for a McIlwain-type tissue chopper and cut into 250- to 400-μm thick slices.

4. Transfer slices to a petri dish using a cut, fire-polished Pasteur pipet filled with dissecting medium. Place 1 ml of interface culture medium in each well of a 6-well, 35-mm multiwell culture plate with 1 ml interface culture medium and place a Millicell-CM insert in each of them. Place up to four confetti (from step 1) on each insert with curved fine forces so that each piece of confetti perfectly adheres to the Millicell-CM membrane.

5. Under the dissecting microscope, select the best slices (criteria: intact morphology, clear cell body layer, homogeneous thickness) and place them on the confetti by aspirating them into a cut, fire-polished Pasteur pipet. Remove the excess medium so they are not covered by culture medium. Place the culture plate into a humidified CO_2 incubator equilibrated with 5% CO_2 at 36°C for 4 days (to recover from the explantation) and then transfer them to a humidified 5% CO_2 incubator set at 33°C (for better quality and more reliable cultures).

6. Change the culture medium after 24 hr and then every 3 to 4 days by replacing the 1 ml of medium present in each of the wells of the multiwell plates. To manipulate slice cultures or transfer them into a recording chamber, simply hold the membrane confetti (with the tightly adhering interface slice culture) with fine tweezers.

Contributors: Beat H. Gähwiler, Scott M. Thompson, and Dominique Muller

UNIT 1.8

Patch-Pipet Recording in Brain Slices

Patch-clamp recording in brain slices provides a powerful approach for investigating the intrinsic electrical properties of neurons and glia and analyzing synaptic interactions between neurons. For troubleshooting patch-pipet recording see Table 1.8.1 at the end of this unit.

BASIC PROTOCOL 1

THE BLIND TECHNIQUE

Materials (see APPENDIX 1 *for items with* ✓)
  ✓ Extracellular artificial cerebrospinal fluid (aCSF) solution
    Carbogen gas mix (95% O_2/5% CO_2)
  ✓ Intracellular patch-pipet solution

  Experimental chamber (either immersion or interface type) for holding brain slices during recording
  Basic patch-clamp setup and associated equipment, including amplifier (headstage and pipet holder), manipulator, and computer (UNIT 1.2)
  Patch pipets (UNIT 1.1)
  Dissecting microscope

1. Place brain slice in the experimental chamber and perfuse with oxygenated (with carbogen gas mix) extracellular aCSF at a rate of ∼1 ml/min (see UNIT 1.3). Fill a patch pipet with intracellular patch-pipet solution and place it in the pipet holder.

2. Apply slight positive pressure, ~20 mm Hg, to the back of the pipet via a piece of tubing attached to the side port of the pipet holder, either by mouth or by depressing the plunger of a 5-ml syringe by ~0.2 ml (to prevent particles in the bath solution entering and fouling the pipet tip when the pipet is lowered into the bath solution). Maintain pressure by closing a valve or three-way tap.

3. Lower pipet into the oxygenated aCSF in the experimental chamber and position it (with the aid of a dissecting microscope at low magnification) over the area of interest, but not touching the slice.

4. Monitor pipet resistance on an oscilloscope or computer screen by applying small (5 to 10 mV), brief (10 ms), positive or negative voltage commands to the voltage-clamped patch pipet at ≥ 10 Hz. Determine the current response using Ohm's law: pipet resistance = test pulse command voltage/size of current deflection.

   For example, with a 10-mV test pulse and an amplifier gain of 1 mV/pA, a 2-V deflection on the oscilloscope screen indicates a pipet resistance of 5 MΩ. Smaller deflections mean higher pipet resistance; larger deflections imply lower pipet resistance.

5. Prior to movement into the slice, apply strong (~100 mm Hg) constant pressure to the back of the pipet, either via a 5-ml syringe (depress ~2 ml) or mouth (blow hard, as if blowing up a balloon).

6. Move pipet into the slice, maintaining high pressure to blow away debris on the slice surface, creating a clear path for the pipet tip. Once pipet is in the slice, reduce positive pressure (between 20 and 100 mm Hg are used at this stage for patch pipets with resistances between 2 and 10 MΩ, higher pressures for smaller-tipped, higher-resistance pipets) so as not to blow away the cell of interest.

7. Slowly (at a few µm/sec) move pipet axially (preferred) or vertically through the slice while watching for a sudden increase in pipet resistance, signified by a decrease in the test-pulse current by 20% to 50% within a few seconds, indicating contact with a cell. When this occurs, immediately release the positive pressure. There should be a further decrease in the test-pulse current, indicating formation of a gigaohm seal. If this does not occur, discard the pipet and make another attempt with a fresh pipet.

8. Apply light suction by mouth until a high-resistance (>1 GΩ) seal is formed. While the seal is forming, hold the patch pipet at a holding potential of approximately -60 mV (this can aid seal formation). If a seal does not form within ~1 min, or requires excessive suction, discard the pipet and try again.

9. Once a gigaohm seal is formed, withdraw pipet by ~5 to 10 µm. Compensate capacitive transients in response to the test pulse at high gain (≥ 50 mV/pA) with little or no filtering of the current signal (low-pass filter set at ≥ 10 KHz) using the fast capacitance compensation controls on the patch-clamp amplifier.

10. Make recordings in the cell-attached mode, obtain a whole-cell recording the patch membrane by rupturing the patch membrane with brief pulses of mouth suction (see *UNIT 1.2*), or acquire excised patch recordings in the outside-out or inside-out configurations (see Hamill et al., 1981).

BASIC PROTOCOL 2

THE DIC TECHNIQUE

Movement of the patch pipet through the brain slice and guidance of the tip of the pipet onto a cell soma or process can be achieved under visual control using a high-power upright microscope equipped with water-immersion DIC optics. Increased resolution can be obtained by using infrared illumination and an infrared-sensitive video camera and viewing the image on a video monitor.

Materials *(see APPENDIX 1 for items with ✓)*

✓ Extracellular artificial cerebrospinal fluid (aCSF) solution
Carbogen gas mix (95% O_2/5% CO_2)
✓ Intracellular patch-pipet solution

Experimental chamber with glass bottom, with sides tapered to provide access of pipets angled at ~15° to 20° to the horizontal
Microscope: upright, preferably of "fixed-stage" type, with DIC or Hoffman optics and high-aperture (>0.7) water-immersion lens and condenser (e.g., Zeiss or Olympus)
Infrared band-pass filter, maximum transmittance near 775 nm and a bandwidth of ~50 nm at half amplitude (e.g., Omega Optical PIN 770 WB 40)
Infrared-sensitive video camera (e.g., Newvicon C2400-07)
Black-and-white video monitor
Pipet manipulator, with movement capability of ≤1 μm, motorized remote-control preferred (e.g., Sutter)
Patch pipets (UNIT 1.1)

1. Place brain slice in the experimental chamber and perfuse with oxygenated extracellular aCSF (see UNIT 1.3). Lower the water-immersion lens of the microscope into the bath solution and focus it on the surface of the slice.

2. Adjust microscope optics for best resolution providing Kohler illumination as follows:

    a. fully close down the microscope field diaphragm and focus the condenser so that the image of the field diaphragm is in focus when the microscope is focused on the surface of the slice,

    b. make sure that the condenser is centered correctly,

    c. make sure that the polarizer is in a position where the image appears darkest,

    d. fully offset the DIC prism above the objective, and

    e. close down the condenser diaphragm to increased contrast when viewing the image through the microscope eyepieces.

3. *For viewing on a video monitor:* Fully open the condenser diaphragm when viewing the image with a video camera., switch the light path to the camera port of the microscope, and move the infrared filter into position in the light path.

4. Select a cell soma or process to be patched and position it in the middle of the field of view. Withdraw the lens from the bath solution without altering the position of the slice.

5. Fill a patch pipet with intracellular patch-pipet solution and place it in the pipet holder, apply slight pressure to the back of the pipet (see Basic Protocol 1, step 2), and lower the pipet into solution and position it above the slice over the selected cell or process. Monitor the pipet resistance in response to a test pulse on an oscilloscope or computer screen (see Basic Protocol 1, step 4).

6. Prior to movement into the slice, adjust the positive pressure (between 20 and 100 mm Hg are used at this stage for patch pipets with resistances between 2 and 10 MΩ, higher pressures for smaller-tipped, higher-resistance pipets) applied to the patch pipet. Maintain pressure by closing a valve or three-way tap.

7. Slowly (at a few μm/sec), under visual control, advance the patch pipet though the slice toward the selected cell soma or process, maintaining pressure to push cells and debris away from the advancing pipet tip. Touch the membrane of cell or process to be patched, seen as

the slight dimpling of the membrane or the small displacement of the process. Immediately release positive pressure, noting a decrease of the test-pulse current by approximately one-third (indicating that formation of a high-resistance seal will occur).

8. Apply slight negative pressure by mouth until a high-resistance (>1-GΩ) seal is established. While the seal is forming, hold the patch pipet at a holding potential of approximately −60 mV (this can aid seal formation). If seal formation does not occur within ∼1 min or requires excessive suction, discard the pipet and try again.

9. Once a gigaohm seal is formed, withdraw pipet a little so that it does not deform the cell or process. Compensate capacitive transients in response to the test pulse (see Basic Protocol 1, step 9).

10. Make recordings in the cell-attached mode, obtain a whole-cell recording of the patch membrane by rupturing the patch membrane with brief pulses of mouth suction (see UNIT 1.2), or use other variants of the patch-clamp technique (see Hamill et al., 1981).

References: Blanton et al.,1989; Stuart et al., 1993

Contributor: Greg Stuart

Table 1.8.1 Troubleshooting Guide for Patch-Pipet Recording

Problem	Possible cause	Solution
No test pulse response	No contact of pipet solution with silver wire in pipet holder	Fill pipet with more solution or use longer silver wire
	Bath ground not in place	Put bath ground in bath
	No voltage command going to patch amplifier or amplifier switched off	Connect or switch on voltage command
	Pipet blocked, e.g., by air bubble	Tap pipet to release bubbles or use a new pipet
Smaller than expected or fluctuating test pulse response	Pipet tip partially blocked	Tap pipet to release bubbles or use a new pipet
Variable DC offset	Poor chloriding of silver wire in pipet holder and/or bath ground	Re-chloride silver wire and/or bath ground
No pressure in system	Leak in tubing	Repair tubing
	Pipet holder not airtight	Replace O-rings
No flow of solution from pipet tip	No pressure	Check for leaks (see above)
	Pipet blocked by air bubble	Tap pipet to release bubbles or use a new pipet; if blockage repeatedly occurs, refilter intracellular solution or make fresh
Optics poor	Object not Kohler illuminated	Check and adjust (see Basic Protocol 2)
	Polarizers not crossed	Rotate polarizer until image is darkest
	Lens or optics dirty	Clean
	Air bubble under objective	Withdraw objective and refocus

continued

Table 1.8.1 Troubleshooting Guide for Patch-Pipet Recording (*continued*)

Problem	Possible cause	Solution
Seals form slowly or not at all	Pipet dirty	Replace pipet
	Cell membrane not clean due to loss of pressure or flow of solution from pipet tip	See above
	Cells unhealthy	Take new slice, reslice, or choose different cell to patch
	Patching the wrong type of cell	If patching under visual control, patch only cells that look smooth and can easily be dimpled by the patch pipet tip
Seals form quickly, but cells are sick	Patching unhealthy cells near the slice surface	Get new slice, reslice, or choose different cell to patch
Cannot break in to obtain whole-cell recording	Too much suction used to get seal	Use less suction
	Loss of pressure	Check for leaks
	Cells unhealthy	Take new slice, reslice, or choose different cell to patch

UNIT 1.9

Single-Channel Recording

STRATEGIC PLANNING

Electrophysiology Setup

The basic electrophysiology setup is essentially identical to that used for the whole-cell voltage clamp (*UNIT 1.2*). The setup must be equipped with a microscope, vibration isolation table, micromanipulator, and additional items mentioned below. Since the mechanical operations leading to a gigaseal are the same in both whole-cell recording and single-channel recording, the required resolution of <1 μm for visualization and positioning is the same, and mechanical stability is essential. Because a low noise level is necessary to detect single-channel currents, extra care must be taken to remove instrumental ground loops that pick up 60-Hz noise. It is critical that metallic components be grounded and that the setup be shielded. A Faraday cage around the setup is usually necessary, although a small metal shield surrounding the preparation on the microscope stage also works well.

Amplifier

Patch-clamp amplifiers can be used either for whole-cell recording (*UNIT 1.2*) or for single-channel recording. If the amplifier will be used only for single-channel recording, the primary consideration is circuit noise. Features such as whole-cell transient cancellation circuitry and series resistance compensation are not necessary, but the amplifiers with the lowest noise usually will have these features. When comparing noise specifications of different amplifiers, the bandwidth should always be taken into account. The Axopatch 200B (Axon Instruments) has somewhat lower open-circuit noise than other commercially available patch-clamp amplifiers.

Electrode Holder

An electrode holder should meet the specifications described in UNIT 1.2. Pressure and suction are still applied during some of the steps, but a pressure meter is less important if no outside-out patches are planned. Pressure readings are most useful during the break-in step of patch clamping, and this step is not used for creating cell-attached or excised inside-out patches.

Patch Electrodes

Patch electrodes (UNIT 1.1) are fabricated, coated, and fire-polished. As emphasized in UNIT 1.2, clean glass surfaces and filtered filling solutions are critical for gigaseal formation. Because of dirt accumulation on the glass surface, electrodes should be used within a day of fabrication. The tip size will determine the area of the patch of membrane through which current is recorded. This may be an important consideration, depending on the membrane density of channels. Simultaneous channel openings make certain kinds of kinetic analysis difficult). Smaller tip sizes will reduce the number of openings by reducing the number of channels in a patch. If the channel density is too low, openings may be infrequent, leading to poor statistical properties in the data analysis. In this case, a larger electrode tip will isolate larger patches that contain more channels, so that more openings will be seen. These considerations must be balanced against the ease of seal formation, which tends to have an optimal tip size that may vary from preparation to preparation.

Many different kinds of glass are available for patch electrode fabrication, and the choice of glass is important for low-noise recording. Lower noise can be obtained with electrodes fabricated from aluminosilicate glass compared to borosilicate glass; flint glass is worse. Quartz electrodes have better noise performance, but fabrication of patch electrodes from quartz is very difficult (UNIT 1.1). Even with borosilicate glass, noise can be reduced appreciably by using thick-walled capillaries, small tip diameters, and low immersion depths.

Electrode coating with Sylgard (184 silicone elastomer, Dow Corning) is critical for noise reduction. It is also important for examining rapid kinetic processes for which a wide instrumental bandwidth is needed. To obtain the lowest possible noise levels, the Sylgard coating should extend to within 50 to 100 μm of the electrode tip. It is especially important to fire polish the electrode tip after coating and curing the Sylgard, because the uncured resin spreads to the tip of the electrode, preventing seal formation. However, even with this precaution it is common to experience difficulty in sealing with Sylgard-coated patch electrodes, and the ratio of elastomer to hardener is critical in minimizing this difficulty. As with whole-cell recording (UNIT 1.2), it is important to coat the barrel of the patch electrode to a sufficient height to prevent any water from contacting the glass above the coating.

Preparation

Cells can be studied either in cultures or in slices (see UNITS 1.3 & 1.10). As with whole-cell voltage clamp recording (see UNIT 1.2, Basic Protocol 1), a naked cell membrane is essential for seal formation. In contrast to whole-cell recording, large cells and elaborate morphology present no problem in single-channel recording. However, with very small cells there is a problem in the cell-attached configuration because the gating of a large channel will generate enough membrane current to change the voltage inside the cell. Single-channel current steps will then be followed by slow relaxations, not unlike those seen when a vesicle forms at the electrode tip after patch excision.

Patch Electrode Solution

The composition of the patch electrode filling solution will depend on what type of channel is to be studied and which patch-clamp configuration is to be used. It is vital to keep the patch orientation in mind when designing both the patch electrode solution and the experimental bathing solution (see below). One or both of these solutions must supply the permeant ion for the channel of interest. Other unwanted channel currents can be blocked by adding appropriate pharmacological agents. Many of the current separation strategies in whole-cell recording (*UNIT 1.2*) are relevant to the isolation of specific single-channel currents. In general, solution compositions vary depending on the preparation and channel type. The comments here provide some useful guidelines, but investigators should consult published reports on channels and cells of interest for solutions tailored for specific applications.

Cell-attached and inside-out

Similar considerations go into the design of patch electrode solutions for cell-attached and inside-out patches. The patch electrode solution is in contact with the extracellular membrane face, so the solution should mimic an extracellular environment. Patch integrity and recording noise are largely insensitive to substitutions among standard monovalent cations and anions. The presence of both Ca^{2+} and Mg^{2+} (1 to 2 mM each) is often important for patch integrity, and this must be tested on a case-by-case basis.

Outside-out

For outside-out patches, the patch electrode solution comes in contact with the intracellular membrane face so the solution should mimic an intracellular environment. It is standard practice to use a chelator to maintain submicromolar Ca^{2+} levels. Many of the guidelines for design of the patch electrode solution in whole-cell recording (*UNIT 1.2*) are relevant to studies in outside-out patches.

Physiological Bathing Solution

To keep the tissue healthy, it is maintained in a physiological bathing solution during preparation and during the setup and formation of the gigaseal. The decision of whether to use the physiological bathing solution during actual recordings will depend on the objectives of the experiment (see Experimental Bathing Solution).

Experimental Bathing Solution

Cell-attached

In many experiments with cell-attached patches, the physiological bathing solution (see above) can be used. Because the bathing solution has no direct contact with either membrane face in a cell-attached patch, manipulation of this solution has virtually no value in the isolation of current through a specific channel type. However, the bathing solution will influence the resting membrane potential, which, together with the pipet holding potential, determines the driving force across the membrane. If the cell membrane potential is not independently measured it can be estimated for the selected bathing solution. To reduce or eliminate the resting membrane potential, the experimental bathing solution can be replaced with isotonic KCl. However, this usually leaves the preparation unsuitable for further recording.

Outside-out

As with cell-attached patches, the physiological bathing solution can often be used in recordings from outside-out patches. However, in this patch configuration the bathing solution is at the outer membrane surface, so it can provide permeating ions. Although ion substitution (for the purpose of isolating currents through specific channels of interest) is common in such experiments, an effort is usually made to maintain an environment that mimics the extracellular milieu. Millimolar Ca^{2+} is usually important; high Na^+ and low K^+ are preferred, but are usually less important.

Inside-out
Conversely, the experimental bathing solution must mimic an intracellular environment for inside-out patches. Low Ca^{2+} is normally maintained with a chelator. Channels often run down (stop opening) after formation of an inside-out patch, and ATP is often effective in preventing this. In other situations enzymes must be added to prevent rundown. Guidelines for the prevention of rundown of membrane currents during whole-cell recording (UNIT 1.2) may be useful in this regard. As noted with outside-out patches, ion substitution in the bathing solution is a common practice with inside-out patches.

Data Acquisition System
Single-channel data can be acquired either in long continuous stretches or in short segments, depending on the nature of the experiment. For experiments on stationary channel properties, long continuous recordings can be made with a tape recorder. A modified video recorder interfaced to a pulse-code modulator is a cheap commercially available system that works well in most situations (Bezanilla, 1985; French and Wonderlin, 1992). The data acquisition rate of these systems is set by the manufacturer to a predetermined value, which is fast enough for most purposes, but may make the system unsuitable for investigating very rapid processes. Alternatively, data can be acquired through an interface into a computer (French and Wonderlin, 1992). However, this can require a large amount of disk space, and a backup system with large capacity may be needed.

Software for Data Acquisition and Analysis
There are many methods of analyzing single-channel data and virtually all require a computer and specialized software. For analysis of stationary channel properties a computer program will be needed for continuous acquisition of data to a disk drive. Even if the data is collected with a tape system (see Data Acquisition System, above), the data will then have to be read into a computer for analysis. It is important to make sure that the acquisition software saves data in a format that is easily read by the analysis software. For voltage-gated currents, data acquisition will have to be synchronized to voltage pulses, and the software discussed in UNIT 1.2 for whole-cell recording will be appropriate.

A variety of general-purpose single-channel analysis computer programs are available (French and Wonderlin, 1992). The software package pClamp (Axon Instruments) has considerable analysis capability for use with PCs. The program TAC (Instrutech) can be used with a MacIntosh. The considerations in selecting the most appropriate software package for analysis of single-channel data are complex and varied and depend strongly on the type of analysis planned. Because of the specialized nature of single-channel analysis, many laboratories have developed their own software, which tends to be idiosyncratic and difficult to use.

BASIC PROTOCOL 1

PATCH FORMATION FOR SINGLE-CHANNEL RECORDING

The setup, electrode positioning, and gigaseal formation for single-channel recording are the same as in whole-cell recording (UNIT 1.2), up to the point where a cell-attached patch is attained.

Materials

Cells or tissue slice (see UNITS 1.3 & 1.10)
Patch electrode solution (see Strategic Planning)
Physiological and/or experimental bathing solution (see Strategic Planning)

Coated patch electrode (see Strategic Planning and UNITS 1.1 & 1.2)
Electrophysiology setup (see Strategic Planning and UNIT 1.2)

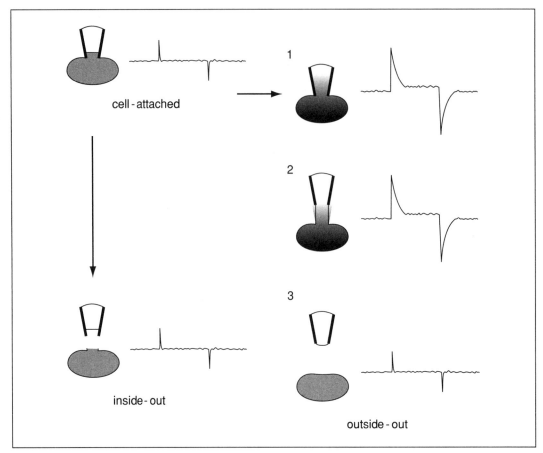

Figure 1.9.1 Patch electrode configurations are shown along with the current responses to a small test pulse (typically 10 msec at 10 mV). The cell-attached patch (top) is the common starting point for all single-channel recordings. Experiments can be performed in this configuration, or an excised patch can be prepared. A simple rapid backward motion of the patch pipet leads to excision in the inside-out configuration, as shown on the left. When a slow backward motion follows break-in, as shown in a sequence of steps on the right, an outside-out patch is formed. In these two excised patch configurations, the orientation of the membrane is opposite.

1. Set up and position the electrode (see UNIT 1.2, Basic Protocol 1, steps 1 to 5). Establish gigaseal formation (see UNIT 1.2, Basic Protocol 1, step 6). Monitor the electrical resistance with a test pulse (shown to the right for each stage of recording in Fig. 1.9.1).

   When studying very small channel currents or rapid gating processes, seal resistances of 10 GΩ or higher may be desired to reduce noise. Reducing the electrode capacitance is also important for low-noise single-channel recording, especially when a wide bandwidth is desired. It may be appropriate to impose a minimum acceptable seal resistance or a signal-to-noise ratio for the experimental objective at hand. For large conductance or slowly gating channels the seal resistance and capacitance are less critical, and channel currents may be clear enough for study in cell-attached patches even with a seal resistance <1 GΩ. However, whenever patch excision is planned, seal resistances >5 GΩ will be necessary.

2. Perform electrode transient cancellation (see UNIT 1.2, Basic Protocol 1, steps 7 and 8) if voltage-gated channels are to be studied using voltage steps (see Basic Protocol 2). Use the faster controls (C-fast and τ-fast on the EPC-7) to remove most if not all of the transient. Recording in cell-attached patches may proceed at this point.

When data are acquired at a fixed voltage, there is no need to remove capacitance transients accompanying voltage changes (Fig. 1.9.1). If a small slow component of transient current remains visible after making these adjustments, further adjustment of the whole-cell transient cancellation circuitry can be made (C-slow and G-series on the EPC-7).

3. For inside-out patches change to an appropriate experimental bathing solution prior to patch excision (if the gigaseal was formed in a physiological bathing solution; see Strategic Planning; Horn and Patlak, 1980; Hamill et al., 1981).

4a. *Excise inside-out patch:* Pull the electrode back using a somewhat sharp movement to excise the patch cleanly. Alternatively, gently tap the micromanipulator while the electrode has a cell-attached patch at the tip.

   The movement of the patch electrode away from the cell should not be so slow as to allow the cell membrane to be stretched because this might lead to the formation of a vesicle at the electrode tip. Vesicles can also form spontaneously after excision, and channel records should be checked periodically for the rounding of gating transients that indicates a vesicle. After a vesicle forms, the patch can sometimes be salvaged by drawing the pipet tip out of the bathing solution and quickly returning it.

4b. *Excise outside-out patch:* Break in using strong suction (see UNIT 1.2, step 9 of Basic Protocol 1). Slowly draw the patch electrode away from the cell (a considerable distance because membranes stretch; see Fig. 1.9.1). Watch the current response to the test pulse for a marked reduction in the amplitude and time constant (Fig. 1.9.1).

BASIC PROTOCOL 2

DATA ACQUISITION AND PULSE SEQUENCES

Continuous Acquisition

When stationary kinetic processes are to be studied, data can be acquired continuously. The data is simply read into the acquisition device after adjusting the holding potential to the desired value, adjusting the filtering cutoff frequency (f_c), and selecting the sampling interval. The length of time for data sampling will depend on how frequently the channels open and how many channel events will be needed for the type of analysis planned. In general, more channel openings translates into higher quality data because of the statistical nature of single-channel analysis.

Pulsed Acquisition

To study voltage-gated channels or voltage-dependent kinetic processes, the voltage can be changed in a stepwise fashion. This is generally done under computer control, and the software developed for whole-cell recording is suitable for this purpose (see UNIT 1.2). The basic process of applying pulses and acquiring the data is identical to whole-cell recording, but the analysis can differ substantially. Transient cancellation (see Basic Protocol 1, step 2) is important with pulse-mode acquisition of single-channel data because capacitance charging during the voltage change can obscure rapid kinetic processes. As with whole-cell recording, this charging transient occurs while the voltage is changing, and so channel currents appearing during this time must be interpreted with cognizance of the fact that the voltage has not yet settled (UNIT 1.2). The precision of timing of voltage pulses in patches has been discussed by Sigworth and Zhou (1992). Generally, voltage settling times in patches are much more brief (e.g., ~100-fold) than those in whole-cell recording.

A cell membrane may contain many channel types, but single-channel patches can be found by random trial. Such patches provide an opportunity to study this channel in isolation in a mode that is free of space-clamp errors and in which the dynamics of the point-clamp are very favorable (see UNIT 1.2). Averages of many single-channel traces from such patches are therefore equivalent to very high quality macroscopic currents, to which the pulse sequences and analysis strategies discussed in UNIT 1.2 can be applied very effectively.

If the charging transient is very large and cannot be removed with the fast transient cancellation circuitry, it usually means one of two things. First the electrode may have been poorly coated with Sylgard, so that the solution in the recording chamber is somehow making contact with glass above the coating. Second, the excision of the patch may not be complete, so that the electrode tip still has some access to the cell interior.

Reference: Jackson, 1993

Contributor: Meyer B. Jackson

UNIT 1.10

Acute Isolation of Neurons from the Mature Mammalian Central Nervous System

It is of the utmost importance that the brain be removed rapidly, yet gently, and that the tissue be handled gently without squeezing. The time from decapitation of the animal to when the slices are introduced to the oxygenated saline should be minimized; with practice this can be done in <5 min.

BASIC PROTOCOL

ACUTE ISOLATION OF NEURONS FROM THE MATURE MAMMALIAN CENTRAL NERVOUS SYSTEM

Materials (see APPENDIX 1 for items with ✓)

    100% ethanol
✓ PIPES saline
    100- to 300-g rat
✓ Proteinase K solution
✓ Trypsin solution

    Tissue chopper (e.g., Stoelting, McIlwain from Brinkman) or slicer
    25-ml spinner flask (Bellco)
    100% oxygen tank and regulator
    60-mm glass petri dish cover
    Guillotine (e.g., Braintree Scientific)
    ~14-cm scissors
    14-cm fine-tipped rongeurs
    Fine-tipped forceps
    Spatula with tapered end
    Size 0 sable-hair paint brush
    Scalpel handle and no. 22 blade
    Pasteur pipet cut to have a wide mouth

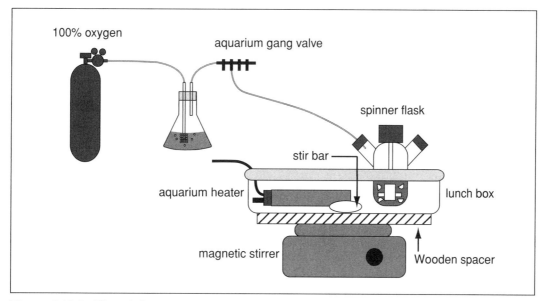

Figure 1.10.1 Dissociation apparatus. The temperature-controlled chamber is constructed from a polyethylene box (~14 × 25 × 5–cm, with clear bottom) and a submersible aquarium heater (Hagen; thermal compact, 6 in. [15 cm], 50 W). Up to four spinner flasks may be inserted through holes cut in the top of the lunch box. The holes should be centered on the edge of the metal plate forming the top of the stirrer. The box should be insulated from the surface of the magnetic stirrer to prevent heat transfer from the stirrer motor to the box; placing it on a 2-cm-thick rectangular wood frame is effective.

Magnetic stirrer (e.g., Cole Parmer)
Heating box fabricated from aquarium heater, gang valve, and polyethylene
  (e.g., Tupperware) box (see Fig. 1.10.1)

1. Prepare the tissue slicer or chopper. Clean new blade with 100% ethanol, then wash with water.

2. Fill the spinner flask with 15 ml PIPES saline and bubble with 100% O_2 for ~30 sec. Move the inlet tube above the level of the solution and continuously superfuse with O_2.

3. Place ~20 ml PIPES saline in a 30-ml beaker (to collect brain after removal) and ~5 ml in a 60-mm glass petri dish cover (to hold the slices after chopping). Place both in the −20°C freezer until ice begins to form. Bubble the ice-cold PIPES saline in the beaker and petri dish for ~30 sec with O_2.

4. Anesthetize the rat and, once it is deeply sedated, swiftly decapitate with a guillotine. Firmly grasp the scalp between thumb and forefinger and, with the scissors, cut through the scalp below the fingers, parallel to the skull surface and in the caudal direction.

5. Break open the occipital crest (the most caudal extension of the skull visible on the dorsal surface) by nipping the slight protrusion with rongeurs, making a hole of ~2-mm diameter. Insert scissors through the hole in the rostral direction. Cut along the midline, keeping the tips of the scissors just under the cranium until past the olfactory bulbs. Again insert scissors through the hole in the occipital crest and cut to the left and right of midline.

6. Remove the flaps of skull with rongeurs (both flaps should come off as single pieces). Pour a few milliliters of ice-cold PIPES saline over the cortex. If the meninges have not been removed, lift away with fine-tipped forceps. Cut through the spinal cord with the fine end of the spatula. Holding the head in the palm of the hand, tilt the rostral end up, insert

spatula under the brain, and gingerly lever the whole brain into beaker of ice-cold PIPES saline.

7. Block off the portion of the brain from which slices are to be made. Cut slices at ∼500 μm with a tissue chopper or slicer or by hand. Use a paint brush wetted with saline to transfer slices into the petri dish of ice-cold PIPES saline.

8. Cut slices into ∼2 × 2–mm pieces with a scalpel blade, removing as much white matter as possible. Transport slices into the spinner flask using a Pasteur pipet cut to have a wide mouth. Gently stir slices with the magnetic stirrer so they are just carried up by the currents.

9. Heat the water in the heating box to 30°C. Once this temperature is reached (∼5 min), add one 100-μl aliquot of proteinase K solution to the spinner flask. Incubate 5 min.

10. Discard the saline in the spinner flask, making sure that slices are not lost in the process. Wash slices with 2 to 3 ml fresh PIPES saline. Add 15 ml PIPES saline and one 500-μl aliquot of trypsin solution. Incubate 30 min at 30°C.

11. Discard the solution in the spinner flask and wash slices with 2 to 3 ml PIPES saline. Add ∼40 ml PIPES saline to the flask and allow it to come to room temperature. Oxygenate every change of solution ∼30 sec by bubbling; then move the oxygen inlet tube above the solution.

12. Place one slice in a 1.5-ml microcentrifuge tube with ∼0.5 ml fresh PIPES saline (do not oxygenate). Triturate the slices (draw into the pipet and then gently expel, avoiding the creation of bubbles) with a fire-polished Pasteur pipet (∼1 mm diameter), then by two to four passages through a pipet with a diameter of ∼0.2 mm, and then a few more passages through the larger-bore pipet.

13. Decant the solution into the recording chamber, allow a few minutes for the cells to settle, and then slowly replace the PIPES saline with an appropriate recording solution.

Reference: Kay and Wong, 1986

Contributors: Alan R. Kay and David J. Krupa

UNIT 1.11

Patch-Clamp Recording from Neuronal Dendrites

BASIC PROTOCOL 1

DENDRITIC RECORDING

The most effortless recordings of the highest quality come from fat, smooth, low-contrast dendritic processes. In fact, the more easily visible and high-contrast a dendrite is, the less likely it belongs to a healthy neuron. This highlights the importance of good IR-DIC optics in allowing one to seek out the less obvious processes. The adjustment of the IR-DIC optics to provide maximum contrast and penetration into tissue is essential (*UNIT 1.8*). The authors have found dichroic IR filters (e.g., Chroma D770/40) to be better than low-pass RG9 filters. Likewise, the Newvicon IR-sensitive video camera is superior to most CCD cameras, although the Hamamatsu C7500-50 comes close and is much smaller in size. An improved IR-DIC image pays dividends in electrophysiology results by allowing identification of healthy, relatively low-contrast processes, which may lie deep in the slice.

Materials *(see APPENDIX 1 for items with ✓)*

    Brain slice (see Support Protocol)
- ✓ Extracellular solution
- ✓ Internal pipet solution, cell-attached (for K^+ currents)
- ✓ Internal pipet solution, whole-cell

    Experimental chamber (see Fig. 1.11.1B)
    Upright microscope with IR-DIC optics (e.g., Zeiss Axioskop 2 FSplus, Olympus BX-51), 40× to 63× water-immersion objective, IR-sensitive video or CCD camera (e.g., Dage Instruments Newvicon, Hamamatsu C7500-50), and high-contrast monitor (e.g., Dage Instruments HR120)
    34-G platinum wire semicircle held with nylon threads (see Fig.1.11.1B)

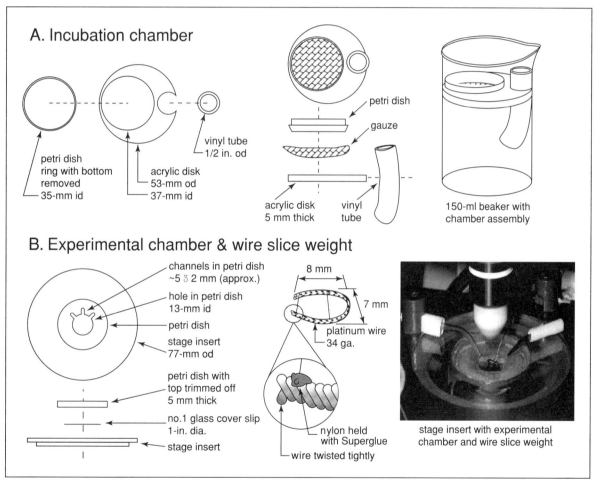

Figure 1.11.1 Schematics for apparatus construction. (**A**) Incubation chamber for warming slices. The incubation chamber can consist of a 150-ml beaker fitted with an acrylic ring that accommodates a 35-mm plastic petri dish. The bottom of the petri dish is removed (e.g., with a Dremel tool), giving a plastic rim that fits within the acrylic ring. A 2 × 2–in. piece of gauze is secured between the plastic rim and the acrylic ring to form a submerged surface to support the incubating slices. (**B**) Experimental chamber and platinum wire weight used to hold slice in place during recordings. The photo of the complete experimental assembly shows bath inlet and outlet tubes, as well as the ground electrode, arranged in the appropriate petri dish channels.

Patch pipets (*UNIT 1.1*), preferably coated with Sylgard and fire-polished with a microforge (e.g., Narishige MF 830) just before use
5-ml syringes with tubing appropriate for pipet
Flexible needle (e.g., Kwick-fill, World Precision Instruments)
Electrophysiology setup (*UNIT 1.2*) with high-resolution electromechanical, hydraulic, or piezoelectric micromanipulator (e.g., Sutter MP-225)
Pressure gauge (e.g., Dwyer Magnelic 2050)

1. Secure a brain slice in an experimental chamber filled with extracellular solution with the "good side" up (see Support Protocol). Weigh down the slice using a 34-G platinum wire formed into a semicircle encompassing the slice, with one or two nylon threads spanning the wire that will cross the slice and hold it down (Fig. 1.11.1B).

2. Visualize slice on an upright microscope with IR-DIC optics, IR-sensitive video or CCD camera, and high-contrast monitor, and identify a viable dendrite of interest (*UNIT 1.8*) by finding processes that are plump, low-contrast, and smooth, and that have a "ground-glass" appearance. Choose a dendrite larger than the diameter of the pipet tip itself (1 to 2 μm) and ensure that it belongs to the neuron type of interest, i.e., that it possesses the expected morphology and is continuous with a soma of expected location and morphology.

3. Select a patch pipet, preferably one coated with Sylgard and fire-polished just before use—a small tip to make it possible to patch smaller processes (down to ∼1 μm) or a larger tip to maximize the current amplitudes that can be recorded in cell-attached voltage clamp, reducing series resistance in whole-cell recordings. In general, unless isolating single channels, use the largest pipet tip practical for the dendrite to be patched.

4. Back-fill the pipet tip with appropriate internal pipet solution (depending on the purpose of the experiment or the ion currents to be isolated) by immersing the tip in the solution and applying suction through a 5-ml syringe connected to the pipet end with tubing. Back fill the barrel of the pipet about halfway using a syringe and flexible needle.

5. Attach pipet to the micromanipulator so that pipet is angled at 30° to 45° from the horizontal plane. Apply steady positive pressure (10 in. H_2O or 18 mmHg) to side port using tubing connected to a pressure gauge.

6. Advance the pipet into the fluid meniscus formed under the microscope objective and locate the pipet tip in the center of the video screen. Monitor pipet resistance in voltage-clamp mode by applying repetitive 5-mV commands (a 1- to 2-μm tip size produced by fire polishing yields resistance of 5 to 10 MΩ).

7. Lower pipet near surface of slice. Confirm location of dendrite and position pipet so that it is displaced in the x axis (perpendicular to long axis of dendrite) a small distance away from the process. Apply a stronger positive pressure to the pipet, ∼30 to 40 in. H_2O (or 55 to 75 mm Hg), and lower pipet into slice, observing a pressure wave of internal solution flowing from the pipet tip.

8. Slowly advance the pipet tip toward the dendrite in both z (perpendicular to surface of slice) and x axes. If necessary, maneuver the pipet around other large dendrites or somata. Position pipet tip so that it is just above (z axis) the dendrite and centered on it in the x axis and lower pipet onto dendrite slowly, depressing surface of cell in to a dimple, resulting from the pressure of internal solution flow. Move the tip, as necessary, in small increments in the x axis to keep it centered on the process and prevent it from moving away under the flow of solution.

9. Once the dimple is seen, release positive pressure on the pipet and apply a small amount of negative pressure while monitoring pipet resistance, which will gradually increase to >1 GΩ as the seal forms.

10. If the seal does not form adequately, apply stronger negative pressure to pull more membrane into the pipet tip. If a gigaohm seal doesn't form within seconds to several minutes, abandon the recording and attempt with another dendrite.

    A high gigaohm seal that forms almost instantly after release of positive pressure can occur with nonviable "ghost" processes.

11. After the seal has formed, reposition the pipet by withdrawing it slightly in the *x*-axis direction, as removing the flow of internal solution from the pipet often causes the surrounding tissue to recoil somewhat.

12a. *For recording in cell-attached voltage-clamp mode:* Adjust capacitive transients using the amplifier and begin data collection.

12b. *For recording in whole-cell mode:* Set the voltage command to −60 mV to prevent depolarization of the cell to near 0 mV upon break-in. Rupture the patch by applying strong negative pressure. Switch to current-clamp mode and apply small repetitive current steps (e.g., 100 pA to produce voltage excursions <10 mV) to monitor series resistance.

13b. Gently apply several puffs of positive pressure to dislodge membrane fragments back into the cell, taking care to produce a brief, small pressure wave visible in the cell to avoid clogging the tip (too little pressure) or ruining the gigaohm seal (too much pressure).

14b. After observing a decline in the series resistance to 10 to 20 MΩ, adjust series resistance and capacitance compensation on the amplifier and begin data collection, periodically monitoring series resistance to keep it minimized.

BASIC PROTOCOL 2

SIMULTANEOUS DENDRITIC AND SOMATIC RECORDING

1. Using the same criteria described above (see Basic Protocol 1, steps 1 and 2), identify a viable neuron of correct morphology with a clearly visible soma and apical dendrite. If necessary for somata that are more than one or two cell layers deep in a stratum (e.g., stratum pyramidale in hippocampus), block the tissue prior to slicing such that the dendrites are oriented "down" into the slice, with the somata closer to the surface (see Support Protocol).

2. Establish dendritic recording first, using the better micromanipulator (for this more difficult recording) and the same methods described in the remaining steps of Basic Protocol 1. Optionally, insert the somatic recording pipet into the bath and position it near the soma with only a small amount of positive pressure before approaching the dendritic site to lessen the chances of jostling the dendritic pipet while positioning the somatic pipet in the bath.

3. Form the somatic recording using the same methods described in Basic Protocol 1.

    In whole-cell mode significant depolarization may be observed at the dendritic site when the cell is approached by a somatic pipet ejecting K^+-based internal solution under pressure; this phenomenon cannot be avoided. Note that the cell will repolarize after the seal has formed.

SUPPORT PROTOCOL

PREPARATION OF BRAIN SLICES

One of the most important areas of variability in dendritic recording is slice health, particularly when mature animals are used. Slice health tends to decline significantly after about 3 weeks of age in rats, yet neuronal ion channel properties in pyramidal neurons have not reached their adult configuration until 5 to 6 weeks. IR-DIC imaging has made it all too clear that brain slices of adult animals can have dismally unhealthy-appearing neurons, even though extracellularly recorded field responses (e.g., population spikes) are present. Nonetheless, careful attention to the process of brain slicing will yield tissue with many healthy neurons and dendrites worthy of patch-clamping.

This method produces brain slices of high quality with viable dendritic processes. The central difference from more conventional methods involves intracardiac perfusion of a low Na^+ and Ca^{2+} solution to clear red blood cells from the brain (which diminish the quality of IR-DIC) and to achieve partial replacement of the brain extracellular milieu with a solution that minimizes Na^+ and Ca^{2+} loading into cells (diminishing the deleterious effects of hypoxia-induced excitotoxicity and Na^+ overload). Cooling the brain also mitigates excitotoxicity. Because the process of perfusion cools the brain less quickly than rapid decapitation and immersion into cold saline, it is important to chill the perfusate as much as possible and perfuse only as long as necessary to achieve wash-out of cerebral blood. Also described is a process for blocking the brain to produce brain slices with the optimal orientation of the apical dendrites from CA1 hippocampal pyramidal neurons.

Materials (*see* APPENDIX 1 *for items with* ✓)

    Anesthetic (e.g., 21 mg/kg ketamine, 3.2 mg/kg xylazine, and 0.7 mg/kg acepromazine given i.p.; pentobarbital; or volatile anesthetics)
    Experimental animal (e.g., rat)
✓ Perfusion solution, ice cold, oxygenated
    Carbogen (95% O_2/5% CO_2), for oxygenation
    Vetbond (cyanoacrylate glue, 3M)
✓ Extracellular solution, oxygenated

    Absorbent underpad (Fisher)
    No. 10 scalpel
    Large dissecting forceps, toothed
    Surgical scissors
    Large hemostat
    18- and 25-G hypodermic needles
    Intravenous (i.v.) perfusion set and delivery system (e.g., 60-ml syringe reservoir with appropriate tubing)
    Pyrex dish, round, 3-in. diameter with 1-in. sides
    Iridectomy scissors, carbide-edged
    Bone rongeurs
    Weighing minispatula with a 30° bend in the blade
    Filter paper disk, 2-in. diameter
    Petri dish, chilled
    Vibrating tissue slicer (e.g., Vibratome 1500; Electron Microscopy Sciences OTS 4000), preferably with glass (fabricated with a Ralph-type knife maker using 6-mm glass strips, Ted Pella) or synthetic sapphire blades
    Incubation chamber: 150-ml Pyrex beaker fitted with an acrylic ring and 35-mm-i.d. plastic petri dish to support gauze platform (see Fig. 1.11.1A)
    35°C water bath

Cutting blade, preferably glass (Pelco knife maker, Ted Pella) or artificial sapphire
(Delaware Diamond Knives)
Dumont tweezers

1. Anesthetize the experimental animal. Position the animal in a supine position on a surface covered with absorbent underpad, securing the upper limbs with laboratory tape so the abdomen is well exposed. Confirm that anesthesia is sufficiently deep by testing for the lack of lower limb retraction from a foot pinch.

2. Make a midline incision with a no. 10 scalpel through the skin from the upper thorax to the mid-abdomen. Grasp the abdominal wall with large, toothed dissecting forceps, and use surgical scissors to make a cut below the sternum through the abdominal wall and rib cage. Extend cuts in a U-shaped pattern to create a flap of the chest wall and retract with a large hemostat to completely expose the heart.

3. Insert an 18-G hypodermic needle connected to an i.v. perfusion set into the apex of the heart, advancing the tip just enough to enter the left ventricle. As soon as the needle is inserted into the heart, use scissors to make a cut into the right atrium, which will allow exsanguination during perfusion.

4. Perfuse with ice-cold oxygenated perfusion solution at a rate of 10 to 15 ml/min via an appropriate delivery system, using a gravity-fed delivery system such as a 60-ml syringe reservoir placed at a height of ∼24 in. above the animal to feed the i.v. perfusion set. Control the flow rate using the thumbwheel of the perfusion set. Keep the solution as close to 0°C as possible by running the perfusion tubing through an ice water bath. Continue perfusion until the fluid leaving the right atrium no longer resembles blood, but is pinkish or mostly clear (<90 to 120 sec).

5. Remove the head with surgical scissors and plunge it into ice-cold perfusion solution in a 3-in.-diameter Pyrex dish with 1-in. sides. Maintain the dish on ice throughout dissection to keep solution and brain cold.

6. Make a midline scalpel incision through the scalp. Using carbide-edged iridectomy scissors, cut through the midline of the skull, starting at the posterior aspect. Remove the skull flaps with bone rongeurs to expose the cerebral hemispheres.

7. Make anterior and posterior cuts through the brain in the coronal plane (Fig. 1.11.2A). Using a weighing minispatula with a 30° bend in the blade, sever the brainstem and dislodge the brain from the skull. Let the freed brain rest in cold perfusion solution for a minute, and then use a scalpel to make a mid-sagittal incision to separate the brain into two hemispheres (Fig. 1.11.2B).

8. Remove the left hemisphere (leaving the right hemisphere in the cold solution) and place medial-side-down on a 2-in.-diameter filter paper disk in a chilled petri dish. Orient the hemisphere so the anterior end faces forward and the posterior end faces toward the dissector.

9. Make a scalpel incision through the dorsal aspect of the hemisphere, angled rightward at 30° from the vertical. Transfer hemisphere to the vibratome stage using a weighing spatula so that the 30° cut surface becomes the base on which the brain rests, the posterior end faces forward, and the ventral aspect of the brain facing up and tilted to the right (Fig. 1.11.2C). Affix the cut surface to the stage with Vetbond. Fill slicing chamber with ice-cold perfusion solution.

The blocking procedure optimizes the angle of the apical dendrites within the tissue slice for CA1 hippocampal pyramidal neurons. The desired result is for the dendrites to course towards the surface of the slice ("up") along their length. This increases the likelihood of making distal apical recordings because the dendrite will be closer to the surface of the slice rather than disappearing

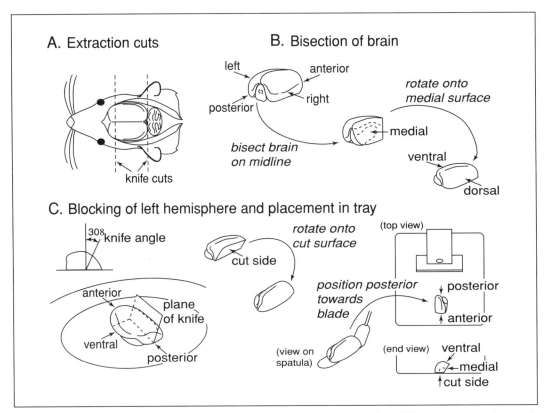

Figure 1.11.2 Procedure for blocking brain to produce optimal orientation of CA1 hippocampal pyramidal neuron apical dendrites. (**A**) Knife cuts used to extract brain. (**B**) Separation of cerebral hemispheres. (**C**) Blocking left hemisphere and placement in tiss-

> into the middle of the slice. A different orientation may be preferred for dual somatic/dendritic recordings (see Basic Protocol 2). Also, because the hippocampus is a curved structure, not all slices cut using this procedure will have the apical dendrites in the correct orientation. Finally, a different blocking procedure is required for preparing neocortical slices, and will depend on the cortical region being studied.

10. To slice the right hemisphere, repeat the blocking process, but angle the scalpel incision through the hemisphere leftward at 30°. Slice one hemisphere at a time, leaving the unsliced hemisphere in cold perfusion solution until ready for use (recommended). Alternatively, slice both hemispheres simultaneously.

11. Prior to perfusion, fill the incubation chamber (Fig. 1.11.1A) with extracellular solution and place in a 35°C water bath. Bubble the solution with carbogen via a 25-G hypodermic needle connected to tubing from the carbogen source. Set the carbogen flow at a level that will not agitate the slices.

12. Adjust the angle of the cutting blade to 20° to 25° from horizontal. Set the vibration amplitude high enough that slicing produces no evidence of shearing or stretching of tissue. Minimize the speed of the blade advance as much as practical while still allowing the overall sectioning process to occur in a reasonable period of time (e.g., 5 to 7 mm/min).

13. Make a first pass into the hemisphere, sectioning away unnecessary tissue until the desired tissue plane comes into view. Using Dumont tweezers, peel away the meningeal membranes adhering to the cortical surface, as they tend to resist cutting and can deform and stretch the tissue during passage of the blade.

14. Cut subsequent tissue slices at 350- to 400-μm thickness, taking care to keep the bath solution at 4°C or less. Arrange cut slices at the bottom of the bath with the just-cut surface face-up to expose the "good" side of the slice and save much time during the recording process.

15. After all slices have been cut but before removing them from the bath, block further as needed to remove unwanted tissue, such as any remaining portion of the thalamus and brain stem, allow for more durable and easy-to-handle slices

16. Using a dropper constructed from the blunt end of a Pasteur pipet, immediately transfer the slices to the incubation chamber (step 11) and incubate 10 min to 1 hr (experimentally determined) with carbogen bubbling to improve slice quality, possibly by rapidly restoring the activity of homeostatic ion pumps. Transfer the chamber to room temperature for 1 hr before recording, while still oxygenating.

Reference: Moyer and Brown, 1998

Contributors: Nicholas P. Poolos and Terrance D. Jones

CHAPTER 2
Neurochemistry/ Neuropharmacology

The interdisciplinary nature of contemporary neurochemistry and neuropharmacology is reflected in the range of protocols included in Chapter 2. UNIT 2.1 describes the scintillation proximity assay, which eliminates the step of separating bound from free radioligand, and thus has several advantages over conventional techniques (including lower cost, virtual elimination of the large volume of radioactive waste generated by conventional separation techniques, and higher sample throughput). As described in this unit, the principles of the scintillation proximity assay can be broadly applied to analysis of signal transduction pathways, making this an extremely valuable technique for neurochemists and neuropharmacologists.

The enzyme-linked immunosorbent assay (ELISA) remains a widely used tool for quantitation of analytes in biological materials. UNIT 2.2 describes an ELISA for the measurement of amyloid β-peptides. This unit emphasizes some of the challenges encountered in sample preparation, perhaps the most problematic aspect of an ELISA. The principles described in UNIT 2.2 may be generally applicable to analyte extraction and quantitation under denaturing conditions.

UNITS 2.3, 2.5 & 2.15 detail methods to assess the consequences of insults to the central nervous system. These units, together with materials presented in Chapter 4, will enable investigators to employ a multidisciplinary approach to study "life-and-death" processes in the central nervous system. UNIT 2.3 details methodologies to assess cell viability in neuron culture. Neuropharmacologists invariably encounter compounds that produce remarkable effects on a model system in vitro, but fail to elicit the predicted actions in vivo. This failure is often attributed to the inability of a compound to cross the blood brain barrier. UNIT 2.15 describes several approaches for measuring the penetration of xenobiotics (ranging from small molecules to peptides and proteins) into the rodent central nervous system. In UNIT 2.5, methods are described for the measurement of oxygen radicals and lipid peroxidation in neural tissues. These methods provide a means of assessing the extent of oxidative injury both in vitro and ex vivo.

The ability to monitor alterations in the levels of transmitters, hormones, and drugs induced by environmental manipulation, drug treatment, or engineered changes in the genome has become an integral component of contemporary neuroscience. An overview of microdialysis covering both historical and practical aspects is presented in UNIT 2.6. Protocols for microdialysis studies in the two most widely used laboratory species, rats and mice, are presented in UNIT 2.7, while UNIT 2.8 contains protocols for microdialysis in nonhuman primates. Protocols describing the analysis and quantitation of neurotransmitters and related substances collected in dialysates are found in UNIT 2.9.

UNITS 2.10 & 2.11 provide detailed protocols for implementing radioligand binding studies. While conventional radioligand binding remains an extraordinarily useful tool for neuroscientists, such approaches generate large quantities of liquid waste, and the separation of bound from free radioligand can be time consuming. The uptake and release of neurotransmitters is an enduring area of investigation. The recent cloning of uptake sites for a variety of neurotransmitters (e.g., GABA, glycine, serotonin, and dopamine) has invigorated studies on transmitter reuptake and release related to both drug action and disease. UNIT 2.11 provides detailed protocols on this topic.

In *UNIT 2.12*, protocols are provided for measurement of anion movement in a variety of preparations ranging from synaptoneurosomes (sometimes referred to as microsacs) and primary cultures to brain slices. In the accompanying *UNIT 2.13*, measurement of cation movement in primary cultures with fluorescent dyes is described.

UNITS 2.4 & 2.14 focus on measurement of additional signal-transduction molecules. The importance of these molecules to fundamental cellular processes transcends a particular organ system, and because of the central role assumed by these biosignaling molecules, these protocols may ultimately be among the most widely used in this manual. *UNIT 2.14* presents protocols for measurement of cAMP and inositol phosphates. These protocols are readily adapted to a variety of preparations ranging from cell cultures to acutely prepared brain slices. Measurement of nitric oxide (NO), an unstable gas with a half-life of several seconds, presents unique challenges to the experimenter. *UNIT 2.4* provides multiple protocols for the measurement of NO and NO synthase that can be adapted to a variety of experimental situations.

Contributor: Phil Skolnick

UNIT 2.1

Scintillation Proximity Assay

Recent advances in simplifying the measurement of biomolecules have had a significant impact on both basic science and biotechnology by improving reliability, reducing costs, and increasing the speed of bioassays. The scintillation proximity assay (SPA) technique provides a mechanism to distinguish bound from free radioactive molecules without the need for a separation step such as filtration or washing. The scintillant-embedded surfaces can be coated with acceptor molecules directly or with reactive groups, which then covalently bind the ligand or molecular complex of interest. Most applications have been designed to work in a standard 96-well plate, reducing reagent costs, minimizing radioactive waste, and lending itself to automated pipetting and detection. Table 2.1.8 summarizes SPA reagents and example applications, and Table 2.1.9 provides references relevant to additional applications. Both tables are located at the end of the unit.

CAUTION: Radioactive materials require special handling; all supernatants must be considered radioactive waste and disposed of appropriately.

NOTE: All culture incubations are performed in a 37°C, 5% CO_2 incubator unless otherwise specified. All solutions and equipment coming into contact with living cells must be sterile, and aseptic technique should be used accordingly.

BASIC PROTOCOL 1

SATURATION ANALYSIS OF [^{3}H]5-HT BINDING TO THE 5-HT$_{1E}$ RECEPTOR

Materials (see APPENDIX 1 for items with ✓)

    LM(tk-) cells expressing 5-HT$_{1E}$ receptor (Kahl et al., 1997)
✓ 50 mM Tris·Cl, pH 7.4, ice-cold
    Bovine serum albumin (BSA)
✓ Saturation assay buffer, room temperature
✓ 20 mg/ml wheat germ agglutinin (WGA) SPA beads
✓ 1 mM 5-hydroxytryptamine (unlabeled 5-HT)
✓ 1 μM [^{3}H]5-hydroxytryptamine ([^{3}H]5-HT)

Teflon pestle/glass tissue homogenizer (e.g., Thomas)
250-ml conical centrifuge tubes (e.g., Corning)
Beckman refrigerated tabletop centrifuge, or equivalent
Beckman J2-21M centrifuge and JA-14 rotor with 250-ml bottles, or equivalent
13 × 100–mm disposable polypropylene tubes
96-well white polystyrene microtiter plates (e.g., clear-bottom from Corning or opaque bottom from Perkin Elmer Life and Analytical Sciences)
Single and multichannel pipettors
Microtiter plate shaker (e.g., Labline Instruments)
Self-adhesive microtiter plate sealers
Liquid scintillation cocktail (e.g., Ready Protein$^+$, Beckman)
Liquid scintillation counter (e.g., Packard Tricarb) and glass scintillation vials
Microtiter plate scintillation counter (e.g., Perkin Elmer Life and Analytical Sciences Trilux or TopCount)

1. Thaw frozen cell pellet (6×10^9 cells) in 50 ml of 50 mM Tris·Cl, pH 7.4, 4°C. Resuspend using four to five strokes with a Teflon pestle/glass homogenizer connected to an overhead motor. Dilute resuspended cell pellet to 250 ml with 50 mM Tris·Cl, pH 7.4, 4°C. Centrifuge the homogenate in 250-ml conical centrifuge tubes for 15 min at $200 \times g$, 4°C.

2. Remove supernatant gently and save on ice. Add 250 ml chilled 50 mM Tris·Cl, pH 7.4, to the pellet and rehomogenize with the Teflon/glass homogenizer. Centrifuge in a 250-ml conical tube for 15 min at $200 \times g$, 4°C. Save supernatant and repeat the homogenization a third time.

3. Combine the supernatants from all three extractions. Centrifuge 40 min at $30,000 \times g$, 4°C, using a Beckman JA-14 or equivalent rotor. Discard the supernatant.

4. Resuspend pellet in 25 ml ice-cold 50 mM Tris·Cl, pH 7.4, using the Teflon/glass homogenizer. Perform a protein determination (e.g., Bio-Rad or BCA) using BSA as standard. Quick-freeze 0.5- to 1-ml aliquots in liquid nitrogen and store at −80°C.

5. Label twelve 13 × 100–mm polypropylene tubes with final assay concentrations of [^{3}H]5-HT, ranging from 0.5 to 75 nM (1.6-fold dilutions, 75, 46.7, 29.3, 18.3, 11.4, 7.2, 4.8, 2.8, 1.8, 1.1, 0.68, and 0.43 nM) as shown in Table 2.1.1. Label tube 13 as "Unlabeled 5-HT."

Table 2.1.1 Preparation of [^{3}H]5-HT Mix Dilutions

Tube	[^{3}H]5-HT (final; nM)	[^{3}H]5-HT (3× mix; nM)^a	Stock used for dilution (nM)	Vol. dilution stock (μl)	Vol. assay buffer (μl)
1	75	225	1000 (stock)	360	1240
2	46.7	140	225	1000	600
3	29.3	87.9	140	1000	600
4	18.3	54.9	87.9	1000	600
5	11.4	34.2	54.9	1000	600
6	7.2	21.6	34.2	1000	600
7	4.8	14.4	21.6	1000	600
8	2.8	8.4	14.4	1000	600
9	1.8	5.4	8.4	1000	600
10	1.18	3.54	5.4	1000	600
11	0.68	2.04	3.54	1000	600
12	0.43	1.29	2.04	1000	600

^{a}The working concentrations (3× mix) are prepared at 3× the desired final incubation assay concentration since 50 μl of each these mix dilutions will be added to the well in a total volume of 150 μl/well.

6. Add 1240 µl of saturation assay buffer to tube 1, 600 µl of saturation assay buffer to tubes 2 to 12, and 4985 µl of saturation assay buffer to tube 13. Add 15 µl of 1 mM unlabeled 5-HT to tube 13 (to determine nonspecific binding), yielding a 3 µM mix.

7. Prepare 1600 µl of a 225 nM [^{3}H]5-HT dilution (3× desired final concentration) by adding 360 µl of the 1 µM stock to tube 1. Prepare 1.6-fold dilutions of [^{3}H]5-HT by adding 1000 µl from tube 1 to the 600 µl of saturation assay buffer in tube 2, mix by vortexing, and continue this dilution scheme for tubes 3 to 12, according to Table 2.1.1.

8. Thaw aliquot(s) of membranes from step 4 (keeping on ice), and prepare 5000 µl of diluted membranes at 0.1 µg protein/µl in saturation assay buffer. Dilute 3000 µl of the 20 mg/ml WGA SPA beads stock into 3000 µl of saturation assay buffer to yield 6000 µl of a 10 mg/ml solution.

   Support Protocol 1 describes a procedure for optimizing the amount of protein for an experiment. Support Protocol 2 provides a procedure for determining the optimal amount of WGA SPA beads.

9. Using a multichannel pipettor, add 50 µl of saturation assay buffer to all wells in Rows A, B, and C and 100 µl of saturation assay buffer to all wells in Rows G and H (to determine nonproximity effects) of a 96-well microtiter plate. Add 50 µl of unlabeled 5-HT to Rows D, E, F (to determine nonspecific effects).

10. Add 50 µl of [^{3}H]5-HT dilution to the wells of a column (one column for each of the twelve dilutions from Table 2.1.1) using a multichannel pipettor reservoir. In addition, in duplicate, add 50 µl of each [^{3}H]5-HT dilution to 10 ml of liquid scintillation cocktail to determine the total amount of [^{3}H]5-HT added to each well.

11. Add 50 µl of diluted membranes to all wells in Rows A to F. Cover the plate with a polystyrene plate lid and shake for 1 min at setting 7 on a microtiter plate shaker. Incubate 30 min at room temperature.

12. Shake or stir the 10 mg/ml WGA SPA beads (see step 8) to homogeneity, then add 50 µl to all wells. Seal plate with adhesive plate sealer. Incubate 1 to 2 hr (optimized using Support Protocol 3) at room temperature with intermittent shaking (1-min shake every 15 min).

13. Read plate in a microtiter plate scintillation counter normalized for ^{3}H, according to the manufacturer's specifications, at a 1-min counting time per well. Count scintillation vials in liquid scintillation counter at 1 min per well.

14. Determine the total number of femtomoles of radioligand added to each well by dividing the dpm results from scintillation vials, for each concentration, by the specific activity of the radioligand determined, as follows, given a stated activity of 123 Ci/mmol (=123 pCi/fmol):

$$\frac{123 \text{ pCi}}{\text{fmol}} \times \frac{2.22 \text{ dpm}}{\text{pCi}} = 273.1 \text{ dpm/fmol}$$

$$\text{sample dpm} \times \frac{\text{fmol}}{273.1 \text{ dpm}} = \text{sample fmol}$$

15. Determine the number of femtomoles of [^{3}H]5-HT in the total binding and nonspecific-binding wells. Divide the cpm results from the microtiter plate scintillation counter by the instrument efficiency to obtain results expressed as dpm. Divide the dpm results by the specific activity of the ligand for the total and nonspecific bound fmol. Subtract nonspecific binding from total binding to determine specific fmol bound. Divide each set of data by the amount of protein used per well (e.g., 0.005 mg) and convert to pmol/mg.

16. Determine the free ligand concentration by subtracting the specific fmol bound from the total number of fmol added to the well. Divide by the assay volume (150 µl) to yield free concentration (nM).

17. Plot the data with the free ligand concentration (nM) on the x axis and the bound ligand (pmol/mg) on the y axis. Determine the binding parameters (K_d and B_{max}) using nonlinear regression analysis.

SUPPORT PROTOCOL 1

OPTIMIZATION OF MEMBRANE PROTEIN CONCENTRATION

1. Prepare cell membranes, reagents, and assay plates (see Basic Protocol 1, steps 1 to 10), except use a single concentration of [^{3}H]5-HT per well (5 nM final).

2. Prepare 1.5-fold serial dilutions of membrane in saturation assay buffer from a 1.3 mg/ml stock. Add 50 µl of each dilution. Shake for 1 min at setting 7 on a microtiter plate shaker, and incubate 30 min at room temperature.

3. Add 50 µl of 10 mg/ml WGA SPA beads prepared in saturation assay buffer. Incubate 2 hr at room temperature with intermittent shaking, and quantify radioactivity in a microtiter plate counter (see Basic Protocol 1, steps 12 and 13).

4. Plot data as shown in Figure 2.1.1.

   The data in Figure 2.1.1 indicate that 5 µg is an acceptable amount occurring within the linear signal response curve.

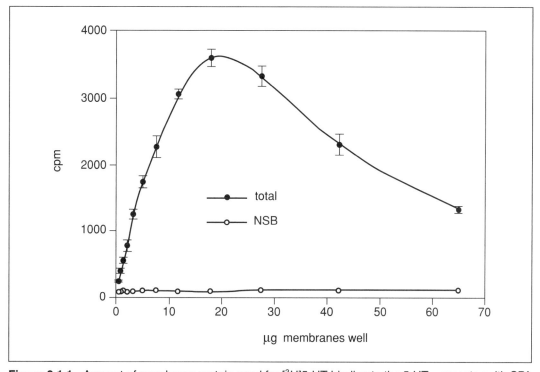

Figure 2.1.1 Amount of membrane protein used for [^{3}H]5-HT binding to the 5-HT$_{1E}$ receptor with SPA in the absence (total; solid circles) or presence (NSB; open circles) of excess unlabeled 5-HT using techniques described in Basic Protocol 1 and Basic Protocol 2. Higher amounts of membrane protein (>15 µg/well) will saturate the limiting amount of WGA SPA beads (0.5 mg) and result in a reduced signal due to non-SPA bead bound receptor.

SUPPORT PROTOCOL 2

OPTIMIZATION OF AMOUNT OF WGA SPA BEADS

1. Prepare cell membranes, reagents, and assay plates (see Basic Protocol 1, steps 1 to 11) except use a single concentration of [^{3}H]5-HT per well (5 nM final). Shake for 1 min at setting 7 on a microtiter plate shaker and incubate 30 min at room temperature.

2. Prepare various concentrations of WGA SPA beads, in saturation assay buffer, from an 80 mg/ml stock (500 mg beads/6.25 ml saturation assay buffer) such that the amount added to the respective wells is 4.0, 3.5, 3.0, 2.5, 2.0, 1.5, 1.0, 0.75, 0.5, 0.25, 0.125, 0.0625 mg in 50 µl, with the highest concentration (4 mg/well) to be added from the 80 mg/ml stock directly. Add 50 µl of each concentration to the appropriate wells.

3. Incubate 2 hr at room temperature with intermittent shaking, and quantify radioactivity in a microtiter plate counter (see Basic Protocol 1, steps 12 and 13).

4. Plot data as shown in Figure 2.1.2, with specific binding determined as the difference between total and nonspecific binding.

   In the data presented in Figure 2.1.2, 0.5 mg of WGA SPA beads was selected as the optimal amount when using 5 µg of this membrane protein preparation.

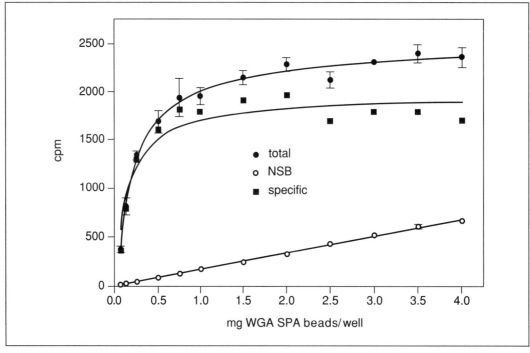

Figure 2.1.2 Determination of optimal WGA SPA bead concentration in a receptor/radioligand binding assay. In this example, [^{3}H]5-HT was incubated with receptor-enriched membranes prior to the addition of various concentrations of WGA SPA beads. Total binding is represented by solid circles and specific binding is represented by squares. Nonspecific binding (NSB; open circles) was determined in the presence of 1 µM unlabeled 5-HT. Since the amount of SPA beads used is also driven by economics, this data demonstrates that 0.5 mg of WGA SPA beads is adequate for capture of nearly all of the specific binding. Higher amounts of WGA SPA beads/well result in increased costs without increased specific signal.

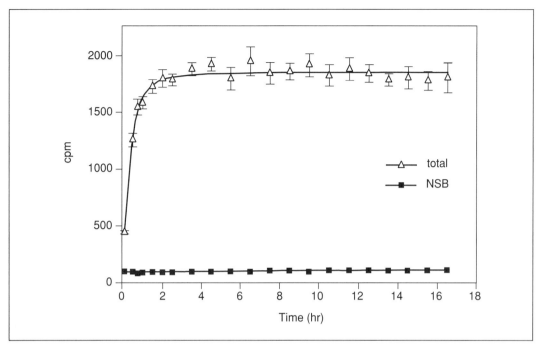

Figure 2.1.3 The incubation time following the addition of WGA SPA beads is a critical parameter in the SPA techniques for receptor/radioligand binding. Here [^{3}H]5-HT is incubated with 5 μg of receptor-enriched membranes followed by an incubation with WGA SPA beads for the indicated amount of time. No appreciable signal from total binding (triangles) or nonspecific binding (NSB; squares) occurs after a 2-hr incubation. The primary incubation of radioligand and membranes was also investigated and determined to be optimal at 30 min (not shown).

SUPPORT PROTOCOL 3

OPTIMIZATION OF INCUBATION TIME

1. Prepare cell membranes, reagents, and assay plates (see Basic Protocol 1, steps 1 to 11), except use a single concentration of [^{3}H]5-HT per well (5 nM final). Shake for 1 min at setting 7 on a microtiter plate shaker and incubate 30 min at room temperature.

2. Add 50 μl of 10 mg/ml WGA SPA beads prepared in saturation assay buffer and shake plate (see Basic Protocol 1, step 12). Immediately count for the first time point. After counting, shake the plate until the next time point.

3. Plot data as shown in Figure 2.1.3.

   These data demonstrate that a 2-hr incubation with WGA beads is adequate to obtain the maximum signal.

BASIC PROTOCOL 2

PHARMACOLOGICAL PROFILE FOR THE 5-HT$_{1E}$ RECEPTOR USING SPA

Materials *(see APPENDIX 1 for items with ✓)*

　　Serotonergic compounds (stock solutions), e.g.:
✓ 10 mM 5-hydroxytryptamine
✓ 10 mM α-methyl-5-hydroxytryptamine

Table 2.1.2 Stock Solutions of Serotonergic Compounds[a]

Compound[b]	Mix concentration	Final assay concentration	10 mM stock (μl)	Saturation assay buffer (μl)
5-HT	30 μM	10 μM	3	997
α-Me-5-HT	3 mM	1 mM	300	700
2-Me-5-HT	300 μM	100 μM	30	970
5-CT	300 μM	100 μM	30	970

[a]Different concentration ranges are used to cover a complete competitive response for each compound.

[b]Abbreviations: 5-HT, 5-hydroxytryptamine; α-Me-5-HT, α-methyl-5-hydroxytryptamine; 2-Me-5-HT, 2-methyl-5-hydroxytryptamine; 5-CT, 5-carboxyamidotryptamine.

✓ 10 mM 2-methyl-5-hydroxytryptamine
✓ 10 mM 5-carboxyamidotryptamine
  1-ml volume/well 96-well polypropylene microtiter plate
  50-ml conical disposable polypropylene tubes

1. Prepare membranes containing the expressed human 5-HT$_{1E}$ receptor (see Basic Protocol 1, steps 1 to 3).

2. Use a 96-well polypropylene microtiter plate (deep-welled, i.e., with a 1-ml capacity per well) to prepare the dilutions for four serotonergic compounds to be tested (compound 1 in row A, compound 2 in row B, and so on). Add the appropriate volumes of saturation assay buffer and 10 mM compound stock solutions to the wells in column 12 as indicated in Table 2.1.2.

3. Add 750 μl of saturation assay buffer to columns 1 to 11 for rows A, B, C, and D in the deep-well plate. Prepare four-fold dilutions in the deep-well plate by pipetting 250 μl from column 12 into the 750 μl contained in column 11. Mix by pipetting up and down several times. Continue the same dilution scheme for the remainder of the columns to yield twelve dilutions of each compound per row, with the highest concentrations in column 12 and the lowest concentrations in column 1.

4. Prepare the following solutions.
   a. Prepare 16 ml of 15 nM [^{3}H]5-HT in a 50-ml disposable conical tube by adding 240 μl of the 1 μM stock to 15.760 ml of saturation assay buffer.
   b. Prepare 16 ml of diluted membranes at 0.1 μg/μl in saturation assay buffer.
   c. Dilute 7.50 ml of the 20 mg/ml WGA SPA beads stock into 7.5 ml of saturation assay buffer to yield 15 ml of a 10 mg/ml solution (optimized using Support Protocol 2).
   d. For nonspecific binding, prepare a 3 μM unlabeled 5-HT mix by adding 15 μl of 1 mM unlabeled 5-HT to 4.985 ml of saturation assay buffer in a 13 × 100-mm polypropylene tube.

5. To test four compounds (two compounds per plate, with four replicates per concentration) with total binding and nonspecific controls, prepare three assay microtiter plates (not deep-welled; same as used in Basic Protocol 1) labeled plate 1, plate 2, and plate 3. Add 50 μl of the unlabeled compound dilutions to the assay plates as in Figure 2.1.4.

6. Add 50 μl of saturation assay buffer to columns 1 to 12 for rows A and B on assay plate 3 for total binding. Add 50 μl of 3 μM unlabeled 5-HT to columns 1 to 12 for rows C and D on assay plate 3 for nonspecific binding.

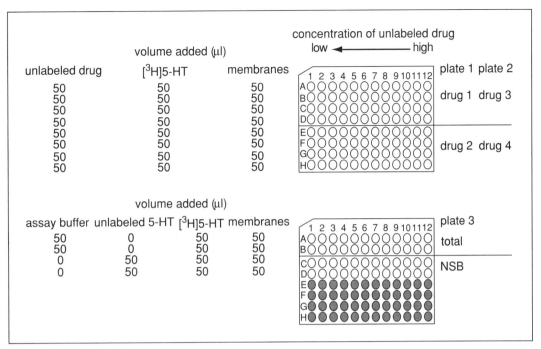

Figure 2.1.4 Plate setup for competitive binding analysis (four compounds).

7. Add 50 μl of 15 nM [^{3}H]5-HT to all assay and control wells. In addition, if determination of total activity added is desired, add 50 μl of the 15 nM [^{3}H]5-HT to 10 ml of scintillation cocktail in triplicate.

8. Add 50 μl of diluted membranes to all assay and control wells. 9. Cover the plates with polystyrene plate lids and shake for 1 min at setting 7 on a microplate shaker. Incubate for 30 min at room temperature.

9. Shake or stir the 10 mg/ml WGA SPA beads to uniformity, then add 50 μl to all assay and control wells. Seal plates with adhesive plate sealer. Incubate 1 to 2 hr (time optimized using Support Protocol 3) at room temperature with intermittent shaking (1-min shake every 15 min).

10. Read plates in a microtiter plate scintillation counter normalized for ^{3}H, according to the manufacturer's instructions, at a 1-min counting time per well. Count scintillation vials (if prepared) in a liquid scintillation counter at 1 min per well.

11. Average the four replicates at each concentration for the four compounds (assay plates 1 and 2). Average the 24 replicates for total binding (assay plate 3, rows A and B) and the 24 replicates for the nonspecific binding in the presence of excess unlabeled 5-HT (assay plate 3, rows C and D). Determine the standard deviation for each calculation.

12. Plot SPA cpm on the y axis versus the concentration of the test compound on the x axis. Determine the IC_{50} for each compound using nonlinear regression analysis as described in UNIT 7.5, and plot the resulting predicted curve fit.

13. Determine the equilibrium dissociation constant (K_i) for each compound using the Cheng-Prusoff equation (Cheng and Prusoff, 1973):

$$K_i = \frac{IC_{50}}{1 + \frac{[L^*]}{K_d}}$$

where $[L^*]$ is the concentration of radioligand and K_d is the dissociation constant for the radioligand.

BASIC PROTOCOL 3

RAPID MEASUREMENT OF cAMP ACCUMULATION IN CHO CELLS STABLY EXPRESSING THE MUSCARINIC M_1 RECEPTOR

Materials (see APPENDIX 1 for items with ✓)

    CHO cells (ATCC CCL-61) stably expressing the human M_1 muscarinic receptor (Felder et al., 1989)
✓ DMEM/F-12/10% FBS medium with HEPES
✓ PBS
    Trypsin/EDTA solution: 0.05% trypsin/0.53 mM tetrasodium EDTA (e.g., Life Technologies)
    Enzyme (trypsin)-free cell dissociation solution (Specialty Media)
✓ Dilution buffer
    1 mM 3-isobutyl-1-methylxanthine (IBMX; Sigma) *or* 10 µM RO20-1724 (Calbiochem) in dilution buffer
    Carbachol (Research Biochemicals)
    Lysis buffer: 1% (w/v) dodecyltrimethylammonium bromide in 50 mM sodium acetate, pH 5.8
    500 µM adenosine 3':5'-cyclic monophosphate (cAMP), sodium salt (Sigma)
    50 mM sodium acetate buffer, pH 5.8
✓ Beads/antibody/[^{125}I]cAMP mixture

    75-cm^2 and 225-cm^2 tissue culture flasks (Costar)
    96-well microtiter plates (white plate; clear bottom; Fisher or Costar)
    Microtiter plate sealing tape (Wallac)
    Microtiter plate scintillation counter (e.g., Wallac Trilux or Packard TopCount)

1. Passage CHO cells in DMEM/10% FBS medium (containing HEPES) in 75-cm^2 tissue culture flasks. Split the cells 1:25 twice a week as follows.

    a. Wash with 10 ml PBS, then add 2 ml trypsin/EDTA solution and incubate 2 min.

    b. Add 8 ml DMEM/F12/10% FBS medium with HEPES.

    c. Detach cells from the flask surface by pipetting up and down.

    d. Transfer cells to a tube and centrifuge 5 min at 500 × *g* in a tabletop centrifuge, room temperature. Resuspend cells in 10 ml warm DMEM/F12/10% FBS medium with HEPES.

    e. Perform a cell count and seed into new 75-cm^2 flasks at 25× dilution.

Table 2.1.3 Carbachol Dilution Series

	Stock conc. used for dilution	Vol. dilution (μl)	Vol. saturation assay buffer (μl)	Dilution conc. (2× final)	Final conc. in 96-well plate
1	10 mM stock	20	980	200 μM	100 μM
2	200 μM from 1	316	684	63.2 μM	31.6 μM
3	200 μM from 1	100	900	20 μM	10 μM
4	63.2 μM from 2	100	900	6.32 μM	3.16 μM
5	20 μM from 3	100	900	2 μM	1 μM
6	6.32 μM from 4	100	900	632 nM	316 nM
7	2 μM from 5	100	900	200 nM	100 nM
8	632 nM from 6	100	900	63.2 nM	31.6 nM
9	200 nM from 7	100	900	20 nM	10 nM
10	63.2 nM from 8	100	900	6.32 nM	3.16 nM
11	20 nM from 9	100	900	2 nM	1 nM

2. For the split prior to collecting cells for the assay, seed cells at a 25× dilution into 225-cm^2 flasks and grow to between 50% and 90% confluence.

3. Aspirate medium and wash cells with 10 ml PBS. Add 10 ml of enzyme-free dissociation solution per 225-cm^2 flask. Incubate 10 min at 37°C.

4. After cells are released from surface of flask, add 10 ml dilution buffer, mix gently, and transfer to a centrifuge tube, saving a 100-μl aliquot for counting. Centrifuge for 10 min at 500 × g at room temperature in a tabletop centrifuge. Count cells in the saved aliquot.

5. Aspirate supernatant and resuspend cell pellet to the appropriate concentration (2000 to 10,000 cells per well for receptors coupled to adenylate cyclase through G_s, 200,000 to 500,000 for receptors weakly coupled to G_s or stimulating adenylate cyclase through other mechanisms) in dilution buffer containing either 1 mM isobutylmethylxanthine (IBMX) or 10 μM RO20-1724 (to inhibit phosphodiesterase activity) to generate adequate levels of cAMP to fall on the standard curve.

6. Dilute a 10 mM stock of the muscarinic receptor agonist, carbachol, in water as shown in Table 2.1.3, and add 50 μl of the appropriate dilutions to corresponding wells of a 96-well microtiter plate as shown in Figure 2.1.5.

7. Add 20 μl of 500 μM cAMP stock to 2.48 ml of 50 mM sodium acetate to make a 4 μM dilution. Serially dilute according to the schedule shown in Table 2.1.4, and add 50 μl of each cAMP standard to the corresponding wells of the 96-well plate (Fig 2.1.5).

8. Add 50 μl of cell suspension (from step 5) to appropriate wells (Fig. 2.1.5) to initiate the reaction and place the 96-well plate into a 37°C incubator or a water incubator set at 37°C in 3 mm of water. Incubate 30 min, then terminate the reaction by adding 50 μl of lysis buffer to each well and pipetting up and down three times.

9. Add 50 μl of beads/antibody/[^{125}I]cAMP mixture to each well and mix by pipetting up and down three times. Cover plate with sealing tape, shake on vortex plate shaker (e.g., Genie) for 30 sec on level 4, and incubate at 4°C overnight.

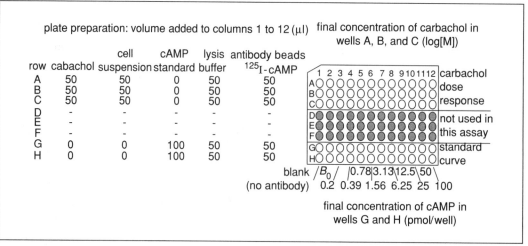

Figure 2.1.5 Plate setup for cAMP assay. A standard curve is run on each plate and can be run with either duplicate or triplicate samples. The wells labeled blank in positions G1 and H1 contain all the assay ingredients but not cAMP or antibody. The wells labeled B_0 in positions G2 and H2 contain all the assay ingredients but not a cAMP standard. Sodium acetate assay buffer should be used to substitute for these components.

10. Quantify the radioactivity in a microtiter plate scintillation counter. Calculate the average cpm for each of the duplicates. Calculate the percent bound for each standard and sample using the following formula:

$$\% \ B/B_0 = \frac{\text{cpm} - \text{nonspecifically bound cpm}}{B_0 \ \text{cpm} - \text{nonspecifically bound cpm}} \times 100$$

11. Generate a standard curve by plotting %B/B0 versus log of standard cAMP concentrations. Read unknowns directly from the standard curve.

Table 2.1.4 cAMP Standard Curve

Dilution no.	Serial dilutions	cAMP (pmol/well)	Pipet 50 µl into positions indicated
12	1000 µl stock + 1000 µl buffer	100	G12 and H12
11	500 µl dilution 12 + 500 µl buffer	50	G11 and H11
10	500 µl dilution 11 + 500 µl buffer	25	G10 and H10
9	500 µl dilution 10 + 500 µl buffer	12.5	G9 and H9
8	500 µl dilution 9 + 500 µl buffer	6.25	G8 and H8
7	500 µl dilution 8 + 500 µl buffer	3.13	G7 and H7
6	500 µl dilution 7 + 500 µl buffer	1.56	G6 and H6
5	500 µl dilution 6 + 500 µl buffer	0.78	G5 and H5
4	500 µl dilution 5 + 500 µl buffer	0.39	G4 and H4
3	500 µl dilution 4 + 500 µl buffer	0.2	G3 and H3
2	500 µl dilution 3 + 500 µl buffer	0.1	G2 and H2
1	1000 µl buffer	0	G1 and H1

BASIC PROTOCOL 4

TIME-COURSE ANALYSIS OF RHINOVIRUS 3C PROTEASE USING SPA

Materials (see APPENDIX 1 for items with ✓)

✓ Protease assay buffer
Stop solution: 10% (w/v) orthophosphoric acid
✓ 50 nM biotinylated ^{125}I-labeled rhinovirus 3C protease substrate stock solution
50 μM purified rhinovirus 3 C protease stock (store at −80°C; Birch et al., 1995)
✓ 20 mg/ml streptavidin SPA beads, 4°C

13 × 100–mm disposable polypropylene tubes
96-well white polystyrene microtiter plates (e.g., clear bottom from Corning or opaque bottom from Perkin Elmer Life and Analytical Sciences)
Single and multichannel pipettors
Microtiter plate shaker (e.g., Labline Instruments)
Self-adhesive microplate sealers
Microtiter plate scintillation counter (e.g., Perkin Elmer Life and Analytical Sciences Trilux or TopCount)

1. Label three 13 × 100–mm disposable polypropylene tubes as "substrate," "enzyme," and "SPA beads." Add 5467 μl protease assay buffer to the "substrate" tube and 2976 μl of protease assay buffer to the "enzyme" tube. Add 4800 μl of stop solution to the "SPA beads" tube.

2. Prepare the following solutions.

    a. Prepare a dilution of biotinylated ^{125}I-labeled substrate by adding 33 μl of the 50 nM stock to the "substrate" tube (final working concentration of 0.3 nM).

    b. Prepare a dilution of the rhinovirus 3C protease enzyme by adding 24 μl of the 50 μM stock to the "enzyme" tube (final working concentration of 0.4 μM).

    c. Prepare 6000 μl of 4 mg/ml streptavidin SPA beads by adding 1200 μl of the 20 mg/ml stock to the "SPA beads" tube.

3. Add the following solutions to the indicated wells and columns.

    a. Add 50 μl of protease assay buffer to all wells in rows E, F, G, and H of a 96-well microtiter plate (Fig. 2.1.6).

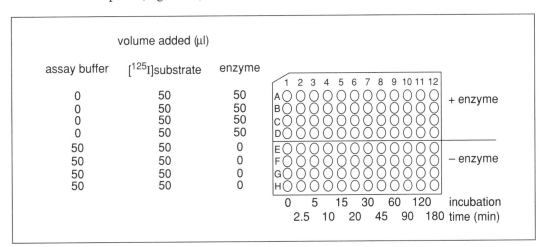

Figure 2.1.6 Plate setup for enzymatic time course.

b. Add 50 µl of 4 mg/ml streptavidin SPA beads from step 2c (0.2 mg) to each of the wells in column 1 to serve as the zero time point (enzyme will be inactivated once added to this well).

c. Add 50 µl of diluted [^{125}I]substrate (from step 2a) to all 96 wells.

d. Add 50 µl of diluted enzyme (from step 2b) to each well in columns 2 to 12. Shake 20 sec on a plate shaker on setting 7. Start timer to begin incubation at room temperature.

e. Add 50 µl of 4 mg/ml streptavidin SPA beads (from step 2c) to columns 2 to 12 after the incubation times (2.5, 5, 10, 15, 20, 30, 45, 60, 90, 120, and 180 min). Shake 20 sec after each addition.

4. Seal plate with adhesive plate seal. Incubate 10 min at room temperature. To avoid settling effects, shake 1 min on plate shaker prior to counting. Read the plate in a microtiter plate scintillation counter normalized for ^{125}I, according to the manufacturer's specifications, at 1-min counting time per well.

5. Plot the time on the x axis versus the SPA cpm on the y axis, obtained for both the enzyme-containing wells and the nonenzyme wells. Use the mean of the four replicates with error bars expressed in SD units.

BASIC PROTOCOL 5

UPTAKE OF [^{14}C]$_P$-HYDROXY-LORACARBEF USING CYTOSTAR-T PLATES

Materials (see APPENDIX 1 for items with ✓)

Human adenocarcinoma cell line Caco-2 (obtained from Dr. J. Fogh at the Research Unit of Memorial Sloan-Kettering Cancer Center, Rye, NY)
✓ DMEM/F12/10% FBS medium without HEPES
Trypsin/EDTA solution: 0.05% trypsin/0.53 mM tetrasodium EDTA (e.g., Life Technologies; store at 4°C)
✓ 4 mg/ml rat tail collagen, type II
0.5% (v/v) acetic acid
Ammonium hydroxide
✓ 20 mM [^{14}C]p-hydroxy-loracarbef
✓ Flux buffer
✓ PBS

75-cm^2 sterile tissue culture flasks (sterile, tissue culture treated; Costar)
Cytostar-T microtiter plate (sterile, tissue culture treated; Amersham Biosciences)
Multichannel pipettor
8-channel aspirator
Self-adhesive microplate seals
Microtiter plate scintillation counter (e.g., Perkin Elmer Life and Analytical Sciences Trilux or TopCount)

1. Passage Caco-2 cells in DMEM/F12/10% FBS medium without HEPES in 75-cm^2 flasks, changing the medium every other day. Split the cells 1:10 every 2 weeks as follows.

   a. Wash with 10 ml PBS, then add 2 ml trypsin/EDTA solution and incubate 2 min.

   b. Add 8 ml DMEM/F12/10% FBS medium without HEPES.

   c. Detach cells from the flask surface by pipetting up and down.

d. Transfer cells to a tube and centrifuge 5 min at 400 × g (1200 rpm in a Beckman tabletop centrifuge), room temperature. Resuspend cells in 10 ml cold DMEM/F12/10% FBS medium without HEPES.

e. Perform a cell count and seed in two new 75-cm^2 flasks at ∼2 × 10^6 cells/flask.

2. Coat plates with a thin layer of rat tail collagen type II as follows.

    a. Prepare a 1/10 v/v dilution of 4 mg/ml rat tail collagen, type II, in 0.5% acetic acid.

    b. Add 50 μl of the collagen suspension to each well of the Cytostar-T microtiter plate using a sterile repeating multichannel pipettor.

    c. Immediately aspirate the collagen from the wells using an 8-channel aspirator.

    d. Place the plate (without the lid) in a hood with an open beaker containing 20 ml of ammonium hydroxide and allow it to incubate under an inverted ice bucket for 45 min with the hood blower turned off and the hood face shield closed.

    e. After 45 min, turn blower back on and carefully remove ice bucket. Allow fumes to dissipate prior to removal of plate.

    f. Add 200 μl of sterile water to all the wells. Place plate in tissue culture incubator and incubate 24 hr prior to seeding plates.

3. Between passages 29 and 48 (to minimize the possibility of cell lines changing properties over time), remove cells from the flask by trypsinization as described in step 1, substeps a and b, and add 8 ml of DMEM/F12/10% FBS medium without HEPES to the trypsinized cells. Centrifuge 5 min at 300 × g (1000 rpm in a Beckman tabletop centrifuge), room temperature.

4. Remove supernatant and resuspend cells in 10 ml fresh medium. Count cells in cell counter to determine the density of cells.

5. Aspirate water from wells of coated plates (step 2f). Seed plates (one for each temperature) at ∼6000 cells per well by adding 200 μl of an appropriate dilution of cells/medium, filling row A wells with medium only (unseeded control).

6. Incubate cells to confluency (which should occur the following day). Allow cells to differentiate (exhibiting a "cobblestone" appearance when viewed under a light microscope) 12 to 14 days post-confluency, with a change of medium every other day.

7. Prepare dilutions of [^{14}C]p-hydroxy-loracarbef in flux buffer using the 20 mM stock solution as shown in Table 2.1.5. Aspirate medium from wells of Cytostar-T plates after the incubation described in step 6.

8. Add 125 μl of each [^{14}C]p-hydroxy-loracarbef dilution according to the plate setup in Figure 2.1.7 for Rows A, B, C, and D on all three plates. Place a lid on each plate and incubate 2 hr at 4°C, room temperature, or 37°C.

9. Aspirate medium from all wells. Add 125 μl PBS to each well. Seal plate with adhesive plate seal. Read plate in a microplate scintillation counter normalized for ^{14}C according to the manufacturer's specifications at 1-min counting time per well.

10. Average the cpm determinations for each concentration, taking into account row A (medium only) to identify any problems with the radioligand adhering to the plastic or collagen coating in the wells. Plot concentration of [^{14}C]p-hydroxy-loracarbef on the x axis versus the averaged cpm results on the y axis to determine the optimum transport temperature.

Table 2.1.5 Preparation of [^{14}C]p-Hydroxy-Loracarbef Dilution Series

Column	Final conc. (mM)	Stock conc. used for dilution	Vol. stock	Vol. flux buffer (μl)
12	20	20 mM	27.4 mg solid	3740
11	15	Column 12	2740 μl	917
10	12.5	Column 11	2657 μl	531
9	10	Column 10	2188 μl	547
8	7.5	Column 9	1735 μl	578
7	5	Column 8	1313 μl	656
6	2.5	Column 7	969 μl	969
5	1.25	Column 6	938 μl	938
4	0.625	Column 5	875 μl	875
3	0.3125	Column 4	750 μl	750
2	0.156	Column 3	500 μl	500
1	0	Flux buffer only	—	—

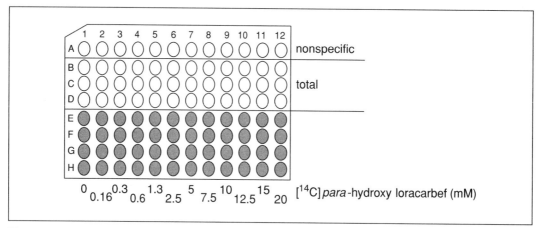

Figure 2.1.7 Plate setup for dipeptide transport in Cytostar-T microplate.

BASIC PROTOCOL 6

MUSCARINIC M$_1$ RECEPTOR-MEDIATED [γ-^{35}S]GTP BINDING: AN SPA APPROACH USING A SPECIFIC ANTI-G PROTEIN ANTIBODY

This assay provides a quantitative measurement of agonist-induced G-protein activation in membranes derived from CHO cells that stable express the human M1 muscarinic receptor (see Fig. 2.1.8)

Materials (see APPENDIX 1 for items with ✓)

Receptor membrane preparation from CHO cells expressing recombinant human M$_1$ muscarinic receptor (PerkinElmer Life and Analytical Sciences)
[γ-^{35}S]GTP (stable aqueous solution; PerkinElmer Life and Analytical Sciences)
Oxotremorine-M (Sigma)
Dimethyl sulfoxide (DMSO)

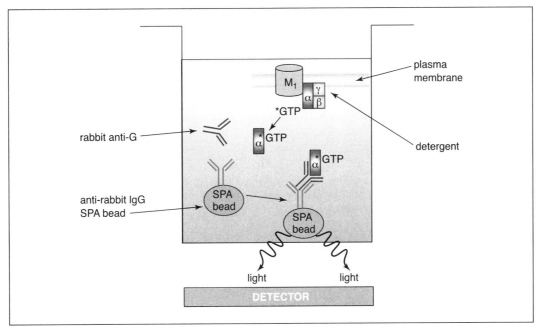

Figure 2.1.8 Antibody capture assay for [γ-^{35}S]GTP binding.

- ✓ Muscarinic assay buffer
  Compound dilution buffer: add 2 ml DMSO to 48 ml muscarinic assay buffer; prepare fresh each day
- ✓ [γ-^{35}S]GTP in ligand solution
  10% (v/v) Nonidet P-40 (NP-40; Roche), 4°C
- ✓ Rabbit anti-Gα$_{q/11}$ antibody, diluted
  SPA PVT antibody-binding beads, anti-rabbit reagent (Amersham Biosciences; store according to manufacturer's recommendations)

50-ml disposable polystyrene tubes
10-ml disposable glass tubes
96-well deep-well polypropylene microtiter plate (1 ml/well capacity)
Tissue homogenizer (e.g., PowerGen 700, Fisher)
50-ml polystyrene reagent reservoirs
Single and multichannel pipettors
96-well polystyrene microtiter plate (white, with clear bottom; Corning Costar)
Self-adhesive microtiter plate seals
Microtiter plate shaker (e.g., Vortex Genie, Scientific Industries)
Tabletop centrifuge (e.g., AccuSpin 1R, Fisher) with microtiter plate carrier
Microtiter plate scintillation counter (e.g., Perkin-Elmer Life and Analytical Sciences Trilux)

1. On the day of the assay prepare the assay ingredients as follows.
   a. Label five 50-ml tubes as "M_1 receptor," "compound dilution buffer," "[γ-^{35}S]GTP," "NP-40," and "Gα$_q$." Label one 10-ml glass tube as "oxo-M."
   b. Rapidly thaw aliquots of the receptor membrane preparation from CHO cells expressing the M_1 receptor and the [γ-^{35}S]GTP. Once defrosted, keep on ice for the remainder of the experiment.
   c. Weigh out 1 mg of oxotremorine-M (mol. wt. 332.3) and dissolve in 0.31 ml DMSO to prepare a 10 mM stock in the "oxo-M" tube.

Table 2.1.6 Preparation of Test Compound (Oxotremorine) Dilution Series

Column	[Oxotremorine] (final conc. in assay)	[Oxotremorine] in 4× mix[a]	Stock used for dilution	Vol. dilution stock (μl)	Vol. compound dilution buffer (μl)
1	0	0	0	0	960
2	100 μM	400 μM	10 mM	40	960
3	30 μM	120 μM	Column 2	195	455
4	10 μM	40 μM	Column 3	195	390
5	3 μM	12 μM	Column 4	195	455
6	1 μM	4 μM	Column 5	195	390
7	0.3 μM	1.2 μM	Column 6	195	455
8	100 nM	400 nM	Column 7	195	390
9	30 nM	120 nM	Column 8	195	455
10	10 nM	40 nM	Column 9	195	390
11	3 nM	12 nM	Column 10	195	455
12	1 nM	4 nM	Column 11	195	390

[a]The working concentrations (4× mix) are 4× the desired final concentration since 50 μl of the working concentration will be added to the well in a total assay volume of 200 μl/well.

2. Use a 96-deep-well microtiter plate to prepare the dilutions of oxotremorine-M (to yield 11 dilutions of oxotremorine-M with the highest concentration in column 2 and the lowest concentration in column 12; column 1 contains buffer only) as follows:

    a. Add the appropriate volume of compound dilution buffer to columns 1 through 12 for rows A, B, and C as indicated in Table 2.1.6.

    b. Prepare a four-fold dilution by pipetting 40 μl of 10 mM oxotremorine-M into column 2 of rows A, B, and C.

    c. Prepare a 1/2 log dilution by pipetting 195 μl from column 2 into the 455 μl contained in column 3. Mix by pipetting up and down several times and dispose of the pipet tips.

    d. Continue the same dilution scheme for the remainder of the plate.

3. Dilute receptor membranes preparations (M_1, M_3, or M_5) by adding 11 ml of muscarinic assay buffer to the tube marked M_1 and pipetting 250 μl of the concentrated stock (as obtained from the manufacturer) into the buffer. For M_2 or M_4 receptors, add 1 μM GDP (22 μl of 1 mM GDP to 11 ml of buffer and membrane). Homogenize the mixture for 5 to 10 sec using a PowerGen 700 or equivalent tissue homogenizer. Store on ice until needed.

4. Referring to the recipe for [γ-^{35}S]GTP ligand solution prepare ∼5.4 ml of [γ-^{35}S]GTP ligand solution in an appropriately labeled tube and store on ice until needed (final assay ^{35}S label concentration 500 pM).

5. Prepare the following dilutions.

    a. Dilute 10% NP-40 to 3% (v/v) using muscarinic assay buffer (∼5 ml final volume).

    b. Referring to the recipe for diluted rabbit anti-G$\alpha_{q/11}$ antibody dilute the anti-G$\alpha_{q/11}$ antibody 1:200 with muscarinic assay buffer for a total volume of 3 ml.

    c. Dilute 500 mg of anti-rabbit SPA PVT antibody-binding beads with 20 ml of muscarinic assay buffer. Mix well by shaking.

6. Pour the contents of the M_1 tube (prepared in step 3) into a polystyrene reagent reservoir. Using a multichannel pipettor, add 100 μl of diluted membrane preparation to columns 1 to 12 of rows A, B, and C of a clear-bottom white 96-well microtiter plate (Fig. 2.1.9).

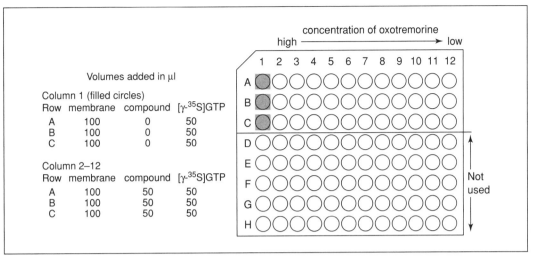

Figure 2.1.9 Plate setup for [γ-^{35}S]GTP assay.

7. Using a multichannel pipet, transfer 50 μl from each well of the deep-well plate containing the oxotremorine-M dilutions (prepared in step 2) to the corresponding well of the clear-bottom white microtiter plate (step 6), mixing well by pipetting up and down several times. Change pipet tips after each transfer. Incubate for 30 min at room temperature.

8. Pour the [γ-^{35}S]GTP solution (prepared in step 4) into a reagent reservoir. Transfer 50 μl of this solution to each well in rows A, B, and C of columns 1 to 12 of the clear-bottom white microtiter plate. Mix thoroughly by pipetting up and down several times. Change pipet tips after each transfer. Incubate the contents of the plate for 30 min at room temperature.

9. Pour the contents of the 3% NP-40 tube (prepared in step 5a) into a reagent reservoir. Transfer 20 μl of 3% NP-40 (final concentration 0.27%) to each well, mixing thoroughly by pipetting up and down several times (this stops the reaction). Change pipet tips after each addition. Incubate the contents of the plate for 30 min at room temperature.

10. Pour the diluted $G\alpha_{q/11}$ antibody (prepared in step 5b) into a reagent reservoir. (For M_2 or M_4 receptors use $G\alpha_{i-3}$ antibody.) Transfer 20 μl of $G\alpha_{q/11}$ antibody to each well, mixing thoroughly by pipetting up and down several times. Change tips after each addition. Incubate the plate for 60 min at room temperature.

11. Shake or stir the SPA bead suspension (prepared in step 5c) to uniformity. Transfer 50 μl of the suspension to each well. Seal the plate with adhesive plate seal and shake the covered plate at setting no. 7 on a Vortex Genie on the microtiter plate shaker for 10 to 15 sec. Incubate the plate for 3 hr at room temperature.

12. Centrifuge the plate 10 min at $1000 \times g$, room temperature, in a tabletop centrifuge equipped with a microtiter plate carrier. Quantify radioactivity in a microtiter plate scintillation counter normalized for [^{35}S] according to the manufacturer's specifications, with 1-min counting time per well.

13. Average the counts in column 1 across rows A to C to obtain the basal (no-compound) level of activity. To obtain the specific counts, subtract the average basal count from the total counts for each well.

14. Average the results for column 2, rows A, B, C. This value is the maximal response obtained by the highest concentration of oxotremorine-M.

15. Divide the value for each well (obtained in step 13) by the value obtained for the maximal stimulation (step 14) and multiply by 100 to obtain the % maximal stimulation for each compound concentration.

16. Plot the % maximal stimulation (y-axis) versus the negative log of the concentration of oxotremorine-M (x-axis). Fit the concentration-response curve using a sigmoidal nonlinear regression with variable slope using Prism or other appropriate software.

BASIC PROTOCOL 7

MINIATURIZATION OF RECEPTOR-RADIOLIGAND INTERACTIONS USING 384-WELL FlashPlates

Materials (see APPENDIX 1 for items with ✓)

✓ Miniature SPA assay buffer
✓ 1.5 mM S (−)-propranolol
✓ 500 nM [^{125}I]iodocyanopindolol
  Membrane preparation (2.0 μg/ml) from HEK293 cells expressing β_2 adrenergic receptor (Perkin Elmer Life and Analytical Sciences)

13 × 100–mm disposable polypropylene tubes
FlashPlates (Perkin Elmer Life and Analytical Sciences, store at 4°C):
  96-well Basic (uncoated) FlashPlate
  96-well WGA FlashPlate
  384-well Basic (uncoated) FlashPlate
  384-well WGA FlashPlate
Multichannel pipettors and corresponding reagents reservoirs
Gamma counter (e.g., ICN Micromedic)
Microtiter plate shaker (e.g., Vortex Genie, Scientific Instruments)
Self-adhesive microtiter plate seals
Microplate scintillation counter (e.g., Perkin Elmer Life and Analytical Sciences Trilux or TopCount)

1. Label 11 tubes with the concentrations of S(−)-propranolol to be used, ranging from 0.0169 nM to 1000 nM (3-fold dilutions, 1000, 333, 111, 37, 12.3, 4.1, 1.4, 0.46, 0.15, 0.051, 0.0169). Add 1797 μl of miniature SPA assay buffer to tube 1 and 600 μl of miniature SPA assay buffer to tubes 2 to 11. Add 3.6 μl of 1.5 mM S(−)-propranolol to tube 1. Mix. Remove 1.2 ml from tube 1 and place in tube 2. Mix and repeat this 3-fold dilution scheme as shown in Table 2.1.7.

2. In addition, prepare the following.

   a. Prepare 6000 μl of 0.15 nM [^{125}I]iodocyanopindolol by adding 1.8 μl of 500 nM stock to 5998.2 μl of miniature SPA assay buffer in a tube labeled "Hot."

   b. Prepare 3000 μl of diluted membranes (0.025 μg/μl) by adding 37.5 μl of the 2.0 μg/μl membrane stock to 2963 μl of miniature SPA assay buffer in a tube labeled "Membranes."

   c. Label four different FlashPlates: 96-well uncoated plate, 96-well WGA-coated plate, 384-well uncoated plate, and 384-well WGA-coated plate.

3a. *To determine a concentration-response curve in 96-well FlashPlates:* Add 10 μl of the corresponding S(−)-propranolol dilution (from step 1) to wells 2 to 12 in rows A, B, C and D (see Fig. 2.1.10A) using a multichannel pipettor to determine a concentration-response curve.

Table 2.1.7 Preparation of Unlabeled S(−)-Propranolol Hydrochloride Dilution Series

Dilution number[a]	Solution to be diluted	Volume measured	Volume of miniature SPA assay buffer
12	1.5 mM[b]	3.6 μl	1797 μl
11	Dilution 12	1.2 ml	600 μl
10	Dilution 11	1.2 ml	600 μl
9	Dilution 10	1.2 ml	600 μl
8	Dilution 9	1.2 ml	600 μl
7	Dilution 8	1.2 ml	600 μl
6	Dilution 7	1.2 ml	600 μl
5	Dilution 6	1.2 ml	600 μl
4	Dilution 5	1.2 ml	600 μl
3	Dilution 4	1.2 ml	600 μl
2	Dilution 3	1.2 ml	600 μl
1	0	Assay buffer only	600 μl

[a] Dilution number corresponds to column number in a 96-well microtiter plate.
[b] See APPENDIX 1.

3b. *To determine a concentration-response curve in 384-well FlashPlates*: Add 10 μl of S(−)-propranolol dilution with the multichannel pipet as follows: dilution 1 goes into wells A3, A4, B3, B4; dilution 2 goes in wells A5, A6, B5, B6; and so on (see Fig. 2.1.10B). For both plates, prepare $n = 4$ replicates for each S(−)-propranolol concentration.

4. Add 10 μl of miniature SPA assay buffer to wells in column 1 (rows A to D) for the 96-well FlashPlates and wells A1, A2, B1, and B2 for the 384-well FlashPlates to determine maximum binding in the absence of competitor.

5. Add 10 μl of [^{125}I]iodocyanopindolol dilution to all wells (final concentration 50 pM) using a multichannel pipettor. In addition, add 10 μl of [^{125}I]iodocyanopindolol dilution to each of two 12 × 75–mm tubes and determine in duplicate the total amount of [^{125}I]iodocyanopindolol added to each well in a gamma counter.

6. Add 10 μl of diluted membranes to all wells. Seal the plate with a self-adhesive plate seal and mix 1 min at setting no. 7 on a Vortex Genie microtiter plate shaker. Incubate 4 hr at room temperature.

7. Quantify the radioactivity in the plate wells in a microplate scintillation counter normalized for [^{125}I] at 1 min counting time per well. If using a Microbeta Trilux, normalize the protocol for a solid-bottom plate (read using top photomultiplier tube only), to achieve noncoincidence counting.

    Two protocols will be needed, one for each plate density.

8. Average the four replicates for total binding on each plate (A1, B1, C1, D1 for 96-well plates and A1, A2, B1, B2 for 384-well plates). For nonspecific binding (NSB), average the four replicates on each plate corresponding to the highest concentration of S(−)-propranolol tested (wells A12, B12, C12, D12 for 96-well plates; wells A23, A24, B23, B24 for 384-well plates). Determine the specific binding for each well by subtracting NSB from the signal and dividing the result by the total binding signal minus NSB as shown in the equation below:

    % specific bound = (cpm − NSB)/(max cpm − NSB) × 100

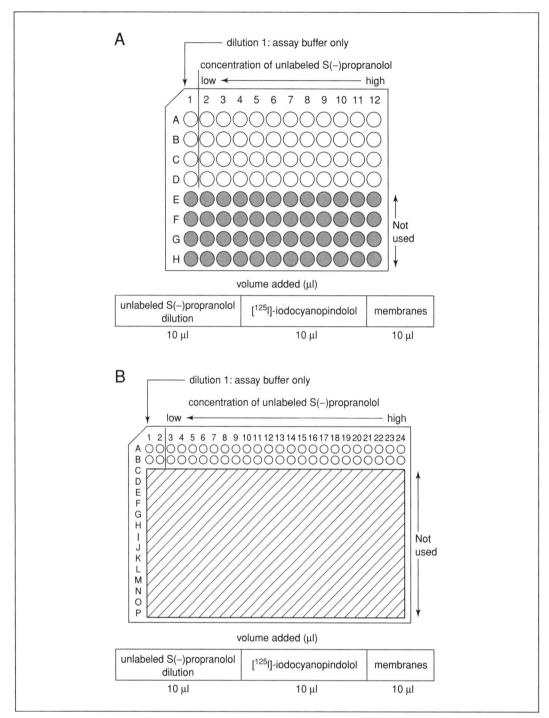

Figure 2.1.10 Plate setup for competitive inhibition of [^{125}I]iodocyanopindolol binding to β_2 adrenergic receptor membranes using (**A**) 96-well microplate, and (**B**) 384-well microplate.

Plot % specific bound on the *y*-axis versus the log concentration of S(−)-propranolol on the *x*-axis. Determine the IC_{50} value for each compound using a four-parameter nonlinear regression analysis.

References: Cook, 1996; Picardo and Hughes, 1997

Contributors: Steven D. Kahl and Christian C. Felder

Table 2.1.8 Summary of Scintillation Proximity Assay Reagents and Example Applications[a]

Reagent	General applications	Published examples	References
Scintillation proximity assay beads			
Anti-mouse	RIA, capture assays	Integrin/fibronectin, acyclovir, IGF-1, PTH, FKBP-12	Pachter et al. (1995); Tadepalli and Quinn (1996); Su et al. (1997); Graziani et al. (1999); Frolik et al. (1998)
Anti-rabbit	RIA, capture assays, [γ-^{35}S]GTP binding	cAMP, IL-5, PTH, muscarinic receptors, 5HT$_{2c}$	Frolik et al. (1998); Hancock et al. (1995); Banks et al. (1995); DeLapp et al. (1999); Cussac et al. (2002)
Anti-sheep	RIA, capture assays	apoA-I, rantidine, CETP	Hanselman et al. (1997); Fenwick et al. (1994); Lagrost et al. (1995)
Protein A	RIA, capture assays	RAS/neurofibromin, GP IIb/IIIa, nNOS	Skinner et al. (1994); Alderton et al. (1998); Bednar et al. (1998)
Wheat germ agglutinin	Receptor binding, [γ-^{35}S]GTP binding	IP$_3$, 5HT$_{1E}$, CCR3, α3 fucosyltransferases, muscarinic receptors	Patel et al. (1997); Dairaghi et al. (1997); Kahl et al. (1997); Hood et al. (1998)
Streptavidin	Enzyme, protein-protein, peptide binding	CD28, PPARγ, p34$_{cdc2}$ kinase, HCMV, HDAC, RAF/MEK/ERK, GSK3β, DNA gyrase, CRF binding protein	Jenh et al. (1998); Nichols et al. (1998); Spencer-Fry et al. (1997); Baum et al. (1996); Nare et al. (1999); McDonald et al. (1999); Peterson et al. (1999); Fowler et al. (2000); Gevi and Domenici (2002); Kahl et al. (1998)
Polylysine	Receptor binding, kinase	TGFα binding, PKA	Holland et al. (1994); Mallari et al. (2003)
Glutathione	Capture of GST fusion proteins	—	
Metal-chelating	His-tagged fusion proteins, measure IP$_3$	NK1, PDGF receptors	Liu et al. (2003)
Uncoated yttrium silicate (Ysi)	Inositol phosphate measurement	Muscarinic receptors	Brandish et al. (2003)
Uncoated polyvinyl toluene (PVT)	ATPase activity	E1 helicase	Jefferey et al. (2002)
FlashBlue GPCR	Receptor binding	MC4, MCH	Baudet et al. (2003)
FlashPlates[b]			
Uncoated/sterile	Receptor, enzyme, protein-protein, peptide binding, cellular uptake and release, [γ-^{35}S]GTP binding	Endothelin, [γ-^{35}S]GTP binding, thymidine uptake, CCR5 receptor, histone acetyltransferase, CCR3 receptor	Holland et al. (1994); Major (1995); Watson et al. (1998); Brown (1996); Brown et al. (1997); Bosse et al. (1998); Turlais et al. (2001); Dillon et al. (2003)

continued

Table 2.1.8 Summary of Scintillation Proximity Assay Reagents and Example Applications[a], continued

Reagent	General applications	Published examples	References
Anti-rabbit	RIA, capture assays	Thromboxane, cAMP, estrogen receptor	Brown (1996); Brown et al. (1997); Kariv et al. (1999); Allan et al. (1999)
Anti-mouse	RIA, capture assays	Estrogen receptor	Allan et al. (1999)
Protein A	RIA, capture assays	TGFα binding	Komesli et al. (1998)
Streptavidin	Enzyme, protein-protein, peptide binding	MAP kinase, HCV helicase, DNA replication	Brown (1996); Brown et al. (1997); AlaouiIsmaili et al. (2000); Earnshaw and Pope (2001)
Wheat germ agglutinin	Receptor binding	Somatostatin-2	Birzin (2002)
Myelin basic protein	Tyrosine kinases	MAP kinase	Brown (1996); Brown et al. (1997)
Metal-chelating	His-tagged fusion proteins		—
ScintiStrips/ScintiPlates[c]			
Uncoated	Receptor, enzyme, protein-protein, peptide binding	c-src kinase, estrogen receptor	Haggblad et al. (1995); Nakayama et al. (1998); Braunwalder et al. (1996a,b)
Anti-mouse	RIA, capture assays, receptor binding	α2 Adrenergic	Sen et al. (2002)
Anti-rabbit	RIA, capture assays		—
Streptavidin	Enzyme, protein-protein, peptide binding	Thyroid hormone receptors, estrogen receptors, c-src kinase	Nakayama et al. (1998); Carlsson and Haggblad (1995); Carlsson et al. (1997)
Cytostar-T plates[d]			
Uncoated/sterile	Cellular uptake and release, calcium flux, apoptosis, transporters	Thymidine uptake, adriamycin release, mRNA levels, glutamate receptors, annexin-V measurement, AMPA receptor, GlyT	Harris et al. (1996); Graves et al. (1997a,b); Cushing et al. (1999); McMurtrey et al. (1999); Smith et al. (2000); Bonge et al. (2000); Williams et al. (2003)

[a] Abbreviations: AMPA, α-amino-3-hydroxy-5-methyl-4-isoxazolepropionic acid; apoA-I, apolipoprotein A-1; cAMP, cyclic adenosine monophosphate; CCR3, CC chemokine receptor 3; CETP, cholesteryl ester transfer protein; CRF, corticotropin releasing factor; FKBP-12, 12-kDa form of FK506 binding protein; GlyT, glycine transporter; GSK3β, glycogen synthase kinase 3 beta; GST, glutathione-S-transferase; HCMV, human cytomegalovirus; HCV, hepatitis C virus; HDAC, histone deacetylase; HIV, human immunodeficiency virus; $5HT_{2c}$, 5-hydroxytryptamine 2C receptor; IGF-1, insulin-like growth factor-1; IL-5; interleukin-5; IP_3, inositol 1,4,5-trisphosphate; NK1, neurokinin 1; NOS, nitric oxide synthase; PDGF, platelet-derived growth factor; PKA, protein kinase A; PPARγ, peroxisome proliferator-activated receptor γ; PTH, parathyroid hormone; RIA, radioimmunoassay; TGFα, transforming growth factor α.

[b] Product of Perkin-Elmer Life and Analytical Sciences (http://www.perkinelmer.com).

[c] Product of Wallac (http://www.perkinelmer.com).

[d] Product of Amersham Biosciences (http://www.amershambiosciences.com).

Table 2.1.9 References Relevant to Other Applications of SPA (also see Table 2.1.8)

Application	References
Imaging/miniaturization	
HIV-RT	Ramm (1999)
HCV RNA polymerase	Zheng et al. (2001)
Serine/threonine kinase	Sorg (2002)
P56(lck) kinase	Beveridge et al. (2000)
General reviews	
Receptor binding assays	Alouani (2000); Carpenter et al. (2002)
Comparisons with other technologies	Park et al. (1999); Kowski and Wu (2000); Sills et al. (2002); Ohmi et al. (2000); Wu (2002); Mandine et al. (2001)

UNIT 2.2

Methods for Sample Preparation for Direct Immunoassay Measurement of Analytes in Tissue Homogenates: ELISA Assay of Amyloid β-Peptides

BASIC PROTOCOL 1

ISOLATION AND HOMOGENIZATION OF TISSUES FOR ANALYSIS OF AMYLOID β-PEPTIDES

Materials (see APPENDIX 1 for items with ✓)

   Experimental animal or tissue sample
✓ Guanidine buffer, ice-cold
✓ Casein buffer with protease inhibitors, ice-cold

   Tube rotator (e.g., Cole-Palmer Roto-Torque)
   Refrigerated centrifuge
   MultiScreen-BV 1.2 μm Durapore PVDF, preassembled 96-well filtration plates (Millipore)
   96-well polypropylene eluant-capture plates (Fisher Scientific)
   Millipore MultiScreen Filtration System vacuum manifold

1. Dissect tissue from animal, obtain wet weight, and mince with scissors in ice-cold guanidine buffer. Alternatively, snap-freeze the tissue on dry ice and either store at either −20°C or −80°C (temperature required for stability assessed by immunoassay) or homogenize as rapidly as possible after removal.

2. Homogenize the tissue 1:10 (w/v) in ice-cold guanidine buffer in polypropylene tubes. Set the power and duration of the homogenizer to cause some foaming and generate a smooth homogenate (30 sec at 50% power for most Polytron-type homogenizers; more power and several 30-sec disruptions for organs with more connective tissue). Keep the sample at or below room temperature.

3. Rotate the capped sample through 360° on a tube rotator for 3 to 4 hr or even overnight for highly polymerized analytes (e.g., amyloid β peptide from plaque) at 4°C. Optimize rotation time for each analyte after assay development. Store sample at 4°C until analysis.

4. Dilute samples at least 1:10 in ice-cold casein buffer containing protease inhibitors, the dilution ratio depending on the final dilution employed in the capture well to yield the desired range of analyte concentrations. Bracket a 0.5 M concentration of guanidine during assay development to establish the maximum guanidine tolerance of the analyte capture efficiency.

5a. *To remove particulates by centrifugation:* Centrifuge for 20 min at 15000 × g, 4°C.

5b. *To remove particulates by filtration:* Add samples from step 4 to a Millipore MultiScreen 96-well filtration plate, vacuum-filter using the Millipore MultiScreen Filtration System vacuum manifold, and collect filtrates in polypropylene 96-well eluant-capture microtiter plates placed underneath the MultiScreen plates in the vacuum filtration device.

6. Proceed with analysis (Basic Protocol 2) or freeze the sample in liquid nitrogen or in a low-temperature freezer (evaluation of analyte stability is advised).

BASIC PROTOCOL 2

EXAMPLE OF ELISA TECHNIQUE FOR THE ANALYSIS OF AMYLOID PEPTIDES Aβ(1-x) AND Aβ(1-42) FROM BRAIN HOMOGENATES

Two capture and detection antibody pairs, which recognize an epitope within the center of the Aβ molecule and in the C-terminal region, are useful for measuring all forms of Aβ and also for quantitation of the highly amyloidogenic species Aβ(1-42). Figure 2.2.1 illustrates the typical "sandwich" ELISA configuration.

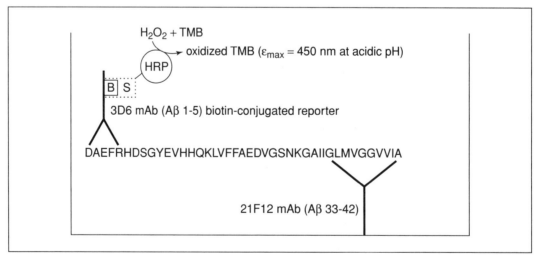

Figure 2.2.1 Schematic of a typical ELISA configuration for measurement of Aβ(1–42) in guanidine extracts of human brain tissue. The capture antibody is electrostatically adsorbed to the ELISA plate. The biotin-conjugated reporter antibody (B) binds streptavidin horseradish peroxidase–conjugate (S-HRP). The bound HRP and H_2O_2 forms a colored dye for quantitation of bound reporter mAbs in each ELISA plate well.

Materials *(see APPENDIX 1 for items with ✓)*

    Capture antibodies:
        266, prepared by immunizing mice with human Aβ peptide fragment 13-28
            (fragment 17-26 substitute available from Biosource International)
        21F12, prepared by immunizing mice with human Aβ peptide fragment 33-42
            (Seubert et al., 1992; Innogenetics N. V.)
- ✓ Plate coating buffer
- ✓ Plate blocking buffer
- ✓ Wash buffer

    Samples to be analyzed for amyloid β-peptides (see Basic Protocol 1)

    Amyloid β-peptide standards (for preparation see manufacturer's recommendations; casein carrier for low concentrations, filtration through 0.45-μm unit, and ultralow temperature storage of aliquots in polypropylene cryovials or microcentrifuge tubes are recommended)

    270 μg/ml de-salted, biotinylated reporter (detection) antibody 3D6 (Innogenetics; follow manufacturer's directions for use). Biotinylate 1.2 mg/ml 3D6 (or other reporter antibody) using EZ-Link Sulfo-NHS-LC-Biotin (Pierce) according to the manufacturer's instructions.

    Streptavidin–horseradish peroxidase conjugate (Amersham)

    TMB substrate kit (Pierce; store at 4°C; once opened, use for no longer than 1 to 2 months)

    Stop reagent: 2 N H_2SO_4

    Polystyrene high-protein-binding ELISA plates (optimize for assay; coefficients of variation for IgG bound between wells should be <5%, and background ±0.005 AU from mean) Plate sealers

    Multichannel pipettor (recommended)

    Spectrophotometer accommodating microtiter plates

1. Coat ELISA capture plates as follows:
   a. Dilute capture antibodies 266 and 21F12 in plate coating buffer to a concentration of 15 μg/ml and 5 μg/ml, respectively.
   b. Add 100 μl of these antibody solutions (1.5 μg and 0.5 μg/well, respectively) to the wells of their respective microtiter plates.
   c. Seal plates with standard film plate seals and place in the refrigerator at 2° to 4°C overnight or optimize time and temperature of incubation for each assay.
   d. The next day, remove the coating buffer by inverting the plates into a sink. Shake out the majority of the liquid in the plate, and blot any residual liquid off by inverting and tapping the plates onto paper towels.

   It is worthwhile evaluating plates from more than one vendor during assay development, because the optimal plate for a particular assay may vary.

   The optimal concentration of diluted capture antibody must be determined for each assay system. Precoated high-protein-binding 96-well plates can accept anywhere from 0.2 to 10 μg of antibody per well. The optimal concentration of antibody will yield minimal variability of the signal 5-fold above and 5-fold below a given concentration of antibody coating solution, determined by performing the ELISA with standards over the range of concentrations.

2. Using a multichannel pipettor for convenience, add 200 μl plate blocking buffer to the plates and allow to stand at room temperature for 1 hr. Remove the plate-blocking buffer in the same manner as the coating solution (see step 1d). At this point, place unused plates in a desiccated container and store at 4°C for future use (monitor for performance). For plates that are going to be used immediately, remove as much liquid as possible by blotting

onto absorbent line-free paper towels. Wash the blocked antibody-coated plates once with 300 μl wash buffer per well, discarding the buffer and blotting on lint-free paper towels as before.

3. Add 100 μl per well of diluted samples (1:10 or greater dilution depending on analyte concentration and maximum guanidine tolerance) to be analyzed, at least in duplicate, as well as dilutions of amyloid β-peptide standards (1000, 500, 250, 125, 63, 31, 16, and 0 pg/ml in this assay) to the capture plates. Seal the plates with high-quality plate sealers to avoid evaporation during the capture phase. Capture overnight at 4°C, or for 3 to 4 hr at 37°C (suggested capture times are only a guideline; time course pilot experiments recommended).

4. After capture of the analyte is completed, discard the solution and wash the wells with 3 cycles of 300 μl wash buffer/well. Discard the final wash.

5. Dilute the 270 μg/ml de-salted, biotinylated reporter (detection) antibody 3D6 1:1500 in casein buffer (optimized to yield minimal variability of the signal 250-fold above and 250-fold below a given concentration of detection antibody). Add 100 μl of the resulting solution to each well of the assay plate. Incubate 1 hr at room temperature. After the incubation, discard the well contents and wash the plates with 3 cycles of 300 μl wash buffer/well.

6. Dilute streptavidin–horseradish peroxidase conjugate 1:1000 in casein buffer. Add 100 μl of this diluted reporter conjugate to each well of the assay plates and incubate for 1 hr at room temperature. After the incubation, discard the well contents and wash the plates with 4 cycles of 300 μl wash buffer/well.

7. Add 100 μl of reporter reagent TMB substrate (1:1 TMB:H_2O_2; from Pierce TMB substrate kit) to each well of the assay plates. Incubate the plates, in the dark and away from any drafts, for the appropriate period of time, e.g., ∼15 min for the total Aβ and ∼30 min for Aβ(1–42).

   TMB turns blue on oxidation, and it is usually useful to monitor the reaction progress by eye. With some experience, the investigator will be able to judge when sufficient signal has been generated, without developing so much chromophore that the OD readings exceed the linear range of the spectrophotometer.

8. Stop the TMB reaction by adding 100 μl of stop reagent (2N sulfuric acid) to each well of the assay plates (oxidized TMB turns yellow at acidic pH), and read absorbance at 450 nm. Subtract readings at 650 nm from the OD readings at 450 nm to compensate for well-to-well variations in OD of the plate matrix/geometry.

9. Analyze data using appropriate software (e.g., Softmax Pro 2.6; Molecular Devices Corp.) supplied with the plate-reading spectrometer or spectrofluorimeter. For example, subtract background, generate standard curves, and calculate unknowns from the standard curve.

References: Crowther, 1995; Gervay and McReynolds, 1999

Contributors: Paul A. Hyslop and Mark H. Bender

UNIT 2.3

Assessment of Cell Viability in Primary Neuronal Cultures

Methods described in this unit are routinely used for cultured primary neurons plated at a density of 2×10^5 cells/ml and grown in 24-well culture vessels with 400 μl bathing medium/well.

Table 2.3.1 Selected Methods for Assessing Neuronal Cell Viability

Guiding principles	Assay
Morphology	Light microscopy, electron microscopy
Membrane permeability	
Marker exclusion	Propidium iodide, ethidium homodimer, trypan blue, nigrosin
Marker retention	^{51}Cr, LDH, immunostaining
Electrophysiology	Membrane potential, action potential generation, synaptic potentials
Ionic homeostasis	
Intracellular free Ca^{2+}	Fura-2, indo-1, fluo-3
Lysosomal pH	Acridine orange, neutral red
Metabolism	
Anabolic rate	Protein or RNA synthesis
Substrate utilization	Glucose utilization or lactate accumulation
Esterase activity	Fluorescein diacetate, fluorescein-AM, calcein-AM, bis-carboxyethylcarboxyfluorescein-AM
Energy state	ATP:ADP ratio
Mitochondrial function	
Succinate dehydrogenase	MTT
Mitochondrial membrane potential	Rhodamine 123
DNA damage	TUNEL, gel electrophoresis

All volumes and amounts can be adapted for cells in other culture vessels (i.e., 96-well, 48-well, 6-well, 35-mm dish, or 60-mm dish) or at other densities. Each of these methods should yield similar results. For a summary of additional methods for assessing neuronal cell viability see Table 2.3.1. Table 2.3.2, at the end of this unit, provides a troubleshooting guide.

To ease interpretation, results from these methods can be expressed as percent cell death rather than in the units obtained directly from the assay:

$$\text{percent cell death} = \frac{(e_i \text{ result} - \text{mean sham result})}{(\text{mean FK result} - \text{mean sham result})} \times 100$$

Here, e_i refers to the experimental result in Figure 2.3.1, which illustrates a typical 24-well plate set up. This calculation requires each experiment to have (1) sham-treated (S) conditions

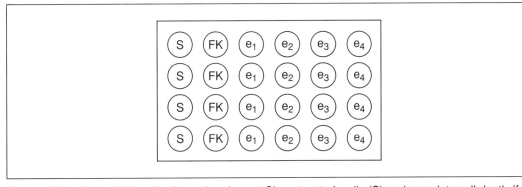

Figure 2.3.1 Typical 24-well culture plate layout. Sham treated wells (S) and complete cell death (full kill, FK) are used to normalize level of cell death in experimental wells (e_{1-4}).

as well as (2) a condition of complete cell death (full-kill or FK). In general, 24-hr exposure to 10 μM ionomycin will produce 100% cell death. In cultures containing both neurons and glia, many treatments affect only the neurons. In this case, selectively killing all of the neurons allows the results to be expressed as percent neuronal death. For 14-day-old cortical cultures obtained from embryonic day 15 mice, 24 hr exposure to 300 μM N-methyl-D-aspartate (NMDA) will kill all the neurons and none of the glia.

BASIC PROTOCOL 1

ASSAY OF CELL VIABILITY BY MEASUREMENT OF MITOCHONDRIAL ACTIVITY (MTT REDUCTION)

Materials (see APPENDIX 1 for items with ✓)

    0.5% (w/v) 3-(4,5-dimethylthiazol-2-yl)-2,5-diphenyltetrazolium bromide (MTT; Sigma-Aldrich) in PBS and filtered (0.2 μm) to remove insoluble debris: store light-protected at 4°C for <1 month
    ✓ PBS
    Cells in 24-well culture vessel, 400 μl bathing medium/well, including wells for sham treatment and full-kill conditions (see Fig. 2.3.1)
    ✓ Lysis buffer
    Microplate spectrophotometer

1. Remove the bathing medium from each well of the 24-well plate containing the cells. Add 100 μl of 0.5% MTT solution (0.1% final; calibrated in this assay for mixed cultures in culture medium containing neurons growing on a confluent bed of astroglia) to each well and return the culture dish to a 37°C incubator for 4 hr.

2. Add 400 μl lysis buffer (i.e., volume equal to original medium volume) and incubate overnight in a humidified incubator at 37°C or until the formazan precipitate is completely solubilized as determined by light microscopy.

3. Measure the absorbance of the culture plate at 570 nm (formazan) and 630 nm (non-formazan-related absorbance) with a microplate spectrophotometer. If desired, convert the A_{570}/A_{630} ratios into percent cell death as described in the unit introduction.

BASIC PROTOCOL 2

ASSAY OF CELL VIABILITY BY MEASUREMENT OF LACTATE DEHYDROGENASE RELEASE INTO THE BATHING MEDIUM

Materials

    Cells in 24-well culture vessel, 400 μl bathing medium/well
    LDH enzyme standard (Sigma)
    0.15 mg/ml NADH solution in potassium phosphate buffer (see APPENDIX A1 for potassium phosphate buffer), make fresh daily
    10 mM sodium pyruvate in potassium phosphate buffer (see APPENDIX A1 for potassium phosphate buffer), store up to 1 month at 4°C

    Empty 96-well culture plate
    Plate-reading spectrophotometer able to make kinetic readings at 340 nm (e.g. Kinetic Microplate reader, Molecular Devices)
    Multichannel pipettors

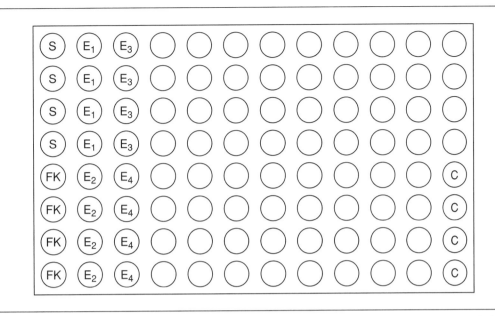

Figure 2.3.2 Typical 96-well plate setup for LDH assay. Sham treated wells (S), complete cell death (full kill, FK), experimental wells (E_{1-4}), control enzyme (C). Note that multiple 24-well plates can be assayed simultaneously.

1. Place 25 μl medium (50 μl for sparse cultures) from each culture well into quadruplicate wells of a 96-well test plate, including quadruplicate wells for sham-treated, full-kill, and control enzyme samples (see Fig. 2.3.2). Add 25 μl LDH standard to the last four wells, i.e., controls, of the test plate. Add 200 μl NADH solution to each well of the test plate.

2. Set the microplate reader for kinetic measurement of OD at 340 nm (every 5 sec for 2 min). Using a multichannel pipettor, add 50 μl of 10 mM sodium pyruvate solution to each well of the test plate and read the plate in the microplate reader immediately.

3. Using the LDH standard, convert raw data (OD/min) into LDH activity (U/ml). If desired, convert data into percent cell death as described in the unit introduction.

ALTERNATE PROTOCOL 1

SIMPLIFIED LDH ASSAY FOR CELLS GROWN IN 96-WELL PLATES

Additional Materials (also see Basic Protocol 2)

  0.3 mg/ml NADH solution in potassium phosphate buffer (see APPENDIX A1 for potassium phosphate buffer), make fresh daily

1. Add 100 μl of 0.3 mg/ml NADH solution directly to each well of cultured cells.

2. Set up the microplate reader to measure absorbance at 340 nm. Add 25 μl of 10 mM potassium pyruvate to each well and begin reading OD immediately.

3. Record a range of LDH standards (lower than the sham wash condition to greater than the full kill condition) on a separate plate (25 μl/well, in 4 wells). Using the LDH standard data, convert the rate of sample absorbance change to LDH units as described in Basic Protocol 2, step 3.

BASIC PROTOCOL 3

ASSAY OF CELL VIABILITY BY COLLOIDAL DYE (TRYPAN BLUE) EXCLUSION

Materials

0.4% (w/v) trypan blue stain (Sigma-Aldrich)
Cells in 24-well culture vessel, 400 µl bathing medium/well, including wells for wash-treated shams

1. Add 20 µl of 0.4% trypan blue stain (0.02% final concentration) to each well and incubate for 30 min at 37°C.

2. Examine cells under bright-field optics at a magnification of 200× (20× objective, 10× ocular). Using a hand-held counter, count at least 300 total cells per well, i.e., live cells plus dead cells (dark blue), within 2 hr of staining, or see Support Protocol to preserve stained cells. Count the same region of each well regardless of treatment conditions (treatment-blinded counting recommended).

3. Quantify cell death by dividing the number of dead cells counted per well by the total number of cells per well.

$$\text{cell death \%} = (\text{number of stained cells/total number of cells}) \times 100$$

Use wash-treated shams to determine baseline cell death. To determine mean cell death specifically due to each condition, subtract the mean baseline cell death for each condition.

ALTERNATE PROTOCOL 2

RAPID TRYPAN BLUE STAINING

Additional Materials (*also see Basic Protocol 3; see* APPENDIX 1 *for items with* ✓)
✓ PBS

1. Add 25 to 100 µl of 0.4% trypan blue (0.1% final concentration) to each well and incubate for 5 min at room temperature.

2. Remove stain with three washes of PBS.

3. Examine cells under bright-field optics. Count cells within 2 hr of staining (dark blue cells are nonviable), or see the Support Protocol to preserve stained cells.

BASIC PROTOCOL 4

ASSAY OF CELL VIABILITY USING PROPIDIUM IODIDE

NOTE: The conditions cited for this protocol were determined empirically for cortical neurons grown in 24-well culture plates and 400 µl bathing volume. The optimal dye concentration and incubation time may depend on cell type and culture conditions, determined in each culture paradigm by comparing the difference in sham conditions and total kill under different propidium iodide concentrations and incubation times.

Materials

0.1 mg/ml propidium iodide (Molecular Probes)
Cultured cells in clear 6- to 96-well culture plates, including wells for sham treatment and full-kill conditions

Fluorescence multiwell plate reader with a detector that can read plates from underneath (e.g., Cytofluor Series 4000, Perseptive Biosystems)
530 ± 25 nm excitation filter and a 645 ± 40 nm emission filter

1. Add 0.1 mg/ml propidium iodide to culture medium to a final concentration of 5 μg/ml (i.e., 20× dilution). Return culture to incubator for 1 to 2 hr. If possible incubate cells for sham-treated and total cell death conditions on the same plate, or at least for the same length of time.

2. Using a fluorometric multiwell plate reader equipped with a 530 ± 25 nm excitation filter and a 645 ± 40 nm emission filter, determine the fluorescence of each well at desired time points. Alternatively, count stained cells manually on a fluorescence microscope (see Alternate Protocol 3).

3. Express cell death as emission at 617 nm or convert the data into % cell death as described in Basic Protocol 3, step 3.

    Care must be taken to observe if dead cells detach from the vessel surface, as this could result in an underestimation of cell death. It is helpful to confirm the correlation between microscopically observed epifluorescent staining and fluorometrically determined fluorescence units.

ALTERNATE PROTOCOL 3

LIVE/DEAD STAINING WITH STANDARD EPIFLUORESCENCE MICROSCOPY

Materials

1.5 mg/ml fluorescein diacetate and 0.46 mg/ml propidium iodide mixture
Epiillumination fluorescence microscope (e.g., Nikon Diaphot equipped with 450 nm excitation filter, 520 nm barrier)

1. Add 1/100 vol of 1.5 mg/ml fluorescein diacetate/0.46 mg/ml propidium iodide mixture to the cultures. Wait 5 min.

2. Examine stained cells with a standard epiillumination microscope and score green-yellow cells as live and cells with red nuclei as dead.

3. Photograph or fix cells within 1 hr, or see the Support Protocol to preserve stained cells.

SUPPORT PROTOCOL

PRESERVATION OF STAINED CELLS

Materials (see APPENDIX 1 for items with ✓)

Stained cells (see Basic Protocols 3 and 4 and Alternate Protocols 2 and 3)
✓ PBS
✓ 4% (w/v) paraformaldehyde in PBS, ice cold

1. Remove stain by washing three times (or until medium is relatively clear), each time with 750 μl PBS. Aspirate all medium and add 250 μl ice-cold 4% paraformaldehyde to fix the cells. Incubate 30 to 60 min at room temperature.

2. Wash away paraformaldehyde with four washes of PBS. Store culture plate up to 1 month at 4°C.

References: Favaron et al., 1988; Hansen et al., 1989; Koh and Choi, 1987; Sattler et al., 1997; Slater et al., 1963; Trost and Lemasters, 1994; Wroblewski and LaDue, 1955

Contributors: Howard S. Ying, Frank J. Gottron, and Dennis W. Choi

Table 2.3.2 Troubleshooting Guide for Assessing Cell Viability in Primary Neuronal Cultures

Problem	Possible cause	Solution
MTT assay		
No signal in healthy cells	MTT stock degraded	Make up fresh MTT stock
	Spectrophotometer is not set up properly	Set machine to read absorbance at 570 nm
Large signal in dead cells	Cells not dead	Assess cell death by an alternate method (trypan blue staining)
	Something else causing product formation	Rerun controls to see if treatments have endogenous reducing activity
High variability (cell death % SEM > 15%)	Variable culture volume	Normalize volumes before adding MTT solution
	Light path partially blocked	Clean bottom of plate, ensure proper placement in machine
Poor correlation with trypan blue staining	Different injuries cause different temporal profiles of MTT activity	Examine time course of injury development
LDH assay		
No signal in standard	NADH or pyruvate degraded	Make up fresh NADH or use new aliquot of pyruvate stock
	Plate reader not set up correctly	Adjust plate reader for kinetics mode
	LDH standard not hydrated	Wait 10 min for lyophilized LDH standard to dissolve fully
No signal in samples	No LDH in medium sample	Stain cells with trypan blue; LDH should correlate with trypan blue staining
	Frozen medium samples not warmed to room temperature	Place frozen sample in 37°C incubator for 1 hr before assay
	Very old frozen medium samples	LDH activity slowly decreases over time; repeat assay immediately after sampling
High variability (cell death % SEM > 15%)	Condensation or scratches on plate	Clear light path by wiping or changing plates
	Foamy or cloudy sample	Remove bubbles or spin down debris before adding sample to assay plate
	LDH residue from previous assay	Clean assay plate after each use and discard if visible residue is present before use
	Inaccurate sampling	Sample 50 or 100 μl of medium to increase volume and activity; avoid scraping cell layer with sampling pipet
	Apoptotic death	Wait several hours longer (24 to 48 hr is enough for cortical neurons) for all LDH to be released into the bathing medium
Low values	Old NADH or pyruvate	Make up new solutions
	Cold solutions or samples	Warm all materials to room temperature before use
	Long delay between NADH and pyruvate additions	Run entire assay quickly to avoid NADH degradation
High background	Microbial contamination of solutions	Store stock solutions at 4°C and warm only amount needed each time
	Serum in medium	LDH assay is best performed in defined medium

Problem	Possible cause	Solution
Trypan blue assay		
No staining	Serum in medium	Wash out serum with defined medium prior to staining or increase trypan blue concentration to 0.1%
	Not stained long enough	Place cells at 37°C and increase staining duration; however trypan blue may become toxic if incubated too long
High background	Trypan blue not well washed prior to fixation	Thoroughly wash out trypan blue before paraformaldehyde fixation
Too few cells stained (<100% staining in complete cell death condition)	Dead cells too degraded to recognize	Stain at earlier time point
	Dead cells detached	Look for floating debris; stain at earlier time point
Too many cells stained (>10% staining in sham condition)	Excessive staining time or trypan blue concentration	Reduce staining time or trypan blue concentration
High variability (cell death% SEM > 15%)	Cells not evenly dispersed across culture vessel	Try to maximize dispersion when plating cells
	Dead cells not evenly distributed; sampling error in counting	Ensure uniform insult delivery or increase counting sample
Propidium iodide assay		
Low fluorescence	Low dye concentration	Optimize dye concentration using a range of dye concentrations
	Wrong optical filter	Check optical filters
	Dye is bleached	Store dye stocks at 4°C protected from light
	Low level of cell death	Measure cell death in culture through independent means to confirm death
	Incomplete dye penetration	Incubate culture longer
	Cells detach upon death	Use alternative assay such as LDH release
	Fluorimeter malfunction	Check stained cultures under epifluorescent microscope
High background	Too much dye	Optimize dye concentration
	Fluorimeter malfunction	Adjust gain on fluorimeter
Too few cells stained (<100% staining in complete cell death condition)	Dead cells detached	Look for floating debris; stain at earlier time point; may need to use alternative assay such as LDH release
Too many cells stained (>10% staining in sham treated condition)	Dye is toxic	Optimize dye concentration and duration of incubation with dye
High variability (cell death% SEM > 15%)	Cells not evenly dispersed across culture vessel	Try to maximize dispersion when plating cells
	Dead cells not evenly distributed	Ensure uniform drug delivery (some insults are enhanced in the center of the well)
	Variable culture volume	Ensure consistent culture volume before adding dye

UNIT 2.4

Measurement of NO and NO Synthase

BASIC PROTOCOL 1

CHEMILUMINESCENT DETECTION OF NO

Materials

    Acidified KI or NaI for nitrite reduction (see Support Protocol 1) *or* acidified VCl_3 for nitrite/nitrate reduction (see Support Protocol 2)

    NO gas cylinder and stainless steel regulator (100 to 200 ppm NO in N_2, for preparation of NO gas standards) and N_2 gas cylinder and stainless steel NO gas regulator (Matheson Gas Products) 10 µM sodium nitrite

10 µM sodium nitrate
Samples for analysis
Antifoam B (Sigma)

    Chemiluminescence detection system consisting of:
        Glass purge vessel with gas-tight Swagelock connections (Fig. 2.4.1; available from Sievers Instruments as no. ASM03292; reflux component required only for nitrite and nitrate measurements)
    Helium gas tank and regulator
    Acid trap with gas bubbler, containing 1 M NaOH (required only for nitrite and nitrate measurements)
    Chemiluminescence detector for NO_x (e.g., Sievers Instruments Model 280A)
    Vacuum pump
    Recirculating heated water bath (required only for nitrite and nitrate measurements)
    Gas-sampling tube (glass, with stopcocks on either end and a septum for extraction with a gas-tight Hamilton syringe)
    Teflon/silicone septa for purge vessel
    Data collection system (i.e., chart recorder, integrator, computer)
    Ringstand, clamps, Swagelock connectors, Teflon ferrules, and stainless steel or PFA connection tubing
10-, 25-, 50-, and 100-µl gas-tight Hamilton syringes

1. Assemble the purge vessel with 5 ml of distilled water for detection of NO in samples, or as described in Support Protocol 1 (using acidified KI or NaI catalyst) for nitrite. For nitrite/nitrate detection, assemble the purge vessel with 5 ml of acidified VCl_3. Connect the purge vessel to a helium tank and begin gassing through the porous glass frit at a rate that causes the liquid level to approach the top of the purge vessel (∼10 ml/min). When purging using catalysts in mineral acids (see Support Protocol 1 or 2), bubble the outflow of the purge vessel first through a trap, fitted with a gas bubbler, containing sufficient 1 M NaOH to fully submerge the gas bubbler. Connect the trap, or the purge vessel itself if a trap is not used, to the needle inflow valve of the chemiluminescence detector.

2. Turn on the vacuum pump [to both introduce sample (NO) into the reaction chamber and evacuate gases that quench light generation by NO_2*; <200 torr] and photomultiplier tube (PMT). Turn on the gas supply to the ozone generator of the chemiluminescence detector.

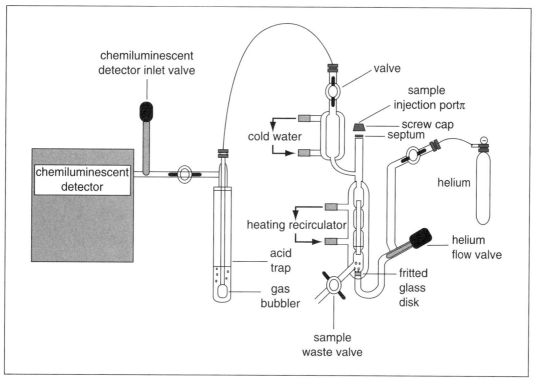

Figure 2.4.1 A gas purge vessel that can be utilized for analysis of NO, NO/nitrite, and NO/nitrite/nitrate by chemiluminescence detection.

3. After stabilization of the ozone gas pressure, turn on the ozone generator and wait several minutes until the PMT output stabilizes, measured with a chart recorder, or more preferably, using an integrator or a computer and data-analysis system (e.g., Maclab from CB Sciences or equivalent system from ADInstruments).

For assay of NO

4a. Under a chemical fume hood, flush a gas-sampling tube with a standard mixture of NO gas in N_2. Preflush a gas-tight Hamilton syringe with N_2 gas, and use the syringe to withdraw the desired volume of the NO gas mixture.

5a. Optimize the sample inflow rate to the chemiluminescence detector via adjustment of the inlet needle valve to the one giving the sharpest NO peak upon repeat injection of 10-μl volumes of 100 ppm NO (or alternative standard mixture), using a range of inlet valve settings.

6a. Calibrate the chemiluminescence detector with appropriate standards. Use triplicate injections of standards to confirm linearity and reproducibility of peak heights and areas. Determine NO concentration based on the universal gas constant of 22.4 liters/mol or 0.0446 mol/liter (e.g., at 100 ppm, NO concentration is 4.46×10^{-12} mol/μl of gas, resulting in a standard curve ranging from 2 to 20 μl of 100 ppm NO gas representing 8.92 to 89.2 pmol of NO).

7a. Inject experimental samples (volumes in the 1- to 100-μl range for the setup in Fig. 2.4.1; larger purge vessel for larger volume samples), using a gas-tight Hamilton syringe through a septum on top of the apparatus and measure, in duplicate or triplicate; quantify relative to the calibration curve based on either peak height or area.

For assay of nitrite (plus NO)

4b. To convert nitrite to NO, use acidic iodide solution in the purge vessel, prepared as described in Support Protocol 1. Confirm reductant efficiency every ~10 samples by re-injecting a nitrite standard. When required, replace with fresh reducing agent and confirm by testing response to a single standard.

5b. Optimize the sample inflow rate to the chemiluminescence detector via adjustment of the inlet needle valve to the one giving the sharpest NO peak upon repeat injection of 10-µl volumes of 10 µM sodium nitrite standard (i.e., each containing the equivalent of 100 pmol NO). Ensure that the injected standard is reaching the acidic iodide reductant.

6b. Prepare a calibration curve with appropriate nitrite standards (e.g., 1-, 10-, and 100-µl injections of 10 µM nitrite will give 10, 100, and 1000 pmol, respectively). Use duplicate or triplicate injections of standards to establish linearity and reproducibility of peak heights or areas.

7b. Measure experimental samples (unknowns) by injection in duplicate or triplicate and quantify relative to the calibration curve based on either peak height or area.

For assay of nitrate and nitrite (plus NO)

4c. To convert nitrate and nitrite to NO, use 5 ml of acidic vanadium chloride (prepared as described in Support Protocol 2) and 100 µl Antifoam B in the purge vessel.

5c. Turn on the cold tap water circulation to the purge-vessel condenser. With the screw cap off and the outlet valve closed to the acid trap, adjust the helium purging to give a slow gentle bubbling of the acidic vanadium reductant through the acid trap and into the chemiluminescence detector. Replace the screw cap and open the purge-vessel outlet to the acid trap.

6c. Set the heating recirculating water bath to 86°C and confirm security of plumbing connections to the purge vessel as the temperature rises to prevent dilation and dangerous loosening of the tubing with heating. Once the final temperature in the purge valve is achieved, readjust the helium flow to ensure small bubbles and gentle purging of the acidic vanadium reductant.

7c. Prepare a calibration curve with nitrate and nitrite standards (e.g., 1-, 10-, and 100-µl injections of 10 µM nitrite to give 10, 100, and 1000 pmol, respectively). Use duplicate or triplicate injections of 10 µM nitrate followed by injection of identical concentrations of nitrite. Replace the catalyst if the nitrate peak is ≤90% of the nitrite peak.

8c. Measure experimental samples (unknowns) by injection in duplicate or triplicate and quantify relative to the calibration curve based on either peak height or area. Prevent sample overflow due to foaming by deproteinizing protein-rich samples (e.g., sera, plasma) as described in Support Protocol 3.

SUPPORT PROTOCOL 1

PREPARATION OF A REDUCING AGENT FOR CONVERSION OF NITRITE TO NO

Additional Materials (*also see Basic Protocol 1*)
Glacial acetic acid
KI or NaI
Antifoam B (Sigma)

1. Isolate the purge vessel by closing off the helium inflow valve, outlet valve, and waste valve. Remove the purge vessel septum and screw cap and add 4 ml of glacial acetic acid. Adjust the helium tank regulator to 1 psi and carefully open the helium inlet to the purge vessel. Let the acetic acid purge for ~10 min to fully deoxygenate.

2. Weigh 50 mg of KI or NaI and dissolve in 1 ml distilled water (as iodide will be present in excess for nitrite reduction, its concentration is noncritical). Add the 1 ml of solution to the deoxygenated acetic acid in the purge vessel and purge with helium for an additional 10 min. Add 10 µl Antifoam B to the purge vessel and replace the septum (Teflon-side-down) and screw cap to seal the purge vessel.

3. Open the valves connecting the purge vessel to the acid trap and the valves connecting the acid trap to the chemiluminescence detector. Adjust the helium flow to bring the acetic acid/iodide solution to a level approaching the top of the purge vessel.

SUPPORT PROTOCOL 2

PREPARATION OF NITRATE REDUCING AGENT

Materials

VCl_3 (Aldrich)
Concentrated HCl

50-ml volumetric flask (dry)
Equipment for refluxing (optional): three-necked flask, condenser, nitrogen bubbler, thermometer, and electric heating mantle
Filter paper

1. Weigh 0.5 g of VCl_3 into a dry 50-ml volumetric flask. Add distilled water to the flask, followed by 4 ml concentrated HCl, and then distilled water to the 50-ml fill line. Stopper the volumetric flask and invert several times.

    CAUTION: Vanadium (III) chloride (VCl_3) is toxic and corrosive. Handle with care. Store refrigerated in full vials with airtight Teflon closures to prevent oxidation.

    The solution will turn blue, but not all VCl_3 will dissolve.

2. *Optional:* To reduce contamination of VCl_3 with nitrite and nitrate, thereby attaining maximal sensitivity for NO detection and reducing the time needed to obtain a stable baseline chemiluminescence signal on the day of assay, reflux the 50 ml of acidic VCl_3 solution in a fume hood for 8 hr at 112° to 114°C under a stream of nitrogen in a 100-ml three-neck flask fitted with a condenser, nitrogen bubbler, and thermometer, heated with an electric mantle.

3. Once cool, filter the acidic VCl_3 through filter paper and store up to 1 month at 4°C. Before use, remove precipitated VCl_3 by filtration.

SUPPORT PROTOCOL 3

DEPROTEINATION OF SAMPLES FOR NITRATE/NITRITE ASSAY

Whether foaming becomes a significant problem depends on sample injection volume as well as sample protein concentration. A convenient protocol, described below, employs ethanolic precipitation and neither interferes with chemiluminescence detection nor affects baseline nitrite/nitrate levels.

Materials

   100% ethanol (absolute ethanol; anhydrous)
   Sample for analysis

1. Chill absolute ethanol to −20°C. Transfer ≤0.5 ml of sample (e.g., plasma) to a 1.7-ml microcentrifuge tube, add an equal volume of cold ethanol, vortex, and place at 4°C for 15 min.

2. Microcentrifuge 5 min at maximum speed, 4°C. Transfer the supernatant to a clean microcentrifuge tube, seal, and refrigerate up to 1 week at 4°C until assay for nitrite/nitrate levels.

BASIC PROTOCOL 2

NITRITE AND NITRITE/NITRATE ASSAY BY THE GRIESS METHOD

Materials (see APPENDIX 1 *for items with* ✓)

   ✓ Griess reagents A and B
      100 μM sodium nitrite standard stock solution (dissolve 69 mg $NaNO_2$ in 10 ml H_2O)
      100 μM sodium nitrate standard stock solution (dissolve 85 mg $NaNO_3$ in 10 ml H_2O)
   ✓ Nitrate reductase buffer
   ✓ 200 mM Tris·Cl, pH 7.6
      Nitrate reductase (from *Aspergillus*; Sigma N-7265)
      L-lactic dehydrogenase (from rabbit muscle; Sigma, L-2500)
      500 mM sodium pyruvate

   96-well microtiter plates (flat-bottom or round-bottom)
   Repeating pipettor (e.g., Repipettor from Eppendorf) and 5-, 1.25-, and 0.5-ml reservoir tips (e.g., Combitips from Eppendorf)
   Microtiter-plate reader with 540- or 550-nm filter (maximum absorbance, 546 nm)

1. Clarify samples [e.g., tissue extracts, plasma, urine, cell culture medium, and solutions from in vitro nitric oxide synthase (NOS) enzymatic assays] containing cells or insoluble matter by microcentrifugation at maximum speed for 5 min prior to assay.

2. Prepare working Griess reagent by mixing equal parts of A and B; store up to 2 to 3 days in a tightly sealed tube at 4°C

For nitrite assay

3a. Prepare nitrite standard solutions (e.g., 0, 2.5, 5, 10, 20, 40, and 60 μM) from stock and add 100 μl to triplicate wells in a 96-well plate. On the same plate add 100-μl volumes of unknowns and appropriate sample blanks (e.g., culture medium that has never contained cells) to triplicate wells.

4a. With a 5.0-ml repeating pipettor, deliver 100 μl of working Griess reagent to all wells. Quantify the pink to purple color developed within seconds as A_{540} or A_{550} in each well using a microtiter plate reader.

5a. Subtract blank absorbance from absorbances of experimental samples and calculate the nitrite concentration from the standard curve (e.g., absorbances in the range of 0.100, 0.200, and 0.400 for 10, 20, and 40 μM nitrite, respectively; a software package such as Softmax from Molecular Devices is recommended).

For nitrite/nitrate assay

3b. Prepare nitrite standards (e.g., 0, 12.5, 25, 50, 75, and 100 μM) from stock and add 25 μl to triplicate wells in a 96-well plate. Similarly, prepare nitrate standards from stock at molar concentrations identical to those for nitrite and add 25 μl to a second set of triplicate wells. On the same plate add 25-μl volumes of unknowns and appropriate sample blanks to triplicate wells.

4b. With a 1.25-ml repeating pipettor, deliver 25 μl of nitrate reductase buffer to all wells.

5b. Dilute nitrate reductase in 200 mM Tris·Cl, pH 7.6, to a final concentration of 0.1 U/ml. With a 1.25-ml repeating pipettor, deliver 25 μl of the diluted nitrate reductase to all wells. Manually rock the microtiter plates to gently mix. Incubate microtiter plates 30 min at 37°C to achieve complete reduction of nitrate to nitrite.

6b. Add 20 μl L-lactic dehydrogenase (LDH) suspension to 500 μl of 500 mM sodium pyruvate. With a 0.5 ml repeating pipettor, deliver 10 μl of LDH/pyruvate to all wells. Incubate 15 min at 37°C to obtain complete oxidation of NADPH to NADP.

7b. With a 5.0-ml repeating pipettor, deliver 100 μl of working Griess reagent to all wells. Quantify the pink to purple color developed within seconds as A_{540} or A_{550} in each well using a microtiter plate reader.

8b. Confirm complete conversion of nitrate to nitrite in samples by demonstrating that the plotted standard curves for nitrite and nitrate are superimposable, subtract blank sample absorbance values from all unknowns, and extrapolate nitrite/nitrate concentrations using a combined standard curve (use of a software package such as Softmax from Molecular Devices recommended).

If complete nitrate reduction is not verified, then the assay results are not valid. The most likely reason for a failed assay is that nitrate reductase or LDH has degraded. Alternatively, a grossly incorrect solution may have been prepared and used. Repeat the assay with fresh solutions and enzymes.

BASIC PROTOCOL 3

L-CITRULLINE ASSAY FOR NITRIC OXIDE SYNTHASE (NOS) ACTIVITY

CAUTION: Radioactive materials require special handling; all column eluates must be considered radioactive waste and disposed of appropriately.

Materials *(see APPENDIX 1 for items with ✓)*
✓ Assay buffer, ice-cold
✓ Assay mix 1 and 2, ice-cold
4 mM -N^G-nitro-L-arginine, ice-cold
Test compounds in H_2O, ice-cold
20 mM EGTA, ice-cold
20 mM EGTA/20 mM N^G-nitro-L-arginine, ice-cold
Stop buffer: 50 mM HEPES, pH 5.5, containing 5 mM EDTA, ice-cold

Assay tubes: 12 × 75–mm disposable borosilicate glass test tubes
Pre-equilibrated Dowex AG 50X-8 columns *or* Dowex AG 50X-8 resin slurry pre-equilibrated with stop buffer (see Support Protocol 4)

1. Prepare NOS samples (see Support Protocol 6). Arrange sample assay tubes in duplicate or triplicate. Include two sets of blanks: an isotope blank and an enzyme blank.

2. Place test tubes in racks and cool the entire rack on ice. Prepare reaction mixtures (total volume 100 µl) as follows.

   a. Add 20 µl enzyme sample to duplicate or triplicate sample tubes, then add 25 µl assay buffer. For isotope blanks, replace the enzyme sample with an additional 20 µl of assay buffer.

   b. Add an additional 25 µl of either assay mix 2 (for samples) or 4 mM-N^G-nitro-L-arginine (a NOS inhibitor for enzyme blanks).

   c. If test compounds are to be examined, add them in a 5-µl volume in distilled water; alternatively add 5 µl distilled water to all samples.

   d. In samples where iNOS is specifically to be determined, add 5 µl of 20 mM EGTA to one set of duplicate tubes and 5 µl of 20 mM EGTA with 20 mM L-N^G-nitro-arginine to the other set of duplicate tubes. Initiate the reaction by the adding 25 µl assay mix 1.

3. Incubate 10 min at 37°C. Establish linearity of L-citrulline synthesis in the assay by examining two time points for the transformation of arginine into NO and L-citrulline.

To separate L-citrulline from L-arginine by Dowex chromatography

4a. Terminate reactions by placing test tubes in an ice-water bath and adding 1 ml of ice-cold stop buffer to each tube.

5a. Apply each resulting mixture to a pre-equilibrated Dowex column and collect the flowthrough in a scintillation vial. Wash each column with 1 ml water and add the effluent to the initial flowthrough.

To separate L-citrulline from L-arginine by centrifugation

4b. Terminate reactions by placing test tubes in an ice-water bath and adding a mixture of 1 ml ice-cold stop buffer and 0.5 ml Dowex resin that was pre-equilibrated with stop buffer. Incubate for 5 min.

5b. Centrifuge 10 min at 10000 × g, 4°C or allow the resin to settle for 10 min. Withdraw an aliquot of the supernatant and count as described in step 6, taking into account the fact that only a fraction of the L-citrulline was removed for counting.

6. Add 4 ml scintillation fluid to counting vials, shake well, and count in a liquid scintillation counter in the dpm mode. Count 25 µl of assay mix 1 (in triplicate) to determine the amount of total radioactivity added to each assay tube. Monitor the recovery of radiolabeled -citrulline by adding a known amount of L-[^{3}H]-citrulline labeled with ^{14}C where L-[^{3}H]arginine is used as a precursor, or L-[^{3}H]arginine when L-[^{14}C]arginine is used. Count samples of the flowthrough in dual-label mode.

7. Determine protein concentration for each sample tested.

8. Use the following formula to calculate the fraction of L-arginine conversion.

$$Con = (FT - BG)/TA$$

where Con is the fraction of L-arginine converted to L-citrulline, FT is the dpm count in the Dowex column flowthrough, BG is the dpm count in the enzyme blank (N^G-nitro-L-arginine-containing tubes), and TA is the dpm count in the 25-µl aliquot of assay mix 1 (see step 6).

9. Calculate the enzyme activity using the fraction of L-arginine converted (Con, calculated in step 8), total L-arginine concentration in the reaction mixture, and sample protein content

(step 7). First, estimate the amount of L-arginine converted according to the following equation.

$$MAC = AC \times V \times Con$$

where MAC is the amount of L-arginine converted (in moles), AC is the final L-arginine concentration (molarity), V is the assay volume (in liters), and Con is the fractional L-arginine conversion determined above.

10. Once MAC is determined, calculate enzyme activity as follows:

$$MAC \times (mg\ protein)^{-1} \times min^{-1}$$

where mg protein is the protein content of the assay tube and min represents the incubation duration in minutes.

SUPPORT PROTOCOL 4

PREPARATION AND REGENERATION OF CATION-EXCHANGE COLUMNS

Materials

Dowex AG 50X-8, H^+ or Na^+ form (100 to 200 mesh recommended; Bio-Rad)
0.5 N HCl
0.5 N NaOH
Stop buffer: 50 mM HEPES, pH 5.5, containing 5 mM EDTA, ice-cold

500-ml glass beaker
pH test paper (e.g., Color pHast; EM Science)
Disposable columns (e.g., polystyrene, 0.4 × 4.0–cm; Bio-Rad)

1. Calculate the amount of resin needed for the desired number of columns with a bed volume of 0.5 ml, allowing for ~15% loss, and use an 8- to 10-fold excess (v/w) of acid or base solution. For example, place 25 g of resin in a 500-ml beaker, add 250 ml of 0.5 N HCl, and stir 2 hr at room temp. Allow resin beads to settle, then decant light particles and acid solution (expected pH close to 1).

2. Wash resin several times, each time adding 100 ml water, stirring for 2 to 3 min, allowing particles to settle, and decanting the supernatant and light particles. Check the pH with test paper and stop the washings when the pH is between 5 and 6. Add 250 ml of 0.5 N NaOH and stir 2 hr at room temp.

3. Allow resin beads to settle, then decant light particles and NaOH solution (expected pH close to 14). Repeat step 2. Stop the washings when the pH is at or below 8.

4. Pack columns by pouring 0.5 ml of resin (~0.42 g) into columns using a pipet with an orifice wide enough to allow easy flow of resin beads. Equilibrate columns by passing 5 ml of stop buffer through them before beginning the citrulline assay.

   To prepare and equilibrate resin for separation of -citrulline from -arginine by centrifugation (see Basic Protocol 3, steps 4b and 5b), wash resin twice with 5 ml stop buffer per 1 ml bead volume (~0.42 g of beads). Adjust pH to 5.5 by adding stop buffer, if required. Use slurry directly in step 4b of Basic Protocol 3.

5. Regenerate columns by adding 1 ml distilled water to each column. Check flow, looking for clogged columns. Treat clogged columns either by applying pressure to the inlet (using house compressed air or a large disposable syringe) or by applying vacuum to the outlet.

6. Wash columns with 4 ml of 0.5 N HCl. Check effluent pH (which should be ∼1). Wash with 4 to 8 ml water, until pH is in the range of 5 to 6. Wash with 4 ml of 0.5 N NaOH. Check effluent pH (which should be ∼14). Wash with 4 to 8 ml water until pH is ≤8.

7. Equilibrate columns by passing 5 ml of stop buffer through them. To avoid faulty separation, discard Dowex resin after four cycles.

SUPPORT PROTOCOL 5

PURIFICATION OF RADIOLABELED L-ARGININE

Materials

Stop buffer: 50 mM HEPES, pH 5.5, containing 5 mM EDTA, ice-cold
Radiolabeled precursor: L-[^{3}H]arginine or L-[^{14}C]arginine (NEN Life Science Products, Amersham, or Sigma)
High-water-capacity scintillation cocktail
0.5 M NH_4OH, ice-cold
2% ethanol solution

Pre-equilibrated Dowex AG 50X-8 columns (see Support Protocol 4)
12 × 75–mm glass test tubes

1. Equilibrate a Dowex AG 50X-8 column (∼0.2 ml resin) with 5 ml ice cold stop buffer (see Support Protocol 4). Dissolve radiolabeled precursor in at least 125 μl stop buffer (to allow for duplicate counting of 10 μl or more of the samples) to yield 10,000 to 20,000 dpm per 100 μl. Apply 100 μl of this precursor solution to the column.

2. Place a scintillation vial under the column and elute the column with 1 ml stop buffer, then continue elution with 1 ml distilled water.

3. Add 4 ml of high-water-capacity scintillation cocktail to the counting vial, mix well, and count. Measure radioactivity in the dpm mode, thus correcting for differences in efficiency.

   The ratio of sample to scintillation cocktail is determined by the nature of the resulting mixture. Only clear solutions or homogeneous gels should be counted.

4. Calculate percent L-arginine in the original sample using the following equation:

$$[(\text{total dpm} - \text{eluate dpm})/(\text{total dpm})] \times 100$$

   where total dpm is the dpm of the radioactive L-arginine sample applied to the column and eluate dpm is the dpm in the column effluent. If samples are <99.0% pure, purify L-[^{3}H]arginine or L-[^{14}C]arginine as described below.

5. Equilibrate a Dowex AG 50X-8 column with 5 ml of ice-cold stop buffer (see Support Protocol 4). Dissolve the desired amount of radioisotope (typically 25 μCi of L-[^{3}H]arginine) in 0.5 ml ice-cold stop buffer and apply to the equilibrated Dowex column.

6. Wash column with 6 ml ice-cold distilled water, then elute precursor with 4 ml ice-cold 0.5 M NH_4OH into a 12 × 75–mm glass test tube. Check radioisotope purity weekly (recommended). Transfer eluate to 1-ml microcentrifuge tubes and lyophilize. Store at −20°C.

7. Reconstitute radiolabeled arginine using an appropriate volume of 2% ethanol. For optimal volume, consider using 0.1 μCi per assay tube prepared in Basic Protocol 3 (also see assay mix 1 recipe). Determine radiolabel concentration by liquid scintillation counting.

SUPPORT PROTOCOL 6

NITRIC OXIDE SYNTHASE (NOS) PREPARATION

Measurement of L-citrulline formation by purified (or partially purified) NOS does not require any of the processing described below.

NOTE: All the following procedures should be performed at 0° to 4°C.

Materials (see APPENDIX 1 for items with ✓)

Cells or tissue of interest
✓ PBS (optional)
✓ Homogenization buffer with and without 1 M KCl
✓ Assay buffer with and without DTT

Rubber policeman *or* Teflon cell scraper (if preparing samples from cultured cells in Petri dishes or flasks, respectively)
Tissue homogenizer (e.g., Polytron from Brinkmann) *or* sonicator
High-speed centrifuge
Dowex AG 50X-8 columns (Na^+-form; Bio-Rad): prepare as in Support Protocol 4 but use 20 to 50 mesh resin and pre-equilibrate in assay buffer without DTT

For cells

1a. Wash cultured cells free from culture medium using either PBS or homogenization buffer. Repeat twice.

2a. Harvest cells from cultures by scraping with a rubber policeman or Teflon scraper and transfer to centrifuge tubes of a size determined by the volume, using 2 ml of buffer (or less if possible) for each 100-mm-diameter tissue culture dish or 75-cm^2 tissue culture flask.

3a. Homogenize cells on ice using a Polytron or by three to six 10-sec cycles of sonication at 100 W. Monitor cell viability by the trypan blue exclusion test for cell viability, if desired.

4a. Centrifuge homogenate 20 min at 20,000 × g, 0° to 4°C. Retain the supernatant for the study of cytosolic NOS activity and the pellet for measurement of membrane-associated NOS. To reduce the contamination of membrane-associated enzyme with soluble enzyme, resuspend the pellet in homogenization buffer containing 1 M KCl, incubate 5 min at 4°C, and centrifuge once more.

5a. Determine protein concentration in pellets. Resuspend pellets in assay buffer to yield a solution containing 2 to 3 mg protein per ml.

6a. To remove endogenous L-arginine from supernatants, apply to Dowex AG 50X-8 columns (prepared as in Support Protocol 4 except with 20 to 50 mesh Dowex and equilibrated with DTT-free assay buffer). Collect flowthrough in test tubes cooled on ice. Store supernatants and pellets 2 months or longer at −80°C.

For tissues

1b. Weigh tissue to be processed and place in a centrifuge tube containing homogenization buffer [typically 1:5 to 1:20 (w/v) ratio of tissue to buffer; determined by the NOS activity level].

2b. Homogenize tissue using a Polytron or similar apparatus (time and power level depending on tissue source; e.g., ~10 sec at a setting of 6 or 7 for brain tissue, longer homogenization and higher power for muscle).

3b. Proceed as in steps 4a to 6a.

Reference: Nathan and Xie, 1994

Contributors: Ben Avi Weissman and Steven S. Gross

UNIT 2.5

Measurement of Oxygen Radicals and Lipid Peroxidation in Neural Tissues

BASIC PROTOCOL 1

SALICYLATE TRAPPING AND HPLC ASSAY OF HYDROXYL RADICAL

Materials

HPLC-grade solvents (EM Science):
    Acetonitrile
    Tetrahydrofuran
    Water
    Ethanol
    30% (v/v) methanol
Monochloroacetic acid (Aldrich)
3,4-Dihydroxycinnamic acid (3,4-DHCA; Aldrich)
Phenylalanine
Mannitol
2,3- and 2,5-Dihydroxybenzoic acid (2,3- and 2,5-DHBA; Aldrich)
Salicylate (Aldrich)
Tissue samples of interest: ~500-mg pieces, dissected and immediately frozen in liquid nitrogen, and stored at $-80°C$
Deferoxamine (Ciba)

Sonicator fitted with a microprobe tip (e.g., VirSonic 60 Ultrasonicator; Virtis Company)
Refrigerated centrifuge (e.g., Tomy MTX 150 equipped with a TMA-11 rotor)
250-μl autosampler vials (e.g., Sun Brokers)
High-performance liquid chromatograph (HPLC) equipped with:
    Autosampler (e.g., Perkin-Elmer ISS-100) with 150-μl vials
    Refrigerated constant-temperature circulating water bath capable of cooling to 4°C
    HPLC pump capable of delivering a flow rate of 1.0 ml/min at pressures up to 4000 psi (e.g., Waters 510 LC)
    Biophase ODS analytical HPLC column (4.6 mm × 250 mm, 5-μm particle diameter, BAS)
    Electrochemical detector (e.g., Bioanalytical Systems)
    Programmable UV absorbance detector (e.g., Applied Biosystems 785A)
    Data acquisition software (e.g., Waters Maxima 820)

1. Combine the following and stir until salts are dissolved:

   200 ml HPLC-grade acetonitrile (20%)
   100 ml of HPLC-grade tetrahydrofuran (10%)
   14.15 g monochloroacetic acid (1.4%)
   HPLC-grade water to 1000 ml.

   Degas just prior to use under vacuum for 10 min.

2. Prepare a 10 mg/ml solution of internal standard by weighing 10 mg 3,4-DHCA into 1 ml ethanol. Vortex to mix. Dilute 100 μl ethanolic 3,4-DHCA solution into 9.9 ml HPLC-grade water (final 100 μg/ml). Freeze 100-μl aliquots in microcentrifuge tubes at $-80°C$ for up to 6 months.

3. On the day of the experiment, combine the following to make extraction medium and vortex to mix:

   100 mg phenylalanine (final 1.0%)
   50 mg mannitol (final 0.5%)
   10 mg deferoxamine (final 0.1%)
   100 μl 100 μg/ml 3,4-DHCA (final 1 μg/ml)
   mobile phase to 10 ml.

4. Prepare separate 1 mg/ml stocks of 2,5-DHBA and 2,3-DHBA, and a 10 mg/ml stock of salicylate, all in HPLC-grade water. Combine the following and freeze the external standard in 20-μl aliquots at $-70°C$.

   100 μl each 2,5-DHBA and 2,3-DHBA solutions (final 10 μg/ml each)
   1 ml salicylate solution (final 1 mg/ml)
   8.8 ml HPLC-grade water.

5. Dilute 100 μl of a thawed aliquot of external standard into 9.90 ml extraction medium (final concentrations 1 μg/ml 3,4-DHCA, from the extraction medium; 0.1 μg/ml each 2,5-DHBA and 2,3-DHBA; and 10 μg/ml salicylate).

6. Weigh each piece of frozen tissue in a tared 1.5-ml microcentrifuge tube and add 0.5 ml extraction medium. (As a general rule, use 100 μl extraction medium per 100 mg tissue.) Sonicate 20 to 30 sec at a setting of 4 in a sonicator fitted with a microprobe tip. Keep tube on ice.

7. Freeze the milky solution on dry ice and then centrifuge twice for 5 min each at $18,700 \times g$ (14000 rpm in a TMA-11 rotor), 4°C (sample does not fully precipitate in the first centrifugation). Place ∼100 μl supernatant in a 250-μl autosampler vial.

8. Equilibrate a 5-μm Biophase ODS analytical HPLC column with mobile phase at a flow rate of 1 ml/min. Monitor elution with an electrochemical detector set at 650 mV and a programmable UV absorbance detector set at 295 nm.

9. Inject 50 μl sample (from step 7) onto the column using an autosampler. Flush the autosampler with 30% (v/v) methanol between runs. Use electrochemical detection to monitor 3,4-DHCA and use UV detection to monitor salicylate.

10. Inject 50 μl external standard (from step 5) onto the HPLC column to use later to determine individual response factors used in the calculations. Use electrochemical detection to monitor 2,3- and 2,5-DHBA (elution of 3,4-DHCA, 2,5-DHBA, and 2,3-DHBA at ∼4.8, 6.1, and 6.9 min, respectively; elution of salicylate at ∼15 min).

    NOTE: Contact of the salicylate-containing dialysate with metal surfaces connecting the dialysis tubing can lead to the artifactual production of DHBA (Montgomery et al., 1995).

11. Determine the area of the chromatographic peaks (electrochemically for DHBAs and 3,4-DHCA and spectrophotometrically for salicylate) using Waters Maxima 820 software.

12. Determine the unit response factors (RFs) for DHBAs and salicylate based on detection of the external standard solution (containing DHBAs, DHCA, and salicylate), using the equation $RF_x = (A/\mu g)_x/(A/\mu g)_{DHCA}$, where x denotes 2,5-DHBA, 2,3-DHBA, or salicylate; A is peak area; and μg is the amount of standard present.

13. Calculate the tissue concentration (TC) of each component in the unknown sample from the equation $TC_x = [(A_x/A_{DHCA}) \times [DHCA] \times V \times RF_x]/m$, where [DHCA] is the molar concentration of DHCA, V is the volume of the extract in liters, and m is the mass of the tissue in mg.

14. Determine the ratios (R_n) of 2,5-DHBA and 2,3-DHBA (n) to salicylate (S) using the equation $R_n = TC_n/TC_S$.

BASIC PROTOCOL 2

SPECTROPHOTOMETRIC ASSAY OF LIPID-CONJUGATED DIENES

NOTE: Commercially available solvents in this protocol can be used without further purification.

Materials *(see APPENDIX 1 for items with* ✓ *)*

Rat liver, fresh or frozen
✓ Sucrose/PBS/EDTA solution (containing ≥1 mM EDTA)
Methanol
Chloroform
2:1 (v/v) chloroform/methanol
Oxygen-free nitrogen
Cyclohexane, spectrophotometric grade

40-ml graduated, heavy-walled, stoppered centrifuge tubes
40° to 50°C water bath
Spectrophotometer with 1-cm-path-length cuvette

1. Prepare a microsomal fraction of rat liver by homogenizing 2 to 3 g tissue sample in 5 ml sucrose/PBS/EDTA solution and centrifuging 5 ml 20 min at 50,000 g, 4°C. Discard the final supernatant fraction and any fat clinging to the walls of the centrifuge tube.

2. Transfer sedimented microsomes and ~6 ml methanol to a 40-ml graduated, heavy-walled, stoppered centrifuge tube and adjust the volume to 7 ml with methanol. Proceed through step 7 without overnight delay.

3. Add 14 ml chloroform, seal with a stopper, and mix thoroughly. Allow the mixture to stand ~10 min at room temperature with occasional mixing. Centrifuge 10 min at ~260 × g, room temperature, to sediment the insoluble material.

4. Decant the supernatant fraction into another 40-ml graduated, stoppered centrifuge tube. Adjust the final volume of the lipid extract to 30 ml with 2:1 chloroform/methanol. Add 10 ml water. Mix gently by inversion only to avoid forming emulsions. Centrifuge 10 min at ~260 × g, room temperature.

5. Aspirate the upper methanol/water phase and any fluffy material at the interface between the two phases. Place the tubes in an ice bath for 5 min, centrifuge as in step 4, and aspirate any remaining methanol/water droplets.

6. Transfer a 2-ml aliquot of the chloroform phase to a clean tube, place in a water bath at 40° to 50°C, and remove all traces of chloroform under a stream of oxygen-free nitrogen. Dissolve the dried-down lipid in several milliliters of fresh chloroform, transfer the solution to a clean tube, and dry down again.

7. Dissolve the extracted chloroform-free lipid in 3 ml cyclohexane and record the absorbance spectrum from 300 to 220 nm against a cyclohexane blank (extinction coefficient at 235 nm is 2.52×10^4 $M^{-1}cm^{-1}$). Correct all absorbance measurements to a uniform base of 1 mg lipid/ml cyclohexane.

BASIC PROTOCOL 3

HPLC ASSAY OF VITAMIN E

Materials

    HPLC-grade water
    HPLC-grade methanol (EM Science)
    Pyridine (Mallinckrodt)
    Sodium perchlorate (Sigma)
    Helium
    α-Tocopherol (vitamin E; Aldrich)
    Experimental animal
    HPLC-grade ethyl acetate (EM Science)

    47-mm, 0.5-μm Teflon filter (Type FH; Millipore)
    Sonicator fitted with a microprobe tip (e.g., VirSonic 60 Ultrasonicator; Virtis Company)
    Refrigerated centrifuge (e.g., Tomy MTX 150 equipped with a TMA-11 rotor)
    250-μl autosampler vials (e.g., Sun Brokers)
    12 × 75–mm polypropylene culture tubes (e.g., Falcon)
    1.5-ml amber autosampler vials (e.g., National Scientific)
    High-performance liquid chromatograph (HPLC) equipped with:
        Autosampler (e.g., Perkin-Elmer ISS-100)
        Refrigerated constant-temperature circulating water bath capable of cooling to 4°C
        Inertsil C8 column (3.0 × 250 mm, 5-μm particle size; Metachem Technologies)
        Inertsil C8 guard column (10 × 4.3 mm, 5-μm particle size; Metachem Technologies)
        3-mm precolumn inlet filter (Rheodyne model 7335)
        HPLC pump (e.g., Waters 510 LC pump or any pump capable of delivering a flow rate of 1.0 ml/min at pressures up to 4000 psi)
        Electrochemical detector with a glassy carbon electrode and Ag/AgCl reference electrode, or any equivalent detector capable of oxidation at 700 mV (e.g., Waters 460)
        Scanning fluorescence detector (Waters 470) or other detector capable of reading emission at 340 nm and excitation at 290 nm
        Data acquisition system (Waters Millennium 2010 version 2.1 or equivalent)

1. Add 50 ml HPLC-grade water to 950 ml HPLC-grade methanol. Add 1 ml pyridine and 3.33 g sodium perchlorate and stir until the salts are dissolved. Vacuum filter the mobile phase through a 0.5-μm Teflon filter and degas for 40 min with helium at a flow rate of 100 ml/min.

2. On the day of experiment, prepare a 10 mg/ml quality control stock solution A (QC-A) by accurately pipetting 10 μl (10 mg) vitamin E into 990 μl ethyl acetate. Vortex and store on ice in the dark.

3. Prepare a 25 μg/ml quality control stock solution B (QC-B) by diluting 10 μl QC-A into 3990 μl ethyl acetate.

4. Prepare a 2.5 μg/ml quality control sample (QC-C; apparent concentration 10 μg/ml.) by adding 200 μl QC-B to 1800 μl ethyl acetate. Split the sample into three autosampler vials for use in analysis of the unknown samples.

   Because samples are extracted in four times their volume of ethyl acetate, the apparent concentration of vitamin E in the QC samples is four times greater when used for quantitation purposes.

5. On the day of experiment, prepare a 10 mg/ml calibration standard stock solution A (calibration A) by accurately pipetting 10 μl (10 mg) vitamin E into 990 μl ethyl acetate. Vortex and store on ice in the dark until needed to prepare the calibration standards.

6. Prepare a 25 μg/ml solution B (calibration B; apparent concentration 100 μg/ml.) by diluting 10 μl calibration A into 3990 μl ethyl acetate.

7. In 1.5-ml microcentrifuge tubes, prepare a set of 6 calibration standards ranging from 1.0 to 50 μg/ml apparent concentration (e.g., 1, 2.5, 5.0, 10, 25, and 50 μg/ml). Transfer prepared standards to 1.5-ml autosampler vials.

For tissue samples

8a. Remove tissue samples from animals. Cut into 50 to 500 mg pieces. Immediately freeze in liquid nitrogen and store up to 6 months at −80°C.

9a. Place frozen tissue into a tared microcentrifuge tube, weigh, and homogenize in 4 vol (v/w) ethyl acetate, e.g., for tissues weighing 25 mg, add 100 μl ethyl acetate. Maintain the samples on ice throughout the procedure.

10a. Sonicate tissues 15 to 20 sec on ice at a setting of 4 in a sonicator fitted with a microprobe tip. Centrifuge homogenates 5 min at 18700 × g (14,000 rpm in a TMA-11 rotor), 4°C. Transfer ∼100 μl supernatant (or standards) to a 250-μl autosampler vial.

For plasma samples

8b. Divide into 300-μl aliquots. Immediately freeze in liquid nitrogen and store up to 6 months at −80°C.

9b. Thaw an aliquot of plasma on ice, place in a 12 × 75–mm polypropylene culture tube, and add 1.2 ml (4 vol) ethyl acetate. Maintain the samples on ice throughout the procedure.

10b. Sonicate plasma 15 to 20 sec on ice at a setting of 4 in a sonicator fitted with a microprobe tip. Mix by vortexing. Centrifuge homogenates 5 min at 18700 × g (14,000 rpm in a TMA-11 rotor), 4°C. Transfer ∼1 ml supernatant to a 1.5-ml amber autosampler vial. For standards, place 100 μl into 250-μl autosampler vials.

11. Equilibrate an Inertsil C8 column equipped with an Inertsil C8 guard column and 3-mm precolumn inlet filter, employing an isocratic mobile phase (from step 1) at a flow rate of 0.7 ml/min (back pressure <3000 psi expected). For fluorescence detection, monitor eluate with excitation at 290 nm and emission at 340 nm. For electrochemical detection, monitor the oxidative electrochemical response of vitamin E with a glassy carbon electrode (filter set at 2), at a potential of 0.7 V versus a Ag/AgCl reference electrode.

    To obtain a stable baseline, the mobile phase may be recirculated overnight with the column in place and the electrochemical and fluorescence detectors turned on. The mobile phase may be continuously recirculated during analysis of samples and does not have to be replaced until there is significant degradation of the signal-to-noise ratio of a known solution standard or a significant change in retention time for vitamin E. After each run, the system may be set to a flow rate of 0.2 ml/min, the electrochemical detector set on standby mode, and the mobile phase continuously recirculated until needed.

12. Inject 5-μl aliquots of sample(s) and standards onto the HPLC system and collect spectral data (e.g., using the Waters Millennium Chromatography System, version 2.1). Flush the autosampler between runs with 95:5 methanol/water. Inject a quality control sample at the first position, at the midpoint of the run, and at the last position of each run (vitamin E elutes at ∼8.2 min).

13. Construct a calibration curve by plotting peak heights versus theoretical concentrations of the calibration standards. Perform linear regression, including an intercept value of zero. Report concentrations of vitamin E in units of μg/ml.

BASIC PROTOCOL 4

HPLC ASSAY OF GLUTATHIONE

In addition to vitamin E, another important cellular antioxidant that responds to LP is the reduced form of glutathione (GSH). Reduction in tissue concentrations of GSH can be a reflection of either an increase in oxidative stress (i.e., increased superoxide and consequent dismutation to H_2O_2) or an increase in oxidative damage (i.e., LP).

Materials (see APPENDIX 1 for items with ✓)

✓ 0.15 M sodium acetate, pH 7.0
  Methanol, reagent grade (EM Science)
  Reduced glutathione (GSH)
  Oxidized glutathione (GSSG)
  10 mM HCl
✓ 5 mg/ml *o*-Phthalaldehyde (OPA solution)
✓ 100 mM sodium phosphate buffer, pH 7.0
  Tissue of interest: 20- to 40-mg samples stored at $-70°C$
✓ 25 mM sodium phosphate buffer, pH 6.0, ice cold
✓ 100 mM Tris·Cl, pH 8.5
  25 and 50 mM dithiothreitol (DTT; Sigma)
  2 mM *N*-ethylmaleimide (NEM; Sigma)
  2.5% (w/v) 5-sulfosalicylic acid (SSA; Sigma)

0.5-μm Teflon filter (Type FH; Millipore)
Sonicator fitted with a microprobe tip (e.g., VirSonic 60 Ultrasonicator; Virtis Company)
Refrigerated centrifuge (e.g., Tomy MTX 150 equipped with a TMA-11 rotor)
High-performance liquid chromatograph (HPLC) equipped with:
    Autosampler with a heater/cooler to maintain constant sample compartment temperature at 4°C, and carousel capable of holding up to 96 autosampler vials (e.g., Waters 717)
    Analytical C18 reversed-phase ODS column (25 cm × 4.6 mm; 5-μm particle size)
    RP18 guard column
    HPLC pump (e.g., Waters 510 LC pump or any pump capable of delivering a flow rate of 1.0 ml/min at pressures up to 4000 psi)
    Scanning fluorescence detector or other detector capable of reading emission at 420 nm and of excitation at 340 nm (e.g., Waters 470)
    Data acquisition system (Waters Millennium 2010, version 2.1 or equivalent)
1-ml autosampler vials containing low-volume inserts with springs and caps (Waters)

Table 2.5.1 Glutathione Standard Solutions

Total	GSH (nmol)	0.01	0.03	0.1	0.3	1	5
	GSH (mM)	0.01	0.03	0.1	0.3	1	5
	GSSG (mM)	0.005	0.015	0.05	0.15	0.5	2.5
Oxidized	GSH (nmol)	0.005	0.015	0.05	0.15	0.5	2.5
	GSH (mM)	0.01	0.03	0.1	0.3	1	5
	GSSG (mM)	0.005	0.015	0.05	0.15	0.5	2.5

1. Add 200 ml of 0.15 M sodium acetate, pH 7.0, to 800 ml methanol. Vacuum filter through a 0.5-μm Teflon filter and degas for 30 min with helium at 50 ml/min. Store up to 24 hr at 4°C.

2. Prepare a 5 mM GSH stock solution by dissolving 15.4 mg GSH in 10 ml of ice-cold 10 mM HCl. Prepare a 2.5 mM GSSG stock solution by dissolving 15.3 mg GSSG in 10 ml of ice-cold 10 mM HCl. Prepare both solutions immediately prior to use and maintain on ice.

3. Prepare two sets of standards, one for total GSH and one for GSSG, as shown in Table 2.5.1. Derivatize a 200-μl aliquot of each standard with 200 μl of 5 mg/ml OPA solution for 1 min at room temperature, then dilute to 800 μl with 100 mM sodium phosphate buffer, pH 7.0. (Note that GSSG produces 2 mol GSH by reduction.)

4. Place 20 to 40 mg tissue in a tared microcentrifuge tube and add 300 μl ice-cold 25 mM sodium phosphate buffer, pH 6.0. Homogenize by sonication for 15 sec at a setting of 3 in a sonicator fitted with a microprobe tip. Keep tubes on ice throughout sonication. Centrifuge 12 min at $18700 \times g$ (14000 rpm in a TMA-11 rotor), 4°C.

5a. *For measurement of total GSH:* Mix 100 μl supernatant with 50 μl of 100 mM Tris·Cl, pH 8.5, and 100 μl of 25 mM DTT.

5b. *For measurement of oxidized GSH:* Mix 100 μl supernatant with 50 μl of 100 mM Tris·Cl, pH 8.5, and 50 μl of 2 mM NEM for 1 min on ice. Then add 50 μl of 50 mM DTT.

6. Incubate 30 min on ice. Precipitate samples by adding 250 μl of 2.5% SSA and centrifuging 6 min at $18,700 \times g$, 4°C. Derivatize a 200-μl aliquot of supernatant with 200 μl of 5 mg/ml OPA solution for 1 min at room temperature, then dilute to 800 μl with 100 mM sodium phosphate buffer, pH 7.0. Use immediately or store in the dark for up to 12 hr at 4°C.

7. Equilibrate a C18 reverse-phase column using an isocratic mobile phase (from step 1) at a flow rate of 1.0 ml/min. Monitor eluate with a fluorescence detector using excitation at 340 nm and emission at 420 nm.

8. Using 1-ml autosampler vials, inject a 100-μl sample and collect fluorescence data (e.g., using the Waters Millennium Chromatography System, version 2.1; elution of derivatized glutathione in 5.0 to 5.5 min.). Flush the column with 100% methanol between runs to elute nonpolar compounds. Alternatively, inject up to five consecutive samples before flushing the column with methanol and re-equilibrating.

9. Inject 20-μl aliquots of standards onto the HPLC column and collect fluorescence data.

10. Determine the quantity of glutathione in a sample (from ∼0.01 to 5 nmol for 20 to 50 mg of brain tissue) by comparing the peak area of the derivatized sample to a standard curve of known amounts of derivatized glutathione.

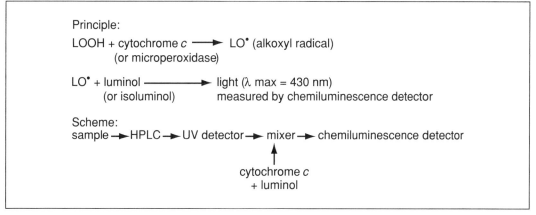

Figure 2.5.1 HPLC-chemiluminescence measurement of lipid hydroperoxides (PCOOH, PEOOH, fatty acid-OOH).

BASIC PROTOCOL 5

HPLC-CHEMILUMINESCENCE ASSAY OF LIPID HYDROPEROXIDES

Figure 2.5.1 presents the principle and scheme for an HPLC-chemiluminescence (HPLC-CL) detection method for selective and sensitive (1 pg) measurement of the levels of lipid hydroperoxides (LOOHs), including phospholipid hydroperoxides, in which the peroxidized PUFA is still attached to the phosphate head group (e.g., phosphatidylcholine hydroperoxide or PCOOH; phosphatidylethanolamine hydroperoxide or PEOOH), or de-esterified fatty acid hydroperoxides (e.g., arachidonic acid-OOH, linoleic acid-OOH). See Miyazawa et al., 1987 and Zhang et al., 1994a,b.

Significant drawbacks of this method include the cost of the HPLC-CL system and the lack of a suitable internal standard. Additionally, certain chemiluminescence peaks may be rather broad, because they are made up of a mixture of similar, but not identical, hydroperoxide species, thus broadening the chromatographic peak.

Materials *(see APPENDIX 1 for items with ✓)*

   HPLC-grade chloroform (EM Science)
   HPLC-grade methanol (EM Science)
   40 mM monobasic potassium phosphate
   25 mg/ml phosphocholine (PC; Avanti Polar Lipids)
   25 mg/ml phosphoethanolamine (PE; Avanti Polar Lipids)
   Methylene blue
   1% (v/v) butylated hydroxytoluene (BHT; Sigma)/methanol
   Tissue samples, dissected and immediately frozen in liquid nitrogen, stored at −80°C
   0.15 M saline: 0.9% (w/v) NaCl in HPLC-grade water
   Nitrogen
   2:1 and 1:9 (v/v) chloroform/methanol, freshly prepared
✓ Chemiluminescent (CL) cocktail

   UV light source
   Tared 4-ml amber glass vials with caps
   Sonicator fitted with a microprobe tip (e.g., VirSonic 60 Ultrasonicator; Virtis Company)
   Refrigerated centrifuge (e.g., Tomy MTX 150 equipped with a TMA-11 rotor)

Reacti-therm heating module/evaporating unit (Pierce)
High-performance liquid chromatograph (HPLC) equipped with:
    3-ml Supelclean LC-Si column (Supelco)
    SIL-LC-Si analytical HPLC column (25 × 4.6–mm, 5-μm particle size; Supelco)
    Guard column (e.g., Supelguard LC-Si; Supelco)
    HPLC pump capable of delivering 1 ml/min at pressures up to 4000 psi
      (e.g., Waters 600)
    Autosampler thermostatted to 4°C and containing a carousel capable of holding
      up to 96 autosampler vials (e.g., Waters 717)
    Postcolumn mixing tee (e.g., ISCO 500D syringe pump; ISCO)
    Chemiluminescence (CL) detector (e.g., Jasco 825-CL; Jasco)
    Data acquisition system (Waters Millennium 2010, version 2.1, or equivalent)
Autosampler vials containing low-volume inserts with springs and caps (Waters)

1. To prepare mobile phase combine the following in order:

    100 ml HPLC-grade chloroform
    880 ml HPLC-grade methanol
    20 ml of 40 mM monobasic potassium phosphate.

   Degas under vacuum for 10 min before use.

2. Prepare individual standards by adding 1 ml (25 mg) PC 18:2 or PE to 25 ml methanol containing 0.1 mM methylene blue as a photosensitizer. Place reaction mixture in a 200-ml beaker cooled with ice water, fix a light source 50 cm above the surface of the reaction mixture, and photoirradiate for 6 hr (for preparation of PCOOH and PEOOH standards by photooxidation; see Miyazawa et al., 1987 and Zhang et al., 1994a,b).

3. Pass the reaction mixture through a 3-ml Supelclean LC-Si column using methanol as an eluant to remove the methylene blue. Recover phospholipid and its oxidized products, including the hydroperoxides, in the methanol extract.

4. Determine the hydroperoxide content using the xylenol orange assay (see Support Protocol). Maintain the resulting hydroperoxides at −20°C in capped 4-ml amber glass vials until use (up to 6 months).

5. For all standards [e.g., 15(S)-HPEPE, 13(S)-HPODE, or 9(S)-HPODE available from Biomol Research Laboratories] make a 100 μM stock solution in 1:9 chloroform/methanol and dilute appropriately for construction of a standard curve.

6. On the day of experiment, add 20 μl of 1% BHT/methanol to 9.98 ml methanol (0.002% BHT), dispense into 340-μl aliquots in 1.5-ml microcentrifuge tubes, and maintain at 4°C. Weigh a 10- to 40-mg sample of tissue and place in the microcentrifuge tube containing 340 μl of 0.002% BHT/methanol. Sonicate 15 sec at 4°C at a setting of 3 using a sonicator fitted with a microprobe tip.

7. Add 660 μl chloroform and vortex for 1 min. Add 400 μl of 0.15 mM saline and vortex for 1 min. Centrifuge 8 min at 18700 × g (14,000 rpm in a TMA-11 rotor), 4°C. Remove the lower solution using a Pasteur pipet and transfer it to a tared 4-ml amber glass vial.

8. Dry under a mild stream of nitrogen at 30°C using a Reacti-therm heating module/evaporating unit at a low setting. Immediately recap the vials when the samples are totally evaporated (∼20 min). I

9. Weigh capped vials and determine total lipid content (in mg) by subtracting the tared weight of the empty vial. Store capped vials overnight at −80°C. Calculate the yield (mg lipid/mg original tissue) and resuspend each sample in 100 μl of 2:1 chloroform/methanol per milligram total lipid.

10. Equilibrate an SIL-LC-Si analytical HPLC column equipped with a guard column using mobile phase (from step 1) at a flow rate of 1 ml/min. Verify that the autosampler compartment is cooled to 4°C. Use an autosampler flush solution of 1:9 chloroform/methanol. Install a postcolumn mixing tee to mix the column effluent with CL cocktail flowing at 1.0 ml/min through the CL detector. Before running the assay, run the mobile phase and postcolumn CL cocktail until background levels on the CL detector stabilize. Use a gain of 100× to keep the background signal energy emission between 0.1 and 0.2.

11. Construct standard curves using 20-μl injections of four concentrations (e.g., ranging from 0.25 to 2 μM) of the photooxidized PCOOH and PEOOH (step 5). Collect data [e.g., using the Waters Millennium Chromatography System, version 2.1; fatty acid-OOH (FFAOOH), PEOOH, and PCOOH elute at ∼3, 4.5, and 7 min, respectively]. Construct a calibration curve by plotting peak areas versus theoretical concentrations of the component in the standard. Use a linear fit with intercept = 0.

12. Inject a 20-μl aliquot of sample onto the equilibrated column and monitor the chemiluminescent signal. Quantify unknowns by peak area using the standard curves. Determine concentrations of PEOOH, PCOOH, and FFAOOH in units of pmol/mg tissue.

SUPPORT PROTOCOL

XYLENOL ORANGE DETERMINATION OF HYDROPEROXIDE CONTENT IN STANDARDS

This procedure is used in conjunction with the HPLC-chemiluminescence assay of lipid hydroperoxides (Basic Protocol 5). It can be performed either on the day of experiment or later, using a stored sample. See Jiang et al., 1991 for more details.

Materials

Degassed water
Ferrous ammonium sulfate (FAS)
Argon
Methanol
Butylated hydroxytoluene (BHT; Sigma)
Concentrated (18 M) H_2SO_4
Xylenol orange (Aldrich)
Standards to be tested (see Basic Protocol 5)
Spectrophotometer and cuvettes

1. Add 1 ml degassed water to 9.8 mg FAS (final 250 μM) in a microcentrifuge tube. Purge with argon by bubbling through a Pasteur pipet and vortex. Transfer solution to a 125-ml container and add the following:

   9 ml water
   90 ml methanol (final 90%)
   88.12 mg BHT (final 4 mM) and stir
   139 μl concentrated H_2SO_4 (final 25 mM) and 8.52 mg xylenol orange (final 100 μM) and stir.

   Store up to 1 week at 4°C.

2. For each standard to be tested, prepare triplicate samples containing 900 μl xylenol orange reagent and the amounts of standard, 90% methanol, and water indicated in Table 2.5.2. Include a triplicate blank. Vortex all samples. Incubate 30 min at room temperature.

Table 2.5.2 Samples for Xylenol Orange Determination

Standard (μl)	90% methanol (μl)	H$_2$O (μl)
5	95	0
10	90	0
20	80	0
45	50	5
90	0	10
0 (blank)	100	0

3. Measure the absorbance at 560 nm. Average all triplicates, subtract the blank value from each sample value, and calculate the amount of hydroperoxide using the extinction coefficient (4.3×10^4 M^{-1}cm^{-1} at 560 nm).

BASIC PROTOCOL 6

THIOBARBITURIC ACID ASSAY OF MALONDIALDEHYDE

Materials (see APPENDIX 1 for items with ✓)

    Biological sample: 1.0 to 2.0 mg/ml membrane protein *or* 0.1 to 0.2 mM lipid phosphate
✓ TCA/TBA/HCl solution
    Boiling water bath
    Spectrophotometer and cuvettes

1. Combine 1.0 ml biological sample with 2.0 ml TCA/TBA/HCl solution and mix thoroughly. Prepare a blank containing 1.0 ml reagents minus lipids in place of sample. Heat 15 min in a boiling water bath.

2. Cool samples to room temperature and centrifuge 10 min at $1000 \times g$, room temperature.

3. Determine the absorbance of the sample at 535 nm against the blank, and calculate the malondialdehyde concentration of the sample using an extinction coefficient of 1.56×10^5 M^{-1}cm^{-1}.

BASIC PROTOCOL 7

HPLC ASSAY OF MALONDIALDEHYDE USING UV DETECTION

Materials (see APPENDIX 1 for items with ✓)

    Tris base
    HPLC-grade water (EM Science)
    HPLC-grade acetonitrile (EM Science)
    0.1 mM H$_2$O$_2$
    1 N HCl
    Malonaldehyde bis(diethylacetal) (also known as 1,1,3,3-tetraethoxypropane or TEP; Aldrich)
✓ 10 mM potassium phosphate buffer, pH 7.0
    Frozen tissue samples: dissected and immediately frozen in liquid nitrogen, stored at $-80°$C

✓ 40 mM Tris·Cl, pH 7.4
✓ 10 mM ferrous ammonium sulfate (FAS) solution
 0.02% (v/v) butylated hydroxytoluene (BHT; Sigma) in acetonitrile

  47-mm, 0.5-μm Teflon filter (Type FH; Millipore)
  250-ml glass-stoppered flask
  37° and 50°C water baths
  10-ml volumetric flask
  UV spectrophotometer (e.g., Beckman model DU-7500 Diode Array)
  Sonicator fitted with a microprobe tip (e.g., VirSonic 60 Ultrasonicator; Virtis Company)
  Refrigerated centrifuge (e.g., Tomy MTX 150 fitted with a TMA-11 rotor)
  1-ml autosampler vials containing low-volume inserts with springs and caps (Waters)
  High-performance liquid chromatograph (HPLC) equipped with:
    Autosampler configured with a thermostatted sample compartment held at 4°C
      and a carousel capable of holding up to 96 autosampler vials (e.g., Waters 717)
    Zorbax SB-CN analytical HPLC column (4.6 × 25 cm, 5-μm particle size;
      MAC-MOD Analytical)
    Zorbax SB-CN guard column (4.0 × 12.5 mm, 5-μm particle size) fitted with
      Peek-encapsulated frit-gasket (MAC-MOD Analytical)
    HPLC Pump capable of delivering 0.9 ml/min at pressures up to 4000 psi (e.g.,
      Waters 600)
    Programmable UV absorbance detector (e.g., Applied Biosystems 785A)
    Data acquisition system (Waters Millennium 2010 version 2.1 or equivalent)

1. Dissolve 3.63 g Tris base in ~800 ml HPLC-grade water and adjust pH to 7.0 with 0.1 N HCl. Vacuum filter through a 0.5-μm Teflon filter. Add 50 ml HPLC-grade acetonitrile and bring total volume to 1 liter with water. Keep refrigerated until needed. Prepare fresh daily. At time of assay, degas for 10 min, add 100 μl of 0.1 mM H_2O_2 per liter mobile phase (final 10 μM), and stir.

2. Add 1 ml of 1 N HCl to 99 ml HPLC-grade water in a 250-ml glass-stoppered flask and stir. Add 1 mmol TEP, replace the stopper, and stir to dissolve. Wrap Parafilm around the stopper to hold it firmly in place (to prevent loss of MDA while heating), and place the flask in a 50°C water bath for 60 min. Cool to room temperature. Store the resulting 10 mM stock solution of MDA for ≥1 month at 4°C.

3. Prepare a 100 μM MDA standard solution in a volumetric flask by diluting 10 μl of 10 mM stock solution to 10 ml using 10 mM potassium phosphate buffer, pH 7.0. Measure the absorption of the 10 μM MDA standard solution at λ_{max} (267 nm) in a UV spectrophotometer. Use the molar extinction coefficient (31,800 $M^{-1}cm^{-1}$) to verify the concentration and purity of the solution.

4. Dilute with 10 mM potassium phosphate buffer, pH 7.0, as necessary to generate the standard curve with at least five MDA standards linear for 20-μl injections of solutions ranging from 1.0 to 20 μM (e.g., 1, 2, 5, 10, and 20 μM).

5. Place 10 to 40 mg frozen tissue into a tared microcentrifuge tube, weigh, and add 10 vol (v/w) of 40 mM Tris·Cl, pH 7.4. Sonicate 15 to 20 sec on ice at a setting of 4 using a sonicator fitted with a microprobe tip. Proceed immediately to the extraction.

6. *Optional:* Split homogenates into two equal volumes in separate tubes. Add 10 mM FAS solution to one tube at a final concentration of 100 μM and immediately vortex to measure spontaneous MDA formation (tube without FAS) and induction by iron.

7. Incubate homogenates 30 to 60 min in a 37°C water bath except for basal samples (i.e., no incubation and no MDA detection).

8. Add an equal amount (as used for homogenization) of 0.02% BHT in acetonitrile at 4°C and vortex to deproteinize the samples. Centrifuge 10 min at 18,700 × g (14,000 rpm in a TMA-11 rotor), 4°C. Retain the supernatants for the MDA assay. Transfer ~100 μl supernatant to 1-ml autosampler vials containing low-volume inserts.

9. Equilibrate a Zorbax SB-CN analytical HPLC column equipped with a Zorbax SB-CN guard column at 4°C. Flush the cooled autosampler with 5:95 (v/v) acetonitrile/water. Use an isocratic mobile phase (from step 1) at 0.9 ml/min and monitor effluent with a programmable UV absorbance detector set at 267 nm.

10. Inject 20-μl aliquots of standards and samples and acquire data (MDA elutes at ~4.0 min).

11. Construct a calibration curve by plotting peak heights versus theoretical concentrations of the component in the standard. Fit the data using a best-fit least squares linear regression forced through the origin.

12. Report concentrations of MDA in units of nmol/g tissue. If appropriate, subtract the values of spontaneous MDA formation from those induced by iron (step 6; Kwon and Watts, 1963).

ALTERNATE PROTOCOL

HPLC ASSAY OF TBA-MALONDIALDEHYDE ADDUCT USING FLUORESCENCE DETECTION

Additional Materials (*also see Basic Protocol 7; see* APPENDIX 1 *for items with* ✓)

  0.9% (w/v) NaCl
✓ TBA/acetate/DTPA solution
  5% (v/v) butylated hydroxytoluene (BHT) in 99% (v/v) ethanol

0.2- and 0.45-μm filters (Millipore)
Water bath capable of maintaining temperatures up to 100°C
150 × 4.6–mm Inertsil ODS-2 analytical HPLC column (MetaChem Tech)
50 × 4.0–mm guard column (MetaChem Tech)
HPLC pump capable of delivering 1 ml/min
Fluorescence detector with excitation at 515 nm and emission at 553 nm

1. Add 300 ml HPLC-grade water to 700 ml HPLC-grade acetonitrile. Filter through 0.45-μm filter. Degas under vacuum for 10 min before use.

2. Prepare MDA standards (see Basic Protocol 7, steps 2 to 4).

3. Prepare a 10% w/v tissue homogenate (see Basic Protocol 7, step 5), but use 0.9% NaCl in place of 40 mM Tris·Cl.

4. Combine the following and incubate 45 min at 95°C:

    10 μl homogenate
    1.0 ml TBA/acetate/DTPA solution
    10 μl 5% BHT in 99% ethanol

   and cool to room temperature under a stream of tap water. Filter through a 0.2-μm filter.

5. Equilibrate a 150 × 4.6–mm Inertsil ODS-2 column equipped with a 50 × 4.0–mm guard column using an isocratic flow of mobile phase (from step 1) at 1.0 ml/min at ambient temperature. Monitor effluent using a fluorescence detector with excitation at 515 nm and emission at 553 nm. Inject 10-μl aliquots of sample and standards and acquire data (the signal of interest produced by the TBA-MDA complex).

6. Create a calibration curve by plotting the integrated peak areas against concentrations of derivatized TEP standard (see Basic Protocol 7, step 11). Determine concentrations in unknowns using the calibration curve. Express values as nmol MDA/mg or g tissue wet weight or per mg protein.

BASIC PROTOCOL 8

IMMUNOASSAY OF 8-ISOPROSTAGLANDIN $F_{2\alpha}$

Materials

Sample: cell cultures or tissues stored at $-80°C$
HPLC-grade solvents (EM Science):
    Absolute ethanol, ice cold
    Hexane
    Ethyl acetate
    Methanol
100 mM sodium phosphate buffer, pH 7.2, prepared with deionized H_2O
8-Isoprostane EIA Kit (8-iso-$PGF_{2\alpha}$; Cayman Chemical)

Polytron tissue homogenizer
C18 Sep-Pak columns (Waters)
Vacuum centrifuge
Microplate reader (e.g., Techan SLT Spectra) with microplate analysis software (e.g., DeltaSOFT II; Biometallics)

1. Weigh a ~50-mg sample and homogenize 20 sec in 2 ml ice-cold HPLC-grade absolute ethanol using a Polytron tissue homogenizer. Allow homogenate to stand 5 min at 4°C, and then microcentrifuge 10 min at maximum speed, 4°C, to pellet precipitated proteins. Collect supernatant, dilute with 8 ml of 100 mM sodium phosphate buffer, and vortex.

2. Prepare a C18 Sep-Pak column by rinsing with 5 ml absolute ethanol followed by 5 ml deionized H_2O. Pass sample through the prepared C18 Sep-Pak column. Rinse the cartridge with 5 ml H_2O followed by 5 ml HPLC-grade hexane. Discard both rinses. Elute 8-iso-$PGF_{2\alpha}$ with 5 ml HPLC-grade ethyl acetate containing 1% (v/v) HPLC-grade methanol.

3. Divide purified sample into five 1.5-ml microcentrifuge tubes and lyophilize them, uncapped, in a vacuum centrifuge (~1 hr) to remove all traces of solvent. Reconstitute with 1 ml EIA buffer from an 8-isoprostane EIA kit.

4. Place 50-µl aliquots in a 96-well plate for EIA analysis. Include standards supplied in the kit, prepared according to the manufacturer's recommendations. Incubate 18 hr at room temperature.

5. Develop EIA for 15 min in the dark on a plate shaker with Ellman's reagent. Read in a microplate reader and construct a four-point standard curve using microplate analysis software.

6. Use the standard curve to determine the concentrations in the samples. Calculate the estimated amount of 8-iso-$PGF_{2\alpha}$ in the sample using the equation 8-iso-$PGF_{2\alpha}$ (pg) = concentration (pg/ml) × vol reconstituted sample (1 ml).

References: Cini et al., 1994; Halliwell and Gutteridge, 1991; Janero, 1990

Contributors: Edward D. Hall and Paula K. Andrus

UNIT 2.6

Overview of Microdialysis

The technique of microdialysis enables the monitoring of extracellular neurotransmitters, neuropeptides, and hormones in the brain and periphery. This unit establishes the groundwork for the basic techniques of preparation, conduct, and analysis of dialysis experiments in rodents and subhuman primates (UNITS 2.7–2.9). Although the methods described are those used for monitoring CNS function, they can be easily applied with minor modification to other organ systems.

The microdialysis technique has disadvantages including: limited time resolution (≥ 1 min; more typically, 10-min collection periods are employed) in comparison to voltammetry (<1 min); creation of an area around the probe in which all solutes capable of crossing the probe membrane are depleted; changes in the neurochemical milieu that may affect basal levels and/or the pharmacological responsiveness of the substance under study; possible diffusion of low-molecular-weight solutes contained in the perfusate into the interstitial space, artificially changing the level or activity of the analyte of interest (e.g., failure to provide appropriate concentrations of Ca^{2+} or Na^+ ions in the perfusate dramatically altering basal levels of neurotransmitters); size limitation to areas large enough and long enough to surround the microdialysis probe; and a fractional representation in the analyte collected by microdialysis of the actual extracellular concentration.

At usual perfusate flow rates (e.g., 1 µl/min), the ratio, or extraction fraction, between the actual extracellular concentration of an analyte and the dialysate concentration of that same analyte is typically less than 40% (known as relative recovery or probe efficiency). Experimental manipulations that modify these neurochemical processes may also modify extraction fraction and confound interpretation of changes in dialysate concentration, mirroring changes in extracellular concentration and reflecting changes in extraction fraction and/or extracellular concentration. Thus, the study of the diffusional and/or convective processes underlying analyte sampling by microdialysis (or solute delivery by the microdialysis probe) has been developed (e.g., quantitative microdialysis) to determine in vivo extraction fraction and the proportional relationship between dialysate and extracellular concentrations of an analyte under study. The theoretical and empirical findings to date are discussed in detail in the following three sections.

DESCRIPTION OF THE MICRODIALYSIS PROCESS

All microdialysis probes are composed of a length of tubular dialysis membrane through which a solution, usually devoid of the analyte of interest, is constantly perfused. The dialysis membrane is semipermeable and typically limited to small molecules with molecular masses less than 20,000 Da. Once the probe is inserted into tissue, substances at higher concentration at the outside surface of the dialysis membrane slowly flow through the length of the dialysis probe into the outflow tubing where it can be collected for subsequent analyte quantification. See Figure 2.6.1 for a representation of a microdialysis probe.

As low-molecular-weight solutes are removed from the external medium by the microdialysis probe, a zone around the probe becomes partially depleted of solute. Information about the difference in concentration between the undisturbed tissue and the dialysate and the size of the area of analyte depletion (penetration distance) will aid in the interpretation of the results from a microdialysis experiment by providing a measure of probe efficiency that makes it possible to calibrate the concentration of the analyte in the dialysate with respect to its extracellular concentration and to identify the anatomical site from which the microdialysis sample is collected (i.e., to determine how far away from the probe it is).

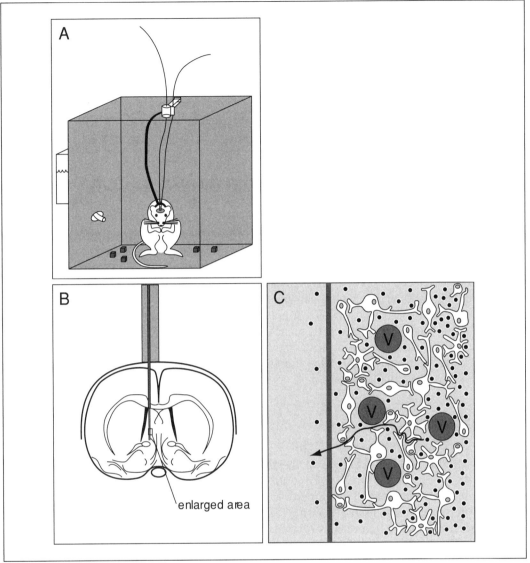

Figure 2.6.1 Representation of the "inside view" of a microdialysis probe. The microdialysis probe, which consists of an inflow and outflow tubing (**A**) separated by tubing made of dialysis membrane, is implanted surgically into a specific area within the brain (**B**). The enlarged view (**C**) illustrates the complex composition of the fluid through which analytes (black dots) must diffuse to get to the microdialysis probe. The presence of impermeable cells such as blood vessels (V) reduces the fluid volume surrounding the probe and increases the diffusional path (arrow) of analytes moving toward the probe. The net effect is a decreased diffusivity in this phase.

The underlying process driving analyte transport during microdialysis is generally accepted to be diffusion. The parameters of the experiment, including flow rate, inflow tube length, and tube diameter and the chemical characteristics of ethanol, may contribute to the occurrence of ultrafiltration. The presence of ultrafiltration during microdialysis (not usually expected) can be identified by determining if fluid volume is lost during dialysis, and additional mathematical considerations (not considered below) are necessary to account for ultrafiltration during analyte transport. An analysis of the slope of the concentration gradient of an analyte from the "undisturbed" tissue into the perfusate, the radial distance over which the concentration gradient occurs, and the net change in concentration of the analyte (extraction fraction) over the course of the concentration gradient can be used to describe the transport of analyte during microdialysis. It is important to note that the concentration gradient is influenced not only by the physical attributes of the analyte (i.e., molecular weight, hydrophobicity, and tertiary structure), but also by the physical composition of the medium in which transport occurs. Therefore, a complete characterization of the transport of an analyte during microdialysis involves a description of the concentration gradient in the external medium (e.g., tissue), the concentration gradient across the membrane, and the concentration gradient across the annular width of the microdialysis probe.

EMPIRICAL METHODS FOR DETERMINING IN VIVO RELATIVE RECOVERY

Estimates of in vivo probe efficacy (extraction fraction, E_d) are important for two purposes: (1) knowledge of in vivo E_d provides a way of accurately converting dialysate levels to extracellular concentration and, thus, serves as a calibration tool; and (2) when E_d is low, changes in the in vivo E_d parallel changes in the resistance of tissue to analyte diffusion, possibly reflecting underlying differences in the metabolism and clearance of the analyte (Justice, 1993; Cosford et al., 1996; Vinson and Justice, 1997). It is recognized that the main source of resistance to diffusion during microdialysis arises from the external medium and not the dialysis membrane. Several groups have developed empirical methods to directly determine in vivo E_d (Jacobson et al., 1985; Lonnroth et al., 1987, 1989; Larsson, 1991), described below. In addition, the use of in vitro assays to estimate in vivo E_d is considered.

Difference Method

The difference method described in UNIT 2.7 for determining extracellular concentrations of analytes will also yield an estimate of the in vivo E_d. It is the slope of the linear regression that describes the dialysate concentration of the analyte under study as a function of experimenter-controlled variations in perfusate analyte concentration. The net difference between the perfusate concentration and the dialysate concentration of the analyte is assumed to result solely from the gain or loss of analyte by diffusion across the microdialysis probe membrane. The direction of diffusion (into or out of the probe) is dictated by the undisturbed tissue concentration and could be either a gain of analyte (if the perfusate concentration is lower than the tissue concentration) or a loss of analyte (if the perfusate concentration is greater than the tissue concentration). In this approach, referred to as the variation in concentration method (Justice, 1993; Parsons and Justice, 1994), analysis is accomplished by perfusing several concentrations of the analyte of interest through the dialysis probe (C_{in}) in random order and determining the net change in perfusate concentration after dialysis ($C_{in} - C_{out}$), where C_{out} is the dialysate concentration for each C_{in}. The slope defined by the linear regression equation describing $C_{in} - C_{out}$ (net change) as a function of C_{in} (starting perfusate concentration) provides the proportional difference between dialysate concentration and the undisturbed tissue concentration of analyte (C_{ext}, the "undisturbed" tissue concentration). This proportion, the extraction fraction, describes the transport of analyte to and from the probe and is defined mathematically by:

$$E_d = \frac{C_{out} - C_{in}}{C_{ext} - C_{in}}$$

Also see UNIT 2.7 for more details on the application of this technique under steady-state and transient conditions and Bungay et al., 1990, for more detailed mathematical description of E_d.

Group comparison of E_d values can easily be made under steady-state conditions by collecting all data necessary for calculation of the linear regression equation in each subject. Therefore, each E_d is obtained for each subject and can be treated as any other dependent variable. Under transient conditions, the data necessary for calculation of the linear regression equation are collected from a group of subjects. In this case, E_d is obtained for each group and can be compared using standard regression statistics that test for parallelism of slopes. Group differences in E_d would suggest that probe efficacy varies by group and that differences in analyte dialysate levels reflect something more than a change in extracellular concentration (Justice, 1993).

The difference technique is easy to apply and analysis of in vivo E_d data is straightforward. Moreover, when group differences in E_d are observed, it is possible to use the E_d as a calibration tool to correct dialysate data to obtain extracellular concentrations and to test hypotheses concerning group differences in extracellular concentration. Theoretical consideration of the factors affecting E_d suggests that changes in E_d parallel changes in the rate of analyte clearance. Therefore, changes in E_d can provide indirect measures of the rate of analyte clearance.

Flow Rate Method

The flow rate method was one of the first empirical methods designed to determine extracellular concentration by determining the relationship between perfusate flow rate, the active area of the membrane, and the mass transfer coefficient and then extrapolating to the case of zero flow rate (Jacobson et al., 1985). The dialysate concentration at zero flow rate is expected to be equal to the extracellular concentration of undisturbed tissue. The equation:

$$C_{out}/C_{ext} = 1 - \exp(-K_0 A/F)$$

is written to describe E_d, where C_{out} is the concentration in the dialysis probe, C_{ext} is the concentration of analyte in undisturbed tissue, K_0 is the mass transfer coefficient, A is the active area of the microdialysis probe membrane, and F is the perfusion flow rate. This equation can be related to the mathematical model of E_d described above. Here $K_0 A$ is equivalent to $1/(R_d + R_m + R_e)$, and the entire expression is equivalent to E_d (Bungay et al., 1990).

The above equation describes a nonlinear regression to a set of dialysate data collected at varying flow rates. The nonlinearity is not surprising given that the source of greatest resistance will vary by flow rate. The source of maximum resistance to transport of the analyte when E_d is low will lie in the external solution, while the source of maximum resistance to transport of the analyte when E_d is high will lie in the membrane. This is a limiting factor for the use of this technique since it does not yield a single best fit as in linear regression (Parsons and Justice, 1994). Furthermore, most of the data should be obtained at very low flow rates (<1 mM/min) to increase the accuracy of the regression analysis in its most dynamic range. Unfortunately, low flow rates require long collection periods to obtain appropriate sample volumes, limiting the technique's application. (see Stahle et al., 1991; Menacherry et al., 1992).

Internal Standard Method

The internal standard method measures E_d by measuring the loss of an internal standard from the perfusate (Larsson, 1991; Scheller and Kolb, 1991). Extraction fraction is given as the relative loss of the standard (RL) as shown by:

$$RL = E_d = \frac{C_{in} - C_{out}}{C_{in}}$$

where C_{in} is the starting concentration of the internal standard and C_{out} is the dialysate concentration of the internal standard. The internal standard is usually a radiolabeled version of the analyte of interest. The assumption is that perfusion of the internal standard across the dialysis membrane is not limited regardless of the amount of unlabeled analyte present. This assumption has received support from in vitro studies in which increasing concentrations of the analyte of interest in the test solution did not change RL (Scheller and Kolb, 1991). With this method it is necessary to show that the labeled standard is diffusionally similar to the unlabeled analyte (e.g., has a similar diffusion coefficient through the membrane and similar clearance mechanisms in tissue), and one must avoid high concentrations that would tend to make the membrane the limiting site of diffusion in the overall diffusional path (Le Quellec et al., 1995; Lonnroth and Strindberg, 1995).

In Vitro Recovery Method

When the primary resistance of the membrane is greater than the resistance of the external medium, then no difference between in vitro recovery and in vivo recovery is expected. Circumstances in which this might occur are found in microdialysis protocols that use very slow flow rates and/or long dialysis fibers, or diffusionally resistant membranes (e.g., when high-molecular-weight substances are sampled or membranes with very small pore sizes are used). Unfortunately, use of a highly resistant membrane is typically not possible, because E_d becomes so low that the detectability of the analyte is compromised (Bungay et al., 1990). Slow perfusate flow rates yield nearly 100% recovery but require a relatively long collection period, or very small sample volume, which may compromise other aspects of the experiment (Menacherry et al., 1992). In vitro recovery methods may become more advantageous as more sensitive detection sampling and holding methods become available.

FACTORS AFFECTING IN VIVO RECOVERY

During microdialysis, analytes pass through a semipermeable membrane from the extracellular fluid (ECF) into a perfusate that is collected over a predetermined time and volume. Because the membrane is semipermeable only some solutes, namely, low-molecular-weight solutes, will be recovered. The concentration of analyte collected by this method will represent only a fraction of the ECF concentration. Consideration of factors determining the amount of analyte recovered (described below) during the design phase of a microdialysis experiment will greatly enhance the success of that experiment (see Strategy for Determining Experimental Design) by assuring detectable levels of analyte in each dialysate sample. In addition, an understanding of these factors reveals the mechanisms that contribute to the stability of recovery during a microdialysis experiment; a major assumption underlying the application of conventional microdialysis methods is that analyte recovery is at steady state throughout the sampling period.

Flow Rate

Relative recovery of an analyte (concentration of analyte per sample) is inversely proportional to the perfusate flow rate (Johnson and Justice, 1983; Tossman et al., 1986; Wages et al., 1986; Alexander et al., 1988; Benveniste, 1989). Thus, as the flow rate decreases the concentration of analyte in each sample increases. Furthermore, at lower flow rates the net depletion of solutes around the probe decreases. For analytes with low extracellular concentration or low diffusivity, reducing the flow rate will increase the relative recovery and, therefore, increase the probability of obtaining a detectable concentration of analyte in each sample.

At extremely low flow rates (<0.1 µl/min) it is possible to reach near 100% recovery of an analyte, in which case the dialysate concentration of that analyte would equal the ECF analyte concentration (Van Wylen et al., 1986; Menacherry et al., 1992; Smith et al., 1992). The flow rate necessary to achieve 100% relative recovery depends on both the diffusional characteristics of the analyte under study as well as the membrane type and length. For many analytes, low flow rates are not necessary to achieve detectable concentrations in dialysates. However, for situations in which long times and/or low volume sample handling are possible, slow flow rates offer the advantage of providing direct measurements of extracellular concentration.

The absolute recovery of an analyte (amount of analyte per sample) is proportional to the perfusate flow rate, up to flow rates of 2 µl/min (Wages et al., 1986; Benveniste, 1989). Thus, for analytical assays in which the total amount of analyte, rather than concentration, is measured (e.g., radioimmunoassay) it may be advantageous to use higher, rather than lower, flow rates (up to 2 µl/min) to achieve a sufficient quantity of sample for detection. However, at these higher flow rates, probe efficiency is reduced (probe efficiency is equivalent to relative recovery), and the probe may be more sensitive to changes in the diffusional characteristics of the analyte in tissue.

Microdialysis Probe Membrane Properties

Recovery is proportional to the membrane surface area of the probe (Hamberger et al., 1983; Johnson and Justice, 1983; Sandberg and Lindstrom, 1983; Ungerstedt, 1984; Tossman et al., 1986; Kendrick 1989, 1990), assuming no change in the homogeneous nature of the tissue surrounding the probe. More often than not, the length of the microdialysis probe is limited by the size of the structure under study.

Probe membrane material may also affect analyte recovery (Ungerstedt, 1984; Kendrick 1989, 1990; Hsiao et al., 1990; Mason and Romano, 1995). The membrane materials currently used in microdialysis probes are regenerated cellulose (Cuprophan from Gambro AB), polyacrylonitrile (PAN), and polycarbonate-ether (proprietary to CMA/Microdialysis). Kendrick (1989, 1990) determined the in vitro recovery for a large number of analytes using probes representative of each type of material and found a great degree of variation in recovery among probe types. These results could be accounted for by differences in the diffusional properties of the analyte in the membrane (see also Mason and Romano, 1995). In vivo, however, relative recovery may be more a function of the diffusional characteristics of the analyte in tissue than in the membrane (Bungay et al., 1990).

Hsiao et al. (1990) compared the in vitro and in vivo recovery of acetaminophen, DOPAC, hydroxyvalproic acid (HVA), and 5-hydroxyindoleacetic acid (5-HIAA) among the three membrane types and found large, significant differences in the in vitro relative recovery. In contrast, no consistent differences in recovery were found in vivo, suggesting that membrane material may have little impact on in vivo recovery under conditions in which the maximum diffusional resistance resides in the tissue, as is typically true for low-molecular-weight solutes. However, as molecular weight increases or if hindrance to analyte diffusion in the membrane increases (e.g., sticky compound; Kendrick, 1989, 1990) significant differences in probe efficiency (relative recovery) as a function of membrane material would be predicted, since the greatest source of diffusional resistance to the analyte would be found in the membrane (Bungay et al., 1990).

Analyte Properties

Reports of relative recovery for different analytes collected by microdialysis vary considerably. This is due, in part, to differences among laboratories in the microdialysis procedure used (i.e., flow rate, membrane length). However, substantial differences in recovery as a function of molecular weight and hydrophobicity can also be shown. By far, the most complete analysis of recovery by analyte type was performed by Kendrick (1989), who determined the in vitro relative recovery of more than 40 analytes under similar conditions (2 µl/min flow rate using the same probe immersed in a solution maintained at 37°C). By analyte class, relative recovery was highest for amino acids (33% to 40%), followed by monoamines (22% to 30%) and then neuropeptides (1.5% to 24%).

Kendrick (1989) observed a strong negative linear relationship between molecular weight and the log percent recovery, suggesting that one explanation for the differences in relative recovery among analytes may be molecular weight. This is not surprising given the inverse relationship between molecular weight and the diffusion coefficient. The lowest relative recoveries were observed for neuropeptides that had greater hydrophobic properties ("sticky"), suggesting that, for these analytes, hindrance may be a significant impediment to their collection by microdialysis. Dialysate levels obtained from sticky analytes also respond slowly to changes in the external concentration of the analyte (particularly decreases) and generally require considerably longer periods of equilibration to achieve a steady level of recovery (Kendrick, 1989; Thompson et al., 1995). (Note that recovery may be concentration dependent for sticky analytes, confounding the interpretation of experimental manipulations on dialysis data.)

Temperature

Temperature directly impacts on recovery by its effects on the diffusion coefficient; a 1% to 2% increase in the diffusion coefficient is observed for every degree Celsius increase in temperature (Bard and Faulkner, 1980), with the impact of temperature on relative recovery decreasing as recovery approaches 100% (Alexander et al., 1988). Analyte recovery evaluated at room temperature (23° to 25°C) underestimates in vivo analyte recovery. If the goal of an in vitro assay is to correspond as closely as possible to the in vivo condition, it is necessary to assess recovery at the expected in vivo temperature (typically 37°C).

Tissue Factors

Diffusion in tissue is typically slower than in an aqueous solution, due, in great part, to the reduced fluid volume and increased diffusional path (or tortuosity) characteristic of tissue (Nicholson and Rice, 1986; Amberg and Lindefors, 1989; Lindefors et al., 1989; Benveniste and Huttemeier, 1990). Diffusion in tissue may be slowed further by analyte binding to cell surface proteins along the diffusional path (Rice et al., 1985). This means in vivo recovery of an analyte should be less than in vitro recovery of that analyte in a stirred solution held at 37°C, perhaps because, in tissue, diffusion of an analyte is also influenced by the rate of analyte clearance (Bungay et al., 1990; Morrison et al., 1991; Dykstra et al., 1992). As the rate of analyte clearance increases, resistance to diffusion decreases. Diffusivity in tissue, then, is somewhat greater for substances, like neurotransmitters, that are rapidly cleared from the extracellular space.

Of course, the extent to which relative recovery, or probe efficiency, during in vivo microdialysis is affected by tissue factors depends on in which medium (tissue, membrane, or perfusate) the greatest resistance to diffusion occurs. When relative recovery is low, most low-molecular-weight hydrophilic substances will meet the greatest diffusional resistance in tissue. In these cases, changes in the physiological characteristics of tissue (i.e., fluid volume, tortuosity, hindrance, and rate of analyte clearance) that affect diffusional resistance will modify probe efficiency in a given microdialysis experiment. The problem is twofold. First, decreases in recovery in vivo as a result of reduced fluid volume, increased tortuosity, and/or increased hindrance in tissue may reduce the dialysate levels of the analyte below the sensitivity of the analytical detection device. Second, experimental treatments (e.g., lesions, drug treatments) or procedures (e.g., implantation of the microdialysis probe) that affect these physiological characteristics of the tissue may lead to changes in probe efficiency. Group differences in probe efficiency and changes in probe efficiency over the course of the microdialysis sampling period would tend to complicate or confound the interpretation of differences in dialysate concentration of the analyte.

PERFUSATE COMPOSITION

The perfusate is typically composed of low-molecular-weight substances, mainly ions, in concentrations similar to those found in the extracellular fluid, and they diffuse into the tissue in the same way that the analyte of interest diffuses into the microdialysis probe. Furthermore, any solute in the extracellular fluid that is small enough to pass through the dialysis membrane, and is in a lower concentration than that in the perfusate, will diffuse into the perfusate and be depleted from the extracellular fluid surrounding the probe. Ideally, then, the perfusate should contain the exact concentration of all solutes that diffuse through the probe membrane and are found in the extracellular fluid, except of course that solute meant to be sampled. In practice, however, this is neither possible nor pragmatic. Instead, microdialysis perfusates are most often made with the ionic composition and pH of plasma (Ringer's solutions) or artificial cerebrospinal fluid (aCSF or mock CSF).

Maintaining proper ionic composition is critical in the evaluation of neurochemical events. The ionic milieu of neuronal extracellular fluid differs from plasma in both concentration and lability (Bradbury, 1979; Wood and Wood, 1980). Numerous reports have shown that small differences in perfusate Ca^{2+} concentration produce marked effects on basal neurotransmitter levels (Westerink et al., 1988; Moghaddam and Bunney, 1989; Osborne et al., 1991). Moreover, changing the concentration of K^+ in the perfusate is a standard technique used to induce neuronal release of many neurotransmitters (Spanagel et al., 1990; Parsons and Justice, 1992; Thompson et al., 1995). Interestingly, Obrenovitch et al. (1995) showed that the concentration of ions in aCSF may act to "buffer" experimentally induced changes in the ionic composition of neural tissue (i.e., spreading depression). In this case, increasing the K^+ concentration in the perfusate was necessary to permit the development of spreading depression in neural tissue. These effects are not surprising given the importance of ion concentration and distribution in the regulation of neuronal activity and emphasize the importance of maintaining appropriate ionic concentrations.

Information regarding the issue of solute depletion around the probe is limited. The magnitude of solute depletion is directly related to the flow rate, such that as flow rate increases solute depletion increases. Sam and Justice (1996) used this relationship to directly test the effect of increasing the solute depletion (by increasing the flow rate) on basal extracellular DA concentration in the rat striatum. They found that changes in the amount of total solute depletion around the microdialysis probe did not, in fact, affect basal extracellular DA concentrations. More research is necessary to determine how well this finding generalizes to other neurochemicals. It is known that the extracellular concentrations of many analytes are regulated by other neurochemicals in the interstitial space. Thus, during microdialysis, there could be depletion of a substance(s) that in turn regulates the analyte under study.

Finally, it should be noted that the perfusate can be used to deliver agents into tissue, as well as to provide preservative agents (e.g., antioxidants) or other substances that aid in the recovery of the analyte once it has diffused into the perfusate (e.g., bovine serum albumin or antibodies). Microdialysis probes can be used to locally administer either a continuous or pulse infusion of drugs. One advantage of this delivery method over microinjections is that the drug is delivered by measurable parameters of diffusion (and possibly ultrafiltration). As a result, precise calculations of the "dose" and area affected can be determined (Bungay et al., 1990).

Preservatives are usually not diffusable through the membrane, thus eliminating the possibility of adding an undesirable compound to the tissue. Bovine serum albumin, which is too large to pass through most microdialysis membranes, has been used to reduce the adsorption of sticky analytes to the probe tubing and collection vials (Kendrick, 1989). Antibodies have also been used for similar purposes (Lambert et al., 1994). Some consideration of the impact of adding these substances on the subsequent analyte detection is often necessary.

Ascorbic acid is frequently included in perfusates when monoamines, particularly DA, are being collected to reduce analyte degradation after collection. Ascorbic acid readily diffuses through most microdialysis membranes and is known to impact on the neurochemical milieu, but it is also found endogenously. Its perfusate concentration should be limited to no more than that found in tissue (~0.2 mM in interstitial fluid). Ascorbate is also electroactive and typically has a short retention time. Because of the high concentration of ascorbate, however, samples containing ascorbate will typically have a long solvent front. Therefore, standards containing known concentrations of DA with ascorbate, or any protective agent, should be tested before adopting it into the experimental protocol.

STRATEGY FOR DETERMINING EXPERIMENTAL DESIGN

A significant advantage of the microdialysis method is that a variety of analytical techniques (e.g., electrochemical detection, radioimmunoassay, and mass spectrometry; also see UNIT 2.9) can be used for the separation and quantification of the analyte of interest. The challenge in designing and implementing a microdialysis experiment is to collect a sample with both sufficient volume and concentration of analyte to permit separation and detection by the analytical technique employed, while satisfying the experimental design specified by the research question. Therefore, the first step in designing a dialysis experiment is to obtain information regarding the analytical sensitivity and minimum volume of dialysate necessary to isolate and quantify the analyte(s) of interest. Familiarity with the physicochemical properties of the analyte of interest (e.g., molecular weight, lipophilicity) and its distribution, concentration, and clearance from the tissue of interest will greatly facilitate subsequent experimental design.

The following strategy is recommended for investigators first setting up microdialysis, or for those establishing procedures for the collection of substances not already described in the literature.

1. Determine the sensitivity of the analytical equipment and the minimum sample volume required for the handling of physiologically relevant levels of analyte.

2. Determine if any loss of the analyte might occur *following* its diffusion into the microdialysis probe. This can be assessed by perfusing a known concentration of the analyte of interest through the inflow/outflow tubing into a collection vial. The concentration of analyte in the collection vial should be the same as the starting concentration. A physiologically relevant concentration(s) of the analyte of interest should be used, and any additional equipment, such as a liquid switch, that the analyte might contact should be tested in this way. If significant loss of analyte is found, then the source of the loss can be identified by systematically testing each component of the dialysis setup. Loss of analyte to the tubing, collection vial, or any other piece of equipment through which the perfusate flows is often the result of surface adsorption. However, degradation of the sample by enzymes (from a bacterial source) or physicochemical interactions (e.g., oxidation) with metal or tubing may also contribute to sample loss.

3. Evaluate the sample storage method. Two issues should be considered. First, some analytes will degrade rapidly at room temperature (e.g., monoamines). Therefore, optimal collection and storage conditions must be determined. Frequently, reducing the temperature of the sample once it is collected or including a protective agent in the collection vial or perfusate will increase analyte stability. For example, the addition of ascorbic acid to the perfusate or perchloric acid to the collection vial will protect catecholamines from oxidative degradation. Second, because microdialysis samples tend to be very small (in the microliter range), evaporation of the sample may occur over long collection periods. In this case, it would be advisable to seal the collection vials during sample collection.

4. Consider the length and diameter of the inflow/outflow tubing. The combination of long, narrow tubing and higher flow rates may produce considerable back pressure in the probe, resulting in ultrafiltration ("sweating") of the perfusate. The net impact on sample collection will be a reduction in relative recovery. In addition, it is important to know the "dead volume" for each dialysis setup, as this will affect the calculated time course of a given neurochemical response. For example, if the capacity of the outflow tubing is 10 μl and the perfusion rate is 2.0 μl/min, there will be a 5-min lag in measuring the response to a given manipulation.

5. Choose a probe membrane. The molecular weight cutoff should be considerably larger than the analyte of interest yet small enough to maintain the semipermeable nature of the membrane. Several types of membranes are used commercially, and at least one report has shown that some membrane materials may be better suited for the collection of analytes that prove to be sticky, e.g., hydrophobic neuropeptides; see Kendrick (1990) and *UNIT 2.7*.

6. Determine the maximum length of active membrane accessible to the tissue under study. The goal here is to increase the surface area through which dialysis occurs. The surface area of a microdialysis probe is increased by increasing the length, and not the diameter, of the probe. In tissue, the maximum length of the probe is generally dictated by the size of the region in which the probe will be implanted. When probe length is not limited by tissue size, the length should be great enough to produce maximum or near maximum recovery. The effect of probe length on analyte recovery can be easily determined in vitro.

7. Choose a perfusate that is compatible with the organ system and analyte being measured. As a rule, perfusion fluids should be isoosmotic. Remember that diffusion during microdialysis is bidirectional, so that low-molecular-weight solutes in the perfusate will diffuse out of the probe and into tissue.

8. Determine the longest collection period that will still permit hypothesis testing. Typically, the longer the collection period the slower is the flow rate necessary to yield a sufficient volume for analytical detection.

9. Choose a flow rate that will yield sufficient volume for the desired collection period. Slower flow rates will increase the relative recovery of the analyte of interest.

10. Use an in vitro assay to determine if the flow rate and collection period selected yield a sample with a detectable concentration of analyte (use an estimate of actual extracellular tissue concentration).

11. Consider an in vivo pilot procedure to assure adequate sample recovery before investing in a larger experiment.

Reference: Bungay et al., 1990

Contributors: Toni S. Shippenberg and Alexis C. Thompson

UNIT 2.7

Microdialysis in Rodents

NOTE: For surgical procedures in this unit all surgical instruments must be sterile, and procedures are to be performed in a sterile manner unless otherwise specified.

BASIC PROTOCOL 1

PREPARATION OF DIALYSIS PROBE/GUIDE CANNULA ASSEMBLY: CONSTRUCTION OF A CONCENTRIC PROBE (DESIGN 1)

This assembly has been designed to twist apart, enabling reuse, and it also locks into place and does not require external adhesives to secure it during the actual experiment. A number of probes are made at one time because ~10% will be discarded because of leaks or breaks. The authors typically prepare probes no more than 4 days prior to intended use.

Materials

Epoxy, rapid drying
70% ethanol

36-G stainless steel tubing
Jeweler's pliers
Fine forceps
Dremel tool (optional; Small Parts)
Dissecting microscope
Sandpaper (optional)
Dialysis fiber (regenerated cellulose hollow fiber; Fisher): prepared according to the manufacturer's directions and maintained in a clean, dry environment
Stainless steel tubing or pin small enough to fit inside dialysis fiber
Microbore tubing to fit over 28-G internal guide cannula
28-G internal cannula
40°C oven
Vernier calipers
29-G stainless steel tubing (optional)

1. Cut 36-G stainless steel tubing into 33-mm lengths by scoring with a razor blade. Grasp the tube (relative to score marks) distally with fine jeweler's pliers and proximally with fine forceps. Gently bend the tube back and forth until it breaks (a dremel tool with cutting disk can also be used for this purpose). Examine the tube with dissecting microscope, and remove rough edges and/or burrs by using a dremel tool or by gently rubbing with sandpaper.

2. Prepare dialysis fiber bundle according to manufacturer's instructions. Cut one fiber to a length 2 to 4 times greater than that of the dialyzing surface in order to have sufficient material to work with. Gently grasp the cut fiber with forceps and dip one end into a small bead of epoxy. Insert stainless steel tubing or pin, which will serve as a holder, into the epoxied end. Tape the ends of the fiber bundle to cardboard while drying. Store the dried bundle in a container up to 6 months to 1 year (or longer) at room temperature. Handle fibers with forceps only.

3. Cut microbore tubing into ~40-mm lengths. Cleanse the 36-G stainless steel tubing and 28-G internal cannula by sonicating in 70% ethanol. Dry in an oven at 40°C.

4. Using a dissecting microscope and forceps, insert 3 to 4 mm of the microbore tubing over the top of the 28-G internal cannula (stretching the tubing using forceps) Bend the tubing to form a right angle to the top of the cannula and pierce with a pin by inserting the pin into the lumen of the microbore tubing at an angle and allowing the tip to enter the internal cannula.

5. Insert the 36-G tubing into the microbore tube via the pinhole used to introduce the microbore tubing over the internal cannula. Thread the tubing through the internal cannula until it protrudes the desired length from the cannula tip (e.g., if the active length of the membrane will be 2.0 mm, then the desired length will be 2.0 mm). Check length using vernier calipers. Place a large bead of epoxy over the microbore tubing so that the 36-G entry point is fully covered. Allow assembly to dry. Optionally, insert a 29-G stainless steel tube over the 36-G tubing to protect the assembly during storage.

6. Cut the dialysis membrane to a length 0.5 mm longer than that desired for the active surface. Measure from the epoxied end. Spread a bead of epoxy around the dialysis membrane at a point ∼0.2 mm above the desired active area to form an O ring.

7. After the epoxy has dried, gently insert the dialysis membrane into the internal cannula so that the O ring forms a loose seal. Apply a small bead of epoxy between the internal cannula and the O ring. Allow to dry. Using a razor blade, cut the membrane from the pin (∼0.2 mm from the bubble).

8. Apply an additional coat of epoxy to the epoxied tip. Inspect assembly under a dissecting microscope for nicks and to ensure that the desired length of the dialysis membrane is free from epoxy. Perfuse the probe prior to use to test for leaks and to assess probe recovery or probe efficiency (see Basic Protocol 4).

9. To recycle the dialysis assembly, remove the dialysis membrane from the assembly by holding the probe at the epoxied end with one hand while twisting the internal guide assembly with the other (this separates the probe into two sections). Insert a thin wire into the internal cannula/dialysis section to dislodge the membrane. Clean the internal cannula with 70% ethanol and then reinsert the 36-G inflow/outflow section as described above (step 5). Apply epoxy to reseal the connection. Insert a newly prepared membrane into the assembly.

ALTERNATE PROTOCOL 1

CONSTRUCTION OF A SIDE-BY-SIDE PROBE (DESIGN 2)

Figure 2.7.1 illustrates a side-by-side probe assembly that employs silicate glass instead of stainless steel tubing. Although the assembly is typically used only once, all materials except the membrane can be recycled.

Preparation of a loop-type probe (Ungerstedt et al., 1982) is achieved with minor modification of procedures described here and in Basic Protocol 1. Dow 50 cellulose tubing (o.d. 0.25 mm) and 23-G stainless steel tubing can be employed. Both ends of the cellulose tubing are glued inside the stainless steel cannula, and the tubing is folded into a loop. Prior to surgical implantation, a stylet is inserted into the loop to add rigidity.

Materials

Polyimide resin
Epoxy, extra-fast sealing
Superglue
70% ethanol

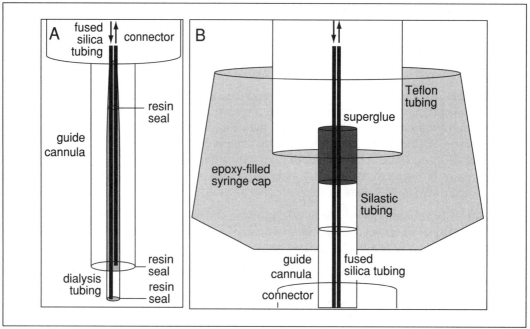

Figure 2.7.1 Diagram of side-by-side microdialysis probe described in Alternate Protocol 1 (Design 2) for in vivo preparations in which the microdialysis probe is inserted through a guide cannula in the rat. (**A**) Two pieces of fused silica tubing are threaded through a connector that is fitted with a short piece of Silastic tubing attached to the inlet tubing and a guide cannula of the appropriate length. The lengths of fused silica tubing exiting the cannula are adjusted so that the tips of the outflow and inflow tubing are separated by the distance that defines the active area of the membrane. Further, the two pieces of silica are placed so that the tip of the outflow tubing is flush with the end of the guide cannula. (**B**) Enlarged view: Once the silica is properly positioned, superglue is applied to the Silastic tubing to permanently fix the silica in place. The guide cannula is then removed and saved for later implantation into the rat. The final step in constructing the dialysis end of the probe involves slipping a tube of dialysis membrane over the end of the two pieces of silica until the sealed end of the dialysis tubing is flush with the inflow silica tubing. The membrane is then attached to the silica with polyimide sealing resin, and all areas of the membrane not intended to support dialysis are sealed with a thin coating of the resin. To complete the probe for in vivo use, the length of silica tubing that runs from the connector to the pump is covered with Teflon tubing, and the junction of the connector, Silastic tubing, and Teflon tubing is stabilized by covering the area with a small tube (e.g., made from a cut-off syringe cap) and filling with fast-drying epoxy. (Adapted from Wages et al., 1986.)

Dialysis fiber (regenerated hollow fiber; Spectrum): prepared according to the
  manufacturer's directions and maintained in a clean, dry environment
Push-pull perfusion connector to fit 26-G guide cannula (Plastics One)
26-G guide cannula, concave
24-G Silastic tubing
Fused silica tubing (o.d. 100 μm, i.d. 0.40 μm; Polymicro)
Teflon tubing
Plastic syringe caps
Needle with plastic hub
Dissecting microscope
26-G needle

1. One day prior to probe construction, cut dialysis fibers into 10- to 15-mm lengths. Seal one end with a plug of polyimide resin or epoxy (as small as possible to avoid adding extra length or width to the probe and let dry 24 hr (see Basic Protocol 1, step 2).

2. Modify a push-pull perfusion connector by removing all but the central tube. Attach connector to 26-G guide cannula. Fit a small piece of 24-G Silastic tubing onto the connector inlet tube.

3. Cut a length of fused silica tubing (long enough to extend from the animal's head to the top of the cage and out from there to a pump or collection vial) to be used as the inflow tube. Cut a second length of tubing (outflow tube) ∼10 cm shorter than the first so there is no confusion as to which is inflow/outflow. Thread both lengths of silica tubing through the connector/cannula.

4. While viewing probe with a dissecting microscope, adjust the length of silica tubing protruding from the cannula so that the longest piece of silica (inflow) extends (from the tip of the cannula) the desired length of the active membrane (e.g., 2.0 to 3.0 mm for rat caudate, depending on the desired active surface of the probe), making sure that the shorter piece (outlet) of silica is flush with the tip of the guide cannula. Secure silica into the Silastic tubing with superglue. Avoid applying glue to the stainless steel cannula, to enable reuse of the connector. Remove guide cannula.

5a. *For use with anesthetized animal:* Cover silica tubing with Teflon tubing and secure with tape at Silastic tubing interface. Leave 2.5 cm of distal inflow line exposed.

5b. *For use with a freely moving preparation:* Fit Teflon tubing over inflow and outflow lines. Allow 2.5 cm of silica tubing to protrude from the Teflon tubing so that it can be inserted into the syringe pump (inflow) or collection vial (outflow). Fit plastic syringe cap, cut to form a tube, over the connector/Silastic tubing connection at a 45° angle. Fill cut off syringe cap with epoxy so that the Teflon tubing and Silastic tubing joint as well as the Silastic tubing/connector tubing joints are coated. Avoid applying epoxy to plastic components of the connector.

The tubing setup described here serves to protect the silica from the strain of motions made by an awake animal. It is critical that the epoxied area include the metal prong of the connector so that the finished product cannot move when tugged. This will reduce breakage when the animal is connected to a spring or other swivel system.

6. Using a dissecting microscope to make sure that there are no glue or resin droplets that would impede insertion into the metal cannula, carefully fit the dialysis membrane (from step 1) onto the two ends of silica tubing protruding from the connector. Seal membrane/silica joint with polyimide resin (or epoxy). Using a 26-G needle as a spatula, coat the silica shaft, silica/membrane joint, and the nondiffusible length of the membrane. Leave the desired length of active membrane free of resin.

7. Attach a plastic-hubbed needle to the inflow line. Thread 2.5 cm of silica tubing which extends from the Teflon tubing into a 26-G needle. Apply epoxy to the needle/silica joint and slide the two pieces back and forth several times to move epoxy into the needle shaft. Finally, slide the silica tubing through the needle and into the Teflon tubing, making sure not to glue shut the opening of the silica tubing. Let dry.

8. Prior to use, remove any remaining glycerol from dialysis membrane by soaking in 70% ethanol for ∼5 min. Perfuse the probe for several minutes with distilled water or perfusion fluid (flow rate 2 μl/min) and check for leaks or air bubbles inside the membrane.

ALTERNATE PROTOCOL 2

CONSTRUCTION OF A HORIZONTAL PROBE

This type of probe can be used for bilateral sampling of discrete nuclei in the brain and, with minor modification, for blood and peripheral tissue. Because of its flexibility and the lack of a rigid guide cannula, it is well suited for dialysis studies in spinal cord.

Materials

Epoxy
Dialysis fiber (o.d. 340 μm Amicon): prepared according to the manufacturer's directions and maintained in a clean, dry environment
Tungsten wire, 125-μm diameter, prestraightened
Sandpaper
Stainless steel tube, 0.65-mm o.d.
Dissecting microscope

1. Cut dialysis fiber to desired length (calculated by determining the width of the structure to be sampled as well as the distance to the skull), avoiding excessive handling of fiber which will close pores.

2. Insert the tungsten wire into the fiber so that one end of the wire extends ~2 mm beyond one end of the dialysis fiber. This provides rigid support of the fiber. Bevel ends of wire with sandpaper to ease insertion. Secure other end of fiber inside stainless steel tube (cut to 1 cm in length) so that 3.0 mm extends over the dialysis fiber. Crimp slightly so that fiber and wire are resistant to pulling. Glue fiber to outer metal tube with epoxy.

3. Measure ~10 mm from the tip of the wire and mark with a waterproof pen (during surgery, the wire will be pulled until this mark rests at the temporal bone). Using a stereotaxic atlas for rat (Konig and Klippel, 1963; Pellegrino and Cushman, 1979; Paxinos and Watson, 1986), mouse (Slotnick and Leonard, 1975), or monkey (Snider and Lee, 1961), calculate the exact distance between the temporal bone and the areas to be dialyzed. Using this distance, epoxy the fiber, leaving areas to be dialyzed unexposed.

4. Allow fiber to dry and then check for leaks or tears using a dissecting microscope (see Basic Protocol 4). Store dust-free and undisturbed.

BASIC PROTOCOL 2

IMPLANTATION AND TETHERING OF DIALYSIS PROBE/GUIDE CANNULA ASSEMBLY IN THE RAT: IMPLANTATION OF CONCENTRIC OR LOOP PROBE/GUIDE ASSEMBLIES

This protocol describes the techniques used for stereotaxic implantation of guide cannula assemblies into discrete regions of the rat brain for short- or long-term measurements of neurochemicals by microdialysis in the awake animal. This provides support and protection of the more delicate dialysis probe and a means to introduce the probe without the trauma of a just-completed surgery. Tethering systems for the freely moving animal are also covered. For illustration of general setup, see Figure 2.7.2. The procedures described are those approved by the National Institutes of Health/National Institute on Drug Abuse (NIH/NIDA) Animal Care and Use Committee. For methods of inserting horizontal probes, see Alternate Protocol 3; for implantations in other species, see Basic Protocol 3 for the mouse and UNIT 2.8 for nonhuman primates.

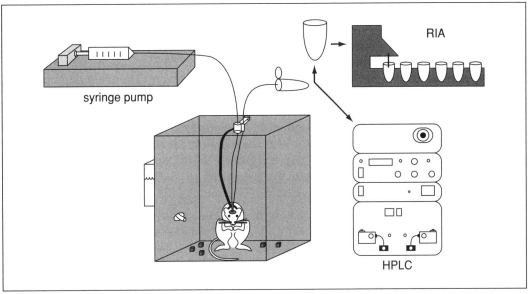

Figure 2.7.2 Sequence of steps in performing microdialysis in the rat. The microdialysis method involves constructing a microdialysis probe that can be implanted surgically into tissue (e.g., brain tissue). The probe consists of an inflow and outflow tube separated by a tube made of dialysis membrane (not visible in this figure). Perfusate, typically artificial cerebrospinal fluid, is pushed through the probe at a slow flow rate (<3 μl/min), and, through the process of diffusion across the dialysis membrane, analytes in the extracellular space are collected into the perfusate and then pushed through the outflow tubing into a collection vial (microcentrifuge tube). The collected sample is then analyzed by whatever method is most appropriate for detection of the particular analyte(s) under study (e.g., HPLC or RIA; see *UNIT 2.9*). Different detection methods will require different amounts of sample handling.

Materials

Rat
Sodium pentobarbital
Atropine sulfate
Betadine disinfectant (povidone-iodine)
Dental acrylic (e.g., Plastics One)

Stereotaxic atlas for rat (e.g., Paxinos and Watson, 1986)
Stereotaxic frame equipped with carrier for mounting the guide cannula or dialysis probe/guide assembly (David Kopf Instruments)
Guide cannula assembly (David Kopf Instruments)/dialysis probe (see Basic Protocol 1 or Alternate Protocol 1)
Scalpel
Serrefine forceps or hemostats
Spatula
Dental or hand drill
Jeweler's forceps and screwdriver
3/16-in. bone screws, 0-80 stainless steel (Small Parts)
Liquid swivel and tether (Instech recommended; also see Alice King Chatham Medical Arts, CMA/Microdialysis, Harvard Apparatus, Spalding Medical Products)

1. Prior to surgery, autoclave or chemically sterilize all surgical instruments. Check probe (if one is to be implanted) for leaks and air bubbles. Ensure that the active length of the probe extends the appropriate distance from the guide cannula tip.

2a. *To determine the interaural line:* Before preparing or mounting the rat, push the ear bars of the stereotaxic frame together symmetrically so that the tip of the guide cannula or probe rests between the tips of the ear bar. Note that the anterior-posterior (AP), lateral (L), and dorsal-ventral (DV) scales of the instrument are defined as the instrument zero. Add or subtract the instrument readings from the respective coordinates obtained from the stereotaxic atlas to determine the coordinates to be used for implantation.

2b. *To use bregma as the reference point:* Anesthetize rat, mount in the stereotaxic apparatus so that the head pivots in the DV but not lateral planes, and expose the skull (steps 2 to 8, below). Lower the cannula so that it is centered on the bregma (midline intersection of the coronal and sagittal sutures). Read the AP, L, and DV coordinates from the scale and add or subtract from the stereotaxic coordinates. Perform surgical procedures in a sterile manner unless otherwise specified.

3. Shave head of animal prior to mounting in the stereotaxic instrument. Gently restrain the rat and inject intraperitoneally (i.p.) with sodium pentobarbital (male, 60 mg/kg; female, 40 mg/kg; 1.0 ml/kg). Administer atropine sulfate (0.05 mg/kg) to relieve respiratory distress. Check level of anesthesia by observing breathing (should be abdominal) and test pain reflexes by pinching footpad with forceps. If animal still responds after 10 min, administer a supplemental dose of anesthetic (e.g., 0.01 ml/250-g male rat; 0.005 ml/female rat).

4. To mount the animal in the ear bars of the stereotaxic apparatus, lock one of the ear bars at the ∼0.5 mm position. Loosely grasp the head, neck, and upper body of the rat from above with one hand and with the free hand maneuver one ear bar into the auditory meatus. Move the ear up and down while gently pushing the ear bar until a popping sound is heard, indicating insertion to the tympanic membrane. Keeping the animal in the same position and maintaining pressure so that the animal's position in the first ear bar is not altered, gently insert the second ear bar. Loosen bar to enable movement in both the DV and L planes. Gently push the second ear bar into the auditory canal while moving it up and down. When no further movement is possible, lock the ear bar in place so that movement is only possible in the DV plane.

5. Read the coordinates of each ear bar and then center the animal between the bars. First, loosen the fittings holding each ear bar in place, while maintaining the ear bar placement manually. Applying equal pressure to both ear bars, move the bars together to adjust the ear bar coordinates. Finally, retighten the fittings to secure the placement of the ear bars. Next, gently pinch open the mouth of the rat and insert the upper incisors over incisor bar. Adjust nose clamp.

6. Disinfect the head using Betadine. Make a midline incision using the point of the scalpel blade to pierce the skin. After piercing the skin, move the blade parallel to the skin so that the whole cutting surface of the blade is used. Make the incision with one stroke and cut to the skull. Retract skin with serrefine forceps or hemostats.

7. With Q-tips or cotton swabs, scrape away fascia and connective tissue. Dry the entire area with a cotton sponge or Q-tips. Apply hydrogen peroxide to facilitate drying of the skull to allow visualization of bregma and proper adherence of dental acrylic (step 11)

8. Place dental or hand drill with a burr sized appropriately for the diameter of the guide cannula or probe assembly (so that it will not catch on the bone) on an electrode carrier or manipulator that has been modified to accommodate the width of drill. Adjust AP and L coordinates (step 2a or 2b). Drill until the dura is punctured, using an up-and-down motion, being careful not to damage the meninges and underlying brain tissue. When using a hand drill first mark the stereotaxic coordinates by inking a stainless steel stylet

that has been mounted into an electrode carrier and calibrate the burrs to extend only through to the dura. Apply pressure with gauze or a Q-tip, cauterize, or apply a gelatin sponge (Gelfoam, Upjohn) to reduce bleeding.

9. Drill three additional holes around the cannula/probe hole. Using jeweler's forceps and a screwdriver, insert 3/16-in. bone screws into the holes (not to extend to the dura).

10. Attach the guide cannula (precut so that it will not extend over the active length of the membrane) or probe into the stereotaxic holder and straighten. Center over the drill hole. Slowly lower into the target region until the cannula or probe base just rests on the skull. Remove skin retractors and dry skull again if necessary.

11. Apply dental acrylic to the base of the guide cannula or probe, the three screws, and the exposed skull. Pour the powder onto and over the skull and add the liquid component drop by drop until the mixture is wet but not loose. Make sure the acrylic extends to the muscle. Gently place the skin over the acrylic so that it will adhere to it when the acrylic is dry. If a tethering system is to be used (step 14), also embed the flat head of a slotted screw in dental acrylic firmly in place on top of the skull, using a guide carrier placed over the screw. Position it as far as possible from the cannula and screws and place the dental acrylic around the cannula, bone screws, and tether screws and attach the protective spring.

12. Allow acrylic to dry and carefully remove the holder. Smooth any jagged edges by applying additional dental acrylic. Leave the probe in situ for some hours prior to the commencement of experiments (see Basic Protocol 5) to allow for recovery from trauma resulting from the introduction of the probe or the probe/cannula assembly into the tissue, During this period, the perfuse the probe with physiological fluid.

13a. *If an anesthetized animal is to be used:* Either keep the animal in the stereotaxic apparatus or remove it. Use a heating blanket and thermometer to maintain body temperature.

13b. *If a guide cannula has been implanted:* Insert the calibrated, perfused probe and commence experiments when neuronal release of neurotransmitter has been demonstrated (see Basic Protocols 4 and 5).

13c. *If experiments will commence at a later time:* place the animal in its cage and provide warmth until consciousness is regained. House animals individually to avoid damage to the implant unless a protective dummy cannula, which cannot be eaten by other rats, is employed.

14. For a freely moving preparation, use the head block tethering system. Tape the inflow and outflow tubes onto the flat head of a slotted screw embedded in dental acrylic in step 11 or insert through the center of the spring. Attach the head mount system with a tether clamp to a liquid swivel and balancing arm. Secure probe inflow and outflow tubings to the wire guide, which pulls the liquid swivel/balance arm (see Fig. 2.7.2). Determine the dead volumes of the swivel inflow and outflow tubes since dead space will introduce a time lag.

ALTERNATE PROTOCOL 3

IMPLANTATION OF A HORIZONTAL PROBE

Materials (also see Basic Protocol 2; see APPENDIX 1 for items with ✓)
✓ 1× artificial cerebrospinal fluid (aCSF) or Ringer's solution
Epoxy

Electrode carriers, modified to position probe in lateral axis
Horizontal dialysis probe (see Alternate Protocol 2)
Stereotaxic ear bars (David Kopf Instruments), modified to raise position of the head in the apparatus
2.5-mm drill bit
22-G stainless steel tubing (cannula) and removable plastic caps (e.g., Plastics One)
Polyethylene (PE) tubing to fit 22-G tubing

1. Mount an electrode carrier that has been modified to manipulate a probe in the lateral axis onto the stereotaxic apparatus. Fix the stainless steel portion of the dialysis probe to the carrier so that it extends laterally. Straighten the probe and then carefully remove carrier.

2. Anesthetize rat and mount in the stereotaxic instrument (see Basic Protocol 2, steps 2 to 5). Use modified ear bars that elevate the head above the stereotaxic instrument to enable clear access and insertion of the probe through the temporal bone. Disinfect head; make incision and expose skull and temporal bones by retracting the skin, muscle, and fascia.

3. Separate temporal muscle from the bone. Use sutures to tie off muscle, thereby reducing bleeding. Pack the muscle with saline-saturated gauze and keep moist throughout surgery. Drill holes (using a 2.5-mm drill bit) bilaterally in the temporal bones at a depth appropriate for the target structure (determined by the stereotaxic atlas employed). Puncture dura.

4. Position probe carrier on stereotaxic arm with the tungsten wire pointed at the exposed dura. Slowly introduce the probe tip (best done using a second electrode carrier mounted on the other arm of the stereotaxic apparatus) into the brain until ~10 mm of the tip extends through the temporal bone on the other side (e.g., use mark as reference point; see Alternate Protocol 2, step 3).

   The second carrier should be modified so that putty or bone wax can be inserted onto it. The putty or bone wax on the carrier is positioned so that the stylet will penetrate it as it exits from the skull. The putty is then used to pull the probe through the brain to the 10-mm mark. This is accomplished by loosening the first carrier to enable use of the second carrier.

5. Once the probe is positioned, tighten the first electrode carrier. Disconnect the tungsten wire from the probe by clipping the wire. Gently retract the remainder of the wire. Check probe for patency by attaching PE tubing to the steel cannula and perfusing the probe with 1× aCSF or Ringer's solution (see Basic Protocol 4), ensuring that the perfusate fluid exits the tube, which indicates that there is no break, leak, or occlusion of the probe.

6. Glue the free end of the probe into the 22-G stainless steel cannula tubing using epoxy. Bend tubing to allow it to rest comfortably around the skull. Check probe patency again using PE tubing (as in step 7).

7. Cement the stainless steel tubes on either side of the probe to the bone with dental cement. Insert bone screws into the skull to aid anchoring of the cannula or probe assembly. Apply dental cement over skull. To prevent damage of the probe (very important), place the exposed fiber in a protective covering (e.g., PE tubing or a pipet tip).

8. Free temporal muscle from its retracted position and suture the skin around the headpiece assembly. Cover the ends of the cannula with removable plastic caps until dialysis studies commence.

BASIC PROTOCOL 3

SURGICAL IMPLANTATION OF DIALYSIS PROBE/GUIDE CANNULA ASSEMBLY IN THE MOUSE

Materials

    Anesthetic: urethane (80 to 85 mg/kg, i.p.; nonsurvival) or sodium pentobarbital (80 to 85 mg/kg, i.p.; survival; supplemented with methoxyflurane if necessary)
Dental acrylic

    Stereotaxic frame (David Kopf Instruments) adapted for use with small rodents (e.g., Kopf mouse adapter set)
Stage to elevate body of mouse to level of ear bars (e.g., inverted pipet-tip box)
Electrode carrier (David Kopf Instruments)
Instruments for small animal surgery
Stereotaxic axis for mouse (e.g., Slotnick and Leonard, 1975)
Guide cannula (CMA Microdialysis)/probe assembly
Bone screws
Dental or hand drill
Slotted peg for tether attachment

1. Anesthetize mouse by injecting intraperitoneally (i.p.) with urethane (1.6 g/kg; 100 mg/ml; a nonsurvival anesthetic with a 4- to 8-hr duration of action). Wait 30 to 60 min. Shave top of head.

2. Insert mouse into stereotaxic frame equipped with Kopf mouse adapter with the mouse level with the bars and the head straight and able to pivot up and down but not from side to side, first placing the head of the mouse into the serrated ear cups that replace the standard ear bars. Lock one ear cup so that it is stationary and place the temporal mandibular joint of the mouse firmly against it. Move the second ear cup onto the opposite joint and tighten the ear cup fittings. For positioning and subsequent surgery, elevate the animal with a small stage.

3. Using a pair of blunt forceps, insert the upper incisors over the incisor bar and gently close nose bar by screwing it lightly in place, allowing the mouse to breath properly.

4. Make a midline incision to expose the skull. Scrape the skull clean of fascia. Make sure the skull is dry so that bony landmarks can be clearly visualized (see Basic Protocol 2, step 7). Adjust head to achieve a flat skull position by rotating the head using the movable cheek and nose bar until the DV coordinates of bregma and lambda are within 0.1 mm (see Basic Protocol 2 for explanation of stereotaxic coordinates).

5. Use a correction factor for determining stereotaxic coordinates. For each animal, divide the actual distance between lambda and bregma by the lambda-bregma distance of the stereotaxic atlas (3.6 mm; Slotnick and Leonard, 1975; Matochik et al., 1994). Multiply the correction factor by the atlas AP, L, and DV coordinates to correct the coordinates.

6. Once the coordinates have been marked on the skull, drill a hole using a dental drill or dremel tool held in a miniature pin vice. For the freely moving preparation, drill two additional screw holes for mounting of bone screws (avoid penetration of the dura when screwing into the skull).

7. Place prestraightened guide cannula into the electrode holder and lower it into the target region. Apply dental acrylic around the cannula (and screws). For anchoring a tethering system (see step 8b), also attach the small peg from the system with dental acrylic.

8a. *For anesthetized mouse:* Insert probe and perfuse (see Basic Protocol 5) for a minimum of 2 hr.

8b. *For freely moving mouse:* Anchor the tether (e.g., Plastics One) by inserting a 0.010-inch looped wire into the small peg and clamping it to a 25-G liquid swivel and balancing arm. Lock the wire into the peg by placing a wire sleeve over it.

BASIC PROTOCOL 4

MICRODIALYSIS IN VITRO

In vitro microdialysis is used for calibrating the microdialysis probe (see Basic Protocol 1) by determining an in vitro measure of probe efficiency to assure that the integrity of the probe remains constant over repeated use, and that probe efficiency is constant over a range of analyte concentrations (i.e., probe efficiency should be concentration-independent), as well as to correct dialysate values for variation in probe efficiency. Another common use for in vitro assays is to evaluate the microdialysis procedure that will be applied in vivo and/or to evaluate the impact of particular changes in the microdialysis procedure on relative recovery, particularly if the substance to be collected has not previously been sampled by microdialysis. Finally, the in vitro method can be applied to ex vivo tissue preparations to provide estimates of interstitial analyte concentrations.

Materials (see APPENDIX 1 for items with ✓)

Analyte of interest (e.g., dopamine, dynorphin, or acetylcholine; usually in powdered form)

✓ Perfusate, e.g., 1× artificial cerebrospinal fluid (aCSF) or Ringer's solution (see UNIT 2.6 for discussion of perfusates)

Microdialysis probe (see Basic Protocol 1 and Alternate Protocols 1 and 2) with holder stand (commercially available or constructed using 1.5-ml microcentrifuge tubes in which the opening has been reduced by insertion of a shortened 10-ml pipet tip, resting in the middle of the tube, which had a groove is etched in the side of the tube with a Dremel tube)
Microsyringe pump and glass Luer-Lok gas-tight microsyringes (Hamilton)
Dissecting microscope
Collection vials (e.g., 0.2-ml microcentrifuge tubes)

1. Prepare in vitro test solution by making standard dilutions of the analyte of interest in the perfusate and spiking the test solution with a known concentration of analyte. Pour test solution into appropriate container (large relative to the probe to avoid damaging the probe and to provide an "infinite" source of analyte in solution relative to what is removed by microdialysis) and incubate at 37°C with shaking.

2. Prepare microsyringe pump (several for more than two to three solutions). Set flow rate and secure syringe and needle containing perfusate onto the pump. Turn pump on and ensure that all air is out of the needle tip and hub. Leave pump on until the probe is attached.

3. Prepare probe by soaking 2 to 3 min in 70% ethanol. Attach a connector to the pump end of the inflow line of the probe so that there is an airtight connection between the probe and the syringe on the pump. Fill another syringe with perfusate and attach it to the probe via the connector just added.

4. Watching the membrane tip of the probe under a dissecting microscope, slowly push perfusate through the probe. Observe the membrane filling up with perfusate and note the flow of perfusate out of the outflow lines. Check for leaks (if noted, discard) and air

bubbles (dislodge by gently prodding the probe while continuing to apply the same amount of pressure or by gently shaking as when shaking down a mercury thermometer). Once the probe has been adequately filled and is free of air bubbles, attach it to the syringe pump, carefully filling the connector with perfusate to avoid introducing air.

If fluid does not appear within the membrane, cut off a small piece of the inflow line from the top (pump end) as sometimes this end becomes blocked. If air bubbles return to the membrane whenever pressure is decreased, check the outflow line integrity and try cutting off the tip of the outflow line to resolve the problem. Use great care when tightening the inflow line-syringe connection. Too much pressure can rupture the membrane.

5. Place probe in the middle of the test solution container. Make sure that it does not accidentally come in contact with the container walls as this decreases the surface available for dialysis and leads to erroneous results. Collect samples into collection vials immediately and continuously, as desired. Estimate equilibration time by collecting several samples until a constant concentration of analyte is achieved across consecutive dialysate samples. Take samples from the test solution to validate the actual beaker concentration.

6. Store samples, standards, and test solution samples at $-80°C$ immediately following collection, but quantify (e.g., by HPLC-EC; see UNIT 2.9) as soon as possible after collection to optimize detection and avoid contamination (e.g., substances leaching out of the collection vials into the sample).

7. Using the analytical method of choice, quantify the concentration of analyte in the samples and in the test solution. Calculate probe efficiency as the percent relative recovery of the analyte collected from the samples according to the following equation:

$$\text{Probe efficiency} = \frac{C_{sample}}{C_{solution}} \times 100$$

where C_{sample} and $C_{solution}$ are the concentration of analyte in the sample and test solution, respectively.

BASIC PROTOCOL 5

IN VIVO MICRODIALYSIS

Materials (see APPENDIX 1 for items with ✓)

✓ Perfusate, e.g., 1× artificial cerebrospinal fluid (aCSF)
Rat or mouse with implanted dummy cannula (see Basic Protocols 2 or 3)

Microsyringe pump and glass Luer-Lok gas-tight microsyringes (Hamilton)
Liquid swivel and tether system (e.g., Instech recommended; Alice King Chatham Medical Arts, CMA/Microdialysis, Harvard Apparatus, Spalding Medical Products)
Testing cage
Microdialysis probe (see Basic Protocol 1 or Alternate Protocol 1)
Collection vials
Dissecting microscope

1. Prepare microdialysis solutions and equipment as follows
    a. Prepare perfusate in a large clean flask.
    b. Prepare microsyringe pumps (see Basic Protocol 4, step 2), fill syringes with perfusate, and attach syringes.

c. Prepare probe. Be sure the inflow and outflow lines are long enough to reach the pump (inflow) and provide easy and unobtrusive removal of the sample (outflow). Check that the inflow and outflow lines are protected against strain and torque, as well as chewing, by means of a tether and liquid swivel system mounted on the testing cage (see Basic Protocols 2 and 3). Attach a connector to the pump end and/or swivel end of the inflow line of the microdialysis probe so that the inflow line can be connected to the syringe on the pump or to the swivel. If nonanesthetized animals are being studied, connect the inflow line directly to a swivel. Use another piece of tubing to connect the swivel to the syringe pump.

2. Attach inflow line from the syringe pump to the swivel and ensure good perfusate flow through the swivel (see Basic Protocol 4, step 4).

3. Attach inflow tube of the probe to the syringe pump or swivel. Place probe securely into a vial filled with distilled water or perfusate until ready to insert the probe into the animal. After the probe is connected to the syringe pump or swivel, initially using a flow rate higher (i.e., twice as high) than that to be used during experiments to make sure the perfusate flows freely through all probes. Adjust pump to the appropriate flow rate and collect perfusate in a collection vial to check that all probes are providing the same and appropriate volume, determined by weighing the collection vial before and after a "collection" period.

4. Remove the dummy cannula from rat or mouse guide cannula and slowly insert the probe into the animal. For nonanesthetized animals, hold the animal (wrapped in a towel, if necessary) in one hand while inserting the probe into the guide cannula. Alternatively, lightly anesthetize the animal with an inhalational anesthetic (e.g., isoflurane).

5. After the probe is inserted, attach the tether to the animal (previously acclimated to the cage and tether system as well as to being handled) and place it in the testing cage. For nonanesthetized animals, be sure the testing cage contains the equipment necessary to tether the animal safely. For anesthetized animals, monitor body temperature to ensure that it is maintained during the experiment.

When testing hypotheses about behavior, the design of the testing cage and the location of the microdialysis probe outlet line require some consideration, so that samples can be removed without confounding concurrent behavioral measurements. Ideally, sample is collected directly into the analytical device used to quantitate the analyte under study. Alternatively, the outflow line must exit the testing cage in such a way that the experimenter can remove the collection vial without disturbing the animal.

6. Following probe implantation, allow equilibration and recovery of the tissue from the physical damage rendered by probe implantation (typically ~2 hr in anesthetized preparations and 6 to 10 hr in an awake preparation). Reduce the flow rate of perfusate until 1 to 2 hr prior to sample collection for a long recovery period).

7. Collect samples, prepare standards and standard dilutions, and perform experimental manipulations (e.g., inject drug or begin behavioral test). Remove and store samples immediately following their collection to help prevent sample degradation. Seal collection vials quickly to avoid evaporation.

8. Quantify using analytical method of choice (*UNIT 2.9*). Evaluate data (see Support Protocols 1 and 2). Demonstrate that baseline samples were at steady state by determining that basal dialysate levels remain constant over several collection periods.

The first time a particular microdialysis procedure is applied, the investigator should demonstrate that the dialysate levels of analyte were in fact of neuronal origin, by observing the result of coperfusion of substances that specifically modify neuronal release of the analyte (e.g., tetrodotoxin, K^+, Ca^{2+} chelators).

SUPPORT PROTOCOL 1

IN VIVO DETERMINATION OF EXTRACELLULAR CONCENTRATION

Quantitative microdialysis, in contrast to conventional dialysis, provides an unbiased estimate of extracellular concentration. One method, put forth by Lonnroth and referred to as the difference method, is presented here, and it has also been adapted for non-steady-state conditions (Olson and Justice, 1993). See UNIT 2.6 for additional methods.

Materials (see APPENDIX 1 for items with ✓)

✓ Perfusate, e.g., 1× artificial cerebrospinal fluid (aCSF) or Ringer's solution
  Analyte of interest

1. Prepare 10 ml each of three or more solutions of perfusate spiked with varying concentrations (straddling the expected extracellular concentration) of the analyte of interest. If desired, include a 0 nM analyte solution to compare the data to that generated from a conventional microdialysis experiment.

2. Prepare microdialysis equipment and test animals through equilibration (see Basic Protocol 5, steps 1 through 6).

3. Under steady-state conditions (demonstrated in a pilot study), change the perfusate to that containing one concentration of the analyte of interest. Allow time for equilibration of the new perfusate to occur (e.g., >5 min depending on the diffusional characteristics of the analyte) and then collect dialysate samples, in duplicate if possible.

4. Replace the perfusate with one containing a different concentration of the analyte and repeat the equilibration/collection procedure. Repeat until samples have been collected for at least three different perfusate concentrations of the analyte.

5. Under transient conditions, assign each animal to a particular perfusate concentration of the analyte, and begin perfusing this concentration at the start of the experiment (after a suitable probe equilibration period; see Basic Protocol 5, step 6). The animal will receive this perfusate for the entire experiment. If necessary, provide fresh perfusate at regular intervals to assure that the analyte does not degrade within the syringe.

6. Store dialysate sample, standards, and representative samples of each perfusate C_{in}. Quantitate dialysate levels of the analyte (see Basic Protocol 4, step 7). Evaluate data by deriving C_{ext} (see Support Protocol 2).

SUPPORT PROTOCOL 2

DETERMINING EXTRACELLULAR CONCENTRATION (C_{ext})

This support protocol describes analysis of data from the quantitative dialysis experiments detailed in Support Protocol 1.

Determination of C_{ext} Under Steady-State Conditions

At the end of a steady-state experiment, each animal will have dialysate levels (C_{out}) for each of three or more different starting perfusate concentrations (C_{in}) of the analyte of interest. To determine the extracellular concentration (C_{ext}), the C_{in} is regressed against the gain or loss of the analyte during dialysis ($C_{in} - C_{out}$). The resultant linear regression equation can be resolved to determine the C_{ext}. Taking the general form for a linear regression, one has

$y = mx + b$, where y is the dependent measure ($C_{in} - C_{out}$), x is an independent variable (C_{in}), m is the slope, and b is the y-intercept. Rearranging the equation to solve for x when y equals zero yields C_{ext} [$C_{in}(x)$ when $C_{in} - C_{out}(y)$ equals zero], $x_{y=0} = -b/m$.

A graphical resolution can also be used by plotting $C_{in} - C_{out}$ (y axis) against C_{in} (x axis) and drawing the best fit line through each point. The x value at the intersection of the regression line with $y = 0$ is the C_{ext}. These calculations are carried out for each animal to determine the estimated C_{ext} values.

Determination of C_{ext} Under Transient Conditions

At the end of a transient experiment, the net gain or loss of analyte from the dialysate ($C_{in} - C_{out}$) can be determined for each subject at each time point. The C_{in} for each animal is known (manipulated independent variable). A linear regression analysis of C_{in} versus $C_{in} - C_{out}$ of data from all animals in each experimental group will yield an estimate for extracellular concentration at each time point. Follow the same graphical or mathematical procedure described above for the calculation of extracellular concentration under steady-state conditions. The dependent measure for each animal is the difference between the starting concentration of the analyte and the dialysate concentration of the analyte. A calculation of C_{ext} can only be made on a group of data, and thus, for each group, only a single estimate of C_{ext} is obtained. It is possible to propagate the error around the slope and y-intercept to obtain a conservative estimate of the error surrounding C_{ext}. A possible alternative application might be to subdivide the animals in each experimental group into smaller matched groups of three or four. The single dependent measure of C_{ext} would then be calculated from this small cohort. The group statistic would then be made on four or more C_{ext} values obtained from each smaller subgroup.

Group comparisons (or pre- versus posttreatment comparisons) can be made by analysis of covariance (ANCOVA), where the C_{in} is a manipulated covariate. However, the proper application of ANCOVA demands that the regression lines between groups or times lie in parallel. The slope defined by the linear regression of the data is, in fact, the in vivo extraction fraction (E_d) or in vivo relative recovery (see UNIT 2.6). Unfortunately, it is very possible that the lines will not be parallel, particularly when relative recovery is low (<50%). In this case, the data could be transformed to force parallelism of the slopes prior to analysis by covariance. Alternatively, the cohort method described above would simplify the statistical analysis to the type described for steady-state conditions.

References: Hsiao et al., 1990; Imperato and Di Chiara, 1984; Kendrick, 1990; Maidment et al., 1989; Pettit and Justice, 1991; Robinson and Whishaw, 1988; Zetterstrom et al., 1984

Contributors: Alexis C. Thompson and Toni S. Shippenberg

UNIT 2.8

Microdialysis in Nonhuman Primates

BASIC PROTOCOL 1

PREPARATION OF DIALYSIS PROBE

Microdialysis probes can be constructed over 3 days as described in this protocol (also see UNIT 2.7; Wang et al., 1990; Saunders et al., 1993), twenty to thirty probes at one time. Testing

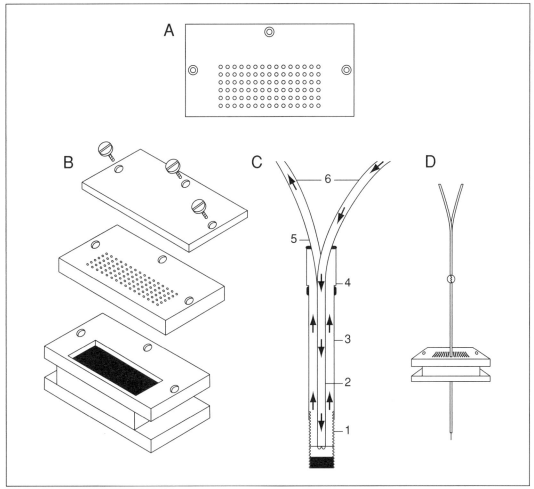

Figure 2.8.1 Diagrammatic representations of the guide assembly and dialysis probe. (**A**) Spacing of the guide holes into six rows of 15 holes (<1 mm in diameter) with the holes spaced 1 mm apart. (**B**) View of the guide holder, guide block, and cap which attach together with plastic screws. (**C**) Schematic view of the dialysis probe: 1, AN 69 dialysis membrane; 2, fused silica inner barrel; 3, fused silica outer barrel; 4, plastic sleeve tubing (PE 50); 5, fused silica barrel of outlet tube; 6, inlet and outlet tubing. (**D**) Example of probe positioned through the guide assembly.

all new probes may take a day or two. The basic concentric design of the probe is shown in Figure 2.8.1C. The two most critical features of the probe to be considered are the overall length of the probe and the length of the exposed membrane. The probe dimensions described in this protocol are for a standard probe size and have been used for multiple subcortical targets.

Materials *(see APPENDIX 1 for items with ✓)*

    Epoxy
    Superglue
    5% to 10% methanol
    Small bags of $CaCl_2$
  ✓ Monkey artificial cerebrospinal fluid (aCSF), prepared fresh for experiment
  ✓ Neurotransmitter standards, e.g., dopamine (DA), aspartate, glutamate, and
       γ-aminobutyric acid (GABA), 37°C
    5% to 10% methanol or 3% to 4% formalin

MRI scan of monkey brain (see Support Protocol 1)
Fused silica tubing (o.d. 425 μm, i.d. 320 μm for outer barrel of probe; o.d. 166, i.d. 101 μm for inner barrel of probe)
Polyethylene (PE) 10 and 50 tubing
25-G needle, blunt
50-ml syringe
35° and 37°C ovens
Dialysis membrane tubing (e.g., Filtral 20, AN 69, Hospal Medica)
Dissecting microscope
1-ml Hamilton syringe; gas-tight
Microinfusion pump
Blue connectors (Bioanalytical Systems)
4-O suture thread (optional)
700-μl (7 × 40–mm) conical-bottom amber glass vials

1. Measure length of probe using a brain atlas or MRI images of the monkey brain from the surface of the brain to the targeted region. Add another 2 cm to the measurement to allow for the guide holder and working space needed.

2. Measure distance from the dorsal border to the ventral surface of the structure to determine the length of the dialysis membrane. Add an additional 5 mm or more to make the membrane easier to work with.

3. To construct the outer barrel of the probe, using a new scalpel blade, cut an 8-cm length of fused silica tubing (o.d. 425 μm, i.d. 320 μm) and a 3-cm piece of PE 50 tubing. Insert the 8-cm piece of silica tubing into the PE 50 tubing, leaving an exposed barrel of 6.5 cm. Use a small drop of non-water-soluble epoxy to glue and seal the parts together. Air dry a minimum of 12 hr.

4. To construct the inner barrel of the probe, with a new scalpel blade, cut a 15-cm piece of fused silica tubing (o.d. 166 μm, i.d. 101 μm) and a 15-cm piece of PE 10 tubing. Insert the 15-cm silica barrel into the PE 10 tubing, leaving 10 cm of the silica barrel exposed. Glue and seal the two segments using Superglue. Air dry a minimum of 12 hr.

5. To construct the outlet tube or port, using a new scalpel blade, cut a 5-cm length of fused silica tubing (o.d. 166 μm, i.d. 101 μm) and a piece of PE 10 tubing 25 cm long. Insert the silica tube into the 25-cm-long piece of PE 10 tubing. Make sure the fused silica barrel extends out of the tubing 1.5 to 2.0 cm. Glue and seal the two segments with Superglue. Air dry a minimum of 12 hr.

6. Flush each unit (from steps 3, 4, and 5) with 5% to 10% methanol to cleanse and ensure unrestricted flow by attaching PE 50 tubing to a blunted 25-G needle on a 50-ml syringe containing 5% to 10% methanol and pushing slowly through. Discard or rebuild units with leaks or blockages.

7. Follow the methanol flush with an air flush. Place the three units in an oven at 37°C and allow to dry 12 to 15 hr.

8. To assemble the three units of the probe, carefully insert the fused silica barrel of unit 2 (from step 4) into the PE 50 tubing of unit 1 (from step 3) until 4 to 6 mm of the inner barrel is seen protruding out of the distal end of the outer barrel. Insert the protruding portion of the fused silica tubing of unit 3 (from step 5) into the PE 50 sleeve adjacent to unit 2 with the fused silica of unit 3 and unit 2 almost touching each other.

9. Place a thick drop of epoxy at the Y junction of units 1, 2, and 3. Hang the assembly from tape attached to a shelf edge to prevent the glue drop from coming into contact and

sticking to anything other than units 1, 2, and 3. Dry assembled probes 8 to 10 hr at room temperature.

10. Prepare the dialysis membrane by cutting short pieces (12 to 15 mm) of the dialysis membrane tubing with a sharp blade at a 45° angle. Handle membranes carefully with blunt plastic forceps, avoiding humid conditions and excessive touching with fingers. Wash hands frequently or wear surgical gloves.

11. Seal one end of the membrane with Superglue as follows. Leave a big drop of Superglue in a small weighing boat for nearly 30 min to allow the glue to thicken but not harden. Gently hold the membrane with tweezers on one end and place the other end to be sealed in the droplet. Allow capillary action to cause the Superglue to slowly enter the membrane. Place membrane fragments in a weighing boat containing small bags filled with $CaCl_2$. Cover and allow to dry completely (allow 8 hr) in a cool place. Do not allow the membranes to come in contact with the $CaCl_2$.

12. Cut the exposed end of the inner silica barrel (unit 2) to the desired membrane length minus 1 mm. Flush the assembled units with 5% to 10% methanol through both the inlet and outlet to identify possible leaks and blockages.

13. Cut the open end of the piece of dialysis membrane tubing at a 45° angle to the desired membrane length plus 5 mm (e.g., 8 to 10 mm). Trim the sealed end and leave ∼1 mm of the glue plug.

14. Slip the open end of the membrane over the protruded portion of the inner barrel (unit 2). While viewing the assembly with a dissecting microscope, make sure the membrane fits inside the outer barrel. Seal the joint of the membrane with the outer barrel with Superglue. Be careful to minimize the amount of glue on the membrane. Place the completed dialysis probe in a 37°C oven and let dry 24 hr.

15. Flush with distilled water to cleanse the membrane and to ensure unrestricted flow through the probe. Store the probe in 5% to 10% methanol or 3% to 4% formalin.

16. Make up fresh monkey aCSF supplemented with neurotransmitter standards; 37°C. Place probe in an airtight beaker with aCSF/standards at 37°C.

17. Fill 1-ml gas-tight Hamilton syringe with monkey aCSF without neurotransmitter standards. Place into microinfusion pump. Connect the inlet tubing of the probe to the 1-ml Hamilton syringe controlled by a precision microinfusion pump using a blue connector to make the connection between the PE tubing of the probe and the microsyringe.

18. Start pump and run several minutes at a rate of 3 to 5 µl/min. Check for leaks. Examine all connections carefully. Stop leaks either by replacing blue connector or by tying 4-O suture thread around the blue connector to tighten the seal around the connector and PE tubing.

19. Place outlet tubing into glass vial for sample collection. Perfuse probe 2 hr at a flow rate of 1 µl/min with monkey aCSF without neurotransmitter standards. Change collection vial every 25 min for each probe. Collect three successive 25-min samples (25 µl) and subject to chromatographic analysis by HPLC, with electrochemical (EC) or fluorescence detection methods (see UNIT 2.9; Saunders et al., 1993; Kolachana et al., 1996).

20. Calculate recovery efficiency of the dialysis probe (≥30% recovery rates for 80% of probes expected) from the mean recovery from three collections, expressed as percentage recovery with respect to the neurotransmitter concentration in the aCSF solution outside the probe (see UNIT 2.6). Save only probes with recovery values greater than 30% for in vivo microdialysis.

21. Store probe/membrane assemblies by immersing for long periods in 5% to 10% methanol or 3% to 4% formalin or for short periods in distilled water. Periodically check the fluid level to ensure the membrane is continuously submersed. Protect membranes from drying out and replace any membranes that dry out.

BASIC PROTOCOL 2

ATTACHMENT OF GUIDE ASSEMBLY TO THE MONKEY SKULL

The guide assembly is composed of three pieces of plastic: the guide holder, guide block, and cap. The guide block is made of a polycarbonate plastic block ($15 \times 30 \times 4$ mm) and consists of six rows of small holes (<1 mm in diameter and 1 mm apart center to center), with 15 holes in each row. Thus, 90 mm^2 of tissue is accessible (in the axial plane) under the guide. The guide block rests securely on the guide holder, a block of plastic ($15 \times 30 \times 8$ mm) in which a 7×16–mm rectangular opening is cut somewhat off center. In addition, along the outside on three sides a 3-mm-wide ledge is etched out. The ledge enables the guide holder to be fixed to the skull with bone cement. The guide block is attached to the guide holder using three screws. A cap of thin clear plastic is attached to the top of the guide block to cover the guide holes (Fig. 2.8.1B). One or more assemblies may be attached during a single surgery. If additional guide assemblies are desired these can be added at a later surgery. These assemblies can be made in most machine shops.

Materials

6- to 12.5-kg rhesus monkey (*Macaca mulatta*)
Ketamine
Isoflurane
Ringer's solution
Bone cement

Guide holder
Intubation tube
Venous catheter
i.v. setup
Vital sign monitor and pads
MRI-compatible stereotaxic instrument, equipped with manipulators
37°C heating blanket
Surgical instruments for aseptic neurosurgery
Guide block, plastic cap, and screws
Marking pen, sterile
Drill, hand-held or dental, 14 to 20 mm × 7 to 9 mm
Rongeurs (optional)
Plastic I-shaped anchors

1. Determine appropriate skull coordinates for guide assembly attachment through MRI scanning (see Support Protocol 1).

2. Sedate the monkey with 10 mg/kg ketamine intramuscularly (i.m.), and insert intubation tube in trachea. Achieve surgical levels of anesthesia using isoflurane gas (1% to 3% to effect). Place venous catheter into the leg vein and attach i.v. setup to allow delivery of Ringer's solution throughout the surgery. Attach monitoring pads for vital sign monitor.

3. Place the monkey in the same MRI-compatible stereotaxic instrument as used during scanning in an orientation identical to that during MRI scanning (see Support Protocol 1, step 1; Saunders et al., 1990).

4. Wrap the animal in a heating blanket (37°C). Attach vital sign monitor and observe vital signs (temperature, heart and respiration rates) for the duration of the surgery. Arrange surgical instruments for aseptic neurosurgery.

5. Following sterile surgical techniques, make a longitudinal incision in the scalp extending from the brow caudally to the back of the head. Retract the subcutaneous tissue and temporalis muscles to expose the surface of the skull. Thoroughly scrape and clean the skull to remove all loose tissue.

6a. *For sampling subcortical or deep cortical regions:* Hold the center of the guide block together with the guide holder (plastic cap remains unattached) over the center of the targeted region using a stereotaxic manipulator. Keep the bottom surface of the guide assembly parallel to the horizontal plane of the monkey's brain (stereotaxic horizontal plane). Place the center of the guide over the center of the target structure. Use the coordinates obtained by MRI scanning to position the guide assembly over the targeted site to maximize the number of probe sites within the target region.

6b. *For sampling a cortical area within a sulcus:* Align a specific row(s) of the guide block 1 mm medial or lateral to the sulcus. Also align the angle of the guide assembly parallel with the bank of the sulcus to be sampled. Ensure that the probe runs down the middle of the cortical tissue comprising one or both banks.

7. With a sterile marking pen, draw the position of each guide holder for each targeted area on the skull using the coordinates derived from the MRI scan. Make a small opening in the dorsal cranium using a small drill. Enlarge the craniotomy using rongeurs or a small drill to 14 to 20 mm × 7 to 9 mm, a size slightly larger than the six rows of holes in the guide holder assembly.

8. Reposition the guide assembly above the craniotomy. Check that the opening aligns appropriately with the guide block. Repeat the craniotomy procedure for each guide holder assembly to be attached. Trim the guide block to allow for additional space when placing more than one holder (up to four possible), leaving a 1-mm ledge minimum around the guide block.

9. Determine the position of plastic I-shaped anchors and prepare anchor holes as follows. With a drill and small-bur bit, make a small slit (width of the I × 3 to 4 mm) in three to four places spaced around a single guide holder or four to six places around multiple guide holder assemblies. Insert the foot of an I anchor through a slit and then turn 90°. Repeat for each I anchor used. Place the guide assembly in the appropriate position.

10. Place small amounts of bone cement between I anchor and guide holder held in the appropriate position. Place the bone cement of the proper consistency (not very runny—mixture of powder and liquid) in successive layers between the I and the guide holder. Carefully fill in the ledge of the guide holder and close all openings between the bone and the guide holder, but do not allow the cement to flow onto the top of the exposed dura.

11. Close the skin with sutures to fit loosely but to cover all the bone alongside the bone cement. If necessary, trim skin with scissors. Keep the monkey warm and turn off the anesthesia. Once the monkey attempts to right itself, place it in its home cage.

12. Allow the monkey a postsurgical recovery period of 10 to 14 days. During the initial postoperative period use antibiotics and analgesics under the guidance of an attending veterinarian. Clean the area in, under, and around the guide assembly twice a week by removing the guide block and flushing the block and guide with sterile saline and dilute hydrogen peroxide. Submerge the guide block in peroxide for ~10 min. Place an ophthalmic antibiotic ointment in the rectangular well of the block.

SUPPORT PROTOCOL 1

DETERMINATION OF SKULL COORDINATES BY MRI

To determine the proper place to attach the specially designed guide assemblies to the animal's skull (see Basic Protocol 2), a combination of MRI and stereotaxic procedures is used (Saunders et al., 1990; Kolachana et al., 1994).

Additional Materials (also see Basic Protocol 2)

    Xylazine
    Container for transporting monkey
    MRI-compatible stereotaxic frame (e.g., see Saunders et al, 1990)
    GE/Signa 1.5 T MRI scanner or equivalent

1. Sedate the monkey using a mixture of ketamine (10 mg/kg, i.m.) and xylazine (3 mg/kg, i.m.). Transport the sedated monkey in a closed container to a GE/Signa 1.5 T MRI scanner. Place the monkey into an MRI-compatible stereotaxic frame. Align the monkey in the frame in the MRI scanner so that all planes (coronal, sagittal, and horizontal) of the scanner are parallel to those of the monkey in the stereotaxic instrument (for details see Saunders et al., 1990 and Kolachana et al., 1994).

2. Collect MRI images in the coronal and horizontal planes using a spoil grass sequence with a repetition time (TR) of 25 msec, an echo time (TE) of 6 msec, and a flip angle of 30°. Use a field of view of 11 cm with four excitations. Collect 60 interleaved images with a 1-mm thickness. Check the monkey regularly to ensure it remains sedated over the course of the ~45 min study.

3. Make measurements using the image analysis capabilities of the MRI scanning unit (see Saunders et al., 1990 and Kolachana et al., 1994 for details of these measurements and reference points). From the images, use the midsagittal sinus as a medial-lateral reference point, the cortical surface as the dorsal-ventral reference, and the vitamin E site in the ear bars as the rostral-caudal reference. Make the measurements in each possible plane to double-check accuracy.

4. Perform measurements to obtain coordinates to outline the targeted brain site for dialysis and guide the accurate placement of the guide holders on the skull (see Basic Protocol 2) as described below.

   a. Measure in millimeters from the midsagittal sinus the medial and lateral limits of the target structure.

   b. Measure in millimeters from the top of the cortex above the target structure to the dorsal and ventral limits of the target structure. Make several similar measurements along the rostral-caudal extent of the targeted structure.

   c. Measure in millimeters from the vitamin E in the ear bars to the rostral and caudal limits of the desired site for dialysis.

5. Return the monkey to its home cage and allow it to awake.

SUPPORT PROTOCOL 2

POSTOPERATIVE VERIFICATION OF GUIDE ASSEMBLY POSITION BY MRI

Additional Materials (*also see Basic Protocol 2 and Support Protocol 1*)

    Vitamin E capsules
    Soft clay
    Fused silica tubing (o.d. 425 μm, i.d. 320 μm)

1. Cut thirty 3-cm-long pieces of fused silica tubing for each guide assembly attached. Remove vitamin E from a capsule and fill the pieces of silica tubing with vitamin E by placing one end into the vitamin E puddle and allowing capillary action to fill the tubes. Seal both ends by rubbing across soft clay.

2. Just prior to scanning, sedate the monkey using a mixture of ketamine (10 mg/kg, i.m.) and xylazine (3 mg/kg, i.m.). Position the monkey in the MRI-compatible stereotaxic instrument (see Support Protocol 1, step 1).

3. Position an individual vitamin E-filled silica tube in each hole of two rows (e.g., rows 2 and 4) in the guide holder. Push the tube down to the dura but do not penetrate. Repeat for all guide holders. Carefully transport the monkey to the MRI center. Take care not to break any of the silica tubes.

4. Rescan using the same preoperative MRI protocol and collect images in the coronal, horizontal, or sagittal planes (see Support Protocol 1, step 2).

5. Remove animal from scanner and remove the vitamin E-filled tubes by pulling them straight up out of the guides. Remove the animal from the stereotaxic frame, take back to the lab, and place in its home cage.

6. Examine the images and visualize the vitamin E-filled tubes above the brain and targeted regions. Choose the holes or row of holes within the guide holder that will provide the best positioning of the dialysis probes within the targeted region.

BASIC PROTOCOL 3

IN VIVO MICRODIALYSIS

Materials (*see APPENDIX 1 for items with* ✓)

    ✓ Monkey artificial cerebrospinal fluid (aCSF)
    Ketamine hydrochloride
    Monkey with attached dialysis probe guide assembly (see Basic Protocol 2)
    Isoflurane
    Ophthalmic antibiotic ointment

    Dialysis probes (see Basic Protocol 1)
    700-μl (7 × 40–mm) conical-bottom amber glass vials
    1-ml Hamilton syringe, gas-tight
    Microinfusion pump
    MRI-compatible stereotaxic instrument
    25-G needle
    25-G stainless steel tubing (o.d. 0.020 in., i.d. 0.010 in.; Hypo tube, Small Parts)
    Blue connectors (Bioanalytical Systems)
    Polyethylene (PE) 10 tubing
    4-O suture thread (optional)

1. Determine the experimental contingencies before beginning the microdialysis, e.g., the neurotransmitter to be examined, the time resolution necessary, pharmacological manipulation(s), the dialysate sample size, the flow rate, dosages required if there is a drug manipulation, behavior tasks to be assessed, and controls (see UNIT 2.6 and Chapter 3).

2. Determine the best hole within the guide holder based on the postoperative MRI (see Support Protocol 1). Chose dialysis probes to be used for in vivo dialysis; use probes with in vitro recovery rates >30%. Allow two probes per site to provide backup in case of probe breakage during insertion.

3. Measure the distance to the ventral limit of the target site from the postsurgical MRI (see Support Protocol 2).

4. Label glass vials to be used to collect dialysate with name of target site and sample number to be used during the experiment.

5. Determine the dead volume of the tubing system from the 1-ml syringe on the infusion pump to the collection vial mathematically by calculating from the length and inner diameter of the tubing or empirically by measuring the volume collected in the time it takes a drop of ink to move from the microsyringe to the probe outlet.

6. Check the flow on the liquid switches. Using a syringe with 25-G needle with blue connector attached, force water through each inlet while working switch to ensure unimpeded flow and no leaks.

7. Make up fresh, particle-free, preferably degassed aCSF the morning of the experiment and keep at room temperature. If desired, add ascorbic acid (3.5 to 5 mg of ascorbate per 250 ml aCSF suggested) to prolong the life of recovered DA in dialysate stored at $-70°C$ for several weeks or months.

8. Divide aCSF into aliquots and modify with pharmacological agents (e.g., neurotransmitter agonists or antagonists) according to experimental design. Fill and label individual airtight 1-ml Hamilton syringes with appropriate room temperature (to avoid formation of bubbles) aCSF. Place in the microinfusion pump. Color-code each syringe, liquid switch, and probe inlets and outlets for convenience.

9. Lightly sedate the monkey with ketamine hydrochloride (10 mg/kg, i.m.). Install intubation tube and insert venous catheters. Start slow drip of Ringer's solution. Lightly anesthetize animal with 1% to 2% (v/v) isoflurane (as needed to achieve effect). Secure the monkey in the stereotaxic instrument.

10. Wrap the animal in a heating blanket for the 8 to 12 hr of a typical dialysis experiment. Connect to a vital sign monitor and observe throughout.

11. Remove the cap and the guide block. Clear the guide holes with a 25-G needle. Clean the well area of fluids and obvious puruloid fluid.

12. *Important:* Immediately before dialysis and every 3 days (at the most), clean the guide holder assembly by removing the guide block and flushing the block and guide with sterile saline and dilute hydrogen peroxide. Submerge the guide block in peroxide for ~10 min. Place an ophthalmic antibiotic in the rectangular well of the block.

   On non-dialysis days, sedate the monkey slightly with ketamine and remove the guide holder. Clean the individual holes and place in hydrogen peroxide. Flush the guide block with sterile saline and with a dilute Betadine/saline (50:50, v/v) solution. If the animal is running a fever or if the infection appears severe, consider administering antibiotics. Discuss this with the facility veterinarian. In general, infection associated with guide holders does not seriously threaten the overall health of the animal or the carrying out of an experiment.

13. Attach the guide block to the guide holder. Pierce the dura by advancing through the selected guide hole a sterile 5-in.-long 25-G stainless steel tubing that has been cut at an angle to form a sharp piercing end. Mark the steel tubing with a marker pen to help control advancement just past the pia mater of the cortex.

14. Cover the animal (except its head) with a surgical drape to keep the site clean. Measure and mark the dialysis probe to indicate the limit of the distance the probe will be advanced through the guide holder to be positioned in the targeted area.

15. Carefully and slowly advance the dialysis probe to the predetermined depth in the target area. Once the probes are in place do not touch them or unnecessarily pull or stretch the inlet and outlets of the probe. Secure the probe in relation to the tissue in such a way that there is minimal risk for movement.

16. Connect 1-ml Hamilton syringes on the infusion pumps to the liquid switches using blue connectors and PE 10 tubing. Connect the outlet port of the liquid switch, using the blue connectors and PE 10 tubing, with the inlet tube of the dialysis probe. Place the outlet tube of the dialysis probe into a collection vial.

17. Turn on infusion pump and set flow rate to 2 to 3 μl/min. Check each individual tubing system of each probe for leaks. Repair leaks by replacing the blue connector or by using 4-O suture thread to tie around the blue connector and PE 10 tubing to seal off the leak. Be careful not to impede flow through the tube by overtightening the suture. If the leak is at a joint, replace the probe.

18. Reduce aCSF flow rate to required rate for experiment (1 to 2 μl/min). Start 2-hr stabilization period (to overcome this initial confounding release of neurotransmitters as a result of the damage inflicted). Set timer for changing of collection vials according to experimental design and transmitter assay capabilities. Alternatively, position the probes 24 hr in advance (see Alternate Protocol 1 for description of dialysis in the awake monkey).

19. Start collecting baseline samples (three to five samples) to establish baseline levels before behavioral or pharmacological manipulation. After baseline samples have been collected, administer pharmacologically active substance to the animal. Turn the liquid switch to change aCSF or infuse systemically via leg vein.

20. Reset the liquid switches to end pharmacological manipulation and restart normal aCSF. Continue to change collection vials at regular intervals according to experimental design and transmitter assay capabilities.

21. At the end of the experiment turn off the infusion pump and disconnect tubing. Carefully remove the probes by lifting straight up. Flush probes with distilled water to prevent deposition of salts in the tubing which may render them useless, and then stored in appropriate solution (see Basic Protocol 1, step 21).

22. Flush the guide holder and block with sterile saline, and place ophthalmic antibiotic ointment into the well. Attach guide and cap.

23. Disconnect gas anesthesia. Remove the monkey from the stereotaxic holder. Continue to keep the monkey warm until it is awake and attempts to right itself. Place the animal back in its home cage.

24. Place the collected samples with external standard aliquots into $-70°$C freezer until ready for assay. Assay samples for neurochemical of interest and compare over time with pharmacological manipulations capable of determining effects of pharmacological manipulation on neurochemical dynamics within a discrete brain structure.

25. Check the monkey regularly over the next 24 to 48 hr to make sure recovery is uneventful. Check with the facility veterinarian if there are any problems.

26. At the end of all dialysis sessions, verify probe placement (see Support Protocol 3).

SUPPORT PROTOCOL 3

ANATOMICAL VERIFICATION OF THE PROBES

All microdialysis experiments should be concluded by determining the localization of the microdialysis probes in the brain. Monkeys are precious research subjects, and when repeated dialysis experimentation with a monkey is under consideration, a reasonable alternative to histological verification to confirm probe placement is postdialysis MRI (see Support Protocol 2) to confirm correct location of the dialysis probe tracks. Probe penetration track marks within the target area are clearly seen on an image. Monkeys have relatively large brain structures, which allow reuse of some of the best probe tracks based on previous experimental results. When all target sites have been completely utilized the monkey is killed by drug overdose and transcardially perfused with fixative. The brain is removed, cryoprotected, and frozen, and 50-μm-thick coronal sections cut. An interrupted 1 in 5 series mounted on glass slides and stained with thionin for histological examination and reconstruction of the probe tracts and placements is recommended.

ALTERNATE PROTOCOL

MODIFICATIONS FOR USING IN VIVO MICRODIALYSIS IN THE AWAKE MONKEY

The type of probes used, their placement, and the actual experimental procedures for microdialysis in the awake animal are, for the most part, the same as described for the sedated animal (see Basic Protocol 3). The biggest difference is that the monkeys must be trained to sit quietly in some sort of apparatus. This procedure may also be used in conjunction with behavioral testing to associate neurochemical dynamics with cognitive behavior.

Additional Materials (also see Basic Protocol 3)

Primate chair, which allows the monkey to sit comfortably and reach out if necessary, but not reach its head
Stainless steel post
Steel straps
Bone cement
No. 8 screws
Steel rod
Crossing rod
4 × 8–in. thermoplastic sheet
Nuts

1. Train the monkey to sit in a restraining primate chair for several hours, depending on the nature of the experimental procedure.

2. Attach guide holder assemblies to the skull as described (see Basic Protocol 2) and perform postoperative MRI to confirm placement of the guide holders (see Support Protocol 2).

3. Use a second surgery to attach a stainless steel post to the monkey's head (see Basic Protocol 2). Use steel straps, no. 8 stainless steel screws, and bone cement to fix the steel post to posterior part of skull. Also add no. 8 small screws in front of the previous bone

cement implant and one immediately behind the caudal guide holder assembly (in front of the steel post).

4. After a short recovery period (1 to 2 weeks), re-adapt the monkey to the chair and regain other behavioral performance, if necessary.

5. Using the steel post implanted in step 3, attach a steel rod from the post to a crossing rod fixed to the chair to keep the animal's head in a stationary position.

6. Place the monkey in a comfortable position in the chair. Adapt the monkey, using several timed steps, to having its head fixed for longer periods until it is comfortable with the duration of the experimental period.

7. Run mock experimental sessions (including penetrating the dura but not advancing to the site, positioning the pump, placing collection vial, and changing vials according to schedule) on several occasions before the actual experiment.

8. Place the monkey in the chair and fix its head. Remove the guide block and clean the guide holder (see Basic Protocol 3, step 12). Replace the guide block.

9. Chose the probes to be used for in vivo dialysis. Allow two probes per site to provide backup in case of probe breakage during insertion.

10. Measure and mark the dialysis probe to indicate the distance the probe should be advanced through the guide holder to be positioned in the targeted area. Pierce the dura by advancing through the selected guide hole a sterile 5-in.-long stainless steel tubing that has been cut at an angle forming a sharp piercing end. Carefully and slowly advance the dialysis probe to the predetermined depth in the target area.

11. Connect all tubing to infusion pumps, liquid switches, and collection vials. Perform experiment according to design.

12. Disconnect all tubing and leave probes in place overnight. Cover the probes with a small protective hat formed from a 4 × 8-in. thermoplastic sheet. Align small holes in the cap with the bolts left in the bone cement. Fix in place by tightening nuts to the bolts. Place animal back in its home cage.

13. The next day replace the monkey in the chair. Remove the protective hat and reconnect the tubing to the syringes and collection vials. Perform the experiment. At the end of the second day remove the probes and clean the guide holder assemblies. Place the animal back in its home cage for 1 week before repeating experimental procedures.

14. Assay all samples for neurochemical of interest. Compare overflow levels over time related to behavioral events so as to associate behavior and/or performance to neurochemical events.

References: Kolachana et al., 1994; Saunders et al., 1990

Contributors: Richard C. Saunders, Bhaskar S. Kolachana, and Daniel R. Weinberger

UNIT 2.9

Detection and Quantification of Neurotransmitters in Dialysates

The protocols below afford detection limits in the low nanomolar range for catecholamines and indoleamines (HPLC-EC), acetylcholine (HPLC-EC), and amino acids (HPLC-fluorescence). For troubleshooting HPLC-EC detection systems see Table 2.9.1 at the end of the unit.

BASIC PROTOCOL 1

DETECTION AND QUANTIFICATION OF DOPAMINE AND SEROTONIN BY HPLC/EC

Materials (see APPENDIX 1 for items with ✓)

✓ Mobile phase buffer I *or* II
  HPLC-grade H_2O
  85% phosphoric acid or 0.5 M NaOH
  Microdialysis samples to be analyzed (see UNITS 2.7 & 2.8)
✓ 10 μM monoamine standard working solutions and dilutions bracketing those estimated for the samples

Filtration apparatus:
  Borosilicate glass vacuum flask
  Fritted filter support and spring clamp
  0.22-μm filter discs (e.g., Millipore)
  Vacuum source
Ultrapure helium with regulator and Teflon tubing
HPLC/EC system consisting of:
  HPLC dual-piston pump equipped with pulse dampener
  3-μm-particle-size octadecylsilane (ODS) column: for mobile phase I, 100-mm length × 3.2-mm i.d. (Bioanalytical Systems); for mobile phase II, 150-mm length × 3-mm i.d. (ESA)
  10-ml gas-tight HPLC priming syringe (e.g., Kloehn or Waters)
  Manual injector (20-μl sample loop/gas-tight HPLC glass microsyringe) *or* autoinjector
  ESA Coulochem II multielectrode electrochemical detector equipped with guard cell (Model 5020) and microdialysis cell (Model 5014B)
  Pen recorder and integrator or equivalent computer software program

1. Prepare 1 to 2 liters of either mobile phase buffer I *or* II using glassware that has been cleaned and rinsed with HPLC-grade water. Check the glassware and HPLC-grade water for electroreactive species that may coelute with the analytes to be measured.

2. Adjust the pH of mobile phase buffer I to 3.25 with 85% phosphoric acid or adjust the pH of mobile phase buffer II to 5.25 with 0.5 M NaOH. Filter solution through 0.22-μm filter under vacuum, then completely degas by passing a gentle stream of ultrapure helium through the solution for 10 min at a pressure of 5 to 10 psi, using Teflon tubing.

3. Purge any air bubbles from the mobile phase inlet lines and pump using a priming syringe. Assemble the HPLC/EC system, attaching the mobile phase buffer to the inlet lines, set the flow rate to 0.6 to 0.9 ml/min (producing a back pressure of ∼2800 psi in the authors' system), and allow several hours for equilibration of the mobile phase and column.

4. Adjust cell potentials to the following:

   Guard cell = 325 to 375 mV
   Electrode 1 = 125 to 175 mV
   Electrode 2 = 225 to 275 mV.

5. Adjust electrode sensitivity to the following:

   Electrode 1 = 10 to 100 mA
   Electrode 2 = 5 to 50 nA.

6. Inject 20 μl of 10 μM monoamine standard working solution to determine retention times.

7. Thaw microdialysis samples. Inject 20 µl of sample manually or using an autoinjector and run HPLC/EC. Inject 20 µl of monoamine standard solutions (in concentrations bracketing those estimated for the samples) prior to, during, and at the end of each series of samples to determine whether chromatographic conditions (e.g., sensitivity and retention times) have changed during the course of the run. If desired, also include an internal standard with a known retention time and concentration for quality-control purposes. Inject mobile phase or solvent between runs to clean the injector/loop.

8. Determine dialysate concentrations of amines by comparing peak heights with those of known concentrations of external standards. Perform a linear regression analysis of the standards and use the resulting linear regression equation to determine actual concentrations in samples.

BASIC PROTOCOL 2

DETECTION AND QUANTIFICATION OF ACETYLCHOLINE BY HPLC/EC

Materials *(see APPENDIX 1 for items with ✓)*

    20% (v/v) nitric acid
    30% (v/v) acetic acid
    HPLC-grade H_2O
✓ Mobile phase buffer III
    50% methanol in HPLC-grade H_2O and 100% methanol
    5.0 mg/ml sodium dodecyl sulfate (SDS)
    Microdialysis samples to be analyzed
✓ Acetylcholine/choline working standards

HPLC/EC system consisting of:
    HPLC dual-piston pump equipped with pulse dampener
    C18 analytical column (100-mm length × 3.0-mm i.d.; Chrompack, ESA, or Bioanalytical Systems)
    Reversed-phase guard columns (10-mm length × 2.0-mm i.d.; Chrompack or Bioanalytical Systems)
    Solvent-saturation reversed-phase column (100-mm length × 4.6-mm i.d.; Chrompack)
    Immobilized enzyme reactor (IMER: 10-mm length × 3.0-mm i.d.; Chrompack) prepared with acetylcholinesterase and choline oxidase (see Support Protocol)
    Dummy IMER (Chrompack)
    Amperometric electrochemical detector equipped with platinum electrode and Ag/AgCl reference electrode (Bioanalytical Systems) and guard cell
    Column holders
    50-µl injection loop
    Pen recorder and integrator or equivalent computer software program
Filtration apparatus:
    Borosilicate glass vacuum flask
    Fritted filter support and spring clamp
    0.45-µm filter discs (e.g., Millipore)
    Vacuum source
Ultrapure helium with regulator and Teflon tubing

1. Set up the HPLC system without the columns and detector cells. Clean and thoroughly passivate the pump (remembering to remove columns and detector cells from flow) by

flushing with the following solutions in the order indicated:

50 ml 20% nitric acid
50 ml 30% acetic acid
100 ml HPLC-grade H_2O
100 ml mobile phase buffer III.

2. Filter 1 liter of mobile phase buffer III through 0.45-μm filter discs under vacuum, then degas by passing a gentle stream of ultrapure helium through the solution for 10 min at a pressure of 5 to 10 psi, using Teflon tubing (also see Basic Protocol 1, step 1 annotation).

3. Connect the C18 analytical column and flush with 100% methanol for 20 min at a flow rate of 1.0 ml/min. Disconnect the column, then flush the pump for 20 min with 50% methanol in HPLC-grade water. Reattach the C18 analytical column and flush again for 20 min with 50% methanol. Check the outflow for signs of sludge, which indicate that additional flushing is necessary.

4. Repeat the washings in step 3 using plain HPLC-grade water in place of the methanol solutions.

5. Filter and degas a solution of 5 mg/ml SDS (see step 2). Load the C18 column with SDS by flushing with this solution for 20 min at a flow rate of 1.0 ml/min (∼1100 to 1200 psi) to convert the reversed-phase column into a cation-exchange column.

6. Disconnect column and flush pump again for 20 min with HPLC-grade water. Reconnect column and flush system for an additional 25 min with HPLC-grade water. Test for cleanliness by collecting samples from the output of the column and adding mobile phase. If the system is clean, the outflow should be clear.

7. With column still in place, replace HPLC-grade water with mobile phase buffer III and equilibrate for 60 min at 0.6 ml/min. Disconnect column, then clean column screens using HPLC-grade water and cotton-tip swabs.

8. Detach the C18 analytical column and connect the solvent-saturation reversed-phase column between the pump and injector (without the guard cell, analytical column, IMER, or electrochemical detector). Rinse with 40 column volumes of mobile phase.

9. Connect the guard and analytical columns but leave out the IMER and use the dummy IMER (a metal cylinder 1.10 cm in length). Rinse the guard column and the analytical column with 40 column volumes of mobile phase buffer III. Finally, put in the real IMER and rinse with 20 column volumes of mobile phase buffer III (complete system set up: solvent-saturation column, injector, guard column, analytical column, IMER, and electrochemical detector).

10a. *To use the column immediately:* Connect the detector and equilibrate the system with mobile phase buffer III for at least 1 hr at 0.6 ml/min.

10b. *To shut down the system overnight:* Keep the IMER connected to the system at a low flow rate (0.2 ml/min).

10c. *To shut down the system for a prolonged time:* Remove the IMER and replace it with a dummy. Rinse the analytical and guard column with HPLC-grade water followed by methanol. Keep the IMER wet with mobile phase and do not allow it to dry out. Store at 4°C.

11. Inject 50 μl of 1.0 μM acetylcholine standard and 50 μl of 1.0 μM choline standard to determine retention times.

12. Thaw microdialysis samples. Inject 50 μl of sample at 0.6 ml/min using a 50-μl loop and run HPLC/EC. Inject 50 μl of 1.0 μM acetylcholine/choline standards, and standards

bracketing concentrations estimated for the samples, prior to, during, and at the end of the run of each test sample. Inject mobile phase or solvent between runs to clean the loop.

13. Determine dialysate concentrations of acetylcholine (see Basic Protocol 1, step 8).

ALTERNATE PROTOCOL

HPLC/EC DETECTION OF ACETYLCHOLINE USING ACETYLCHOLINE/CHOLINE ASSAY KIT

Additional Materials *(also see Basic Protocol 2; see* APPENDIX 1 *for items with* ✓ *)*

    40% and 100% acetonitrile
✓ Mobile phase buffer IV
    Microdialysis samples for analysis
✓ Acetylcholine/choline working standards

    Acetylcholine/choline assay kit (Bioanalytical Systems) including:
        Analytical column
        Postcolumn immobilized enzyme reactor (IMER)
        Platinum electrode and Ag/AgCl reference electrode
        Column holder, threaded body and union
        Precolumn immobilized enzyme reactor (IMER) with choline oxidase catalyst
            (30-mm length × 4.6-mm i.d.; Bioanalytical Systems; optional)
        20-or 50-μl injection loop

1. Passivate system (see Basic Protocol 2, step 1). Cleanse the pump by flushing it at a flow rate of 2.0 ml/min, first for 25 min with 40% acetonitrile, then for 55 min with 100% acetonitrile, and finally for 25 min with 40% acetonitrile.

2. Filter 1 to 2 liters of mobile phase buffer IV through a 0.22-μm filter under vacuum. Flush the system for 1 hr with filtered mobile phase buffer IV at 1.0 ml/min.

3. Install the analytical column and flush for 15 min with mobile phase buffer IV at 1.0 ml/min. Install the postcolumn IMER and flush overnight with mobile phase buffer IV at 1.0 ml/min, to equilibrate.

4. Using acetylcholine/choline standard working solutions, determine retention times and detection limits. If it is desirable to reduce the height of the choline peak, install precolumn IMER and flush for 15 min prior to use with mobile phase buffer IV at 1.0 ml/min.

5. Thaw microdialysis samples. Inject 30 or 60 at 1.0 ml/min using a 25- or 50-μl loop and run HPLC/EC. Inject 30 or 60 μl of 50 to 100 nM acetylcholine/choline standard working solution prior to, during, and at the end of the run of each test sample. Inject mobile phase or solvent between runs to clean the loop.

6. Determine dialysate concentrations of acetylcholine (see Basic Protocol 1, step 8).

SUPPORT PROTOCOL

PREPARATION OF IMMOBILIZED ENZYME REACTOR (IMER) FOR DETECTION OF ACETYLCHOLINE

Additional Materials *(also see Basic Protocol 2)*

    Acetylcholinesterase (Sigma)
    Choline oxidase (Sigma)
    0.1 M KOH

Immobilized enzyme reactor (IMER: 10-mm length × 3.0-mm i.d.; Chrompack), unloaded
500-μl injector loop

1. Replace 50-μl sample injector loop with 500-μl loop.

2. Filter an adequate volume of mobile phase buffer III (see Basic Protocol 2, step 2). Disconnect the C18 analytical column and store in mobile phase buffer III at 4°C. Also disconnect the EC cell. Connect the IMER in place of the analytical column and flush with mobile phase buffer III at a flow rate of 0.05 ml/min.

3. Mix 160 U acetylcholinesterase in 248.96 μl mobile phase buffer III and 80 U choline oxidase in 248.8 μl mobile phase buffer III. Inject the enzymes into the injector loop at a flow rate of 0.05 ml/min and wait 1 min. Adjust the flow rate to 0.6 ml/min, then equilibrate the IMER with the mobile phase for 1 hr.

4. Disconnect the IMER from the injector. Check the pH of the 0.1 M KOH solution. Inject the KOH solution into the injector loop and check the pH at the outlet to make sure that it is the same as the initial pH of the 0.1 M KOH solution.

5. Check the pH of some HPLC-grade water. Inject HPLC-grade water into the injector loop and check the pH at the outlet. Repeat injection until the pH after the loop is the same as the initial pH.

6. Reinstall the 50-μl injector loop. Reconnect the analytical and guard columns. Flush mobile phase buffer III through the system for 20 min at a flow rate of 0.6 ml/min.

7. Connect the IMER to the analytical column, but do not connect the IMER output to the detector. Flush for 20 min at 0.6 ml/min.

8. Turn on the cell and set to +500 mV.

9. Equilibrate the system overnight at a flow rate of 0.2 ml/min and an initial current of 50 nA. Gradually decrease the current as background current falls. Before use, increase the sensitivity of the system by decreasing the current setting to 1.0 nA. Store the enzyme reactor at 4°C, wetted with mobile phase and not allowed to dry out.

BASIC PROTOCOL 3

DETECTION OF AMINO ACIDS IN MICRODIALYSATES BY HPLC AND FLUOROMETRIC LABELING WITH o-PHTHALDIALDEHYDE

Amino acids in microdialysates can be readily separated using reversed-phase high-performance liquid chromatography (RP-HPLC). This protocol describes the fluorimetric detection of primary amino acids treated with o-phthaldialdehyde (OPA) in the presence of 2-mercaptoethanol at basic pH in an aqueous solution. Retention times range from ∼2 min for aspartate to 25 min for γ-aminobutyric acid. The fluorescence detector can measure picogram quantities of the OPA-derivatized amino acids as they elute from the column.

Materials (see APPENDIX 1 for items with ✓)

✓ Mobile phase buffer V
✓ OPA stock reagent
✓ Perfusate buffer used in microdialysis experiment (e.g., Ringer's solution or aCSF)

Filtration apparatus:
- Borosilicate glass vacuum flask
- Fritted filter support and spring clamp
- 0.22-μm filter disks (e.g., Millipore)
- Vacuum source

HPLC system consisting of:
- Glass low-volume inserts (Waters)
- 10-ml gas-tight HPLC priming syringe (e.g., Kloehn or Waters)
- Chromatography pump: Waters 510 or comparable pump capable of delivering flow rate of 1.0 ml/min at pressures up to 4000 psi
- Injector: e.g., Waters 717 autosampler or manual injector
- C_{18} column: Alltech Adsorbosphere 287095
- Fluorescence detector: Waters 420 or comparable detector equipped with 338- and 450-338-nm excitation filter and 450-nm emission filter
- Data acquisition software: Waters Millennium or equivalent

1. Prepare working dilution of the OPA reagent by mixing 1 part OPA stock solution with 1 part of the perfusate buffer used in the microdialysis experiment (24 hours before use for best reactivity). Pass through a syringe filter before use. Avoid exposing the working dilution of OPA reagent to direct light.

2. Adjust the pH of the mobile phase buffer V to 5.88 (at 25°C) with H_3PO_4. Filter and degas solution through 0.22-μm filter under vacuum. Perform these steps before each chromatographic run.

3. Purge any air bubbles from the mobile phase inlet lines and pump using a priming syringe. Assemble the HPLC system and column, attaching the mobile phase buffer to the inlet lines, set the flow rate to 1 ml/min, and then allow the column to equilibrate with mobile phase for at least 30 to 60 min. Turn on the fluorescence detector 30 min before the beginning of the run to warm up the lamp.

4. Add 1 part working OPA reagent (diluted in step 1) to 2 parts microdialysate sample or standard and mix by pipetting up and down several times. Incubate for 2 min at room temperature protected from exposure to direct light. Inject 2 μl of this sample/OPA mix into the HPLC system and run for 30 min (optimally at a low flow rate, e.g., 1 μl/min).

   Derivatization and injection of the sample can be automated through incorporation of an autosampler (e.g., Waters 717) into the HPLC system. Collection of microdialysates into Waters glass low-volume inserts, which are compatible with both the CMA/200 microsampler and Waters 717 autosampler carousels, reduces the risk of contamination and sample loss. For microdialysate samples collected at 20-min intervals at a flow rate of 1 μl/min, an autosampler would be programmed to perform step 4 as follows: add 10 μl of working OPA reagent to 20 μl of sample, mix a volume of 20 μl three times, and inject a volume of 20 μl (leaving a residual dead volume of 10 μl of derivatized sample in the glass low-volume inserts/collection tube).

5. Determine dialysate concentrations of amino acids by comparing peak areas with those of known concentrations of external standards. Perform a linear regression analysis of the standards and use the resulting linear regression equation to determine actual concentrations in samples.

References: Lunte and Lunte, 1979; Pettit and Justice, 1991; Tossman et al., 1986

Contributors: A.K. Pani, Toni S. Shippenberg, Christian Heidbreder, and Michael Graham Espey

Table 2.9.1 Troubleshooting Guide for HPLC/EC Detection System

Problem	Possible cause	Solution
Fluctuations in pressure	Air in check valves	Purge pump with freshly degassed mobile phase to remove air from check valves. In the case of Bioanalytical Systems pump or equivalent, switch inlet check valves between the two heads of the pump; if pressure drop occurs on opposite pump head, clean (e.g., sonicate) or replace check valves.
Low pressure	Leaks or bad column, as indicated by mobile phase salt residue that collects at valves or unions as well as by occasional spikes that occur in chromatogram	Check/replace fittings or seals
High pressure	Clogged frits, injection ports, or guard columns; also dirty or too-tightly-packed column	Open fittings, beginning at detector, and work backwards toward pump. At some point the pressure will drop, pinpointing location of clog.
Pump stops	Clog in flow path	Open fittings, beginning at detector, and work backwards toward pump. At some point the pressure will drop, pinpointing location of clog.
Wavy baseline (wide irregular, positive and negative fluctuations)	Bad column or inadequate column equilibration	Replace or re-equilibrate column
Baseline drift	Variations in room temperature or chemical contamination	Passivate system with 6 M nitric acid
Peaks that occur in only one direction	Grounding problems or alterations in the circuit in which the HPLC is connected	Check and repair electrical connections
Very fast spikes that occur in both the positive and negative direction	Air trapped in detector and/or pump (more periodic spiking)	Inspect the compartment housing the reference electrode and tap to remove any entrapped bubbles. Inspect reference electrode for air bubbles on its tip and replace if necessary. Loosen cell clamp of working electrode to allow mobile phase to run from the gasket, clearing trapped air. If this does not alleviate the problem, degassing and purging of the system is necessary. A pulse dampener will reduce pulsations from the pump and is highly recommended.
Decrease in sensitivity	Mobile phase, electrodes, or column may be old	Inject fresh standards to see if problem persists. If sensitivity is still low, use new mobile phase and replace columns, reference, or working electrodes and/or repolish the working electrode.
High background current	Can occur immediately after an electrode is polished or replaced; may also be consequence of mobile phase degradation, worn seals, scratched electrodes, and/or chemical contamination.	Wait 1 to 2 days until reconditioning of the electrode has occurred; prepare freshly degassed mobile phase; replace seals or rods; ensure that detector is operating at the minimum potential for the analyte of interest; and/or passivate the system. If problem persists, check column by connecting pump directly to analytical cell. If background current drops despite high pulsation, clean or replace column.

continued

Table 2.9.1 Troubleshooting Guide for HPLC/EC Detection System, continued

Problem	Possible cause	Solution
Split peaks	Poor injector alignment or old column	Check injector alignment or age of column
Insufficient sensitivity		Use microbore HPLC technique.[a] Because sample volume collected is substantially less than that of conventional HPLC, the perfusate flow rate used during dialysis can be reduced. This increases probe efficiency, resulting in greater recovery of analyte through the probe.

[a]Several detailed reviews of the equipment and protocols used for this technique are available (Church and Justice, 1989; Pettit and Justice, 1991).

UNIT 2.10

Saturation Assays of Radioligand Binding to Receptors and Their Allosteric Modulatory Sites

Generally, there are two classes of radioligand-receptor binding assays: equilibrium and nonequilibrium. Nonequilibrium binding assays (e.g., association and dissociation assays) are used to measure the kinetics of ligand interaction with a receptor. In contrast, equilibrium binding assays (e.g., competition and saturation assays, Fig. 2.10.1) are conducted under conditions where the amount of ligand bound to a receptor has reached steady-state levels for given concentrations of ligand and receptor. Analysis of the data acquired from a saturation assay allows the ligand binding affinity (K_d) and the maximum number of binding sites (B_{max}) to be determined.

BASIC PROTOCOL 1

SATURATION ANALYSIS OF LIGAND BINDING TO THE BENZODIAZEPINE SITE

Materials *(see APPENDIX 1 for items with ✓)*

  Rat
  ✓ 320 mM sucrose (optional), 0° to 4°C
  ✓ 50 mM Tris citrate buffer, pH 7.4, 0° to 4°C
  ✓ 10 mM flunitrazepam
  ✓ 100 nM [^{3}H]flumazenil (i.e., [^{3}H]Ro15-1788) or other radioligand (e.g., agonists [3H]flunitrazepam, [3H]diazepam, and [3H]zolpidem, or the partial inverse agonist [3H]Ro 15-4513)
  Scintillation fluid (e.g., Cytoscint, ICN Biomedicals)

  Dissection tools
  Tissue homogenizer (e.g., Polytron; Brinkmann)
  Sorvall RC-5C+ centrifuge and SS-34 rotor (DuPont), or equivalent
  12 × 75–mm disposable borosilicate glass test tubes, *or* 1 ml × 96-well polystyrene microtiter plates
  Glass fiber filter strips (grade 32; Schleicher & Schuell)
  Filter manifold (M-24R; Brandel)
  Liquid scintillation spectrometer (e.g., Beckman LS 6500) and scintillation vials

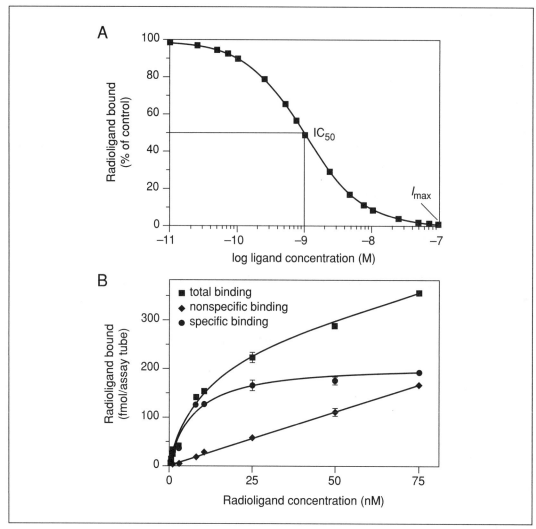

Figure 2.10.1 The two representative types of equilibrium binding assays: the competition assay (**A**) and the saturation assay (**B**). In the competition assay (A), increasing concentrations of unlabeled ligand compete with a fixed concentration of radioligand for the binding site, resulting in a sigmoidally shaped decrease in the amount of radioligand specifically bound. The competition assay determines the concentration of unlabeled ligand that inhibits 50% of the radioligand binding (IC_{50}; 1 nM), the efficacy of radioligand binding inhibition (I_{max}; 100%), and the presence of cooperativity (or multiple binding sites: the Hill coefficient, n_H ; 1.1). The saturation assay (B) employs increasing concentrations of radioligand to bind to a fixed amount of receptor. The total and nonspecific radioligand bindings are determined directly, while the specific radioligand binding is derived by subtracting the nonspecific binding from the total. Fitting a rectangular hyperbola to the binding data by nonlinear regression analysis allows the K_d (6.5 nM) and B_{max} (209 fmol/assay) to be determined.

1. Decapitate animal under AAALAC-approved conditions. Rapidly remove the brain and weigh. (For further dissection, first place the brain into 320 mM sucrose at 0° to 4°C to cool and firm it.)

2. Place the brain in a 125-ml Erlenmeyer flask and add 50 vol of 50 mM Tris citrate buffer at 0° to 4°C (e.g., if brain weight is 1.5 g, add 75 ml of buffer). Homogenize for 10 sec in a Polytron or similar homogenizer. Place the homogenate in centrifuge tubes and centrifuge 20 min at 20,000 × g, 0° to 4°C.

Table 2.10.1 Design of Saturation Assay of [³H]Flumazenil Binding to the Benzodiazepine Receptor

Tube number	Tissue	[³H]Flumazenil	50 mM Tris citrate buffer	10 μM flunitrazepam
1–2	50 μl	0.5 nM	400 μl	–
3–4	50 μl	0.5 nM	350 μl	50 μl
5–6	50 μl	1 nM	400 μl	–
7–8	50 μl	1 nM	350 μl	50 μl
9–10	50 μl	2.5 nM	400 μl	–
11–12	50 μl	2.5 nM	350 μl	50 μl
13–14	50 μl	5 nM	400 μl	–
15–16	50 μl	5 nM	350 μl	50 μl
17–18	50 μl	7.5 nM	400 μl	–
19–20	50 μl	7.5 nM	350 μl	50 μl
21–22	50 μl	10 nM	400 μl	–
23–24	50 μl	10 nM	350 μl	50 μl

3. Pour off the supernatant, retain the pellet, and add another 50 vol ice-cold buffer. Resuspend the pellet in the buffer with the Polytron, then centrifuge again for 20 min at 20,000 × g. Repeat for a total of five washes. Keep the homogenate chilled at 0° to 4°C at all times.

4. Label tubes (or use 96-well polystyrene microtiter plates for convenience in manipulating multiple samples) with six concentrations of [³H]flumazenil, ranging from 0.5 to 10 nM (0.5, 1, 2.5, 5, 7.5, and 10 nM), as indicated in Table 2.10.1. At each radioligand concentration, use two tubes for determining total binding and two tubes for determining nonspecific binding. Add the following to the labeled tubes:

   400 μl assay buffer to the total binding tubes,
   350 μl buffer to the nonspecific binding tubes, and
   50 μl of 100 μM flunitrazepam to the nonspecific binding tubes only.

5. Prepare a 100 nM stock of [³H]flumazenil by adding 8.1 μl from the source vial to 1.091 ml assay buffer, then dilute this stock to obtain the range of working concentrations to be used in the assay (5 to 100 nM; 400 μl of each) as indicated in Table 2.10.2 to give a final concentration of radioligand of 0.5 to 10 nM. Add 50-μl aliquots of each working concentration of [³H]flumazenil to the appropriate experimental test tubes (Table 2.10.1).

6. Resuspend the cell pellet from step 3 in 10 vol assay buffer (which should yield a concentration of ∼2 μg protein/μl). Add 50 μl of this receptor preparation to each assay tube (Table 2.10.1) to start the binding assay. Incubate 60 min at 25°C (room temperature).

Table 2.10.2 Preparation of Working Concentrations of [³H]Flumazenil

[³H]Flumazenil concentration	Volume of 100 nM [³H]flumazenil stock	Volume of assay buffer
100 nM	8.1 μl[a]	1091 μl
75 nM	300 μl	100 μl
50 nM	200 μl	200 μl
25 nM	100 μl	300 μl
10 nM	40 μl	360 μl
5 nM	20 μl	480 μl

[a] Volume of [³H]flumazenil taken from source vial.

7. Deposit the contents of the tubes onto glass fiber filter strips using the filter manifold. Wash twice with 5 ml assay buffer, 0° to 4°C.

8. Remove the filter disks and place in scintillation vials. At this time, set up the standards for each working concentration of radioligand by adding 50-μl aliquots of each radioligand concentration to duplicate scintillation vials. Add 4 ml scintillation fluid to each vial, seal, and shake for 1 hr (or let stand overnight). Place the vials into a liquid scintillation counter and count each vial for 2 min.

9. Determine the actual concentration of radioligand added to each tube by dividing the radioligand concentration in each standard tube by the assay volume [e.g., $(260.4 \times 10^{-15}$ mol/assay tube$)/(500 \times 10^{-6}$ liter/assay tube$) = 0.52$ nM]. Determine the amount of radioligand specifically bound to the receptor by subtracting the nonspecific binding data from the appropriate total binding data.

10. Plot the data (in terms of fmol/assay tube) with the free ligand concentrations (obtained by subtracting the amount of radioligand specifically bound from the actual amount of radioactivity added to each tube) on the x axis and the amount of specifically bound ligand on the y axis (this yields a rectangular hyperbola). Determine the radioligand affinity for the receptor (K_d) and the maximal density of binding sites (B_{max}) by fitting a curve to the data using nonlinear regression (preferred method). Alternatively, linearize the data in a Scatchard-Rosenthal plot (Scatchard, 1949; Rosenthal, 1967).

BASIC PROTOCOL 2

SATURATION ANALYSIS OF LIGAND BINDING TO THE GABA$_A$ RECEPTOR

Materials (see APPENDIX 1 for items with ✓)

    Rat
✓ 50 mM Tris citrate buffer, 0° to 4°C
✓ 10 mM γ-aminobutyric acid (GABA)
    [^{3}H]Muscimol (NEN; 30 Ci/mmol)
✓ 100 μM unlabeled muscimol
✓ 0.03% polyethylenimine

    Dissection tools
    Tissue homogenizer (e.g., Polytron; Brinkmann)
    Sorvall RC-5C+ centrifuge and SS-34 rotor (DuPont), or equivalent
    12 × 75–mm disposable borosilicate glass test tubes *or* 1 ml × 96-well microtiter plates
    Glass fiber filter strips (grade 32; 2 1/4 × 12 1/4–in; Schleicher & Schuell)
    Filter manifold (M-24R; Brandel)
    Liquid scintillation spectrometer (e.g., Beckman LS 6500) and scintillation vials

1. Decapitate rat under AAALAC-approved conditions. Remove the brain and weigh. Place the brain in a flask and add 50 vol of 50 mM Tris citrate buffer, 0° to 4°C. Homogenize for 10 sec in a Polytron, then centrifuge 20 min at $20,000 \times g$, 0° to 4°C.

2. Discard the supernatant, resuspend the pellet in 50 vol of 50 mM Tris citrate buffer, and centrifuge again. Repeat this washing procedure five times. After the last centrifugation, pour off the supernatant, seal the centrifuge tube with its pellet using Parafilm, and freeze overnight at $-70°$C.

3. The next day, thaw the pellet, resuspend it in 50 vol of 50 mM Tris citrate buffer, and wash twice more. *Optional:* Incubate 15 min at 37°C, with or without GABAase and appropriate cofactors (process may degrade receptors; use carefully).

4. Label tubes with sixteen concentrations of [^{3}H]muscimol (a GABA$_A$ receptor agonist), ranging from 0.5 to 500 nM, as indicated in Table 2.10.3. Perform both total and nonspecific radioligand bindings in duplicate. Add the following to the labeled tubes:

   175 µl of 50 mM Tris citrate buffer to the total binding tubes,
   125 µl buffer to the nonspecific binding tubes, and
   50 µl of 10 mM GABA to the nonspecific binding tubes only.

5. Prepare a 2000 nM working stock of [^{3}H]muscimol by adding 84 µl radioligand from its source vial to 1.316 ml assay buffer [(2000 × 10^{-9} M/liter) × (30 × 10^{-6} Ci/liter) ×

Table 2.10.3 Design of the Saturation Assay of [^{3}H]Muscimol Binding to the GABA$_A$ Receptor

Tube number	Tissue	[^{3}H]Muscimol	50 mM Tris citrate buffer	1 mM GABA
1–2	50 µl	0.5 nM	175 µl	–
3–4	50 µl	0.5 nM	125 µl	50 µl
5–6	50 µl	1 nM	175 µl	–
7–8	50 µl	1 nM	125 µl	50 µl
9–10	50 µl	2.5 nM	175 µl	–
11–12	50 µl	2.5 nM	125 µl	50 µl
13–14	50 µl	5 nM	175 µl	–
15–16	50 µl	5 nM	125 µl	50 µl
17–18	50 µl	10 nM	175 µl	–
19–20	50 µl	10 nM	125 µl	50 µl
21–22	50 µl	15 nM	175 µl	–
23–24	50 µl	15 nM	125 µl	50 µl
25–26	50 µl	25 nM	175 µl	–
27–28	50 µl	25 nM	125 µl	50 µl
29–30	50 µl	40 nM	175 µl	–
31–32	50 µl	40 nM	125 µl	50 µl
33–34	50 µl	50 nM	175 µl	–
35–36	50 µl	50 nM	125 µl	50 µl
37–38	50 µl	75 nM	175 µl	–
39–40	50 µl	75 nM	125 µl	50 µl
41–42	50 µl	100 nM	175 µl	–
43–44	50 µl	100 nM	125 µl	50 µl
45–46	50 µl	150 nM	175 µl	–
47–48	50 µl	150 nM	125 µl	50 µl
49–50	50 µl	200 nM	175 µl	–
51–52	50 µl	200 nM	125 µl	50 µl
53–54	50 µl	300 nM	175 µl	–
55–56	50 µl	300 nM	125 µl	50 µl
57–58	50 µl	400 nM	175 µl	–
59–60	50 µl	400 nM	125 µl	50 µl
61–62	50 µl	500 nM	175 µl	–
63–64	50 µl	500 nM	125 µl	50 µl

Table 2.10.4 Preparation of Working Concentrations of [³H]Muscimol

[³H]Muscimol concentration	Volume of 2000 nM [³H]muscimol	Volume of 100 μM muscimol	Volume of 50 mM Tris citrate buffer
5000 nM	200 μl	6 μl	–
4000 nM	200 μl	4 μl	–
3000 nM	200 μl	2 μl	–
2000 nM	84 μl[a]	–	1316 μl
1500 nM	150 μl	–	50 μl
1000 nM	100 μl	–	100 μl
750 nM	75 μl	–	125 μl
500 nM	50 μl	–	150 μl
400 nM	40 μl	–	160 μl
250 nM	25 μl	–	175 μl
150 nM	15 μl	–	185 μl
100 nM	10 μl	–	190 μl
50 nM	5 μl	–	195 μl
25 nM	2.5 μl	–	197.5 μl
10 nM	1 μl	–	199 μl
0.5 nM	0.5 μl	–	199.5 μl

[a]Volume of [³H]muscimol taken from source vial.

$(1.4 \times 10^{-3}$ liter$) \times (1 \times 10^{-6}$ liter$/1 \times 10^{-6}$ Ci$) = 84$ μl] (Table 2.10.4). Dilute this stock to create the 5 to 1500 nM working concentrations of [³H]muscimol. (In cases where multiple binding sites are expected, expand the range of radioligand concentrations to at least eight per binding site.)

For labile ligands that may be destroyed in the binding assay, it may be necessary to add antioxidants, use a cocktail of protease inhibitors (Hanley, 1985), or perform the assay in low-light conditions.

6. Add 200-μl aliquots of the undiluted 2000 nM working stock to the tubes labeled 3000, 4000, and 5000 nM. Then add 2, 4, and 6 μl, respectively, of a 100 μM solution of unlabeled muscimol to these tubes. After the dilutions are made, add 25-μl aliquots of each working concentration of [³H]muscimol to all tubes to give specific activities of reduce the specific activity of [³H]muscimol of 20, 15, and 12 Ci/mmol.

7. Resuspend the pellet from step 3 in 10 vol assay buffer, and initiate the assay by adding 50 μl of the receptor preparation to all the assay tubes. Incubate 90 min at 0° to 4°C.

8. Terminate the reaction by rapid filtration under vacuum (>25 in. Hg) using glass fiber filter strips pretreated with 0.03% polyethylenimine (~20 ml per strip) for ~10 to 15 min. Wash samples twice with 3-ml aliquots of assay buffer at 0° to 4°C. Place the filters in scintillation vials along with 4 ml scintillation cocktail, and shake for 1 hr.

9. Count each vial in a liquid scintillation counter for 3 min. Calculate specific binding by subtracting the nonspecific binding from the total binding.

10. After counting, multiply all the count values for each of the three highest concentrations (standards, total, and nonspecific binding) by a factor equivalent to the amount of isotopic dilution: i.e., for the 300 nM concentration, multiply all the count values by 1.5 (for

300/200), and multiply the 400 and 500 nM count data by 2 and 2.5, respectively. Determine the K_d and B_{max} values for this data set by nonlinear regression analysis, expressing the results as fmol/250 µl assay volume.

If the count values are too low to reliably distinguish from background, double the total volume of each assay tube (i.e., 100 µl + 50 µl + 350 µl = 500 µl).

11. Plot the data to yield a rectangular hyperbola or attempt to linearize by a Scatchard-Rosenthal transformation (particularly effective in detecting deviations from linearity implying the presence of multiple radioligand binding sites).

References: Bennett and Yamamura, 1985; O'Brien, 1986

Contributor: Anthony Basile

UNIT 2.11

Uptake and Release of Neurotransmitters

CAUTION: Radioactive materials require special handling; all supernatants must be considered radioactive waste and disposed of accordingly.

NOTE: All culture incubations are performed in a humidified 37°C, 5% CO_2 incubator unless otherwise specified.

BASIC PROTOCOL 1

STUDY OF UPTAKE AND RELEASE OF DOPAMINE IN INTACT ATTACHED CELLS EXPRESSING THE RECOMBINANT DOPAMINE RECEPTOR

Materials (see APPENDIX 1 *for items with* ✓)

 Cells stably expressing the recombinant transporter (see Eshleman et al., 1995; contact authors at *janowsky@oshu.edu*)
✓ Uptake buffer
 Drug stock solutions to be tested (at concentrations 10× desired final levels)
 50 µM mazindol (Research Biochemicals) in uptake buffer (prepare from 10 mM stock in 0.1 N HCl)
 200 nM [^{3}H]dopamine (40 to 60 Ci/mmol; NEN Life Sciences)
✓ PBS, ice-cold
 3% (w/v) trichloroacetic acid (TCA)
✓ Release buffer: uptake buffer, without calcium and ice-cold

 24-well tissue culture plates
 25°C water bath
 Scintillation vials and cocktail
 Software for analyzing radioligand binding data (e.g., GraphPad Prism or EBDA-Ligand)

1. Plate cells at a density of 15,000 cells per cm^2 of well surface in 24-well plates (enough wells for triplicate data points) and incubate 3 days or to ~100% confluence. (If using cells that transiently express the recombinant transporter, e.g., COS-7 African green monkey kidney cells, grow them for 48 hr after transfection.) Remove medium by aspiration, add 400 µl uptake buffer to each well, and place plates in a 25°C water bath.

2a. *For uptake assay:* Add 50 μl of test drugs, 50 μl of 50 μM mazindol (5 μM final concentration), or 50 μl of solvent (blank) to triplicate wells and preincubate 10 min (or up to 2 hr if indicated by preliminary experiments) at 25°C. Include a set of three wells that contain only [^{3}H]dopamine and cells (to determine total uptake) and wells that contain 5 μM Mazindol or another potent and selective uptake blocker (to define nonspecific uptake).

2b. *For release assay:* Proceed to load cells with [^{3}H]dopamine (step 3) without adding drugs.

3. Add 50 μl of 200 nM [^{3}H]dopamine to each well for a final concentration of 20 nM to initiate uptake with the following considerations: to determine drug IC$_{50}$ values, use 20 nM [^{3}H]dopamine and conduct the assays in the linear portion of the time curve, before steady state is reached; and to determine K_m and V_{max} values, use nonlabeled dopamine at appropriate concentrations to dilute the specific activity of [^{3}H]dopamine. Incubate for the appropriate period of time (determined in pilot time curve studies; varies among transfected cell lines) and temperature (varies the rate of uptake) to achieve steady state.

For uptake assay

4a. Remove solution by aspiration to terminate assay and place plates on ice. Rapidly wash each well twice, each time with 1 ml ice-cold PBS. Immediately add 0.5 ml of 25°C 3% TCA to each well and let plates stand 10 min at 25°C.

5a. Transfer the TCA to scintillation vials, add scintillation cocktail, and determine radioactivity in the TCA by conventional liquid scintillation counting. Alternatively, add scintillation fluid directly to the wells after the cells have been rinsed and use a microbeta plate reader for determination of remaining radioactivity.

For release assay

4b. Rinse the cells (loaded to steady-state levels with ^{3}H-labeled dopamine) twice, each time with 1 ml ice-cold release buffer. Add sufficient ice-cold release buffer for a 0.5 ml final volume in each well, and then add test drugs. Place plate in a 37°C water bath to initiate release.

> *This rinse step is crucial for the removal of any [^{3}H]dopamine that has not been taken up or that has leaked from the cells. The assay is sensitive to the time period of the rinses, which can result in large standard errors for triplicate determinations, and some cells may respond better if room temperature buffer is used. In addition, some cell types (including C6 rat glioma cells) that are used to express the recombinant dopamine transporter will spontaneously leak [^{3}H]dopamine. This problem can be circumvented by using [^{3}H]1-methyl-4-phenylpyridinium (MPP$^+$), a potent neurotoxin; NEN Life Sciences; Johnson et al., 1998) instead of [^{3}H]dopamine as the substrate. COS-7 cells transiently expressing the recombinant human dopamine transporter appear to leak [^{3}H]dopamine very slowly, and thus allow the determination of drug-induced inhibition of spontaneous release, as well as the stimulation of release.*

> *The release assay may also be run at room temperature, depending on the cell type, [^{3}H]substrate, and rinse buffer temperature (room temperature rinse buffer may best be followed by a room temperature or 37° incubation). For determination of the effects of drugs on [^{3}H]substrate release, a time point should be used at which the drug has not caused the release of all of the loaded [^{3}H]substrate. Thus, the $t_{1/2}$ value—i.e., the time point at which a drug has caused the release of 50% of loaded [^{3}H]substrate—should be determined, and dose-response curves for the drug should be conducted at that time point (Johnson et al., 1998).*

5b. Remove solution by aspiration to terminate release assay. Immediately add 0.5 ml of 25°C 3% TCA to each well and let plates stand 10 min at 25°C. Transfer the TCA to scintillation vials, add scintillation cocktail, and determine radioactivity remaining in the cells by liquid scintillation counting.

6. Analyze data using a nonlinear curve-fitting program.

ALTERNATE PROTOCOL 1

STUDY OF DOPAMINE UPTAKE IN DETACHED CELLS

Additional Materials (*also see Basic Protocol 1*)

    HEK-hDAT cells or other poorly adherent cells expressing the dopamine transporter
    0.05% (w/v) polyethylenimine
    0.9% (w/v) NaCl

    150-mm tissue culture plates
    Cell scrapers
    96-well arrays of assay tubes or microvials appropriate for use with cell harvester
    Whatman GF/C glass fiber filters (or equivalent) appropriate for use with cell harvester

1. Grow HEK-hDAT cells on 150-mm plates until confluent. Prepare cells for uptake by gently washing plates twice, each time with 7 ml PBS. Remove the last wash, then add 2.5 ml uptake buffer per plate (appropriate resuspension volume must be determined for each cell line) and place in 25°C water bath for 5 min. Scrape cells gently with a cell scraper and triturate gently with a 5-ml pipet. Pool cells from up to 6 plates in a reservoir suitable for pipetting.

2. Pipet 350 µl uptake buffer into each of the assay tubes or microvials set up in the 96-well array. To triplicate sets of wells add 50 µl of test drugs, 50 µl of 50 µM mazindol (5 µM final concentration), or 50 µl of solvent (blank).

3. Add 50 µl of the cell suspension from step 1 and allow to preincubate 10 min at 25°C. Add 50 µl of 200 nM [^{3}H]dopamine for a final concentration of 20 nM to initiate uptake. Incubate 2 to 20 min (depending on the cell line) until the linear portion of the uptake time course is reached (linearity determined in pilot time curve and temperature experiments).

   For the determination of drug IC_{50} values, 20 nM [^{3}H]dopamine is commonly used, and the assays are conducted in the linear portion of the time curve, before steady state is reached. To determine K_m and V_{max} values, nonlabeled dopamine is used at appropriate concentrations to dilute the specific activity of [^{3}H]dopamine (e.g., 100 nM to 10 µM for a K_m value of 1 µM).

4. Presoak glass fiber filters ≥30 min in 0.05% polyethylenimine. Terminate uptake during the linear portion of the uptake time course (2 to 20 min, depending on the cell line) by harvesting in a cell harvester onto the polyethylenimine-soaked glass fiber filters. Wash filters twice, each time with 4 ml ice-cold 0.9% NaCl, or wash for 3 sec with 0.9% NaCl on an automated harvester, taking care not to disrupt the integrity of the cells during harvesting.

5. Transfer filters to scintillation vials, add scintillation cocktail, and determine radioactivity in cells by liquid scintillation counting. Analyze data by nonlinear curve-fitting program (e.g., GraphPad Prism or EBDA-Ligand).

BASIC PROTOCOL 2

STUDY OF DOPAMINE UPTAKE IN SYNAPTOSOMES

NOTE: Conduct experiments with fresh tissue and perform all steps for tissue preparation at 0° to 4°C.

Materials (*see APPENDIX 1 for items with* ✓)

    Rats or other suitable animals
    0.32 M sucrose, ice-cold

✓ Uptake buffer
Drug stock solutions to be tested (at concentrations such that desired final levels can be achieved by adding 50 µl per well)
50 µM mazindol (Research Biochemicals) in uptake buffer (prepare from 10 mM stock in 0.1 N HCl)
200 nM [^{3}H]dopamine (40 to 60 Ci/mmol; NEN Life Sciences)
0.9% NaCl
0.05% (w/v) polyethylenimine

Glass Potter-Elvehjem tissue homogenizer with Teflon pestle
96-well arrays of assay tubes or microvials appropriate for use with cell harvester
Whatman GF/C glass fiber filters (or equivalent) suitable for use with cell harvester
Scintillation vials and cocktail
Software for analyzing radioligand binding data (e.g., GraphPad Prism or EBDA-Ligand)

1. Dissect striata from brain immediately after sacrifice of animal by decapitation. Homogenize striatal tissue in 5 ml ice-cold 0.32 M sucrose using ten strokes of a glass Potter-Elvehjem homogenizer with a Teflon pestle at 1000 rpm. Centrifuge the homogenate 10 min at $1000 \times g$, 0° to 4°C.

2. Decant the supernatant (S1) and centrifuge it for 20 min at $10,000 \times g$, 0° to 4°C. Using the Potter-Elvehjem homogenizer, resuspend the resultant pellet (P_2) as above in 20 vol of 0.32 M sucrose (on a v/w basis with respect to the original wet weight of the tissue). For alternative methods see Gray and Whittaker, 1962.

3. Pipet striatal synaptosomes (100 µg protein) and uptake buffer into each of the assay tubes or microvials set up in the 96-well array for a volume of 400 µl. To triplicate sets of wells add 50 µl of test drugs, 50 µl of 50 µM mazindol (5 µM final concentration), or 50 µl of solvent (blank). Preincubate 10 min at 37°C prior to addition of [^{3}H]dopamine (volume 450 µl).

4. Add 50 µl of 200 nM [^{3}H]dopamine for a final concentration of 20 nM (final volume 500 µl) to initiate uptake and incubate for the appropriate period of time at 37°C within the linear portion of the uptake time course.

5. Presoak glass fiber filters ≥30 min in 0.05% polyethylenimine. Terminate uptake during the linear portion of the uptake time course by harvesting in a cell harvester onto the polyethylenimine-soaked glass fiber filters. Wash filters twice, each time with 4 ml ice-cold 0.9% NaCl, or wash for 3 sec with 0.9% NaCl on an automated harvester.

6. Transfer filters to scintillation vials, add scintillation cocktail, and determine radioactivity in cells by liquid scintillation counting. Analyze data by nonlinear curve-fitting program.

BASIC PROTOCOL 3

STUDY OF DOPAMINE RELEASE FROM SYNAPTOSOMES

Materials (see APPENDIX 1 for items with ✓)

Rats or other suitable animals
✓ Uptake buffer
200 nM [^{3}H]dopamine (40 to 60 Ci/mmol; NEN Life Sciences)
0.32 M sucrose
Release buffer: uptake buffer without calcium ($CaCl_2$)
Test drugs in release buffer
0.05% (w/v) polyethylenimine
0.9% (w/v) NaCl, 4°C

Glass Potter-Elvehjem tissue homogenizer with Teflon pestle
Refrigerated centrifuge
Whatman GF/C glass fiber filters (or equivalent) appropriate for use with cell harvester
Scintillation vials and cocktail
Software for analyzing radioligand binding data (e.g., GraphPad Prism or EBDA-Ligand)

1. Dissect striata from brain immediately after sacrifice of animal by decapitation. Place in 1.8 ml uptake buffer, mince with scissors, and preincubate for 5 min with shaking at 37°C. Add 200 μl of 200 nM [^{3}H]dopamine for a final concentration of 20 nM to initiate uptake and incubate 10 min at 37°C.

2. Add 20 ml of 0.32 M sucrose, transfer mixture to a centrifuge tube, and sediment by centrifuging 5 min at 1000 × g, 4°C. Decant the supernatant, add 10 ml of 0.32 M sucrose to the pelleted mince, and then homogenize using ten strokes of a glass Potter-Elvehjem homogenizer with a Teflon pestle.

3. Centrifuge the homogenate at 1000 × g for 10 min. Decant the supernatant into a new centrifuge tube and centrifuge 20 min at 10,000 × g, 4°C. Resuspend pellet in 0.32 M sucrose, centrifuge 20 min at 10,000 × g, 4°C, decant supernatant, and resuspend pellet in 1 ml 0.32 M sucrose per rat brain used.

4. Add 0.9 ml of a solution test drug in release buffer (or release buffer alone as a blank) to 0.1 ml of the suspension of synaptosomes for a final volume of 1.0 ml. Incubate 3 min at 37°C to allow release, then quench the reaction by addition of 4 ml of 4°C 0.9% NaCl.

   Spontaneous release for these striatal P2 preparations is less than 1% of the drug-induced release. Results can be expressed as the percent of the amount of [^{3}H]dopamine loaded. One of the releasing amines such as amphetamine or methamphetamine can be included as a positive control.

5. Harvest synaptosomes in a cell harvester onto glass-fiber filters that have been soaked ≥30 min in 0.05% polyethylenimine. Wash filters twice, each time with 4 ml ice cold 0.9% NaCl, or wash for 3 sec with 0.9% NaCl on an automatic harvester.

6. Transfer filters to scintillation vials, add cocktail, and determine radioactivity by liquid scintillation counting. Analyze data by nonlinear curve-fitting program.

ALTERNATE PROTOCOL 2

DETECTION OF UPTAKE OR RELEASE OF DOPAMINE BY HPLC WITH ELECTROCHEMICAL DETECTION (HPLC-EC)

Additional Materials (*also see Basic Protocol 1; see* APPENDIX 1 *for items with* ✓)

Dopamine·HCl (unlabeled)
✓ Solvent for dopamine standards
3% (w/v) trichloroacetic acid (TCA)
✓ HPLC mobile phase

HPLC microvials (Hewlett-Packard)
HPLC/EC system (see UNIT 2.9 for more detail) including:
  Reversed-phase C18 column (3-μm particle size; ESA HR-80)
  ESA Coulochem electrochemical detector

1. Conduct uptake and/or release as described for cells (see Basic Protocol 1), using up to 10 or 100 μM unlabeled dopamine·HCl instead of [^{3}H]dopamine, depending on both the cell density and the capacity of the cells for uptake of dopamine.

2. Prepare HPLC standards by dissolving 2.47 mg dopamine·HCl (2 mg of dopamine as the free base) in 10 ml of solvent for dopamine standards, then using 3% TCA to make dilutions of standards in the experimental range. Transfer a 100-μl aliquot of each dilution into an HPLC microvial.

3. Transfer each TCA solution from uptake or release assay (see Basic Protocol 1, step 5a or 5b) into a 1.5-ml microcentrifuge tube, then microcentrifuge 10 min at maximum speed, 4°C. Transfer a 100-μl aliquot of each supernatant into an HPLC microvial.

4. Inject 5 to 25 μl of each TCA extract and standard dilution into an HPLC/EC apparatus with reversed-phase C18 column at a flow rate of 0.9 ml/min at room temperature (see UNIT 2.9; dopamine peak at ~7 to 10 min).

5. Measure dopamine with ESA Coulochem electrochemical detector by oxidation at 0.4 V followed by reduction at −0.35 V (UNIT 2.9). Convert dopamine peak height to the desired units (e.g., ng dopamine/well) by comparison with the standard curve constructed using known amounts of dopamine (UNIT 2.9).

ALTERNATE PROTOCOL 3

USING A SUPERFUSION APPARATUS FOR TIME SAMPLING

Materials (see APPENDIX 1 for items with ✓)

Rats
✓ Modified Krebs-HEPES buffer
200 nM [^{3}H]dopamine (40 to 60 Ci/mmol; NEN Life Sciences)
Test drugs: transporter inhibitors
✓ 20 mM K^+/low Na^+ buffer
Transporter substrates, e.g.:
    S-(+)-Amphetamine (Research Biochemicals)
    (+)-Methamphetamine (Research Biochemicals)
    Tyramine (Sigma)
0.2 N HCl
✓ Krebs-bicarbonate buffer
Pargyline (Research Biochemicals)
Despiramine (Research Biochemicals)
Fluoxetine (Research Biochemicals)
1 N NaOH

McIlwain tissue chopper (Brinkmann)
Superfusion apparatus including 12 superfusion chambers, tubing, and pump (Brandel SF12)
Glass fiber filter discs
Scintillation vials and cocktail
95% O_2/5% CO_2 gas mixture in cylinder, with regulator and tubes
Electrical pulse generator (Brandel ES12955)

1. Sacrifice rats, remove brains, and place on ice. Dissect striata and nucleus accumbens, chop into 225 × 250–μm strips using a McIlwain tissue chopper, and resuspend in 5 ml of modified Krebs-HEPES buffer by trituration through a plastic transfer pipet.

For chemical stimulation of vesicular release

2a. Wash tissue three times in 25°C modified Krebs-HEPES buffer, each time by allowing strips to settle in the buffer, then removing the buffer and adding fresh Krebs-HEPES buffer at 37°C. Resuspend in 2 ml of 37°C modified Krebs-HEPES buffer and add [^{3}H]dopamine to a 15-nM final concentration. Incubate 15 min at 37°C.

A serotonin uptake inhibitor such as fluoxetine can be added to prevent uptake of [^{3}H]dopamine into serotonergic terminals.

3a. Remove incubation buffer and resuspend tissue in modified Krebs-HEPES buffer at 40 mg tissue per 12 ml buffer (based on original wet weight) and distribute in 275-µl aliquots between glass fiber filter discs into chambers of a superfusion apparatus.

4a. Superfuse buffer at 37°C, with (added to the buffer throughout the experiment) or without added transporter inhibitor as appropriate, over the tissue at a flow rate of 0.5 ml/min for 25 min to establish a low, stable baseline release, collecting superfusates at 2-min intervals into scintillation vials.

5a. While continuing to collect superfusates at 2-min intervals, superfuse 20 mM K^+/low Na^+ buffer over the tissue for 2 min, which will cause vesicular release of [^{3}H]dopamine. Return the inflow to modified Krebs-HEPES buffer for 10 min to allow release to return to baseline level. Then, to measure release of [^{3}H]dopamine via the transporter, superfuse with transporter substrates such as S-(+)-amphetamine, (+)-methamphetamine, or tyramine dissolved in modified Krebs-HEPES buffer. Again, return the inflow to buffer for 10 min to allow release to return to baseline levels. Wash the superfusion apparatus extensively after experiments with drugs (e.g., GBR-12909) that may adhere to the superfusion chambers.

6a. Expose tissue to 0.2 N HCl for 45 min to extract the remainder of the radioactivity from the sample, collecting the effluent into scintillation vials. Collect tissue into the final scintillation vial with the filter discs. Add scintillation cocktail and measure released radioactivity by liquid scintillation counting.

7a. Analyze data. Express data as radioactivity released (predominantly [^{3}H]dopamine in this case) above baseline during the collection interval as a fraction of total radioactivity contained in the tissue at the beginning of the release interval, or as the percent of radioactivity expressed by the control stimulus.

For electrical stimulation of vesicular release

2b. Preincubate the slices (step 1) from the same brain region of two animals for 15 min at 37°C in 2 ml Krebs-bicarbonate buffer saturated with 95% O_2/5% CO_2, in the presence of 10 µM pargyline, 0.5 µM desipramine, and 1 µM fluoxetine to block monoamine oxidase and other biogenic amine transporters. Add [^{3}H]dopamine to a final concentration of 100 nM and incubate 30 min in the dark at 37°C.

3b. Transfer the slices to twelve low-volume (200-µl) chambers of a superfusion apparatus. Place 3 to 5 mg of striatal slices or 2 to 4 mg of nucleus accumbens slices between glass fiber filter discs within each chamber, making sure that there are no air bubbles in the chambers.

4b. Begin superfusion at a rate of 1 ml/min at 37°C. After 60 min of presuperfusion, begin collecting the effluent in 5-min fractions.

5b. Induce the release of radiolabeled amines by electrical field stimulation (120 bipolar pulses, 2 msec long, 20 mA, 0.5 Hz) using a constant current stimulator. Stimulate 20 min after the 60 min presuperfusion period. Follow with a second stimulation 50 min later. Add test drugs 20 min before the second stimulation.

In experiments in which the interaction between two or more drugs is studied, start the first treatment at the beginning of the superfusion and the second treatment 20 min before the second stimulation. Keep both drugs present throughout the rest of the superfusion.

6b. At the end of the superfusion, remove the slices with their bottom filters from the chamber, solubilize in 1 N NaOH, and transfer to a scintillation vial. Add scintillation fluid to all vials and measure released radioactivity by liquid scintillation counting.

7b. Analyze data. Calculate the fractional basal release (percentage of the amount of radioactivity in the tissue at the beginning of that collection) as the amount of radioactivity present in one 5-min sample just before the first and second electrical stimulation. Express the electrically induced release as the sum of the increased fractional release above base line in the four fractions after the start of stimulations. Compare the baselines and stimulated releases in the presence and absence of the drug to asses its effect. Compare the ratios for control slices and drug-treated slices in each experiment.

BASIC PROTOCOL 4

EXAMINATION OF THE DOPAMINE TRANSPORTER WITH RADIOLIGANDS

There is evidence to support the hypothesis that drugs for characterizing drug interactions with recognition sites on the transporters bind to different sites. Assays involving binding to the dopamine transporter in whole cells, as well as in membranes, are discussed in this unit. As with establishing any initial binding assay, time course, temperature dependency, protein concentration, and filter binding should be determined with the membrane preparation and filtration conditions employed (see Support Protocol and UNIT 2.10).

Materials (see APPENDIX 1 for items with ✓)

Cells expressing the dopamine transporter (see Eshleman et al., 1995; contact authors at *janowsky@oshu.edu*)
✓ PBS
Lysis buffer: 2 mM HEPES/1 mM EDTA (ice-cold)
0.32 M sucrose
Test drugs
400 to 600 pM [^{125}I]3β-(4-iodophenyl)tropane-2β-carboxylic acid methyl ester ([^{125}I]RTI-55; 2200 Ci/mmol; NEN Life Sciences)
50 μM mazindol
0.9% NaCl, ice-cold
✓ Uptake buffer
Unlabeled 3β-(4-iodophenyl)tropane-2β-carboxylic acid methyl ester (RTI-55; Research Triangle Institute)

150-mm tissue culture plates
Cell scrapers
Refrigerated centrifuge
Polytron tissue homogenizer (Brinkmann)
Whatman GF/C glass-fiber filters (or equivalent)
Scintillation vials and cocktail
Software for analyzing radioligand binding data with nonlinear curve fitting programs, e.g., GraphPad Prism or EBDA-Ligand; the Cheng-Prusoff correction (Cheng and Prusoff, 1973) for deriving K_i values from experimentally determined IC_{50} values

1. Grow cells expressing the dopamine transporter to 80% to 100% confluency on 150-mm diameter tissue culture plates. (If using cells that transiently express the recombinant transporter, e.g., COS-7 cells, grow them for 48 hr after transfection.)

2. Remove medium and wash cells with 25°C PBS. Add 7 ml ice-cold lysis and allow cells to lyse for 10 min at 4°C. Scrape plates, transfer to a centrifuge tube, and centrifuge 20 min at 31,000 × g, 4°C.

3. Resuspend pellet of cell membranes in 0.32 M sucrose (volume depending on expression levels of the transporter; e.g., 24 ml for a confluent 150-mm plate of HEK-hDAT cells expressing the dopamine transporter at a density of 12 pmol/mg protein) using a Polytron tissue homogenizer for 10 sec at setting 7.

   The volume of resuspension medium is determined empirically so that no more than 10% of total radioactivity added is bound to the membranes.

4. For each experimental point, mix a 50-μl aliquot of cell membranes (containing ∼15 and 50 μg protein for C6 and COS-7 cells expressing the transporter, respectively), and the test drug. Preincubate for 10 min (typically) prior to the addition of radioligand.

5. Add 25 μl of 400 to 600 pM [^{125}I]RTI-55 (for a 40 to 60 pM final concentration) for a final volume of 250 μl. Include samples containing 5 μM mazindol in place of the test drug to determine nonspecific binding. Incubate 90 min at 25°C (or an incubation time/temperature that allows ligand binding to come to equilibrium; see Support Protocol).

6. Terminate the assay by filtration on glass fiber filters. Wash filters twice, each time with 4 ml ice-cold 0.9% NaCl. Determine radioactivity on the filters by conventional liquid scintillation counting.

   If a positively charged radioligand is used, filters should be soaked for ≥30 min in 0.05% (w/v) polyethylenimine.

   The conditions described above may need to be adjusted for specific ligands; if a ligand has low specific activity and high affinity, a larger incubation volume may be needed to achieve a reasonable amount of bound radioligand. The total amount bound should not exceed 10% of total radioactivity added to an assay so that pseudo first-order equations may be used to fit the data.

   To determine the affinity and density of binding sites (K_d and B_{max} respectively) for [^{125}I]RTI-55—as well as other ligands such as [^{3}H]CFT and [^{3}H]cocaine—experiments can be conducted by diluting the specific activity of the radioligand with increasing concentrations of the unlabeled drug (see UNIT 2.10).

7. *For saturation isotherms:* Prepare an initial 160 nM solution of RTI-55. Prepare 13 tubes for serial dilutions by adding 0.4 ml uptake buffer to each tube. Add 0.6 ml of the RTI-55 solution to the first tube, mix, and then add 0.6 ml from the first tube to the second tube. Continue until there are 14 concentrations of RTI-55 with concentrations ranging from 0.24 to 160 nM. Add 25 μl of buffer, 50 μM mazindol (5 μM final), or the RTI-55 concentrations to assay tubes or wells in triplicate (48 assay tubes total). Proceed with binding as in steps 4 to 6 above.

   For other ligands the range of concentrations should optimally span 0.1 to 10 times the estimated K_d.

8. Analyze data using appropriate software. Transform the resulting competition-curve data to saturation data by calculating the change in the specific activity of the radioligand for every concentration of unlabeled ligand.

SUPPORT PROTOCOL

ESTABLISHING INITIAL BINDING-ASSAY PARAMETERS

As with establishing any binding assay, the time course, temperature dependency, protein concentration, and filter binding should be determined with the membrane preparation used.

Time course and temperature dependence
Materials (for materials, see Basic Protocol 4)

1. Pipet 150 μl uptake buffer into every tube or well. Add 25 μl buffer to half of the tubes and 25 μl of 50 μM mazindol (5 μM final concentration) to the other half. Finally, add 50 μl membrane preparation to every tube.

2. At time points varying from 120 to 1 min from harvesting time (usually at least 12 time points, with wells for total and nonspecific binding at every time point in duplicate), add 25 μl of 400 to 600 pM [^{125}I]RTI-55 to wells (40 to 60 pM final concentration). At time 0, terminate incubation (see Basic Protocol 4, step 6).

   Alternatively (to avoid exposing tissue to experimental conditions for long periods of time before testing), add tissue to all tubes at the same time and harvest tubes at indicated times after addition. Use a single well filtration device in this case.

3. Conduct time courses at 0°, 15°, 25°, and 37°C to determine equilibrium times for each possible experimental temperature. (Saturation and competition binding should be conducted at an equilibrium time point.)

4. If equilibrium is not reached in this amount of time, the experiment should be repeated with longer intervals. If equilibrium is reached rapidly, the experiment should be repeated with more early time points. See Table 2.11.1 for descriptions of assay conditions for commonly used radioligands. (Optimal and appropriate incubation time, temperature, and protein concentration can differ significantly across laboratories).

Table 2.11.1 Incubation Conditions for Binding of Selected Dopamine Transporter Radioligands

Ligand[a]	Buffer[b]	Incubation time/temp.	Reported K_d values (nM; approx.)[c]
[^{3}H]cocaine	Tris/NaCl	20 min/ 20°C	360 (Kennedy and Hanbauer, 1983)
	Tris/NaCl	1 hr/4°C	19.2, 1120 (Madras et al., 1989a)
	Tris/KCl/NaCl	1 hr/4°C	210, 26400 (Schoemaker et al., 1985)
	Sodium phosphate/Tris or bicarbonate/Tris	20 min/25°C	46 to 175 (Eshleman et al., 1993)
[^{3}H]CFT (WIN 35,428)	Krebs/HEPES (uptake buffer)	20 min/on ice	10 (A. Janowsky, unpub. observ.)
	Tris/NaCl	2 hr/on ice	4.7, 60 (Madras et al., 1989b)
	Krebs/Tris/HEPES	2–3 hr/4°C	10 (Pristupa et al., 1994)
	Sucrose/sodium phosphate	1 hr/20°C	3.6, 1300 (Eshleman et al., 1995)
[^{3}H]GBR-12935	Krebs/HEPES (uptake buffer)	1 hr/25°C	1 (A. Janowsky, unpub. observ.)
	Tris/NaCl	45 min/25°C	1 (Janowsky et al., 1986)
	Sodium phosphate/Tris or bicarbonate/Tris	45 min/on ice	4 to 14 (Eshleman et al., 1993)
	Sucrose/sodium phosphate/NaCl	90 min/20°C	1 (Eshleman et al., 1995)
[^{3}H]mazindol	Tris/KCl/NaCl	1 hr/0°C	1.8 (Javitch et al., 1984)
	0.9% NaCl	2 hr/on ice	14.4 (Shimizu and Prasad, 1991)
[^{3}H]methylphenidate	Tris/NaCl	30 min/4°C	235 (Schweri, 1985)
[^{125}I]RTI-55	Tris/NaCl/sucrose phosphate	100 min/22°C	0.066, 1.5 (Little et al., 1993)

[a]CFT (WIN 35428), 2β-carboxymethyl-3β-(4-fluorophenyl)tropane; GBR-12935, 1-[2-diphenylmethoxy]ethyl-4-(3-phenylpropyl)-piperazine; RTI-55, 3β-(4-iodophenyl)tropane-2β-carboxylic acid methyl ester.

[b]If buffer is not listed in APPENDIX 1, see literature citations in rightmost column for full buffer compositions.

[c]Values separated by commas indicate that high- and low-affinity binding sites were reported.

Determining optimal protein concentration
1. Prepare cell membranes (see Basic Protocol 4, steps 1 to 3).

2. Resuspend membranes in a minimal volume (1 or 2 ml) of 0.32 M sucrose (see Basic Protocol 4, step 3), and make serial dilutions of membranes. Determine protein concentration at each dilution.

3. Conduct binding (see Basic Protocol 4, steps 4 to 6), with duplicate determinations of total and nonspecific binding at each protein concentration. The dilution of membranes selected for use in subsequent experiments should bind no more than 10% of the added radioligand and should be in the area of the curve in which specific binding increases linearly with increasing protein concentration.

Determining filter binding
Materials (for materials, see Basic Protocol 4)
1. In triplicate, pipet 200 μl of uptake buffer, 25 μl of buffer or 50 μM mazindol (5 μM final), and 25 μl of 400 to 600 μM [^{125}I]RTI-55 (40 to 60 pM final concentration), for a final volume of 250 μl (also see Basic Protocol 4, steps 4 and 5). Incubate for the optimal time and at the optimal temperature determined in the steps above for time course and temperature dependence (see Basic Protocol 4, step 6).

2. Terminate the reaction by harvesting over glass fiber filters treated in an identical manner as will be used for membrane binding assays (see Basic Protocol 4, step 6).

   If there is substantial filter binding, filters can be presoaked in a variety of substances, such as polyethylenimine or bovine serum albumin, and then tested for filter binding.

Intact cell binding
1. Grow attached cells until confluent in 24-well plates. Remove medium and wash with 1 ml uptake buffer.

2. Add buffer, radioligand ([^{125}I]RTI-55), and test drugs or solvent (blank) to a final volume of 0.5 ml per well of 24-well plate. Incubate reaction 2 to 3 hr at 4°C as described in Pristupa et al. (1994). Optimize conditions to prevent entry of radioligand into the cells.

3. Terminate binding by pouring off the buffer and washing plates twice with PBS. Extract radioactivity into TCA and count (see Basic Protocol 1).

References: Eshleman et al., 1995; Rudnick, 1997

Contributors: Aaron Janowsky, Kim Neve, and Amy J. Eshleman

UNIT 2.12
Measurement of Chloride Movement in Neuronal Preparations

BASIC PROTOCOL 1

UPTAKE/EFFLUX OF RADIOISOTOPIC Cl⁻ IN SYNAPTONEUROSOMES

NOTE: Prior to starting this assay, have all solutions and ^{36}Cl⁻ mixtures prepared and have the filter apparatus ready, because tissue and ^{36}Cl⁻ are added to the tubes in a timed fashion.

Materials *(see APPENDIX 1 for items with ✓)*

    PEI assay buffer: $^{36}Cl^-$ flux assay buffer containing 0.05% (v/v) polyethyleneimine
✓ $^{36}Cl^-$ flux assay buffer
    $^{36}Cl^-$ (~10 to 20 mCi/g, in HCl form; NEN Life Sciences)
    10× γ-aminobutyric acid (GABA) or other agonist in $^{36}Cl^-$ flux assay buffer
    Synaptoneurosomes from individual brain regions (see Support Protocol 1)
    Picrotoxin assay buffer, ice cold: $^{36}Cl^-$ flux assay buffer containing 100 μM picrotoxin
    Water-miscible scintillation fluid

    2.4-cm glass fiber filters (Whatman GF/C or Schleicher & Schuell no. 30)
    1-liter vacuum flask with Tygon tubing and vacuum pump (do not use a vacuum line)
    Single-place stainless steel filter holder (Hoefer Pharmacia Biotech)
    Gas drying jar (containing Drierite or equivalent)
    7-ml polyethylene scintillation vials
    12 × 75–mm glass test tubes
    Shaking water bath, 30°C (uptake) or room temperature (efflux)
    Filter forceps (not pointed)
    Repeat pipettor with 1-liter total volume and 1- to 10-ml dispensing range

1. Prepare the filtration apparatus as follows.
   a. Spread out 2.4-cm glass fiber filters in a petri dish, and pour enough PEI assay buffer into the dish to cover the filters. Use sufficient filters for triplicate assays and a triplicate blank.
   b. Set up a 1-liter vacuum flask with a single-place stainless steel filter holder. Connect a gas drying jar between the vacuum pump and the vacuum flask to protect the pump.
   c. Arrange 7-ml polyethylene scintillation vials (equal to the number of filters) in a rack.

For uptake

2a. Arrange up to 36 total 12 × 75–mm glass test tubes in triplicate (eleven samples and a blank) in a test tube rack. Add any additional drugs (e.g., antagonists), if desired, and 150 μl $^{36}Cl^-$ flux assay buffer (or an amount that will result in a final assay volume of 250 μl) to each tube. Add 200 μl $^{36}Cl^-$ flux assay buffer to triplicate tissue blanks, and place at the end of the row. Place the rack in a 30°C water bath.

3a. In a separate tube, prepare a 5× master solution containing 5 μCi/ml $^{36}Cl^-$ (0.25 μCi final per assay) and 5 times the appropriate concentration of GABA or other drug of interest in Cl^- flux assay buffer (typically 1 to 500 μM for 1× GABA concentrations).

4a. At 30-sec intervals, with a stopwatch in view, add 50 μl synaptoneurosomes (gently mixed before adding; final ~0.5 mg protein in 250 μl) to each tube, mix by gentle vortexing, and immediately place tube back in the 30°C water bath. Do not add synaptoneurosomes to the tissue blanks. Save a 10-μl aliquot of synaptoneurosomes for a standard protein analysis (e.g., Lovrien and Matulis, 1995), if desired.

5a. Incubate tubes with gentle shaking 20 min from the time they are placed in the water bath. At 18 min, turn on vacuum pump to prepare it for filtering. Using filter forceps, place a wet filter on the filter holder and close the holder. Make sure pump is pulling a steady vacuum.

6a. When the first tube is at exactly 20 min, with a stopwatch in view, add 50 μl of the $^{36}Cl^-$ mixture (containing drugs) from step 3a to the first tube to initiate uptake. Mix quickly by gentle vortexing. Exactly 5 sec later, use a repeat pipettor to add 5 ml ice-cold picrotoxin assay buffer to the tube, and quickly pour the entire mixture over the filter under vacuum (to terminate uptake).

7a. Repeat rinse two more times, making sure each wash has completely cleared the filter before adding the next. Place filter in a scintillation vial.

8a. Place a fresh filter on the filter holder, and repeat steps 6a and 7a every 30 sec until all samples have been filtered. After the last assay tube, repeat steps 6a and 7a for the tissue blanks. Proceed to step 9.

For efflux

2b. Remove a 10-μl aliquot of synaptoneurosomes for a standard protein assay (e.g., Lovrien and Matulis, 1995), if desired. Add 5 μCi $^{36}Cl^-$ per ml synaptoneurosomes and incubate 60 min at 4°C to load the synaptoneurosomes with $^{36}Cl^-$. Prepare a separate tube containing 1 μCi $^{36}Cl^-$ in 200 μl $^{36}Cl^-$ flux assay buffer as a tissue blank.

3b. During the 60-min incubation, arrange 12 × 75–mm glass test tubes for triplicate samples and blanks in a test tube rack. Add 2.64 ml $^{36}Cl^-$ flux assay buffer to each tube (total volume 3 ml).

4b. Add 300 μl $^{36}Cl^-$ flux assay buffer or 10× GABA (typically 1 to 500 μM for 1× GABA concentrations) or agonist to appropriate tubes. Place assay tubes in a gently shaking water bath at room temperature. Turn on vacuum pump to prepare it for filtering.

5b. Add 60 μl of $^{36}Cl^-$-loaded synaptoneurosomes or tissue blank to an assay tube (1/50 dilution; final 0.6 mg protein) to initiate efflux. Vortex gently and return tube to water bath. Using filter forceps, place a wet filter on the filter holder and close the holder. Make sure the pump is pulling a steady vacuum.

6b. After exactly 2 min (or desired time), use a repeat pipettor to add 5 ml ice-cold picrotoxin assay buffer to the tube, and quickly pour the entire mixture over the filter under vacuum to terminate efflux of $^{36}Cl^-$.

7b. After the first wash has completely cleared the filter, wash the filter once more with 5 ml ice-cold picrotoxin assay buffer. Place the filter in a scintillation vial.

8b. Working quickly, repeat steps 5b to 7b for each assay tube. Proceed to step 9.

To save time, triplicates can be staggered at 30-sec intervals during a 2-min incubation (e.g., at $t = 0$, 30, and 60 sec).

9. Add ∼4 ml water-miscible liquid scintillation fluid to each vial, and shake well, and wait for at least 1 hr before counting. Place vials in a scintillation counter and count each sample for at least 5 min.

10. Calibrate counter to give data as dpm. Subtract the average of the three tissue blanks (typically ∼500 dpm for uptake assays) from each of the sample averages (triplicates). Convert dpm to nmol/mg protein, if desired, using the specific activity of $^{36}Cl^-$ (in mCi/g or MBq/g) adjusted for Cl^- molarity in the assay buffer, and the conversion factor for dpm (1 mCi = 37 MBq = 2.22×10^9 dpm). Adjust for the amount of protein determined in the protein assay.

11. Subtract the values for the basal samples (no agonist) from samples containing agonist to determine the $^{36}Cl^-$ uptake stimulated by the agonist.

BASIC PROTOCOL 2

UPTAKE OF RADIOISOTOPIC Cl⁻ IN PRIMARY CULTURES

NOTE: Prior to assay, cultured cells should be kept at 37°C until needed. Keep the $^{36}Cl^-$ flux assay buffer at room temperature and the wash buffers ice cold. Prepare all solutions ahead of time.

Materials *(see APPENDIX 1 for items with ✓)*

$^{36}Cl^-$ (~10 to 20 mCi/g in HCl form; NEN Life Sciences)
✓ $^{36}Cl^-$ flux assay buffer
γ-Aminobutyric acid (GABA) or agonist
Picrotoxin assay buffer, ice cold: $^{36}Cl^-$ flux assay buffer containing 100 μM picrotoxin
Cultured neurons plated on coverslips, grown in modified essential medium (MEM) or equivalent for 4 to 8 days
0.2 N NaOH
0.2 M HCl
Water-miscible scintillation fluid

35-mm plastic petri dishes
Forceps
Blotting or tissue paper
20-ml liquid scintillation vials

1. Make a 2 μCi/ml solution of $^{36}Cl^-$ in $^{36}Cl^-$ flux assay buffer. Subdivide the solution and add GABA (typically 1 to 500 μM), GABA agonist, or other drugs of interest to separate aliquots, keeping one aliquot for basal uptake levels. Keep chilled.

2. Dispense 2 ml of $^{36}Cl^-$ solution into the appropriate 35-mm plastic petri dishes (one dish for three replicate coverslips in step 3). Pour 1 liter ice-cold picrotoxin assay buffer into each of two 1-liter beakers, with constantly stirring.

3. Remove coverslips containing cultured neurons from the medium with a pair of forceps, and rinse in $^{36}Cl^-$ flux assay buffer for 2 to 4 min at room temperature. Take one coverslip at a time and gently blot dry on blotting or tissue paper. Place in the appropriate petri dish containing $^{36}Cl^-$ (with or without drugs) for 5 sec (or other time during linear uptake of up to 30 sec).

4. Rapidly transfer the coverslip to the first beaker containing ice-cold picrotoxin assay buffer to terminate uptake; leave for 3 sec. Transfer the coverslip to the second beaker containing ice-cold picrotoxin assay buffer and leave for 7 sec.

5. Drain the coverslip on blotting paper and transfer it to a 20-ml scintillation vial. Add 0.5 to 0.8 ml of 0.2 N NaOH to each vial to solubilize the protein. Let stand 1 to 2 hr (or heat 20 min at 60°C to hasten protein digestion).

6. Neutralize the basic extract by adding an equivalent volume of 0.2 M HCl. Remove a 50- to 100-μl aliquot for determination of protein concentration by standard procedures (e.g., Lovrien and Matulis, 1995). Add 10 ml water-miscible scintillation fluid. Count in a liquid scintillation counter for at least 5 min per vial.

7. Express data as dpm. If desired, convert dpm to nmol/mg protein as described (see Basic Protocol 1, step 10). To obtain net $^{36}Cl^-$ uptake, subtract the basal values (no GABA or agonist present) from the agonist-stimulated values.

BASIC PROTOCOL 3

EFFLUX OF Cl⁻ IN SYNAPTONEUROSOMES MEASURED WITH FLUORESCENT DYES AND PHOTOMETRY

Materials (see APPENDIX 1 for items with ✓)

    Synaptoneurosomes (see Support Protocol 1)
✓ SPQ/MQAE loading buffer
    2 M 6-methoxy-N-(sulfopropyl)quinolinium (SPQ) *or* 1 M
       N-(6-methoxy-quinolyl)acetoethyl ester (MQAE; both from Molecular Probes; store
       in small aliquots at $-20°C$)
✓ SPQ/MQAE wash buffer, ice cold
✓ Low-Cl⁻ buffer, room temperature and 37°C
    10 mM pentobarbitone
    2.5 mM nigericin in ethanol
    3.5 mM tributyltin acetate in ethanol
    10.5 M KSCN
    1.75 mM valinomycin in ethanol
    Fluorescence spectrophotometer (Hitachi F 2000 or F 4000 or equivalent) equipped with
       thermosetting, magnetic stirring, and a 1-ml quartz cuvette

1. Pellet 20 to 50 mg synaptoneurosomes by centrifuging 5 min at $1000 \times g$, 4°C. Remove supernatant, then add 2 ml SPQ/MQAE loading buffer (∼10 mg/ml final synaptoneurosome concentration) to centrifuge tube and resuspend pellet by gentle trituration.

2. Add 5 µl of 2 M SPQ or 1 M MQAE per ml SPQ/MQAE loading buffer (final 10 mM and 5 mM, respectively), and incubate 30 to 45 min in a 37°C incubator or water bath with gentle stirring.

3. Centrifuge the suspension 5 min at $1000 \times g$, 4°C, and pour off the supernatant. Resuspend the pellet by gentle trituration in ice-cold SPQ-MQAE wash buffer and keep the suspension on ice. Use immediately or store 3 to 4 hr on ice.

4. Place an appropriate amount (typically 150 to 250 µl) of the synaptoneurosome suspension into a microcentrifuge tube, microcentrifuge ∼10 sec at maximum speed, and aspirate the supernatant.

5. To wash away extrasynaptoneurosomal dye and Cl⁻, resuspend the pellet in 0.5 ml room temperature low-Cl⁻ buffer and centrifuge as in step 4.

6. Carefully aspirate the supernatant and resuspend the pellet in 350 µl of 37°C low-Cl⁻ buffer in a 1-ml quartz cuvette. Place the cuvette in a fluorescence spectrophotometer at 37°C with stirring (critical).

7. Monitor fluorescence at an excitation wavelength of 350 nm and an emission wavelength of 445 nm for SPQ, or at an excitation wavelength of 355 nm and emission wavelength of 460 nm for MQAE.

8. Monitor baseline fluorescence (which shows an upward drift because Cl⁻ is still leaking out from the synaptoneurosomes) for 1 to 2 min, then add a drug of interest, e.g., 3.5 µl of 10 mM GABA (100 µM final) and monitor the signal for ∼2 min. Add 3.5 µl of 10 mM pentobarbitone (100 µM final) and monitor signal again.

9. At the end of the experiment, produce the maximal fluorescence (F_{max}) by adding 1 µl of 2.5 mM nigericin (7 µM final) and 1 µl of 3.5 mM tributyltin acetate (10 µM final). When the fluorescence signal stabilizes, produce the minimum fluorescence (F_{min}) by adding 1 µl of 1.75 mM valinomycin (5 µM final) and 5 µl of 10.5 M KSCN (150 mM final).

10. Calculate the intracellular Cl^- concentration before and after the addition of drugs using the Stern-Volmer equation and the K_q constant obtained from the calibration of dye (see Support Protocol 3).

SUPPORT PROTOCOL 1

PREPARATION OF SYNAPTONEUROSOMES

Materials (see APPENDIX 1 for items with ✓)

✓ Synaptoneurosome preparation buffer
Animal for sacrifice

Nylon monofilament mesh with 147-µm pore size (Small Parts)
Swinnex filter holders (Millipore)
10-µm Mitex filters (Millipore)
10- and 60-ml plastic syringes
15-ml glass tissue grinder (e.g., Duall, Kontes)
12 × 75–mm glass test tubes

1. To prepare a nylon mesh filter unit (one per gram tissue), cut three layers of nylon mesh into a circle to fit a Swinnex filter holder. Place the three layers on the filter holder and squirt with water to make the nylon stick. Place filter cap on holder. Place a 10-µm Mitex filter into each holder (two per gram tissue).

2. Place a 60-ml plastic syringe on each nylon filter unit, and a 10-ml plastic syringe on each Mitex filter unit. Push a few milliliters of synaptoneurosome preparation buffer through the filter units to wet the filters. Place a clean 50-ml centrifuge tube under each filter unit for collecting filtrate.

3. Sacrifice animal by decapitation (preferably without anesthetics) and rapidly dissect the desired brain region. Dissect gray matter free from white matter, if possible. Keep tissue chilled (on ice) and all reagents on ice.

4. Quickly place tissue in a tared plastic weigh boat, weigh the tissue, and use two razor blades to slice tissue into small pieces, if required for large brain regions.

5. Add 7 ml ice-cold synaptoneurosome preparation buffer per gram of tissue to the weigh boat, and pour the entire contents into a 15-ml glass tissue grinder. Homogenize tissue pieces with five to seven strokes to form a thick mixture. Do not overhomogenize. Pour the homogenate into a 50-ml centrifuge tube. Rinse homogenizer with 30 ml ice-cold synaptoneurosome preparation buffer and add to homogenate.

6. Pour the diluted homogenate into the 60-ml syringe attached to the pre-wetted nylon filter unit. To initiate and end gravity filtration, provide a little pressure to the top of the syringe with the palm of the hand (not the syringe plunger).

7. *Optional:* Pour ∼10 ml of the filtered homogenate into the 10-ml syringe on one Mitex filter unit, and use a plunger to gently push the filtrate through. Repeat with another 10 ml using the same filter unit, and then with 10-ml aliquots in the second filter unit.

Do not use >20 ml per Mitex filter. If excessive force is necessary, the filter may be clogged. Use a fresh filter instead.

8. Centrifuge filtrate 15 min at 1000 × g, 4°C. Discard the supernatant, and resuspend the pellet gently by pouring 5 ml buffer onto the pellet and pouring the entire contents into a 12 × 75–mm glass test tube. Vortex gently to suspend the pellet into a homogenous mixture. Do not overvortex.

9. Pour the contents into a clean 50-ml centrifuge tube and add 25 ml ice-cold synaptoneurosome preparation buffer. Centrifuge 15 min at 1000 × g, 4°C. Discard the supernatant and resuspend the pellet as described in step 8. To obtain ∼10 mg protein/ml, use 1.2 ml resuspension volume per gram tissue (in step 4). Keep final suspension on ice and use within 1 hr.

BASIC PROTOCOL 4

EFFLUX OF Cl^- IN PRIMARY CULTURES MEASURED WITH FLUORESCENT DYES AND PHOTOMETRY

Additional *(also see Basic Protocol 3; see* APPENDIX 1 *for items with* ✓ *)*
 Rat cerebellar granule cells cultured on
 ✓ Poly-L-lysine-coated glass coverslips
 10 mM GABA
 10 mM pentobarbitone
 4-ml quartz spectrophotometer cuvette with coverslip holder

1. Using forceps, transfer coverslip containing cultured rat cerebellar granule cells to a 3- or 6-cm petri dish containing SPQ/MQAE loading buffer and rotate the dish to rinse off the culture medium.

2. Transfer the rinsed coverslip to another small petri dish containing 2 ml SPQ/MQAE loading buffer. Add 10 μl of 2 M SPQ or 1 M MQAE (final 10 mM and 5 mM, respectively) and mix by gently rotating the dish. Incubate 30 to 45 min in a 37°C incubator.

3. Rinse coverslip once in low-Cl^- buffer to wash away excess dye and extracellular Cl^-. Add 2 ml low-Cl^- buffer to a 4-ml quartz cuvette with a coverslip holder. Mount the coverslip in the cuvette, place the cuvette in a fluorescence spectrophotometer, and stir gently at 37°C to ensure even distribution of added components.

4. Monitor SPQ fluorescence at an excitation wavelength of 350 nm and emission wavelength of 445 nm. For MQAE, use an excitation wavelength of 355 nm and emission wavelength of 460 nm. Record baseline fluorescence (showing an upward drift because of some Cl^- leakage out from the cells) for ∼1 min, then add drug(s) of interest, e.g., 20 μl of 10 mM GABA (10 μM final) and then 20 μl of 10 mM pentobarbitone (100 μM final).

5. Add 5.6 μl of 2.5 mM nigericin (7 μM final) and 5.7 μl of 3.5 mM tributyltin acetate (10 μM final) and monitor fluorescence until it reaches a maximum value (F_{max}). Then add 5.7 μl of 1.75 mM valinomycin (5 μM final) and 28.6 μl of 10.5 M KSCN (150 mM final) to produce the minimum fluorescence (F_{min}).

   This procedure may cause cells to detach from coverslips, in which case the actual Cl^- concentration cannot be calculated. However, changes in fluorescence can be used as a relative measure of Cl^- flux (e.g., increase in fluorescence from basal or % increase as compared to the signal generated using tributyltin).

6. Calculate the intracellular Cl^- concentration before and after the addition of drugs using the Stern-Volmer equation and the K_q constant obtained from the calibration of dye (see Support Protocol 3).

BASIC PROTOCOL 5

UPTAKE OF Cl⁻ IN THE ACUTE BRAIN SLICE MEASURED WITH FLUORESCENT DYE AND EPIFLUORESCENCE IMAGING

CAUTION: Protect eyes from UV light. Use the shield that attaches to the microscope and do not look directly at the blue light beam.

Materials *(see APPENDIX 1 for items with ✓)*

  MEQ-loaded acute brain slice (see Support Protocol 2)
✓ Ringer's solution, oxygenated
  50 to 1000 µM γ-aminobutyric acid (GABA) or other agonist in Ringers solution
  Plexiglas imaging chamber (Fig. 2.12.1) with outflow and inflow tubes (∼1.7-mm-o.d.;
    ∼1.2-mm i.d.), flow rate restrictor (in line with tubing), and switchable valve
    (chamber can be constructed by any instrument shop to fit stage)
  Upright epifluorescence microscope (Nikon Optiphot or equivalent) equipped with:
    10× objective
    UV water-immersible 40× objective (Olympus or equivalent)
    Mercury lamp
    335-nm excitation filter
    360-nm dichroic beam splitter
    440-nm emission filter (100-nm band pass)
    Silicon-intensified target (SIT) camera or charge-coupled device (CCD)
      intensified camera
  Platinum-wire holder containing threads of nylon stocking glued with Superglue ("harp")
  Image analysis computer system and video monitor

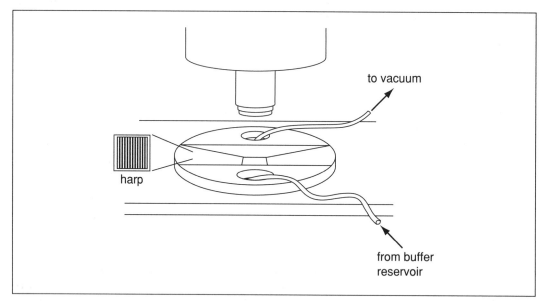

Figure 2.12.1 A typical brain slice superfusion (imaging) chamber. The chamber is made of Plexiglas with two wells for filling and removing buffer from the chamber. The wells are connected to the central chamber by a thin tunnel to allow the buffer to fill the central chamber. The slice is placed in the central chamber and the harp is placed on top of the slice to keep it from moving. A single inflow tube is shown here. For treatment of dye-loaded slices with GABA or other agonists, a separate reservoir is used, and a switch valve is used to alternate between perfusion media.

1. Using the polished (large) end of a Pasteur pipet to pick up and deliver slices, place an MEQ-loaded acute brain slice in a Plexiglas imaging chamber containing oxygenated Ringers solution. Perfuse with Ringers solution at 1.5 to 2 ml/min.

2. Orient the slice for imaging with an upright epifluorescence microscope using $10\times$ and $40\times$ objectives in transmission mode, and anchor the slice with a platinum-wire holder (harp). Continue to perfuse the slice 20 min at room temperature before starting the experiment.

3. Using the $40\times$ objective, illuminate the slice with UV light and quickly find a suitable area to be imaged. Keep the illumination to <5 sec, if possible. Adjust the gain and background (using the software) so that the brightness intensity is within a working dynamic range. With the imaging software, acquire an image using an average of ≥ 16 frames (1 frame = 1/30 sec). Close the shutter. If desired, mark each cell to be used for data collection by drawing a box on the part of the cell to be measured.

   Depending on the type of detection camera, neutral density filters can be placed in the illumination path to help reduce photobleaching.

4. Turn the data logging function on using the software and illuminate the slice for ≤ 1 sec to save brightness intensity values within the drawn boxes. Repeat to obtain three to four basal time points. If desired, with the imaging software, acquire an image using an average of ≥ 16 frames (1 frame = 1/30 sec).

5. Add antagonists (if included in the experiment) and wait at least 15 min. Apply 50 to 1000 µM GABA (or other agonist) to the perfusion chamber via a separate inflow reservoir connected to a valve that can be switched to deliver different perfusion buffers. To deliver drugs more quickly, add a $100\times$ concentrated solution directly to the perfusion chamber as the valve is switched.

6. At a desired time, illuminate the cells in the presence of GABA (agonist). Open the shutter for ≤ 1 sec for each data point. If an acquired image is desired, use a frame averaging function (≥ 16 frames) and save the image to disk. Save the data files containing the brightness intensity values from each cell marked for analysis.

7. If washout is desired, switch the valve back to normal Ringers solution (without drug) and perfuse at least 30 min.

8. Save the data files. Analyze data (optical density units) with the imaging software or another spreadsheet program. Subtract background values from both basal and drug-induced values. Calculate the change in fluorescence (F) using the equation, $\Delta F = [1 - (F/F_b)] \times 100$, where F_b is the basal fluorescence value. To convert the change in fluorescence to a change in intracellular $[Cl^-]$ if calibration of the dye is performed see Support Protocol 4.

ALTERNATE PROTOCOL

UPTAKE OF Cl⁻ IN THE ACUTE BRAIN SLICE MEASURED WITH FLUORESCENT DYE AND CONFOCAL IMAGING

CAUTION: Protect eyes from the UV light. Use the shield that attaches to the microscope and do not look directly at the blue light beam.

Additional Materials (*also see Basic Protocol 5*)

Confocal microscope attached to an upright microscope (Nikon Optiphot or equivalent) equipped with:
  $10\times$ objective
  UV water-immersible $40\times$ objective (Olympus or equivalent)

Air table and air tank
UV laser with a 364-nm excitation line
400-nm barrier filter
Optical disk recorder (Panasonic OMDR LQ 3031 or equivalent)

1. Stabilize and orient a brain slice on an upright confocal microscope as described for epifluorescence (see Basic Protocol 5, steps 1 and 2).

2. Follow the confocal microscope manufacturer's instructions to set the proper laser output intensity. Choose an intensity that provides adequate illumination without producing too much background signal or photobleaching of the dye.

   The extent of photobleaching can be measured by illuminating a slice continuously and recording the rate of loss of brightness intensity. A loss of brightness intensity of less than 6% over 20 sec is acceptable.

3. Before illuminating slice, choose parameters within the imaging software for desired slit width for the confocal aperture (15 to 25 µm is a workable range; the smaller the slit width, the more confocality) and desired emission filter (~400 nm works well).

4. Illuminate the slice using a UV laser line (~364 nm) and a dichroic mirror, to exclude all but the 364-nm laser line. Find and mark cells for imaging and data collection as for epifluorescence (see Basic Protocol 5, step 3).

5. Acquire basal fluorescence data as described for epifluorescence (see Basic Protocol 5, step 4). Optimally, send frames directly to optical disk using an optical memory disk recorder (OMDR; faster than saving images to a hard drive).

6. Superfuse with GABA (or other drugs) and illuminate the field at a desired time as described for epifluorescence (see Basic Protocol 5, steps 5 and 6). For very rapid application of GABA or other drugs (versus bath perfusion), place a picospritzer pressure ejection pipet (or equivalent) to within 40 µm of a neuron to be imaged. Apply a brief 0.5- to 2-sec "puff" of GABA and record images with a 250-µsec acquisition time every sec for at least 10 sec. Record images to an optical disk (see step 5) l for rapid data acquisition.

7. Proceed with washout and analysis as for epifluorescence (see Basic Protocol 5, steps 7 and 8).

SUPPORT PROTOCOL 2

PREPARATION OF MEQ-LOADED ACUTE BRAIN SLICES

NOTE: Use distilled, deionized water throughout.

Materials *(see APPENDIX 1 for items with ✓)*

6-methoxy-*N*-ethylquinolinium chloride (MEQ; Molecular Probes)
N_2 tank
Sodium borohydride
Ethyl acetate
Anhydrous $MgSO_4$
✓ Ringer's solution, oxygenated, ice cold and room temperature
Brain tissue for slicing (fresh from 12- to 21-day-old rat pups)

12×75–mm glass test tubes
0.5- to 1.0-ml glass microvial with conical insert
Mesh supports (glue a piece of crinoline, obtained at any fabric store, to a plastic ring to make a floatable support that fits an 80-ml beaker)
Vibrating tissue slicer (Vibratome or equivalent)

1. Dissolve 5 mg MEQ in 0.1 ml water in a 12 × 75–mm glass test tube. Place under a stream of N_2 for 1 min. Prepare 20 to 50 µl of a 12% (w/v) solution of sodium borohydride in water, and quickly add it to the MEQ solution (dark red in color). Return to the stream of N_2 for 30 min (solution turns yellow).

2. Dilute the mixture with 0.5 ml water and 0.5 ml ethyl acetate. Vortex 30 sec and then allow the aqueous and organic layers to separate. Remove the organic (top) layer with a pipet and transfer it to a fresh test tube.

3. Re-extract the aqueous dihydro-MEQ solution with another 0.5 ml ethyl acetate. Mix vigorously and allow the layers to separate. Remove the organic layer and add it to the extract in the fresh test tube. Add ∼100 mg anhydrous $MgSO_4$ and let stand 5 min. To determine if the reduction was successful, quickly shine a shortwave UV lamp on the extract. If it fluoresces green, it has been reduced. If it fluoresces blue, there is still considerable MEQ present.

4. Transfer the organic layer to a 0.5- to 1.0-ml glass microvial with a conical insert, and evaporate the ethyl acetate with a stream of N_2. Store the dried dihydro-MEQ in a tightly closed microvial under N_2 up to 2 days at $-20°C$, or up to 2 weeks at $-70°C$.

5. Place a mesh support in an 80-ml glass beaker containing 40 ml oxygenated room-temperature Ringers solution, and continue bubbling 95% O_2/5% CO_2 into the beaker.

6. Prepare acute brain slices from areas of interest such as hippocampus or cerebellum (e.g., see UNIT 1.3). Slice brain tissue at 200 to 300 µm in ice-cold oxygenated Ringers solution using a vibrating tissue slicer. Transfer each slice to the mesh support with the blunt end of a Pasteur pipet. Do not let slices touch one another. Continue bubbling O_2/CO_2 into the beaker without disturbing the slices, and incubate at room temperature for ∼15 min.

7. For each 5 mg MEQ in step 1, add 15 µl ethyl acetate to the dried dihydro-MEQ extract (step 5). Mix well. Add 12 µl resuspended dihydro-MEQ to the beaker containing slices. Incubate 30 min at room temperature.

8. Rinse slices by carefully transferring them one by one (using the blunt end of a Pasteur pipet) to a second mesh support in an 80-ml glass beaker containing 60 to 80 ml fresh oxygenated Ringers solution at room temperature.

9. Continuously bubble slices with oxygen for at least 15 min prior to transferring them to an imaging chamber. Use slices within 5 hr (older slices will leak some MEQ and cells may become unhealthy).

SUPPORT PROTOCOL 3

CALIBRATION OF FLUORESCENT DYE IN SYNAPTONEUROSOMES AND CULTURED CELLS

Materials (see APPENDIX 1 for items with ✓)

    2.5 mM nigericin in ethanol
    3.5 mM tributyltin acetate in ethanol
    SPQ- or MQAE-loaded synaptoneurosomes (in 1-ml quartz cuvette; see Basic Protocol 3, step 7) or cultured cells (in 4-ml quartz cuvette; see Basic Protocol 4, step 4)
  ✓ Low-Cl^- buffer, supplemented with varying NaCl and gluconate
    2 M NaCl
    10.5 M KSCN
    1.75 mM valinomycin in ethanol

1. Deplete intrasynaptoneurosomal Cl⁻ by adding 1 μl of 2.5 mM nigericin (7 μM final) and 1 μl of 3.5 mM tributyltin acetate (10 μM final) to a 1-ml quartz cuvette containing 350 μl SPQ- or MQAE-loaded synaptoneurosomes in low-Cl⁻ buffer. If using a 4-ml cuvette, add 5.6 μl of 2.5 mM nigericin and 5.7 μl of 3.5 mM tributyltin acetate to 2 ml dye-loaded synaptoneurosomes to achieve the same final concentrations. Measure fluorescence to obtain the maximal fluorescence, F_{max}.

2. Make four cumulative additions of 2 M NaCl up to a maximum of 50 mM (e.g., final concentrations of 5, 10, 20, and 30 mM in one experiment and 25, 30, 40 and 50 mM in the next) to the stirring cuvette while monitoring real-time fluorescence. To achieve the first concentration series in a 350-μl assay in a 1-ml cuvette, add 2 M NaCl as follows: 0.875 μl, 0.875 μl, 1.75 μl, and 1.75 μl. For a 2-ml assay in a 4-ml cuvette, add 5 μl, 5 μl, 10 μl, and 10 μl. To achieve the second concentration series in a 350-μl assay, add 2 M NaCl as follows: 4.38 μl, 0.87 μl, 1.75 μl and 1.75 μl. For a 2-ml assay in a 4-ml cuvette, add 25 μl, 5 μl, 10 μl, and 10 μl.

3. For a 350-μl assay, add 5 μl of 10.5 M KSCN (150 mM final) and 1 μl of 1.75 mM valinomycin (5 μM final) to quench the fluorescence of the dye. For a 2-ml assay, add 28.6 μl 10.5 M KSCN and 5.7 μl of 1.75 mM valinomycin to achieve the same final concentrations. Measure fluorescence to obtain the lowest fluorescence intensity, F_{min}.

4. Calculate the total quenchable signal (F_0; equivalent to F_t in Engblom and Kerman, 1993): $F_{max} - F_{min} = F_0$.

5. At each [Cl⁻], determine F_{Cl} as the fluorescence at that [Cl⁻] minus F_{min}. Then determine F_0/F_{Cl} for each [Cl⁻]. Calculate the K_q (in M⁻¹) according to the Stern-Volmer equation, $F_0/F_{Cl} = 1 + K_q[Cl^-]$. Alternatively, determine the K_q from the slope of the line obtained by plotting F_0/F_{Cl} versus [Cl⁻].

    The reciprocal of K_q (M^{-1}) is the concentration of Cl⁻ that quenches the fluorescence by 50% (also referred to as K_{sv} in Engblom and Kerman, 1993).

6. Monitor the fluorescence of the synaptoneurosomes. Determine F_{max} and F_{min} as defined above and calculate the Cl⁻ concentrations using the Stern-Volmer equation and the K_q constant obtained from the calibration.

SUPPORT PROTOCOL 4

CALIBRATION OF FLUORESCENT DYE IN BRAIN SLICES

This protocol is modified from Support Protocol 3 for the calibration of MEQ-loaded brain slices. Calibration of the dye is only necessary if an estimation of the intracellular change in Cl⁻ concentration is desired. It should be determined whenever a change in assay procedures occurs (e.g., MEQ concentration, temperature).

Additional Materials (*also see Support Protocol 3; see APPENDIX 1 for items with* ✓)

MEQ-loaded brain slice (see Support Protocol 2)
✓ Low-Cl⁻ Ringer's solution, with 200 μM nigericin/100 μM tributyltin acetate
✓ Low-Cl⁻ Ringer's solution, with 200 μM nigericin/100 μM tributyltin acetate containing varying NaCl and gluconate concentrations
✓ Low-Cl⁻ Ringer's solution, with 150 mM KSCN/25 μM valinomycin

1. Superfuse an MEQ-loaded brain slice in low-Cl⁻ Ringer's solution, with 200 μM nigericin/100 μM tributyltin acetate (maximally effective concentrations for slice penetration) for 30 min. Record the fluorescence intensity with a 1-sec exposure, obtaining the maximal fluorescence, F_{max}.

2. Every 10 min, add low-Cl^- Ringer's solution, with nigericin, tributyltin, and three increasing Cl^- concentrations (20 to 125 mM), and record the fluorescence intensity.

3. Add low-Cl^- Ringer's solution, with 150 mM KSCN/25 μM valinomycin. Wait 10 min and record fluorescence intensity, obtaining the minimal fluorescence, F_{min}, which should be close to zero.

4. Calculate the total quenchable signal (F_0) by subtracting fluorescence in the presence of KSCN from fluorescence in the low-Cl^- buffer: $F_0 = F_{max} - F_{min}$. Plot F_0/F_{Cl} versus [Cl^-] to obtain a straight line. Calculate the slope, which is equal to the Stern-Volmer constant, K_q (M^{-1}).

5. Convert fluorescence values (basal or any other sample) obtained in Basic Protocol 5, step 8, to Cl^- concentrations by substituting F and F_b for F_{Cl} in the Stern-Volmer equation. The difference in [Cl^-] values is the net change in intracellular [Cl^-]. Estimate the change in intracellular Cl^- induced by an agent exposed to the slice using the Stern-Volmer equation and K_q obtained from the calibration experiment (expected to be larger in response to drug addition in brain slices than in cell suspensions).

Contributors: Rochelle D. Schwartz-Bloom, A. Christine Engblom, Karl E.O. Åkerman, and Jon R. Inglefield

UNIT 2.13

Measurement of Cation Movement in Primary Cultures Using Fluorescent Dyes

STRATEGIC PLANNING

A critical consideration in performing these experiments is the recording hardware. The light source will typically be an arc lamp, but may be a laser. Commonly used arc lamps contain either mercury or xenon. Xenon is preferred because it has a relatively flat intensity spectrum, such that a similar amount of light is emitted over a wide range of wavelengths. Lasers are least flexible in terms of the number of emitted wavelengths, but offer the greatest control over illumination intensity. Argon-filled lasers generate both ultraviolet (UV) and visible spectral lines that are convenient for emission ratio and single-wavelength cation indicators.

For excitation ratio dyes, the method of wavelength selection is important. The most popular option currently used is a computer-controlled, motor-driven wheel that holds barrier filters of the desired wavelength. These devices allow switching between filters within ~0.1 sec, which limits the maximal rate of acquisition of ratios to >0.3 sec. A more expensive but more flexible option is a computer-controlled monochromator that allows selection of any wavelength between 270 to 650 nm. This device switches between wavelengths approximately ten times more quickly than the fastest filter wheel and offers an infinite number of wavelengths without the purchase of additional barrier filters. Some commercially available systems also use a pair of fixed monochromators. The light from each is selected using a rotating chopper, and is combined using a light guide. This also offers fast wavelength changing.

The addition of neutral density filters between the wavelength selector and the sample, and the ability to change the amount of filtering, is important to optimize recording conditions. It is necessary to empirically determine the amount of filtering required, based on the intensity of the illumination device, the sensitivity of the dye to bleaching, and the dynamic range of the recording device.

Light, once filtered and attenuated, enters the microscope (upright or inverted) by the epifluorescence adapter and is directed to the sample by a dichroic mirror and objective. It is important to note that both the neutral density filters and the objectives need to be made of quartz glass if they are to be used with UV dyes, because normal glass absorbs UV light. Samples are mounted in a recording chamber that ideally can be perfused with solutions of interest during experiments.

The emitted fluorescence is collected by the objective and passes through a dichroic mirror that separates the illumination wavelengths from the emitted fluorescence. The light then typically continues to the video adapter of the microscope. Emitted light may pass through a single barrier filter or the combination of an additional dichroic mirror and multiple barrier filters, depending on whether excitation or emission ratios are being collected, respectively.

The final important component is the detector. This may be a camera or a photomultiplier tube (PMT). Cameras offer spatial resolution, so that images may be acquired. This is clearly invaluable if conclusions are to be drawn about the localization of ion changes, and also offers the advantage of recording from multiple cells in a microscopic field, which increases productivity. However, cameras are significantly more expensive than PMTs, require more sophisticated software to operate, and may be less sensitive to low levels of light. They are relatively impractical for emission ratio dyes because of the difficulty in balancing the sensitivity of the two camera detection systems. The limited light sensitivity of the less expensive cameras and the consequent need to integrate a number of video frames to generate a suitable image may also become a rate-limiting step in data acquisition.

For many simple applications, light detection with a PMT is sufficient, sensitive, fast, and relatively inexpensive. A PMT will provide a measure of the total amount of light that reaches the detector surface. To restrict the signal to a single neuron, a circular or rectangular diaphragm can be placed in the light path, along with an eyepiece or video camera on the detector side of the diaphragm to ensure correct alignment with the sample. This is a straightforward approach to limiting light collection to a single neuron.

To detect hardware-related problems, it is helpful to have a fluorescent, nonbiological standard. Many investigators use uranium glass, which provides a bright emission at >510 nm when illuminated with wavelengths between 330 and 410 nm. A range of fluorescent beads is also available from a number of suppliers.

BASIC PROTOCOL

MEASUREMENT OF [ION$^+$]$_i$ USING RATIOMETRIC DYES

The properties of the dye dictate whether excitation ratios (the ratio of emitted fluorescence at two different excitation wavelengths) or emission ratios (the ratio of fluorescence intensities at two different emitted wavelengths) are measured. It is necessary to optimize recording conditions for each dye to provide an adequate signal with minimal bleaching or dye loss.

Materials (see APPENDIX 1 *for items with* ✓)
  ✓ 1 mM dye solution, acetoxymethyl (AM) ester
  ✓ HEPES buffer, with and without 5 mg/ml BSA
  Primary cultures of neurons prepared on no. 1 glass coverslips
  Drugs for stimulating ion changes (optional): e.g., 1 mM glutamate, glycine, or
    N-methyl-D-aspartate (NMDA) in HEPES buffer

  35-mm disposable plastic culture dishes
  Fluorescence recording system (see Strategic Planning)

1. Add the appropriate volume of 1 mM dye solution (optimal dye concentration determined empirically; typically 1 to 5 μM) to 1 ml BSA-supplemented HEPES buffer. Immediately vortex vigorously, and pour into a 35-mm disposable plastic culture dish.

2. Carefully rinse no. 1 coverslips containing cultured cells by exchanging the culture medium at least three times with an equal volume HEPES buffer. Pick up a coverslip with fine forceps and place it in the culture dish containing dye solution. Put the culture dish in a dark incubator at 37°C (incubation time determined empirically, typically 15 to 60 min).

3. Carefully rinse coverslips twice by exchanging the buffer solution with fresh HEPES buffer. Optionally, return cells to the dark incubator for 30 min or less to facilitate additional dye de-esterification.

4. Mount the coverslip in the recording chamber and select suitable neurons well separated and healthy (i.e., phase bright, with intact processes and no obvious signs of vacuolization or process fragmentation) for recording data. Start the recording software and perform the desired experiment (e.g., stimulate of neurons by superfusing drug solutions through the recording chamber for <1 min at 5 to 20 chamber vol/min).

   It is usually possible to use morphological criteria to distinguish neurons from nonneuronal cells on a coverslip. In addition, neurons, but not glia, express N-methyl-D-aspartate (NMDA) receptors. Thus, when measuring $[Ca^{2+}]_i$, a change in fluorescence in response to addition of 100 μM NMDA combined with 10 μM glycine can be used to confirm that a given cell is, in fact, a neuron.

   Until the characteristics of a culture system and drug response are established, it can be helpful to have a range of drug solutions that can be used to test the responsiveness of the recording system.

5. At the end of the experiment, move the coverslip to view an area that does not contain neurons, and record the level of background fluorescence, typically <25% of the lowest fluorescence value measured. Discard the coverslip.

6. Perform data analysis offline. Subtract values for background fluorescence from raw fluorescence intensity values at each wavelength. Calculate ratios by dividing the fluorescence intensity at one wavelength by that of the other. Convert the ratios to $[ion^+]_i$ using separately determined calibration parameters (see Support Protocols 1 to 4).

CALIBRATION FOR RATIOMETRIC DYES

Calibration for conversion of fluorescence intensity or fluorescence ratios to ion concentrations of single-wavelength dyes is done on a cell-by-cell basis and is described in the Alternate Protocol. Calibration of ratiometric dyes can be achieved either by measuring ratios in dye solutions in a cell-free environment, or by manipulating the ionic milieu of the dye inside cells to measure the extremes of the fluorescence ratios (R_{min} and R_{max}), which, assuming the affinity of the dye is known, can be used to calculate the relationship between the ratio and the ion concentration.

The advantages of the first approach are that the ion concentrations can be effectively clamped and that the approach does not require an assumption of the affinity of the dye for the ion. The major disadvantage is that the intracellular environment (specifically the viscosity and protein concentration) may have a significant impact on the properties of the dye. In this case, calibration parameters measured in a cell-free environment may not provide a reasonable approximation of the properties of the dye inside the cell, and the resulting ion concentration estimates will be of little value.

The difficulty associated with dye calibration within cells (second approach) is that the dye signal must be measured in both completely ion-free and ion-saturated environments. Achieving these two states can be difficult, especially for Ca^{2+} because neurons have very effective Ca^{2+} buffering mechanisms. Moreover, conditions that result in high levels of intracellular Ca^{2+} may also represent a departure from the usual state of the cell (cells typically swell, for example), so that the calibration condition does not faithfully mimic the recording conditions. In most cases, these problems are hard to resolve unequivocally, and the resulting concentrations of ions should be regarded as estimates rather than absolute values. Empirically, studies have suggested that Fura-2 and Magfura-2 can be calibrated in a cell-free environment, while Indo-1 and SBFI require calibration within cells.

SUPPORT PROTOCOL 1

CELL-FREE CALIBRATION OF Ca^{2+} DYES

Materials (see APPENDIX 1 for items with ✓)

✓ 1 mM dye solution
✓ Ca^{2+} calibration solutions

Clean coverslips (without neurons)
Curve-fitting software (e.g., Prism, GraphPad Software)

1. Mark a clean coverslip with a permanent marker to provide an object to focus on. Mix the dye of interest with a range of buffered Ca^{2+} calibration solutions (ten to fifteen concentrations in a range between $0.1 \times K_d$ and $10 \times K_d$, including 0 mM and a supersaturated ion concentration, e.g., 1 mM; see Tables 2.13.1 and 2.13.2). Prepare equal volumes of Ca^{2+} calibration solutions without dye for background determination. Put a small drop (30 to 50 μl) of the dye solution on the coverslip adjacent to the focus mark.

Table 2.13.1 Properties of Commonly Used [ion$^+$]$_i$ Indicators

Dye	λ_{ex} (nm)[a]	λ_{em} (nm)[a]	K_d[b] Ca^{2+} (μM)	Mg^{2+} (mM)	Na^+ (mM)
Fura-2	340/380	510	0.22	>5	–
Indo-1	345	405/490	0.25	>5	–
BTC	405/480	525	14	–	–
Fluo-3	488	535	0.32	–	–
Calcium green 1 N	488	535	0.19	–	–
Calcium green 2 N	488	535	0.57	–	–
Calcium green 5 N	488	535	3.3	–	–
Magfura-2	340/380	510	50	1.5	–
Magfura-5	340/380	510	6.5	2.6	–
Magindo-1	345	405/490	29	2.7	–
Magnesium green	488	535	4.8	0.9	–
SBFI	340/385	510	–	–	18

[a]Wavelengths reflect commonly available filter sets that are typically used with these dyes. More optimal settings may be obtained with the use of a monochromator. λ_{ex}, excitation wavelength; λ_{em}, emission wavelength.

[b]Values for dye affinity were taken from Grynkiewicz et al., 1985; Harootunian et al., 1989; Minta et al., 1989; Haugland, 1993; Iatridou et al., 1994; Zhao et al., 1996.

Table 2.13.2 Ca^{2+} Standard Solutions for Dye Calibration

Desired [Ca^{2+}] (nM)	Vol. Ca^{2+}/EGTA solution (ml)	Vol. 100 mM EGTA (ml)
0	0.000	10.000
20	0.735	9.265
50	1.656	8.344
100	2.841	7.159
200	4.425	5.575
300	5.435	4.565
400	6.135	3.865
500	6.649	3.351
750	7.485	2.515
1000	7.988	2.012
1500	8.562	1.438
2000	8.881	1.119

The concentration of dye should be titrated to approximate the signal obtained from loaded cells under the same recording conditions. This will almost inevitably be a higher concentration than is used to load the cells. Typically, 10 to 50 μM dye is required.

The acetoxymethyl ester of the dye is not used for cell-free calibration, as membrane permeability is not an issue, and the dye will not become de-esterified.

2. Measure fluorescence values at each ion concentration. Subtract background fluorescence from the signal at each wavelength, and determine the fluorescence ratio at each ion concentration. Plot the data (ratios) on the *x* axis, and Ca^{2+} concentrations on the *y* axis. Use curve-fitting software to determine the values of K, R_{min}, and R_{max}.

The value K is a constant given by $\beta \times K_d$, where K_d is the affinity of the dye for the ion, and β is the ratio of the fluorescence intensity of the free form of the dye at low and saturating ion concentrations (i.e., S_{f2}/S_{b2}; Grynkiewicz et al., 1985). Note that using the method presented here does not require the assumption of an affinity of the dye because K is determined from the curve fitting. The actual ratio values will be instrument-specific; however, the resulting shape of the curve should be the same, regardless of the instrument used.

3. Convert experimentally determined ratios (see Basic Protocol) to [Ca^{2+}]$_i$ using the equation:

$$[Ca^{2+}]_i = K \times \frac{R - R_{min}}{R_{max} - R}$$

SUPPORT PROTOCOL 2

IN SITU CALIBRATION OF Ca^{2+} DYES

Materials *(see APPENDIX 1 for items with ✓)*

✓ Intracellular buffer
✓ 1 mM dye solution, acetoxymethylester (AM) of dye to be calibrated
  BSA
  1 mM antimycin in methanol

1 mM nigericin in ethanol
100 mM 2-deoxyglucose (2-DG) in water
✓ 100 mM EGTA
Primary culture of neurons prepared on no. 1 glass coverslips
1 mM carbonyl cyanide p-trifluoromethoxyphenyl hydrazone (FCCP) in methanol
1 mM Ca^{2+} ionophore (ionomycin or 4-bromo-A23187; Molecular Probes) in dimethyl sulfoxide (DMSO)
140 mM $CaCl_2$ in H_2O
35-mm tissue culture disk

1. Supplement 1 ml intracellular buffer with 5 µM dye of interest, 5 mg/ml BSA from powder, 1 µM antimycin, 5 µM nigericin, 1 mM 2-DG, and 1 mM EGTA, and add to a 35-mm tissue culture dish. Place a coverslip containing cultured neurons in the dish and incubate 45 min at 37°C. Rinse twice with 1 ml intracellular buffer.

2. Mount coverslip in the recording chamber in intracellular buffer containing 1 µM antimycin, 1 mM 2-DG, 750 nM FCCP, 5 µM nigericin, and 1 mM EGTA. Add new solutions directly to the chamber and mix gently by triturating with a pipet to avoid damaging fragile neurons.

3. Add 10 µl 1 mM Ca^{2+} ionophore per milliliter intracellular buffer (final 10 µM) and allow the signal to reach equilibrium, which should be a ratio somewhat lower than the starting value. Record this ratio (R_{min}).

4. Add sufficient 140 mM $CaCl_2$ to raise the Ca^{2+} to 5 mM. Add additional aliquots of 10 µM ionophore until the ratio stops increasing. Record this ratio (R_{max}).

   This approach sometimes does not result in dye saturation, as evidenced by higher experimental ratio values than R_{max} values determined using this calibration method. If this occurs, the addition of a high concentration of glutamate (~100 µM with 30 µM glycine) will add an additional Ca^{2+} load and may raise the R_{max} determined from this calibration approach.

5. Move chamber to a cell-free area and measure background at each wavelength. Subtract background from signal. Repeat in several cells to obtain mean values of R_{min}, R_{max}, and β.

   β is the ratio of the fluorescence intensity of the free form of the dye at low and saturating ion concentrations (i.e., S_{f2}/S_{b2}; Grynkiewicz et al., 1985). K is determined by $\beta \times K_d$, where the latter value is taken from the literature (see Table 2.13.1).

6. Use these values to convert experimentally determined ratios (see Basic Protocol) to $[Ca^{2+}]_i$ using the equation in Support Protocol 1, step 3.

SUPPORT PROTOCOL 3

IN SITU CALIBRATION OF Na^+ DYES

The principal dye for measuring $[Na^+]_i$ is SBFI, for which in situ approaches appear to be necessary (Harootunian et al., 1989). SBFI calibration is performed as described for in situ calibration of Ca^{2+}-sensitive dyes (see Support Protocol 2), except that it is usually feasible to calibrate the dye at several values of $[Na^+]_i$.

Materials (see APPENDIX 1 for items with ✓)

✓ HEPES buffer
1 mM SBFI acetoxymethyl ester (SBFI AM; Molecular Probes) in dimethyl sulfoxide (DMSO)
BSA
Primary culture of neurons prepared on no. 1 glass coverslips

✓ Na$^+$-free HEPES buffer
  1 mM gramicidin D in DMSO
  HEPES buffer containing 5, 10, 50, 100, and 140 mM Na$^+$ (prepare by mixing HEPES buffer with Na$^+$-free HEPES buffer)

1. Supplement 1 ml HEPES buffer with 5 μM SBFI-AM and 5 mg/ml BSA, and add to a 35-mm tissue culture dish. Place a coverslip containing cultured neurons in the dish and incubate 60 min at 37°C.

2. Rinse cells twice with 1 ml HEPES buffer. Mount coverslip in a recording chamber and perfuse with Na$^+$-free HEPES buffer for 10 min.

3. Add 5 μM gramicidin D to the perfusion solution. Sequentially perfuse the cell with HEPES buffer containing increasing Na$^+$. Measure R for each solution when the signal reaches a new equilibrium. Continue until there is no further increase in R. Move the coverslip to a neuron-free region and measure background.

4. Subtract background ratios from ratios at each dye concentration. Plot against [Na$^+$] and fit the curve to derive R_{min}, R_{max}, and K. Repeat in several cells to obtain a mean value.

   β is the ratio of the fluorescence intensity of the free form of the dye at low and saturating ion concentrations (i.e., S_{f2}/S_{b2}; Grynkiewicz et al., 1985). K is determined by $\beta \times K_d$, where the latter value is taken from the literature (see Table 2.13.1).

5. Use these values to convert experimentally determined ratios to ion concentrations (see Support Protocol 1, step 3).

SUPPORT PROTOCOL 4

CELL-FREE CALIBRATION OF Mg^{2+} DYES

Magfura-2 can be calibrated by the cell-free method, as described for Fura-2 (see Support Protocol 1). In this case, intracellular buffer (see APPENDIX 1) containing 0.1 to 50 mM MgCl$_2$ is used for the calibration solution, and the pentapotassium salt of the dye is used from a 1 mM aqueous solution. Additionally, it is usually necessary to add 20 μM EGTA to these calibration solutions to prevent contamination of the Magfura-2 signal with Ca^{2+}. However, it is not necessary to buffer the free Mg^{2+} concentration at the millimolar concentrations used in this approach. Measuring ratios at a range of Mg^{2+} concentrations is followed by curve fitting to generate values for R_{min}, R_{max}, and K, which are then used for calculating free Mg^{2+} concentrations. Using cell-based approaches to calibrate Magindo-1 is impractical because of the difficulty in loading high millimolar concentrations of Mg^{2+} into neurons.

ALTERNATE PROTOCOL

CALIBRATION AND MEASUREMENT USING NONRATIOMETRIC [Ca^{2+}]$_i$ Dyes

Dye loading and data collection are performed as for ratiometric dyes (see Basic Protocol). However, because the dye intensity is related to both ion and dye concentrations, and because the dye concentration will vary from cell to cell, calibrating responses in each cell is the only reliable way of converting these data to ion concentrations.

Although nonratiometric dyes require more laborious calibration procedures, they are otherwise simple to use and do not require wavelength selectors or multiple detectors. The most widely used nonratiometric dyes are used for [Ca^{2+}]$_i$ measurements, and include Fluo-3 and various forms of calcium green (see Table 2.13.1). Fluo-3 has proven to be useful because the fluorescence intensity increases up to 40-fold upon saturation with Ca^{2+}. In addition, these dyes do not require UV optics, and are quite widely used in laser-based confocal microscopy applications.

Materials (also see Basic Protocol; see APPENDIX 1 for items with ✓)

✓ Ca^{2+}-free HEPES buffer, containing 1 mM EGTA
1 mM Ca^{2+} ionophore (ionomycin or 4 Br-A23187; Molecular Probes) in dimethyl sulfoxide (DMSO)
100 mM $CaCl_2$ in H_2O

1. Load cells and perform experiment as for ratiometric Ca^{2+}-sensitive dyes (see Basic Protocol, steps 1 to 5).

2. At the end of the experiment, perfuse the chamber with Ca^{2+}-free HEPES buffer containing 1 mM EGTA. Stop the perfusion when the solution change is complete. Add 10 μl of 1 mM Ca^{2+} ionophore per ml of chamber volume (10 μM final) and allow the fluorescence intensity to stabilize. Record fluorescence (F_{min}).

   Neurons tend to become fragile when challenged with ionophores. Stopping the perfusion will help prevent dislodging the cell(s) under study. Also, adding the ionophore directly to the chamber will reduce the amount of ionophore needed, and will avoid contaminating the perfusion system with an agent that could adversely impact experiments.

3. Add a sufficient volume of 100 mM $CaCl_2$ to the recording chamber to bring the concentration to 5 mM. Add ionophore in 10 μM increments until the fluorescence intensity has reached a stable maximum. Record fluorescence (F_{max}). Record background fluorescence from a region of the coverslip that does not contain neurons, and discard the coverslip.

4. Perform data analysis offline. Subtract background fluorescence from each fluorescence value. Determine $[Ca^{2+}]_i$ from fluorescence by the equation:

$$[Ca^{2+}]_i = K_d \times \frac{F - F_{min}}{F_{max} - F}$$

where K_d is the affinity of the dye (see Table 2.13.1).

Reference: Mason, 1993

Contributor: Ian J. Reynolds

UNIT 2.14

Measurement of Second Messengers in Signal Transduction: cAMP and Inositol Phosphates

BASIC PROTOCOL 1

DETERMINATION OF CYCLIC AMP CONCENTRATION IN CULTURED CELLS

Materials (see APPENDIX 1 for items with ✓)

Adherent or suspension cells
✓ Incubation buffer, 37°C

Incubation buffer containing 1 mM isobutylmethylxanthine (IBMX): prepared using 100 mM IBMX in DMSO (IBMX stock stored at −20°C), 37°C
Test agents
0.2 N HCl
10 and 5 N NaOH
10 mM cAMP in H_2O (store in small aliquots at −20°C)

✓ TE cAMP buffer

✓ ~60 μg/ml protein kinase A (PKA; a cAMP-dependent protein kinase)
0.5 μCi/ml [^{3}H]cAMP (50 Ci/mmol) in TE cAMP buffer (prepare 10 ml immediately before use; keep at 4°C; total [^{3}H]cAMP 5 μCi)

✓ Charcoal suspension

24- or 48-well microtiter plates
12 × 75–mm glass tubes

1a. *For adherent cells:* The day before the assay, plate cells at 80% confluency in 24- or 48-well microtiter plates. Before beginning the assay, aspirate culture medium and wash each well twice with incubation buffer warmed to 37°C.

1b. *For nonadherent cells:* The day of the assay, add 500 μl cell suspension, containing $0.5–1 \times 10^6$ cells, to each well of a 24- or 48-well microtiter plate.

2. Add 500 μl prewarmed incubation medium containing 1 mM IBMX to each well. Incubate 10 min at 37°C.

3. Add agents to be tested to each well; prepare triplicate wells for each agent or condition to be tested and include control wells (incubated without drug additions). Incubate 30 min at 37°C or as otherwise determined to be appropriate.

4. Aspirate medium and add 500 μl of 0.2 N HCl to each well to lyse the cells. Incubate 30 min at room temperature.

5. Transfer lysates to microcentrifuge tubes and microcentrifuge 2 min at top speed, room temperature. Neutralize the clarified lysates with 10 μl of 10 N NaOH.

6. From 10 mM cAMP stock, prepare 1 ml of 320 nM cAMP in TE cAMP buffer in a 12 × 75–mm glass tube. Prepare six serial 1:1 dilutions from the 320 nM cAMP solution by mixing 200 μl of each dilution with 200 μl of TE cAMP buffer. Label tubes 5 nM, 10 nM, 20 nM, 40 nM, 80 nM, 160 nM, and 320 nM cAMP.

7. Label duplicate microcentrifuge tubes as follows: B (blank), 0, 0.25 pmol, 0.5 pmol, 1 pmol, 2 pmol, 4 pmol, 8 pmol, and 16 pmol. To tubes labeled B and 0, add 50 μl TE cAMP buffer. To the remaining tubes labeled 0.25 pmol, 0.5 pmol, 1 pmol, 2 pmol, 4 pmol, 8 pmol, and 16 pmol, add 50 μl of 5 nM, 10 nM, 20 nM, 40 nM, 80 nM, 160 nM, and 320 nM cAMP, respectively.

8. Add the following:

    a. 50 μl neutralized lysates (from step 5) to each tube.

    b. 100 μl TE cAMP buffer to blank samples (B). To all other samples add 100 μl of ~60 μg/ml PKA.

    c. 50 μl [^{3}H]cAMP to all samples.

    Mix briefly and incubate ≥2 hr at 4°C, allowing the [^{3}H]cAMP and the lysate cAMP to compete for binding to PKA, a cAMP-dependent protein kinase.

9. Add 100 μl charcoal suspension (to remove free cAMP) chilled to 4°C. Vortex and set on ice for 1 min. Spin in microcentrifuge 2 min at top speed. Transfer 200 μl of the

supernatant to a scintillation vial, add scintillation fluid, and quantify radioactivity in a scintillation counter.

10. Add 200 μl of 5 N NaOH to microtiter plate wells (step 5) to dissolve protein adhering to plate. Determine protein concentration for each sample.

11. Construct a calibration curve (with a negative slope) of cAMP concentrations using the formula:

$$[^3H]cAMP\ (cpm) = a \times \log[\text{standard concentration}] + b$$

12. Calculate the concentration of cAMP for experimental samples by extrapolation from the calibration curve. Express the results as pmol cAMP (per well)/μg protein (per well).

BASIC PROTOCOL 2

MEASUREMENT OF [^{3}H]CYCLIC AMP ACCUMULATION IN SYNAPTONEUROSOMES AFTER PRELABELING WITH [^{3}H]ADENINE

Materials (see APPENDIX 1 for items with ✓)

    2 Hartley guinea pigs (male, ~200 g) or rats (male, ~150 g)
✓ Krebs-Henseleit (KH) buffer, bubbled with 95% O_2/5% CO_2 immediately before use and kept on ice
    KH buffer containing 10 μg/ml adenosine deaminase and 1 μM [^{3}H]adenine (10 to 15 Ci/mmol)
    KH buffer containing 10 μg/ml adenosine deaminase (no label)
    0.15 mM cAMP in 5.6% (v/v) trichloroacetic acid (TCA)
    Cation-exchange resin (AG 50 W-4X 200–400 mesh, H^+ form; Bio-Rad)
    0.2 M NaOH
    0.2 M HCl
    Dry alumina
    0.1 M imidazole·HCl, pH 7.3

    Glass/glass homogenizer
    Test tubes prewarmed to 37°C
    Poly-Prep columns (Bio-Rad; 0.8 × 4 cm, 2-ml bed volume)
    Scintillation fluid and vials

1. Prepare nonfiltered synaptoneurosomes (~20 samples) as follows (for greater detail see Hollingsworth et al., 1985): Dissect the cerebral cortex from two guinea pigs (or rats) and place in ice-cold Krebs-Henseleit (KH) buffer bubbled with 95% O_2/5% CO_2. Homogenize tissues in a glass/glass homogenizer using 7 ml of cold KH buffer for each ~1 g of tissue. Add another 28 to 30 ml cold KH buffer and centrifuge 10 min at 1000 × g, 4°C.

2. Resuspend pellet in 10 ml ice-cold KH buffer containing 10 μg/ml adenosine deaminase and 1 μM [^{3}H]adenine. Incubate 30 min at 37°C while bubbling with 95% O_2/5% CO_2.

3. Centrifuge 10 min at 1000 × g, and resuspend pellet in 20 ml ice-cold KH buffer. Centrifuge again and resuspend the pellet in 10 ml of KH buffer containing 10 μg/ml adenosine deaminase. Incubate 30 min at 37°C while bubbling with 95% O_2/5% CO_2.

4. Distribute 1-ml aliquots of preincubated homogenate to prewarmed tubes (37°C). Incubate an additional 10 min at 37°C and add desired test agents. Incubate 10 min at 37°C.

5. Centrifuge samples 5 min at 12,000 × g, and discard supernatants. Terminate the reaction by adding 1.2 ml of 0.15 mM cAMP in 5.6% trichloroacetic acid to each tube and vortexing. Store TCA extracts at 4°C while the columns are prepared.

6. *Cation-exchange columns:* Prepare a 50% (v/v) slurry of cation-exchange resin in distilled water. Pour 3 ml into Poly-Prep columns. Prior to first use, perform three regeneration cycles on each column, with a single cycle consisting of sequential washes with 4 ml of 0.2 M NaOH, 4 ml H_2O, 6 ml of 0.2 M HCl, and 4 ml H_2O. Perform a single regeneration cycle before each subsequent use.

7. *Alumina columns:* Add 1.2 g dry alumina to Poly-Prep columns prefilled with H_2O, avoiding bubble formation. Do not let columns dry out. Before use, wash columns with 8 ml of 0.1 M imidazole·HCl, pH 7.3.

8. Centrifuge samples (from step 5) 5 min at 12,000 × g. Transfer 50 µl of supernatant to scintillation vials for determination of total radioactivity incorporated into the tissue. Add scintillation fluid and determine radioactivity using a scintillation counter.

9. Apply the remainder of the supernatant onto equilibrated cation-exchange columns. Discard flowthrough and wash column with 1 ml H_2O. Discard flowthrough and then wash with another 2 ml H_2O. Discard flowthrough. Place each cation-exchange column over an alumina column. Add 3 ml H_2O to each cation-exchange column, making certain that the eluate drips into the alumina columns.

10. Remove the cation-exchange columns. Wash each alumina column with 1 ml of 0.1 M imidazole·HCl and discard flowthrough. Place alumina columns over scintillation vials and elute [^{3}H]cAMP with 4 ml of 0.1 M imidazole·HCl. Transfer 1 ml of each eluate to another set of vials or test tubes.

11. Record absorbance of a UV blank containing 0.1 M imidazole·HCl at 258 nm. Also record triplicate absorbance measurements for 0.15 mM cAMP in 0.1 M imidazole·HCl (UV_{abs} standard). Measure the absorbance of the 1 ml of eluate prepared in step 10 (UV_{abs} sample).

12. Mix the remaining eluate from step 10 with scintillation fluid and analyze by liquid scintillation spectroscopy.

13. For each sample, record: total radioactivity incorporated (from step 8), recovery measured as cold cAMP UV absorbance (in step 11; UV_{abs} sample), and [^{3}H]cAMP (step 12).

$$\% \text{ conversion} = \frac{[^3H]cAMP \text{ (cpm)} \times UV_{abs} \text{ (standard)} \times 5.555}{\text{total (cpm)} \times UV_{abs} \text{ (sample)}}$$

The calculation described results in a % conversion value that represents the fraction of the radioactivity incorporated by the cells that has been processed into [^{3}H]cAMP. The conversion factor (5.555) corrects for the amount of [^{3}H]cAMP cpm counted: (3 out of 4 ml, or 1.33) × % (100)/total counts correction factor (50 µl out of 1.2 ml counted, or 24).

BASIC PROTOCOL 3

MEASUREMENT OF PHOSPHOINOSITIDE TURNOVER IN SYNAPTONEUROSOMES

Materials (*also see Basic Protocol 2; see* APPENDIX 1 *for items with* ✓)
    [^{3}H]Inositol (10 to 20 Ci/mmol)
    Nonfiltered synaptoneurosomes (see Basic Protocol 2, step 1)

- ✓ Krebs-Henseleit (KH) buffer, bubbled with 95% O_2/5% CO_2 immediately before use and kept on ice
- 200 mM LiCl
- 6% (v/v) trichloroacetic acid (TCA)
- Anion-exchange resin (AG 1-X8, formate form; Bio-Rad)
- 60 mM ammonium formate/5 mM $Na_2B_4O_7 \cdot 10\ H_2O$ (column washing solution; store at room temperature)
- 200 mM ammonium formate/100 mM formic acid (IP_1 elution buffer; store at room temperature)
- 400 mM ammonium formate/100 mM formic acid (IP_2 elution buffer; store at room temperature)
- 1 M ammonium formate/100 mM formic acid (IP_3 elution buffer; store at room temperature)
- ✓ Lipid cleaning solution (LCS)
- Chloroform

5-ml propylene tubes
1.5-ml microcentrifuge tubes
Poly-Prep columns (Bio-Rad; 0.8 × 4 cm, 2-ml bed volume)

1. Resuspend to the synaptoneurosome preparation in 12 to 14 ml KH buffer and add 200 μCi [^{3}H]inositol. Distribute 300 μl suspension into 5-ml propylene tubes. Incubate tubes 1 hr at 37°C with agitation to avoid sedimentation of the synaptoneurosomes.

2. Add 20 μl of 200 mM LiCl to each tube. Mix and incubate 10 min at room temperature. Add test agents and bring the final incubation volume to 400 μl with KH buffer. Incubate tubes 90 min at 37°C with agitation to prevent sedimentation of synaptoneurosomes.

3. Transfer the contents of each tube into 1.5-ml microcentrifuge tubes. Centrifuge 1.5 min at low speed. Remove supernatant and wash pellet with 1 ml fresh KH buffer. Centrifuge 1.5 min at low speed. Remove buffer and add 1 ml of 6% trichloroacetic acid (TCA). Vortex to disrupt the pellet. Centrifuge 2 min at high speed.

4. Prepare a slurry of 50 g anion-exchange resin in 50 ml distilled H_2O. Add 0.85 ml of slurry to Poly-Prep columns. Apply supernatant from step 3 to the top of columns without disturbing resin. Discard flowthrough and wash columns five times with 3 ml distilled H_2O and once with 8 ml column washing solution.

5. Place columns on top of scintillation vials and add two successive 1-ml aliquots of IP_1 elution buffer to each column. Collect the [^{3}H]IP_1 (inositol monophosphate) fraction. Add 4 ml scintillation fluid and quantify radioactivity.

6. *To analyze other inositol polyphosphates (optional):* Wash columns with 8 ml IP_1 elution buffer. Place columns on top of fresh scintillation vials and add two 1-ml aliquots of IP_2 elution buffer. Collect the resulting [^{3}H]IP_2 (inositol bisphosphate) fraction. Wash columns with another 8 ml IP_2 elution buffer. Collect the [^{3}H]IP_3 (inositol trisphosphate) fraction by eluting with two 1-ml aliquots of IP_3 elution buffer. Add scintillation fluid to each fraction and quantify as in step 5.

7. Process the TCA pellet from step 3 as follows: add 0.5 ml LCS and 0.5 ml chloroform. Vortex to disrupt pellet and shake tubes 10 min at room temperature. Centrifuge 15 to 20 sec at 5000 × g. Aspirate and discard aqueous upper phase, and transfer 200 μl of the organic phase to scintillation vials. Allow chloroform to evaporate at room temperature, add scintillation fluid, and quantify radioactivity as in step 5.

8. For each sample, record the cpm of [^{3}H]IP$_1$ (from step 6) and the cpm for the lipid fraction (from step 7). Calculate a turnover value as follows:

$$\text{turnover} = \frac{[^3\text{H}]\text{IP}_1 \text{ (cpm)}}{[^3\text{H}]\text{lipids (cpm)} + [^3\text{H}]\text{IP}_1 \text{ (cpm)}} \times 10,0000$$

The turnover value obtained represents the ability of the tissue to convert inositol incorporated into the phospholipid fraction into a water-soluble negatively charged form, IP$_1$. Under different stimuli, this process is mediated by the activation of phospholipase C.

Reference: Schulz and Mailman, 1984

Contributor: Fabian Gusovsky

UNIT 2.15

In Vivo Measurement of Blood-Brain Barrier Permeability

BASIC PROTOCOL 1

BLOOD-BRAIN BARRIER INFLUX MEASUREMENT: IN SITU BRAIN PERFUSION PROCEDURE FOR RATS

Materials *(see APPENDIX 1 for items with ✓)*

✓ Krebs-bicarbonate buffer, 37°C
  Perfusate solution: test compound(s) dissolved in Krebs-bicarbonate buffer, 37°C
  Laboratory rats, either gender, 200 to 350 g
  Anesthetic (as specified by the approved animal care protocol)
  Saline: 0.9% (w/v) NaCl (sterile)
  Tissue solubilizer (e.g., Soluene 350; Packard) or 2 N NaOH, for radiolabeled test compounds only

Vials (scintillation vials, 7 to 20 ml, with caps, can be used when working with radiolabeled compounds)
1-, 10-, 20-, and 50-ml syringes
22-G hypodermic needles with the beveled end cut off (see Support Protocol 1)
Warming pads (e.g., Deltaphase Isothermal Pad; Braintree Scientific), preheated to 37°C
Syringe infusion pump (e.g., Pump 22; Harvard Apparatus)
Animal fur shaver
Surgical instruments (e.g., Harvard Apparatus), including:
    4.5-in. (~11.4-cm) operating scissors, sharp or blunt
    Disposable scalpels
    4-in. microdissecting forceps, full or strongly curved
    5.5-in. Olsen-Hegar needle holder
    3.5-in. Hartman hemostatic forceps
    6-in. probe with eye (eye probe)
    3-in. Vannas spring scissors, straight
    Tissue forceps (straight) or bone rongeurs
    Inox no. 7 sharp-point forceps

Suture, USP size 4.0

Electrocautery (e.g., Change-A-Tip low-temperature power handle and disposable sterile tips; Aaron Medical Industries), 1100° to 1300°F

Microvessel arterial clamp (Kleinert-Kutz microvessel clip; blades 6 mm × 1 mm, curved; 11 mm in length; Pilling Weck Surgical)

Perfusion cannula (see Support Protocol 2)

PE50 polyethylene tubing, 0.58 mm (0.023 in.) i.d., 0.965 mm (0.038 in.) o.d. (e.g., Intramedic Clay Adams; Becton Dickinson Labware)

22-G stainless steel connectors (see Support Protocol 1)

Rodent guillotine

Filter paper

50°C water bath, optional

1. Label and record the individual weights of an appropriate number of empty vials (six vials for each rodent in this protocol) for collecting brain tissue (one vial for every discrete brain region per rodent).

2. Fill a 1-ml syringe with prewarmed Krebs-bicarbonate buffer and attach a 22-G hypodermic needle with the beveled end cut off. Keep the syringe and its contents warm by wrapping it in a preheated warming pad.

3. Fill a 10- or 20-ml syringe with prewarmed perfusate solution (containing the test compound) and attach a 22-G hypodermic needle with the beveled end cut off. Keep the syringe and its contents warm by wrapping it in the preheated warming pad.

4. Set up a syringe infusion pump within easy reach of the surgical area. Program the pump for the desired infusion rate (step 16).

5. Anesthetize a 200- to 350-g laboratory rat with an appropriate anesthetic as approved by the IACUC. Administer sufficient anesthetic (with or without supplemental doses) to maintain proper anesthesia for a surgical procedure of ≤1 hr (see *APPENDIX 2B*).

6. Place anesthetized rat upside down on its back on another warming pad with its head toward the surgeon and its tail away from the surgeon. Shave fur from its neck and shoulder areas using an animal fur shaver.

7. Midway between the trachea and esophagus and the shoulder, make an ~2-cm-long skin incision with the operating scissors or disposable scalpel, running parallel to the trachea and esophagus (Fig. 2.15.1). Using 4-in. curved microdissecting forceps, gently tease away the subcutaneous fat and muscle to expose the right common carotid artery (left CCA for left-handed investigators; Fig. 2.15.1).

8. Using the forceps, gently tease away the white nerve running along the length of the CCA. Expose the CCA to reveal the bifurcation of the external and internal carotid arteries (ECA and ICA; Fig. 2.15.2). If necessary, remove the hyoid bone using surgical shears to expose the ECA.

9. Using 4.0 suture (and the Olsen-Hegar needle holder and Hartman hemostatic forceps, if necessary), ligate shut the ECA rostrally (towards the snout) at the point where the ECA undergoes additional bifurcation (Fig. 2.15.3). Use an electrocautery to cauterize small arterial branches (superior thyroid and occipital arteries) that originate from the ICA and ECA (Figs. 2.15.2 and 2.15.3). If necessary, insert the curvature of the forceps under the occipital artery to lift it upwards slightly for cauterization.

10. Locate the bifurcation of the ICA where the pterygopalatine artery branches from the ICA (Fig. 2.15.3). Carefully cauterize the pterygopalatine artery. Using 4.0 suture, place a loose ligature around the CCA caudally (towards the tail) before the bifurcation into

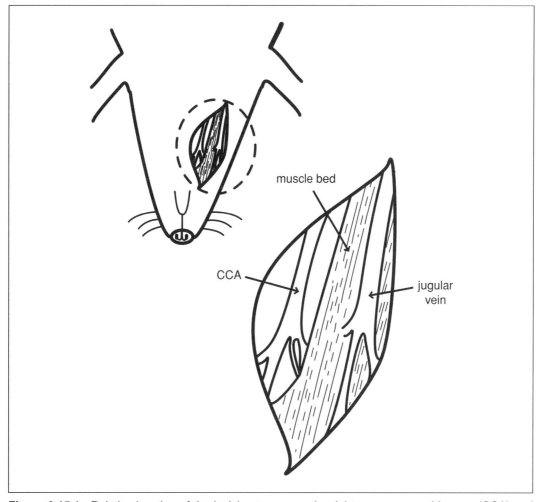

Figure 2.15.1 Relative location of the incision to expose the right common carotid artery (CCA) and jugular vein of a rat. The ventral view of a rat's head and shoulder area is shown with an enlarged view of the incision area.

the right ECA and ICA. Place the ligature at the point of the arrow identifying the right CCA (Fig. 2.15.3). Do not tighten the ligature yet.

Incomplete cauterization of the pterygopalatine is the major cause of surgical failure.

11. Place a microvessel arterial clamp around the ECA at the point where it branches from the CCA. Be certain that the clamp does not impede blood flow through the CCA.

12. Using the 1-ml syringe (step 2) connected to the PE50 tubing of the cannula using the cut 22-G needle, fill a perfusion cannula with Krebs-bicarbonate buffer. Catheterize the ECA with the buffer-filled perfusion cannula. Insert the cannula such that the tip points towards the caudal end. Carefully release the microvessel arterial clamp. Advance the tip of the cannula up to, but not past, the bifurcation point where the ECA branches off the CCA.

If the beveled cannula (cannula configuration 1) is used, a 25-G needle should be used to puncture and create a hole in the ECA. If the 25-G needle epoxied to PE50 tubing (cannula configuration 2) is used, no prior needle puncture is needed. Alternatively, the Vannas spring scissors can be used to make a small cut in the ECA.

Advancing past the bifurcation point may disturb cerebral blood flow.

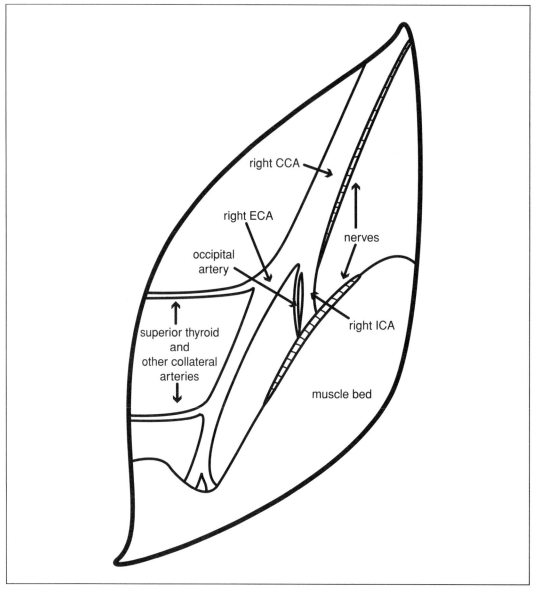

Figure 2.15.2 Detailed anatomy of the branches of the right CCA as viewed through the incision depicted in Figure 2.15.1. CCA, common carotid artery; ECA, external carotid artery; ICA, internal carotid artery.

13. Using 4.0 suture, tie one or two ligatures around the cannula to anchor it in place. Do not tie the ligatures so tightly as to occlude the cannula.

14. Place the 10- or 20-ml syringe containing the perfusate solution (step 3) in the infusion pump. Connect the syringe to a length of PE50 polyethylene tubing that can extend to the implanted perfusion cannula. Attach a 22-G stainless steel connector and fill the PE50 tubing with perfusate solution from the syringe.

15. Connect the free end of the fluid-filled PE50 tubing to the perfusion cannula. Tighten the caudal ligature around the CCA (step 10). *Quickly* open the chest cavity and sever the cardiac ventricles with a pair of 4.5-in. (∼11.4-cm) operating scissors. *Immediately* turn on the pump to start the perfusion.

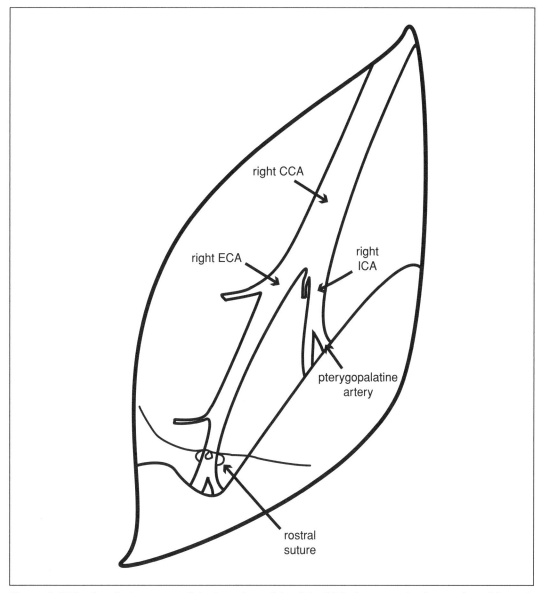

Figure 2.15.3 Detailed anatomy of the branches of the right CCA after cauterization as viewed through the incision depicted in Figure 2.15.1. The rostral suture of the ECA is also indicated. CCA, common carotid artery; ECA, external carotid artery; ICA, internal carotid artery.

16. Perfuse the cerebral hemisphere for 30 to 180 sec at a rate of 5 ml/min for <5 min (recommended; determined in pilot studies from the linear phase of the influx clearance versus time graph; see step 23). Retain a 0.25-ml aliquot of the perfusate for test compound concentration analysis, and store up to 1 month at −20°C.

17. Stop the perfusion and quickly decapitate the rat using a rodent guillotine. Using a pair of tissue forceps or bone rongeurs, remove the entire brain (forebrain and cerebellum) from the skull. Place the brain in a disposable petri plate containing a piece of filter paper moistened with sterile saline. Place the petri plate on ice so the brain dissection is performed under cold conditions.

18. Using a pair of Inox no. 7 sharp-point forceps, remove meningeal membranes and associated surface blood vessels. Grossly dissect the cerebellum and the left (unperfused;

pinkish) hemisphere from the right (perfused; blanched or whitish) hemisphere. Set aside the cerebellum and left hemisphere for proper disposal.

19. Grossly dissect the right hemisphere into anatomically defined brain regions: frontal cortex, occipital cortex, posterior cortex, striatum, hypothalamus and thalamus, and hippocampus (for details see Ohno et al., 1978).

20. Use a lint-free tissue to wick away any excess adherent moisture from the brain regions. Place each piece of brain tissue in a preweighed vial (step 1) and determine the weight of the tissue. Store up to 1 month at −20°C.

21. Analyze the perfusate aliquot (step 16) using an appropriate test compound analysis protocol, e.g., gas chromatography–mass spectrometry (GC/MS) or high performance liquid chromatography–mass spectrometry (HPLC/MS).

22a. *For radiolabeled compounds in tissue:* Digest brain tissue with an appropriate amount of a tissue solubilizer as specified by the manufacturer's instructions. Alternatively, digest tissue with 1 to 2.5 ml (5 vol per gram of tissue) of 2 N NaOH and incubate 3 hr in a water bath at 50°C. Vortex vigorously. Add an appropriate volume of scintillation fluid that is chemically compatible (e.g., Hionic Fluor; Packard Instruments, if using NaOH) with the tissue solubilizer so the sample forms a clear solution (with no emulsion). Quantify the radioactivity using an appropriate counting protocol in a scintillation counter.

22b. *For nonradiolabeled compounds in tissue:* Follow the tissue extraction and analysis procedures as specified by the test compound analysis protocol.

23. Calculate influx clearances, corrected for test compound present in the cerebrovascular blood volume, using the equation:

$$\text{Cl}_{in} = \frac{q_{tot} - (V_{vasc} \times C_{pf})}{T \times C_{pf}}$$

where Cl_{in} represents the influx transfer clearance (ml/min/g wet brain weight), q_{tot} is the observed total amount of test compound in the brain region (g compound/g wet brain weight) and represents the observed test compound entrapped within the parenchyma and cerebrovasculature, V_{vasc} represents the vascular blood volume (ml/g wet brain weight), C_{pf} is the compound perfusate concentration (g compound/ml), and T is the time (sec or min) of perfusion.

Influx clearance is a parameter that measures test compound penetration across the BBB into brain parenchyma (Takasato et al., 1984; Smith et al., 1992; Adkison and Shen, 1996). Equation 2.15.1 assumes negligible efflux (which may occur by passive diffusion or efflux transporters) and no metabolism during the short duration of the perfusion. Literature values of V_{vasc} may be used (e.g., see Smith et al., 1988, for comprehensive tabular data on inulin regional distribution volumes in rat brain). Alternatively, V_{vasc} is determined by perfusion with an intravascularly retained marker (such as [^{3}H]inulin; see Support Protocol 3).

If performing pilot studies to determine the appropriate length of time for perfusion, graph the Cl_{in} versus time of perfusion. Select the duration of perfusion from the linear portion of this graph, which typically occurs between 0 and 180 sec of perfusion.

24. Calculate the apparent cerebrovascular permeability-surface area product (PA) using the equation:

$$PA = -F[\ln(1 - \frac{\text{Cl}_{in}}{F})]$$

where F is cerebral perfusion fluid flow.

PA is a parameter that takes into account the effects of cerebral perfusion fluid flow. F is an important factor for highly lipophilic compounds where BBB permeability is rate limited by perfusion fluid flow (see Smith, 1989, for an excellent theoretical and practical overview). It is calculated from the brain uptake of [3H]diazepam and is $\sim 5 \times 10^{-2}$ ml/sec/g or 3 ml/min/g (Takasato et al., 1984; Smith et al., 1992) for an externally determined perfusion rate of 5 ml/min. Alternatively, F may be experimentally determined (see Support Protocol 4).

Equation 2.15.2 assumes that the test compound does not bind to proteins contained in the perfusion fluid. If the perfusion fluid contains albumin or other compound-binding protein components, then PA should be corrected for the unbound fraction of test compound (f_u) that is available for BBB permeation. In this case, the correction requires multiplying the right-hand side of Equation 2.15.2 by $1/f_u$ to obtain PA′, as shown in the equation below. In the case of an albumin- or protein-free perfusate, $f_u = 1$ and PA = PA′.

$$PA' = -(F/f_u)[\ln(1 - \frac{Cl_{in}}{F})]$$

ALTERNATE PROTOCOL

BLOOD-BRAIN BARRIER INFLUX MEASUREMENT: SIMPLIFIED IN SITU BRAIN PERFUSION FOR RATS

1. Perform the experimental setup and animal preparation as described above (see Basic Protocol 1, steps 1 to 6), but use a 10-, 20-, or 50-ml syringe for the perfusate solution in step 3.

2. Perform the surgical procedure as in Basic Protocol 1, steps 7 to 9. Do not cauterize the pterygopalatine artery, but suture the CCA as in step 10.

3. Proceed as in Basic Protocol 1, steps 12 and 13, but cannulate the CCA (instead of the external carotid artery) for perfusion. Insert the cannula such that the tip points rostrally (towards the head). Do not advance the tip past the bifurcation point where the ECA branches off the CCA.

4. Perfuse the cerebral hemisphere (Basic Protocol 1, steps 14 to 16), but perfuse for 20 to 180 sec at a rate of 10 ml/min using a 10-, 20-, or 50-ml syringe for the perfusate.

5. Harvest tissue and analyze the tissue and perfusate for the test compound (steps 17 to 24).

BASIC PROTOCOL 2

BLOOD-BRAIN BARRIER INFLUX MEASUREMENT: IN SITU BRAIN PERFUSION PROCEDURE FOR MICE

Materials *(see APPENDIX 1 for items with ✓)*

✓ Krebs-bicarbonate buffer, 37°C
  Perfusate solution: test compound(s) dissolved in Krebs-bicarbonate buffer, 37°C
  Laboratory mice (adult weight, either gender)
  Anesthetic (as specified by the approved animal care protocol; APPENDIX 2B)

Vials (scintillation vials, 7 to 20 ml, with caps, can be used when working with radiolabeled compounds)
1- and 10-ml syringes
22-G hypodermic needles with the beveled end cut off (see Support Protocol 1)

Warming pads (e.g., Deltaphase Isothermal Pad; Braintree Scientific), preheated to 37°C
Syringe infusion pump (e.g., Pump 22; Harvard Apparatus)
Animal fur shaver
Surgical instruments (e.g., Harvard Apparatus; see Basic Protocol 1)
    4.5-in. (∼11.4-cm) operating scissors, sharp or blunt
    Disposable scalpels
    4-in. microdissecting forceps, full or strongly curved
    5.5-in. Olsen-Hegar needle holder
    3.5-in. Hartman hemostatic forceps
    Tissue forceps (straight) or bone rongeurs
    Inox no. 7 sharp-point forceps
Suture, USP size 4.0
Perfusion cannula configuration 1 or 2 (see Support Protocol 2), but made with a 26-G (not 25-G) hypodermic needle
PE50 polyethylene tubing, 0.58 mm (0.023 in.) i.d., 0.965 mm (0.038 in.) o.d. (Intramedic Clay Adams; Becton Dickinson Labware)
22-G stainless steel connectors (see Support Protocol 1)
Sharp surgical shears or rodent guillotine

1. Label and record the individual weights of an appropriate number of empty vials (six vials for each rodent in this protocol) for collecting brain tissue (one vial for every discrete brain region per rodent).

2. Fill a 1-ml syringe with prewarmed Krebs-bicarbonate buffer and attach a 22-G hypodermic needle with the beveled end cut off. Keep the syringe and its contents warm by wrapping it in a preheated warming pad.

3. Fill a 10-ml syringe with prewarmed perfusate solution and attach a 22-G hypodermic needle with the beveled end cut off. Keep the syringe and its contents warm by wrapping it in the preheated warming pad.

4. Set up a syringe infusion pump within easy reach of the surgical area. Program the pump for an infusion rate of 2.5 ml/min.

5. Anesthetize a laboratory mouse with an appropriate anesthetic as approved by the IACUC. Administer sufficient anesthetic (with or without supplemental doses) to maintain proper anesthesia for a surgical procedure of ≤1 hr.

6. Place anesthetized mouse upside down on its back on another warming pad with its head towards the surgeon and its tail away from the surgeon. Shave fur from its neck and shoulder areas using an animal fur shaver.

7. Midway between the trachea and esophagus and the shoulder, make an ∼1- to 2-cm-long skin incision with the operating scissors or disposable scalpel, running parallel to the trachea and esophagus (Fig. 2.15.1).

8. Using a pair of 4-in. curved microdissecting forceps, gently tease away subcutaneous fat and muscle to expose the right CCA (left CCA for left-handed investigators) and the bifurcation of the right ECA and ICA (Fig. 2.15.2).

9. Using a 4.0 suture (and the Olsen-Hegar needle holder and Hartman hemostatic forceps, if necessary), ligate shut the ECA rostrally (towards the snout) after the bifurcation of the CCA into the right ECA and ICA (see Fig. 2.15.3 for anatomical reference).

   It is not necessary to ligate the pterygopalatine artery in the mouse.

10. Using 4.0 suture, place a loose ligature around the CCA caudally (towards the tail) before the bifurcation of the right ECA and ICA (at the point of the arrow identifying the right CCA; Fig. 2.15.3). Do not tighten the ligature yet.

11. Using the 1-ml syringe (step 2) connected to the PE50 tubing of the cannula using the cut 22-G needle, fill a perfusion cannula with Krebs-bicarbonate buffer.

12. Place the 10-ml syringe containing the perfusate solution (step 3) in the infusion pump. Connect the syringe to a length of PE50 polyethylene tubing that can extend to the implanted perfusion cannula. Attach a 22-G stainless steel connector and fill the PE50 tubing with perfusate solution from the syringe. Connect the free end of the fluid-filled PE50 tubing to the perfusion cannula.

13. Catheterize the CCA caudally (before the bifurcation) with the buffer-filled perfusion cannula (with the tip of the cannula pointing towards the head). Using 4.0 suture, tie one or two ligatures around the cannula to anchor it in place. Do not tie the ligatures so tightly as to occlude the cannula.

14. *Quickly* open the chest cavity and sever the cardiac ventricles with a pair of 4.5-in. operating scissors. *Immediately* turn on the pump to start the perfusion.

15. Perfuse the cerebral hemisphere for 30 to 180 sec at a rate of 2.5 ml/min for <5 min (recommended; determined in pilot studies from the linear phase of the influx clearance versus time graph; see Basic Protocol 1, step 23). Retain a 0.25-ml aliquot of the perfusate for test compound concentration analysis and store up to 1 month at $-20°C$.

16. Stop the perfusion and quickly decapitate the mouse using a pair of sharp surgical shears or a rodent guillotine. Harvest tissue and analyze as described for the rat (see Basic Protocol 1, steps 17 to 24).

SUPPORT PROTOCOL 1

CONSTRUCTION OF PE50-STAINLESS STEEL CONNECTORS AND CUT 22-G HYPODERMIC NEEDLES

Materials

Metal file
22-G stainless steel disposable hypodermic needles
5.5-in. (14-cm) Olsen-Hegar needle holder (e.g., Harvard Apparatus)
Hartman hemostatic forceps, 3.5 in.

1. Using a metal file, score the stainless steel tubing near the beveled point of a 22-G stainless steel disposable hypodermic needle, being careful not to file all the way through the tubing.

2. Holding the luer-lock end of the needle with a pair of 5.5-in. Olsen-Hegar needle holders and the other end of the needle with Hartman hemostatic forceps, snap the needle cleanly into two pieces along the score mark without bending the needle.

3. *Optional:* Construct a PE50-stainless steel connector by repeating the process at the end near the plastic luer-lock hub.

4. If necessary, file off any burrs left on the cut end(s) of the tubing.

SUPPORT PROTOCOL 2

CONSTRUCTION OF IN SITU PERFUSION CANNULAS

Configuration 1 Cannulas

Materials

> PE50 polyethylene tubing, 0.58 mm (0.023 in.) i.d., 0.965 mm (0.038 in.) o.d. (e.g., Intramedic Clay Adams; Becton Dickinson Labware)
> Forceps and scissors
> Boiling water bath

1a. Cut a 15-cm length of PE50 polyethylene tubing. Using a pair of forceps, immerse one end of the PE50 tubing into a beaker of boiling water for 10 to 30 sec until tubing is pliable.

2a. Remove the immersed end and quickly stretch the tubing to elongate it to create a narrower bore for easier catheterization. Allow tubing to cool.

3a. Using a pair of scissors, cut the elongated length of the PE50 tubing at an ∼30° angle to give the tubing a sharp bevel.

Configuration 2 Cannulas

Materials

> 5.5-in. (14-cm) Olsen-Hegar needle holder (e.g., Harvard Apparatus)
> Metal file
> 20- and 25- (for rats) or 26-G (for mice), 1-in. (2.5-cm) stainless steel hypodermic needles
> PE50 polyethylene tubing, 0.58 mm (0.023 in.) i.d., 0.965 mm (0.038 in.) o.d. (e.g., Intramedic Clay Adams; Becton Dickinson Labware)
> Epoxy adhesive

1b. Using a 5.5-in. Olsen-Hegar needle holder and a metal file, cut a 25- or 26-G 1-in. stainless steel hypodermic needle from the plastic luer-lock syringe hub (see Support Protocol 1). Discard the plastic hub.

2b. Using the needle holder and metal file, cut a 20-G hypodermic needle from the plastic luer-lock syringe hub. Retain the plastic hub. Alternatively, remove the epoxy cementing the needle to the hub using a razor blade, and then extricate the entire needle from the plastic, luer-lock syringe hub.

3b. Cut a 15-cm length of PE50 polyethylene tubing. Insert the cut end of the 25- or 26-G needle into the PE50 tubing, leaving 1.5 cm of the needle exposed. Glue the needle to the tubing using epoxy adhesive.

4b. Slide the 20-G plastic luer-lock syringe hub over the needle-tubing connection. Use epoxy adhesive to affix the plastic hub to the tubing and allow the epoxy to cure overnight.

SUPPORT PROTOCOL 3

EXPERIMENTAL DETERMINATION OF INTRAVASCULAR CAPILLARY VOLUME USING RADIOLABELED INULIN

Calculation of the mass of the test compound in the cerebrovasculature is based on knowledge of the concentration of test compound in the perfusate and the volume of the cerebrovascular bed, V_{vasc}. To experimentally determine V_{vasc} for animal models other than healthy, adult male rats, an in situ brain perfusion using the nonpenetrable BBB tracers [^{3}H]inulin or [^{14}C]methoxyinulin will provide good measurements of cerebrovasculature capillary bed volume.

Additional Materials (*also see Basic Protocols 1 and 2*)

Perfusate solution: 0.5 µCi/ml [^{3}H]inulin (American Radiolabeled Chemicals) or 0.3 µCi/ml [^{14}C]methoxyinulin (American Radiolabeled Chemicals) dissolved in Krebs-bicarbonate buffer (see APPENDIX 1), 37°C

1. Perform in situ brain perfusion of rats (see Basic Protocol 1, steps 1 to 22b) or mice (see Basic Protocol 2, steps 1 to 16), using a perfusate solution containing [^{3}H]inulin or [^{14}C]methoxyinulin.

2. Calculate the vascular blood volume (the cerebrovascular capillary bed volume) using the following equation:

$$V_{vasc} = q_{tot} / C_{pf}$$

where V_{vasc} is the vascular blood volume (ml/g wet brain weight); q_{tot} is the observed total test compound amount in the brain region (g compound/g wet brain weight), representing the observed compound entrapped within the lumen of the cerebrovasculature; and C_{pf} is the [^{3}H]inulin or [^{14}C]methoxyinulin perfusate concentration (g/ml).

SUPPORT PROTOCOL 4

EXPERIMENTAL DETERMINATION OF CEREBRAL PERFUSION FLUID FLOW USING RADIOLABELED DIAZEPAM

Cerebral perfusion fluid flow (F) (which differs from the externally controlled perfusion rate) is an important factor for highly lipophilic compounds where BBB penetration is often limited by the rate of perfusion fluid flow (see Smith, 1989). F is determined experimentally by in situ perfusion with radiolabeled diazepam or from literature estimates. For adult male rats and an externally controlled perfusion rate of 5 ml/min, F is $\sim 5 \times 10^{-2}$ ml/sec/g or 3 ml/min/g (Takasato et al., 1984; Smith et al., 1992). Diazepam is highly lipophilic and penetrates across the BBB quite rapidly. Therefore, it is assumed that all observed brain tissue diazepam is localized in the brain parenchyma. No correction for intravascularly entrapped tracer is required.

Additional Materials (*also see Basic Protocols 1 and 2*)

Perfusate solution: 0.5 µCi/ml [^{3}H]diazepam (American Radiolabeled Chemicals) or 0.3 µCi/ml [^{14}C]diazepam (American Radiolabeled Chemicals) dissolved in Krebs-bicarbonate buffer (see APPENDIX 1), 37°C

1. Perform an in situ brain perfusion and tissue harvest for rats (see Basic Protocol 1, steps 1 to 22b) or mice (see Basic Protocol 2, steps 1 to 16), using a perfusate solution containing radiolabeled diazepam.

2. Calculate cerebral perfusion flow using the following equation:

$$F = \frac{q_{tot}}{T \times C_{pf}}$$

where F is the cerebral perfusion flow rate in ml/min/g wet brain weight, q_{tot} is the observed total test compound amount in the brain region (g compound/g wet brain weight), T is the total time of the externally controlled perfusion, and C_{pf} is the concentration of radiolabeled diazepam in the perfusate (g/ml).

SUPPORT PROTOCOL 5

USE OF THE IN SITU PERFUSION TECHNIQUE FOR MECHANISTIC ASSESSMENT OF CARRIER-MEDIATED OR SATURABLE TRANSPORT

The in situ brain perfusion technique readily lends itself to mechanistic studies of BBB transport (Smith et al., 1987; Takada et al., 1991; Stoll et al., 1993). Variations in test compound perfusate concentration will facilitate a discernment of saturable transport. Adding postulated competitive transport inhibitors to the perfusion fluid will assist in understanding the substrate specificity of the transport system.

Additional Materials (also see Basic Protocols 1 and 2)

Perfusate solution: variable concentrations of a test compound with or without a competitive inhibitor dissolved in Krebs-bicarbonate buffer (see APPENDIX 1), 37°C

1. Perform an in situ brain perfusion for rats (see Basic Protocol 1, steps 1 to 22b) or mice (see Basic Protocol 2, steps 1 to 16), with Krebs-bicarbonate buffer (without test compound) for 15 sec to flush the cerebrovasculature of blood-borne endogenous competitive substrates. Quickly switch to the perfusate solution (only one solution per animal) containing variable concentrations of a test compound with or without a competitive inhibitor.

2. Determine Cl_{in}, the influx transfer clearance (ml/min/g wet brain weight), using Equation 2.15.1. Then transform the observed Cl_{in} data to the mass transport influx parameter, J_{in} (g compound/min/g wet weight) using the equation:

$$J_{in} = Cl_{in} \times C_{pf}$$

where C_{pf} is the test compound perfusate concentration (g/ml). Use this equation when the BBB permeability mechanism is carrier-mediated with saturable transport.

When the BBB permeability mechanism is subject to competitive inhibition, the observed Cl_{in} data will decrease significantly compared to control (perfusate containing only the test compound).

3. Fit the resulting J_{in} data to the Michaelis-Menten equation:

$$J_{in} = \frac{V_{max} \times C_{pf}}{K_M + C_{pf}} + J_{in(ns)}$$

where V_{max} is the maximal rate of saturable transport (g compound/min/g wet brain weight), K_M is the dissociation constant of the transporter-test compound complex (g/ml), and $J_{in(ns)}$ represents mass transport influx due to diffusion.

If saturable transport is observed, a graph of J_{in} versus C_{pf} will be parabolic. If no saturable transport is observed, the graph will be linear.

BASIC PROTOCOL 3

BLOOD-BRAIN BARRIER INFLUX MEASUREMENT: INTRAVENOUS ADMINISTRATION/MULTIPLE TIME POINT PROCEDURE FOR RATS

NOTE: This procedure involves rodent survival surgery. Sterile surgical procedures must be employed, as specified by prevailing animal care regulations. All surgical instruments and consumable items used for surgery must be sterile.

Materials

Laboratory rats, either gender, 200 to 350 g
Anesthetic, antibiotic, and other required medications for rodent survival surgery and rapid sacrifice (as specified by the approved animal care protocol; APPENDIX 2B)
Betadine (povidone/iodine solution)
Saline: 0.9% (w/v) NaCl (sterile), 37° and 4°C
10 U/ml heparinized saline, sterile
✓ Krebs-bicarbonate buffer, containing test compounds, 37°C
Tissue solubilizer (e.g., Soluene 350; Packard) or 2 N NaOH, for radiolabeled test compounds only
Hydrogen peroxide, 30% (e.g., Sigma)

Warming pads (e.g., Deltaphase Isothermal Pad; Braintree Scientific), preheated to 37°C
Animal fur shaver
Surgical instruments (e.g., Harvard Apparatus), including:
    4.5-in. (~11.4-cm) operating scissors, sharp or blunt
    Scalpel
    4-in. (10-cm) microdissecting forceps, full or strongly curved
    3.5-in. Hartman hemostatic forceps
    6-in. (15-cm) eye probe
    3-in. (7.5-cm) Vannas spring scissors, straight
    Dumont no. 7 microdissecting forceps
    5.5-in. Olsen-Hegar needle holder
    Inox no. 7 sharp-point forceps
Sterile drapes
Sterile sutures with and without needle, USP size 2.0 and 4.0
Jugular vein catheter (see Support Protocol 6)
22-G hypodermic needles with the beveled end cut off (see Support Protocol 1)
1-, 3-, and 5-ml syringes
22-G stainless steel wire, sterile
Vials (scintillation vials, 7 to 20 ml, with caps, can be used when working with radiolabeled compounds)
Rodent guillotine
PE50 polyethylene tubing, 0.58 mm (0.023 in.) i.d., 0.965 mm (0.038 in.) o.d. (Intramedic Clay Adams; Becton Dickinson Labware)
22-G stainless steel connectors (see Support Protocol 1)
Tissue forceps or bone rongeurs
Filter paper
50°C water bath, optional
1.5-ml microcentrifuge tubes with caps

1. Anesthetize a 200- to 350-g laboratory rat with an appropriate anesthetic as approved by the IACUC. Administer sufficient anesthetic (with or without supplemental doses) to maintain proper anesthesia for a surgical procedure of ≤1 hr. Administer appropriate antibiotic to the animal, as approved by the IACUC.

2. Place anesthetized rat right side up (lying on its belly) on a warming pad. Shave the fur from its back, between the shoulder blades, using an animal fur shaver. Wash the surgical area with Betadine solution. Using a pair of 4.5-in. operating scissors, make a small incision (0.3 cm long) between or slightly ahead of the shoulder blades.

3. Reposition the rat so that it is upside down (lying on its back) on the warming pad with its head located towards the surgeon and its tail away from the surgeon. Shave the fur from its neck and shoulder areas. Wash surgical area with Betadine solution and cover area with sterile drapes, leaving only the surgical site exposed.

4. Using a scalpel, make a 2- to 3-cm-long incision (Fig. 2.15.1) over the skin site where pulsation of the right jugular vein (left jugular vein exposed for left-handed investigators) can be observed. Using a pair of 4-in. curved microdissecting forceps, bluntly separate the subcutaneous and muscle tissues to expose a section of the jugular vein (superficially located under the skin, purple-bluish in color, disappearing under the chest muscle towards the heart).

5. Using the forceps, carefully clean the vessel of connective tissue from the point of the chest muscle penetration (towards the heart) to a point where it bifurcates into two smaller vessels. Using a sterile 4.0 suture without needle (and Hartman hemostatic forceps, if necessary), ligate the vein shut at the point of the bifurcation (towards the snout).

6. Subcutaneously tunnel a 6-in. eye probe from the incision in the back (between the shoulder blades) to the incision in the neck and shoulder area. Insert a jugular vein catheter into the eye of the probe (Fig. 2.15.4A). Pull the probe out so that the catheter is

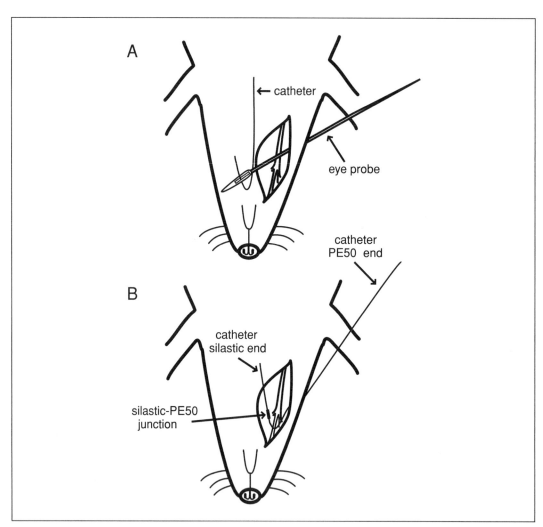

Figure 2.15.4 (**A**) Placement of the subcutaneously tunneled eye probe with the catheter threaded through the probe slit. (**B**) Orientation of the catheter in the jugular-vein incision for determination of the appropriate length for the silastic tubing end. In the final orientation of the catheter, the silastic tubing end is inserted into the jugular vein up to the point of the silastic-PE50 junction. The curved "U" of the catheter will lie comfortably within the incision site.

subcutaneously tunneled under the skin, such that the catheter end extends out from the incision on the back and the silastic tubing end extends from the jugular vein incision (Fig. 2.15.4B).

7. Position the catheter so that it lies comfortably in the jugular vein incision site. Cut the silastic tubing such that the free end is ~0.5 cm short of the midline of the front paws and the silastic-PE50 junction rests over the jugular vein (tubing typically no more than 2 to 3 cm in length; see Fig. 2.15.4B). Make a beveled cut at the end of the silastic tubing.

8. Connect the PE50 tubing of the catheter to a 22-G hypodermic needle with the beveled end cut off, affixed to a 3-ml syringe containing prewarmed sterile saline. Fill the catheter with sterile saline to eliminate air (important in avoiding an air embolism).

9. Place two or three more pieces of 4.0 suture under the jugular vein, towards the heart. Using a pair of 3-in. Vannas spring scissors, make a small, nicking incision in the jugular vein. Be careful not to cut through the vein.

10. Insert one point of a pair of Dumont no. 7 microdissecting forceps into the jugular vein incision. Slide the beveled end of the catheter along the probe into the incision. Continue inserting and guiding the catheter into the vein until the cut end of the vein covers the juncture of the silastic and PE50 tubing.

11. To test for catheter patency, inject 0.3 ml sterile saline and then slowly pull back on the syringe plunger to withdraw blood.

    If no blood is forthcoming, then slightly pull the catheter back out from the vein and test for catheter patency again. If still no blood is forthcoming, remove the catheter, trim down the length of the silastic tubing and re-implant.

12. Clear the catheter of blood by flushing with 0.5 ml sterile saline. Anchor the catheter to the vein by tying the 4.0 sutures (step 9) securely (but not too tightly) around the vein and the PE50 tubing (not the silastic tubing).

13. Suture the catheter to the muscle bed using an Olsen-Hegar needle holder and sterile 4.0 suture with needle. Close the internal incision by suturing the muscle and subcutaneous tissues with 4.0 sutures. Close the skin incision using a sterile 2.0 suture with needle. Alternatively, use wound clips or staples to close the incision.

14. Place anesthetized animal right side up (on its belly). Suture the exteriorized catheter to the skin and then close the incision on the back using 2.0 suture. Trim the exteriorized PE50 tubing so that no more than 2.5 to 3 cm are exposed. Plug the open end of the catheter with a sterile piece of 22-G stainless steel wire.

15. Apply topical antibiotics to all incision sites as directed by the approved animal care protocol and allow animal to recover from the anesthetic using approved postoperative care protocols. Slowly flush the jugular vein catheter with 0.5 to 1.0 ml sterile 10 U/ml heparinized saline each day to maintain patency. Allow at least 24 hr to elapse between the surgery and experimental study.

16. Label and record the individual weights of an appropriate number of empty vials (six vials for each rodent in this protocol) for collecting brain tissue (one vial for every discrete brain region per rodent). Label an appropriate number of 1.5-ml microcentrifuge tubes with caps for the number of desired blood samples per rat.

17. Fill a 1-, 3-, or 5-ml syringe (depending on the dosing volume) with prewarmed test compound solution for each animal to be studied. To each syringe attach a 22-G hypodermic needle with the beveled end cut off. Keep the syringe and its contents warm by wrapping it in a preheated warming pad.

18. Prepare an appropriate number of 1-ml syringes filled with 1 ml prewarmed sterile saline, e.g., for each rat, prepare one syringe for every time point of blood collection plus one syringe to be used after compound dosing (to flush the catheter) and one extra syringe (for extra measure).

19. Prepare an appropriate number of empty 1-ml syringes for blood collection, depending on the number of blood samples obtained per rat and the number of rats in the experimental group. If plasma is desired, rinse the syringes with sterile heparin solution (1000 to 5000 U/ml) and allow to dry.

20. Prepare sufficient medications for rapid sacrifice of the rat (as approved by the IACUC; e.g., an intravenous bolus dose of a fast acting barbiturate, followed by intravenous administration of isotonic potassium chloride). Set up a rodent guillotine for brain harvesting. Transfer the catheterized rat to an appropriate animal holding cage.

21. Connect one end of a 20-cm-long piece of PE50 tubing to a 1-ml saline-filled syringe using a 22-G hypodermic needle with the beveled end cut off. Fill tubing with sterile saline and connect the free end of the saline-filled PE50 tubing to the exteriorized jugular vein catheter using a 22-G stainless steel connector. Gently flush catheter with 0.3 ml saline.

22. Withdraw a predose blood sample (blank; time zero) by gently pulling back the syringe plunger so that the luer-lock needle hub is filled with blood. Attach a fresh syringe (step 19) to the luer-lock hub and collect a 0.1- to 0.15-ml blood sample (larger sample volumes will cause a reduction in hematocrit with potential alterations in BBB permeability pharmacokinetics).

23. Remove the sample syringe, transfer sample to the prepared vial (step 16), and reattach the first syringe (filled with saline and possibly mixed with blood) to the needle hub. Flush the PE50 line and refill it with 0.3 to 0.5 ml sterile saline. Replace the initial syringe with a fresh syringe containing sterile saline.

24. Attach the test compound dosing syringe (step 17) to the PE50 tubing. Administer the dose intravenously over 30 to 60 sec. Remove the dosing syringe and attach a saline-filled syringe to the PE50 line. Flush the residual dose intravenously with saline and refill the tubing with sterile saline. Retain a 0.25-ml aliquot of the test compound solution for later analysis. Store the aliquot up to 1 month at $-20°C$.

25. Withdraw blood samples at predetermined time intervals, e.g., 0 (predose, blank), 1, 2.5, 5, 10, 15, 20 min, and every 10 min thereafter for the duration of the study (study period is typically determined by the compound's preliminary pharmacokinetics). Withdraw the last blood sample ∼30 to 60 sec before sacrifice, quickly anesthetize the rat, and decapitate it.

26. Using a pair of tissue forceps or bone rongeurs, remove the entire brain (forebrain and cerebellum, and optionally brainstem) from the skull. Place the brain in a disposable petri plate containing a piece of filter paper moistened with sterile saline. Place the petri plate on ice so the brain dissection is performed under cold conditions.

27. Using a pair of Inox no. 7 sharp-point forceps, remove meningeal membranes and associated surface blood vessels. Grossly dissect the forebrain, cerebellum, and separate the left hemisphere from the right hemisphere. Grossly dissect both hemispheres into anatomically defined brain regions: frontal cortex, occipital cortex, posterior cortex, striatum, hypothalamus and thalamus, and hippocampus (see Ohno et al., 1978 for details).

28. Use a lint-free tissue to wick away any excess, adherent moisture from the brain regions. Place each piece of brain tissue (forebrain regions plus cerebellum and brainstem) in

a preweighed vial (step 16) and determine the weight of the brain tissue. Store up to 1 month at −20°C.

29. Analyze each blood sample, tissue sample, and dose aliquot (step 24) for the test compound as defined by the appropriate analysis protocol.

30a. *For radiolabeled test compounds in tissue:* Digest each tissue sample with an appropriate amount of a tissue solubilizer as specified by the manufacturer's instructions. Alternatively, digest tissue with 1 to 2.5 ml (5 vol per gram of tissue) of 2 N NaOH and incubate 3 hr at 50°C. Vortex vigorously. Add an appropriate volume of chemically compatible scintillation fluid (e.g., Hionic Fluor; Packard Instruments, if using NaOH). Sufficient scintillation fluid is added when the sample forms a clear solution (with no emulsion formation). Quantify the radioactivity using an appropriate counting protocol in a scintillation counter.

30b. *For nonradiolabeled test compounds in tissue:* Follow the tissue extraction and analysis procedures as specified by the test compound analysis protocol.

30c. *For radiolabeled test compounds in blood:* Transfer an appropriate volume of blood sample to a 20-ml scintillation vial with cap. For every 100 μl of whole blood, add 300 μl of 30% H_2O_2 to decolorize the sample for liquid scintillation counting. Cap the samples and incubate the sample overnight at room temperature, followed by an additional 2 hr in a 45°C water bath (to destroy any residual H_2O_2). Vortex vigorously. Add an appropriate volume of chemically compatible scintillation fluid so the sample forms a clear solution (with no emulsion).

30d. *For nonradiolabeled test compounds in blood:* Follow the blood extraction and analysis procedures as specified by the test compound analysis protocol.

31. Calculate the area under the curve (AUC) for the concentration of test compound in the blood versus time from time 0 to the time of sacrifice. The AUC is calculated by the linear trapezoidal rule:

$$\text{AUC} = \sum_{0}^{n} \frac{1}{2}(C_n + C_{n+1})(t_{n+1} - t_n)$$

where C_n is the nth blood sample test compound concentration at time t_n, C_{n+1} is the next successive concentration at time t_{n+1}, and t is the observed time of collection for the blood sample.

32. Using the brain and blood data, calculate an apparent unidirectional (influx) transfer coefficient, $K_{in(app)}$, for BBB penetration into brain parenchyma by linear regression analysis using the equation:

$$\frac{q_{\text{tot}}}{C_T} = K_{in(app)} \left[\frac{\int_0^T C\, dt}{C_T} \right] + V_{\text{vasc}}$$

where q_{tot} is the observed total test compound amount in the brain region (g compound/g wet brain weight) and represents the observed compound entrapped within the parenchyma and cerebrovasculature, C_T is the test compound blood concentration at the time of sacrifice (T), $\int_0^T C\, dt$ represents the AUC from time 0 to T (from step 31), and V_{vasc} represents the calculated vascular blood volume (ml/g wet brain weight).

Equation 2.15.9 is in the form of a straight line with a slope of $K_{in(app)}$ and a y intercept of V_{vasc} (Ohno et al., 1978; Rapoport et al., 1980; Smith et al., 1988; Smith, 1989). This mathematical analysis assumes that: (1) the test compound is not metabolized by either peripheral or brain tissues; (2) the series of studied rats have similar values of $K_{in(app)}$ and V_{vasc}; (3) the pharmacokinetic influx rate microconstant (k_{in}) is substantially greater than the pharmacokinetic efflux rate microconstant (k_{out}); and (4) radiotracer efflux from brain tissue is negligible at early times after dosing (i.e., there is unidirectional transfer of tracer from the cerebrovascular capillary bed into brain tissue). The validity of the unidirectional transfer assumption can be assessed by inspecting a plot of data derived from Equation 2.15.9 for curvilinearity; significant radiotracer efflux from brain tissue will produce data that systematically fall below those values predicted by the initial linear regression line). For comparative purposes for V_{vasc}, Smith et al. (1988) contains comprehensive tabular data on inulin regional brain distribution volumes.

SUPPORT PROTOCOL 6

CONSTRUCTION OF JUGULAR VEIN CATHETERS

CAUTION: Chloroform is a toxic irritant and mild carcinogen; take suitable precautions and work in a fume hood.

Materials

Chloroform
PE50 polyethylene tubing, 0.58 mm (0.023 in.) i.d., 0.965 mm (0.038 in.) o.d.
  (e.g., Intramedic Clay Adams; Becton Dickinson Labware)
Silastic tubing, 0.50 mm (0.020 in.) i.d., 0.965 mm (0.038 in.) o.d. (e.g., Silastic Medical Grade; Dow Corning)

1. Cut a 15-cm length of PE50 polyethylene tubing and a 5-cm length of silastic tubing. Immerse one end of the silastic tubing into a beaker of chloroform for ~20 to 60 sec. Quickly remove from the chloroform and slip over one end of the PE50 tubing. Allow chloroform to evaporate.

2. Wrap the PE50 tubing in a U shape around a glass Pasteur pipet so the U of the PE50 tubing is ~1 to 1.5 cm from the juncture of the silastic and PE50 tubing. Using one pair of forceps to hold the silastic end and another pair (or the experimenter's hand) for the PE50 end, immerse the Pasteur pipet into a beaker of boiling water to cover the U shape for 10 to 30 sec.

SUPPORT PROTOCOL 7

CAPILLARY DEPLETION METHOD

The protocols in this unit assume that binding or endocytosing of test compounds by cerebroendothelial cells is negligible. However, certain physiochemical properties may predispose the xenobiotic to cerebrocapillary binding and/or endocytosis. Additional experimental methods, e.g., a capillary depletion method (Triguero et al., 1990), may be performed to discriminate mechanistically between actual penetration across the BBB into brain parenchyma versus capillary binding or endocytosis. This method is performed on harvested brain tissue using any of the BBB influx protocols described in this unit (see Basic Protocols 1, 2, or 3 or the Alternate Protocol).

Additional Materials (*also see Basic Protocols 1, 2, and 3; see* APPENDIX 1 *for items with* ✓)

✓ Krebs-bicarbonate buffer, 4°C
26% (w/v) dextran solution (average MW 74,000), 4°C
Glass homogenizer with glass pestle, prechilled

1. Follow the procedures as described (see Basic Protocol 1, steps 1 to 20; Alternate Protocol; Basic Protocol 2, steps 1 to 16; or Basic Protocol 3, steps 1 to 28), but keep all brain tissues cold after harvesting and for the remainder of the protocol by placing all items on ice. For Basic Protocol 3, collect one blood sample immediately prior to sacrifice.

2. Transfer brain tissue to a prechilled glass homogenizer. Add 2 vol (based on tissue weight) ice-cold Krebs-bicarbonate buffer. Homogenize the tissue with eight to ten up-and-down strokes of the pestle. Add an equal volume of ice-cold 26% dextran solution. Homogenize the tissue with three additional strokes as rapidly as possible (within a minute or so).

3. Retain a 1-ml aliquot of the homogenate for test compound analysis and transfer remaining homogenate to a preweighed 50-ml tube. Store the aliquot up to 1 month at −20°C. Centrifuge homogenate 15 min at 5400 × g, 4°C, in a swinging-bucket rotor.

4. Carefully separate the supernatant (containing the test compound liberated from brain parenchyma and the cerebrovascular lumen) from the pellet (containing red blood cells, brain nuclei, and brain microvessels containing bound or endocytosed test compound) by pipetting.

5. Record the weight of the pellet fraction. Determine the concentration of test compound in the whole homogenate, supernatant, pellet, and perfusate aliquot as defined by the appropriate test compound analysis protocol.

6. Calculate the volume of distribution, V_d, a measure of test compound distribution in each of the homogenate, supernatant, and pellet fractions using the equation:

$$V_d = \frac{M_F}{C}$$

where M_F, the test compound mass in the measured fraction, represents the compound amount per gram of brain tissue in the aliquot of the homogenate, pellet, or supernatant fractions, and C is the compound concentration in the perfusate (g/ml) for Basic Protocols 1 and 2, or the compound concentration in the blood sample immediately prior to sacrifice for Basic Protocol 3.

The V_d of the pellet fraction represents the compound bound and/or endocytosed in the brain cerebrovessels. The V_d of the supernatant is a composite value consisting of the V_d for test compound localized in brain tissue and the V_d for test compound in the cerebrovascular lumen. The fraction of the supernatant V_d attributable to luminally entrapped test compound can be determined by in situ perfusion using an intravascular marker, such as radiolabeled inulin (see Support Protocol 3) coupled with the capillary depletion technique. The supernatant V_d for inulin is entirely derived from tracer resident in the cerebrovasculature.

BASIC PROTOCOL 4

BLOOD-BRAIN BARRIER EFFLUX MEASUREMENT: THE BRAIN EFFLUX INDEX METHOD

There is an emerging interest in the structure and function of efflux transporters at the BBB (Kakee et al., 1996; Kusuhara et al., 1997; Hosoya et al., 1999), e.g., P-glycoprotein, a well-known efflux transporter expressed at the BBB and other organ sites, exports lipophilic molecules from brain endothelial cells. There are circumstances in which a compound has sufficient lipophilicity for BBB penetration, yet fails to accumulate in brain tissue in pharmacologically sufficient amounts. One likely cause may be the active export of the test compound by BBB efflux transporters. This protocol employs a traditional pharmacokinetic approach to quantitation of blood-brain barrier drug efflux. Because the amounts of injected test compounds are small in this protocol, a highly sensitive analytical detection method is required. Radiolabeled test compounds are frequently used. Suggested reference compounds include ^{3}H- or ^{14}C-labeled inulin or carboxyinulin. If the radiolabeled form of a test compound is to be used, an isotopic form different from the inulin reference must be selected.

Materials

    Laboratory rats (200 to 350 g, either gender)
    Anesthetic (as specified by the approved animal care protocol; *APPENDIX 2B*)
    2% (w/v) lidocaine/0.001% (w/v) epinephrine (commercially available as Xylocaine 2% with epinephrine 1:100,000; J.A. Webster)
    ✓ ECF buffer, containing test and reference compounds, 37°C
    Saline: 0.9% (w/v) NaCl (sterile)
    Tissue solubilizer (e.g., Soluene 350; Packard) or 2 N NaOH

    Scintillation vials, 7 to 20 ml, with caps
    Stereotaxic atlas (e.g., Paxinos and Watson, 1986, for rats)
    5-µl syringe (e.g., Hamilton Gas-Tight removable needle microliter syringe; Hamilton Company), with 28-G, 2-in. (5-cm), point-style no. 4 (12° bevel) needle, 0.18 mm (0.007 in.) nominal i.d., 0.36 mm (0.014 in.) nominal o.d. (may need to be custom fabricated by Hamilton Company)
    Rodent stereotaxic frame
    Animal fur shavers
    Warming pads (Deltaphase Isothermal Pad; Braintree Scientific), preheated to 37°C
    Scalpel
    Drill and small drill bit, e.g., 1/32-in. (0.8-mm) bit inserted in a Dremel unit or affixed to a manual handle
    Surgical shears
    Cerebrospinal fluid (CSF) collection unit (see Support Protocol 8)
    Olsen-Hegar needle holder (e.g., Harvard Apparatus)
    Rodent guillotine, optional
    Tissue forceps or bone rongeurs
    Filter paper
    Inox no. 7 sharp-point forceps (e.g., Harvard Apparatus)
    50°C water bath, optional

1. Label and record the individual weights of an appropriate number of empty scintillation vials (three vials for each rodent in this protocol) for collecting brain tissue (one vial for every discrete brain region per rodent).

2. Using a stereotaxic atlas, identify the target brain region for microinjection (Par 2 region of the frontal cortex recommended; see Kusuhara et al., 1997; Hosoya et al., 1999).

3. Mount a 5-µl syringe on the arm of a rodent stereotaxic frame. Weigh a laboratory rat and anesthetize it with an appropriate injectable anesthetic. Shave its scalp using an animal fur shaver. Place animal in the rodent stereotaxic frame, with the animal's trunk on top of a preheated warming pad to maintain normothermia.

4. Using a scalpel, make an ∼1-cm incision along the anterior posterior axis to expose the skull. Blot the skull dry using a lint-free tissue, gauze, or a cotton-tipped applicator. Apply ∼0.5 ml of 2% lidocaine/0.001% epinephrine to the exposed skull and skin. After a minute or so, blot the skull dry again.

5. Identify the bregma (a reference point where the coronal and sagittal sutures intersect) and mark it with the tip of a sharp pencil. Align the needle tip of the microsyringe with the pencil mark defining the bregma. Use the micromanipulator of the stereotaxic frame

to lower the needle tip so that it just touches the pencil mark on the skull. Record the anterior and lateral stereotaxic coordinates.

6. Based on the published coordinates of the brain target region (obtained from the stereotaxic atlas, e.g., Paxinos and Watson, 1986), calculate the new micrometer coordinates (for the Par 2 region in this case 0.2 mm anterior, 5.5 mm lateral with respect to the bregma, and 3.1 mm deep from the dura mater.

7. Raise the microliter syringe and manipulate the micromanipulator so that the needle is over the target brain region. Lower the needle tip so that it nearly touches the skull over the target brain region. Mark the needle spot on the skull with a pencil mark. Raise the microliter syringe and swing the micromanipulator arm to the side.

8. Use a drill and small drill bit to gently drill a small hole through the skull, taking care not to puncture the dura mater.

    A change in resistance will occur as the drill penetrates the skull.

9. Load the syringe with 1 µl injectate solution without disturbing the mounted position of the syringe. Retain a 10-µl aliquot of the injectate for later analysis; store up to 1 month at −20°C.

10. Swing the micromanipulator arm back into place over the target brain region. Lower the needle until it nearly touches the dura mater. Record the vertical coordinate. Based on the published vertical coordinate of the brain target region, calculate the new vertical micrometer coordinate.

11. Slowly lower the microliter syringe to the new vertical coordinate. Gently wick away minor bleeding with a lint-free tissue or cotton-tipped applicator, but be sure not to disturb the needle.

12a. *For a single test compound:* Slowly administer 0.5 to 0.6 µl injectate into the brain tissue by depressing the microliter syringe plunger over a period of 2 sec. Let the needle remain in the injection site for 10 sec. Slowly withdraw the needle by carefully raising the syringe using the micromanipulator. Allow the animal to remain positioned in the stereotaxic unit for a period of time (as specified by the experimental design) before harvesting CSF and brain tissue.

12b. *For substrate inhibition studies:* Align two syringe needles for the same brain region target site. Preadminister a 50-µl volume of injectate containing the competitor substance in isotonic ECF buffer over a period of 30 sec (slow administration is essential to avoid traumatic brain damage) using a 50-µl gas-tight microliter syringe (Hamilton Company). After an appropriate period of time (as determined by experimentation; may range from 5 to 30 min), administer 0.5 to 0.6 µl of test compound.

    Pharmacokinetic analysis of the apparent efflux rate constant of the test compound of the brain, K_{eff}, requires the sacrifice of multiple rats at different time intervals following the microinjection (a minimum of three rats per time point). It is suggested the experimental design include a minimum of four to five time points after injection, e.g., 0, 5, 10, 20, 30 min, or longer, if necessary.

NOTE: Steps 13 to 15 should be performed as quickly and reliably as possible because test compound efflux is time dependent.

13. Use a pair of surgical shears to carefully cut the skin over the back of the skull and outstretched neck of the anesthetized rat. Use a scalpel to cut the neck muscle anchored at the base of the skull. Using the scalpel, gently scrape the cut neck muscle downwards (caudally) to expose the white alanto-occipital membrane that covers the cisterna magna. Control excessive bleeding by infiltrating the muscle and skin tissue with 2%

lidocaine/0.001% epinephrine, using a 25-G hypodermic needle and 1-ml hypodermic syringe.

14. Hold a CSF collection unit with an Olsen-Hegar needle holder. Advance the needle through the membrane to a depth where the needle bevel just punctures through the membrane. Do not penetrate of the cerebellar tissue

15. Hold the CSF collection unit steadily in place with one hand and pull back on the syringe plunger with the other hand. Be careful to aspirate CSF (100 to 150 µl) uncontaminated with blood. Transfer CSF to a 1.5-ml microcentrifuge tube with cap and store up to 1 month at −20°C.

16. Remove the animal from the stereotaxic frame and quickly decapitate it using sharp surgical shears or a rodent guillotine. Using a pair of tissue forceps or bone rongeurs, remove the entire brain (forebrain and cerebellum) from the skull. Place the entire brain in a disposable petri plate containing a piece of filter paper moistened with saline. Place the petri plate on ice so that the brain dissection is performed under cold conditions.

17. Using a pair of Inox no. 7 sharp-point forceps, remove meningeal membranes and associated surface blood vessels. Harvest the cerebellum and both hemispheres (ipsilateral and contralateral). Use a lint-free tissue to wick away any excess, adherent moisture from the brain region tissue. Place tissue in a preweighed vial (step 1) and determine the weight of the tissue. Store up to 1 month at −20°C.

18a. *For radiolabeled test and reference compounds in tissue:* Digest brain tissue with appropriate amount of a tissue solubilizer as specified by the manufacturer's instructions. Alternatively, digest tissue with 1 to 2.5 ml (5 vol per gram of tissue) of 2 N NaOH and incubate 3 hr at 50°C. Vortex vigorously. Add an appropriate volume of chemically compatible scintillation cocktail (e.g., Hionic Fluor; Packard Instruments, if using NaOH) until the sample forms a clear solution (with no emulsion). Quantify the radioactivity using a dual-label counting protocol (for concurrent ^{3}H- and ^{14}C-labeled test and reference compounds) in a scintillation counter.

18b. *For radiolabeled test and reference compounds in CSF:* Pipet 100 µl of CSF into a 7- or 20-ml scintillation vial with cap. Add an appropriate volume of chemically compatible scintillation cocktail until the sample forms a clear solution with no emulsion. Quantify the radioactivity using a dual-label counting protocol (for concurrent ^{3}H- and ^{14}C-labeled test and reference compounds) in a scintillation counter.

19. Calculate the brain efflux index (BEI) using the equation:

$$\text{BEI} = 1 - \left(\frac{M_{\text{TB}}}{M_{\text{RB}}}\right)\left(\frac{M_{\text{RI}}}{M_{\text{TI}}}\right)$$

where M_{TB} is the mass of the test compound in the brain, M_{RB} is the mass of the reference compound in the brain, M_{RI} is the mass of the reference compound injected, and M_{TI} is the mass of the test compound injected.

20. Multiply the BEI by 100 to convert the fraction into the percentage of efflux. Calculate the percentage retained in the brain as $100 - \%\text{BEI}$. Determine the apparent efflux rate constant of the test compound from the brain, K_{eff}, as the slope of the plot of $\ln(100 - \%\text{BEI})$ versus time. Alternatively, calculate K_{eff} as the slope $\times$ 2.303 of the plot of $\log_{10}(100 - \%\text{BEI})$ versus time.

SUPPORT PROTOCOL 8

CONSTRUCTION OF CEREBROSPINAL FLUID COLLECTION UNITS

Materials

    5.5-in. (14-cm) Olsen-Hegar needle holder (e.g., Harvard Apparatus)
    Metal file
    22- and 25-G stainless steel disposable hypodermic needles
    Silastic tubing, 0.50 mm (0.020 in.) i.d., 0.965 mm (0.038 in.) o.d. (e.g., Silastic Medical Grade; Dow Corning)
    Suture, USP size 4.0
    1-ml syringe

1. Using a 5.5-in. Olsen-Hegar needle holder and a metal file, cut a 25-G stainless steel disposable hypodermic needle from its plastic luer-lock hub. Save the needle and discard the hub.

2. Cut a 10-cm length of silastic tubing and slide one end over the cut end of the 25-G needle. Securely tie the tubing onto the needle using 4.0 suture.

3. Using the needle holder and metal file, cut a 22-G stainless steel hypodermic needle in half. Save the end still connected to the plastic luer-lock hub. Properly discard the cut sharp needle.

4. Insert the cut end of the 22-G needle (with hub) into the free end of the silastic tubing. Insert a 1-ml syringe into the plastic luer-lock hub.

References: Boje, 1995; Kakee, et al., 1996; Smith, 1989; Smith et al., 1987, 1988; Zlokovic, 1995

Contributor: Kathleen M.K. Boje

CHAPTER 3
Behavioral Neuroscience

Behavioral neuroscience is the study of the neural mechanisms mediating normal and abnormal behaviors. Behavior is the final output of the nervous system at the level of the whole organism. Discrete animal behaviors are used to test hypotheses about anatomical, neurochemical, neurophysiological, and genetic substrates of behavior. Appropriate animal behavior paradigms provide the critical functional tools for translational research, moving discoveries at the molecular level toward clinical applications. The study of user-friendly behavioral tasks that model specific symptoms of neuropsychiatric disorders is a critical phase in the testing of new psychotherapeutic treatments in vivo.

The protocols presented in Chapter 3 are designed to serve both as a solid introduction for the novice and as a guide to specific methods for the established behavioral neuroscientist. To facilitate their use in the functional analysis of new compounds for the treatment of neuropsychiatric disorders, an attempt has been made to present only protocols that are well-validated, quantitative, easily replicated, rapid, and simple. Nonetheless, hands-on training and/or collaboration with a good behavioral neuroscience laboratory with expertise in the chosen behavioral tasks is highly recommended.

This chapter highlights many of the best major protocols developed in the long and illustrious history of behavioral neuroscience and presents protocols in common use for both rats and mice. Quantitative measures of developmental milestones in mice and rats, relevant to modeling many neurodevelopmental diseases, are described in UNIT 3.1. Classic methods for measuring spontaneous motor activity are described in UNIT 3.2. Exploratory locomotion underlies almost all behavioral output, and is considered a specific measure of mesolimbic dopaminergic pathway activation. Motor coordination and balance tests, described in UNIT 3.3, can be used in rodent models of neurodegenerative diseases such as Parkinson's disease and amyotrophic lateral sclerosis.

In UNIT 3.4, a protocol for measuring food intake over a set period of time is detailed. A simple but very versatile procedure for measuring conditioned flavor aversions, in which administration of a drug following ingestion of a novel food or solution can suppress subsequent intake of the food or solution (conditioned taste aversion), is presented in UNIT 3.13. The associative learning is extremely robust, and provides a sensitive and widely used behavioral index of drug side effects.

Methods for quantitating male and female sexual receptivity and activity in rats are described in UNIT 3.5. Social behaviors in rats and mice are well documented in the ethological literature. Continuing on the topic of sexual behaviors, protocols for measuring parent-infant interactions are presented in UNITS 3.6. Infant rodent pups emit vocalizations that elicit parental responses. Active parental behaviors, described in UNIT 3.6, include carrying, licking and nursing the pups, nest building, and quiescent contact in the nest.

Stress paradigms, including restraint, footshock, social isolation, maternal separation, and sleep deprivation, are described in UNIT 3.7. Behavioral despair tasks sensitive to antidepressant drugs are described in UNITS 3.8 & 3.9. The Porsolt swim test, learned helplessness, and the tail suspension test are applied to both rats and mice. Well-characterized animal models of anxiety-related behaviors, including the traditional Geller-Seifter conflict test, social interaction, light/dark exploration, and the elevated plus maze, are presented in UNIT 3.10.

Established protocols for evaluating cognitive processes in mice and rats are presented in UNITS 3.11 & 3.12; spatial learning tasks, including the Morris water maze and radial maze tasks, are described. An emotional memory task, cued and contextual fear conditioning, is explained in UNIT 3.12. Sensorimotor reflexes and gating processes abnormal in schizophrenia, including measures of prepulse inhibition and habituation of the startle response, are described for rats in UNIT 3.14. Latent inhibition, described in UNIT 3.15, is a task that measures the ability to ignore irrelevant stimuli and models aspects of attentional deficits in schizophrenia.

The protocols collected in Chapter 3 are designed to provide a comprehensive overview of the behavioral neuroscience armamentarium. Most of these protocols can be applied to the behavioral phenotyping of transgenic and knockout mice.

Contributor: Jacqueline N. Crawley

UNIT 3.1

Assessment of Developmental Milestones in Rodents

At birth, the rat is capable of some specific activities—principally, suckling, rolling over, and vocalizing. However, its movements are uncoordinated and seemingly random, its tactile sensitivity is not fully developed, and its ear canals and eyes remain closed until several days after birth. Postnatal development consists mainly of the continuation of processes begun earlier. See Table 3.1.1 for typical ages for the appearance of developmental milestones.

STRATEGIC PLANNING

The main strategic planning issue to be considered is selection of items for the test battery. The selection of a test item should be based on the focus of the inquiry and time considerations, taking into account the various confounding factors (e.g., handling, time of day, behavior definition and scoring, litter size and nutritional effects, postnatal maternal influences, and statistical considerations) associated with developmental testing, which may increase with more time-consuming extensive test batteries.

One convention is to divide the tests into those that assess sensory development and those that assess motor development, although clearly, most of these tests involve both sensory and motor function. Examples of tests representative of sensory functions are: negative geotaxis, cliff avoidance, placing responses, and the startle response. Examples of motor tests are: righting reflexes, crossed extensor reflex, rooting reflex, grasp reflex, bar holding, and the screen tests.

BASIC PROTOCOL 1

PHYSICAL LANDMARKS OF RODENT DEVELOPMENT

Materials

Experimental subject (rat or mouse)
Nontoxic indelible marker or equipment for animal tattooing
Data sheet (see Fig. 3.1.1)
Holding cage: 43 × 27 × 15–cm standard plastic animal cage
Wood-chip bedding

Table 3.1.1 Typical Ages for Appearance of Developmental Milestones

Measure	Average age for response (days)	Range (days)
Physical landmarks		
Pinnae detachment	15	10–20
Eye opening	13	7–17
Incisor eruption	7	5–10
Fur development	11	3–15
Reflexes		
Surface righting	5	1–10
Air righting	18	16–21
Negative geotaxis	7	3–15
Cliff avoidance	8	2–12
Visual placing	15	11–18
Forelimb/hindlimb placing[a]	5	1–10
Vibrissa placing response	9	5–15
Auditory startle	15	11–21
Tactile startle	15	3–20
Crossed extensor reflex[a]	3	1–10
Rooting reflex[a]	2	1–15
Grasp reflex[a]	7	3–15
Bar holding	14	10–21
Level screen test	8	5–15
Vertical screen test	19	15–21
Locomotor behavior		
Elevation of the head	12	9–21
Elevation of the forelimbs and shoulders	7	5–15
Pivoting[a]	7	2–17
Crawling[a]	11	7–16
Walking	16	12–21

[a]This behavior either disappears or reduces to ~0 in frequency.

    Waterproof, hospital-grade heating pad
    Dissecting microscope
    Small animal scale

1. Prepare the holding cage by covering the floor of a 43 × 27 × 15–cm standard animal cage with wood-chip bedding and placing one-half of the cage on a waterproof, hospital-grade heating pad. Set the temperature of the heating pad such that the temperature directly over it is ~32°C to 34°C, with the ambient temperature of the cage ~22°C to 24°C, allowing each individual animal to choose a thermal preference.

2. *Randomly* select the animals to be tested from the litter and place them into the heated holding cage. Return test animals to the holding cage until all animals have been tested from that litter. Return all animals to the home cage at the same time.

3. Record the measures in steps 4 and 5 from each pup on a data sheet (Fig. 3.1.1). If the pups are to be repeatedly tested, mark each animal with a unique identification using either a tattoo or a nontoxic, indelible marker (reapply daily as needed).

4. Record the age of the following events:

    a. *Pinnae detachment* (detachment of both pinnae, defined as the complete separation of the pinna from the cranium)

| Date: | Experimenter Name: |

Animal#										
Age										
Sex										
Physical landmarks										
Pinnae detachment										
Eye opening										
Incisor eruption										
Body weight										
Fur development										
Developmental reflexes										
Surface righting*										
Air righting										
Negative geotaxis*										
Cliff avoidance										
Forelimb placing										
Vibrissa placing										
Visual placing										
Auditory startle										
Tactile startle										
Crossed extensor										
Rooting reflex										
Grasping reflex										
Bar holding*										
Level screen										
Vertical screen										

*These represent timed behaviors. The rest are scored as either present (Yes) or absent (No).

Figure 3.1.1 Sample data sheet for developmental testing in rodent pups.

    b. *Fur development* (appearance of dorsal and ventral pigment and fur), and

    c. *Incisor eruption* (observation under a dissecting microscope of the upper and lower teeth).

5. Record the following every day:

    a. *Eye opening* (whether the eyes are open) and

    b. *Body weight* (using a small container on a weighing scale).

BASIC PROTOCOL 2

DEVELOPMENTAL REFLEXES IN RODENTS

In these tests, the dependent variable is the appearance and/or disappearance of several reflexes.

Materials

Experimental subject (rat or mouse)
Nontoxic indelible marker or equipment for animal tattooing
Data sheet (see Fig. 3.1.1)
Holding cage: 43 × 27 × 15–cm standard plastic animal cage
Wood chip bedding
Waterproof, hospital-grade heating pad
Stopwatch
Padding: e.g., foam rubber
Round metal bar thin enough for pup to grasp (∼4- to 7-mm diameter)
Inclined plane: wooden or plastic floor, 30 × 30 cm, set at an angle of 30°, covered with 16-mesh wire screen
Wooden platform ≥30 cm in height
Handheld metal clicker (can be bought from canine training facilities)
Metal wire mesh screens: 30 × 30 cm, 16-mesh

1. Prepare heated holding cage (see Basic Protocol 1, step 1). Randomly select the animals to be tested from the litter and place them into the heated holding cage (see Basic Protocol 1, step 2).

2. Record the measures described in steps 3 to 17 from each pup on a data sheet (Fig. 3.1.1). If the pups are to be repeatedly tested, mark each animal with a unique identification using either a tattoo or a nontoxic indelible marker (reapplied daily, as needed).

3. *Righting reflex, surface righting:* Place the pup gently onto its back and record the time (in sec) for the subject to turn over onto its belly. If the pup takes longer than 60 sec, stop the test, right the pup, and record the measure as 60 sec.

4. *Righting reflex, air righting:* Hold the pup upside-down at a height of 60 cm. Drop the pup onto a padded surface (e.g., foam rubber). Record whether the pup successfully righted itself during the fall.

   For more detailed analysis of responses tested in steps 3 and 4 using video recording systems for frame-by-frame analysis, see Pellis and Pellis (1994) and Vorhees et al. (1994).

5. *Negative geotaxis:* Place each pup on an inclined plane (typically 30° from horizontal) with its head facing downwards. Observe and record whether the pup succeeds in changing its orientation so that its head faces up the incline and record the latency period (in sec) to make the change. If the pup takes longer than 60 sec, stop the test and record 60 sec. Protect young pups from injury, e.g., by using a high-friction surface (such as 16-mesh wire) on the incline and soft padding at the end of the incline.

   Thiessen and Lindzey (1967) have shown that repeated negative geotaxis testing in adults improved performance (i.e., a practice effect). Therefore, it may be necessary to limit repeated testing and/or include other assessments of negative geotaxis.

6. *Cliff avoidance:* Place the pup on the edge of a wooden platform (no less than 30 cm in height) with its nose and forefeet over the edge. Record whether or not the pup moves away from the "cliff" by backing up or by turning sideways (can be assessed before the eyes are open).

7. *Placing response, forelimb/hindlimb placing response:* Suspend the pup in the air by grasping the animal gently around the trunk, making sure that no paws are touching a solid surface. Bring a thin metal bar into contact with the back of its paw. Record whether the pup raises and places one of paws on the surface of the bar.

8. *Placing response, vibrissa placing response:* Gently grasp the pup around the trunk and suspend it in the air. Using a thin metal bar, make contact with the vibrissa. Observe whether, when contact is made with a solid object, the head is raised and the forelimbs extended to grasp the object. Record (yes or no) if this reflex is observed.

9. *Placing response, visual placing response:* Gently suspend the animal by the tail and lower it toward a solid surface (e.g., a tabletop). Observe whether the pup raises its head and extends the forelimbs in a placing fashion. Record (yes or no) if this reflex is observed.

10. *Startle response, auditory startle:* Test for responses to acoustic stimuli in a room with a very low level of ambient noise by operating a clicker at a distance of 25 to 30 cm above each pup. Record whether or not the animal is "startled" (i.e., slight jerk, kicking, squirming, or combinations of all three are typical reactions observed immediately after the presentation of the auditory stimulus). Be careful not to confuse random movements with actual responses to the click.

    More extensive procedures exist using automated equipment (see Davis, 1984; Schnerson and Willott, 1980; Pletnicov et al., 1995; Martinez et al., 2000; also see UNIT 3.14 for more extensive tests of the startle response).

11. *Tactile startle:* Gently apply a puff of air (this can be a puff of the experimenter's breath) to the pup. Record whether the animal is "startled" or not (i.e., exaggerated jumping and running behavior response).

    Tactile sensation can also be assessed using Von Frey hairs (see Marsh et al., 1999; Pitcher et al., 1999; also see UNIT 4.6). This affords a more controlled tactile pressure to a specific body area.

12. *Crossed extensor reflex:* Gently pinch the foot of one hindlimb and observe for opposite hindlimb extension.

13. *Rooting reflex:* Bilaterally stimulate the face region of the pup with the forefingers or two cotton swabs. Record whether the animal exhibits the rooting reflex (i.e., crawling forward and pushing the head in a rooting fashion during the actual stimulation). Do not confuse with a startle response or spontaneous movement.

14. *Grasping reflex:* Stroke the paw with a blunt instrument (examples of blunt instruments that can be used include a wooden toothpick and a blunt metal dissecting tool). Record whether the reflex is present on the basis of whether or not the animal grasps the instrument.

15. *Bar holding:* Gently lift the pup by the trunk and bring it up close to a thin (~4 to 7-mm diameter) metal bar. Allow it to grab hold with its front paws, such that when the animal is released, it is hanging only by its front paws. Immediately start the stopwatch. Record how long the animal holds on, up to a maximum of 10 sec (an adult-like response).

    Be very careful because the animals may jump or fall off the bar suddenly.

16. *Screen test, level screen test:* Place the pup onto a piece of 16-mesh metal screen. Gently drag the pup in a horizontal direction (~10 cm) across the screen by the tail. Record whether the pup can hold onto the screen.

17. *Screen test, vertical screen test:* Place the pup at one end of a wire mesh (16-mesh) screen. Rotate the screen to a vertical position (with the animal's head pointing up) to

assess whether the animal exhibits a climbing response within 60 sec. Score as "yes" or "no".

More extensive description of these methodologies can be found in Anderson and Patrick (1934) and Irwin (1968).

BASIC PROTOCOL 3

DEVELOPMENT OF LOCOMOTOR BEHAVIOR IN RODENTS

An excellent description of these behaviors assessed in this protocol can be found in Altman and Sudarshan (1975).

NOTE: Experiments conducted over an extensive period of time must take into account the possible effects of prolonged maternal deprivation and control for possible body temperature loss.

Materials

Experimental subject (rat or mouse)
Nontoxic indelible marker or equipment for animal tattooing
Data sheet (see Fig. 3.1.2)
Heated holding cage: $43 \times 27 \times 15$–cm standard plastic animal cage
Wood chip bedding
Waterproof, hospital-grade heating pad
Stopwatch
Open field (recommended size $50 \times 50 \times 25$ cm)
Video camera (optional)

1. Prepare heated holding cage (see Basic Protocol 1, step 1). Randomly select the animals to be tested from the litter and place them into the heated holding cage (see Basic Protocol 1, step 2).

2. Using either the videotaping or time-sampling method described above, record the measures described in steps 3 to 7 from each pup on a data sheet (Fig. 3.1.2). If the pups are to be repeatedly tested, mark each animal with a unique identification using either a tattoo or a nontoxic indelible marker (reapply daily, as needed).

3. *Development of quadruped stance, elevation of the head:* Determine the frequency of elevation of the head.

4. *Development of quadruped stance, elevation of the forelimbs and shoulder:* Determine the frequency of elevation of the forelimbs and shoulder (see Altman and Sudarshan, 1975).

5. *Development of quadruped locomotion, pivoting:* Determine the frequency of the pivoting motion, in which the pup makes broad swipes with the paws, producing a "paddling" motion that results in turning, with the pelvis remaining anchored to the ground (most frequently observed by the end of the first week of life).

6. *Development of quadruped locomotion, crawling:* Determine the frequency of the crawling motion, in which the paddling movements of the paws result in animal dragging itself forward or pushing itself backward.

7. *Development of quadruped locomotion, walking:* Determine the frequency of walking (the hindlimbs often slip or are dragged; see Clarke and Still, 1999 for more information about gait analysis).

References: Altman and Sudarshan,1975; Anderson and Patrick,1934; Clarke and Still,1999; Irwin,1968

Contributor: Charles J. Heyser

Date: ____________________ Experimenter Name: ____________________

NOTE: Record the following in each time periord (H= elevation of head; F = elevation of forelimbs P = pivoting; C= crawl; W = walk)

Animal#												
Age												
Sex												

T10												
T20												
T30												
T40												
T50												
T60												
T70												
T80												
T90												
T100												
T110												
T120												
T130												
T140												
T150												
T160												
T170												
T180												
T190												
T200												
T210												
T220												
T230												
T240												
T250												
T260												
T270												
T280												
T290												
T300												

Figure 3.1.2 Sample data sheet for locomotor development in rodent pups using a time-sampling procedure.

UNIT 3.2

Locomotor Behavior

The study of locomotor activation in rodents does not involve the extensive learning or conditioning required by other behavioral tasks, and so this dependent measure is often used as the initial screen for pharmacological effects predictive of therapeutic efficacy of a drug class in humans. As with all behavioral studies, consistency is critical. Such seemingly trivial factors as the attire of the experimenter, time of day of testing, habituation of the animal to human handling, and its acclimation to the behavioral apparatus should be held as constant as possible from one experimental group to the next. It is important to assess the entire time course of a given behavior response as well as a range of doses when evaluating a drug. It is also important to control for the potential influence of conditioning on locomotor activity when animals are subjected to repeated drug treatments (e.g., the exposure to the chamber where injections are given). To minimize inter- and intra-observer scoring differences, test observers at the beginning and end of an experiment for close agreement to avoid drift from a scoring norm.

BASIC PROTOCOL

LOCOMOTOR ASSESSMENT BY DIRECT OBSERVATION USING INTERVAL AND ORDINAL SCALES

The interval scale in observational experiments measures the frequency of occurrence of a behavior, and the ordinal scale reflects intensity and duration of the activity.

Materials

Rat subjects of appropriate strain, sex, and age (see UNIT 3.10)
Soapy water
90% ethanol
Saline for vehicle injection
Drug to be tested

Plexiglas behavioral arena, 40 × 40 × 40–cm
Sponge
Sound-attenuating chamber equipped with lighting, fresh air circulation, and large two-way mirror

1. Paint lines on the underside of floor of the Plexiglas behavioral arena (e.g., sixteen 10 × 10–cm squares optimal for a 40 × 40–cm floor).

2. Place the behavior arena in the sound-attenuating chamber that shields the subject from the presence (sight, sound, and, if possible, scent) of the experimenter. To observe line crosses, install a two-way mirror on one side or on the ceiling of the sound-attenuating chamber, and sit in a position with an unobstructed view of the apparatus floor.

3. Using a sponge, wash the Plexiglas behavior arena with soapy water and then alcohol. Allow the alcohol to evaporate fully before using the apparatus. Clean the behavioral arena thoroughly before each use to eliminate any stimuli left in the chamber by the previous subject.

4. Place the test subject in the arena and allow it to habituate to the test apparatus for 60 min to minimize novelty-induced freezing or exploratory behavior.

5. Inject the subject with saline or vehicle if a drug is to be injected during the experimental portion of the procedure. Leave the animal in the arena for a further 120 min. Return the subject to its home cage.

6. The day after the habituation session, thoroughly clean the apparatus as in step 3. Place the test subject in the arena and record assessments as the subject rehabituates to the behavioral arena during this period, and locomotor activity returns to baseline (see Fig. 3.2.1).

    a. *Interval-scale assessment:* Record the number of line crosses for 60 min. Define a line cross strictly; score a line cross only when both front paws cross a line, i.e., when the rat moves either forward or backward (recommended).

    b. *Ordinal-scale assessment:* Record the behavioral response for 60 min. Score the locomotor behavior produced upon drug administration separately for intensity and duration (0 to 3 each). Multiply the intensity and duration scores to yield a continuum for each behavior from 0 (not present) to 9 (maximally intense and of continuous duration). See operational definitions in Table 3.2.1.

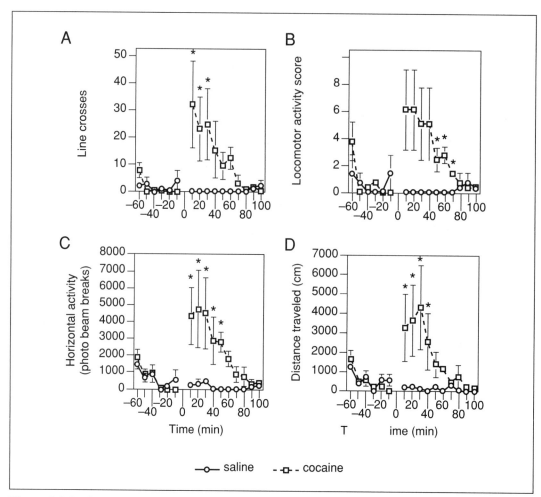

Figure 3.2.1 Comparison of sample data collected using three simultaneous methods of locomotor behavior assessment before and after i.p. injection of saline or 15 mg/kg cocaine. As two trained observers (see Basic Protocol) scored line crosses (**A**) or locomotor activity (**B**), a photocell-based system (see Alternate Protocol) monitored horizontal activity (**C**) and distance traveled (**D**). Asterisks indicate significant differences between cocaine and control groups at a given time point (see Support Protocol for data analysis). Magnitude and duration of locomotor activity recorded by all three assessment methods following injection of cocaine or saline were similar.

Table 3.2.1 Scoring Criteria for Locomotor Activity

Activity level	Score
Intensity	
Inactive	0
Locomotor activity confined to $1/4$ of arena floor	1
Locomotor activity within $\sim 1/2$ of arena floor	2
Extensive locomotor behavior encompassing entire floor	3
Duration[a]	
Inactive	0
Locomoting <20 sec	1
Locomoting 20–40 sec total	2
Locomoting 40–60 sec total	3

[a]Duration of activity within a 60-sec recording interval.

7. Perform the experimental manipulation (e.g., drug injection or introduction of novel stimulus) and record assessments as in step 6 over the time course determined for the experiment.

   Assess a range of doses in order to adequately evaluate the full behavioral effects of a drug. Assess the entire time course of a given behavioral response. Many drug effects last several hours, and the nature of the response can shift dramatically.

8. Return the subject to its home cage, and analyze data (see Support Protocol).

ALTERNATE PROTOCOL

LOCOMOTOR QUANTIFICATION BY PHOTOCELL-BASED SYSTEMS

In contrast to human observers, photocell-based apparati only measure behavior quantitatively, within a relatively narrow range of parameters. When using these systems, it is important to characterize the behavioral response in tandem with direct observation initially, in order to verify that changes in locomotor activity are measured appropriately by the automated system, and to identify any unexpected qualities of the locomotor behavior not discernible from simple photobeam breaks.

Because automated behavioral apparati can be used in conjunction with observer ratings, locomotor activity can be monitored automatically, while a human observer simultaneously scores stereotyped behaviors. Stereotypy is defined as beam breaks that occur when the animal is not locomoting (i.e., repetitive breaks of the same set of photocell beams). Although this measure is limited, as it cannot differentiate among stereotyped behaviors such as head bobbing, scratching, and grooming, it is important to distinguish and separate these repetitive behaviors to obtain an accurate measure of locomotion.

When using photocell-based systems, ensure that the photocells are unobscured prior to each testing session, and calibrate each photocell frequently with the use of a metal rod. This protocol is written as a procedure to assess drug-modified locomotor behavior, but other stimuli may be tested in the same manner.

Additional Materials (*also see Basic Protocol*)

   Photocell-based activity monitor (e.g., Omnitech Electronics, Columbus Instruments, Coulbourn Instruments, Lafayette Instrument, MED Associates, and San Diego Instruments)
   Sound-attenuating chamber, equipped with lighting and circulating fresh air
   Computer system with commercial analytical software (e.g., Omnitech) and digital converter interface
   Printer

1. Position the behavioral arena in the activity monitor and adjust the photocell arrays 4 cm above the floor for rat locomotor activity and spaced at 2.5-cm intervals for measuring horizontal activity. Add a second (or third) row of photocells to measure vertical activity. Place the behavioral arena and activity monitor in the sound-attenuating chamber.

2. Set up the computer and interface according to the manufacturer's instructions and attach the appropriate cables to the activity monitor. Calibrate the system before placing the test subject in the arena. Habituate the subject to the test apparatus (see Basic Protocol, steps 4 and 5).

3. The day following the habituation session, clean the apparatus and place subject in the test arena. Set the computer to record the locomotor response for 60 min to allow the subject

to rehabituate to the behavioral arena. Save the data on computer disk and/or print out the data.

4. Perform the experimental manipulation (e.g., drug injection or introduction of novel stimulus). Set the computer to record the locomotor response over the time course determined for the experiment (may vary according to drug and dosage).

   Assess a range of doses in order to adequately evaluate the full behavioral effects of a drug. Assess the entire time course of a given behavioral response. Many drug effects last several hours, and the nature of the response can shift dramatically.

5. Save the data on computer disk and/or print out the data. Turn off all experimental equipment and return the subject to its home cage. Analyze data (see Support Protocol).

SUPPORT PROTOCOL

DATA ANALYSIS

The following statistical analysis is effective in showing the relationship of the scoring methods to one another. To achieve adequate statistical power in assessing locomotor response to an experimental manipulation, use a sample cohort of eight to twelve rats in each group.

Figure 3.2.1 presents sample data from locomotor behavior tests used to assess the locomotor activity induced by a cocaine or saline injection. The data (time points prior to 0 min in panels A, C, and D) were analyzed with a two-way mixed factors analysis of variance (ANOVA). The between-subjects measure was cocaine or saline injection; the within-subjects measure was time. The analysis revealed a significant main effect of time for horizontal activity $[F(5,20) = 9.36, p < 0.0001]$, distance traveled $[F(5,20) = 5.34, p < 0.0028]$ and (marginally) line crosses $[F(5,20) = 2.34, p < 0.079]$. There were no other significant treatment effects or interactions. Analysis of the data summarized in panel B with the nonparametric Kolmogorov-Smirnov test revealed no significant treatment effect. The outcome of the effect of cocaine or saline injection was the same as for the data described above. The ANOVA analyses revealed a significant main effect of time for horizontal activity $[F(9,36) = 3.86, p < 0.0017]$, distance traveled $[F(9,36) = 2.94, p < 0.01]$ and line crosses $[F(9,36) = 2.31, p < 0.036]$. In addition, there were significant time × injection interactions for line crosses $[F(9,36) = 2.67, p < 0.017;$ panel A], horizontal activity $[F(9,36) = 3.12, p < 0.0071;$ panel C], and distance traveled $[F(9,36) = 2.73, p < 0.015;$ panel D]. Pairwise comparisons were made with Fisher's LSD. Nonparametric Kolmogorov-Smirnov analysis of the scoring of locomotor activity (panel B) revealed a significant treatment effect ($\chi^2 = 19.27, p < 0.0001$). Pairwise comparisons were made with the Mann-Whitney U test.

Reference: Kalivas et al.,1988

Contributors: R. Christopher Pierce and Peter Kalivas

UNIT 3.3

Motor Coordination and Balance in Rodents

Measurement of motor coordination and balance can be used not only to assess the effects of test compounds or other experimental manipulations on mice and rats, but also to characterize the motor phenotype of transgenic or knockout animals.

BASIC PROTOCOL 1

ASSESSING BALANCE USING A ROTAROD

Most untreated control mice perform to criterion (maintaining balance for 60 sec) even at fast rotation speeds (e.g., 44 rpm). Thus, this test is good for quantifying progressive impairments in motor coordination and balance. By contrast, the rotarod is poor at detecting minor deficits or improvements in performance.

Materials

Male or female mice
70% (v/v) ethanol
Rotarod apparatus (e.g., AccuScan Instruments, Biotech Instruments, Columbus Instruments, Letica Scientific Instruments, San Diego Instruments, Stoelting, Ugo Basile Biological Research Apparatus) with appropriate rotating cylinder for mice (diameter ~30 mm) or rats (diameter ~70 mm); see Figure 3.3.1

1. To train the subjects transfer the male or female mice, in their home cages, from the holding room to the experimental room with constant temperature, humidity, and light intensity. Allow mice to habituate to the experimental room for 60 min.

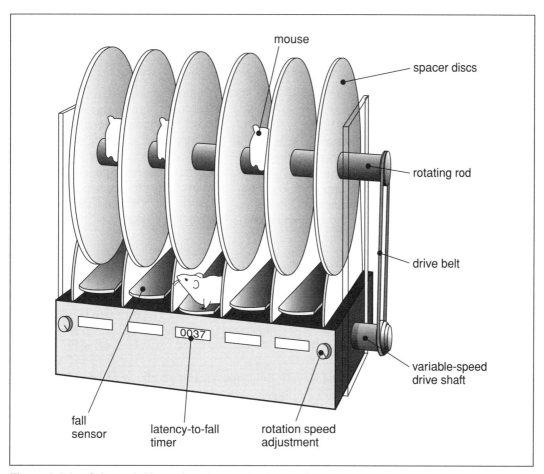

Figure 3.3.1 Schematic illustration of a standard rotarod apparatus for testing motor coordination.

2. Start the rotarod apparatus turning (initially, 24 rpm recommended); if necessary, manually calibrate it so that it is revolving at a consistent speed. Place each mouse individually on the rotating rod for a maximum of 60 sec. Record the latency to fall off the rotarod within this time period. Return mice to their home cages between trials, allowing a 5- to 10-min intertrial interval.

3. Give the mice four trials per day for three consecutive days to attain a stable baseline level of performance (if not achieved, continue training for an additional 2 or 3 days). Clean the apparatus frequently during testing. At the end of each day's training session, return all mice to the holding room, turn off the rotarod, and clean the apparatus thoroughly with 70% ethanol.

4. To test the subjects transfer the mice, in their home cages, from the holding room to the experimental room. Allow the mice to habituate to the experimental room for 60 min. Turn the rotarod on; if necessary, manually calibrate the rotarod so that it revolves at several different predefined speeds (e.g., 5, 8, 15, 20, 24, 31, 33, and 44 rpm; see Carter et al., 1999).

5. Give the mice two trials at each predefined speed level as described in step 2. For animals impaired on the rotarod, use fewer speed levels (omitting the fastest speeds first) and longer intertrial intervals for testing.

6. At the end of the testing session, return mice to the holding room, turn off the rotarod, and thoroughly clean the apparatus with 70% ethanol. Analyze data (see Support Protocol).

BASIC PROTOCOL 2

BEAM WALKING

This protocol measures footslips and latency to traverse the beam, although a number of other parameters of motor coordination and balance can also be measured (e.g., see Table 3.3.1). All apparatuses necessary to conduct this test can be made simply and cheaply in-house and therefore tend not to be available commercially. The protocol describes the apparatus required for mice, with variations noted as appropriate for rats. Two investigators are usually required to carry out this protocol. However, if only latency measurements are required, the test can be performed by one investigator, and the video camera may not be needed.

Table 3.3.1 Example of a Neurological Scoring System for Beam Walking[a]

Score	Performance on the beam
7	Traverses beam normally with both affected paws on horizontal beam surface, neither paw ever grasps the side surface, and there are no more than two footslips; toe placement style is the same as preinjury.
6	Traverses beam successfully and uses affected limbs to aid >50% of steps along beam.
5	Traverses beam successfully but uses affected limbs in <50% of steps along beam.
4	Traverses beam and, at least once, places affected limbs on horizontal beam surface.
3	Traverses beam by dragging affected hindlimbs.
2	Unable to traverse beam but places affected limbs on horizontal beam surface and maintains balance for ≥5 sec.
1	Unable to traverse beam; cannot place affected limbs on horizontal beam surface.

[a]Adapted from the method of Feeney et al. (1982) used to evaluate unilateral lesions of sensory cortex in rats.

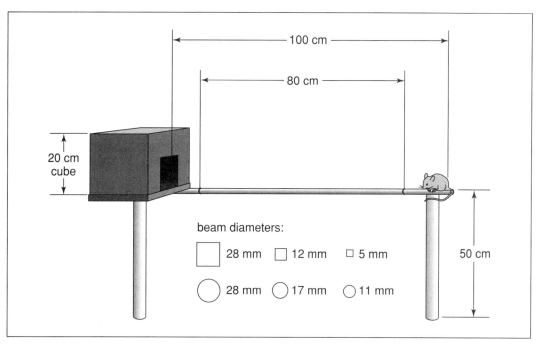

Figure 3.3.2 Schematic illustration of a raised-beam apparatus for testing motor coordination.

Additional Materials *(also see Basic Protocol 1)*

   Beams: 1-m strips of smooth wood with small, medium, and large square, i.e., 5, 12, and 28 mm wide, and round, i.e., 11, 17, and 28 mm in diameter, cross-sections; longer beams, if necessary and beam widths or diameters of 25 to 50 mm for rats or large mice Narrow support stand to hold up the start section of the raised beam (1.5-cm cross-section, 50 cm high; 60 cm high for rats or large mice)
   Goal box (20 cm on each side, with a 4 × 5–cm entrance hole; larger entrance hole for rats or large mice) secured on a narrow support stand (3 cm cross-section, 50 cm high; 60 cm high for rats or large mice)
   Two 60-watt (W) desk lamps
   Video camera, tripod, and blank video tapes

1. Define the middle 80 cm of each beam (the area for taking measurements) by drawing lines 10 cm from the beginning and 10 cm from the end of the beam using a permanent marker. Mount the "start" end of a beam on a narrow support stand on a stable laboratory bench in the experimental room and attach a goal box at the other end (see Fig. 3.3.2).

2. Position two 60-W desk lamps above and to one side of the start of the beam to create an aversive stimulus (bright light) to encourage the mice to traverse the beam to the dark enclosed goal box. If necessary, dim the main light in the test room in order to enhance the contrast of the bright light stimulus.

3. Attach a video camera to a tripod and position it 1.5 to 2 m away from the center of the beam so that the full length of the beam fills the view finder of the video camera. Record all trials on video tape to enable rescoring, if necessary.

4. To train subjects transfer male or female mice, in their home cages, from the holding room to the experimental room [kept constant (with respect to temperature, humidity, and light intensity) for all test sessions]. Allow the mice to habituate to the experimental room for 60 min. Turn lamps on so that they illuminate the start area of the beam. Place a mouse

at the start of the training beam (12-mm-wide square beam recommended for training mice, and the 28-mm-wide square beam for rats) and time the latency to traverse the beam (up to 60 sec).

5a. *For two investigators:* Begin recording with the video camera, with one investigator operating the video camera and recording footslips from the video camera side of the beam and the second investigator placing mice on the beam at the start of each trial, removing them at the end of each trial, recording the latency to traverse the beam, and counting the number of footslips on his or her side of the beam.

5b. *For one investigator:* Repeat each trial, with the animal viewed first from the left side and then from the right, so that footslips can be recorded separately for the two sides of the body.

6. Give the mouse four consecutive trials and return it to its home cage. Clean the apparatus with 70% ethanol (thoroughly, if a mouse urinates or defecates on the beam) and allow alcohol to evaporate fully before reusing the apparatus. Repeat for all additional mice. At the end of the day's training session, return all home cages to the holding room and clean apparatus thoroughly with 70% ethanol. Repeat for a total of three consecutive days to generate a stable baseline level of performance (traversing the 12-mm-wide beam in <20 sec; if not achieved, continue training for an additional 2 or 3 days).

   During the first few trials, some animals may need encouragement (such as an occasional nudge with a finger) to traverse the full length of the beam.

7. To test the subjects transfer the mice, in their home cages, from the holding room to the experimental room. Allow mice to habituate to the experimental room for 60 min.

8. Give mice two consecutive trials on each of the square and round beams. In each case, test first with the widest beam (square ones first) and progress to the narrowest beam. Test with fewer beams when animals are poor at the task. Measure the latency to traverse each beam, and the number of times the left and right hindfeet slip off each beam for each trial. Allow mice up to 60 sec to traverse each beam. Record falls from the beam as failures or allocate a maximum latency of 60 sec for inclusion in latency analyses (see Support Protocol).

9. After completing two trials on each of the six beams, return mice to their home cages. At the end of the testing session, return all home cages to the holding room. Clean apparatus thoroughly with 70% ethanol. Analyze data (see Support Protocol).

BASIC PROTOCOL 3

FOOTPRINT ANALYSIS

Additional Materials *(also see Basic Protocol 1)*

Nontoxic paints in two contrasting colors (e.g., orange and purple)
Open-top runway (for mice: 50 cm long, 10 cm wide, with walls 10 cm high; for rats: 100 cm long, 10 cm wide, with walls 20 cm high), brightly illuminated from above
Enclosed goal box (20-cm square, with a 4 × 5–cm entrance hole)
White paper (A3 size: 29.7 × 42–cm)
Fine paint brushes

1. Set up an open-top runway with an enclosed goal box at one end in the experimental room (with constant temperature, humidity, and light intensity)

2. Allocate an identity number to each mouse. Label the corner of each sheet of white paper with an identity number so that each mouse has its own sheet. Place the sheet of paper

(lengthwise and the sides of the paper folded up to fit the runway) for the first mouse to be tested on the floor of the runway.

3. To train the subjects transfer the male or female mice, in their home cages, from the holding room to the experimental room. Allow the mice to habituate to the experimental room for 60 min.

4. Pour a small quantity of each nontoxic paint into separate petri plates. Hold a mouse by the scruff of the neck and paint the forepaws one color (e.g., orange) and the hindpaws the other color (e.g., purple). Use a different fine paint brush for each color, and use just enough paint to cover the pads of the paws.

5. Immediately place the mouse on the end of the sheet of paper opposite the goal box. Allow the mouse 60 sec to run over the paper to the goal box (Fig. 3.3.3). Immediately place mouse in a clean cage for 5 to 10 min to remove paint from the paws. Remove sheet of paper from the runway. Allow the footprint patterns to dry in a well-aerated room for ≥1 hr before storing.

   During the first few trials, some mice may need encouragement (such as an occasional nudge with a finger) to reach the goal box.

6. Give the mice three consecutive trials on each of three consecutive days to attain stable performance before testing, using a fresh sheet of paper and reapplying paint for each trial. Return mice to their home cages after the final trial, and at the end of the day's training session, return all home cages to the holding room. Clean apparatus thoroughly with 70% ethanol.

7. Test the subjects by repeating steps 3 to 6, except give mice only one trial per week (allow a second trial if the resulting footprint pattern is poor). Analyze data (see Support Protocol).

SUPPORT PROTOCOL

DATA ANALYSIS

To compare two groups of mice (for example, transgenic and wild type, or test compound-treated and control), use two- or three-way analysis of variance (ANOVA), depending upon the number of variables and specific design (e.g., groups × age; groups × age × task difficulty; Winer et al., 1991; Sokal and Rohlf, 1994). In cases of significant interactions, use Sidak's test for series of independent post hoc pair-wise comparisons between groups (Rohlf and Sokal, 1995); Dunnett's test for changes in performance over time against a baseline level within groups; and Tukey's, Newman-Keuls', or Duncan's tests for other multiple nonorthogonal comparisons (Winer et al., 1991; Sokal and Rohlf, 1994). The values used for statistical analyses in the Basic Protocols are described below.

Rotarod

The mean latency to fall for the two trials at each speed level, for each mouse, is analyzed. Any mouse that stays on the rod for the full 60-sec trial is allocated a maximum value of 60 sec for analysis.

Beam walking

Mean scores of the two trials for each beam for each measure (i.e., latency, right hind footslip, left hind footslip) are analyzed. Forepaw slips are typically rare, and so are not analyzed

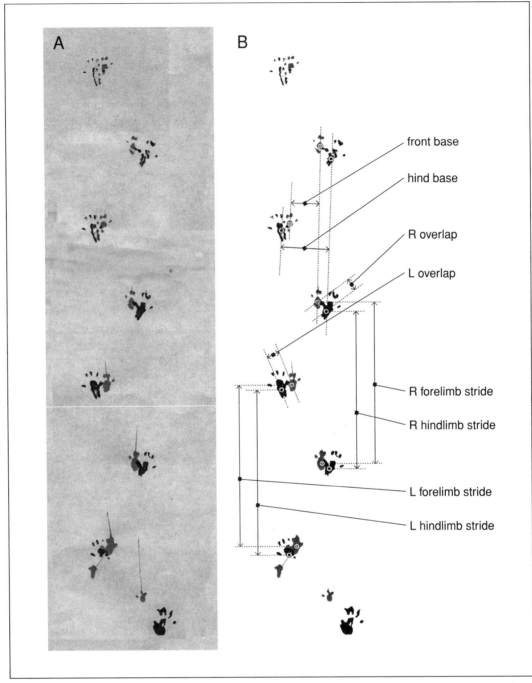

Figure 3.3.3 Illustration of (**A**) footprints and (**B**) the standard measurements that can be collected from them. The darker footprints are those of the hindlimbs. Overlap is the mean of the right and left overlaps, and stride length is the mean of the right (R) and left (L) forelimb or hindlimb strides.

further. As mice become progressively impaired, they can cling tightly onto the beam to prevent themselves from falling. This increases latency to cross; any animal that does not cross within the full 60-sec trial is allocated a maximum value of 60 sec for analysis. Falls are rare in normal mice, but in transgenic animals the total numbers of falls on each test can increase to significant levels as neurological phenotypes become more severe.

Footprint analysis

Four step parameters are typically measured (in centimeters) from the footprint patterns (Fig. 3.3.3) and analyzed. The first, *stride length*, is measured as the average distance of forward movement between each stride. The second, *hind-base width*, is measured as the average distance between left and right hind footprints. This value is determined by measuring the perpendicular distance of a given step to a line connecting its opposite preceding and succeeding steps. The third, *front-base width*, is measured as the average distance between left and right front footprints. The fourth, *overlap between forepaw and hindpaw placement*, is measured as the distance between the front and hind footprints on each side. This is used as a measure of the accuracy of foot placement and the uniformity of step alternation. During normal locomotion in rats and mice, the center of the hind footprint falls on top of the center of the preceding front footprint, so that the overlap value is close to zero. With increasing impairment, the footprint placements become more variable and the distance between front and hind prints on the same side increases. Thus, the overlap—the distance between the centers of the footprints—becomes greater with increasing impairment.

For each step parameter, three values are measured from the middle portion of each runway trial, excluding footprints made at the beginning and end of the trial where the mice initiate and finish movement, respectively. The mean of each set of three values is used in the analysis. In studies of transgenic animals or test compound effects, the data from left and right sides of the body can be pooled. However, for animals with lesions or other unilateral manipulations it is necessary to measure and analyze the two sides of the body independently. In addition, in particular with video recording of stepping, others have recorded the timing (Clarke and Still, 1999), toe spread (Walker et al., 1994), and posture and weight distribution (Miklyaeva et al., 1995) while walking.

Reference: Ossenkopp et al.,1996

Contributors: Rebecca J. Carter, A. Jennifer Morton, and Stephen B. Dunnett

UNIT 3.4

Basic Measures of Food Intake

BASIC PROTOCOL

This protocol generates reliable and stable baseline measures of feeding and is appropriate for experimental manipulations that either enhance (e.g., injecting a peptide into the brain) or suppress feeding (e.g., systemic injections of amphetamine).

Materials

  Rat subjects, preferably males weighing 200 to 250 g (one strain used consistently in a study)
  Solid-pellet or other test diet (e.g., ground chow, high-fat, liquid, or wet mash diets; Purina, Harlan Teklad, Research Diets, Bio-Serv, or equivalent or home made; see Table 3.4.1; maintain same supplier throughout a study)

  Individual rat cages
  Top-loading balance with range of 0 to 600 g, ± 0.1 g
  80-mm-deep $\times$ 210-mm-diameter weighing bowl
  Polycarbonate test cage *or* wire-mesh hanging cage
  Wire grid floor (Lab Products)

Table 3.4.1 Manipulations of Palatability and Caloric Density

Taste manipulation	Additive	Concentration	Base diet/solvent
Negative[a]	Butyric acid	1.75–7.0%	Wet mash
	Acetic acid	1.75–7.0%	Wet mash
	Citric acid	5–20%	Wet mash
	Quinine	0.08–0.2%	Wet mash
Positive	Sucrose[b]	12.5%	Tap water
	Mineral oil[c]	33%	Ground chow
	Crisco[c]	33%	Ground chow
	Saccharin[b]	0.1%	Tap water

[a]Peters et al. (1979).
[b]Grill and Kaplan (1990).
[c]Corbit and Stellar (1964).

    Cardboard spillage pad, cut to cover bottom of cage under wire grid
    500-ml square French glass water bottles
    No. 6 rubber stoppers
    110-mm angled or straight stainless steel sipper tubes (Ancare)
    Food hoppers
    50- to 100-ml calibrated glass drinking tubes or graduated 50-ml polycarbonate centrifuge tubes
    Laboratory timer capable of timing ≥60-min duration to 1-sec resolution
    Data sheets (e.g., see Fig. 3.4.1)
    Test diet containers (e.g., glass petri dish glued to the base of a 4-oz. glass baby food jar with Super Glue)
    Forceps

1. After arrival, house the rats in individual home cages coded with a unique number and experiment code. Identify individual rats as required by the IACUC (e.g., using a permanent marker to numerically code the tail or see Wellman, 1994). Maintain the colony room with constant humidity (60% to 70%) and temperature (23°C) under an appropriate lighting schedule (e.g., lights on from 0600 to 1800 hr). Allow animals continuous access to a rodent pellet diet and tap water.

2. At the same time each day, weigh each rat to the nearest gram in a weighing bowl on a top-loading balance, handling each animal for a few minutes before and after weighing. Record weight and subsequent experimental data on a sheet such as that shown in Figure 3.4.1. Monitor for illness, indicated by a 10- to 15-g weight loss in a 24-hr period or scruffy fur, red material around the eyes, and/or extreme vocalizations upon handling.

3. Prepare test cages (in separate racks or on separate shelves from the housing cages) by readying a drinking tube mount (e.g., see Fig. 3.4.2) and placing a dry cardboard pad in the test cage beneath the wire floor to catch small food particles and waste materials. Keep each test cage clean (no bedding) and dry.

4. In mid to late afternoon of each day, begin setup for trials by assembling drinking tubes. Before presenting the sipper tube and drinking tube to a rat, hold upright and shake sharply to verify that fluid flows from the tip of the sipper tube, and that it does not continue to drip. Extend the inside end of the sipper tube a few millimeters beyond the stopper to ensure reliable water flow. If a rat chews on the edges of a rubber stopper, mount a stainless steel washer on the cage wall and extend the sipper tube through the hole in the washer.

Date: ________________

Experiment Title: ________________________

Experimenter: ________________________

Trial: ________________________

		Food (g)			Water (g or ml)		
Rat	BW (g)	In	Out	Intake	In	Out	Intake
1							
2							
3							
4							
5							
6							
7							
8							
9							
10							

Comments:

____

____

____

____

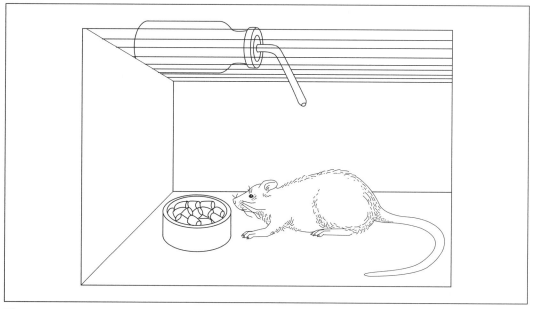

Figure 3.4.1 Sample data sheet for a rat feeding experiment. BW, body weight.

Figure 3.4.2 Rat cage and feeding setup. (Figure courtesy of M. Meagher.)

5. Place the balance near the test cages. Clean and dry the balance pan. Wash hands before handling the diet, and do not use potent perfumes or lotions prior to handling test diets. Weigh the test diets and drinking tubes for each rat in the series to the nearest 0.1 g. Record these values on the data sheet prior to the start of the ingestive test. Weigh out enough diet to last each rat throughout the established length of the test period (e.g., 12 to 15 g of pellets for a 30-min test period).

6. At the same time each day (starting at ~1600 hr, preceding offset of lights and coinciding with a time of day when rats spontaneously consume a 1- to 1.5-g meal in a 30-min period), transfer each rat from its home cage to a test cage, establishing a consistent time interval by staggering the start time for successive rats (e.g., a 2-min interval). Minimize disturbance while providing the food and water, and while the ingestion test is in progress. Test each rat for the same duration.

   The test also can be timed during the period of maximum food intake when the lights go out in the colony room, or the period of minimum intake when the lights come on.

7. Place the diet at the rear of the test cage away from the water source to prevent wetting directly on the wire floor for pelleted diet or in a small container (a petri dish glued with Super Glue to the base of a baby food jar or tuna can) for nonpellet diets. Place the water bottle above or in front of the cage, with the sipper tube extending into the cage interior. Start a timer for the test interval of the first rat (e.g., 30 min).

8. Remove the rat from the test cage and place it in a nearby holding cage. Remove each drinking tube and record its weight to the nearest 0.1 g (or volume, for graduated drinking tubes) on the data sheet. Recover the uneaten portion of the diet from the test cage and place into the weighing pan. Retrieve the cardboard pad and pour the remaining diet spillage into the weighing pan. Remove and discard any fecal material or debris from the pad with forceps. Record the total weight of remaining diet (plus spillage) to the nearest 0.1 g.

   If a rat urinates on the food spillage on the cardboard pad, or a drinking tube leaks onto food spillage on the cage floor, label the pad with cage number and date, set it aside to dry with the leftover diet, and determine the food weight the following day.

9. Return each rat to its home cage. Return the respective feeder and water bottle of each rat to the home cage. Always discard old food and start with fresh diet for each trial: do not reuse test diets already offered to rats. Clean and refill drinking tubes with fresh water prior to the next feeding test.

10. Calculate the total amount of diet and water consumed during the test using initial and final weights.

Reference: Toates and Rowland, 1987

Contributors: Paul J. Wellman and Lance R. McMahon

UNIT 3.5

Sexual and Reproductive Behaviors

Reproductive behaviors are critical for the evolutionary success of an individual. They are contingent upon the action of gonadal steroid hormones within the central nervous system, and this action results in expression of different types of behaviors in males and females. Thus, these behaviors can serve as model systems for investigating steroid-dependent and sexually dimorphic behaviors.

BASIC PROTOCOL 1

ASSESSMENT OF MOUNTS, INTROMISSIONS, AND EJACULATIONS IN MALE RATS

This protocol describes how to conduct tests for the ability of a male rat to display mounts, intromissions, and ejaculations. These behaviors are operationally defined as follows:

Mount: A properly oriented mount occurs when the male mounts the female from the rear and grasps her flanks with his front feet. In many species, mounts are accompanied by multiple pelvic thrusts. Sexually inexperienced males often perform incompetent mounts in which the orientation is wrong, resulting in mounts of the female's head or side. Some experimenters score "correct" and "incorrect" mounts as separate behaviors.

Intromission: An intromission occurs when vaginal penetration is achieved by the male during a mount. Like mounts, intromissions may be accompanied by pelvic thrusting, and therefore it can be difficult to distinguish between the two behaviors unless the observer has a ventral view of the male. For this reason, a mirror positioned under a glass testing arena is recommended.

Ejaculation: At the end of an intromission, or after a series of rapid intromissions, the male ejaculates sperm and seminal fluid into the vagina. When learning to recognize the motor patterns correlated with an ejaculation, perform a vaginal lavage of the female with 0.1 ml water using an eyedropper and view the lavage fluid on a coverslipped slide with low-level brightfield microscopy to determine if sperm are present in the vagina. Ejaculations are marked behaviorally by an intromission in which a longer-lasting thrust occurs, and they are accompanied by characteristic pelvic motor patterns. Ejaculation is always followed by extended (several minutes or more) genital grooming and a period of disinterest in the female, even if she solicits the male, but genital grooming should not be used as the sole indication of ejaculation, as males frequently groom after nonejaculatory intromissions.

It is recommended that the novice observer not rely solely on real-time observation to learn to recognize the motor behaviors, but also videotape experimental test sessions so that behaviors can be reviewed several times if necessary for correct identification. Additionally, recording creates a permanent record of the test session, so that behaviors may be scored in a different way at a later time, and additional behaviors of subsequent interest can be scored. Software programs designed specifically for data collection and analysis of reproductive behavior are described by Rakerd et al. (1985), Holmes et al. (1988), Claro et al. (1990), Weed and Boone (1992), and Mallick et al. (1993). A skilled programmer can also develop customized software.

Many nocturnal animals display the most robust reproductive behavior during the first 2 to 4 hr after the beginning of the dark cycle, and it is recommended that testing occur during this time period. When testing animals during the dark portion of the light/dark cycle, it is important to have no light source other than dim red light (≤ 25 W per bulb). It is important to test the capabilities of the video camera and lens for producing a clear picture under dim red light conditions. Special low-light cameras and lenses can be purchased if necessary. Animals should be moved to the testing room during the light portion of the cycle and the door covered with a light-impermeable baffle (e.g., heavy black plastic) to allow entry and exit into the testing room without exposing animals to light. A two-room testing area (i.e., anteroom and test room) allows for all animals to be brought to the testing area during the light portion of the cycle, while separating odors and sounds from animals in the testing room from animals on hold in the anteroom. For convenience, animals can be kept on a reversed light/dark cycle so that the animals' dark period occurs during the experimenter's day.

Materials

    Female rats 60 to 120 days old, either ovariectomized and steroid primed (see Support Protocols 2 and 5) or gonad intact, one per test
    70% ethanol
    Test males, either castrated and steroid primed (see Support Protocols 1, 3, and 4) or gonad intact
    20-gallon glass aquarium mounted on an open framework with tilted mirror positioned underneath for visualization of ventral side of the animal
    Lamp(s) with dim red light bulb, if testing during dark portion of the light/dark cycle
    Video camera, videotapes, VCR, and monitor
    Pen and data pad *or* computer and appropriate software for data entry and analysis
    Stopwatch

1a. *Using ovariectomized female rats:* To prepare stimulus females that are sexually receptive for the tests, induce estrus by sequential treatment with estradiol and progesterone (see Support Protocol 5). Use stimulus females just once on any given testing day.

1b. *Using gonad-intact, cycling female rats:* Track the estrous cycle by daily monitoring of cytological changes in vaginal smears; rats become receptive ~3 to 4 hr after the preovulatory surge of luteinizing hormone (see Freeman, 1994 for details of cytological monitoring).

2. If using castrated rats as test animals, prepare them for the test by priming with testosterone (see Support Protocols 3 and 4).

3. Set up the testing equipment (aquarium, lamps, camera and VCR, and computer) in a quiet room. Position the camera equipment and lights so that the aquarium and the mirror image underneath it are clearly visible. Ensure that the computer program, if used, is functioning properly.

4. Clean the aquarium with paper towels and soapy water and then with 70% ethanol. Let the ethanol evaporate before beginning the test. Set the correct date and time on the camera/VCR. Place a piece of tape identifying the male subject and stimulus female on the side of the aquarium.

5. Place the test male in the aquarium and allow 10 min for it to acclimate. Start tape recording, place the stimulus female in the cage with the male, and begin timing the test. Do not talk or move about during the test.

6a. *Scoring mounts, intromissions, and ejaculations:* Record these by marking on a scoresheet each time one of these behaviors occurs during the test session, and determine the number of mounts and intromissions associated with each ejaculation.

6b. *Scoring latency time to the first occurrence of a mount, intromission, or ejaculation:* Check the stopwatch upon the first occurrence of each behavior.

6c. *Scoring inter-intromission interval and postejaculatory intromission latency:* Determine the average length of intervals between intromissions and the amount of time between an ejaculation and the next intromission using videotape or computer-based scoring.

7. Conduct tests for a minimum of 10 min, longer, in general, for sexually inexperienced males (a 1-hr test is more than sufficient for most experimental objectives). Terminate a test if it appears that serious harm could come to the male from an aggressive female.

8. At the end of the test, remove the male and stimulus female, and turn off the recording equipment. Before conducting the next test, clean the aquarium as described in step 4.

BASIC PROTOCOL 2

ASSESSMENT OF LORDOSIS IN FEMALE RATS

Lordosis is the reflexive posture assumed by a receptive female in response to appropriate tactile stimulation, such as mounts by a male or pressure on the back, flanks, or anogenital region. In female rats, lordosis is characterized by pronounced arching of the back, with the head and hindquarters elevated, back feet extended, and tail deflected to one side. Lordosis is commonly scored as a lordosis quotient (LQ), in which the frequency of lordosis is scored as a ratio to a fixed number of mounts (usually 10) × 100. For example, if the female lordoses during five out of ten mounts, the LQ would be 50, and it would be inferred that the female is moderately receptive. A 10- to 15-min testing time is usually sufficient to obtain a lordosis quotient.

Materials

Stimulus male 60 to 120 days old, either castrated and steroid primed (see Support Protocols 1, 3 and 4) or gonad intact; one per test
70% ethanol
Test females, either ovariectomized and steroid primed (see Support Protocols 2 and 5) or gonad intact

20-gallon glass aquarium
Video camera, videotapes, VCR, and monitor
Lamp(s) with dim red light bulb, if testing during dark portion of the light/dark cycle

1a. *Using intact male rats:* Use sexually experienced, gonad-intact adult male rats as stimulus males. Use an individual male in only one test per day, if possible.

1b. *Using castrated male rats:* If using castrated male rats as stimulus males, use castrated males (with prior sexual experience) primed with testosterone (see Support Protocols 3 and 4).

2. If using ovariectomized female rats as test animals, prepare them for the test by treating them sequentially with estradiol and progesterone (see Support Protocol 5).

    Without additional information on whether the female is cycling or her level of circulating ovarian steroids, the use of gonad-intact females does not allow ascribing the absence of lordosis to acyclicity or inappropriate amounts of ovarian steroids, or to the inability of the central nervous system to respond to ovarian steroids.

3. Set up testing equipment (see Basic Protocol 1, step 3; mirror not necessary unless scoring male behavior, too).

4. Clean the aquarium (see Basic Protocol 1, step 4). Set the correct date and time on the camera/VCR. Place a piece of tape identifying the female subject and stimulus male on the side of the aquarium.

5. Place the test female in the aquarium and allow 10 min for it to acclimate. Start tape recording, place the stimulus male in the cage with the female, and begin timing the test. Do not talk or move about during the test. Score the frequency of occurrence of lordosis either in real time or later from the videotape.

6. At the end of the test, remove the female and stimulus male, and turn off the recording equipment. Before conducting the next test, clean the aquarium (see Basic Protocol 1, step 4).

SUPPORT PROTOCOL 1

CASTRATION

Two common experimental objectives require castration of an animal. One is to determine whether a particular behavior is dependent on testicular secretions. A second is to "clamp" (hold constant) circulating testosterone levels in experimental animals; this objective requires removal of the source of endogenous testosterone (castration) so that exogenous testosterone can be administered to produce similar levels of circulating steroid in all animals. Castrations should be performed in accordance with individual institutional guidelines and with the Institutional Animal Care and Use Committee (IACUC) guidelines and recommendations for aseptic surgery.

Materials

Male rat
Anesthetic (as recommended by IACUC)
Topical anti-infective: 5% Nolvasan solution or Betadine
Topical antiseptic cream, antibiotic, and analgesic (if recommended by IACUC)

Surgical gloves, mask, and lab coat
Padded bench paper or surgical pads
Heating pad
Shaver
Surgical scissors, operating and iris
Mouse-tooth forceps
Gauze
3-0 surgical silk and suturing needles *or* 9-mm wound clips and applicator
Holding cage lined with paper towels

1. Set up an aseptic surgical area. Place a heating pad set on low on the bench and cover with padded bench paper. Perform surgery under conditions approved by the IACUC and wear surgical gloves, mask, and lab coat throughout the procedure.

2. Anesthetize the animal (e.g., using a short-acting inhalant such as methoxyflurane for the <15 min surgery). Determine that the animal is sufficiently anesthetized by pinching the tail or hind foot, or by lightly touching the eyelid to check for lack of reflexive withdrawal or eye blink.

3. Place the animal on its back. Shave the scrotum and clean the area with gauze and Nolvasan or Betadine. The testes may ascend into the body during handling and anesthetic administration; if necessary, palpate the abdomen to make the testes descend.

4. Using surgical scissors, make a 10-mm incision in the skin over one testis. Position the incision laterally so that the anus and penis will not be compromised by either the incision or by wound clips (see Fig. 3.5.1A).

5. Using iris scissors, make a 10-mm incision through the sac in which the testis is enclosed. In castrating an adult, there will be a large fat pad associated with the testes; grasp this with mouse-tooth forceps. In castrating a prepubertal animal, grasp the testis itself with the forceps. Pull the testis and fat pad out of the sac. Take care not to pull out the intestines or the bladder (see Fig. 3.5.1B).

6. Tie a loop of surgical silk around the connective tissues leading to the testis, which include the fat pad, testicular vein, and vas deferens, and knot it tightly several times to prevent heavy bleeding when the testis is removed (see Fig. 3.5.1C).

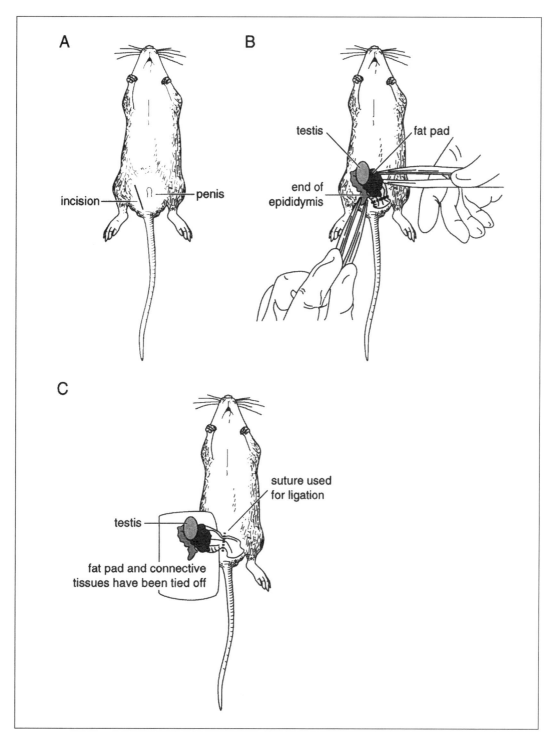

Figure 3.5.1 Castration procedure. (**A**) For castration, a 10-mm lateral skin incision is made in the skin over the testis. (**B**) The testis and associated fat pad (in adults) are retracted with forceps through the sac which contains the testis. The area to be tied off with surgical silk is indicated by the dotted line. (**C**) A loop of surgical silk is tied around the connective tissues leading to the testis, which include the fat pad, testicular vein, and vas deferens. The silk is tightly knotted several times prior to excision of the testis and fat pad.

7. Using operating scissors, cut the connective tissue between the testis and the knot around the connective tissue. Also remove the epididymis (attached to the testis) at this time. Return the ligated connective tissue to the body cavity.

8. Clean the wound gently with gauze and Nolvasan or Betadine. Close the skin with sutures or two or three wound clips. Repeat steps 4 to 8 to remove the other testis.

9. Administer topical antiseptic cream, antibiotic, or analgesic, if recommended by IACUC. Place the animal in the holding cage placed on top of a heating pad set to low. Monitor the animal closely as it recovers from the anesthetic, and when it has regained consciousness, place individually in a home cage (not group-housing). After 3 days remove the wound clips, if desired.

SUPPORT PROTOCOL 2

OVARIECTOMY

The rationale for performing ovariectomy is similar to that for castration—it removes the source of endogenous ovarian hormones so that the level of circulating steroids can be controlled by administration of known doses of exogenous hormone. Ovariectomies should be performed in accordance with individual institutional guidelines and with IACUC guidelines and recommendations for aseptic surgery, anesthesia, and analgesics.

Materials

    Female rat
    Anesthetic (as recommended in IACUC guidelines)
    Topical anti-infective: 5% Nolvasan solution or Betadine
    Antiseptic cream, analgesic, and antibiotic (if recommended by IACUC)

    Surgical gloves, mask, and lab coat
    Padded bench paper or surgical pads
    Heating pad
    Shaver
    Surgical scissors, operating and iris
    Mouse-tooth forceps
    Hemostats
    Gauze
    3-0 surgical silk and suturing needles *or* 9-mm wound clips and applicator

1. Set up an aseptic surgical area. Place a heating pad set on low on the bench and cover with padded bench paper. Perform surgery in accordance with institutional IACUC guidelines and wear surgical gloves, mask, and lab coat throughout the procedure.

2. Anesthetize the animal (e.g., using a short-acting inhalant such as methoxyflurane for the <15 min surgery). Determine that the animal is sufficiently anesthetized by pinching the tail or hind foot, or by lightly touching the eyelid to check for lack of reflexive withdrawal or eye blink.

3. Position the animal on its side and place a finger or thumb parallel to the bottom of the ribs (perpendicular to the spinal cord) so the fingertip is touching the spine. The area under the fingertip (in the angle formed by the ribs and the spine) is where an incision should be made (see Fig. 3.5.2A). Shave this area and clean with gauze and Nolvasan or Betadine.

4. Using surgical scissors, make a 5- to 10-mm incision that bisects the angle formed by the ribs and spinal column. Cut through both the skin and underlying fat. Under the skin

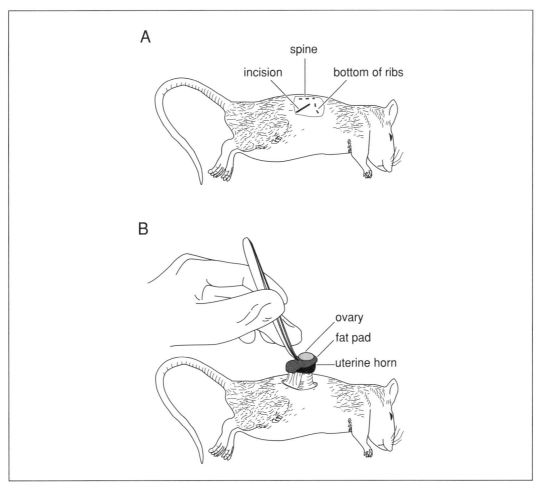

Figure 3.5.2 Ovariectomy procedure. (**A**) The incision made for ovariectomy should be 5 to 10 mm long and bisect the angle formed by the ribs and spinal column (shown as dotted lines). (**B**) Retraction of the forceps inserted ~5 mm through the incision will grasp either the ovary or its associated fat pad. Shown here are the ovary, the fat pad, and the uterine horn to which they are attached. A hemostat should be clamped between the uterine horn and ovary while the suture should be tied off behind the hemostat.

and fat is the muscle layer. Using an iris scissors, make an incision 5 mm long in the muscle.

5. Insert mouse-tooth forceps straight down into the incision in the muscle to a depth of ~5 mm; close the forceps and gently retract. If the incision was correctly placed, the forceps will grasp either the ovary or its associated fat pad. Pull the ovary, fat pad, and the uterine horn to which they are attached out of the opening in the muscle (see Fig. 3.5.2B).

6. Clamp the uterine horn with a hemostat. Making two knots, tie off the uterine horn with surgical silk behind the hemostat, and cut away the ovary and most of the fat pad. Remove the hemostat. Gently feed the ligated uterine horn back into the hole in the muscle with the forceps.

7. Once the horn is returned to the body cavity, clean the wound gently with gauze and antiseptic. Suture the incision in the muscle layer with surgical silk. Close the skin incision with sutures or wound clips. Repeat steps 3 to 7 on the other side to remove the other ovary.

8. Administer topical antiseptic, antibiotic, or analgesic, if recommended by IACUC. Place the animal in a heated holding cage lined with brown paper towels. Monitor the animal closely as it recovers from the anesthetic and for a couple of days after surgery. Do not group-house animals immediately after surgery. After 3 days remove wound clips, if desired.

SUPPORT PROTOCOL 3

TESTOSTERONE TREATMENT USING SUBCUTANEOUS IMPLANT

Administration of hormone to gonadectomized animals is an effective way of controlling circulating levels of hormones in experimental animals. The most common method of testosterone treatment is subcutaneous implantation of a porous capsule containing testosterone. If making capsules as described is not practical, time-release pellets impregnated with a specific amount of steroid hormone can be purchased commercially from Innovative Research. With any method of hormone replacement, it is advisable to assay plasma samples to determine the circulating levels of hormone that are produced by the treatment. The physiological range of circulating testosterone in adult male rats is 1 to 6 ng/ml.

Materials

Silicone type A medical adhesive (Factor II)
Crystalline testosterone (Sigma or Steraloids)
70% ethanol
Saline solution: 0.9% NaCl
Castrated male rat (see Support Protocol 1)
Anesthetic (e.g., methoxyflurane)
Topical anti-infective: 5% Nolvasan solution or Betadine

Silastic medical-grade tubing (Dow Corning 601-285, 1.57-mm i.d., 3.18-mm o.d.)
Two 1-ml syringes and one blunt-ended 16-G needle
Padded bench paper or surgical pads
Surgical gloves, mask, and lab coat
Shaver
Gauze
Surgical scissors
Forceps
9-mm wound clips and applicator

1. Cut Silastic tubing in 20-mm lengths. Using a syringe with no needle, fill one end of the piece of tubing with silicone adhesive to a length of 5 mm. Dry for 24 hr.

2. Working in a hood, fill tubing with testosterone powder using a small funnel made from a Pasteur pipet. Score and break off most of the tip and most of the barrel and fire each end to smooth. Fit funnel into open end of tubing and fill funnel with hormone. Touch capsule/tubing unit to vortex mixer to shake hormone into tubing. Use a straightened large paper clip to tamp hormone into tubing. Pack tubing so that the length of the hormone filling is 10 mm.

3. Fill open end with silicone adhesive using a syringe and 16-G needle with blunted end. Dry for 24 hr.

4. Clean capsule with 70% ethanol and trim ends so that 2 mm adhesive remains on each end. Soak in 0.9% saline for 24 hr prior to implanting in animal.

5. Set up an aseptic surgical area. Perform surgery in accordance with IACUC guidelines and wear surgical gloves, mask, and lab coat throughout the remainder of the procedure.

6. Anesthetize the animal (e.g., using a short-acting inhalant such as methoxyflurane for the <5 min surgery). Shave a small area on the animal's nape. Clean with gauze and Nolvasan or Betadine. Make a 5-mm incision using surgical scissors. Use closed forceps to separate the skin from the muscle layer and create a cavity just large enough to contain two capsules.

7. Insert two capsules (number varies between laboratories) into the cavity and close the incision with a single wound clip. Apply topical antibiotic ointment. House animals individually after surgery and monitor animal until it recovers from anesthesia.

SUPPORT PROTOCOL 4

TESTOSTERONE TREATMENT BY SUBCUTANEOUS INJECTION

Additional Materials *(also see Support Protocol 3; see* APPENDIX 1 *for items with* ✓ *)*

✓ Steroid hormone (testosterone propionate in sesame oil)
Surgical gloves, mask, and lab coat
1-ml syringe with 26-G needle

1. Pinch a fold of skin on the nape of the animal and inject an appropriate quantity of testosterone between the skin and muscle, using a 1-ml syringe and 26-G needle. Be careful not to push the needle through the skin fold and out the other side. Wear surgical gloves, mask, and lab coat when performing injections.

2. Inject the steroid hormone testosterone propionate at a dose of 2 mg/kg (0.07 ml of the 10 mg/ml testosterone for a 350-g rat) daily for a minimum of 7 days. For rats of different body weights alter the concentration to permit an injection volume of ∼0.1 ml.

   Plasma levels of testosterone will peak within 2 to 3 hr. However, plasma testosterone concentrations will remain within physiological range throughout the 24 hr following each daily injection.

SUPPORT PROTOCOL 5

ESTRADIOL AND PROGESTERONE TREATMENT

Ovariectomized females can be brought into estrus by sequential treatment with estrogen and progesterone. This protocol can be used to create stimulus females for tests of male reproductive behaviors, or to determine whether an experimental manipulation interferes with the ability of ovarian hormones to induce receptivity.

Materials *(see* APPENDIX 1 *for items with* ✓ *)*

✓ Steroid hormone (β-estradiol 3-benzoate in sesame oil)
✓ Steroid hormone (progesterone in sesame oil)
Ovariectomized female rat (see Support Protocol 2)

Surgical gloves, mask, and lab coat
Two 1-ml syringes and 26-G needles

1. Wearing surgical gloves, mask, and lab coat, inject the steroid hormone estradiol subcutaneously at a dose of 50 μg/kg (or a standard amount, e.g., 10 μg estradiol to rats in the weight range of 200 to 300 g) as described for testosterone (see Support Protocol 4, step 1).

2. Forty-eight hours after the estradiol injection, inject the steroid hormone progesterone subcutaneously at a dose of 2.5 mg/kg (or a standard amount, e.g., 500 μg progesterone to rats in the weight range of 200 to 300 g).

   The female will become receptive ∼5 hr after the progesterone injection, and will remain receptive for ∼12 to 15 hr thereafter.

3. Wait 4 days after induction of estrus before readministering estradiol if stimulus females will be used again.

References: Cherry, 1993; Meisel and Sachs, 1994; Pfaff et al., 1995; Price, 1993; Smith et al., 1977

Contributors: Cheryl L. Sisk and Leslie R. Meek

UNIT 3.6

Parental Behaviors in Rats and Mice

DESCRIPTION OF PARENTAL BEHAVIORS

Parental behavior in rats and mice encompasses two general categories of responses: active behaviors (e.g., construction of a nest site, carrying pups from one place to another, and licking the pups) and quiescent behaviors, interspersed with active behaviors, when parental rodents cease moving and quietly position themselves over the pups for prolonged periods of time (e.g., nursing behavior). These parental behaviors are defined below.

Nest construction

Nest construction involves carrying of nest material with the mouth, or pushing it with the snout or paws, to the potential nest site, creating a tightly bound mass of material with a depression in the middle in which the subject interacts with the pups.

Retrieval

Retrieval of displaced or scattered pups first involves the subject orienting to and moving towards the pup, often sniffing the pup before gently picking it up with the incisors, carrying it over a relatively long distance back to the nest site, and finally depositing it there. In contrast, when a subject orally repositions pups for short distances within or around the nest, their behavior is sometimes referred to as "mouthing."

Licking of pups

Parental rodents often spend a great deal of time licking neonates, a behavior that cleans the pups, increases their activity, and stimulates elimination of waste from the young. Licking also has long-term effects on the emotional, endocrinological, and sexual development of pups. Furthermore, lactating dams obtain important resources from the pups' urine. Therefore, it is often helpful to separate maternal licking into two categories: licking specifically of the pup's anogenital region and licking of the rest of the pup's body.

Actively hovering over pups

The subject is positioned over some or all of the pups in the nest, but is not quiescent. Instead, the subject is relatively active, performing a variety of activities such as licking the pups, self-grooming, or moving nest material. In lactating rodents, hovering over the litter provides the pups access to the dam's nipples.

Quiescent positioning over pups

In lactating rats, quiescence is typically induced by the suckling provided by a sufficient number (i.e., at least four) of pups. One of the most frequent nursing postures displayed is *kyphosis* (upright crouch; Stern, 1996), which is characterized by the dam standing over the litter with rigidly splayed limbs and a dorsal arch of the spinal column. When the animal adopts a kyphosis posture, it does not lick the pups or engage in other active behaviors. A second common nursing response, lying *supine*, occurs when the animal is stretched out on its back or side with its nipples exposed. Other postures that may also be observed include laying *prone* on top of the litter with no limb support and the subject sitting *hunched* over the litter on its haunches with back limb support but no forelimb support. Nonlactating rats, and probably other nonlactating rodents as well, do not display kyphosis for prolonged periods of time and are very often found over the litter in the supine, prone, or hunched positions (Lonstein et al., 1999).

NOTE: In all cases, animals should be housed under standard laboratory conditions, with an ambient temperature of ∼22°C, relative humidity of 40% to 50%, and a 12:12 light:dark photoperiod. Unless otherwise noted, standard laboratory rat or mouse chow and water should be available ad libitum.

ASSESSMENT OF PARENTAL BEHAVIORS

Two commonly used methods to examine parental behaviors in rodents are periodic (spot-check) observations of the subject's undisturbed behavior with the litter, and continuous observation—either undisturbed or following varying periods of separation between the parent and litter (ranging from 5 min to 4 hr). Test durations can also range from 10 min to a number of hours. A shorter test using continuous observation is sometimes followed by periodic spot-checks throughout the day. The rationale for choosing a particular procedure depends on both the primary objective of the exercise and practical constraints. Procedures are also provided for assessing nest construction and retrieval of separated pups.

BASIC PROTOCOL 1

ASSESSMENT OF PARENTAL BEHAVIORS USING DIRECT PERIODIC SPOT-CHECK OBSERVATION

This procedure is advantageous because it requires little equipment and can provide a "snapshot" of the parent-litter interaction under undisturbed conditions at various periods throughout the day.

Materials

Parental lactating or nonlactating rodent subject
Litter of 1- to 8-day-old pups provided by a lactating "donor" animal (there are usually 6 to 8 pups in a culled litter, with equal numbers of males and females included)

Clear glass or polypropylene home and testing cage (for rats: ~48 × 28 × 16–cm; for mice: 21 × 16 × 13–cm) with wood chips or shavings for bedding (~8 cups for rats, 4 cups for mice). Cotton balls or pads can be used for mice.

Paper data sheets with each minute of the observation separated into 5-, 10-, or 15-sec periods. Each sheet is divided into columns that represent time periods and rows that represent the individual behaviors.

Stopwatch

1. Test the subject in a familiar home cage (recommended) or transfer the subject and the pups into a clean transparent cage (if the subject's home cage is opaque) with fresh nesting material at least 24 hr prior to testing.

2. To begin the test, start the stopwatch and quietly note the subject's behavior without disturbing her or the pups. Record the subject's behavior (e.g., hovering over, nursing, licking, nesting; see unit introduction) on the data sheet every 5, 10, or 15 sec until the conclusion of the test (typically 10, 20, or 30 min). Record each behavior only once within any single interval. Take care not to disturb the subject when exiting the room at the end of testing.

3. Repeat steps 1 and 2 as often as necessary throughout the day.

BASIC PROTOCOL 2

ASSESSMENT OF PARENTAL BEHAVIOR USING CONTINUOUS OBSERVATION

Materials

Parental lactating or nonlactating rodent subject

Litter of 1- to 8-day-old pups provided by a lactating "donor" animal (there are usually 6 to 8 pups in a culled litter, with equal numbers of males and females included)

Clear polypropylene pan cage (for rats: ~48 × 28 × 16–cm; for mice: 21 × 16 × 13–cm) with wood chips or shavings for bedding (~8 cups for rats, 4 cups for mice). Cotton balls or pads can be used for mice.

Humid incubator set at nest temperature (~34°C)

Plastic or glass dish to hold pups

Cotton swabs

Small animal scale

Laptop computer

Data acquisition software that allows continuous recording of events at least once every second; available commercially (e.g., Noldus Observer, Noldus Information Technology; or Behavior Evaluation Strategies and Taxonomies software, Scolari, Sage Publications Software) although a rather simple data acquisition program may be generated in-house by a person with some computer programming experience

Continuous observation using a subject-litter separation

1a. Wearing gloves (recommended), remove pups from subject or surrogate lactating dam. Leave the subject alone in test cage. Place pups in a small bowl or container and place them in a warm (34°C) environment for 5 to 10 min or in a humid 34°C incubator if separating pups for longer periods of 3 to 4 hr.

2a. Just before testing, remove pups from the bowl or incubator, again using gloves to prevent contamination by human odors. Gently hold each pup on its back and swab its anogenital region to stimulate urination. Place the entire litter on the scale and record the total weight to the nearest 0.1 g. Place pups back in the bowl and bring into the testing room.

3a. Turn on computer data acquisition system and prepare program to begin recording.

Each behavior of interest should have a single key on the keyboard representing it (e.g., "R" could be used for retrieval of pups, "L" for licking the pups, etc.). For convenience, each key should be able to be toggled on so that individual keys do not have to be held down by the experimenter for the duration of each behavior.

4a. Quietly remove the top of the test cage and scatter the pups in the cage opposite to where the subject has built a nest, or where the subject is sitting if a nest has not been constructed. Start the computer and press the appropriate key each time that a particular behavior is observed. Stop the program after the necessary amount of time has passed.

5a. Remove the litter from the subject and again weigh the pups all together. Record the weight and return the litter to either the subject or a lactating dam from the colony. Determine the amount of milk that the pups ingested during testing from the difference in litter weight between the beginning and conclusion of testing.

Continuous observation without a subject-litter separation

1b. Begin this test as one would a spot-check undisturbed observation (Basic Protocol 1, step 1). Without disturbing the subject or litter, quietly sit and start the data acquisition software.

2b. Continuously record each behavior displayed by the dam as it occurs by pressing the keys on the keyboard that represent each activity. After the desired period of time, conclude the test and exit the software program. Quietly leave the room without disturbing the subject and litter.

BASIC PROTOCOL 3

ASSESSMENT OF NEST CONSTRUCTION

Materials

Parental lactating or nonlactating rodent subject
Litter of 1- to 8-day-old pups provided by a lactating "donor" animal (there are usually 6 to 8 pups in a culled litter, with equal numbers of males and females included)

Clear glass or polypropylene home and testing cage (for rats: ~48 × 28 × 16–cm; for mice: 21 × 16 × 13–cm) with wood chips or shavings for bedding (~8 cups for rats, 4 cups for mice). Cotton balls or pads can be used for mice.
Two-ply paper towels (16 × 25–cm) shredded into thin strips

1. Re-house subject and pups in the observation cage at least 24 hr before the first observation. If desired, remove pups from the cage immediately prior to performing step 2. Scatter shredded paper towel strips over entire cage.

2. Four to 24 hr later, rate nest on 5-point scale:

   0 = no nest, paper strips still scattered over entire floor of the cage
   1 = poor nest, not all paper strips are used and the nest is flat
   2 = fair nest, all paper is used but the nest is flat
   3 = good nest, all paper is used and the nest has relatively low walls (<5 cm)
   4 = excellent nest, all paper is used and the nest has relatively high walls (>5 cm).

3. If repeated measurements of nest construction are needed, destroy nests after scoring and repeat the procedure as many times as needed.

ALTERNATE PROTOCOL

ASSESSMENT OF NEST CONSTRUCTION—MICE

Materials

    Parental lactating or nonlactating mouse subject
    Clear glass or polypropylene home and testing cage (21 × 16 × 13–cm)
    Cotton balls or pads (e.g., Nestlet squares)

1. Each day place a large number of cotton balls in the bottom of one side of the animals' food bin and place food over the cotton. Allow the animal to pull the cotton through the cage wiring, just as it does for its food, and use the cotton for a nest.

2. The next day, check for the presence of a nest and rate it on the following scale:

    1 = no nest
    2 = saucer shaped nest
    3 = nest with raised sides
    4 = fully enclosed nest.

3. After rating, weigh the nest. Dispose of the nest and allow the subject to create another one, and rate the new nest the next morning. Repeat as necessary.

BASIC PROTOCOL 4

ASSESSMENT OF RETRIEVAL

Materials

    Parental lactating or nonlactating rodent subject
    Litter of 1- to 8-day-old pups provided by a lactating "donor" animal (there are usually 6 to 8 pups in a culled litter, with equal numbers of males and females included)

    Clear glass or polypropylene home and testing cage (rats: ~48 × 28 × 16–cm; mice: 21 × 16 × 13–cm) with wood chips or shavings for bedding (~8 cups for rats, 4 cups for mice), with floor of cage divided by 1-in. high plastic barriers into four equally-sized quadrants to prevent the pups from crawling to the subject. Cotton balls or pads can be used for mice.
    Stopwatch

1. Remove the pups from the subject and leave the subject alone in the test cage for at least 5 min. Scatter the pups in the quadrant of the test cage opposite to the subject's nest or sleeping corner. Start stopwatch and monitor subject's behavior for 10 min.

2. Note the time that the subject picks up the first pup, carries it back to the nest, and deposits it there. Note the time that the last pup is deposited into the nest. Note the number of pups retrieved within the 10-min test.

3. If more information is needed about retrieval of any stray pups that occurs after the 10-min test, use spot checks to note the position of dam and pups each hour after the 10-min retrieval test. Record the number of pups in the nest at each hour.

BASIC PROTOCOL 5

INDUCTION AND ASSESSMENT OF PARENTAL BEHAVIOR IN NONLACTATING RATS AND MICE (SENSITIZATION)

Materials

Nonlactating rodent subject

Unmanipulated lactating rats that can provide 1- to 8-day-old litters of test pups and that can foster any hungry pups (the total number of surrogate lactating dams necessary to provide recently fed pups each morning and feed hungry foster pups is approximately the same as the number of subjects that are actually being sensitized)

Clear glass or polypropylene home and testing cage (for rats: ~48 × 28 × 16–cm; for mice: 21 × 16 × 13–cm) with wood chips or shavings for bedding (~8 cups for rats, 4 cups for mice). Cotton balls or pads can be used for mice.

Stopwatch

1. Re-house subjects individually in clean, clear polypropylene cages.

2. At 48 hr after re-housing, place four to eight 1- to 8-day-old pups in the cage opposite to the corner where the subject sleeps or has nested. Observe the subject's behavior continuously for 15 min. Note if the subject sniffs, retrieves, licks, or hovers over any of the pups.

   Occasionally, a retrieval or mouthing episode may involve the female biting too hard and breaking the skin of the pup. In this case, animals may start cannibalizing the pup. When this or any other injury to the pups occurs, immediately remove the pups from the cage.

3. Because the subjects are not lactating, remove and replace the pups numerous times each day (every 12 or 24 hr) with freshly fed pups obtained from a lactating "donor" dam.

4. Observe the subject's behavior towards pups for 15, 30, or 60 min each morning when the pups are removed and then replaced.

5. If a subject is observed to sniff, retrieve, lick, and hover over all pups during each of two consecutive daily observations, consider it to be fully parental and conclude testing. Record the first day of the two consecutive observations of full parental behavior as the subject's latency to become fully parental.

6. Continue exposing subjects to pups for 14 to 21 consecutive days to sensitize them (time required varies with strains). After this time, terminate testing and consider the remaining subjects not parental.

7. If necessary, assess ongoing parental behaviors in sensitized subjects as in lactating subjects, although nursing behavior will noticeably differ between virgin and lactating rats.

BASIC PROTOCOL 6

HORMONAL INDUCTION OF MATERNAL BEHAVIOR IN VIRGIN RATS

Most females undergoing the following treatment (Bridges, 1984) will typically respond maternally within 24 to 48 hr after the first introduction to pups, with a small minority (20% to 40%) being spontaneously maternal upon first exposure to the neonates.

Materials

Adult female rats
Crystalline estradiol
Crystalline progesterone
Clear glass or polypropylene home and testing cage (for rats: ∼48 × 28 × 16–cm; for mice: 21 × 16 × 13–cm) with wood chips or shavings for bedding (∼8 cups for rats, 4 cups for mice). Cotton balls or pads can be used for mice.

1. Anesthetize and bilaterally ovariectomize adult virgin female rats (see UNIT 3.5, Support Protocol 2). Subcutaneously implant a 2-mm Silastic capsule filled with estradiol in the nape of the subject's neck (UNIT 3.5, Support Protocol 3).

2. Three days later, subcutaneously implant three 30-mm capsules filled with progesterone.

3. Ten days later, reanesthetize the subject and remove the progesterone capsules; do not remove the estrogen capsule.

4. Twenty-four hours after removal of the progesterone capsules, expose subjects to pups as described in the sensitization protocol (see Basic Protocol 5).

ASSESSMENT OF MATERNAL PREFERENCES AND MOTIVATION IN RATS AND MICE

The tests described can be used to test how salient and rewarding the pups or their sensory cues are to the subject. Numerous factors can be investigated using these motivation procedures, including:

1. Parity: compare nulliparous versus primiparous versus multiparous female rats.

2. Maternal experience: compare maternally induced virgins versus maternally naive virgins; compare primiparous females with varying amounts of postpartum experience with pups.

3. Hormonal status: compare ovariectomized virgin females with and without hormonal priming.

4. Deprivation condition: compare females tested after varying amounts of separation from the unconditioned stimulus (pups or pup-related cues).

5. Specificity of unconditioned stimulus: compare females exposed to pups, novel food pellets, Froot Loops, or other neutral stimuli.

6. Sensory features of pups: compare females exposed to pups of different ages, scent characteristics, or temperatures.

7. Sensory capacity of mother: compare females after experiencing a reduction in sensory capacity (e.g., olfactory denervations, etc.).

BASIC PROTOCOL 7

TEST OF THE UNCONDITIONED PREFERENCE FOR PUPS OR PUP-RELATED CUES

Materials

Rat or mouse subject
Pups (six 1- to 3-day-old pups) *or* pup-related olfactory cues (4 ounces of clean nest-material on which pups have been placed for 6 to 12 hr, or onto which pup urine

has been excreted); if nest material is used, use the same and standardized amounts of nest material during all tests

Unconditioned preference chamber (Fig. 3.6.1A): glass or Plexiglas aquarium (90 × 45 × 47–cm) divided into two goal areas (areas A and B; each 15 cm) at either end and two center areas, A1 (adjacent to A; 30 cm) and B1 (adjacent to B; 30 cm)

Clear Plexiglas box (8 × 8 × 8–cm) with perforated sides, top, and bottom (top removable to permit insertion of stimuli)

Glass bowls (∼8 cm diameter × 3 cm high) to hold nest material

Control stimuli: six small pink pencil erasers or control non-pup related olfactory cues (4 ounces of clean nest material on which juvenile animals or adult animals have been placed for a 6- to 12-hr period, or onto which urine has been excreted); if nest material is used, use the same and standardized amounts of nest material during all tests

Data acquisition software that allows continuous recording of events at least once every second; available commercially (e.g., Noldus Observer, Noldus Information Technology; or Behavior Evaluation Strategies and Taxonomies software, Scolari, Sage Publications Software) although a rather simple data acquisition program may be generated in-house by a person with some computer programming experience

1. To prepare pup stimuli, place six 1- to 3-day-old pups into one 8 × 8 × 8–cm Plexiglas box. As control stimuli place six pencil erasers into another Plexiglas box.

2. *Optional:* Prepare pup-related olfactory stimuli by placing 4 ounces of nesting material in a small cage and place pups onto the nesting material for 1 hr. Before and after nest material is removed, express pup urine by anogenital swabbing with a small cotton swab and place the swab into the nest material. Place 4 ounces of clean nest-material for 6 to 12 hr in the cage of another conspecific animal, for example, a female in the diestrous stage of the estrous cycle or a juvenile animal, and then use this nest material during testing as a control stimulus.

3. Place the two stimuli, i.e., test (pup or pup-related) and control in either area A or area B. For half of the test subjects within each group, place pup stimuli in area A. For the other half, place pup stimuli into area B. For the second preference test put pup stimuli into the area opposite to the area in which they were placed on the first test.

4. Place the subjects in the center of the arena between A1 and B1 and conduct a 5-min preference test. Using an event recorder (data acquisition system), continuously record:

    a. The time that subjects spend in each of area A and area B.

    b. The time spent in contact with the Plexiglas cube or the bowl containing nest material.

    c. The time spent sniffing, pawing and digging at the Plexiglas cube or nest material.

    d. The time spent in areas A1 or B1 of the arena.

5. To ensure that responses to either stimulus reflect attraction to or aversion from that particular stimulus, administer additional tests in which each stimulus is tested alone, e.g., pup stimuli versus no stimulus, and a neutral stimulus versus no stimulus.

6. To determine preferences scores, compute the proportion of total test time that subjects spent in area A versus B and in areas A + A1 versus B + B1 (see Fig. 3.6.1A).

    The time spent sniffing the stimulus cubes or digging into the bowls can also be computed and used to assess preference. A proportional measure can be computed for each subject (time in preferred side/total time) and different groups compared by between-group tests. Compare the proportion of animals (number showing a preference/total number) showing a preference in each group using χ^2 tests. Alternatively, assess single groups for their preferences by means of Fisher's exact probability test or paired t tests.

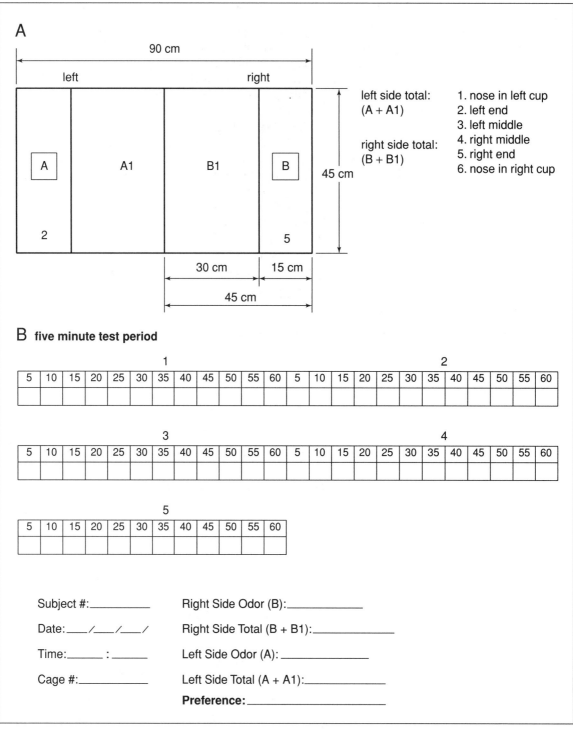

Figure 3.6.1 (**A**) Schematic representation of the unconditioned preference chamber. (**B**) Representative data sheet.

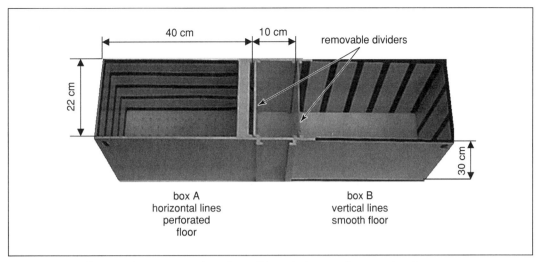

Figure 3.6.2 Schematic representation of conditioned place preference chamber.

BASIC PROTOCOL 8

CONDITIONED PLACE PREFERENCE TESTS

This procedure involves learning processes to examine the reinforcing effects of pup stimulation.

Materials

Pregnant female rat subjects housed individually in polypropylene cages
(∼48 × 28 × 16–cm) with wood chips or shavings for bedding
70% ethanol

Large plastic weighing containers (weigh boats)
Conditioned place preference (CPP) apparatus (Fleming et al., 1994; see Fig. 3.6.2)

1. *Days 0 to 2:* When pregnant females give birth, adjust the litter to contain 8 pups. Leave dams with pups for 48 hr in the home cage.

2. *Day 2:* Wearing gloves, remove pups from the subject dam and place them with a surrogate lactating dam. Leave subject dam alone in the home cage with food and water freely available.

3. *Day 3:* After a 23-hr mother-pup separation, begin the CPP procedure.

4. *Days 2, 4, and 6:* Give subjects a paired-chamber exposure to pups (to control for possible individual preferences for specific chambers). Place six 1- to 3-day-old pups into a weigh boat in the far corner of the paired chamber of the CPP apparatus (either chamber A or B). Present the pups to half of the subjects in chamber A, the other half in chamber B. Make sure the chamber dividers are in place to prevent the subject from leaving the paired chamber.

5. Remove the subjects from the home cage and place them in the appropriate paired chamber with the pups for 1 hr. Observe the subjects' behavior with pups by continuous observation or through periodic spot checks throughout the 1-hr period (see Basic Protocols 1 and 2).

6. Remove subjects from the CPP chamber and return them alone to their home cage. Return pups to lactating surrogate mothers. Clean the CPP apparatus with 70% ethanol and allow it to dry before next use.

7. *Days 3, 5, and 7:* Give subjects a nonpaired chamber exposure. Remove subjects from their home cages and place them in the previously empty (non-pup, either A or B) chamber for 1 hr. Remove the chamber dividers so that subjects can freely explore the CPP apparatus.

8. Observe the subjects' behavior by continuous observation or spot checks and note which chamber they are in. Also note their nonmaternal behaviors such as self-grooming, sniffing cage, eating, drinking, and sleeping. Return subjects to their home cage.

9. To insure that all animals receive 1 hr of daily interaction with pups, place six 1- to 3-day-old pups into the subjects' home cage beginning 1 hr after exposure to the non-paired chamber, and observe their behavior for 1 hr.

10. Remove pups and place with lactating surrogate mother. Clean the CPP apparatus with 70% ethanol and allow it to dry.

11. The next day (day 8), test subjects for their preference in the CPP apparatus. Remove the dividers between central box and chambers A and B to give animal access to both chambers. Place subjects into the central start box and observe their behavior continuously over a 10-min period. Note the frequency of entries into each chamber, the time spent in each chamber, the time spent in each corner of the chambers, and the time spent in the central start box. Remove the subject after testing and return it to the home cage.

12. To analyze the data, compute the duration of time that subjects spent in each chamber as well as the proportion of time that they spent in the paired chambers. Also compute the percentage of subjects that showed more than 50% of their time in the paired chamber

   For example, if an animal spent 7 min in the paired chamber, 2 min in the unpaired chamber, and 1 min in the central box, its preference proportion is 7/(7 + 2) = 0.78 or 78% of the time spent in the two chambers.

BASIC PROTOCOL 9

TESTS OF OPERANT RESPONDING

Materials

Rat test subject housed individually in polypropylene cages ($\sim$48 × 28 × 16–cm) with wood chips or shaving for bedding ($\sim$8 cups)
Kellogg's brand Froot Loop breakfast cereal or similar food reward
1- to 3-day-old rat pups

White noise generator or radio
Modified operant chamber (Fig. 3.6.3)
Computerized system to automate the operant schedule and deliver stimuli according to the specified response schedule
Desk lamps with red light bulbs for nighttime observation

1. Habituate the animals by handling them for 5 min each day for 10 days during the dark phase of the light/dark cycle with room illuminated by a red light bulb. Use a white noise generator or a radio set at low volume to mask extraneous noises in the environment. On each day (to familiarize subjects with handling, food, and environment) place the subjects in an inactivated operant chamber (i.e., a chamber in which pressing the levers does not produce any effect) for 1 hr during which 10 Froot Loops are made available in the dispenser.

2. Place animals on a food deprivation schedule that will achieve and maintain a body weight level of 85% of their baseline body weights, slowly, over $\sim$5 to 7 days. To maintain weight, provide animals with 1 to 4 pellets of standard rodent chow daily, depending on the daily

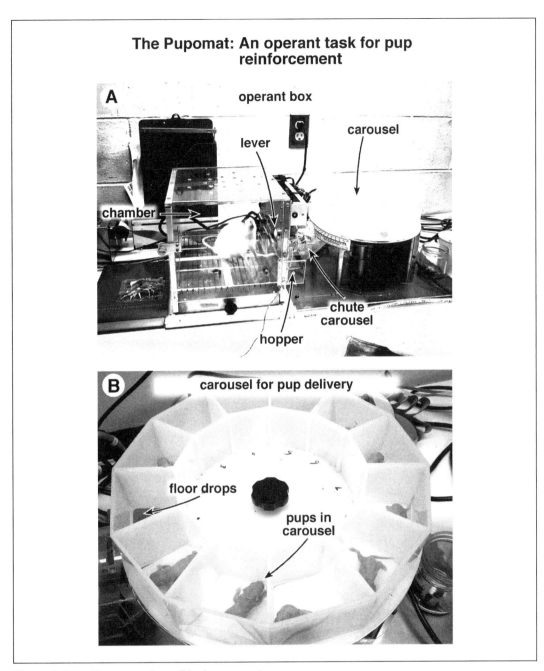

Figure 3.6.3 The operant conditioning apparatus.

recorded body weight. Provide water ad libitum during this time. Once 85% of baseline body weight is achieved, maintain this level of food deprivation during the shaping and conditioning phases of the experiment, which lasts ~8 days.

3. Carry out shaping sessions over an 8-day period with each session lasting ~30 min. Begin each session by placing the subjects in the active operant chamber (i.e., bar presses produce food reward). Carry out all sessions under red light illumination using a white noise generator or radio set at low volume in order to mask any noise made by experimenters or the rotating carousel. Clean the operant box and dispenser with 70% alcohol before each training session.

4. To accomplish shaping, reinforce behaviors that resemble bar-pressing behavior. For instance, reward the following behaviors with a 1/4 Froot Loop each time they occur: rearing in the chamber, rearing on the wall of the levers, sniffing the levers/dispenser, rearing over the levers, or placing forepaws on the levers. Initially reward subjects for behaviors that remotely resemble bar pressing (e.g., rearing anywhere in the chamber), then gradually shift to reinforcing actions most closely resembling a true bar press (e.g., rearing directly over the lever).

5. Continue 30-min shaping sessions until the subject begins to press the bar and retrieve the food reward. Once subjects acquire the ability to bar press independently, place them on an FR-1 reinforcement schedule (i.e., a single Froot Loop is delivered following each bar-press response). Consider subjects to have learned the bar press response once they achieve 50 bar presses within the 30-min testing session. Once subjects reach this criterion, remove them from the operant box, place them back into their cages, and provide food and water ad libitum.

6. Following the last conditioning session on the eighth day, transfer all subjects that meet the conditioning criterion to a large observation cage, and provide two shredded paper towels and six freshly fed pups. Remove subjects that fail to reach the criterion from the study. Observe maternal behavior in the home cage using one of the observation procedures described above.

7. Begin the pup test phase after the maternal behavior observations by removing pups from the cage and placing them into a bowl with nesting material on a heating pad (30°C). Separate subjects from pups for 60 to 120 min before operant tests begin.

8. Transport subjects to the room containing the operant chamber. Place 11 freshly nursed pups (1 to 10 days old) into individual compartments in the 12-compartment carousel, leaving the one compartment over the chute open to the chute but empty.

9. Place the subject into the operant chamber. During the first 10 min of the 30-min session, record the subjects' behaviors using a paper and pencil procedure at 10-sec intervals during which several behaviors (both maternal and nonmaternal) are recorded. Divide data sheets into columns representing time intervals and rows representing individual behaviors. Note the presence or absence of each behavior at each 10-sec interval. Record the following behaviors:

    a. bar pressing: subject rears over one of the two levers, applies pressure on the bar using its forepaws resulting in the delivery of a single pup.
    b. pup retrieval: subject removes pup from the dispenser.
    c. lick pup: subject engages in licking the body of the pup.
    d. anogenital licking: female licks the genital region of the pup.
    e. hover over: rat positions herself over the pups.
    f. self-grooming.
    g. sniffing air.
    h. sniffing magazine.
    i. sniffing the operant box.
    j. sniffing the levers.
    k. settling: subject is inactive, staying in one particular portion of the apparatus.
    l. number of urinations/defecations.

10. Record the frequency of bar presses and pups retrieved for the next 20 min.

11. To avoid the accumulation of pups in the hopper, whenever a group of six pups has been delivered following bar-presses, remove all pups from the operant chamber through a

door in the hopper if pups have not been retrieved out of the hopper. If the pups have been retrieved, remove them through a door in the back of the chamber. Recycle these pups and place them back into the carousel.

12. Place subjects back in their home cage at the end of the 30-min session and immediately give them the pups that had been removed from them before testing began.

13. Employ the same procedures used during the operant chamber pup testing phase (steps 7 to 12) during the extinction phase of the experiment with the exception that bar pressing no longer delivers a pup/Froot Loop reinforcer. Carry out each extinction phase for 5 to 7 days following both the pup testing phase and Froot Loop testing phase. Conduct 8-min behavioral observations of press-presses, grooming, sniffing air, chamber, lever, magazine, and settling. Exclude maternal behaviors.

14. On the final day of testing, do not place pups back in the subject's cage following testing in the operant chamber.

BASIC PROTOCOL 10

USING A T-MAZE EXTENSION OF THE HOME CAGE TO ASSESS MATERNAL MOTIVATION

This test assesses retrieval of pups within a Plexiglas T-maze extension of the subject's home cage (Bridges et al., 1972). It is based on the premise that only very motivated subjects will leave the home cage to locate the pup.

Materials

Rat test subject housed individually in a polypropylene cage (~48 × 28 × 16–cm) with wood chips or shavings for bedding (~8 cups) with 10 × 10–cm hole in wall to accommodate T-maze and which can be kept closed until testing

Neutral stimulus (e.g., small pink pencil erasers)

T-maze (10 cm width; 10 cm height; 33 cm stem length, 37 cm arm length) fitting into hole in subject's home cage (Fig. 3.6.4)

1- to 8-day-old rat pups

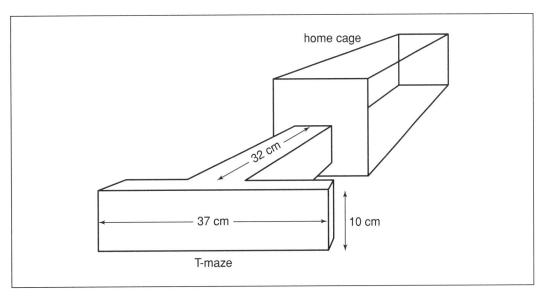

Figure 3.6.4 Schematic representation of T-maze extension of the home cage.

1. Attach the T-maze (to the subject's home cage through a 10 cm × 10–cm hole in the home cage wall that is closed until testing; Fig. 3.6.4). Place a pup and a neutral stimulus in opposite arms of the maze.

2. Expose the opening in the home cage and record subjects' behavior for 10 min. Record the time taken for the subject to emerge into the maze, as well as the time taken for them to locate the pup in the maze arm, and then retrieve the pup back to the home cage.

3. Return subject to home cage and close off the opening to the T-maze. Return stimulus pup to its lactating mother.

References: Fleming and Corter, 1995; Fleming et al., 1989, 1994; Lee et al., 2000; Numan, 1994; Sharpe and Koperwas, 2000

Contributors: Joseph S. Lonstein and Alison S. Fleming

UNIT 3.7

Application of Experimental Stressors in Laboratory Rodents

Stress induction is a critical element in the study of neural and neuroendocrine mechanisms involved in establishing and maintaining a state of stress. Acute and chronic stress can lead to physical and psychiatric pathology, and knowledge of the biologic mechanisms may lead to the development of effective means for relieving significant human suffering and distress.

STRATEGIC PLANNING

Humoral Indices of Stress. Stress hormone secretion from the pituitary and adrenal glands is a gauge of the presence and intensity of the physiological effects of various stressors. Plasma adrenocorticotropic hormone (ACTH) and corticosterone levels are responsive to the onset of stress, and are directly influenced by several hypothalamic releasing factors for ACTH. Hence, a preponderance of studies in the field of stress biology employ plasma ACTH and corticosterone levels as possible indices of differential physiological activation produced by stressor exposure. A stressor dose-response curve of plasma ACTH levels is achieved by increasing the variety, duration of exposure, or strength of stimulation. This profile, taken together with plasma corticosterone levels (a more controversial index of stressor intensity), suggests that graded hormone responses to stress occur with different amounts of physiological stimulation. It is recommended that trial studies be performed to establish appropriate ranges of stimulation for the protocols in this unit.

Choice of Dependent Measure. A commonly quantified consequence of stressor exposure is emotionality in an animal model of anxiety (see *UNIT 3.10*), in which behavioral alarm and withdrawal responses compete with such intrinsic drives as hunger and exploration. Equally popular is the use of stressors to study immunocompetence, algesia, and neural pathway activation, and to model psycho- and histopathologies of every description. Because stressors are robustly effective in a wide variety of paradigms, selection of dependent measures (endpoints) often explores hypotheses related to the physiological or pathophysiological effect of a particular biological system in eliciting the stress response.

Choice of Independent Variable. Three physical stressor procedures and four psychological stressor procedures are described. The choice of stressor will depend on many variables, such as ease of administration; issues regarding ethological, more "natural" stressors; intensity of response; and topography of response desired. In addition, more than one stressor is often applied concomitantly to test the generality of a hypothesized mechanism.

Other Stressors. The stressor protocols described here have broad utility, but by no means constitute an exhaustive listing. Alternative validated rodent stressors reported in the literature include hemorrhage, pharmacological agents, infection, parasitism, cat odor, ultrasonics, drug abstinence, novelty, dietary manipulation, hypothermia, startling noises, hyperbolic exposure, centrifugation, shipping and transfer, anesthesia, and poisoning. The interested reader may wish to pursue some of these other options, although the classical battery of stressor protocols described here fulfills most needs for basic stress tests.

Regardless of the protocol selected, it is important to adhere to the support protocols such as the stress hormone assays and animal calming routines, which are designed to establish and quantify steady state and the stress-induced departure from it.

BASIC PROTOCOL 1

RESTRAINT STRESSOR

Restraint stress (also known as immobilization) produced by confining a naive animal inside a hemicylindrical plastic tube is a frequently employed stressor that stimulates the hypothalamo-pituitary-adrenal axis. Comparative studies of stressor intensity suggest that a prone restraint procedure (fixing the trunk and all four paws splayed flush against a surface with tape) or electric shock applied to the tail are both more severe stressors than this tube restraint procedure.

Although subjects can be tube-restrained over a fairly long interval, peak stressor effect is achieved within minutes and may have subsided by the time of withdrawal from the tube. After an initial acute response the subject habituates rapidly to repeated tube restraint, limiting the utility of this protocol for chronic re-exposures.

Colonies of animals to be used in this protocol must be maintained appropriately so as to provide a nonstress baseline, and rats should be handled prior to testing (see Support Protocols 1 and 2). If desired, stressor efficacy can be quantified independently prior to the experiment (see Support Protocol 3).

Materials

Naive experimental subject (rat or mouse)
Hemicylindrical acrylic restraint tube (Braintree Scientific), 4.5-cm diameter, 12 cm long for a 150- to 300-g rat, or 3-cm diameter, 8 cm long for a 10- to 40-g mouse
Absorbent pad
Timer

1. Place tubes on a hard surface atop a clean, dry absorbent pad to absorb urine. Confine subjects individually and continuously for a period of time, typically 15 to 60 min (duration determined in pilot experiments), in hemicylindrical plastic tubes: allow the subject to walk into the open end of the restraint tube, then insert the slotted keeper so that the animal can neither depart the tube nor contort end-for-end within the enclosure.

2. Collect data from untreated controls before the stress procedure, or in a different location entirely to avoid audible or odor-borne alarm signals that cause stress.

3. Remove the subject at the conclusion of restraint by withdrawing the slotted keeper and allowing the subject to back out of the tube, or grasp the base of the tail and gently withdraw the animal. Wash and thoroughly dry restraint tubes prior to reuse in order to avoid transmission of olfactory cues.

BASIC PROTOCOL 2

ELECTRIC FOOTSHOCK STRESSOR

The suggested shock duration is short and the shock levels are very low by the standards recorded in the literature. This protocol effectively mobilizes the pituitary-adrenal axis, but is not severe enough to cause any known structural harm. If desired, stressor efficacy can be quantified independently prior to the experiment (see Support Protocol 3).

Materials

Experimental subject (rat or mouse), maintained appropriately to provide a nonstress baseline and handled prior to testing (see Support Protocols 1 and 2)
70% ethanol

Multimeter set to read milliamperes (mA)
Shock generator (Coulbourn Instruments) with fuses, output scrambler, overshoot protection circuitry, and a spring-loaded toggle switch (preferable) for shock administration
Covered enclosure with electrified grid floor (Coulbourn Instruments), with bar spacing and diameter of the shock grid appropriately sized for rats or mice
Holding cage

1. Titrate the dose of shock (typically 0.2 to 1.0 mA for a rat weighing 150 to 350 g or 0.15 to 0.6 mA for a mouse weighing 24 to 35 g, depending on the desired effect) using steps 2 to 4 in validation studies.

2. Prior to each test, verify the footshock intensity by setting the shock generator and reading the current input with a multimeter set to read milliamperes. Check footshock intensity periodically between animals to make sure it has not changed.

3. Place subjects on an electrified floor grid within the covered enclosure. Expose subjects to a fixed number of 0.5-sec pulses, typically two per minute over 10 min for a total of 20 shocks. Deliver the shocks with an average intershock interval of 30 sec but with an actual intershock interval that varies randomly from a minimum of 10 to a maximum of 50 sec (variable-interval 30-sec, time-out-10 schedule). Place control subjects in the cleaned shock apparatus for 15 min without administering shocks.

4. Following exposure to shock, place subjects in a holding cage for 15 min in order to avoid aggressive encounters among group-housed animals in the home cage. Treat nonstressed control subjects similarly.

5. After each session, clean all surfaces of the shock delivery apparatus and especially the grid floor with 70% ethanol to eliminate any odors, fecal deposits, and urine.

BASIC PROTOCOL 3

SWIM STRESSOR

Forced swimming at ambient water temperatures is a stressor that activates the hypothalamo-pituitary-adrenal axis and has an anxiogenic effect on behavior.

Materials

Experimental subject (rat or mouse)
Tap water containing 2 mg/liter granular chlorinating concentrate (available from local Pool & Spa store), maintained at $22° \pm 1°C$

Cylindrical water pool (Nalgene): plastic container 45 cm tall × 30 cm in diameter for a rat or 25 cm tall × 15 cm in diameter for a mouse
Thermometer
Stopwatch
Bath-sized terry cloth towels

1. Fill water pool with fresh 22°C chlorinated water to a depth of 30 cm for a rat or 15 cm for a mouse (deep enough to prevent coping strategies or escape facilitated by touching the pool bottom or mounting submerged objects). Place subject in the cylindrical pool. Change water after 30 (at most) sequential swim trials.

2. Time for 90 sec with a stopwatch, then grasp subject gently by the trunk or tail, lift from the water, place into a dry, bath-sized terry cloth towel, and dry the entire body surface by vigorously rubbing the animal inside the towel to allow subsequent testing of the subject without excessive grooming.

3. *Optional:* If desired, elevate the intensity of the stressor by either increasing the duration of swimming or lowering the water temperature.

    Some mice and rats will begin to float after several minutes in a swimming task. This acquired immobility is interpreted as behavioral despair and is employed as an animal model of depression.

4. For controls, take subjects directly from the home cage (nonstressed control) or place them in a dry, unfilled water pool for 90 sec (sham control).

BASIC PROTOCOL 4

SOCIAL ISOLATION STRESSOR

This protocol causes social isolation during which animals have relatively normal auditory and olfactory experiences, but cannot at any time see, touch, or be touched by the colony animals. Aside from their solitary housing, animals to be used in this protocol must be maintained so as to provide a nonstress baseline (see Support Protocol 1). If desired, stressor efficacy can be quantified independently prior to testing (see Support Protocol 3).

Materials

Experimental subjects: same-sex rats or mice from 21 days (weaning) to 3 months of age, preferably from an in-house breeding colony
Opaque polyethylene group rodent cages (Nalgene), 40 × 30 × 10–cm for mice or 50 × 40 × 20–cm for rats
Opaque polyethylene single rodent cages (Nalgene), 30 × 15 × 10–cm for mice or 40 × 20 × 20–cm for rats

1. House social controls in groups (four to six per group for mice, four to ten per group for rats) in group housing cages. House isolated subjects individually in smaller single housing cages. Subdivide each litter in equal proportion into social and isolated groups in order to balance maternal/developmental confounds.

2. House both groups in the same vivarium for 1 to 12 weeks (typically 2 to 4 weeks) isolation. Change bedding regularly and permit isolated subjects to hear and smell the other rats or mice in the colony.

BASIC PROTOCOL 5

RESIDENT/INTRUDER STRESSOR

Social conflict stress involving threat from an aggressive male counterpart is an ethologically derived, species-typical stressor. Resident exposure produces pituitary-adrenal activation; in defeated intruder animals, both ACTH and corticosterone levels remain significantly elevated for several hours following an aggressive encounter. Maximal ACTH activation occurs within 10 min of the initiation of restraint stress, but has a latency of 20 to 40 min following initiation of social conflict. After a brief social defeat, the intruder is continually threatened but prevented from physical contact for an additional period of time. If desired, stressor efficacy can be quantified independently prior to the experiment (see Support Protocol 3).

Materials

Experimental subjects: several mixed-gender pairs of male and female Long-Evans rats (preferred for reliable aggressiveness) or Swiss-Webster male mice, maintained appropriately to provide a nonstress baseline and handled prior to testing (see Support Protocols 1 and 2)

Resident cage, $72 \times 52 \times 35$–cm for rats or $30 \times 20 \times 15$–cm for mice
Wooden hutch (for rats only): two 30×15–cm pieces of plywood joined at right angles
Intruder protective enclosure made of Plexiglas/wire mesh, $19 \times 20 \times 17.5$–cm for rats or $10 \times 15 \times 10$–cm for mice

1. House one pair of subjects in the resident cage over the long term (not less than 4 weeks), typically >12 months. Cover floor of cage with rodent bedding, and for rats, insert a hutch in order to provide odor and object constancy over time (no hutch required for mice).

2. After the resident pairs have littered successfully, perform steps 3 and 4 to select resident males with reliable and robust aggressive responses to intruder rats for experimental use. Keep successful breeding pairs together, one pair per cage, for the long term, but remove offspring at postpartum day 12 in order to avoid tiring the resident male.

3. To initiate social stress, remove the resident female and litter (if any) to a holding cage and introduce an unfamiliar experimental male rat into the home territory of an aggressive male resident (a $\geq$100-g size advantage desirable).

4. Interrupt the ensuing characteristic pattern of olfactory investigation and agonistic offensive and defensive behaviors once a resident attack results in social defeat (display of a submissive/supine posture after being attacked) of the intruder rat.

5. Place the defeated intruder within a protective enclosure and place it back inside the resident cage for 15 min to allow the social conflict to resume and continue without physical contact or injury.

6. Allow each resident to defeat an equal number of subjects (four defeats per resident per day for practical purposes) within each experimental treatment group, in order to control for individual differences in attack intensity. For controls, take subjects directly from the home cage (nonstressed control) or place them in the protective enclosure for 15 min (sham control).

BASIC PROTOCOL 6

MATERNAL SEPARATION STRESSOR

Although perhaps most germane to nonhuman primates, maternal deprivation in laboratory rodents serves as an excellent perinatal stressor in the study of maturation and the biology of separation. Reactivity of the hypothalamo-pituitary-adrenal axis to experimental stressors appears to be permanently deranged by brief, repetitive postnatal maternal separation and handling. Moreover, sensitive measures of arousal in the preweanling rat pup, such as ultrasonic distress call vocalizations, increase acutely and reliably within minutes of maternal separation. Sensitivity of these endpoints to anxiolytic drugs, manual stroking, and restoration of the family unit serve to validate the utility of this naturalistic stressor in the study of developmental stress biology. If desired, stressor efficacy can be quantified independently (see Support Protocol 3).

Materials

Experimental subjects: pregnant in-house dams of the desired strain of rat or mouse, maintained appropriately to provide a nonstress baseline and handled prior to testing (see Support Protocols 1 and 2)

Rodent caging with bedding, preferably wood shavings

Electric heating pad

1. House pregnant dams in standard group cages and monitor daily for parturition. Designate date of birth as day 0. On the first postpartum day, sex the pups using standard measures of genital size and anogenital distance, cull litters to eight pups (usually four females and four males), and place each dam and litter together in a clean cage.

    IMPORTANT NOTE: From this time forward, do not handle the dam and litter in any way or clean their cages until the time of deprivation or testing, in order to minimize disruption of the mother/infant relationship.

2. Prepare a deprivation chamber consisting of a small housing cage containing ∼5 cm of bedding, and place it on an electric heating pad set at 30° to 33°C to allow the pups to partially regulate their body temperature by burrowing for warmth or returning to the surface for cooling. Identify individual pups through a distinguishing pattern of marks with an indelible lab marker.

3. To achieve maternal deprivation stress, remove the pups from their home cage either once or repeatedly over several consecutive days. Place them in the deprivation chamber for periods of 2 to 24 hr. Allow approximately half of each litter to remain with the dam as experimental controls. Replace pups with their dams in the home cage until subsequent deprivations, if applicable.

BASIC PROTOCOL 7

SLEEP DEPRIVATION STRESSOR

Sleep deprivation is most commonly achieved by forced perpetuation of wakefulness. Placement of the animal upon a pedestal, a rotating drum, or a spin-controlled disk within a water bath denies any opportunity for sleep. Although controversy surrounds the use of standard sleep deprivation protocols to achieve selective effects (e.g., rapid eye movement sleep), the utility of this protocol as a stressor appears straightforward. If desired, stressor efficacy can be quantified independently prior to the experiment (see Support Protocol 3).

Materials

Tap water containing 2 mg/liter granular chlorinating concentrate (available from local pool & spa store), room temperature

Experimental subject (mouse or rat), maintained appropriately to provide a nonstress baseline and handled prior to testing (see Support Protocols 1 and 2)

Water pool, with pool floor extending >12.5 cm in all directions from the central point (e.g., a standard polycarbonate rodent housing cage)
Immersible cylindrical pedestals, small (2-cm diameter for mice or 6.5-cm diameter for rats) and large (6-cm diameter for mice or 16-cm diameter for rats)

1. Prepare a watertight pool with laboratory chow and drinking water bottle available ad libitum either from an overhead wire mesh lid or on one of the sidewalls. Flood the pool with chlorinated water (1 cm deep for mice or 2 to 3 cm deep for rats) at room temperature.

2. Fix a cylindrical pedestal to the floor of the pool in the center of the enclosure, with the height of the pedestal such that it emerges 1 cm above the water level of the flooded pool. Place animals initially on the pedestal, and confirm continued access to food and drinking water. Keep subjects in the enclosure for 5 days (may be shortened or lengthened as needed). Monitor diligently as duration of deprivation increases, in order to safeguard animals.

3. Place untreated control animals concurrently in identical, but dry, bedding-lined enclosures; place sham controls (large-diameter pedestals of the sham controls do not induce sleep debt, but mimic other stressful aspects of the small-pedestal condition) in water-filled pools on larger-diameter pedestals.

SUPPORT PROTOCOL 1

COLONY MAINTENANCE

On arrival, animals should be group-housed three to five per cage in standard caging. It is important not to employ filter-top shrouds in experiments involving stressor application, as ammonia buildup within the cage has a deleterious effect on dependent measures sensitive to stressors. Provide subjects with ad libitum access to water and laboratory chow. Place the entire group of animals into a clean cage and replenish water and food supply twice weekly. The colony room must be light- and temperature-controlled (lights on at 0600, off at 1800; temperature $22° \pm 1°C$). Radios tuned to a 24-hr music station provide constant background noise. Any disturbance of the colony environment involving lighting changes, loud noises, or natural calamity (e.g., earthquake) occurring during the acclimation period necessitates a new 7-day acclimation period prior to experimentation.

In order to prepare for an experimental trial, select a day of the week on which the animal technicians are not changing bedding or cages. Beginning when lights are on in the colony room and at least 30 min prior to initiation of testing, move all cages containing experimental subjects to an anteroom adjacent to a procedure room. Match the lights in the anteroom and testing room to the circadian cycle of the colony during the entire period of time that experimental subjects are present. Perform the experimental stressor in the procedure room, separate from the anteroom. Prohibit entry of nonessential personnel into either the anteroom where animals are held prior to testing or the procedure room for the duration of the test.

SUPPORT PROTOCOL 2

RAT HANDLING

This procedure should be implemented in group-housed adult subjects soon after arrival, since isolation housing and advanced age make handling/calming problematic in general, and the risk of biting high. Handle all subjects after acclimation and 24 to 48 hr prior to testing. This handling procedure (or one comparable) is critical to achieving a reliable baseline in nonstressed groups.

Materials

Rats
Laboratory coat
Indelible marking pen
Portable scale
Flat surface

1. Permit new arrivals to acclimate to the colony for 7 days.

    It is important that the individual manipulating the animal on test day is the same one who performed the preliminary handling; this provides familiarity with odors, style of handling, and appearance that distinguish experimental personnel.

2. Holding each subject by the tail, make unique identification marks on the tail with an indelible pen, and record body weight.

3. Place subjects on a smooth tabletop in the company of their cagemates and repeatedly hold each animal by the torso, lift off the surface, and replace. Continue for ~2 min.

4. Grasp the animal by the upper torso so as to shroud the eyes, and hold the upper torso still while turning the lower body in a circular motion rapidly through several revolutions. Alternatively, hold the animal by the base of the tail and place it on a hard surface in order to produce the same pinwheel motion.

    The resulting disorientation prepares animals for close contact with the handler and reduces the tendency to escape, leap, struggle, and otherwise fear the handler.

5. Lift the animals and place them somewhere on the handler's person, such as in the crook of the elbow, in a pocket of the laboratory coat, along the horizontal forearm, or in the hand. Move around the room for a period of ~5 min carrying the animals along.

SUPPORT PROTOCOL 3

ACTH/CORTICOSTERONE DETERMINATIONS

Corollary physiological assays for stress hormones serve the dual purpose of confirming the presumed steady state of experimental controls, and quantifying the response to stressor exposure. Plasma ACTH and corticosterone can be readily assayed using commercially available kits and the accompanying step-by-step instructions.

Materials

Experimental subjects (mice or rats)
50 mg/ml EDTA (ethylenediaminetetraacetic acid, tetrasodium salt)
5000 kallikrein inhibitor units (KIU)/ml aprotinin
ACTH immunometric assay kit (Nichols)
Corticosterone radioimmunoassay kit (ICN Biochemicals)

Indwelling vascular catheter apparatus (see Caine et al., 1993 for implant procedure)
12 × 75–mm centrifuge tubes, ice cold
Tabletop centrifuge
Scintillation counter

1. Draw blood samples from an indwelling vascular catheter line at separate time points from 1100 to 1500 hr, beginning 5 hr into the light phase of the circadian cycle (when endogenous hormone levels are lowest) before and after stressor exposure from subjects kept in a quiet, dimly lit room. Alternatively, collect blood from the trunk following decapitation.

2. Collect 0.3- to 0.5-ml blood samples in tubes on ice containing 2.5 µl of 50 mg/ml EDTA to halt coagulation, and 2.5 µl of 5000 KIU/ml aprotinin to inhibit peptide (i.e., ACTH) degradation.

3. Microcentrifuge samples 4 min at 14,000 rpm, room temperature. Pipet off supernatant plasma for analysis. (Store samples frozen up to 1 year.) Determine plasma ACTH using an immunometric assay kit. Determine plasma corticosterone using a radioimmunoassay kit.

Reference: Selye, 1976

Contributors: Stephen C. Heinrichs and George F. Koob

UNIT 3.8

Rodent Models of Depression: Forced Swimming and Tail Suspension Behavioral Despair Tests in Rats and Mice

Rodents forced to swim in a narrow space from which there is no escape adopt, after an initial period of vigorous activity, a characteristic immobile posture, moving only when necessary to keep their heads above the water. The animals' immobility was hypothesized to show they had learned that escape was impossible and had given up hope. Immobility was therefore given the name "behavioral despair".

BASIC PROTOCOL

BEHAVIORAL DESPAIR TEST IN THE RAT

Materials

    200- to 230-g male Wistar rats (e.g., Elevage Janvier), acclimated for at least 5 days before starting the experiment
    Standard rodent diet
    Treatment solutions (5 ml/kg):
        Drug or test compound dissolved in distilled water (for oral administration) or sterile physiological saline (for intraperitoneal or subcutaneous injections); for water insoluble compounds, disperse in 0.2% (w/v) hydroxypropylmethylcellulose in distilled water (for oral administration) or physiological saline (for intraperitoneal or subcutaneous injections).
        Reference compound (e.g., imipramine hydrochloride; 32 mg/kg for intraperitoneal or subcutaneous injections, 64 mg/kg for oral administration; Sigma-Aldrich or equivalent)
        Vehicle alone (for control)

    Transparent plastic cages (41 × 25 × 15–cm), padded with wood shavings
    Two transparent Plexiglas cylinders (20 cm in diameter × 40 cm high) containing water (25°C) to 13 cm (made in house or obtained from commercial suppliers of Plexiglas material)
    Opaque screens for separating cylinders
    Metric balance (e.g., Sartorius model 1401.001.2), accurate to 1 g
    2-ml syringes for intraperitoneal and subcutaneous injections (e.g., Terumo type BS-025)

23-G × 1-in. (0.6 × 16–mm) needles for intraperitoneal injections (e.g., Terumo)
21-G × 1.5-in. (0.8 × 40–mm) needles for subcutaneous injections (e.g., Terumo)
Luer gastric probes with oval extremity (70 mm long × 1.5 mm wide oval diameter) for oral administration

1. House six rats per cage in 41 × 25 × 15–cm plastic cages padded with wood shavings, and provide free access to standard rodent diet and tap water, except during the test. House the animals at 21°C on a standard (nonreversed) light/dark cycle with illumination from 0700 to 1900. Use six rats per group for a standard experiment (usually five groups: vehicle control, reference compound, and three doses of test substance).

2. Equip the experimental room with white neon ceiling lights (standard lighting). Set up two transparent cylinders separated visually from each other by opaque screens.

3. On Day 1, 25 hr prior to testing, place the animals in the experimental room 60 min before the beginning of the habituation session (during the light phase of the cycle). Immediately place an identifying mark on the tail of each animal with indelible ink. Randomly assign animals to a drug treatment, but give all animals within a cage the same treatment. Make food and water available throughout the experiment. Weigh two animals individually, then place one rat in each of the two cylinders for 15 min (habituation session).

4. Remove the rats from the cylinders, dry them with a cloth towel, and place them into a cage. Weigh a new group of two rats, and place them in the cylinders for 15 min.

5. Treat the first group of two rats with the appropriate treatment (the 24-hr pretest treatment coded to avoid bias in evaluating the animals' behavior) in a volume of 5 ml/kg, and place them back in their home cages. Repeat this sequence until the end of the session. Change the water in the cylinders after every three rats.

6. When the Day 1 session is completed, return the animals to the colony room and provide food and water ad libitum.

7. *Optional:* Administer drug or test substance to the animals 24 and 4 hr prior to the final treatment to provide more stable pharmacological results than a single administration. Administer vehicle to control animals according to the same schedule.

8. Administer the final treatment 30 min (for intraperitoneal or subcutaneous injection) or 60 min (for oral administration) prior to the test in a fixed rotation (A, B, C, etc.) to ensure a regular distribution of the different treatments over time. Maintain the same treatment order during the different phases of the test.

9. Place two animals simultaneously in individual side-by-side cylinders separated by an opaque screen. Observe (one observer for both animals) their behavior for 5 min. Score the duration of immobility by summing the time spent immobile; do not score as immobile those movements necessary to maintain the animal's head above water.

10. Decode scores after all evaluations have been completed to avoid bias. Compare data from treated groups with data from the control group using nonpaired Student's *t* tests (two tailed).

ALTERNATE PROTOCOL 1

FORCED SWIMMING TEST IN THE MOUSE

Materials

20- to 25-g male NMRI (Naval Medical Research Institute) mice (e.g., Elevage Janvier), acclimated for at least 5 days before starting the experiment

Standard rodent diet
Treatment solutions (5 ml/kg):
> Drug or test compound dissolved in distilled water (for oral administration) or sterile physiological saline (for intraperitoneal or subcutaneous injections); for water insoluble compounds, dispersed in 0.2% (w/v) hydroxypropylmethylcellulose in distilled water (for oral administration) or physiological saline (for intraperitoneal or subcutaneous injections).
> Reference compound (e.g., imipramine hydrochloride; 32 mg/kg for intraperitoneal or subcutaneous injections, 64 mg/kg for oral administration; Sigma-Aldrich or equivalent)
> Vehicle alone (for control)

Transparent plastic cages (25 × 19 × 13–cm) padded with wood shavings
Two Plexiglas cylinders (13 cm in diameter × 24 cm high) containing water (22°C) to 10 cm
Metric balance (e.g., Sartorius type 1401.001), accurate to 0.1 g
1-ml syringes (e.g., Terumo type BS-01-T)
25-G × 0.625-in. (0.5 × 16–mm) needles for intraperitoneal injections (e.g., Terumo)
23-G × 1-in. (0.6 × 25–mm) needles for subcutaneous injections (e.g., Terumo)
Luer gastric probes with oval extremity (25 mm long × 1.2 mm wide oval diameter) for oral administration

1. House ten mice per cage in transparent plastic cages (25 × 19 × 13–cm) padded with wood shavings. Provide free access to standard rodent diet and tap water, and maintain a controlled temperature of 21°C and a standard (nonreversed) light/dark cycle with illumination from 0700 to 1900. Use five groups of ten mice per group for a standard experiment (vehicle control, reference compound, and three doses of test substance).

2. Equip the experimental room with white neon ceiling lights (standard lighting). Set up two transparent cylinders separated visually from one another by an opaque screen. Perform experiments during the light phase of the cycle.

3. Place the animals in the experimental room 60 min before the beginning of the experiment (during the light phase of the cycle), and immediately place an identifying mark on the animals' tails with indelible ink. Make food and water available throughout the experiment.

4. Weigh the animals and immediately administer the appropriate treatment in a volume of 5 ml/kg. Wait a predetermined time after administration to begin the test, usually 30 min for intraperitoneal and subcutaneous injections and 60 min for oral administration.

5. Place two animals simultaneously in individual side-by-side cylinders separated by an opaque screen. Observe (one observer for both animals) behavior for the last 4 min of the 6-min test session. Score the duration of immobility by summing the time spent immobile; do not score as immobile those movements necessary to maintain the animal's head above water.

6. Compare data from treated groups with data from the control group using nonpaired Student's t tests (two tailed).

ALTERNATE PROTOCOL 2

TAIL SUSPENSION TEST IN MICE

This procedure has several advantages over the forced swimming procedure (see Basic Protocol and Alternate Protocol 1). First, the situation appears to be less stressful to the experimental animal in that no hypothermia is induced, and the animals, once removed from the experiment, resume normal spontaneous activity immediately. No special post-experimental treatment (rubbing down, maintenance in a warmed environment) is required. The procedure lends itself readily to automation, with a resulting increase in throughput for screening purposes. In addition, it shows a different spectrum of pharmacological sensitivity from the forced swimming procedures, thus providing a complementary approach to the behavioral screening of antidepressant and other psychotropic activity.

Materials

    20- to 25-g male NMRI mice (e.g., Elevage Janvier), acclimated for at least 5 days before starting the experiment
    Standard rodent diet
    Treatment solutions (5 ml/kg):
        Drug or test compound dissolved in distilled water (for oral administration) or sterile physiological saline (for intraperitoneal or subcutaneous injections); for water insoluble compounds, dispersed in 0.2% (w/v) hydroxypropylmethylcellulose in distilled water (for oral administration) or physiological saline (for intraperitoneal or subcutaneous injections).
        Reference compound (e.g., imipramine hydrochloride; 32 mg/kg for intraperitoneal or subcutaneous injections, 64 mg/kg for oral administration; Sigma-Aldrich or equivalent)
        Vehicle alone (for control)

    Transparent plastic cages (25 × 19 × 13–cm) padded with wood shavings
    Automated tail suspension apparatus (e.g., TAILSUSP-1N96; MED Associates) consisting of plastic enclosures (20 × 25 × 30–cm) fitted with a ceiling hook connected to a strain gauge and computer assembly with Windows compatible software
    Metric balance (e.g., Sartorius type 1101601), accurate to 0.1 g
    1-ml syringes (e.g., Terumo type BS-01-T)
    25-G × 0.625-in. (0.5 × 16–mm) needles for intraperitoneal injections (e.g., Terumo)
    23-G × 1-in. (0.6 × 25–mm) needles for subcutaneous injections (e.g., Terumo)
    Luer gastric probes with oval extremity (25 mm long × 1.2 mm wide oval diameter) for oral administration

1. House 10 mice per cage in transparent plastic cages (25 × 19 × 13–cm) padded with wood shavings, and provide free access to standard rodent diet and tap water. Maintain a controlled temperature of 21°C and a standard (non-reversed) light/dark cycle with illumination from 0700 to 1900. Use six groups of 10 mice per group for a standard experiment (vehicle control, two reference compounds, and three doses of test substance).

2. Equip the experimental room with white neon ceiling lights (standard lighting). Place the animals in the experimental room 60 min before beginning the experiment (during the light phase of the cycle), and immediately place an identifying mark on each animal's tail with indelible ink. Remove food and water for the duration of the test.

3. Weigh the animals and immediately administer the appropriate treatment in a volume of 10 ml/kg. Wait a predetermined time after administration to begin the test, usually 30 min for intraperitoneal and subcutaneous injections, and 60 min for oral administration. Administer drugs and test substances to individual animals in a fixed rotation (A, B, C), to ensure a regular distribution of the different treatments over time.

4. Wrap adhesive tape around the animal's tail in a constant position three quarters of the distance from the base of the tail. Suspend the animals by passing the suspension hook through the adhesive tape as close as possible to the tail (1 to 2 mm) to ensure that the animal hangs with its tail in a straight line (to avoid injury).

For nonautomatic observation, the same observer can comfortably observe two animals simultaneously. For an automated procedure, the number of animals measured simultaneously depends on the configuration of the system. Simultaneous observation of six animals during the same measurement period is recommended, with all animals being placed in the apparatus before starting the measurement. The animals should be visually shielded from one another during the test.

5. Observe the animals continuously for 6 min. If an automated testing apparatus is not available, use separate stopwatches for each animal and sum the time spent immobile by each animal over the 6-min observation period. Compare data from treated groups with data from the control group using nonpaired Student's *t* tests.

Reference: Porsolt et al., 1977, 1991

Contributors: Roger D. Porsolt, Genevieve Brossard, Carine Hautbois, and Sylvain Roux

UNIT 3.9

Rodent Models of Depression: Learned Helplessness Induced in Mice

The learned helplessness paradigm has been used to model depression in humans, to identify the neurochemical substrates that may be associated with stressor-induced depressive symptoms in an animal model, and to assess the efficacy of antidepressant treatments.

BASIC PROTOCOL

INDUCTION OF A LEARNED HELPLESSNESS EFFECT IN MICE

Materials

Male or female mice, >60 days of age, housed individually if the procedure is to be carried over >1 day
70% (v/v) ethanol
Drugs to be tested (optional)
Saline or appropriate vehicle
Syringes and needles: 26 G, 3/8 in.

Shuttle boxes (Figure 3.9.1; e.g., Stoelting, Coulbourn Instruments, or see Support Protocol)
Sound-attenuating chambers (ventilated) to house a shuttle box and accessory equipment (∼60 × 60 × 90–cm; make in-house or purchase, e.g., Coulbourn Instruments)
Shock generator and scrambler (or neon bulbs as an alternative to a shock scrambler) (Stoelting or Coulbourn Instruments)
Computer (to control trial delivery and record data) and software (from equipment suppliers; e.g., Coulbourn Instruments) or custom-written

1. Acclimate male or female mice to the laboratory for ≥2 weeks before using them as experimental subjects. House the animals in groups of four until 5 days before testing, and then house them individually.

2. Weigh the animals and bring them to the laboratory ≥1 hr before testing (24 hr is preferable) to diminish the immediate effects of stress associated with transport from the housing area

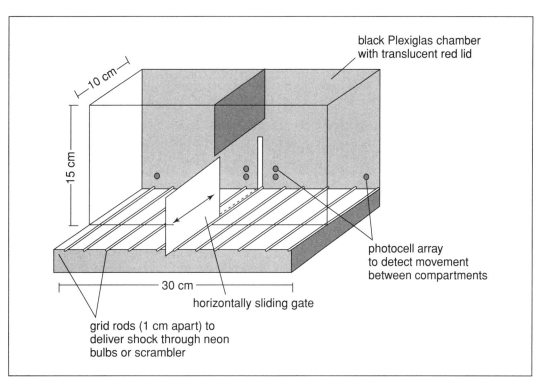

Figure 3.9.1 Schematic depiction of a shuttle box. This apparatus can be used to deliver escapable or inescapable shock and can be used subsequently to test escape performance after an inescapable shock session.

to the test room, as well as the effects of odors that may be present in the hallways. Restrict testing to a relatively narrow window (e.g., 0900 to 1200 hr) to avoid potential impact of diurnal variations of various neurochemicals on performance

3. Quickly place three animals individually into shuttle boxes and strap each box into a sound-attenuating chamber where the mice cannot hear, see, or smell one another. Connect the first and second shuttle boxes to a common shock generator and scrambler.

4a. *For escapable-shock group:* Expose the mouse to 60 shock escape trials allowing intervals of 60 sec between trials. On a given trial, present the foot shock but keep gate separating the compartments closed for the initial 1.5 sec of shock (shorter durations can be used, but this is near optimal), thereby delaying the escape response. Then open the gate (automatically) gently and quietly to avoid startling the animal, permitting the mouse access to adjacent compartment. Terminate the shock and close the gate when the mouse enters the adjacent compartment. Automatically terminate the trial if an escape does not occur within 30 sec.

4b. *For uncontrollable-shock group:* Configure the apparatus so the foot shock in the second shuttle box terminates when the first animal successfully escapes or when 30 sec have elapsed.

4c. *For no-shock group:* Leave mouse in the third chamber undisturbed, omitting any shock treatment (see Anisman et al., 1978 for more details on this triadic design).

5. Measure latency (in seconds) from the time that the gate opens to escape from shock to maintain a record of how much shock each animal received. Return mice to their home cages until subsequent testing, which typically commences 24 hr later. Clean the apparatus thoroughly with 70% alcohol.

6. After a predetermined length of time, place the three animals treated in steps 2 to 5 individually into shuttle boxes, each located in a sound-attenuating chamber, and commence testing 1 min later. During the initial five test trials, synchronize shock onset with opening the gate separating the compartments (i.e., do not impose an escape delay) to reduce the probability of behavioral disturbances being observed in control mice. For each trial, record latencies to escape starting from the time the gate is opened.

7. During the next 25 trials, introduce a 4-sec delay between shock onset and the gate being opened. Terminate the trial when the mouse crosses from one chamber to the next or if the mouse does not escape within 30 sec. Start the next trial 30 to 60 sec later. Maintain the shock intensity at 150 μA; AC, 60 Hz. For each trial, record latencies to escape starting from the time the gate is opened.

8. Average escape latencies over successive blocks of five trials, and then subject these values to a repeated-measures analysis of variance (ANOVA).

    It is important to assess the course of the interference during the trials, as treatments may not influence the overall magnitude of the effect, but may alter the development of the effect during the trials. It is also important to recognize that a treatment may increase response latencies, but without the production of failures to escape from the foot shock. Thus, the frequency of escape failures (or the number of animals that display a preset number of escape failures) must be specified.

ALTERNATE PROTOCOL

ADMINISTRATION OF UNCONTROLLABLE SHOCK

Additional Materials *(also see Basic Protocol)*

Shock chambers (optional; Coulbourn Instruments)

1. Prepare mice for exposure to uncontrolled and no shock as described (see Basic Protocol, steps 1 and 2). Place two animals individually into shuttle boxes or shock chambers, each located within a sound-attenuating chamber. Connect the first shuttle box to a shock generator and scrambler.

2. Initially expose uncontrolled shock mice to 60 inescapable shock presentations of predetermined duration (e.g., 2 sec) or use shock of variable duration averaging 2 sec (e.g., 0.5 sec on some trials, 2 sec on others, 4 sec on others). Apply these at fixed intervals, such as every 60 sec, or administer them at variable intervals averaging 60 sec (i.e., 20 sec intervals between some trials, 80 sec intervals between others).

    On occasion an animal will fail to make a successful response, and thus an arbitrary cut-off for the shock is imposed (e.g., 30 sec, but it can be as short as 10 sec). When using a procedure calling for a fixed number of shock trials of set duration, 60 shock presentations of 2-sec duration are sufficient to induce the interference effect. However, a more pronounced interference is induced with a greater number of shock presentations (e.g., 180).

    If data are collected with respect to activity during the foot shock itself, or during intertrial intervals, then these data can be correlated with later performance (e.g., by Pearson product moment correlation or using more sophisticated regression analyses).

3. Return mice to their home cages until subsequent testing, which typically commences 24 hr later. Clean the apparatus thoroughly with 70% ethanol. Test animals using the shuttle escape procedure as described in Basic Protocol, steps 6 to 8.

SUPPORT PROTOCOL

DESCRIPTION AND FUNDAMENTAL CHARACTERISTICS OF SHUTTLE BOXES FOR ESCAPE TESTING

The apparatus must have absolutely no cracks, ledges, or protrusions that the mouse can use to escape the stressor (see Fig. 3.9.1).

Materials

Black Plexiglas (0.63 cm thick)
Red translucent Plexiglas
Stainless-steel rods (0.32 cm in diameter)
Scrambler (available from virtually all companies that manufacture avoidance/escape equipment) or neon bulbs wired in series
Stainless-steel sheets (0.1 cm thick)
Solenoid to control gate opening and closing
Photocells (e.g., Radio Shack)

1. Construct 30.0 × 10.0 × 15.0–cm shuttle boxes of black Plexiglas (0.63 cm thick) and make the roof of each out of red translucent Plexiglas. Make a slot in one side wall of the shuttle apparatus (~6.2 cm high and 0.25 cm wide) to accommodate a horizontally moveable gate (see Fig. 3.9.1).

   When constructing the chamber in house, it is advisable to have it made up of two portions: a lower portion containing the grid floor, and an upper chamber that contains the gate, photocells, and other components. The lower portion ought to be ~5 cm wider than the upper part. The end walls of the upper portion rest on the back portion of the lower part, and the two are connected by clasps. However, the side walls should be elevated slightly (~0.20 cm) above the grid floor. By separating the upper portion from the grid floor, and by having drip rings located ~0.5 cm from the walls of the lower portion, it is possible to eliminate short circuits produced by urine running against the side walls and contacting two grid bars. In addition, having two separate portions permits easy cleaning of the grid.

2. Use 0.32-cm stainless-steel rods, spaced 1.0 cm apart (center to center) as the floor of each chamber (to permit feces to drop through). Wire the grid floor to receive shock through a scrambler or through neon bulbs wired in series to prevent animals from standing on grids of the same polarity, thereby avoiding or escaping shock. Deliver foot shock (~150 to 300 μA; AC, 60 Hz) to the grid floor through a high-voltage source (~3000 V shock generator) in order to minimize current changes owing to the resistance contributed by the mouse.

3. Line the end walls of each chamber with stainless-steel sheets and connect them to the grid floor to assure that animals will not avoid or escape shock by standing on a single grid rod and leaning on a back wall. The apparatus must have absolutely no cracks, ledges, or protrusions that the mouse can use to escape the stressor.

4. Divide each shuttle box into two compartments with a stainless-steel sheet, with an opening of ~5.0 × 6.0 cm. Include a stainless steel-lined, solenoid-controlled, horizontally sliding gate to cover the passage between the two compartments.

   When open, the gate triggers a lever that reduces the current to the solenoid, thereby preventing it from overheating.

5. Position two photocells on both sides of the gate (1.5 cm); place one photocell ~2.5 cm above the grid floor and the second ~4.5 cm above the floor. Position an additional photocell 2.5 cm from each end wall.

References: Anisman et al., 1978, 1992; Meyer and Seligman, 1976; Weiss et al., 1981

Contributors: Hymie Anisman and Zul Merali

UNIT 3.10

Animal Tests of Anxiety

Animal tests of anxiety are used to screen novel compounds for anxiolytic or anxiogenic activity, to investigate the neurobiology of anxiety, and to assess the impact of other occurrences such as exposure to predator odors or early rearing experiences.

BASIC PROTOCOL 1

GELLER-SEIFTER CONFLICT

Materials

Male or female rats, group housed
Rat food pellets
Drugs to be tested
Saline or other appropriate drug vehicle for control injections

Animal scale
Standard rat operant chamber (Skinner box) with floor of metal bars for delivery of footshock, and with house light, cue light triggered by the response, lever, and a pellet dispenser
Shock generator and scrambler (*UNIT 3.7*)

1. Weigh the rats 1 week prior to training, and start food restriction by giving measured amounts of food to reduce animal weight to ∼80% of free-feeding weight. Place small quantities of the reward pellets that will be used in the Skinner boxes in the home cages during this week to familiarize the rats with this food.

2. Begin initial training by placing a few food pellets in the food hopper, and placing the animal in the test chamber. Initially reward any movement toward the response lever by delivering a food pellet. Once the rat remains in the lever zone, reward only rearing over the bar. Run initial daily shaping sessions for 30 min, but restrict to 10 to 15 reinforcements. Keep the cue light on.

   Train more animals than needed for the experiment, because at all phases of training some rats learn too slowly and have to be discarded. Most rats learn after five 15-min sessions. Those that do not should be given an additional 15-min session in the afternoon; sessions may be prolonged for a further 15 to 20 min if a rat begins responding only towards the end of the session.

3. When the rat has a stable response on continuous reinforcement, switch it to a schedule in which responses are reinforced at random intervals with a mean interval of 20 to 30 sec (a variable-interval, or VI, schedule). Allow daily training sessions to last up to 30 min for several days until response rates stabilize. Alternatively, use a fixed-ratio (FR) schedule, and increase the ratio from day to day from FR1 to FR2 to FR4 to FR8 or FR10 (see Thiebot et al., 1991). Keep the cue light on during training sessions.

4. Begin alternating punished and unpunished schedules clearly distinguished by visual cues, e.g., three components, with each unpunished and punished schedule separated by a short (2-min) time-out period, in which the test chamber is dark (see Hodges et al., 1987). Use a higher reinforcement rate for the punished schedule than the unpunished (often FR1, but can be VI). Start and finish with a nonpunished schedule when there are only two schedules.

5. Gradually introduce footshock in the conflict periods, adjusting the current individually for each rat (typically shock levels from 0.1 to 0.3 mA, but not higher than 0.5 mA) to give appropriate response rates. After obtaining a steady response (about 4 weeks; unpunished response rates varying daily by <10%), maintain steady responding for at least 1 week before drug administration.

6. Habituate rats to weighing and injection procedures (using saline injections) for at least 3 days before initial testing. Inject rats with drug or saline. Use a subject in only one drug session per week. Re-establish response criteria (see step 5) for at least 3 days before performing a second drug test. Randomly allocate rats among the drug groups each week.

BASIC PROTOCOL 2

SOCIAL INTERACTION

In this test the dependent variable is the time that pairs of male rats spend in social interaction, which decreases when rats are anxious. Both bright light and unfamiliarity are used to generate anxiety. Anxiolytic action is best detected in a condition that generates low levels of interaction (e.g., high light, unfamiliar condition); anxiogenic action is best detected in the low light, familiar test condition.

Materials

Adult male or adolescent male or female rats: hooded Lister (recommended for high levels of social interaction), Sprague-Dawley, or Wistar; male rats preferred
Drugs to be tested
Saline or other vehicle for control injections

Single rat cages in low light (50 lux) housing area
Quiet test room away from disturbance with accurately controllable light levels
Video camera with automatic iris
Social interaction test arena: 60 × 60–cm wooden box with 35-cm high walls, and infrared photocells mounted in the walls 4.5 and 12.5 cm from the floor, illuminated either brightly (300 radiometric lux) or dimly (30 radiometric lux)

1. Singly house rats for 5 days before test (to increase levels of social interaction during the test) and weigh each rat daily. Provide food and water ad libitum and maintain a 12-hr light:dark cycle with light onset at least 1 hr before testing.

2. Place the test arena in a quiet room. Mount the video camera vertically above the arena. Assign test partners on the basis of weight (±5 g) and house test pairs singly in adjacent cages. Randomly allocate pairs of rats to test conditions and/or drug treatments. Give both rats in a pair the same drug treatment and use a combined score for each pair.

   In experiments involving lengthy surgical procedures (e.g., central injections or lesions), modify the procedure by testing each operated rat with an unoperated, untreated partner and scoring only the social interaction initiated by the treated rat (e.g., Higgins et al., 1988). This reduces the sensitivity of the test, as each rat's behavior is influenced by that of its partner.

3. Use one or more of the four possible test conditions (low light, familiar arena; low light, unfamiliar arena; high light, familiar arena; high light, unfamiliar arena), depending on the purpose of the experiment. For those allocated to one of the familiar test conditions, place each rat singly in the test arena for a 5- to 10-min familiarization session on 2 consecutive days.

4. Move the racks of rat cages to a dimly lit waiting area adjacent to the test room at least 1 hr before testing. Perform injections of drug or vehicle in the dimly lit area. Test pairs of

rats in an order randomized for experimental treatments, between 0800 and 1300 hr, e.g., for 4.5 to 10 min. Record all sessions on videotape and observe the rats from a television screen in an adjacent room.

5. Record social behavior with a custom-made keyboard (observer blinded to drug or test condition). Depress a key when the behavior starts and release it at the end of the behavior to automatically enter scores into a computer. Score the time spent in active social interaction. If possible, distinguish aggressive (kicking, boxing, wrestling, submitting) and nonaggressive (sniffing, following, and grooming the partner) behaviors. Score passive body contact separately.

Interruption of beams between the photocells provides automated measures of locomotor activity (lower beams) and rearing (upper beams).

BASIC PROTOCOL 3

LIGHT/DARK EXPLORATION

The light/dark exploration test uses the ethological conflict between the tendencies of mice to explore a novel environment and to avoid a brightly lit, open area. Anxiolytic drugs increase the number of transitions between the light and dark compartments and nonanxiolytics do not.

Materials

Male mice of appropriate strain, group housed (adapt 3 weeks to reversed cycle, if used)
Drugs to be tested
Saline or vehicle for control injections

Quiet, darkened test room away from disturbance
Illuminated test chamber, screened from observers (Fig. 3.10.1)

1. Group house the mice at a constant temperature of 21°C with food and water freely available.

2. Bring mice into test room at least 1 hr before testing and protect from external perturbation. Test the mice in the morning for standard light cycle, or in the afternoon, for reversed cycle. Standardize the time of testing to minimize diurnal variation.

3. Inject mice with drug or vehicle and immediately replace in a holding cage for at least 10 minutes to minimize stressful effects of injection. Place mice individually in the middle of the illuminated chamber, facing away from the opening. Allow 5 or 10 min of free exploration and count the number of crossings from the white to the black side, the time spent on each side, and the latency period before first leaving the white side.

    Additional measures of the activity on each side are optional. The observer must not know the drug treatment status of the mouse.

4. Remove any feces, wipe up urine, and wipe down the box with water after each trial. At the end of each test day wipe the apparatus with water and mild laboratory detergent. Avoid using strong-smelling detergent.

BASIC PROTOCOL 4

ELEVATED PLUS-MAZE TEST

This test relies on the inherent conflict between exploration of a novel area and avoidance of its aversive features. It may be used with male or female rats or mice, or male gerbils. It is reliable

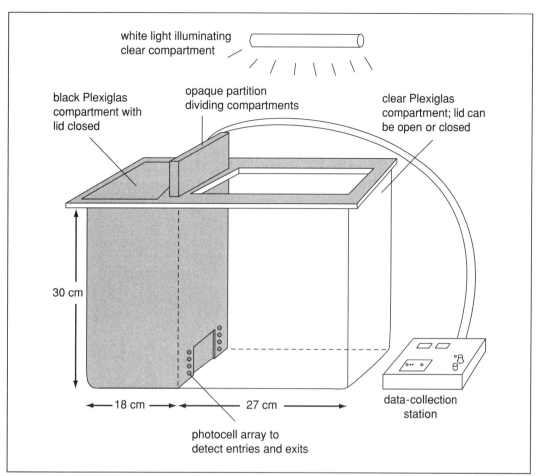

Figure 3.10.1 An example of an open-topped apparatus for measuring light/dark exploration behavior as a model of anxiety-related behaviors (J. Crawley, pers. comm.). Normally made entirely from Plexiglas (although Costall's apparatus has metal walls), it consists of a clear (light) side and a smaller, fully opaque (dark) side separated by a partition with a small opening (7.5 × 7.5 cm or 12 × 5 cm). Typical dimensions are 27 × 27 × 30 cm for the light side and 18 × 27 × 30 cm for the dark side. The light side is illuminated by a 60-W light bulb, giving an illuminance of ∼400 lux; the dark side can either be completely closed to observation or may have an observation window in the lid and be lit by a 60-W red bulb, giving an illuminance of <10 lux. If the dark side is equipped for observation, then the white light on the other side must be carefully directed only at the light side to avoid any light spillage into the dark compartment. Movements from one side to another, or within one side, can be scored from the videotapes or tracked by infrared motion detectors fitted in the walls of the apparatus. Sufficient photobeams must be used to detect movement accurately both in each compartment and through the partition. In order to record only whole-body entries, the infrared beams around the partition and between the two sides of the box must be broken and unbroken in a set order. This test works better if the whole apparatus is raised off the floor, which exposes the mice to fewer distracting stimuli.

in a wide range of strains and in group-housed or isolated animals, but distinct differences are reflected in baseline scores.

Materials

Male or female rats or mice, or male gerbils
Quiet test room away from disturbance, with few features or contents
Video camera and monitor
Elevated plus-maze apparatus (Fig. 3.10.2)
Automated test system or keyboard for scoring behaviors

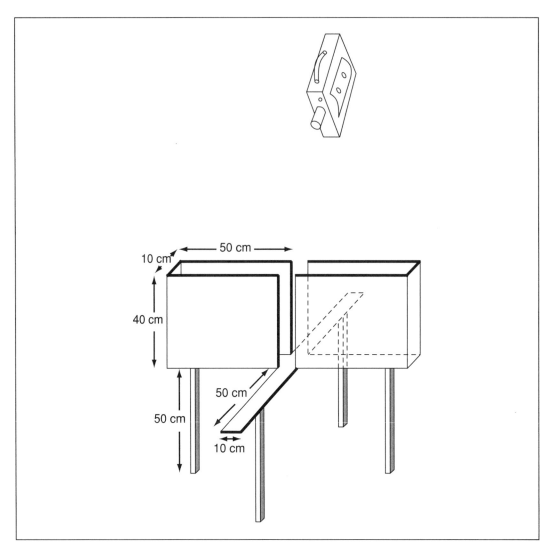

Figure 3.10.2 Elevated plus-maze (sized for rats; central platform is 10 × 10 cm). The maze may be made of wood, metal, or Plexiglas, and consists of four arms in form of a plus: two open arms (50 × 10 cm for rats; 30 × 5 cm for mice) and two arms of the same size, also with an open roof, but enclosed by walls (40 cm high for rats; 15 cm for mice). If Plexiglas is used, the side and end walls should be clear to allow even illumination. The two open arms are opposite each other and converge into a central platform (10 × 10 cm for rats; 5 × 5 cm for mice). A video camera mounted above the maze is used to observe the animal's behavior and record the trials for later additional scoring or checking. (If the observer is in the same room, the animal will be distracted by movement, sound, and odors.)

1. Set up the plus maze in the test room 40 to 60 cm above the floor (depending on the light level in the test room, the definition of the floor, and whether the animal is pigmented) to prevent the rats from clearly seeing the ground and jumping off. Mount the video camera vertically above the maze. Place the video monitor in an adjacent room so the observer does not disturb the animals.

2. Randomly allocate animals to the various drug groups and inject them at the interval appropriate to the route of injection of a particular drug. At least 1 hr before the test, bring the animal to a holding area immediately adjacent to the test room. Keep the holding area quiet.

3. Place the animal in the central platform, facing an open arm. Observe the animal for 5 min during a standardized 3- to 4-hr period. If the animal falls off, exclude its scores. For animals housed under reversed lighting, hold and test them under dim red light. Test animals *once only*, unless the intention is specifically to study the different form of anxiety evoked by a second trial.

4. Use a strict definition of an arm entry (all four paws must enter the arm) and arm exit (both forepaws must leave the arm). Score the following ethological measures (see Rodgers et al., 1995):

   a. Number of entries into open arms;

   b. Number of entries into closed arms;

   c. Time spent in open arms;

   d. Time spent in closed arms;

   e. Time spent in the central square.

   The following optional measures may also provide useful information:

   f. Number of rears (these occur almost exclusively in the closed arms);

   g. Entries into distal portion of open arms;

   h. Head-dipping over sides of open arms;

   i. Stretch-attend, scanning, flat-back posture.

5. Remove the animal and any feces, wipe the maze with a damp cloth, and dry it. Avoid using strong-smelling detergent.

BASIC PROTOCOL 5

DEFENSIVE BURYING

This test situation requires no animal pretraining, with the behavior generally occurring after a single shock-probe experience. Anxiolytic drugs produce dose-dependent suppression of defensive burying, which may be accompanied by an increase in the number of probe shocks. Anxiolytic-induced suppression of defensive burying is not secondary to general motor impairment, associative learning deficits, or analgesia. Anxiogenic drugs increase probe burying.

Materials

Male or female rats of desired strain and age
Drugs to be tested
Saline or other vehicle for control injections

Individual polycarbonate rat cages
Aspen chips
Absorbent bedding material (e.g., cat litter)
Quiet room away from disturbance
Video camera
Shock-probe test apparatus (Fig. 3.10.3)
Multimeter

1. Individually house subjects in polycarbonate cages lined with aspen chips, and maintain on a 12-hr light/dark cycle (lights on at 0700 hr), with food and water freely available. Handle rats briefly in the colony room on each of 3 consecutive days prior to habituation.

2. Place the test chamber in the test room and line it with a 5-cm depth of cat litter bedding. Mount the video camera vertically above the chamber, and place the shock source and the VHS recorder and video monitor in an adjacent room.

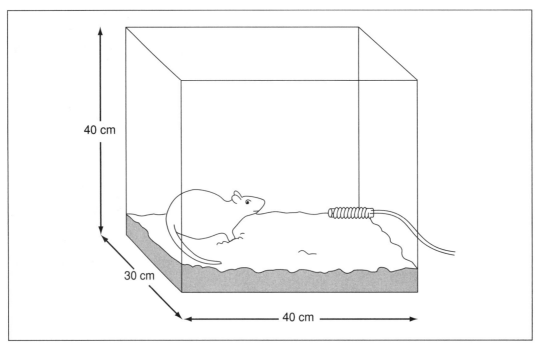

Figure 3.10.3 Shock-probe burying test apparatus. The apparatus consists of a 40 × 30 × 40–cm probe chamber of transparent Plexiglas with an open roof. The floor is covered evenly with a 5-cm layer of bedding material (e.g., odor-absorbent cat litter). In the center of one of the end walls of the chamber, 7 cm above the floor, is a 7-mm-diameter hole through which the shock probe can be inserted and secured with adhesive tape. The inserted portion of the probe is a 0.5 × 6.0–cm Plexiglas rod, helically wrapped along its length with two uninsulated copper wires separated by gaps of ∼3 mm. These wires feed to the outside of the chamber, where they are carefully insulated and connected to a shock source (leads must be long enough to permit the shock source to be placed in a separate room from the test chamber).

3. Habituate individuals or groups of four to six rats by placing them in the test chamber (no shock probe present) for 30 min on each of 4 consecutive days between 0900 and 1700 hr. Do not put the shock probe in the test chamber during these four habituation trials. At the end of each habituation period, return rats to the animal colony.

4. Randomly allocate rats to drug treatment groups and inject at appropriate intervals before the test. Just prior to the test, insert the shock probe ∼6 cm into the Plexiglas chamber and secure in place. Test with a multimeter.

5. Take rat to the test room. Activate the video camera and recorder, and enter the rat's code into the video display. Place rat into the test chamber, facing away from the shock probe (which is not yet electrified). Move to the adjacent room that contains the video monitor, VHS recorder, and shock source.

6. Activate the shock source (set to 1 to 2 mA) only when the rat's snout and/or forepaws make clear and full contact with the probe (otherwise, the rat may receive only a slight shock, to which it will respond less vigorously). After the first shock, keep the probe electrified for 15 min, during which time subsequent probe contacts result in further shocks.

7. Record the following main measures:
   a. Total duration of burying (i.e., the total time each rat sprays bedding material toward the probe using rapid, alternating forward thrusts of its forelimbs; any other form of displacement of material does not constitute burying behavior).

b. Total number of contact-induced shocks (i.e., the number of abrupt, reflexive withdrawals away from the point of probe contact).

c. Score of the behavioral reaction to each shock on the following 4-point scale:

   1 = flinch involving only head or forepaw;
   2 = whole-body flinch, with or without slow ambulation away from the probe;
   3 = whole-body flinch and/or jumping, followed by immediate walking away from the probe;
   4 = whole-body flinch and jump (all four feet in the air), followed by immediate running to the opposite end of the chamber.

d. Duration of immobility, defined as the complete absence of body movement, other than respiration (e.g., standing still or lying on the chamber floor).

8. Calculate a rat's mean reactivity to shock as the sum of its individual reactivity scores divided by the total number of shocks received during the session.

9. After 15 min, switch off the shock source and remove the rat from the test chamber. Test the entire shock circuit with a multimeter between tests to ensure that no change in conductivity has occurred. Use rats on only one occasion in this test.

10. Remove feces from the bedding and smooth to a uniform depth of 5 cm. Carefully clean the shock probe to remove dust or debris that might have accumulated during the test.

BASIC PROTOCOL 6

THIRSTY RAT CONFLICT

This simple procedure involves thirsty rats that are periodically shocked for drinking water. The dependent variable is the number of shocks received by the test animal. The effects of anxiolytics in this procedure (i.e., an increase in shocks received) are extremely robust.

Materials

Male or female rats (5 to 10 per experimental group)
10% dextrose-water solution for water bottle
Drug to be tested (for screening, oral doses of 5, 10, and 50 mg/kg and intraperitoneal doses of 1, 5, and 25 mg/kg)
Active control drug (e.g., 12.5 mg/kg chlordiazepoxide, p.o.)
Vehicle control (saline or other appropriate drug vehicle)
Clear Plexiglas box [1/4 in. thick; $38 \times 38 \times 30$ (h) cm] with stainless steel grid floor and Plexiglas lid
Black Plexiglas compartment [1/4 in. thick; $10 \times 10.5 \times 10.5$ (h) cm]
Water bottle with metal drinking tube
Constant current shock generator with circuit for counting the number of shocks (e.g., Anxio-Meter model 102; Columbus Instruments)

1. Attach the small black Plexiglas compartment to the back wall of the large clear Plexiglas compartment. Make a small opening (5-cm wide × 7.5-cm high) leading from the small compartment to the large compartment. Attach the water bottle (containing 10% dextrose in water) to the outside of the small compartment 3 cm above the grid with the drinking tube extending not more than 2 cm into the box. Attach the shocker leads between the metal drinking tube and the grid floor.

2. Deprive rats of water for 48 hr and food for 24 hr. Inject rats with drug or control solution. Initially include a no-injection control group as well as a vehicle control and an active control to verify that the no-injection controls yield comparable data to the vehicle controls,

after which the no-injection control group is no longer necessary. Randomly allocate rats among the drug groups and use only once.

3. At some experimentally determined time period after drug administration (initially at 0.5 and 2 hr to detect activity; then at peak times of activity determined by dose-response curves), place the rat in the apparatus for up to 5 min to allow it to find the drinking tube and become familiar with its surroundings. If rat does not initiate drinking within 5 min, discard it. Allow the rat to drink unpunished (shock-free) for 25 sec.

   Active compounds should be compared at times of peak activity.

4. Begin alternating punished and unpunished periods in 5-sec intervals, beginning with a punished period in which electric shock is applied (from 0.1 to 0.3 mA, but not higher than 0.5 mA; empirically determined) between the metal drinking spout and the grid floor whenever the rat makes contact with the spout. Record the number of shocks received during each minute of the 5-min test session.

References: Blumstein and Crawley, 1983; Costall et al., 1989; Crawley, 1981; File, 1980, 1993; File and Hyde, 1978; File and Seth, 2003; Howard and Pollard, 1991; Lister, 1987; Pellow et al., 1985; Treit, 1985, 1991; Vogel et al., 1971

Contributors: Sandra E. File, Arnold S. Lippa, Bernard Beer, and Morgen T. Lippa

UNIT 3.11

Assessment of Spatial Memory Using the Radial Arm Maze and Morris Water Maze

BASIC PROTOCOL 1

USE OF RADIAL ARM MAZE TASK TO TEST BASIC WORKING MEMORY

The radial arm maze task has been most extensively used to investigate and is primarily sensitive to impairments in working memory (information only useful to a rat during the current experience with the task).

Materials

   Rats
   Food reward (e.g., 10-mg pellet of chow or sweetened breakfast cereal or chocolate milk)
   Radial arm maze (Fig. 3.11.1), handmade or fully automated (Coulbourn Instruments or Columbus Instruments)

1. Weigh each rat daily throughout training and testing to monitor health and degree of food deprivation. Restrict food available to rat so that its body weight attains 85% of that prior to training. During testing and training, allow rat to gain ~5 g body weight per week. Handle rat to allow it to become comfortable with the experimenter (see APPENDIX 2D). Give food reward in home cage for a few days prior to training in order to acclimate the rat to the reward in a familiar environment.

2. Set up the radial arm maze (see Fig. 3.11.1) in a room that contains various external cues (e.g., a doorway, overhead lights, a noisy radio, and large simple designs on the walls) that are visible to the rat while it is on the maze).

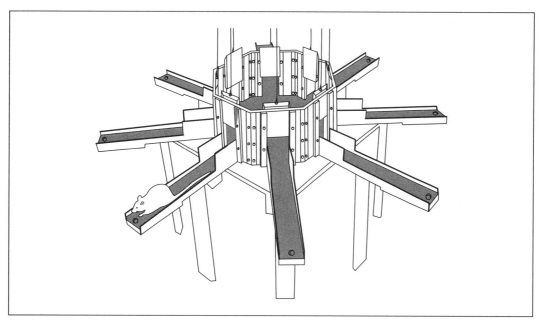

Figure 3.11.1 8-arm radial maze. For rats, the central platform should be ≥45 cm in diameter to accommodate the animal and allow it to turn easily between arms. A Plexiglas wall, 25 cm high, surrounds the central platform. The arms are 87 cm long and 10 cm wide, radiating from the central platform at equal angles. For mice, use shorter (35-cm), narrower (5-cm) arms and a smaller (20-cm) central platform. Each arm has a 5-mm-deep hole 1 cm from the end, which is used as a food cup, and each arm is separated from the center platform by a transparent Plexiglas guillotine door that covers a hole in the Plexiglas wall. The guillotine door can be raised or lowered to allow or prevent entry. The guillotine doors are connected by individual strings to a pulley system that allows the experimenter to open any door from one location within the testing room. Short walls (2 cm high) along the edge of the maze arms prevent the animal from falling off the maze.

3. Place a well-handled pair of rats (preferably cagemates) on the maze at the same time. Spread food rewards around the entire maze to encourage exploration and acclimation (1 or 2 days). On subsequent days, place food only on the arms, then only at the ends of the arms. Finally, place a rat alone on maze and food only in the food cup at end of arms.

   Testing can begin when rat is comfortable being picked up by the experimenter and, when placed alone on the maze, explores without hesitation and without excessive defecation or urination (typically within ~7 days). Run rats in the maze once a day every day (including weekends, ideally) during training and testing.

4. Place food reward at end of each arm before each test session. Place rat on central platform with all guillotine doors closed. Raise all doors simultaneously. Allow rat to enter an arm. Close doors to all other arms. Allow rat time to eat food and to return to central platform. Close door to that arm and confine rat to the central platform area for a set time (from 0 sec to many minutes; 5 sec is ideal to begin).

   Longer waits make the task more difficult to solve—i.e., increase the length of time for which the rat must remember which arms it has entered.

5. Repeat step 4 until all food pellets have been retrieved or until a predetermined length of time has elapsed. Record the following data:

   Which arm the rat entered each time and whether it received a food reward.
   Time elapsed between the beginning of the test session and the rat's obtaining all eight food rewards (determines how fast the rat is making choices and finding the food rewards; an indirect indication of motivation).

Number of correct arm choices: i.e., those that are chosen the first time.
Number of incorrect arm choices: i.e., visitations to the same arm more than once during a single test session (considered a working memory error, normally <15%, within 15 days).

6. Perform data analysis. Performance for all groups is typically expressed as either:

    a. the percentage of correct choices made in each test session in relation to the total number of arms entered,

    b. the absolute number of correct choices made in the first eight to twelve choices of each test session, *or*

    c. the percentage of correct choices made in relation to the number of incorrect choices.

    Present the data (optimally) as a line drawing comparing a performance measure for each group versus daily test sessions. Average data from two or four days of testing into blocks, if desired.

7. When performance is stable and choice accuracy is >85%, administer pharmacological agents or produce lesions to begin studies to assess their effects upon performance.

8. Alternatively, study the effects of drugs or lesions upon the acquisition of performance of this task by administering drugs (or producing lesion) prior to training

ALTERNATE PROTOCOL 1

USE OF RADIAL ARM MAZE TASK TO TEST WORKING VERSUS REFERENCE MEMORY

This protocol allows a disassociation to be achieved between working (see Basic Protocol 1) and reference types of memory (information that is useful across all exposures to the task, i.e., on any day of testing).

1. Train rats in radial arm maze (see Basic Protocol 1, steps 1 to 3).

2. Place food reward at the end of only four arms of the radial arm maze before each test session (the same for a given rat but varying between rats, e.g., arms 1, 3, 6, and 8 every time for rat #1, arms 1, 4, 5, and 7 every time for rat #2, and so on).

3. Place rat on maze with all doors raised and allow it to explore the maze completely and retrieve all food rewards. Remove the rat from the maze for a set period of time (1 hr to 1 day; longer delays make the task more difficult).

4. Bait appropriate four arms of maze for rat. Repeat step 3.

5. Record the following data:

    Number of correct entries into baited arms.
    Number of entries into unbaited arms (reference memory errors).
    Number of reentries into baited arms (working memory errors).
    Time elapsed between the beginning of the test session and the rat's obtaining all available food rewards.

    Repeat steps 2 to 5 until performance is stable and choice accuracy is high (>85%).

6. Perform data analysis. Performance for all groups is typically expressed as:

    a. the percentage of correct entries into baited arms in relation to the total number of arms originally baited (working memory performance), *and*

b. the percentage of entries into arms that are never baited in relation to the number of unbaited arms (reference memory performance).

Present the data (optimally) as a line drawing comparing working and reference performance for each group versus daily test sessions. Average data from two or four days of testing into blocks, if desired.

7. Once the rat makes few, or no, reference or working memory errors (<15%, within 10 days), begin studies with pharmacological challenges or lesions.

BASIC PROTOCOL 2

USE OF MORRIS WATER MAZE TASK TO TEST SPATIAL MEMORY

The time it takes a rat to find a hidden platform in a water pool after previous exposure to the setup, using only available external cues, is a measure of spatial memory. The water maze task is particularly sensitive to the effects of aging (Brandeis et al., 1991). This maze is typically used for rats, but it can be scaled down in size (by ~50%) for use with mice. Using the maze is quite labor-intensive and requires that a tester be present, or nearby, throughout the task.

Materials

Rats
Water maze apparatus (Fig. 3.11.2)
Tracking system and software (Columbus Instruments, HVS Image, San Diego Instruments, or CPL Systems)

1. Set up water maze (see Fig. 3.11.2), positioned in a room with various external cues that are visible to a rat swimming in the pool, e.g., a doorway, overhead lights and camera (if desired), and large simple designs on the walls. Make the water opaque by adding powdered milk or nontoxic white paint to the water. Design the pool so that it can be easily drained on a regular basis.

2. Insert platform into one quadrant of the pool. Place the rat into water with its head pointed towards the side of pool (a different, and randomized, location each day of testing, e.g., north, south, etc).

3. Record time (in seconds) it takes the rat to find the submerged platform. Guide the rat to platform on first few trials if it requires >120 sec.

    A tracking camera, positioned ~200 cm above the center of the pool, can be used to quantify the distance swam on each trial and thereby determine swimming speed when combined with latency measurements. The tracking system can also display swim path and distance and provide additional information on search efficiency and exploration patterns during acquisition and probe trials. This equipment and associated computer software can be obtained from several commercial manufacturers.

4. Allow rat to remain on platform for 10 to 15 sec. Remove rat from pool. Wait 5 min.

5. Release rat into pool (from the same location) with platform in same location. Record time for the rat to find platform. Give each rat four trials on the first day.

6. On second day, insert the platform in same location as on the first day. Release the rat with its head pointed towards the side of the water pool. Record time it takes the rat to find the platform. Give the rat eight to ten trials per day with 5-min intertrial intervals for several days until performance is stable and latency to find the platform is low (<5 to 7 sec).

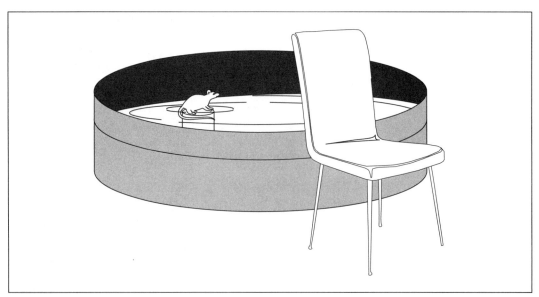

Figure 3.11.2 Morris water maze. The pool should be watertight, 200 cm in diameter and 75 cm deep, and filled with 50 cm water. The actual dimensions of the pool can be varied depending upon the space available to contain it and whether rats or mice are being tested (e.g., for mice, the pool need only be ~100 cm in diameter and 30 cm deep). The water is made opaque by adding nontoxic white paint or powdered milk. A 10-cm circular escape platform should be constructed of a water-resistant material and covered with material (e.g., cloth) that allows the animal to remain on the top when it is submerged. The platform should be made heavy enough to remain upright when submerged or may be attached to the bottom of the pool. The platform should be 48 to 49 cm in height so that it is submerged 1 to 2 cm below the surface. A chair can be positioned near the pool to allow the experimenter easy access and provide an additional cue for the animal. The water temperature should be maintained at ~20°C. It is useful to have an ample supply of towels nearby to dry animals between trials. Also, an incubator can be used to keep them warm between trials.

7. Perform data analysis. Performance is expressed as the average time it takes each rat to find the submerged platform. The data are best presented as a line drawing comparing the latency to find the platform for each group versus daily test sessions. Data from 2 or 4 days of testing can be averaged into blocks.

   Additional analyses utilizing sophisticated computer tracking programs can classify the spatial location of the animal with regard to the platform in order to provide information on the spatial pattern of the rat's search during both the training and testing phases, as well as during the probe trial (see Alternate Protocol 2).

8. When performance is stable, begin studies with pharmacological agents or lesions to assess their effects upon performance.

9. Alternatively, study the effects of drugs or lesions upon the acquisition of performance of this task by administering drugs (or producing lesions) prior to training to assess their effects upon acquisition.

ALTERNATE PROTOCOL 2

USE OF WATER MAZE TASK FOR SPATIAL PROBE TRIAL

The spatial probe trial is used to test the rat's knowledge of the precise location of the platform in the water maze. An accurate direction of the swimming behavior provides evidence that the rat has learned the spatial location of the platform relative to the available external cues (Sutherland et al., 1982).

1. Train rats in the water maze (see Basic Protocol 2, steps 1 to 6).

2. Set up the water pool without a platform. Release the rat with its head pointed towards the side of the water pool. Remove rat after 90 sec.

3. Record time the rat spent in the quadrant that previously contained the platform and calculate as a percentage of total time in pool. If possible (i.e., if using a computer tracking system), also record the percentage of time spent in the other quadrants.

4. Perform data analysis. Express the data as either:

    a. a histogram showing the average amount of time that the rats in each group spent exploring each quadrant of the pool (most typically), *or*

    b. the percentage of time the rat spent exploring the quadrant that had contained the platform in relation to the total time spent exploring the entire pool.

ALTERNATE PROTOCOL 3

USE OF WATER MAZE TASK TO TEST WORKING MEMORY

This protocol has also been referred to as a "reversal test."

1. Train rats in the water maze (see Basic Protocol 2, steps 1 to 6).

2. Release the rat with its head pointed towards the side of the water pool (start position the same each day).

3. Record time it takes the rat to find submerged platform. Allow the rat to remain on platform for 10 sec. Remove the rat from pool and place in holding cage for 15 sec. Move submerged platform to new location.

4. Release the rat from same location as in step 2. Allow the rat to swim for up to 120 sec. Record time it takes for the rat to find platform. Guide the rat to platform if necessary. Allow the rat to remain on platform for 10 sec. Remove the rat from pool and place in holding cage for 15 sec.

5. Repeat step 4 until the rat swims directly and quickly to the platform. Record time on each attempt.

6. Repeat steps 4 and 5 once per day for 4 days, with the platform in a different quadrant of the pool each day (decreasing latency to find the platform expected with each day of testing).

7. Perform data analysis. Performance is expressed as the average time it takes each rat to find the submerged platform at each new location. The data are best presented as multiple line drawings comparing the latency to find the platform for each group and for each location versus daily test sessions. Data from 2 or 4 days of testing can be averaged into blocks.

References: Brandeis et al, 1989; Olton, 1985

Contributor: Gary L. Wenk

UNIT 3.12

Cued and Contextual Fear Conditioning in Mice

BASIC PROTOCOL

ONE-TRIAL CUED AND CONTEXTUAL FEAR CONDITIONING IN MICE

Fear conditioning protocols for both rats and mice are very similar. Many factors such as age, prior experience, and genetic makeup can influence fear conditioning and may produce random or systematic variation. Aging may alter an animal's ability to hear the auditory cue or process features of the context, and early-onset hearing loss is characteristic of some strains of mice in which the ability to hear high frequency pure tones is especially impaired—even in early adulthood. For this reason, the characteristics of the auditory conditioning stimulus are important and high-frequency pure tones should be avoided. A broad-band clicker is employed as the auditory cue in the Basic Protocol. If deafness is suspected, a more thorough examination of auditory function might be pursued (see Crawley, 2000).

Care also must be exercised when animals are to be put through a battery of behavioral tests because order effects or interactions (order-by-treatment or order-by-genotype) may obscure fear conditioning results. The training (day 1) and the testing (day 2) should be performed at the same time of day. To measure contextual learning accurately, the same person must perform both the training and testing.

Materials

    70% ethanol or 4% acetic acid solution for cleaning chambers
    Mice (e.g., C57BL/6J, male or female), typically 8 to 12 per group; 20 to 25 per group in two separate experiments for more subtle effects (e.g., those that are produced by a single gene manipulation)

    Fear conditioning chambers (Figs. 3.12.1 and 3.12.2; also see Paylor et al., 1994; Wehner et al., 1997 for mice, or Rudy and Morledge, 1994 for young rats) made in-house from components described below
    Shock generator and scrambler for administering shocks at various intensities (0.1 to 1.0 mA; e.g., Med Associates, Lafayette Instruments, Columbus Instruments, or Coulbourn Instruments)
    Sound generator that either delivers broad-band clicker sounds or low-frequency tones (e.g., Med Associates, Lafayette Instruments, Columbus Instruments, or Coulbourn Instruments)
    Computer that will run Med PC software and interface with the chambers to allow a timed program (e.g., Med Associates PC for Windows)
    Sound-attenuating chambers (Fig. 3.12.1) equipped with
        Light (24 V, DC)
        Small fan (to provide air circulation and white noise)
        Speaker attached to the sound generator
        Doors that allow viewing of animals through a window while chambers are visually isolated from each other (so two subjects can be trained and tested simultaneously)
    Transparent acrylic contextual conditioning chambers (typically 26 × 21 × 10–cm for mice or young rats; 23.5 × 29 × 19.5–cm for adult rats; see Fig. 3.12.2A) with
        Removable grid floors constructed of stainless steel rods 1.5 mm in diameter and spaced 0.5 cm from center-to-center for mice (Owen et al., 1997a,b;

Wehner et al., 1997) and 1.2 cm center-to-center for young rats (Rudy and Morledge, 1994; 16 stainless steel rods (2.5 mm in diameter) spaced 1.25 cm from center to center (Fanselow, 1990) for adult rats

Screen top

Smooth plastic floor (Fig. 3.12.2B) the size of the grid floor (for altered context testing)

Rectangular plastic wall placed on the diagonal in the chamber (for altered context testing)

Voltmeter and sound meter to check stimulus intensities

Scoring sheets with subject number, date, tester identity, any other relevant information (e.g., genotype or protocol number) and columns for each subject for scoring of freezing behavior in 10-sec blocks (e.g., Fig. 3.12.3)

Automated scoring systems (optional) using, e.g., photo beam interruptions (Bolivar et al., 2001), pressure transducers (Fitch et al., 2002), digital tracking systems (Kim et al., 1993), image subtraction methods (Anagnostaras et al., 2000; Marchand et al., 2003), a head video-tracking system (Moita et al., 2003), and a proprietary motion detection system (Miller et al., 2002) requiring additional equipment, some commercially available (Lafayette Instruments, Med Associates, Actimetrics, Viewpoint Life Sciences).

Day 1: Carry out training procedure

1. Turn on all equipment, and clean all chambers thoroughly with 70% alcohol or 4% acetic acid solution. Verify that the following occurs:

   a. A shock is delivered to the floor grid either with a voltmeter or by touch. Record the intensity of the shock used in every experiment for future comparisons.

   b. The auditory cue is generated. For future comparison, periodically record the background noise level from the fan in addition to the decibel level of the auditory conditioning stimulus.

   c. The cameras are in the proper position and in focus (for automated systems that require the use of cameras).

2. Transport the animals to be conditioned from the colony or holding room to the testing room in individual holding cages. Only transport the number that is equal to the number of conditioning chambers.

3. Place each subject into the conditioning chamber (Fig. 3.12.2) and replace the top of the chamber. Close the sound-attenuating chamber door (Fig. 3.12.1). Keep a record of the specific chamber used for each animal if multiple chambers are used.

4. Activate the computer program. For contextual learning, keep the animal in the chamber for 2 min (phase A) before presentation of the clicker sound for 30 sec (phase B), followed by presentation of the foot shock for 2 sec (phase C). If multiple rounds of conditioning are required, repeat phases A, B, and C. Score animals for freezing during the first 2 min in order to obtain a measure of baseline freezing.

5. Remove animals from the chambers 30 sec after receiving the last shock. Place animals in the transport cages and then return them to their home cages. Clean each chamber very thoroughly and reset the computer.

6. Obtain the next set of animals to be tested and repeat training steps 2 to 5.

Day 2: Test for contextual or cued fear conditioning

7. Turn on the equipment and verify that the auditory cue (no shock) is generated. For automated systems that require the use of cameras, verify that the cameras are in the proper position and in focus.

Figure 3.12.1 Typical equipment used for contextual and cued fear conditioning. (**A**) A sound-attenuating chamber holds the conditioning chamber; the exterior view is shown. (**B**) The inside of the sound-attenuating chamber has a fan, light, and speaker.

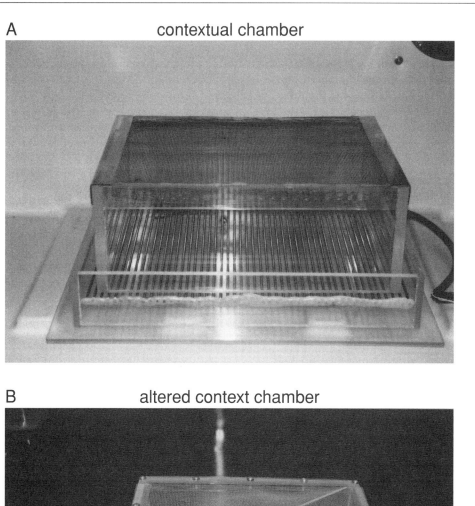

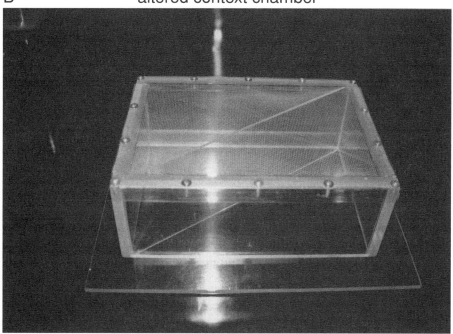

Figure 3.12.2 Typical chambers for the context and altered contexts. (**A**) The contextual conditioning chamber is composed of a wire shock grid and a plastic top. (**B**) The chamber for measurement of freezing to the altered context and the auditory cue (CS) is shown. In order to show the details for this picture, it has been removed from the outer sound-attenuating chamber.

8. Transfer subjects from home cage into transporting cage, with one subject per cage, and transport to testing room (typically 24 hr after training). Place the subject into the same conditioning chamber that was used for training.

9. Activate a silent hand-held timer and examine the animal through the chamber window by looking into the chamber every 10 sec. For consistency, start the first bout of scoring at the same time for all experiments (e.g., start scoring at 20 sec after closing the chamber door). Watch the timer and then look up at the animal just for an instant to determine immediately whether it is freezing (total lack of movement except for respiration) or not. Score the animal as to whether it is freezing (1) or moving (0) for 5 min (i.e., 30 possible freezing scores are recorded; sample scoring sheet shown in Fig. 3.12.3).

10. Remove the animal from the chamber into the transporting cage and return to home cage for 1 hr. Clean the chambers and repeat the testing procedures with remaining subjects (steps 7 to 10).

11. Reset the computer to the proper program for altered context and freezing to the auditory conditioning stimulus (CS). Remove the grid floor of the conditioning chamber and replace it with a smooth plastic floor (Fig. 3.12.2B). Change the configuration of the conditioning chamber from a rectangle to a triangle by adding a plastic wall (colored to make the altered context more distinct, if desired) placed at the diagonal in the chamber. Change the olfactory cues if desired, e.g., by placing a small plastic cup containing a drop of orange extract in the corner of the chamber.

12. Starting with the first animals that were scored on contextual learning, transfer animals into transporting cage and bring into testing room. Place each animal in the altered context chamber. Activate the timed program and score animal every 10 sec as freezing or not freezing for 3 min (i.e., 18 possible scores).

13. At the completion of scoring in step 12, activate the auditory conditioning stimulus for 3 min. Score each animal every 10 sec as freezing or not freezing for 3 min (i.e., 18 possible scores). Remove animals to home cages, clean chambers, and score the next set of animals.

14. Perform calculations, most commonly measured as a percent score, because of the time differences in the scoring for the different types of learning. In this case:

    Context = (the number of bouts of freezing in the context/30 bouts) × 100%
    Altered context (pre-CS) = (the number of bouts of freezing in the altered context/18) × 100%
    Auditory cue (CS) = (the number of bouts of freezing to the auditory CS/18) × 100%.

    Present the number of bouts of freezing for each of the three tests [i.e., context, altered context (pre-CS), and auditory cued (CS)] and analyze the data by analyses of variance (ANOVA) for each of the different measures of freezing. Use a repeated-measures ANOVA when comparing the degree of fear conditioning in the context versus altered context or to the auditory cue. Determine group differences by a between-subjects design.

ALTERNATE PROTOCOL 1

PRE-EXPOSURE EXPERIMENTS

Some genotypes show improved contextual learning after pre-exposure to the context. This is usually done the day before the training day (prior to step 1 of the Basic Protocol). This pre-exposure is usually brief (0.5 to a few minutes) and is thought to facilitate hippocampal function.

Date: | Tester: J.W. | Experiment: #101 | C57BL/6 (B6) compared to DBA/2 (D2) mice

Time:

	Condition	1	2	3	4	5	6	7	8	9	10	11	2	3	4	15	16	17	18	19	20	21	22	23	24	25	26	27	28	29	30	Freezes
1	naïve B6	0	0	1	1	1	0	1	1	0	0	0	0	0	1	1	0	1	0	1	0	0	1	1	0	0	1	1	0	1	1	15
2	naïve B6	1	1	1	0	0	0	1	1	1	0	1	1	1	0	0	0	1	1	0	1	1	0	0	1	1	0	1	1	1	0	18
3	naïve D2	0	0	1	1	0	0	0	1	0	0	0	1	0	1	1	0	0	0	0	0	0	0	0	0	0	0	0	0	1	0	7
4	naïve D2	0	0	0	0	1	0	1	0	1	0	0	0	0	0	0	1	1	1	0	0	1	0	0	0	1	0	0	0	1	0	9
5	naïve B6	0	1	1	1	0	1	1	0	1	1	0	1	1	0	1	0	0	0	0	1	0	1	1	1	0	1	1	1	0	0	17
6	naïve D2	0	0	1	0	0	0	0	1	0	0	0	0	0	0	0	0	1	0	0	0	1	0	0	0	0	1	1	0	0	0	6
7																																
8																																
9																																
10																																
11																																
12																																
13																																
14																																
15																																
16																																
17																																
18																																
19																																
20																																

		Altered Context																		CS																		Alt	CS
		1	2	3	4	5	6	7	8	8	10	11	12	13	14	15	16	17	18	19	20	21	22	23	24	25	26	27	28	29	30	31	32	33	34	35	36		
1	naïve B6	0	0	0	0	0	0	0	0	0	1	0	0	1	0	0	0	0	0	1	1	1	0	1	1	1	0	0	0	1	1	0	1	0	0	1	1	2	11
2	naïve B6	0	0	1	0	0	1	0	0	0	0	0	0	0	0	0	0	0	0	0	1	1	1	1	1	0	0	1	1	1	1	1	0	0	1	1	0	2	12
3	naïve D2	0	0	0	0	0	0	0	0	0	0	0	1	0	0	0	0	0	0	1	0	0	0	1	1	1	1	0	1	0	1	1	1	0	0	0	1	1	10
4	naïve D2	0	1	0	0	0	0	0	0	0	0	0	0	0	0	0	1	0	0	0	1	1	1	0	0	0	1	1	1	1	1	0	0	0	1	0	0	2	9
5	naïve B6	0	0	0	0	0	0	0	1	1	0	0	0	0	0	0	0	0	0	1	1	1	1	0	1	0	1	0	1	1	1	1	1	0	1	1	1	2	14
6	naïve D2	0	0	0	0	0	1	0	1	1	0	0	0	0	0	0	0	0	0	1	1	0	1	1	1	0	1	0	0	1	1	1	0	1	1	1	1	2	13
7																																							
8																																							
9																																							
10																																							
11																																							
12																																							
13																																							
14																																							
15																																							
16																																							
17																																							
18																																							
19																																							
20																																							

Figure 3.12.3 Sample scoring sheet for recording freezing on testing day. Scores for several C57BL/6J and DBA(2J mice are shown. Mice are scored as "0" when they are moving and "1" if they are freezing.

ALTERNATE PROTOCOL 2

DISCRIMINATION PROTOCOL

The dependence of contextual learning on the hippocampal formation may be greater when an animal must discriminate between two contexts: one in which they received the shock and another in which they were placed for the equivalent time but did not receive the shock. The following protocol involves a pre-exposure step but could be performed without pre-exposure to the chambers. The 5-day protocol is modified from Frankland et al. (1998).

Materials (also see Basic Protocol)

Fear conditioning chambers (see Basic Protocol, except that two chambers must be different with respect to as many visual and olfactory cues as possible, and the chambers must be in two separate rooms, A and B)

1. Prepare the testing environment such that the characteristics of the two rooms and the conditioning chambers differ maximally. In room A and chamber A, keep the room dark, use the usual rectangular chamber, and clean chambers with ethanol. In room B and chamber B, turn on room lights, place a circular plastic ring in the conditioning chamber, clean the chamber with detergent, and place a small cup with a few drops of orange extract in the corner of the chamber.

2. On day 1, expose animals for 5 min to each of the two chambers (5 min in A and then 5 min in B).

3. On days 2 to 4, expose animals to chambers A and B, but shock them in one of the two chambers, for a total exposure time of 3 min in *either* chamber with 1 hr between the two exposures. Expose half of the animals to the shock unconditioned stimulus (US) in *either* chamber A or B. Time the shock such that it is delivered 2.5 min after the animal is placed in the chamber and then remove the animal 30 sec later. Monitor and record freezing scores every 10 sec during the first 2.5 min each day in each chamber.

    Each animal must be shocked in the same chamber each day.

4. On day 5, do not shock the animals, but obtain freezing scores in both chambers.

5. Calculate scores and analyze data (see Basic Protocol).

    The expected result is that the animals will freeze more in the context where they were shocked as compared to the other context.

ALTERNATE PROTOCOL 3

CONTEXTUAL LEARNING WITHOUT THE AUDITORY CUE

While it is common practice to evaluate contextual and cued fear conditioning in the same animal, contextual learning can be evaluated without the presentation of the auditory conditioning stimulus. The procedure is the same as the Basic Protocol, except that after exposure to the context the shock is administered without the auditory cue. Scoring is also the same as the Basic Protocol.

ALTERNATE PROTOCOL 4

IMMEDIATE SHOCK CONTROL PROTOCOL

Animals will not show freezing in the context of the experiment unless they are given adequate exposure to the chamber to allow time to make the association between the context and the shock. However, an immediate shock condition can be used to the investigator's advantage when biochemical studies or genetic comparisons are performed because it can be used as a control condition to eliminate the possible effects of the shock from those specific to contextual learning (Young et al., 2000). The same number of shocks is administered as for the normal fear conditioning procedure. Immediately following placement into the chamber, the shocks are delivered consecutively, and the animal is allowed to remain in the chamber for the same duration as in the normal fear conditioning procedure.

Reference: Wehner et al., 1997

Contributors: Jeanne M. Wehner and Richard A. Radcliffe

UNIT 3.13

Conditioned Flavor Aversions: Assessment of Drug-Induced Suppression of Food Intake

BASIC PROTOCOL

Administration of a drug following ingestion of a novel food or solution often suppresses subsequent intake of the new food or solution. This suppression is associative, in that consumption is not suppressed when there is no temporal relationship between consumption and drug administration. Aversion conditioning is extraordinarily robust, acquired with limited trials and long delays between consumption and drug administration, and at doses of the drug that are innocuous under other circumstances. The robust nature of aversion learning has made this procedure a sensitive and widely used behavioral index of drug side effects.

Materials (see APPENDIX 1 for items with ✓)

    Rats (6 to 10 animals per group), individually housed (see UNIT 3.4)
    Tastant solution [e.g., 0.1% or 0.2% (w/v) sodium liter tap water], prepared fresh
✓ Drug solution (e.g., lithium chloride)
    Drug vehicle for control injections (usually sterile H_2O)

Top-loading balance
50- to 100-ml calibrated glass or plastic drinking tubes (or noncalibrated water bottles with flat bottoms for weighing)
Rubber stoppers (size dependent on drinking tubes)
Stainless steel sipper tubes bent at 45° angle
Stopwatch
Disposable syringes and 23- to 25-G needles for injection of drugs or gavage needles for oral administration

1. Deprive animals of water for ~24 hr. Weigh animals to the nearest gram using a top-loading balance.

2. Fill water bottles with fresh tap water or sterile water and twist stoppers firmly in place. Tap bottles gently to displace air from sipper tube. Measure level in calibrated bottles (or weigh noncalibrated bottles) and record as preconsumption (PRE).

3. At intervals of 30 sec (or other specific interval, timed with a stopwatch), place each bottle at a 30° angle on home cage of each water-deprived animal. Allow animals limited access to water (e.g., 20 min). Remove bottles from cages at the same time intervals so that all animals get the same access period.

4. Remove bottle and measure fluid level in calibrated bottle or weigh noncalibrated bottle. Record this measure as postconsumption (POST). Subtract POST from PRE to calculate amount of fluid consumed.

5. Repeat above procedure daily until fluid consumption is stable, ±2 ml, over 3 consecutive water-adaptation days.

6. Perform steps 2 to 4, except fill the bottles with a novel tastant solution (e.g., 0.1% sodium saccharin). Use the same time intervals and fluid-access periods as were used in the water-adaptation phase. Immediately following fluid access, rank the animals in terms of the amount of fluid consumption on this conditioning trial, and assign the animals to the appropriate experimental and control groups, matched for fluid consumption.

7. Within 1 to 15 min following fluid access, remove the animals from their cages, give them the appropriate experimental (e.g., 76 mg/kg lithium chloride) or control (vehicle) injection and return them to their home cages. Inject animals in the same order and at the same time intervals used in placing and removing the bottles from the cages.

8. For 2 to 3 days following this conditioning trial allow animals limited access to tap water as during the water adaptation phase by performing steps 2 to 4, once again filling the bottle with water. Make no injections during these water-recovery sessions (usually two to three sessions, or more if required, to recover baseline levels).

9. Repeat the conditioning/water recovery cycle (steps 6, 7, and 8) until consumption of the tastant solution for the drug-injected subjects is stable (±2 ml over two consecutive conditioning cycles; generally five to ten cycles).

10. Alternatively, use a two-bottle taste aversion design. Follow steps 1 to 9, but once consumption of the tastant has stabilized (±2 ml for two consecutive conditioning trials), give subjects two bottles during the aversion test, one containing the tastant and the other containing water. Use relative consumption of the toxin-associated taste to indicate the degree of the aversion.

References: Barker et al., 1977; Braveman and Bronstein, 1985; Bures et al., 1998; Garcia and Ervin, 1968; Klosterhalfen and Klosterhalfen, 1985; Logue, 1979; Milgram et al., 1977; Revusky and Garcia, 1970; Rozin and Kalat, 1971; Spiker, 1977

Contributors: Anthony L. Riley and Kevin B. Freeman

UNIT 3.14

Measurement of Startle Response, Prepulse Inhibition, and Habituation

The startle response is comprised of a constellation of reflexes elicited by sudden, relatively intense stimuli. It offers many advantages as a behavioral measure of central nervous system (CNS) activity and can be measured in numerous species, including humans, when elicited by acoustic (noise bursts), electrical (cutaneous), tactile (air puff), or visual (light flash) stimuli.

STRATEGIC PLANNING

The materials and procedures needed to study the startle response vary enormously depending on the particular goals of the study, e.g., drug toxicity, auditory system physiology, neurodevelopment, and behavioral genetics. The specific application must determine decisions related to the particular startle apparatus; animal species, strain, sex, and age; stimulus and test session characteristics; and experimental design. Commercially available systems (e.g., the SR-LAB system recommended for these protocols, San Diego Instruments) with interchangeable chambers of different sizes for different species or ages of animals (e.g., mice, hamsters, guinea pigs, rats, ferrets, and marmosets) make it possible to quickly change according to the size of the animal. Rodent studies are the most convenient and efficient for screening drug effects on or assessing the neurobiological mechanisms of startle plasticity.

Gender

Both male and female rats have been used in measures of startle reflex magnitude and plasticity. The effects of rat gender on many variables and drug responses have not been studied systematically, so it is important to match the gender of the rats used previously when studies are replicated or extended.

Age

Startle magnitude and prepulse inhibition (PPI) are influenced by rat age. Prepubescent rats exhibit smaller startle responses and less robust and more variable PPI compared to adults. Aged rats may experience frequency-dependent hearing loss, which can alter startle response characteristics.

Animal Care and Housing

Because of the effects of stress on startle behavior, animal care procedures should be designed to limit exposure of rats to stress. Consideration should be given to the means of shipment from the supplier and the amount of time the animals must spend within confined and poorly ventilated shipping boxes before they reach their home cages.

Isolation or extremely crowded housing can significantly affect startle variables. Animals should be housed two to three per cage, unless the effect of different housing conditions on startle is a test parameter. Animals should be kept in a housing facility offering conditions appropriate for other sensitive behavioral measures, including constant ambient temperature and a regular circadian lighting pattern. Startle is best measured in the dark phase of a rat's circadian cycle, when it is most robust and least variable; effects of circadian cycle on startle plasticity have not been studied completely.

It is important to control the ambient noise level and assess whether sound insulation is needed in the housing area. The sound-attenuated testing room must be separated from any hallway or heavily used thoroughfare. Floor vibrations, such as those triggered by pedestrian traffic or doors, can contaminate the results, as any phasic sensory events perceived by the test subjects can affect startle measures by virtue of prepulse inhibitory or dishabituation effects.

Because of the sensitivity of startle to stressors, begin animal handling procedures within three days of shipment arrival, and at regular intervals thereafter throughout the course of the experiment (see Support Protocol 1). Health checks are particularly important because of the possible impact of infectious processes on the auditory system, as well as the effects of generalized debilitation on reflex responsiveness. Inspect the animals daily, and handle thoroughly (at least 1 min per animal) every 3 to 4 days.

Drug Effects

Startle testing is commonly performed after drug administration via systemic injection, oral drug delivery, drug delivery via chronic indwelling pumps or drug pellets, or intracerebral drug delivery. In the interpretation of the effects of a drug on startle habituation, the time course of the drug effect is a potential confounding factor, as it may produce altered levels of reactivity late in the trial series not evident initially. Such a pattern could indicate either a delayed drug effect per se or a drug effect on habituation.

Data Analysis

The practical steps required for analyzing startle data are dictated by the nature of the data files created by the test system. Assemble all the data from an entire experiment into a single ASCII file for use by a standard statistical package such as SAS, SYSTAT, SPSS, or BMDP, condensing the data by averaging trials for each animal prior to actual analysis. While this step can be accomplished using spreadsheet programs, it is more efficient to use programs that condense the data automatically and rapidly, such as SAS or SYSTAT. The software provided with the San Diego Instruments system combines all the raw data generated from the experiment into one ASCII file using a utility program. The authors reduce these data to the descriptive variables of interest for each animal and sort them using SYSTAT, and conduct inferential statistical analyses using SYSTAT, SPSS, or BMDP (PC systems) or SuperANOVA (MAC systems).

BASIC PROTOCOL 1

BASIC TEST OF ACOUSTIC STARTLE REACTIVITY

The most basic startle experiment assesses the level of reactivity in groups of rats manipulated in a particular manner, such as by drug treatment. For this purpose, startle reactivity is defined as the magnitude of the startle response on either the initial stimulus presentation or over a relatively small number (e.g., 10) of startle trials. Data obtained from Basic Protocol 1 are used to match groups of rats for the prepulse inhibition (PPI) experiment described in Basic Protocol 2.

Both within- and between-subject comparisons are possible in startle experiments. This matching procedure reduces the large subjective variation typical of startle reactivity measures. Because these differences tend to be stable over days or weeks, animals can serve as their own control by being tested on two different days.

Materials

Forty naive male Sprague-Dawley rats weighing 250 to 300 g (typically 8 to 12 per cohort)
Drugs and placebos

Vivarium with 12 hr light/dark cycle (e.g., lights on at 8:00 a.m. and off at 8:00 p.m.)
Sound-attenuated testing room
SR-LAB startle apparatus with digitized electronic signal output (San Diego Instruments) or equivalent; programming of all aspects of the stimuli and test sessions accomplished by the menu-driven SR-LAB software
Calibrated sound level meter (Quest)
Drug delivery system

1. Obtain and prepare animals. Handle all forty rats (see Support Protocol 1) and acclimate them to the vivarium and light cycle for 1 week before testing. Delay startle testing after a surgical procedure for at least 1 week, and continue handling during that postoperative week.

2. Select and define testing parameters, e.g., plan to present one initial and ten subsequent trials of a single acoustic stimulus to each rat, with reference to the following considerations:.

Maintain a continuous background noise level of 65 dB within each startle chamber to provide a consistent acoustic environment and to mask external noises.

Conduct testing during the dark phase of the animals' diurnal cycle, no closer than 1 hr to either light change.

For the acoustic startle stimulus, use a fast-rise-time (<1 msec) burst of noise presented for 40 msec at an intensity of 120 dB. Set the intertrial intervals (ITIs) to average 15 sec, with a range from 8 to 23 sec, resulting in a test session of ∼8 min per rat. Use a variable ITI to minimize habituation of startle (Graham, 1975; Hoffman and Searle, 1968).

3. Calibrate the stimulus delivery and response recording systems. For multiple test stations, calibrate all stations to the same standards and program them to deliver the startle stimuli simultaneously. Measure sound levels with continuous tones and experimental chambers closed. Measure stabilimeter sensitivity with the SR-LAB dynamic startle response calibration unit.

4. Clean and dry the animal enclosure before testing each rat. Weigh and test each of the forty rats, beginning the test session with a 5-min period of acclimation to the background noise.

5. Collect the peak or average response from each rat on each of eleven trials. Record the initial response value separately, and average the remaining ten responses together for each rat.

   Only the initial startle response provides a pure measure of startle reactivity uninfluenced by habituation and sensitization, but single response measures are too variable for most practical purposes. Therefore, the standard practice is to average the responses across blocks of five to ten trials (after the initial response) for each subject.

6. Define four matched groups of ten rats each by sorting the forty startle reactivity values to give each group the same mean and variance. For multiple test stations, include the same number of rats from each station in each group. Ensure that the four groups match by the less contaminated but more noisy measure of the initial startle response.

ALTERNATE PROTOCOL 1

BETWEEN-SUBJECTS TESTS OF STARTLE REACTIVITY

With manipulations such as prenatal exposure to a noxious agent, where animals cannot be tested before the manipulation, use control subjects and a between-subjects design. In selecting manipulations to study the startle reflex, consider factors such as the route of drug administration, the expected drug effect, and dose-response characteristics. Certain drugs produce ceiling or floor effects (i.e., plateau or threshold measurements) in startle testing that complicate data interpretation. In cases where this sort of effect is anticipated, select a range of startle pulse intensities to delimit submaximal to maximal startle magnitude characteristics.

To test the effects of a manipulation on startle reactivity where the expected effect size is likely to have adequate power between subjects carry out the basic test of acoustic startle reactivity (see Basic Protocol 1) with the following alterations:

1. Modify steps 1 and/or 4 as appropriate for the manipulation being studied. For example, if a dose-response study of a drug is the goal, treat each of the rats with the appropriate vehicle or dose at a suitable time before introducing the animal into the chamber in step 4.

2. Modify step 6 by including the independent variable (e.g., vehicle or drug treatment) as a factor in analyses of variance (ANOVA) on the two dependent measures (initial response and average of trials 2 through 11).

ALTERNATE PROTOCOL 2

BETWEEN-SUBJECTS TESTS OF STARTLE HABITUATION

To increase the number of trials to demonstrate adequate levels of habituation, carry out the basic test of acoustic startle reactivity (see Basic Protocol 1) with the alterations described in the steps below.

Materials (see Basic Protocol 1)

1. Modify steps 1 and/or 4 of Basic Protocol 1 as appropriate for the manipulation being studied (see Alternate Protocol 1).

2. Modify step 2 to include more trials (e.g., with acoustic startle stimuli, a total of 121 trials to ensure that adequate habituation occurs).

3. Modify step 6 by including the independent variable (e.g., vehicle or drug treatment) as a between-subjects factor and blocks of trials as a within-subjects factor in ANOVA.

   If the response to the initial trial is excluded from this analysis as being a unique event (see Basic Protocol 1, step 5), the principal dependent measure becomes the repeated blocks of trials. Average the data for each subject as 24 blocks of five trials each or 12 blocks of ten trials each, as dictated by the rapidity of habituation observed (evidenced by a statistically significant effect of trial block in the ANOVA). Treatment-induced changes in startle habituation are evidenced by a statistically significant treatment-by-trials interaction in the ANOVA. Alternatively, assess habituation as the percent decrease in startle reactivity between the first and last blocks of trials.

BASIC PROTOCOL 2

TESTING PREPULSE INHIBITION OF STARTLE

In combination with pharmacological and neurosurgical procedures, this protocol allows for the systematic investigation of the neurochemical and neuroanatomical systems that modulate sensorimotor inhibition. Startle magnitude is reduced when the pulse stimulus is preceded 30 to 500 msec by a weak prepulse. PPI provides an operational measure of sensorimotor gating. Three types of prestimuli used in intramodal studies of sensorimotor gating of acoustic startle are: (1) delivery of a discrete acoustic prepulse several msec before the startle pulse, with an intensity below startle threshold; (2) a variation of this design, the peak-on-pedestal stimulus, in which the prepulse is a continuous elevation of the background noise (the pedestal) beginning at a set interval prior to delivery of the pulse (the peak); and (3) gap inhibition, where a discrete reduction (gap) in background noise precedes the startle pulse by a prescribed interval. In all these cases, the interval length, intensity, and duration of the prepulse significantly influence the amount of PPI.

Additional Materials (also see Basic Protocol 1)

   Apomorphine·HCl for the administration of 0.1, 0.3, and 1.0 mg/kg injections at a
      volume of 1.0 ml/kg, prepared fresh
   Vehicle solution of sterile isotonic saline containing 0.1 mg/ml ascorbic acid, purged of
      oxygen by sparging with nitrogen gas for 10 min, prepared fresh

   Sterile 1-cc syringes
   Sterile 25-G syringe needles

1. Follow the basic acoustic startle reactivity test (see Basic Protocol 1) to define four matched groups of ten rats each.

   Omitting the baseline matching step described in Basic Protocol 1 permits proceeding directly to assessment of PPI in animals previously untested in startle, as in studies of development.

2. Define test parameters as follows:

>A pulse-only trial containing only a 40-msec burst of 120-dB noise;
>Variable ITIs to average 15 sec;
>Three different trial types containing prepulse stimuli (3, 6, or 12 dB above the 65 dB background noise with a duration of 20 msec) preceding the 120-dB pulse stimulus by 100 msec (onset to onset);
>Defined sequence of presentation of the four trial types;
>Recording of the startle response (i.e., the recording window) to begin at the onset of the pulse stimulus;
>Test session beginning with a 5-min acclimation period and then six presentations of the pulse-only trial (these trials will not be used in the assessment of PPI);
>Defined pseudorandom sequence in which the pulse-only trial is presented ten times and each of the three prepulse trial types is presented five times;
>Conclusion with an additional set of five pulse-only trials (for a total of 36 trials).

3. Clean and calibrate the equipment.

4. Assign each of the four matched groups of rats to receive either vehicle or one of the three doses of apomorphine. Bring rats from the vivarium to the laboratory 60 min prior to testing, shielding them from any sounds from startle testing of other animals.

5. Weigh each rat and prepare appropriate syringes for drug administration. Administer the vehicle and apomorphine injections subcutaneously 5 min prior to introducing the rats to the startle chambers.

6. Test the rats using the test session defined in step 2 above. For each rat, define the following descriptive statistics:

>Response amplitude on the first trial;
>Average response magnitude on pulse-only trials 2 to 6 and 32 to 36;
>Average response magnitude in each of the four trial types between trials 7 and 31 inclusively (i.e., ten pulse-only trials and five each of the three prepulse variations).

Analyze the first response and the first block of pulse-only trials as measures of startle reactivity. Analyze the first and last blocks of pulse-only trials together in a repeated-measures ANOVA to assess habituation of acoustic startle across the test session, or use the percent habituation score for this purpose.

7. Using the four values (3, 6, or 12 dB above background) derived from trials 7 to 31 to assess PPI, calculate for each rat:

Percentage score (most widely accepted):

$$PPI = 100 \times \{[\text{pulse-only units} - (\text{prepulse} + \text{pulse units})]/(\text{pulse-only units})\}$$

Difference score:

$$PPI = \text{pulse-only units} - (\text{prepulse} + \text{pulse units})$$

SUPPORT PROTOCOL

RAT HANDLING

Handling techniques vary greatly from laboratory to laboratory. This specialized procedure aids in reducing variation in behavioral results arising from stressful ambient conditions.

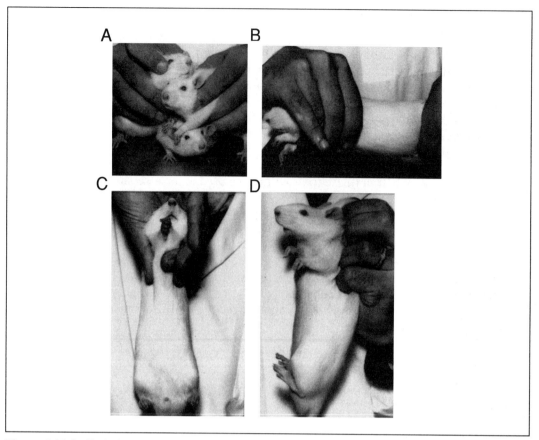

Figure 3.14.1 Technique to reduce acute stress in response to pretest handling of rats (see Support Protocol). (**A**) Hug three animals together, gently. (**B**) and (**C**) Handle an animal individually. (**D**) A fully relaxed rat reacts via a primitive relaxation reflex that leaves the body lax with all four paws extended but limp against the body (not fully achieved in this illustration).

1. Bring animals to the test laboratory, three per cage, within 24 hr of shipment arrival.

2. Place three animals at a time on a large, open table surface, and move them quickly and gently so that they experience a significant amount of contact with each other and with the experimenter's hand.

3. Hug the three animals together between the experimenter's hands gently but firmly (see Fig. 3.14.1A). After the animals become completely calm, usually within 3 min, return them to their cage.

    Kicking or vocalizations suggest the handling is not being performed correctly.

4. Remove each animal individually and place it on the table with one hand covering the head in a gentle fashion to shade any light, and the other hand on its hindquarters to prevent perambulation (see Fig. 3.14.1B).

5. Grip the animal gently between the hands, again firmly but not tightly, holding the rat's head so that the thumb and forefinger support the bottom of the neck and upper chest, and squeeze together the forepaws so that they cross over the underside of the neck and jaw region (to prevent the rat from accidentally biting the experimenter or scratching with the forepaws; see Fig. 3.14.1C). Lift the rat off the table with the hands moving in a massaging fashion.

6. Massage the animal for 2 min: gently twist the torso by rotating each hand in an opposite direction for a total displacement of ∼30°, and then rotate in the opposite direction.

7. Verify that the rat is fully relaxed by lifting it by the scruff of its neck, gripped between the thumb and several fingers and noting its limp body with all four paws extended but lying limp against the body (see Fig. 3.14.1D).

References: Ison and Hoffman, 1983; Swerdlow and Geyer, 1996

Contributors: Mark A. Geyer and Neal R. Swerdlow

UNIT 3.15

Latent Inhibition

Latent inhibition (LI) is considered to index the capacity to ignore insignificant stimuli, and as such has become of increased interest to neuroscientists studying the neural processes underlying stimulus selectivity and competition between conflicting associations, as well as modeling disorders in which such capacity is impaired, such as schizophrenia.

BASIC PROTOCOL 1

LATENT INHIBITION (LI) IN THE CONDITIONED EMOTIONAL RESPONSE (CER) PROCEDURE

Materials

    3- to 4-month-old male Wistar rats

    Operant chambers, e.g., Campden Instruments or any commercially available rodent operant chambers, or any Plexiglas or aluminum boxes constructed according to the following specifications (see Fig. 3.15.1):

        25 cm wide × 23 cm deep × 21 cm high (inner box measures); with a floor made of stainless steel rods (4 mm in diameter, 1 cm apart), 4.5 cm above a sawdust pan

        2.3-cm diameter hole in the middle of the left wall of the chamber, 3.8 cm above the floor (hole center); a square piece of metal, 4 mm behind the hole with a 1-cm diameter hole, concentric to the chamber hole

    Retractable 140-ml bottle fitted with stainless steel twin ball point sleeve (Classic Pet Products), positioned on the outer side of the chamber, 3 mm behind the piece of metal, at a 30° angle.

    Drinkometer (Campden Instruments) connected to the bottle sleeve and the square piece of metal, to record each lick the rat makes (chamber hole covered by a metal lid when bottle not present)

    Shock scrambler connected to the floor of the chamber

    Shock source/generator (connected to the scrambler) set at a 0.5-mA intensity and 1-sec duration

    Light bulbs on the left wall (two on the sides and one in the center) and one on the roof of the chamber

    Sonalert modules (Campden Instruments) located on the roof of the chamber, one for generating tones, and another for generating white noise; pre-exposed, to-be-conditioned stimulus is a 10-sec, 80-dB, 2.8-kHz tone

    Ventilated, sound-attenuating chamber to enclose each chamber.

    Experimental room (separate from the vivarium) for the chambers, dimly lit

An interface (fabricated in-house) connecting the chambers, Sonalert modules, shock sources, and drinkometers to a personal computer equipped with software programs to control stimuli administration and collect data

Software programs for baseline, pre-exposure and conditioning, and testing: written in-house (e.g., in Pascal) or commercially available packages, e.g., ABET (Animal Behavioural Environment Test system run on Windows and interfaced to the operant chambers using the ABET interface system and the ABET computer interface card; Campden Instruments and Lafayette Instruments); WMPC (MED-PC for Windows interfaced to the operant chambers using SmartCard; Med Associates); *or* WINLINC (interfaced to the operant chambers using LABLINC HABITEST LINC; Coulbourn Instruments)

Baseline program: controls house light, side lights, session length, and number of blocks for data recording; ASCII file output includes rat number, time to first lick, total number of licks, and number of licks per block

Pre-exposure and conditioning program: controls house light, side lights, type of intertrial interval (fixed or variable) and its average duration, presence of stimuli (on or off), type of stimuli delivered on each trial (e.g., tone in pre-exposure, tone and shock in conditioning), delay of onset of each stimulus from the beginning of the trial, and their duration and frequency (in the case of flashing lights or white noise)

Test program: controls type of stimulus, number of licks to stimulus onset, stimulus duration, and number of blocks for data recording; creates an ASCII file for each rat including rat number, time to first lick (pre-A period first lick), time to complete the first 50 licks (pre-A period, licks 1 to 50), time to complete 25 licks prior to tone onset (A period, licks 51 to 75), time from tone onset to the third lick (B period, third lick; the third, not the first, lick is measured since sometimes a drop of water between the metal square and the bottle sleeve may be wrongly counted as a lick), time to complete 25 licks after tone onset (B period, licks 76 to 100), total number of licks during the tone, and number of licks per block from tone onset

Standard statistical package (e.g., StatView, SuperANOVA, SPSS, or STATISTICA)

1. Upon the arrival of rats to the vivarium, house them four to a cage on a reversed light cycle (lights on: 7:00 p.m. to 7:00 a.m.). Acclimate rats to the vivarium and to the reversed light cycle for at least 10 days with freely available food and water.

2. Restrict water available to rats in the home cages to 1 hr/day (at least 1 hr after the end of the last behavioral session and at approximately the same time each day). Adjust times of running according to the times of water access (i.e., do not finish baseline for all of the rats before the first daily access to water). Determine the daily times of water access according to the longest day of the LI procedure (pre-exposure and/or conditioning; the exact length will depend on the total number of rats run), and give half of the rats access to water earlier in the day and the other half later in the day.

3. Weigh rats daily or every second day from the onset of water restriction to monitor health and to assure that body weight does not fall below 90% of the first day's weight. Randomly assign cages to the PE and NPE conditions. If using two levels of conditioning, randomly assign the PE and NPE cages to each of two levels of conditioning. Label each rat according to the operant box in which it will be run (e.g., mark the tail with 1 to 4 stripes according to box number; see step 6).

4. Start handling the rats on the first water restriction day and continue for 5 days for ∼2 to 3 min per rat daily at the same time each day in a separate, darkened room adjacent to the vivarium. See UNIT 3.14 for details about rat handling. After handling, replace the rat into its home cage.

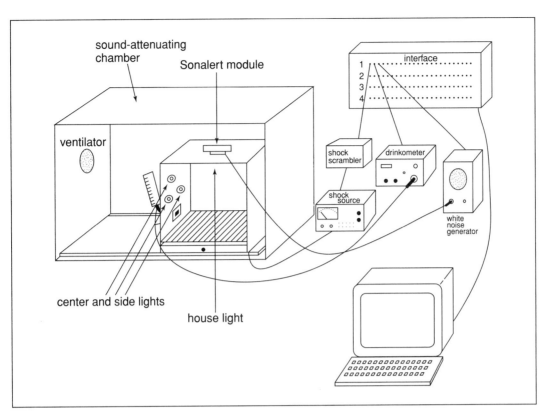

Figure 3.15.1 A schematic drawing of the operant chamber used for LI in the CER procedure.

5. Prepare behavioral scoring sheets for the baseline, rebaseline, and test stages, on which all the data displayed on the screen for each rat will be manually recorded to prevent data loss if the computer records are lost or damaged (see Fig. 3.15.2).

6. Run each cage separately, i.e., in squads of four. Make sure that each rat runs in the same experimental box and that each squad is run at approximately the same time of day throughout the experiment (see step 2). Run pre-exposed and nonpre-exposed squads in an alternating order, so that the running of the two conditions is spread equally throughout the day.

7. On baseline, pre-exposure, conditioning, rebaseline, and test days, transfer rats in squads to the testing facility 30 min prior to the beginning of the behavioral session, and keep them in a separate room (quiet and dimly lit) adjacent to the testing room. Ensure that they are not exposed to the tones while in this room. Alternatively, keep them in a more remote room. Stay close by the testing room throughout each testing day, and check often (without making noise when entering the testing room) that all equipment is functioning properly. At the end of the behavioral session, return the cage to the waiting room for an additional 15 min, and then to the vivarium. At the end of each day of sessions, clean and dry the chambers and test the program of the next day manually (the shock is mild enough to be tested with a finger on the grid).

8. On day 6, begin lick training (*baseline*) in the operant boxes and continue for 5 days. Select appropriate parameters:

    house light (on)
    side lights (off)
    session length (20 min first day, 15 min on subsequent days)
    number of blocks (10)

Investigator:						Date:	
Rat#	Box	Pre A		A Period	B Period		Total Licks
		1st Lick	Licks 1-50	Licks 51-75	3rd Lick	Licks 76-100	

Figure 3.15.2 A behavioral scoring sheet for manual data recording during test in the CER procedure.

Put rats gently in their respective cages. On the first day, place a few drops of water on the inside of the chamber, ~1 cm above the drinking hole, before putting the animal in (this should be done rapidly and efficiently). On the second day, encourage rats that failed to drink on the first day and during the first 7 min of the second day, by gently squeezing the bottle from the outside so that a few drops of water are released, and if this does not help, insert a plastic syringe filled with water in place of the bottle and when the rat approaches the syringe and licks, very slowly withdraw the syringe until it is outside of the chamber and replace the bottle. In the very rare case when a rat fails to drink by the third day, exclude it from the experiment.

Water in the operant chambers is given in addition to the daily ration of 1 hr given in the home cages.

9. On day 11, conduct the *pre-exposure* stage. Remove the bottle and seal the hole with a metal lid. Select proper parameters:

   house light (on)
   side lights (off)
   presence of stimulus ("on" for PE rats, "off" for the NPE rats)
   type of stimulus (tone)
   number of trials (10 or 40)
   intertrial interval (fixed, 50 sec)
   tone duration (10 sec)
   delay of onset from beginning of the trial (0).

   Gently place each rat in its operant chamber. After the session, return rats to the home cage.

10. On day 12, conduct the *conditioning* stage with the bottle removed as in the pre-exposure stage (step 9). Select proper parameters:

    house light (on)
    side lights (off)
    presence of stimulus (on)
    type of stimulus (tone, shock)
    number of trials (2 or 5)
    intertrial interval (fixed, 300 sec)
    tone duration (10 sec)
    delay of tone onset from beginning of the trial (0)
    shock duration (1 sec)
    delay of shock onset from beginning of the trial (10, i.e., shock immediately follows tone offset).

    Gently place each rat in its operant chamber. After the session, return rats to the home cage.

11. On day 13, conduct the *rebaseline*, namely, give rats a drinking session as in last day of baseline. Rats that fail to complete 600 licks can be excluded from the experiment.

12. On day 14, conduct the *test* session. Test rats individually. Select appropriate parameters:

    type of stimulus (tone)
    lick number for tone onset (75)
    duration of tone (300 sec).

    Place each rat in the box and run the test program. Remove the rat, but do not replace the rat in its home cage with other rats that are still to-be-tested (keep it in a separate cage).

13. Merge the individual data files from the rebaseline phase into a single file, and the individual data files from the test into another file for use by a standard statistical package.

14. Perform *t* tests for independent samples (if using one level of pre-exposure and one level of conditioning) or ANOVAs that include the independent variable of pre-exposure (NPE, PE) and number of pre-exposures (if using two levels of pre-exposure) or number of conditioning trials (if using two levels of conditioning). When using two levels of pre-exposure, first compare the two NPE controls, and if they do not differ significantly, merge their data into a single NPE control and analyze the data with one-way ANOVA.

15. Perform *t* tests or ANOVAs on pre-A lick 1, pre-A, and A periods (licks 50 to 75) times as the dependent variable.

16. Perform an additional *t* test or ANOVA using one of three dependent measures of the suppression of licking.

    a. Calculate the logarithmic transformation of times to complete licks 76 to 100 (log B period).

    The log transformation is appropriate because usually the data obtained have a skewed distribution because of the cutoff at 300 sec after tone onset. This measure can be used only if no significant differences were obtained in the analysis of A periods. Longer log B times indicate stronger suppression of drinking. LI is defined as significantly shorter log B times to complete licks 76 to 100 of the PE as compared to NPE rats. The logarithmic transformation of log B lick 1 period can also be used, but this is a less commonly used measure.

    b. Calculate the suppression ratio using A/(A + B).

    Note that this ratio is inversely related to the amount of lick suppression: a high suppression ratio indicates low suppression of drinking, whereas a low suppression ratio indexes high suppression of drinking. Pre-exposure effect is therefore manifested by a decrease in suppression of drinking and an increase in suppression ratio. LI is defined as a significantly higher suppression ratio of the PE as compared to NPE rats. This measure must be used if there are differences in the analysis of A periods.

 c. Determine the number of licks made during tone presentation in 30-sec blocks. In this case, perform an ANOVA with blocks as a repeated measurements factor and independent variable/s as in the ANOVAs/*t* test.

 LI is defined as a higher number of licks of the PE compared to the NPE rats.

17. When a significant main effect of pre-exposure in one-way ANOVA without repeated measurements, or significant interactions with pre-exposure in all the other analyses, are obtained, perform post-hoc comparisons comparing each of the PE groups to its NPE control.

18. As an alternative statistical approach (when LI is expected a priori in some of the conditions, but not in others), perform planned comparisons on any of the dependent measures of the suppression of licking comparing each PE group to its NPE control. If desired, perform planned comparisons for each of the independent variables in addition to that of pre-exposure.

ALTERNATE PROTOCOL 1

ADDING DRUGS, LESIONS, OR OTHER MANIPULATIONS: DISRUPTION AND POTENTIATION OF LI

In combination with environmental, pharmacological, and neurosurgical procedures, Basic Protocol 1 allows for a systematic investigation of the neurochemical and neuroanatomical systems that mediate the capacity of an animal to ignore irrelevant stimuli. A treatment of interest can potentially have one of three effects on LI: it can leave LI intact, disrupt LI, or potentiate LI. In order to disclose no effect or LI disruption, use pre-exposure and conditioning parameters that produce LI in controls (40 pre-exposures and 2 conditioning trials). In order to disclose LI potentiation, use pre-exposure and conditioning parameters that do not produce LI in controls (10 pre-exposures and 2 conditioning trials, or 40 pre-exposures and 5 conditioning trials). Bear in mind that LI-potentiating treatments spare LI under conditions that lead to LI in controls. Therefore, when LI remains intact, one cannot conclude that the treatment is ineffective without first excluding the possibility that it potentiates LI, by retesting it with LI-disrupting parameters.

Alternatively, use designs combining LI-yielding and LI-disrupting parameters (0, 10 and 40 pre-exposures with 2 conditioning trials, or 0 and 40 pre-exposures with 2 and 5 conditioning trials). These designs allow a clear conclusion that a given treatment does not affect LI (i.e., both control and treatment conditions show LI with 40 pre-exposures and 2 conditioning trials, and do not show LI with 10 pre-exposures and 2 conditioning trials or with 40 pre-exposures and 5 conditioning trials), or that it potentiates LI (i.e., both control and treatment conditions show LI with 40 pre-exposures and 2 conditioning trials, but only the treatment condition shows LI with 10 pre-exposures and 2 conditioning trials or with 40 pre-exposures and 5 conditioning trials).

For permanent surgical treatments allow ~2 weeks of recovery prior to the beginning of the behavioral experiment (see Basic Protocol 1, steps 2 to 12). In Basic Protocol 1, step 3, assign lesioned and sham-operated rats to groups in a factorial design consisting of pre-exposure (NPE, PE) and lesion (sham, lesion). Keep in each home cage 4 rats from the same pre-exposure condition (PE or NPE), 2 sham-operated and 2 lesioned. Counterbalance the test boxes so that rats from each condition are run in all boxes.

When the effects of drugs, lesions, or other manipulations of interest (e.g., effects of prenatal or postnatal manipulations, etc.) on LI are assessed, these are included as independent variables in addition to that of pre-exposure. When using two levels of pre-exposure or two levels of conditioning, include number of pre-exposures or number of conditioning trials as an independent variable (this can be applied to all designs presented below). Temporary treatments (systemic or intracerebral injections, reversible lesions, or stimulation) can be administered either in both the pre-exposure and conditioning stages, or confined to one of the stages, depending on the experimental question. Perform Basic Protocol 1 with the following modifications.

1a. *To test whether a drug is active in LI:* Administer it in both pre-exposure and conditioning stages. In Basic Protocol 1, step 3, assign rats to groups in a factorial design consisting of main factors of pre-exposure (NPE, PE) and treatment (vehicle, treatment). If using several drug doses, the different doses are levels of the drug factor; assign rats to groups in a factorial design consisting of pre-exposure × dose, e.g., 2 (NPE, PE) × 3 (3 doses including no-drug).

1b. *To determine the stage at which the drug acts:* Use a drug-no-drug design, i.e., administer the drug in either the pre-exposure stage, the conditioning stage or in both. In this case, the stage of drug administration is an additional independent variable; assign rats to groups in a 2 × 2 × 2 factorial design consisting of pre-exposure (NPE, PE), drug in pre-exposure (vehicle, drug), and drug in conditioning (vehicle, drug). Conduct a separate experiment for each drug dose (or each level of pre-exposure(conditioning).

2. Keep in each home cage rats from the same pre-exposure condition and assign rats within each cage to control and treatment conditions. Counterbalance test boxes so that rats from the same condition are run in different boxes. In Basic Protocol 1, step 4 (handling), accustom rats to the position at which they will be held when given the treatment (e.g., injected i.p., s.c., or intracerebrally).

3. Modify Basic Protocol 1, step 7, depending on the required injection-testing interval. If this is <30 min, bring the squad to the waiting room in the testing laboratory 30 min before the session as described above, and on pre-exposure and/or conditioning days, treat each rat with the appropriate vehicle or drug dose during the 30-min waiting period, at a suitable time before putting the rat into the experimental chamber. If the interval is >30 min (e.g., 60 min for haloperidol), bring the rats into the waiting room, inject them, and keep them in the room for 60 min before putting them into the experimental chambers. If the treatment is stressful (as may be with drugs of certain classes), clean and dry cages between squads; use an odorless detergent. Administer drugs (and other temporary manipulations) only in pre-exposure and/or conditioning stages; rebaseline and test are conducted without treatment.

4. Perform ANOVAs as described in the data analysis portion of Basic Protocol 1 which include the independent variable of pre-exposure, conditioning (if using two levels of conditioning), and depending on the design of the experiment, additional variables such as drug (vehicle versus one or more doses of a drug), lesion (sham versus operated), and others. When significant interactions between pre-exposure and any of the other variables are obtained, perform post-hoc comparisons comparing the PE to the NPE groups within each of the conditions constituting the interaction. When lower level interactions are contained in higher level interactions, perform post-hoc comparisons only for the latter. Alternatively, perform planned comparisons comparing the PE and the NPE groups within each condition.

BASIC PROTOCOL 2

LI IN A TWO-WAY ACTIVE AVOIDANCE PROCEDURE

The advantage of this procedure compared to CER is that it does not require water deprivation, and is shorter. The disadvantage is that manipulations that affect activity may confound the results.

Materials

> 3- to 4-month-old male Wistar rats, housed four to a cage under reversed cycle lighting (lights on: 7:00 p.m. – 7:00 a.m.).
>
> Four commercially available shuttle boxes (e.g., Campden Instruments or Coulbourn Instruments) containing house lights, side lights (on the side walls) and tone/white noise generators, with no barrier between the two compartments of the box, each set in a soundproof shelter equipped with fans. The pre-exposed to-be-conditioned stimulus is a 10 sec, 2.8 kHz, 80 dB tone produced by a Sonalert module located on the roof of the chamber. Shock is supplied to the grid floor by a scrambled shock generator set at 0.5 mA intensity. In-house interface connects the shuttle boxes, Sonalert modules/light sources, and shock sources to a PC computer that is equipped with a software program (programs written in-house, as well as commercially available software packages, e.g., from Campden Instruments, Med Associates, or Coulbourn Instruments, can be used).
>
> Software program: controls the house lights, the side lights, stage (pre-exposure, test), number of trials, the type of intertrial interval (fixed or variable) and its average duration, presence of stimuli (on or off), the type of stimuli delivered on each trial, delay of onset of each stimulus from the beginning of the trial, their duration and frequency (in the case of flashing lights or white noise), number of blocks (blocks of equal duration in pre-exposure and blocks of equal number of trials in test);creates two ASCII files, one for pre-exposure and one for test: the former includes rat number, total number of crossings, and number of crossings per block; the latter includes in addition, total number of escape responses (crossings during shock), total number of avoidance responses (crossings during the stimulus), and number of escape and avoidance responses per block. Escape and avoidance responses terminate the stimuli currently on (namely, escape response terminates the tone and the shock, and avoidance response terminates the tone) and initiate the intertrial interval.

1. Perform Basic Protocol 1, steps 1, 3, 4, 5, 6, and 7 (on pre-exposure and test days).

2. On day 6, conduct the *pre-exposure* stage. Choose parameters:

   > house lights (on)
   > side lights (off)
   > stage (pre-exposure)
   > number of trials (50 or 100)
   > intertrial interval (variable 50 sec, ranging from 1 to 100 sec)
   > stimulus ("on" for the PE rats, "off" for the NPE rats)
   > type of stimulus (tone)
   > delay of onset of stimulus from the beginning of the trial (0)
   > tone duration (10 sec)
   > number of blocks (10).

   Put each rat gently into the box, and return the rats to their home cages after the session.

3. On day 7, conduct the *test* stage. Choose parameters:

   > house lights (on)
   > side lights (off)

stage (conditioning)
number of trials (100)
intertrial interval (variable, 50 sec ranging from 1 to 100 sec)
stimuli (on)
type of stimuli delivered on each trial (tone, shock)
delay of onset of tone from the beginning of the trial (0)
tone duration (11 sec)
delay of onset of shock from the beginning of the trial (10 sec)
shock duration (1 sec; namely, the stimulus remains on with the shock)
number of blocks (10).

Each avoidance trial begins with a 10-sec tone followed by a 1-sec shock, the stimulus remaining on with the shock. If the rat crosses to the opposite compartment during the tone, the tone is terminated and no shock is delivered (avoidance response). A crossing response during shock terminates the tone and the shock (escape response). If the rat fails to cross during the entire stimulus-shock trial, the stimulus and the shock terminate together after 11 sec.

4. Put each rat gently into the box and return the rats to their home cages after the session. Make sure that each squad of rats is run at approximately the same time of day on the pre-exposure and test days.

5. Merge the individual data files from the pre-exposure into a single file, and the individual data files from the test into another single file for use by a standard statistical package such as StatView, SuperANOVA, SPSS, or STATISTICA.

6. Perform *t* test or ANOVA as detailed in step 14 of the data analysis portion of Basic Protocol 1 using the dependent measure of total number of avoidance responses or the number of trials to reach a criterion of four consecutive avoidance responses.

7. Alternatively, use the dependent variable of the number of avoidance responses in ten 10 trial blocks. In this case, perform an ANOVA with blocks as a repeated measurements factor, and independent variables as in ANOVAs/*t* test.

8. When a significant main effect of pre-exposure in one-way ANOVA without repeated measurements or significant interactions with pre-exposure in all the other analyses are obtained, perform post-hoc comparisons comparing each PE group to its NPE control.

9. As an alternative statistical approach, perform planned comparisons on any of the dependent measures comparing each PE group to its NPE control.

In all the analyses, total number of crossings during pre-exposure can be used as a covariate in order to remove the possible confound of activity on avoidance learning.

ALTERNATE PROTOCOL 2

LI IN TWO-WAY AVOIDANCE WITH CONTEXT SHIFT

LI is context-specific, namely, stimulus pre-exposure is effective in retarding its subsequent conditioning only if pre-exposure and conditioning take place in the same context (which in animal experiments typically refers to the apparatus in which the experiment is conducted), and a change in context from pre-exposure to conditioning disrupts LI (Lubow and Gewirtz, 1995).

The LI-with-context-shift procedure is a modification of the LI in two-way avoidance procedure (see Basic Protocol 2) which both produces and disrupts LI within one experimental design. The apparatus and the steps of this procedure are identical to Basic Protocol 2 with 100 pre-exposures, except that the pre-exposure stage is conducted for half of the rats in context A and for the other half in context B.

Additional materials (also see Basic Protocol 2)

Ylang-Ylang oil and cinnamon oil, or any two oils with distinct scents, e.g., rose oil or eucalyptus oil

1. Assign rats to groups in a 2 × 2 factorial design consisting of pre-exposure (0, 100) and context (same, different). Label each rat according to the operant box in which it will be run (e.g., mark the tail with 1 to 4 stripes according to box number). Keep in each home cage, 4 rats from the same pre-exposure and context condition. Counterbalance shuttle boxes within each condition so that rats from the same condition are run in different boxes.

2. Prepare two sets of shuttle boxes. Use two sets of four identical shuttle boxes, one set as context A and the second set as context B. In the latter boxes, cover the door with black and white checkered wallpaper and the grid floor with plyboard. Cover the trays below the grid floors with a thick absorbing paper. House each set of boxes in a different room in the laboratory.

3. Conduct the *pre-exposure* stage for half of the rats in context A and the other half in context B. Before the session, put one drop of Ylang-Ylang oil in the middle of the tray of the shuttle boxes which serve as context A, and a drop of cinnamon oil in the tray below the grid floor of the shuttle boxes which serve as context B.

4. Conduct the *test* for all rats in context A. Before the session, put a drop of Ylang-Ylang oil (without spraying) in the trays of the boxes. At the end of each session, throw away the paper covering the tray and replace with a new one.

   Remember, the same context condition will lead to LI, while a different context condition will lead to LI disruption.

5. Perform an ANOVA as detailed in the data analysis portion of Basic Protocol 2 in which the independent variables are pre-exposure and context (same, different). When significant interactions including pre-exposure and context are obtained, perform post-hoc comparisons comparing the PE to the NPE groups in each of the context conditions. Alternatively, perform planned comparisons comparing the PE and the NPE groups within same and different context conditions.

ALTERNATE PROTOCOL 3

ADDING TREATMENTS IN THE TWO-WAY AVOIDANCE PROCEDURE: DISRUPTION AND POTENTIATION OF LI

When the effects of treatments of interest on LI in two-way avoidance are assessed, include them as independent variables in addition to that of pre-exposure. Assign rats to groups and modify relevant steps as detailed in Alternate Protocol 1. For the context shift procedure, include context as an independent variable. Assign rats to groups in a 2 × 2 × 2 factorial design consisting of main factors of pre-exposure, treatment, and context (same, different). Keep in each home cage rats from the same pre-exposure and context condition, assign rats to treatment and control conditions, and counterbalance test boxes within each condition so that rats from the same condition are run in different boxes. Perform ANOVAs as described in Basic Protocol 2 or Alternate Protocol 2 including additional independent variables such as lesion, drug used in pre-exposure and/or test, and others, depending on the design of the experiment. Since the context shift procedure produces LI and LI disruption within the same experimental design, it can disclose spared or disrupted LI (in the same context condition) as well as persistent LI (in the different context condition). Performing a dose-response with such a design, however, is cumbersome; conduct a separate experiment for each drug dose. Reversible treatments can be confined to pre-exposure or conditioning, or given in both, as detailed in Alternate Protocol 1.

References: Lubow, 1989; Lubow and Gewirtz, 1995; Weiner, 1990, 2000; Weiner and Feldon, 1997

Contributor: Ina Weiner

CHAPTER 4
Preclinical Models of Neurologic and Psychiatric Disorders

Preclinical models of neurologic and psychiatric disorders provide a mechanism both for hypothesis testing at the organismic level and, in the ideal, also provide a mechanism for hypothesis generation that can be investigated at more elementary levels. New insights obtained at the molecular and cellular levels will result in a better understanding of disease processes, and ultimately improved prevention and treatment modalities.

The protocols in UNIT 4.1 describe the use of neurotoxins to produce lesions on dopamine-containing neurons in order to produce a preclinical model of Parkinson's disease. Cerebral ischemia remains a major cause of death and disability, and the protocols in companion UNITS 4.2 and 4.3 detail rodent models of global and focal ischemia, respectively. The techniques described are often used to model the pathophysiology of stroke, the most common among cerebrovascular insults. These techniques may also be useful in evaluating agents that either reduce ischemia-evoked damage or improve recovery following insult. In UNIT 4.13, methods to induce experimental autoimmune encephalomyelitis (EAE) in rodents are presented. While the etiology of multiple sclerosis (MS) remains unknown, EAE has been used to model this disease because of the presence of perivascular inflammatory cell infiltrates and demyelination common to both.

The most common methods of inducing focal ischemia in rodents involve mechanical occlusion of major cerebral arteries as described in UNITS 4.2 & 4.3. However, the majority of stroke cases are attributable to vascular thrombosis, and a model employing platelet aggregation as the primary cause of cerebral ischemia may offer greater fidelity to the human condition. UNIT 4.4 describes such a method for inducing photochemical lesions in the rat cortex through the use of a photoactive dye and laser. The cortical infarcts in this model are, in general, smaller and more reproducible than models involving mechanical occlusion of vessels. Further, this model does not require extensive surgery, and is less invasive than the procedures described in UNITS 4.2 & 4.3.

In UNIT 4.5, a rodent model of percussive closed head injury is described. This type of traumatic brain injury is a leading cause of death and disability, particularly among young adults. The ability to elicit a reproducible, controlled percussive impact on a circumscribed area of the brain is invaluable not only for examining potential treatment modalities, but also for studying the fundamental cellular processes that underlie neurological dysfunction. Peripheral nerve injury caused by trauma, disease and certain toxins can result in neuropathic pain syndromes. Neuropathic pain is generally chronic and refractory to standard analgesics. UNIT 4.6 describes three procedures that are widely used as models of neuropathic pain. This unit also details behavioral assays used to measure the resulting symptoms.

The origin of migraine pain has traditionally been associated with an inappropriate dilatation of the cerebral vasculature. However, more recent theories suggest migraine pain results from a complex series of events, with inflammation of the dural membranes a central component of this process. In UNIT 4.7, a procedure is described to produce dural inflammation in rodents.

The extent of dural inflammation is then quantified by measuring one of the hallmarks of this inflammatory process, plasma protein extravasation.

Neuropathy is one of the most common complications of diabetes. An estimated 50% of diabetics develop some form of peripheral neuropathy, with ~32% suffering from chronic, severe, and unremitting pain. Patients with diabetic neuropathy can exhibit a variety of aberrant sensations, including spontaneous pain (pain in the absence of an overt stimulus) and one or more kinds of stimulus-evoked pain, including allodynia (exaggerated pain sensations to normally nonpainful stimuli) and hyperalgesia (exaggerated pain sensations to normally painful stimuli). These symptoms are often concurrent with a paradoxical loss of stimulus-evoked sensation. UNIT 4.8 describes the streptozotocin (STZ)-induced diabetic rat, among the most commonly employed animal model used to study mechanisms of painful diabetic neuropathy and to evaluate potential therapies.

A basic understanding of the mechanisms involved in pain transmission is important in order to choose the model most appropriate for examining the analgesic properties of specific agents or classes of compounds. Methods for evaluating nociception, including hot-plate, tail-flick, and formalin tests, are presented in UNIT 4.9.

UNIT 4.10 details conditioned place preference often used to assess the positive reinforcing properties of test compounds. The procedures described for both rats and mice in this unit are capable of detecting substances abused by humans. Conditioned place preference procedures are generally less time consuming than other techniques (such as the intravenous self-administration procedure described in UNIT 4.11) used to examine the rewarding properties of compounds. Further, conditional place preference procedures do not require instrumental learning.

The protocols described in UNITS 4.11 & 4.12 are designed to study the mechanisms contributing to the rewarding properties of alcohol. UNIT 4.11 presents a model of intravenous self-administration in mice. The protocol described in this unit may be of particular value in studies aimed at exploring the neural substrates of alcoholism using genetically engineered animals. UNIT 4.12 provides a series of protocols to explore alcohol drinking behaviors in rats, with particular emphasis on the development of amethystic agents.

UNIT 4.14 details two models of Amyotrophic Lateral Sclerosis: an in vitro method using organotypic spinal cord cultures developed to study aspects of the excitotoxic hypothesis of this disease. The second model is a transgenic mouse expressing a mutated human *SOD1* gene. The latter model is especially valuable for screening putative therapeutic agents.

Anxiety disorders are the most prevalent of all psychiatric illnesses. Panic attacks, a hallmark of panic disorder (a subtype of anxiety disorder), can be induced in susceptible individuals by infusion of sodium lactate. UNIT 4.15 describes methods for inducing panic-like responses in rats following intravenous infusion of sodium lactate.

UNIT 4.16 presents an animal model of chronic spontaneous seizures induced by the chemoconvulsant, kainic acid. This treatment protocol reliably induces temporal lobe epilepsy in nearly all subjects with a relatively low mortality rate compared to other models. This modified chemoconvulsant treatment protocol using multiple, low doses of kainic acid is efficient and relatively simple to perform. The properties of the chronic epileptic state induced by this regimen of kainic acid appear similar to those of severe human temporal lobe epilepsy.

Contributor: Phil Skolnick

UNIT 4.1
Preclinical Models of Parkinson's Disease

STRATEGIC PLANNING

Model Selection: Species

Rodent models of Parkinson's disease (PD) are good anatomical and biochemical models of the disease. However, they are not good functional models because the behavioral syndrome only superficially resembles some aspects of idiopathic PD, and the progressive nature of the disease is not reproduced. The complex behavioral repertoire of monkeys provides greater scope for determining the extent of functional recovery.

In nonhuman primates, administration of MPTP can induce a stable parkinsonian syndrome that is remarkably similar to the idiopathic disease (Burns et al., 1983). All primates are susceptible to MPTP toxicity. However, marmosets show considerable motor recovery despite biochemical and histological evidence for extensive damage, so that in the chronic stage only mild akinesia and incoordination of movement remain. The choice of species for specific experiments depends on the cost, availability of the animals, the type of study proposed (some species, like squirrel monkeys, are difficult to train), and the size of the brain required (for certain neuroimaging studies, such as positron emission tomography or PET scanning, the large brain of baboons is preferable). All results and specifications concerning doses and behavior in this unit refer to macaques (Rhesus and Cynomolgus) in the MPTP models and marmosets for the 6-OHDA lesions.

As in primates, MPTP and 6-OHDA are the most common toxicants used to induce dopamine (DA) depletion in rodents. Rats are the most commonly used experimental rodents, but they are rather insensitive to MPTP, requiring high doses and, with most strains, intracerebral administration of 1-methyl-4-phenyl-pyridinium (MPP^+; the toxic metabolite of MPTP). So far, 6-OHDA lesioning continues to be the most popular parkinsonian model in rats.

Lesion

Different lesions yield appropriate animal models to evaluate different therapeutic strategies, such as DA replacement (cell-mediated or pharmacological), neuroprotection, and neuroregeneration, as well as physiopathological mechanisms of neurodegeneration. Complete DA lesions in experimental animals are considered, at least in a neuropathological sense, analogous to end-stage PD. Data obtained from studies using this type of model are useful for the evaluation of therapeutic strategies aimed at DA replacement, as the DA-mediated positive effects will not be complicated by the potential response of remaining host DA neurons. On the other hand, experimental PD models sparing some DA terminals are analogous to less severe stages of the disease and are appropriate for investigating the potential benefit of regenerative approaches (e.g., the use of trophic factors) that require the presence of some residual DA neurons. In the context of tissue transplantation, the relative contributions of DA replacement or graft-induced trophic responses of a particular procedure are probably best addressed when efficacy is compared in both types of model.

Toxin

MPTP

The capacity of MPTP to induce persistent parkinsonism in humans (Davis et al., 1979) and nonhuman primates (Burns et al., 1983) provides the opportunity to investigate both therapeutic approaches and possible pathogenic mechanisms of PD in primate models. Three models can be induced in primates by MPTP administration: (1) the unilateral model, induced by internal carotid artery (ICA) administration; (2) the bilateral model, induced by systemic (i.v., i.m., or s.c.) administration; and (3) the combined overlesioned or bilateral asymmetric model, induced by unilateral ICA plus i.v. administration. Basic Protocol 1 describes procedures for both ICA and systemic administration.

In the acute stage, administration of MPTP in primates produces mydriasis, piloerection, tachycardia, and tachypnea lasting 5 to 15 min. Repeated injections produce involuntary facial (perioral) movements, facial grimacing, retrocollis, and dystonic flexion of the extremities lasting 10 to 30 min. Intake of food and water should be carefully monitored in the acute phase; some animals might require tube feeding and L-DOPA support. In the first 48 hr following ICA administration of MPTP, the animal might present partial motor seizures with occasional secondary generalization; this should be adequately treated with diazepam at 0.5 mg/kg.

Mice exhibit characteristic neuropathological and biochemical signs of DA system damage following systemic administration of MPTP (Basic Protocol 4). Loss of tyrosine hydroxylase (TH)-positive neurons is seen in the substantia nigra pars compacta (SNc) and decreased levels of DA and its metabolites are observed in the striatum. These changes are accompanied by akinesia-depressed spontaneous motor activity (Heikkila et al., 1985; Hallman et al., 1985; Donnan et al., 1987; Sundstrom et al., 1990). Rats are very resistant to MPTP; to produce significant DA damage, the toxic metabolite MPP^+ has to be directly administered into the CNS.

6-OHDA

6-Hydroxydopamine has been used extensively to produce parkinsonism in rodents; however, few studies in primates have used this toxin (Schultz, 1982; Annett et al., 1992), most likely because it requires intracerebral administration.

CAUTION: When preparing, handling, and injecting concentrated MPTP solutions (e.g., those used for i.v. administration), always use a chemical hood, skin protection (gloves, Tyvex laboratory coat), and a mask. Any spilled MPTP should be degraded by spraying with 0.1 N HCl, while excess solution should be mixed 1:1 with 6 M H_2SO_4 and degraded by adding 4.7 mg potassium permanganate per 100 ml. In case of exposure, MAO-B inhibitors block the conversion to MPP^+, thus preventing the toxic effects.

BASIC PROTOCOL 1

COMBINED ICA AND INTRAVENOUS ADMINISTRATION OF MPTP: THE OVERLESIONED (BILATERAL ASYMMETRIC) PRIMATE MODEL

Materials

Sterile saline: 0.9% (w/v) NaCl
10-mg vial of 1-methyl-4-phenyl-1,2,3,6-tetrahydropyridine HCl (MPTP·HCl; Sigma)
Adult macaques (Rhesus and Cynomolgus; Sierra Biomedical or Charles River), capable of interacting well with investigators
Ketamine/xylazine
Isoflurane
Betadine
70% (v/v) ethanol
0.1 M HCl
Peroxide

10- and 60-ml sterile syringes
10- or 30-ml sterile vial
Animal balance (accurate to 0.1 g)
Intravenous (i.v.) extension set
Electric shaver
Alcohol pads
22-G i.v. catheter
Tracheal tube
Surgical table
Absorbent blue benchpads
Stretch gauze and cotton swabs
Sterile surgical tools: scalpel, large forceps, delicate curved forceps, scissors, needle holders, retractors, mosquito hemostats
27-G sterile needle
Infusion pump (fitted for a 60-ml syringe)
3-0 Vicryl with needle
Plastic, transparent millimeter-scale ruler
22-G angiocatheter
Heated water pad
Drapes
2/0 silk suture
Towel clamps

1. Fit a 10-ml sterile syringe with a 20- or 23-G needle and fill with sterile saline. Infuse 1 ml sterile saline into a sterile vial containing 10 mg MPTP·HCl. Aspirate the 1-ml solution back into the syringe to complete the 10-ml volume (final 1 mg/ml MPTP). Transfer the entire 10-ml MPTP solution into a sterile 10- or 30-ml vial.

2. Weigh the animal (within 0.1 g) and determine the total dose, adjusting for the HCl group using a 1.2× conversion factor.

   Particular caution should be exercised when calculating the dose for intracarotid administration. MPTP is toxic during the infusion (i.e., first pass effect) only because it undergoes peripheral conversion to MPP^+, which cannot cross the blood brain barrier. For this route of administration, MPTP should be calculated according to brain size, which tends to be constant over wide body weight ranges, instead of per body weight. The latter can give the false impression that older (heavier) animals are more susceptible to MPTP, when in fact they are receiving larger doses for similar brain sizes. As a general guideline, 2 to 2.5 mg can be used for small (3 to 5 kg) animals, 3 to 3.5 mg for a 6- to 10-kg monkey, and 4 mg for large (>10 kg) animals.

3. Fill a sterile 60-ml syringe with 2 to 4 ml MPTP solution (2 to 4 mg MPTP) to give 0.033 to 0.067 mg/ml MPTP final. Fill the rest of the syringe with sterile saline and attach an i.v. extension set.

4. Anesthetize a macaque monkey with ketamine (10 mg/kg) and xylazine (1 mg/kg) in the home cage. Transfer to the procedure room. Shave the calves and neck region of the animal using an electric shaver. Palpate the calf muscle and identify the saphenous vein. Clean the skin with an alcohol pad. Using a 22-G i.v. catheter, cannulate the vessel and flush with saline. Intubate the animal with a 3.5- to 5-mm (inside diameter) tracheal tube, according to the size of the animal.

5. Transfer the animal to a surgery room, place it on a surgical table with absorbent blue benchpads, and a heated water pad at 37°C. Maintain isoflurane anesthesia at 1.5% with an oxygen flow of ∼0.3 liters/min.

6. Hyperextend the head on the surgical table by placing stretch gauze through the canines and applying gentle retraction. Secure the head in position by tying the gauze to the surgery

table. Cover the animal completely with a series of sterile drapes. Cut out an opening in the drape for the neck incision and secure drapes with towel clamps. Scrub the surgery site with Betadine followed by 70% ethanol.

7. Make a midline incision through the skin of the neck with a sterile scalpel. Using blunt dissection technique, locate and open the carotid sheath, exposing the common carotid artery, internal jugular vein, and vagus nerve (Fig. 4.1.1). Isolate the common carotid artery below the carotid bifurcation.

8. Locate the superior thyroid artery and the external carotid artery and temporarily clamp the vessels on the branch medial to the carotid bifurcation using mosquito hemostats.

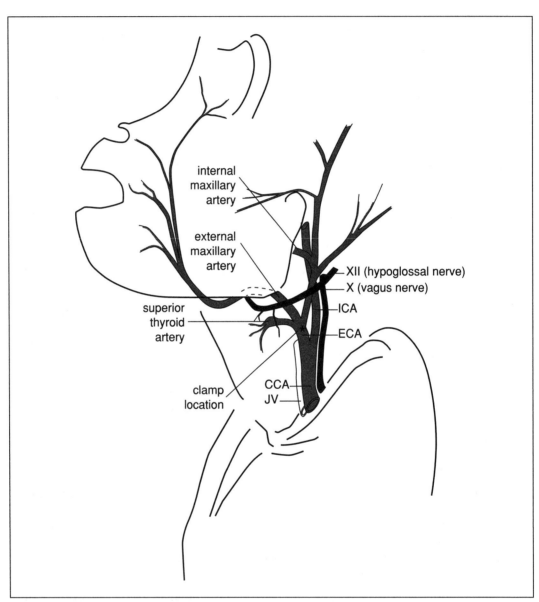

Figure 4.1.1 ICA dissection. The common carotid artery (CCA) at the level of the bifurcation showing the internal carotid artery (ICA), the X and XII cranial nerves, the external carotid artery (ECA), and its first branches: the superior thyroid artery and the external maxillary artery. An asterisk marks the recommended location of the vascular clamp, in the ECA proximal to the superior thyroid artery origin. JV, jugular vein.

Alternatively, clamp both the thyroid superior and external carotid (below the external maxillary branch) arteries using single hemostats distal to the carotid bifurcation. Use only mosquito hemostats with an adjusted clipping mechanism formed by bending handles inside.

9. Attach a 27-G needle to the i.v. extension tubing and drain all the air from the needle by activating an infusion pump for several seconds. Drain all excess MPTP onto a sponge soaked with 0.1 M HCl solution.

10. Insert the 27-G needle (with attached extension set and 60-ml MPTP syringe) into the internal carotid artery in a direction retrograde to the direction of blood flow. Program the infusion pump to deliver 4 ml/min and infuse the entire volume from the syringe (2 to 4 mg MPTP). Remove the vascular clamps (or mosquito hemostat) from the superior thyroid and external carotid arteries. Withdraw the needle from the common carotid artery and apply pressure for 5 min (until bleeding has stopped). Clean the wound with peroxide, close the incision site at the neck with 3-0 Vicryl, and clean the sutured area.

11. Monitor the recovery of the animal and return it to the animal room. Record ipsilateral pupil mydriasis (dilation) by measuring the diameter of both pupils with a plastic transparent millimeter-scale ruler (typically 3 to 5 mm following ICA infusion of MPTP solution, but may not occur even in the presence of a lesion). House animals for 2 weeks before i.v. administration.

    CAUTION: *Whenever possible, keep the animals in a quarantine room for 48 to 72 hr. Animal excreta may contain considerable amounts of unmetabolized MPTP; thus, precautions should be taken to minimize skin contact with animal fur and excreta and to avoid inhalation of bedding dust from MPTP-treated animals. As a general precaution, 0.1 N HCl solution should be used to spray the bedding in monkey cages during the first 48 hr after MPTP administration, and the cages should then be washed with an acidic solution. This ensures that any free MPTP is converted to its nontoxic form (for safety concerns see Yang et al., 1988).*

12. Weigh the animal (within 0.1 g) and determine the total dose of MPTP required (0.3 mg/kg body weight for i.v. administration, recommended), adjusting for the HCl group using a 1.2× conversion factor.

13. Anesthetize the animal with ketamine (10 mg/kg) and xylazine (1 mg/kg) in the home cage. Transfer the monkey to the procedure room. Shave the calves using an electric shaver. Palpate the calf muscle and identify the saphenous vein. Clean the vein with an alcohol pad. Using a 22-G angiocatheter, cannulate the vessel and flush with sterile saline.

14. Fill a sterile 3- to 6-ml syringe with the appropriate volume of 1 mg/ml MPTP solution (0.3 ml/kg or 0.3 mg/kg) and infuse the MPTP into the vein over 3 min. Flush the line with 3 ml sterile saline. Remove the needle and apply pressure to the saphenous vein until bleeding stops.

    See step 11 annotation for safety guidelines.

15. One week after i.v. administration of MPTP, assess the syndrome using a motor rating scale for primates (Table 4.1.1) based on the unified Parkinson's disease rating scale (UPDRS). Monitor response to L-DOPA (see Support Protocol 1) and activity (see Support Protocol 2).

ALTERNATE PROTOCOL 1

SYSTEMIC MPTP LESION IN PRIMATES

The principal advantage of this approach is that the behavioral syndrome closely resembles that of Parkinson's disease in humans. The main drawbacks are that it takes a long time to

Table 4.1.1 Rating Scale for Parkinsonian Primates[a]

Parameter	Score	Degree
Tremor (right arm/left arm)[b]	0	Absent
	1	Occasional or barely detectable (normal for aged), occurring while active
	2	Frequent or easily detectable, occurring while active or at rest
	3	Continuous or intense, occurring while active and at rest
Freezing	0	Absent
	1	Occasional episodes of short duration (<5 sec)
	2	Occasional episodes of moderate duration (6 to 10 sec)
	3	Frequent episodes or episodes of long duration (>10 sec)
Locomotion	0	Uses all four limbs smoothly and symmetrically
	1	Walks slowly (normal for aged), noticeable limp
	2	Walks very slowly and with effort, may drag limb or refuse to bear weight
	3	Unable to ambulate
Fine motor skills (right arm/left arm)[b]	0	Normal function: able to grasp/retrieve small objects, aims accurately, independent use
	1	Reduced ability in grasping/retrieving small objects, independent use, may have reduced aim
	2	Able to grasp small objects rarely, only with assistance, or with great difficulty
	3	Unable to grasp/retrieve small objects
Bradykinesia (right arm/left arm)[b]	0	Quick, precise movements
	1	Mild slowing of movements (normal for aged)
	2	Slow deliberate movements with marked impairment initiating movements
	3	No movements
Hypokinesia	0	Moves freely, alert, responsive
	1	Reduced activity (normal for aged), moves less frequently (without provocation)
	2	Minimal activity, moves with provocation, may have reduced facial expression
	3	Akinetic (essentially no movements)
Balance	0	Requires no assistance for maintaining posture
	1	Requires assistance for standing
	2	Requires assistance for walking, or falls
	3	Face down, or unable to maintain posture
Posture	0	Normal posture, stands erect
	1	Reduced posture (normal for aged), stands with feet apart, knees flexed
	2	Stooped posture, hunched, legs bent
	3	Unable to maintain posture, recumbent
Startle response	0	Immediate, robust response to provocation
	1	Slightly diminished or delayed response, open mouth threat
	2	Minimal response or longer delay, without open mouth threat
	3	No response to provocation
Gross motor skills (right arm/left arm)[b]	0	Normal function, able to grasp/retrieve large objects accurately
	1	Reduced ability/frequency of grasping/retrieving large objects
	2	Great difficulty in grasping/retrieving large objects, rarely used
	3	Unable to grasp/retrieve large objects

[a] According to mean scores in the scale, animals are classified into five stages. Stage 0: A maximum of 5 points is considered normal according to the authors' results in healthy animals. Stage 1: Total of 5–12 points; hemiparkinsonian monkeys do not show axial impairment. Stage 2: Total 12–20 points; mild to moderate bilateral symptoms. Stage 3: Total 21–30 points; moderate to severe bilateral symptoms, but without major systemic consequence. Stage 4: Total >30 points; severely damaged and can be difficult to manage due to feeding difficulties and complications of akinesia; may require DA replacement.

[b] Score each arm separately and add the total.

produce animals with stable and uniform lesions and, if the lesion is too extensive, it produces a considerable mortality. Total dose is not predetermined, but will depend on the highly variable response of each animal to the toxin. Therefore, careful clinical evaluation is mandatory throughout the lesion progression. The toxic effect is cumulative.

Additional Materials *(also see Basic Protocol 1)*

    3-ml syringes equipped with 26-G needles
    Cage with back-squeezing mechanism

1. Prepare MPTP as described (see Basic Protocol 1, step 1). Weigh the animal and determine the appropriate dose (0.33 to 1 mg/kg), adjusting for the HCl group with a 1.2× conversion factor. Fill a sterile syringe with the appropriate volume of MPTP solution.

2. With the monkey in a cage with a back-squeezing mechanism, squeeze the back of the cage to hold the animal and inject the MPTP intramuscularly or subcutaneously into the thigh or arm. For better results, inject twice a week.

    See Basic Protocol 1, step 11 annotation for safety guidelines.

    If the animal shows general symptoms such as feeding difficulty or profound akinesia, stop the injections and reevaluate the motor syndrome (see Table 4.1.1) 2 to 4 weeks later.

3. Assess the syndrome as described (see Basic Protocol 1, step 15).

BASIC PROTOCOL 2

UNILATERAL 6-OHDA LESION IN PRIMATES

6-OHDA uses the catecholamine uptake system to enter catecholamine neurons and kill the cells through oxidative mechanisms. Unilateral injection of 6-OHDA into the medial forebrain bundle of marmosets produces a severe loss of tyrosine hydroxylase-immunoreactive neurons in the ipsilateral SNc and DA depletion of >90% in the dorsal striatum, accumbens, and frontal cortex ipsilateral to the lesion (Annett et al., 1992). Levels of 5-hydroxytryptamine and noradrenaline (NA) are also decreased, but previous administration of NA uptake blockers (e.g., desipramine) limits the non-DA damage. The animals show ipsilateral spontaneous rotation, contralateral apomorphine-induced rotation, reduced spontaneous activity, contralateral sensorimotor neglect, and ipsilateral hand preference with variable impairment of hand skill (Annett et al., 1992). Persistent deficits are observed in animals with >95% DA depletion.

Materials

    Sterile saline: 0.9% (w/v) NaCl
    Adult marmosets (Sierra Biomedical or Charles River), capable of interacting well with investigators
    Ketamine/xylazine
    Isoflurane
    Betadine
    70% (v/v) ethanol
    4 mg/ml 6-hydroxydopamine (6-OHDA), HBr (Sigma) in 0.01% (w/v) ascorbate/0.9% (w/v) NaCl (protect from light)

    Stereotactic frame/tower (David Kopf Instruments)
    Manipulator arm
    Spinal needle
    5-ml syringes
    Intravenous (i.v.) line with 3-way stopcock
    22-G angiocatheter

Animal balance
Electric razor
Alcohol pads
Surgical tape
Tracheal tube
Isoflurane inhalation chamber
Sterile drapes, gauze, and rubber bands
Towel clamps
Tissue forceps
Scissors
Electric cauterizer
Water heating pad
Dremel drill with carbide bur excavating tip
10-μl Hamilton syringes and needle
3-0 Vicryl
2-0 silk sutures

1. Calibrate stereotactic frame by fixing the settings on the ear bars of a stereotactic frame so the bars are ~1 mm apart. Fit the manipulator arm with a 20-G spinal needle and position it on the stereotactic frame. Ensure that the mark readings on both ear bars are identical. Advance the spinal needle into the space between the ear bars. Adjust the position of the needle along the anterior/posterior (A/P) scale to correlate with the plane of the ear bars.

2. Flush an i.v. line with sterile saline using a 5-ml syringe, then prime a 22-G angiocatheter and 3-way stopcock with saline. In the animal room, weigh a marmoset and anesthetize with 10 mg/kg ketamine/1 mg/kg xylazine. Transfer animal to the surgery room.

3. Shave the calves, arms, and head for surgical and intravenous access using an electric razor. Palpate the calf muscle or flex and extend the ankle to visualize the saphenous vein. Clean the shaved area with an alcohol pad. Insert the 22-G angiocatheter into the saphenous vein and flush with saline to ensure patency.

4. Attach the 3-way stopcock with i.v. line to the catheter and start a slow saline drip (1 ml/min) to maintain patency of the vessel. Secure the i.v. line onto the calf muscle using surgical tape. Intubate the animal with an appropriate size tracheal tube and maintain on isoflurane anesthesia by inhalation.

5. Position animal in stereotactic frame
   a. Place the animal in the stereotactic frame and tighten the right ear bar to the frame. Elevate the animal's head and insert the right ear bar into the ear canal.
   b. Position the left ear bar in the animal's left ear. Ensure that the animal's eyes are parallel to the front plane of the frame.
   c. Center the animal's head such that the same setting for both ear bars is achieved. Position the eye bars in the inferior border of the orbits. Set the incisor bar and readjust the eyebars to the inferior orbital rim. Fix the incisor bar at an adequate height.

6. Clean the entire surface of the cranium by scrubbing in an outward circular motion with Betadine scrub followed by 70% ethanol. Spray the stereotactic frame with 70% ethanol. Use sterile gloves for cranium preparation. Cover the animal completely with a heating pad and then cover it with a series of sterile drapes. Cut out an opening in the drape for the craniotomy and secure drapes with towel clamps.

7. Determine the target skin incision site using the baseline A/P coordinates as a guideline. Make a small sagittal incision through the skin and fascia using an electric cauterizer.

Using sterile gauze, retract the skin and fascia to expose the cranial surface. Secure the skin in place with towel clamps. Cover a Dremel drill with sterile drapes and fix in place with sterile rubber bands. Use a carbide burr excavating tip and make a burr hole that exposes the dura.

8. Position the spinal needle over the first target site and manually advance the needle to the surface of the cranium. For marmosets, use the following coordinates with reference to stereotactic zero (Annett et al., 1992), with all five injection sites at A/P +6.5:

    one lateral (3 µl): lateral (L) ±3.2, ventral (V) 7.5
    two central (2 µl each): L ±2.2, V 6.5 and V 7
    two medial (2 µl each): L ±1.2, V 6 and V 7.

9. Insert the needle at the first target site, then touch the electric cauterizer to the needle and withdraw manually. Place a 10-µl Hamilton syringe on the arm holder and infuse the selected volume of 4 mg/ml 6-OHDA at a rate of 0.5 µl/min. Leave the syringe in place for 5 min to avoid overflow. Repeat steps 8 and 9 at the remaining four injection sites.

10. Suture the fascia with sterile 3-0 Vicryl. Close the incision site by subcutaneous suturing of the skin with 2-0 silk sutures. Clean and dry the incision site with sterile gauze and spray with Betadine.

11. Evaluate nigral damage in vivo and post mortem using the methods described for the MPTP model (see Support Protocol 1 and Support Protocol 2). Also evaluate rotational behavior (see Support Protocol 3).

SUPPORT PROTOCOL 1

EVALUATION OF CHANGES IN MOTOR BEHAVIOR IN RESPONSE TO L-DOPA

The effect of L-DOPA can be evaluated using the activity or motor tasks shown in Table 4.1.1. For each animal, the response to saline injection should be used as control.

Materials

Lesioned animal (see Basic Protocol 1, Alternate Protocol 1, and Basic Protocol 2) with stable deficit (~6 weeks)
L-3,4-Dihydroxyphenylalanine methyl ester (M-L-DOPA; Sigma)
Benserazide (Sigma)
Sterile saline: 0.9% (w/v) NaCl

Cage with back-squeezing mechanism
3-ml syringes
26- to 30-G needles

1. Prepare the total dose of M-L-DOPA (typically 5 to 50 mg/kg) and add benserazide at a 1:10 benserazide/M-L-DOPA ratio. Mix and dilute in 1 to 2 ml sterile saline.

2. With a lesioned monkey in a cage with a back-squeezing mechanism, squeeze the cage and inject i.m. into the thigh of the animal using a 3-ml syringe with a 26- to 30-G needle.

3. Rate the response according to the parameters described in Table 4.1.1. Evaluate the response 45 to 60 min after injection and at later time points when duration of the response is relevant. If desired, videotape the animals, particularly when they display abnormal movements that might be difficult to categorize.

SUPPORT PROTOCOL 2

MONITORING ACTIVITY TO ASSESS MPTP-TREATED MONKEYS

Whole-body activity measurement is a useful indicator of DA lesion in MPTP-treated monkeys and correlates with the global motor score. This objective measure approximates akinesia, which is most pronounced in animals with moderate to severe lesions. The baseline motor activity is reduced and the normal pattern disappears after the MPTP lesion. The personal activity monitors (PAM) contain a biaxial piezoelectric sensor that is calibrated to detect threshold activities $>0.024 \times g$ acceleration. Sensitivity and epoch length is programmable. The acceleration signal is sampled and digitally integrated to quantify all activity under the signal curve. The information is converted to a reference scale of data counts or acceleration units (G). The PAM monitor can be inserted into a vest that the animal wears during testing or into a collar around the neck. Activity data are normally acquired over four to five days that should include a weekend.

Materials

Lesioned animal (see Basic Protocol 1, Alternate Protocol 1, and Basic Protocol 2)
Ketamine/xylazine

Personal activity monitors (PAM; ActiTrac, Individual Monitoring Systems)
PAM connector cable
Computer
Nylon cable ties
Vetwrap
Nylon collars/vest

1. Connect a PAM to a computer. Check that the battery level for the PAM is >5.25 volts, and program the sensitivity, epoch length (1 min is adequate for time periods >60 min; to detect short-lived changes, use 30 sec), and start time and date. Enter the animal and test data. Place the PAM in a horizontal position such that the PAM label faces outward. Secure the PAM to the collar using nylon cable ties. Wrap the monitor and cable ties with Vetwrap.

2. Anesthetize a lesioned animal in its home cage with 7 mg/kg ketamine and 1 mg/kg xylazine. Place a nylon collar around the animal's neck and then wrap the buckle with Vetwrap. Record the animal's identification number, the date and time, and the PAM identification number.

3. Begin activity data acquisition 60 min after administration of anesthesia (exact time delay not critical for consistent time periods).

4. Anesthetize the animal as in step 2 and remove the PAM.

5. Connect the PAM to the interface and download information to a data file. Extract 12-hr segments (e.g., 6:00 am to 6:00 pm) and summarize them in a table, normally at the following time points: prelesion baseline, postlesion baseline, and postexperimental treatment. Compare relative values (e.g., percentage with respect to baseline) rather than mean and total counts (because of considerable variability between subjects).

SUPPORT PROTOCOL 3

ROTATIONAL BEHAVIOR AS A MEASURE OF UNILATERAL NIGROSTRIATAL LESIONS

In animals with a unilateral DA lesion (see Basic Protocol 2), there is an imbalance of motor activity such that they usually display spontaneous turning toward the side of the lesion. Administration of indirect DA agonists (e.g., DA-releasing drugs such as D-amphetamine) increases the imbalance and the ipsilateral rotation. Administration of direct DA agonists (e.g., apomorphine) evokes contralateral turning, which is considered to be the result of denervation hypersensitivity of DA receptors in the lesioned side. Rotational behavior in response to DA agonists grossly correlates with the severity of the lesion (it is, in fact, better correlated with asymmetry indices). Quantification of the rotational response can be accomplished in monkeys by videotaping the animals and counting the turns, and in rats by using specific devices called rotometers (Ungerstedt, 1971 a,b).

Suggested compounds and doses include 0.5 to 5 mg/kg D-amphetamine sulfate (Sigma) in saline and 0.025 to 0.25 mg/kg apomorphine·HCl (Sigma) in saline. Avoid amphetamine whenever possible, particularly if tests have to be repeated, as it might increase the mortality rate. For either compound, use the lower end of the range for primates (primates are susceptible to compound effects), the higher end for rats.

BASIC PROTOCOL 3

6-OHDA LESIONS IN RATS

In rats, several models can be induced by unilateral intracerebral stereotactic injection of 6-OHDA into different brain structures. Bilateral application of 6-OHDA is rarely used because of high mortality due to diencephalic damage (adipsia and aphagia); hence, only unilateral models are discussed here.

Complete DA lesion can be induced by unilateral injection of 6-OHDA in the medial forebrain bundle or in the SNc (Ungerstedt and Arbuthnott, 1970; Ungerstedt, 1971 a). These animals demonstrate a characteristic asymmetric motor behavior in response to antiparkinsonian drugs (Ungerstedt, 1971b) that enables distinction between drugs with predominantly DA receptor agonist activity from those with predominantly DA-releasing activity. Systemic administration of L-DOPA or direct DA receptor agonists leads to contralateral rotation (towards the undamaged side), while administration of DA-releasing substances (amphetamine, amantadine) leads to ipsilateral rotation (in the direction of the damaged side). This model is also useful in studies of DA replacement therapy and neuroprotection factors.

Partial lesion models are induced by injection of 6-OHDA in the medial forebrain bundle (MFB) or SNc in smaller doses that leave a number of DA neurons intact (Zigmond and Stricker, 1989). These are useful models for the study of pathophysiological mechanisms and neuroregeneration.

The *selective A-9 lesion model* can be induced by injection of 6-OHDA in the SNc, leaving the ventral tegmental area (VTA or A-10 region) neurons intact (Perese et al., 1989; Thomas et al., 1994), thus reproducing the selective vulnerability of DA neurons in the SNc observed in idiopathic PD. This model is useful to study the contribution by other DA areas to the restoration of the nigrostriatal pathway and the differential effects of protective and repair mechanisms on different DA nuclei.

The *striatal lesion model* is induced by injections of 6-OHDA into the striatum, causing progressive retrograde degenerative changes of the corresponding DA neurons in the SNc

(Sauer and Oertel, 1994; Lee et al., 1996). This is a useful model for pathophysiological, neuroregenerative, and neuroprotective studies.

Materials

    Sprague-Dawley rats, 200 to 250 g
    Isoflurane
    Betadine
    70% (w/v) ethanol
    4 mg/ml 6-hydroxydopamine (6-OHDA, HBr; Sigma) in 0.01% (w/v) ascorbate/saline (protect from light)
    Sterile saline: 0.9% (w/v) NaCl

    Animal balance (accurate to 0.1 g)
    Isoflurane inhalation chamber
    Electric razor
    Stereotactic frame
    Scalpel
    Tissue forceps
    Scissors
    10-μl Hamilton syringes and needles
    Dental drill
    Sutures or staples

1. Weigh a Sprague-Dawley rat (within 0.1 g) and place in an isoflurane chamber until deeply anesthetized (alternative anesthetics: ketamine/xylazine or chloral hydrate). Position the animal in a stereotactic frame and fix the plastic tube connected to the anesthesia machine to the nose of the animal using surgical tape (make sure that the snout bar does not collapse the tube). Maintain isoflurane at ~1.5% with an oxygen flow of 2 to 3 liters/min.

2. Shave the head with an electric razor. Clean the skin with Betadine and 70% ethanol. Perform a midline incision with a scalpel and identify the bregma at the intersection of the coronal and the sagittal sutures.

3. Fill a 10-μl Hamilton syringe with the appropriate 6-OHDA solution. Attach syringe to the holder on the stereotactic frame. Adjust 6-OHDA solutions according to the region of injection:

    For medial forebrain bundle lesion: 2 μg/μl 6-OHDA in saline containing 0.1% (w/v) ascorbic acid
    For SNc lesion: 4 μg/μl 6-OHDA in saline containing 0.02% (w/v) ascorbic acid
    For striatal lesion: 0.4 μg/μl 6-OHDA in saline containing 0.1% (w/v) ascorbic acid
    For A-9 selective lesion: 2 μg/μl 6-OHDA in saline containing 0.02% (w/v) ascorbic acid.

4. Calculate the stereotactic coordinates for injection (Paxinos and Watson, 1986). Some examples of possible injection sites include:

    For medial forebrain bundle lesion: Anteroposterior (A/P) -2.2 mm, mediolateral (M/L) 1.5 mm with reference to bregma; ventrodorsal (V/D) -8.0 mm with reference to dura
    For SNc: A/P -5.4 mm, M/L 2.2 mm with reference to bregma, V/D -7.5 mm with reference to dura
    For striatal lesion: A/P $+0.5$ mm, M/L 2.8 mm with reference to bregma, V/D -4.5 mm with reference to dura

For A-9 selective lesion:
1st lesion: A/P +3.5 mm, M/L 1.9 mm, V/D −7.1 mm with reference to lambda and dura with needle bevel directed rostrally
2nd lesion: A/P +3.5 mm, M/L 2.3 mm, V/D −6.8 mm with reference to lambda and dura.

5. Adjust the incisor bar in the animal until the heights of lambda and bregma skull points are equal. Drill a burr hole at the target site using a dental drill.

6. Lower the needle of the Hamilton syringe through the dura down to the selected depth and start injection at a rate of 0.5 to 1 µl/min. Adjust volumes (and doses) of 6-OHDA according to the region of injection:

   For medial forebrain bundle lesion: 4 µl solution (8 µg 6-OHDA)
   For SNc lesion: 2 µl solution (8 µg 6-OHDA)
   For striatal lesion: 20 µl solution (8 µg 6-OHDA)
   For A-9 selective lesion: 2 µl solution (4 µg 6-OHDA).

   Leave the needle in place for 5 min and withdraw slowly (1 mm/min).

7. Close scalp margins with sutures or staples. Remove the animal from the stereotactic frame and place it in its home cage. Put food on the floor of the cage and monitor the animal's weight for 3 days after surgery. Supplement the diet (e.g., with fruit) if there is >10% weight loss.

8. Evaluate rotational behavior (see Support Protocol 3).

BASIC PROTOCOL 4

MPTP LESION IN MICE

Many different protocols have been used with various doses, routes of administration, and species. An example of the model induced by intraperitoneal administration of MPTP to C57Black mice, using four injections of 10 mg/kg (total 40 mg/kg), is described (Ricaurte et al., 1987).

Materials

C57Black mouse, age 8 to 12 months
1-Methyl-4-phenyl-1,2,3,6-tetrahydropyridine HCl (MPTP·HCl; Sigma)
Isotonic saline

Animal balance (accurate to 0.1 g)
1-ml syringe equipped with 26-G needle

1. Weigh a C57Black mouse (within 0.1 g). Prepare a 1 mg/ml MPTP solution in isotonic saline. Fill a 1-ml syringe, equipped with a 26-G needle, with MPTP solution to provide a dose of 10 mg/kg (e.g., 0.4 ml for 40-g mouse).

2. Hold mouse in dorsal recumbency with left leg immobilized. Insert the needle in the lateral aspect of the lower left abdominal quadrant through the skin and musculature, and immediately lift the needle against the abdominal wall and inject the solution. Repeat the injection three times at 1-hr intervals.

3. Observe animals for any changes in general locomotion, stooped posture and piloerection (see Support Protocol 4).

SUPPORT PROTOCOL 4

MONITORING ACTIVITY IN MPTP-TREATED MICE

Open-field locomotor activity can be measured to evaluate the toxic effect of MPTP in mice and correlates with the content of dopamine in the striatum and nucleus accumbens (Leroux-Nicollet and Costentin, 1986). However, MPTP induces an acute increase in activity in some mouse strains, as the predominant effect in the early stage is inhibition of DA reuptake by MPTP. This effect can be blocked with neuroleptics. In C57 Bl(6 mice, MPTP produces a decrease in baseline activity (~60% in albino and 40% in black mice) that is prevented by administration of monoamine oxidase inhibitors. Amphetamine induces an increase in locomotor activity in normal animals that is absent in MPTP-lesioned animals. Apomorphine induces a decrease in locomotor activity that is not significantly different between normal and MPTP-injected animals.

Measurement of locomotor activity can range from simple observation to sophisticated automated procedures. In general, these instruments use an array of infrared photobeams and reveal the activity of the animal by the number and pattern of beam interruptions. *UNIT 3.2* describes these procedures in greater detail.

Materials

MPTP-lesioned mouse (see Basic Protocol 4)
Dexamphetamine sulfate
Saline: 0.9% (w/v) NaCl
Automated open-field instrument (e.g., Digiscan Animal Activity Monitor; Omnitech Electronics) consists basically of a Plexiglas cage with infrared monitoring sensors at determined distances
1-ml syringe with 26- to 30-G needle

1. Place an MPTP-lesioned mouse in a Digiscan cage and record activity for the selected period of time.

    It is mandatory to perform pre-exposure habituation to get a reliable baseline.

2. Prepare dexamphetamine sulfate at 0.5 mg/ml in saline and place in a 1-ml syringe with a 26- to 30-G needle. Weigh mouse and inject intraperitoneally at 1.5 mg/kg.

3. Place the animal in the Digiscan cage and record data (e.g., horizontal activity, movement time, total distance, average speed, average distance per horizontal movement, and number of movements) for ≥30 min. Perform all experiments at the same time of day as activity varies with circadian cycles.

Reference: Annett et al., 1992

Contributors: Krys S. Bankiewicz, Rosario Sanchez-Pernaute, Yoshitsugu Oiwa, Malgorzata Kohutnicka, Alex Cummins, and Jamie Eberling

UNIT 4.2

Rodent Models of Global Cerebral Ischemia

Several rodent models have been developed (see Table 4.2.1) that mimic the effects of cerebral ischemia on either the whole brain (global ischemia) or in one region (focal ischemia). Two models of global cerebral ischemia are widely used and are discussed in detail in this unit: the

Table 4.2.1 Animal Models of Global and Focal Cerebral Ischemia

Global ischemia	Focal ischemia
Gerbil bilateral carotid artery occlusion (BCAO)[a]	Intraluminal monofilament middle cerebral artery occlusion (MCAO)
Rat 4-vessel occlusion (4-VO)[b]	Endothelin-1 (Et-1) MCAO
Rat 2-vessel occlusion	Permanent MCAO (Tamura)
Neck tourniquet	MCA in combination with common carotid artery (CCA) occlusion in spontaneously hypertensive rats (SHR) rats
	Photochemical MCAO
	Microsphere embolization
	Blood clot embolization
	Arachidonic acid-induced thrombosis

[a]See Basic Protocol 1.
[b]See Basic Protocol 2.

gerbil bilateral carotid artery occlusion model and the rat 4-vessel occlusion (4-VO) model. A troubleshooting guide for these methods (Table 4.2.2) is at the end of the unit.

CAUTION: Formaldehyde and other fixatives are harmful and care should be taken to carry out the fixation procedures over a sink and in a fume hood. Wear safety glasses and gloves at all times.

BASIC PROTOCOL 1

GERBIL BILATERAL CAROTID ARTERY OCCLUSION (BCAO) MODEL TO TEST SYSTEMICALLY ACTIVE NEUROPROTECTIVE AGENTS

The Mongolian gerbil *Meriones unguiculatus*, a small rodent found in dry regions and deserts, is unique in that it has an incomplete circle of Willis, and therefore lacks connections between the carotid and vertebral arteries. Transient forebrain ischemia can be produced in gerbils by bilateral occlusion of the common carotid arteries. As described in this protocol, 5-min BCAO causes a selective pattern of damage in the CA1 hippocampal region that develops over 2 to 3 days, and the effectiveness of compounds in preventing this damage can be assessed.

Materials (see APPENDIX 1 for items with ✓)

    Male Mongolian gerbils, 60 to 80 g (Bantin and Kingman in the U.K. and Harlan in North America recommended for strains consistently lacking basilar artery connections), eight to ten animals per group
    Test compounds
    Inhalational anesthetic (halothane or isoflurane)
    Saline: 0.9% NaCl
✓ 10% buffered formalin
    Pentobarbital or chloral hydrate
    60%, 80%, and 90% industrial methylated spirits (IMS; e.g., Fisher) in distilled H_2O

    Syringes and needles suitable for route of compound administration (e.g., intramuscular, intravenous, or oral)
    Inhalation anesthetic apparatus comprising, e.g., oxygen, vaporizer, tubing, and vacuum trap (International Market Supply)
    Heating pad/blanket

Rectal temperature monitor (RS Components)
Electric hair clippers
Methiolate tincture (Lilly)
Fiber optic light source
Surgical instruments: including scissors, scalpel, and forceps (e.g., John Weiss & Sons)
Vascular clamps (Holborn Surgical Instruments), clean, without sharp edges and with tips that meet and exert sufficient pressure
Silk thread (6/6 braided suture; International Market Supply)
Suture (Ethicon W529 6/0 Mersilk; A.C. Daniels)
Thermacages (Beta Medical & Scientific) or thermostatically controlled incubators
Dram vials, glass scintillation vials, or similar glass vials
Rodent brain matrix, coronal gerbil (ASI Instruments)
Automated tissue processor (e.g., Tissue-Tek VIP 2000 vacuum infiltrator processor; Miles Scientific or Bayer Diagnostics)
Sledge microtome (Leitz 1400)

1. If test compounds are to be administered prior to carotid occlusion, treat animal by the intraperitoneal, subcutaneous, intramuscular, oral (gavage), or intravenous route (via an exposed jugular vein) as dictated by the experimental protocol (see Donovan and Brown, 1995 for techniques for different types of parenteral injection). For compounds with short half-lives, administer several times or via the subcutaneous or intraperitoneal implantation of osmotic minipumps.

2. Anesthetize male adult gerbils with inhalational anesthetic (halothane or isoflurane) to maintain a surgical plane of anesthesia in which the animals are unresponsive to hind-limb pinch applied with fingers and have no withdrawal reflex or motor response to surgery. Place the animal on its back on a thermostatically controlled heating blanket. Insert a rectal temperature probe 2 cm into the colon of the animal and attach to the thermostatic controller that powers the heating pad to maintain the animal's core temperature at 37°C.

   Hypothermia produces neuroprotection in this model, so it is important that the surgery be carried out on a heating pad. After surgery, the animals should be returned to either temperature-controlled cages or incubators, and rectal temperatures must be monitored at regular (30-min) intervals. Ensure temperatures in holding rooms are maintained within normal ranges.

3. Shave the neck using hair clippers and swab area with methiolate. Illuminate the operative area with a fiber optic light source. Expose the ventral surface of the neck, divide the sternohyoid muscles, and locate and expose the carotid arteries. Clear all connective tissue surrounding the artery by running a forceps up and down the length of the artery. Be careful not to damage the vagus or sympathetic nerves that run close to the artery. Do not try to run perpendicular to (i.e., across) the artery as there is a much greater chance of puncturing it.

4. Prepare arteries for clamping by carefully lifting each artery and pulling a silk thread underneath using a fine forceps. Lift each artery in turn using the silk thread and secure clamps for 5 min (shorter, e.g., 3 min, for initial proof of concept studies; longer, e.g., 8 to 10 min, for later, more severe challenges). Treat the sham group identically except for the placing of the clamps.

5. After 5 min (or other allotted time during which the artery turns white), remove the clamps and visually check the patency of the arteries, i.e., the artery turns red after the clamp is removed). Suture the wound in the neck and allow the animals to recover (normally ~10 to 15 min) in thermostatically controlled incubators or thermacages at 28°C to maintain core body temperature of 37°C.

6. If dictated by the experimental protocol, administer test compounds post carotid occlusion by the intraperitoneal, subcutaneous, intramuscular, oral (gavage), or intravenous route (via an exposed jugular vein).

7. Measure body temperature at various intervals (30-min intervals for the first 6 hr following surgery recommended) by manually restraining the animal and inserting a probe ∼1.5 cm into the rectum.

8. *Optional:* Measure the effects of BCAO on locomotor activity 1 to 7 days after surgery (see Support Protocol 2 and *UNIT 3.2*).

9. Five to seven days after surgery (or up to 3 months depending on individual protocols), prepare perfusion apparatus as follows (also see Gerfen, 2006).
    a. Connect the ends of some tubing to two bottles (one containing saline, the other 10% formalin) and place them on a shelf 1.5 meters above a sink.
    b. Place clamps on the tubing to prevent flow of the solutions. Connect the other end of the tubing from each of the bottles to an inlet on a 3-position flow stopcock/tap. Attach a 16-G needle to a piece of tubing connected to the outlet of the tap.

10. Administer an overdose of anesthetic (e.g., 100 mg/kg pentobarbital or 200 mg/kg chloral hydrate, intraperitoneal). Open the thorax and insert the needle into the left ventricle of the heart to allow perfusion of the brain, and make a nick in the heart on the opposite side from the needle insertion to allow the perfusate to be released. Clamp the descending arteries. Release the saline clamp and turn the tap to allow ∼30 ml of saline to flow, washing out any blood. After the perfusate runs clear, turn the tap to switch flow to the other bottle, containing 10% formalin.

    The animal, although now dead, may move as the fixative flows in, but this is expected due to the effects of the fixative. The animal will become stiff when sufficient fixative (∼100 ml) has been infused.

11. Remove the head using scissors or a guillotine. Remove the brain (which should now be clear of blood, firm and white in color) and place in a dram vial, liquid scintillation vial, or similar glass vial containing 10% formalin for 2 to 3 days.

12. Cut the brains into segments using a rodent brain matrix (Fig. 4.2.1), then process the segment containing the hippocampus and embed in paraffin wax (usually carried out using an automated tissue processor) typically by the following processing steps for 6-mm brain segments:

    10% buffered formalin for 60 min
    60% IMS for 50 min
    80% IMS for 50 min
    90% IMS for 50 min
    100% IMS for 50 min
    100% IMS for 50 min
    100% IMS for 50 min
    Xylene for 50 min
    Xylene for 50 min
    Xylene for 50 min
    60°C paraffin wax for 50 min
    60°C paraffin wax for 50 min
    60°C paraffin wax for 50 min.

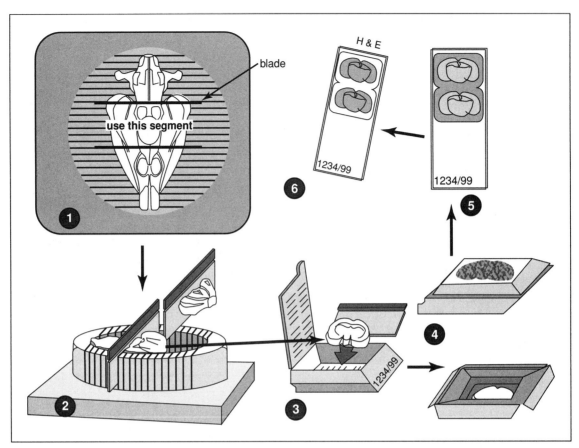

Figure 4.2.1 Illustrations of the steps involved in preparing gerbil brain sections (the process is similar for rat brain). (**1**) Place the brain into a rodent brain matrix and chop into segments with blades. (**2**) Remove the middle segment (which contains the hippocampus). (**3**) Place this segment into a prelabeled cassette and process overnight. (**4**) The next day place the segment in a histology tray and embed using hot wax. Allow to cool and remove from tray. (**5**) Embed the brain in paraffin wax as illustrated; it is then suitable for sectioning on a microtome. Pick up the sections on microscope slides and allow to dry. (**6**) Stain the sections with hematoxylin/eosin as outlined in Support Protocol 1.

13. Take 5-μm sections at 1.5, 1.7, and 1.9 mm caudal to bregma using a sledge microtome (see the gerbil brain atlas for details; Loskota et al., 1974). Pick up sections on microscope slides and allow to dry. Stain the sections with hematoxylin and eosin (Support Protocol 1). Allow the mountant to dry and then clean the slides.

14. Count the neuronal density in the CA1 subfield of the hippocampus using a microscope with grid lines (e.g., a 0.05 mm × 0.05 mm grid). Place the grid in the middle of the CA1 region, directly above the dentate gyrus, and using the 40× lens, count the number of intact pyramidal cells. Count only the cells that are fully intact (i.e., have a nucleus, are round and have a clearly defined border).

15. *Optional:* Use a scoring system to quantify the degree of damage: 0 = intact, 1 = 0 to 25% damage, 2 = 25% to 50% damage, 3 = 50% to 75% damage, and 4 = >75% damage (not as accurate as step 14, and results can vary between investigators).

16. Express results as number of viable cells per mm CA1 hippocampus. Carry out statistical analysis of histological data as appropriate (using ANOVA followed by suitable post hoc test, with p values <0.05 being statistically significant).

ALTERNATE PROTOCOL 1

GERBIL BCAO MODEL TO TEST NEUROPROTECTIVE AGENTS THAT DO NOT PENETRATE THE BRAIN

This protocol allows "proof of concept" tests to be carried out with agents that do not cross the blood-brain barrier. It combines central administration of substances with bilateral carotid artery occlusion.

Additional Materials *(also see Basic Protocol 1)*

Stereotaxic frame (e.g., model 900 from David Kopf Instruments or Bilaney Consultants)
28-G injection cannula (Plastics One)
Electrode holders/cannula holder (e.g., model 1770, 1771 from David Kopf Instruments)
25-µl Hamilton syringe
Infusion pump (World Precision Instruments)
Anesthesia mask (David Kopf Instruments or Bilaney Consultants)
Fine-tipped felt pen
Dental drill (Biotech Instruments) and 0.9-mm steel drill bits/burrs (Interfocus Ltd.)

1. Arrange the 28-G infusion cannula in the needle holder of the stereotaxic device, fill a 25-µl Hamilton syringe with the compound to be infused, and connect to the cannula using appropriate plastic tubing, ensuring there are no air bubbles in the tubing and solution can flow unimpeded. Place the Hamilton syringe on the infusion pump and set the flow rate and volume (typically a flow rate of 1 µl/min and volume of 2 to 3 µl). Activate the pump and check that the pump is delivering the solution by observing the infusion cannula.

2. Anesthetize gerbils and maintain using inhalation anesthetic (halothane or isoflurane). Place the anesthetized gerbil in a stereotaxic apparatus (head holder), insert and tighten the ear bars, and adjust face mask and anesthetic level to maintain deep anesthesia. Insert a rectal temperature probe 2 cm into the colon of the animal and attach to the thermostatic controller that powers the heating pad to maintain the animal's core temperature at 37°C.

3. Ensure the animal is well anesthetized before incising the scalp. With the head of the anesthetized animal held firmly in the stereotaxic device, shave the crown of the head and swab with methiolate. Make a midline incision with a scalpel, extending from the most caudal aspects of the eyes to between the ears.

4. Retract the skin and use some small clips to hold down the edges. Scrape the skin to remove fascia until the skull is dry. Use a bone chisel if necessary to remove membranes and clean the skull surface. Remove excess blood with swabs and surgical spears to make the various intersections of bony plates visible.

5. Locate bregma and move the manipulators until the infusion cannula is just over bregma and note the coordinates (see the gerbil brain atlas; Loskota et al., 1974). Move the manipulators to the desired coordinates from bregma (e.g., hippocampus at coordinates 2 mm AP, ±2 mm lateral and −2.5 mm ventral to bregma). Mark this point with the felt-tipped pen.

6. Using a dental drill, drill small holes in the skull at the positions marked by the felt-tipped pen using a slow drill speed at the beginning and end of the procedure to avoid penetrating the brain. Clean excess debris and blood with surgical swabs.

7. Insert the 28-gauge steel cannula and lower to the required level below the skull. Infuse the drug over a 3-min period. Alternatively, continuously administer the compound or vehicle intracerebrally or intracerebroventricularly via implanted osmotic minipumps attached

to the cannula (implanted at the time of initial surgery in both control and drug-treated animals) for up to 72 hr post-occlusion.

8. Leave the cannula in place for 5 min, and then slowly remove it from the brain using the manipulator to avoid creating a canal through which the drug can travel back up the cannula track. Suture the wounds (muscles and skin), but keep the animal anesthetized.

9. Perform BCAO (see Basic Protocol 1, steps 2 to 5) and fix, section, and examine brain tissue (see Basic Protocol 1, steps 9 to 16).

ALTERNATE PROTOCOL 2

GERBIL BCAO TO INDUCE ISCHEMIC TOLERANCE

For materials, see Basic Protocol 1

1. Perform steps 1 to 3 of Basic Protocol 1. Secure bilateral clamps for 2 min (preconditioning ischemia; see Basic Protocol 1, step 4). Treat the sham group identically, except for the placing of the clamps. Remove the clamps and visually check the patency of the arteries. Suture the wound in the neck and allow the animals to recover on thermostatically controlled blankets or in incubators at 37°C.

2. Two days later, reanesthetize the animals, expose the carotid arteries, and clamp for 3 to 5 min (test ischemia) to produce a more robust tolerant effect. As before, treat the sham-operated group identically, other than the placing of the clamps.

3. Proceed with perfusion fixation and sectioning as outlined in steps 9 to 16 of Basic Protocol 1.

BASIC PROTOCOL 2

USE OF 4-VESSEL OCCLUSION (4-VO) MODEL TO STUDY NEURONAL DEGENERATION AND TEST THE EFFECTS OF NEUROPROTECTIVE AGENTS AGAINST GLOBAL CEREBRAL ISCHEMIA

Materials (see APPENDIX 1 *for items with* ✓)

  Male Wistar rats (this strain required), 280 to 300 g
  Isoflurane
  Antiseptic solutions (e.g., Betadine and 70% ethanol)
  Pentobarbital or chloral hydrate
  ✓ 10% (v/v) buffered formalin

  Inhalation anesthetic apparatus suitable for use with isoflurane (Vetamac)
  Stereotaxic frame (David Kopf Instruments)
  Rat anesthesia mask (David Kopf Instruments)
  Electric hair clippers
  Surgical instruments including scissors, scalpel, forceps (Miltex Surgical Instruments)
  Operating microscope
  Electrocautery unit no. 160-1370 (hand piece with electrocautery needle, 0.5 mm diameter; Tiemann & Co.)
  Wound clips

Atraumatic carotid clasps (constructed according to Support Protocol 3)
PE20 polyethylene tubing (Fisher)
Bulldog artery clamp, Johns Hopkins, straight, 1.5-in. length (Roboz Surgical)
Rectal thermistor (Yellow Springs Instrument)
YSI Temperature Controller Model 73A (Yellow Springs Instrument)
Harvard Compact Infusion Pump Model 975 (Harvard Instruments)
Heat lamp (from local hardware store)
Physitemp temperature controller (Physitemp)
Physitemp type IT-21 tissue implantable thermocouple (Physitemp)
Flow-through swivel (Harvard Instrument)
Masterflex peristaltic pump (Cole Parmer)

1. Anesthetize rat with isoflurane. Place rat in stereotaxic instrument, tighten earbars and adjust face mask to keep animal anesthetized. Shave dorsal surface of neck and swab area with antiseptic solution.

2. Make an ∼1-cm midline incision behind the occipital bone directly over the first two cervical vertebrae. Separate paraspinal muscles from the midline and, with the use of an operating microscope, expose the right and left alar foramina of the first cervical vertebra.

3. Insert the 0.5-mm diameter electrocautery needle through each alar foramen and electro-cauterize both vertebral arteries (which travel within a vertebral canal and pass beneath the alar foramen before entering the posterior fossa) to permanently occlude them (the cautery tip needs to be pushed in and down centrally ∼1 to 3 mm). Hold the cautery button until the tip glows (∼2 to 5 sec, depending on the metal type used in the cautery tips).

   Sometimes smoke rises and/or atlas bone around the foramen turns white.

4. If necessary, soak blood away with a pad or a spear and use wound clips to close the incision. Remove the rat from stereotaxic instrument, but keep the animal anesthetized. Shave ventral surface of the neck and apply antiseptic solution to the skin.

5. Make a midline ventral incision and localize both common carotid arteries. Carefully dissect the arteries free from connective tissue and closely associated nerves. Place the atraumatic arterial clasps loosely around each common carotid artery without interrupting blood flow. Exteriorize the ends of the clasps constructed from the polyethelene tubing. Fix the clasps in place and close the wounds with wound clips. In this case, anchor one end of the small Silastic tubing to one hole of the baby button and exteriorize the tubing that surrounds the carotid artery by passing through the other hole in the button up through the PE90 tubing (see Support Protocol 3).

6. If the object of the study is to deliver drugs systemically (see step 11), implant a small PE20 jugular cannula at the same time the atraumatic carotid clasps are placed. Exteriorize the end of the drug-delivery cannula at the nape of the neck and hold it in place with a wound clip. Return the rats to their cages and allow them to recover from the anesthesia for 24 hr, at which time they should behave like normal animals.

7. On the day when ischemia is to be produced, hold the animals by hand and tighten the carotid clasps to produce 4-vessel occlusion by pulling on each of the Silastic tubes and placing a bulldog arterial clamp on the end of the Silastic tubing precisely where the tubing exits the PE90 tubing that protrudes from the ventral neck incision. Regulate body temperature at 37°C with a rectal thermistor coupled to a heating lamp via a temperature controller (see Basic Protocol 1, step 2 annotation about the importance of temperature control).

8. To regulate the brain temperature at 37°C (especially important when long periods of global ischemia are utilized), insert a Physitemp tissue-implantable thermocouple needle microprobe between the temporalis muscle and the skull at the time of electrocoagulation. Connect the probe to the temperature controller, which is coupled to a heating lamp. Direct the heating lamp to the opposite side of the skull to ensure that the probe stays near 37°C and accurately reflects brain temperature.

9. Observe the animals for their level of consciousness (unconsciousness, unresponsiveness, and lack of the righting reflex are expected if the vertebral arteries are completely cauterized and the blood flow through the carotids is completely halted). If the animal is still conscious retighten the carotid clasps; if this does not produce unconsciousness, it is likely that the vertebral arteries were incompletely cauterized.

10. After the period of ischemia (which could range anywhere from 10 to 30 min), remove the carotid clasps to restore blood flow through the carotids.

    The duration of post-ischemic survival is important because neuronal cell death in this model is delayed. After 10 to 15 min of 4-VO, moderate to extensive hippocampal CA1 neuronal cell death is observed; however it takes ~7 days for complete maturation of neuronal cell death. In contrast, if the period of ischemia has a duration of 30 min, massive CA1 hippocampal neuronal cell death is observed within 72 hr.

    The other important factor here is the selective vulnerability of different neuronal populations. 10 to 15 min of 4-VO is adequate for substantial reduction of hippocampal CA1 neurons. An ischemic period of 10 to 15 min, however, produces a very variable degree of striatal damage. Some animals show very little striatal damage after 10 to 15 min of ischemia. A period of 4-VO between 20 and 30 min produces substantial striatal damage and also causes neuronal cell death in cerebral cortex layers 3, 5, and 6. Cortical damage is not as reliably produced as striatal or hippocampal damage. If the goal of the study is to investigate mechanisms of neuronal cell death, it is important to know that the rate of cell death differs in various neuronal populations. Neurons in the dorsal lateral corpus striatum die within 24 hr, whereas hippocampal neurons in the CA1 layer take at least 3 days to undergo neuronal death.

11. If the objective is to deliver drugs, administer intravenous injections directly through the jugular cannula, or if constant infusion is desired, connect the cannula to the infusion pump via the flow-through swivel (Clemens et al., 1993).

12. Sacrifice the rats with an overdose of anesthetic (e.g., 100 mg/kg pentobarbital or 200 mg/kg chloral hydrate, intraperitoneally). Sever the jugular veins and place a piece of tubing from a peristaltic perfusion pump in the left ventricle of the heart (see Gerfen, 2006). Perfuse the animal with 10% buffered formalin at a rate of 10 ml/min to fix the brain.

13. Remove the brains, block, and prepare for paraffin imbedding and histological staining (see Basic Protocol 1). Evaluate neuronal densities in a manner similar to that used in Basic Protocol 1 for gerbils.

SUPPORT PROTOCOL 1

HEMATOXYLIN AND EOSIN STAINING OF BRAIN TISSUE

Materials (see APPENDIX 1 for items with ✓)

Slides of paraffin-embedded brain tissue (see Basic Protocol 1)
Xylene (Fisher)
70%, 75%, and 100% industrial methylated spirits (IMS; e.g., Fisher) in Coplin jars

Gill's hematoxylin (available as ready-to-use preparation from Surgipath)
✓ 2% (w/v) eosin
DPX mountant (BDH)

Coverslips
Slide trays

1. Place slides in a rack and process as follows.

    Deparaffinize 2 min in xylene; repeat for a total of 3 times
    Immerse 2 min in 100% IMS; repeat for a total of 3 times
    Immerse 2 min in 70% IMS
    Wash 2 min in H_2O to complete rehydration series
    Stain 10 min in Gill's hematoxylin
    Wash 5 min in H_2O
    Immerse 2 min in 75% IMS
    Immerse 2 min in 100% IMS; repeat for a total of 3 times to complete dehydration series
    Immerse 2 min in xylene; repeat for a total of 3 times to clear sections.

2. Coverslip the sections and mount using DPX. Allow the mounted sections to dry and observe at a later period. Store the slides in slide trays at room temperature.

SUPPORT PROTOCOL 2

MEASUREMENT OF LOCOMOTOR ACTIVITY AFTER ISCHEMIA IN GERBILS

Materials

Automated locomotor activity apparatus (e.g., Greenacre; also see UNIT 3.2)
SAS Software for gathering and analyzing data (e.g., Misac Instruments)

1. Connect automated locomotor activity apparatus to a PC and configure software to record photocell interruptions in the apparatus.

2. Carry out bilateral carotid artery occlusion (BCAO) in gerbils (see Basic Protocol 1, steps 1 to 7).

3. At a specified time (e.g., 24, 48, 72, or 96 hr) after surgery, place the animal in an individual compartment of the locomotor activity apparatus. Record the movement/locomotor activity within that compartment using infrared sensors positioned at the side of each compartment. See UNIT 3.2 for more details.

SUPPORT PROTOCOL 3

FABRICATION OF ATRAUMATIC CLASPS FOR RAT 4-VO

Materials

White baby buttons (Blumenthal Industries)
Silastic tubing, 0.012 in.-i.d. × 0.25 in.-o.d. (Dow-Corning)
PE90 polyethylene tubing (Fisher)
5-min epoxy glue (from local hardware store)

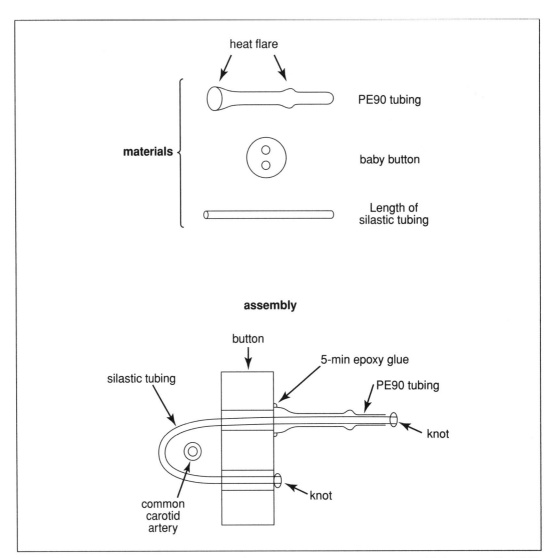

Figure 4.2.2 Materials required and procedure for assembly of atraumatic carotid clasps used in 4-VO.

1. Knot one end of the Silastic tubing. Using 5-min epoxy glue, glue one end of PE90 tubing to anchor it over one hole in the white baby button.

2. Thread the free end of the Silastic tubing through the other hole of the baby button (see Fig. 4.2.2), leaving the Silastic tubing free until surgery, when it is placed loosely around the carotid artery and exteriorized through PE90 tubing over the other hole on baby button.

References: Gill et al., 1987; Kirino, 1982; Kirino et al., 1991; Pulsinelli and Brierley, 1979

Contributors: Michael J. O'Neill and James A. Clemens

Table 4.2.2 Troubleshooting Guide for Rodent Models of Global Cerebral Ischemia

Problem	Possible cause	Solution
Gerbil BCAO model		
Animal is not deeply anesthetized, or dies	Anesthetic level too low/high	Adjust halothane/isoflurane percentage
		Adjust oxygen flow rate
Artery bleeds on dissection or clamping	Artery nicked on exposure	Take greater care on exposure
	Forceps damaged or with sharp edges	Check and replace
	Clamps damaged	Check and clean or replace
High mortality rate	Damage during surgery	Take greater care with surgical technique
	Too long a period of occlusion, i.e., >5 min	Shorten occlusion time
	Compound toxic or dose too high	Check toxicity effects in some nonoccluded animals and reduce dose if necessary
Writhing and irritation	Administered compound irritating the abdomen due to poor solubility or pH	Check that compound is completely dissolved and check pH
		Alter route of administration
No damage in the CA1	Strain and supplier an issue	Check supplier and request information
	Animal may have a basilar artery allowing flow of blood to the brain or some form of intact connections forming a circle of Willis	Transcardially perfuse the animal with charcoal to visualize vasculature and determine if there are connections in the circle of Willis
	Blood getting through carotid arteries upon clamping	Check that clamps are not damaged or that there is no buildup of dirt
	Body temperature drop of 3°C or 4°C	Check temperature data; adjust room temperature and heating pad if required
Compound fails to protect against ischemic damage	Compound not crossing blood-brain barrier or attaining sufficient brain levels	Check ADME[a] data and measure brain levels or administer centrally
	Compound administered too late	Repeat experiment, giving compound at earlier times
	Agent does not protect against ischemic damage	Change pharmacological intervention or use a positive control
Rat 4-VO model		
Excessive bleeding from cervical vertebra	Rupture of vertebral artery because of improper cauterization	Be more gentle during placement of cautery tip into alar foramen so as not to puncture vertebral artery
Rat fails to lose righting reflex	Improper cauterization of vertebral arteries or incomplete occlusion of carotid arteries	Retighten carotid clasps or completely cauterize vertebral arteries
Animal becomes unconscious and loses righting reflex; however, little neuronal damage is observed	Animals are sacrificed too quickly after ischemia	Wait a longer period of time before animals are sacrificed to allow damage to mature
Severe neurological deficits prior to 4-VO	Spinal cord damage from overzealous cauterization of vertebral arteries	Reduce heat used in cauterization procedure
High mortality rate	Damage during surgery or too long a period of occlusion	Take greater care during surgery or shorten occlusion time

[a]ADME: absorption, distribution, metabolism and excretion.

UNIT 4.3

Rodent Models of Focal Cerebral Ischemia

NOTE: Sterilize surgical instruments (i.e., forceps and scissors) using a hot bead sterilizer.

A troubleshooting guide for the methods described in these protocols (Table 4.3.1) is located at the end of the unit.

BASIC PROTOCOL 1

THE INTRALUMINAL SUTURE MODEL OF MIDDLE CEREBRAL ARTERY OCCLUSION (MCAO) TO TEST NEUROPROTECTIVE AGENTS

Materials (see APPENDIX 1 *for items with* ✓)

Male Sprague-Dawley rat, 270 to 300 g
Isoflurane (Abbott Laboratories)
30% oxygen/70% nitrogen
Cyanoacrylate glue
Test compounds of interest (optional)
Anesthetic (e.g., pentobarbital or chloral hydrate)
✓ 10% (v/v) buffered formalin

Small PE20 jugular cannula (polyethylene tubing; Becton Dickinson)
Wound clips (e.g., autoclip; Stoelting) with applier and remover (Fisher Scientific)
Anesthetic vaporizer and flowmeter (e.g., Vetmac)
Dissecting microscope (e.g., stereomaster zoom microscope, Fisher Scientific) with 10× wide-field eye piece and fiber optic light source
Homeothermic temperature system (Harvard Apparatus)
Assorted surgical instruments:
    Animal clippers (Fisher Scientific)
    Microdissecting tweezers (RS-5005, Roboz Surgical Instruments)
    McPherson-Vannas microdissecting scissors (RS-5630; Roboz Surgical Instruments)
    Straight 6-mm microaneurysm clip and 5.75-in. (14.6-cm) applying forceps with lock (Roboz Surgical Instruments)
    Bone-cutting forceps (Roboz Surgical Instruments)
    Microdissecting forceps (Roboz Surgical Instruments)
    Fine straight 11-cm iris scissors (Fine Science Tools)
    Dumont forceps (no. 11297-00 and 11251-30; Fine Science Tools)
    10-cm curved and 14-cm straight hemostatic forceps (Fine Science Tools)
    Fine forceps (no. 111500-10; Fine Science Tools)
    Needle holders (Fine Science Tools)
    Retractors (Fine Science Tools)
Portable hot bead sterilizer (Fine Science Tools)
Laser Doppler monitor (optional; e.g., Perimed Periflux 4001) with probe (Probe 407) and probe miniholder (PH 07-5)
Double-sided tape (Perimed PH 105-3)
Electrocautery unit (e.g., Geiger NY model 100 with light-duty cautery tips, style B; George Tiemann)
5-0 silk black braided suture (Roboz Surgical Instruments)

Intraluminal monofilament occluder (see Support Protocol 1)
Cryostat or microtome for sectioning
Glass microscope slides
Computerized image analysis system (e.g., Optimus 5.2 or ImageProplus; Datacell Software)
16-G needle
Peristaltic pump (e.g., Master-flex, Cole Palmer)

1. To deliver test compounds systemically, implant a small PE20 jugular cannula ∼24 hr prior to MCA occlusion to allow intravenous injections directly through the jugular cannula or an infusion pump via a flow-through swivel (Clemens et al., 1993). Exteriorize the end of the cannula used for test compound delivery at the nape of the neck and secure it in place with a wound clip.

2. Anesthetize a male 270- to 300-g Sprague-Dawley rat with 3% isoflurane in 30% oxygen/ 70% nitrogen using an anesthetic vaporizer and flowmeter. Place the animal on its back so the surgical field can be visualized under a dissecting microscope to avoid severing arteries and nerve. Maintain body temperature at 37°C with a homeothermic temperature system.

3. *Optional but recommended:* Use a laser Doppler monitor to monitor relative cerebral blood flow during the surgical procedure and recovery.

    a. Turn the animal on its side and make a 10 to 15-mm incision in the middle of the ipsilateral temporalis muscle.

    b. Use retractors to spread the muscle and expose the skull between the eye and ear, and dry the bone as much as possible with a cotton swab.

    c. Cover the bottom of the probe miniholder of a laser Doppler monitor with double-sided tape and insert the end of the probe into the holder.

    d. Coat the tape with cyanoacrylate glue, and place the probe either on the temporal surface of the squamous part of the temporal bone, or on the anterior surface of the parietal bone.

    e. Hold the probe firmly and steadily in place for 30 to 60 sec to allow the glue to set. Turn the animal onto its back and continue the procedure.

4. To occlude MCA by intraluminal suture, first make a 25-mm surgical midline incision to expose the left common carotid, external carotid, and internal carotid arteries (Fig. 4.3.1). Carefully dissect the arteries free from surrounding nerves and fascia and use a retractor to help expose the surgical field.

5. Isolate the occipital artery branches of the external carotid artery. Dissect and coagulate them with an electrocautery unit (i.e., melt the arteries with the electrocautery unit until they separate(dissect). Dissect the external carotid artery further distally and coagulate it along with the terminal lingual and maxillary artery branches.

6. Isolate the internal carotid artery and carefully separate it from the adjacent vagus nerve. Ligate the pterygopalatine artery close to its origin with a 5-0 silk suture. Loosely tie two 5-0 silk sutures around the external carotid artery stump, and apply a microaneurysm clip to the external carotid artery near its bifurcation with the internal carotid artery.

7. Make a small puncture opening in the external carotid artery. Make a mark on the suture to monitor how far it is to be advanced. Insert the monofilament (intraluminal suture) through the opening and tighten the silk sutures around the lumen containing the filament.

8. Remove the microaneurysm clip from the external carotid artery and gently advance the monofilament from the lumen of the external carotid artery into the internal carotid artery

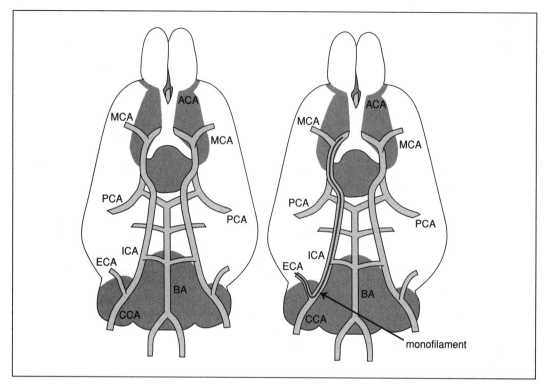

Figure 4.3.1 Illustration of the intraluminal monofilament model of focal cerebral ischemia. Carotid and cerebral arteries in the rat before (**A**) and after (**B**) insertion of the intraluminal monofilament. ACA, anterior cerebral artery; BA, basilar artery; CCA, common carotid artery; ECA, external carotid artery; ICA, internal carotid artery; MCA, middle cerebral artery; PCA, posterior cerebral artery.

for a distance of ~19 to 20 mm beyond the bifurcation of the common carotid artery. If using a laser Doppler monitor, advance the monofilament until the blood flow drops (see Fig. 4.3.2).

The right half of Figure 4.3.1 shows an appropriately inserted intraluminal suture.

9. Tighten the suture around the external carotid artery stump to prevent bleeding. Close the neck incision with surgical wound clips. Leave the laser Doppler probe holder in place to monitor cerebral blood flow at later time points (e.g., during reperfusion or when test compounds are administered), or remove it and close the wound with surgical sutures.

10. Allow the animal to regain consciousness. Check to determine whether the MCA was successfully occluded by placing the animal on a flat surface, lifting it by the base of the tail ~2 in. (5 cm) off the surface, and observing its behavior.

    When the MCA is successfully occluded, the rat will exhibit marked thorax twisting when suspended by its tail. Contralateral forepaw flexure should also be observed. This neurological indicator of damage (referred to as postural asymmetry) is scored as follows: (0) no postural asymmetry (rat reaches down with both forepaws); (1) rat retracts contralateral forepaw; (2) rat twists whole body towards the contralateral side. If these neurological symptoms are not observed, the attempt to occlude the MCA was not successful. In this case, the animal should be excluded from the study.

11a. *For transient occlusion of the MCA:* Withdraw the suture after a predetermined time (typically 2 hr) back to the stump of the external carotid artery.

    When the monofilament is withdrawn, the blood flow should return to baseline levels (Fig. 4.3.2).

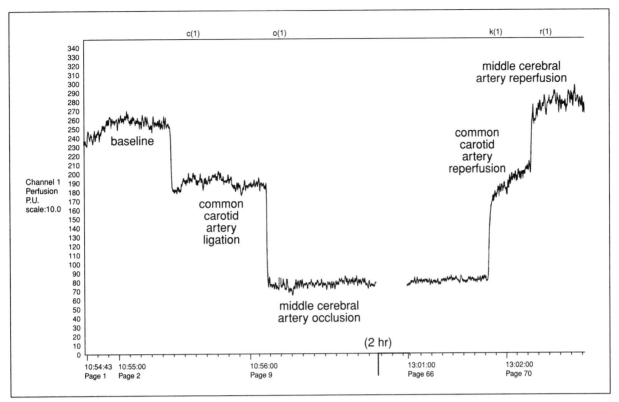

Figure 4.3.2 A laser Doppler monitoring trace of relative cerebral blood flow during intraluminal occlusion of the MCA in a male Wistar rat. The recording from left to right shows baseline flow levels, a drop in blood flow on clamping of the common carotid artery, and a further drop in flow after advancing the monofilament to occlude the MCA. The low blood flow is maintained for 2 hr (note the time and page on the bottom of the trace). The flow rate recovers quickly upon removal of the monofilament and returns to baseline on removal of the carotid artery clamp. P.U., perfusion units. Data from Deborah McCarty, Eli Lilly, Cardiovascular Division.

11b. *For systemic delivery of test compounds:* Make intravenous injections directly through the implanted jugular cannula (from step 1). After experimentation is complete, freeze the brain tissue and section on a cryostat (steps 12 a to 13a) or perfusion fix, process and embed in paraffin wax for sectioning with a microtome (steps 12b to 15b). Stain the sections with cresyl violet (see Support Protocol 4). Immunostain the sections for various proteins (e.g., glial fibrillary acidic protein to visualize astrocytes) if appropriate.

For cresyl violet staining

12a. Sacrifice rat by overdose of anesthetic (e.g., intraperitoneal injection of 100 mg/kg pentobarbital or 200 mg/kg chloral hydrate). Remove brain and freeze on dry ice. Store frozen until sectioning (up to 3 months at $-70°C$).

13a. Cut serial 30-μm sections on a cryostat and harvest sections at eight to fifteen predetermined stereotaxic levels. Choose levels that cover the whole brain so that an exact volume of infarct can be calculated. Mount on glass microscope slides and stain sections with cresyl violet (see Support Protocol 4).

14a. Use commercially available image analysis software to measure the area of damage (mm^2) in each section. Calculate the infarct volume (mm^3) based on the known distance between the sections. Alternatively, view the slides through a microscope and transcribe the region of damage onto brain maps to calculate, the area and volume of damage (see Tamura et al., 1981; Bederson et al., 1986).

For histological procedures requiring fixation

12b. Administer an overdose of barbiturate (step 12a) and sever the jugular veins.

13b. Connect a piece of tubing from a peristaltic infusion pump to a 16-G needle and insert the needle into the left ventricle of the heart. Perfuse brain with neutral buffered formalin at a rate of 10 to 30 ml/min.

14b. Remove brain and prepare for histological evaluation of infarct size (see UNIT 4.2 for histological preparation of fixed brain tissue).

15b. Cut 8-μm serial sections using a sledge microtome (e.g., Leitz). Harvest sections at eight to fifteen predetermined stereotaxic levels.

16b. Perform desired histology and determine infarct volume (step 14a).

SUPPORT PROTOCOL 1

FABRICATION OF INTRALUMINAL MONOFILAMENT OCCLUDERS

The intraluminal monofilament occluder needs to be carefully prepared, and it is recommended that one person always fabricate the occluders to achieve consistent results. Two types of occluders are described below, using two different types of monofilament. The Dermalon blue occluders are prepared as described by Belayev et al. (1996).

Materials

0.1% (w/v) poly-L-lysine (Sigma)
Nylon monofilament:
    Stren monofilament fishing line, 6- to 8-lb test, 0.28-mm diameter (DuPont) with tip rounded by dipping in fingernail polish (e.g., Double Magic Cutex, base coat/topcoat), *or*
    3-0 Dermalon blue nylon monofilament (The Butler Company)
60°C incubator

For Stren monofilament fishing line

1a. Cut fishing line into ~40-mm segments. Examine the tip of the filament under a microscope and make sure the cut is smooth and even.

2a. Round the tip of the filament (to make it shaped like a small, flattened cotton swab) by dipping in fingernail polish.

3a. Let fingernail polish dry. Examine the monofilament under the microscope to make sure that the diameter of the tip is 0.37 to 0.4 mm.

For Dermalon blue nylon monofilament

1b. Cut 3-0 Dermalon blue nylon monofilament into 20-cm lengths and stretch (i.e., tape in stretches to a table or wall) for a few days to remove kinks. Cut into 40-mm segments.

2b. Round the tip of each segment by heating the tip near glowing embers. To accomplish this, set a wooden applicator stick on fire, blow out the flame, and advance the monofilament toward the glowing ember perpendicular to the stick just until the end begins to round.

3b. Examine each monofilament microscopically and evaluate for size (∼250 μm in diameter; final tip diameter ∼0.4 ± 0.02 mm) and appearance (shaped like a small, flattened cotton swab). Discard any filament with rough edges or if it curves to one side.

4. On the day of surgery, dip the monofilament in 0.1% poly-L-lysine and then heat 1 hr at 60°C in an incubator.

BASIC PROTOCOL 2

MIDDLE CEREBRAL ARTERY OCCLUSION (MCAO) USING STEREOTAXIC INFUSION OF ENDOTHELIN 1

Unlike the monofilament (Basic Protocol 1) and Tamura (Basic Protocol 3) models, stereotaxic infusion of endothelin 1 (Et-1) is a fast and noninvasive method to produce MCAO (first described by Sharkey et al., 1993).

Materials (see APPENDIX 1 for items with ✓)

✓ Et-1 solution, prepared fresh prior to each new experiment
Male Sprague-Dawley rat, 280 to 320 g (at least ten animals per group, twelve to fourteen recommended)
Isoflurane (Vet Drug)
Test compounds of interest

28-G steel infusion cannula (e.g., Plastics One, c/o Bilaney Consultants)
Stereotaxic frame (e.g., David Kopf Instruments or Bilaney Consultants)
25-μl Hamilton syringe (e.g., Alltech Associates Applied Science)
FEP tubing (CMA, Biotech Instruments Ltd.)
Infusion pump (e.g., World Precision Instruments)
Rat anesthesia mask (e.g., David Kopf Instruments or Bilaney Consultants)
Rectal temperature monitor and probe (e.g., RS Components)
Heating pad/blanket (e.g., International Market Supply) to which rectal probe can be attached
Electric hair clippers
Surgical instruments (e.g., see Basic Protocol 1)
Large artery clips
Small clips
Bone chisel
Cotton swabs
Surgical spears (e.g., Interfocus)
Dental drill (e.g., Biotech Instruments)
0.9-mm steel drill bit/burr (e.g., Interfocus)
Mersilk 410 W536 Ethicon braided silk sutures (International Market Supply)
Thermacage (Beta Medical & Scientific)

1. Arrange a 28-G steel infusion cannula in the needle holder of a stereotaxic frame as illustrate in Figure 4.3.3 A. Fill a 25-μl Hamilton syringe with Et-1 solution and connect to the cannula using FEP tubing. Ensure that there are no air bubbles in the tubing and that Et-1 is flowing. Attach the Hamilton syringe to an infusion pump and set the flow rate to 1 μl/min and the volume to 3 μl. Press the start button and ensure that the pump is delivering Et-1 by observing the infusion cannula.

2. Anesthetize a male 280- to 320-g Sprague-Dawley rat (weight range important for stereotaxic injection accuracy) with 3% isoflurane in 30% oxygen/70% nitrogen using an anesthetic vaporizer and flowmeter and a rat anesthesia mask. Maintain using an inhalation anesthetic (isoflurane).

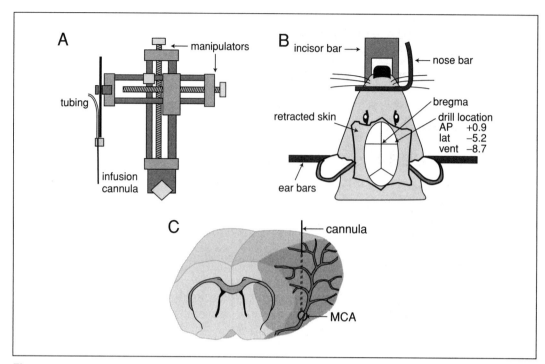

Figure 4.3.3 The endothelin-1 model of MCA. (**A**) The infusion cannula and the manipulators on the stereotaxic frame. (**B**) Rat with head in stereotaxic frame. The bregma and site of infusion of Et-1 are marked. (**C**) Diagram of the brain showing the area of damage after infusion of Et-1 adjacent to MCA.

3. Place anesthetized rat in the stereotaxic apparatus (Fig. 4.3.3B), insert and tighten the ear bars, and adjust rat anesthetic face mask and anesthetic level until animal is deeply anesthetized. Insert a rectal temperature probe 4 cm into the colon of the animal and attach to the thermostatic controller that powers the heating pad to maintain the animal's core temperature at 37°C.

4. *Optional:* Monitor blood pressure by cannulating a vessel or by placing a pressure cuff on the tail. Monitor blood gases by taking regular samples from a cannulated femoral artery (cannulated at the time of MCA surgery, and samples taken until 6 hr after occlusion).

5. With the head of the well anesthetized animal held firmly in the stereotaxic device, shave the crown of the head with electric hair clippers and use a scalpel to make a midline incision from the most caudal aspects of the eyes to between the ears (Fig. 4.3.3B).

6. Retract the skin with large artery clips and use small clips to hold down the edges. Scrape the skin to remove fascia until the skull is dry. Use a bone chisel, if necessary, to remove membranes and to clean the skull surface. Use cotton swabs and surgical spears to remove excess blood that may accumulate.

7. Note the various intersections of bony plates (Fig. 4.3.3B). Locate bregma, move the manipulators until the infusion cannula is just over bregma, and note the coordinates. Move the manipulators to AP = +0.9 mm and lateral = −5.2 mm from bregma. Mark this point with a fine-tipped felt pen.

8. Drill a small (1- to 2-mm) hole in the skull at this location, taking care not to cause damage to the brain. Use a slow speed at the beginning and end of the procedure to minimize the chances of penetrating into the brain. Clean excess debris and blood with surgical swabs.

9. Insert the 28-G steel cannula and lower to 8.7 mm below the skull (Fig. 4.3.3 C). Infuse 10 to 300 pmol Et-1 solution in a 1- to 3-μl volume over a 3-min period (amount and

volume depending on the purpose of the experiment). Leave the cannula in place for 5 min, and then slowly remove it from the brain using the manipulator to avoid creating a canal through which the drug can travel back up the cannula track.

10. Close wounds (muscles and skin) with Mersilk sutures and allow the rat to recover for 3 days (time may vary) in a thermacage at 28°C.

11. Administer test compounds post-occlusion by the intraperitoneal, subcutaneous, intramuscular, oral (gavage), or intravenous route. Compounds with short half-lives may be administered several times or via the subcutaneous or intraperitoneal implantation of osmotic minipumps.

    Administration of compounds prior to occlusion provides the maximum opportunity of neuroprotection. If they provide good protection, the time window of protection should be evaluated. This is best done by moving the time of initiation of treatment forward in 30-min intervals (for example, start treatment 30 min before occlusion, then immediately after occlusion, 30 min after occlusion, and so on).

12. Evaluate the effects of MCAO on functional outcomes (see Support Protocols 2 and 3).

13. Section brain tissue on a cryostat and stain with cresyl violet (see Basic Protocol 1, steps 12a to 14a) or perfusion fix for other histological procedures (see Basic Protocol 1, steps 12b to 16b).

SUPPORT PROTOCOL 2

USE OF HORIZONTAL AND INCLINED BALANCE BEAM TO ASSESS SENSORIMOTOR PERFORMANCE AFTER Et-1 MCAO

Materials

Rat: untreated for training, and then Et-1-treated (see Basic Protocol 2) for testing
Two pieces of wood, 2 cm wide and 1 m long

1. Set up one piece of wood as a horizontal beam 18 in. (46 cm) above a bench. Set up a second piece of wood as an inclined beam with one end touching the bench and the other supported at an angle of 30°.

2. Place an untreated rat on each beam for 1 min and score as follows:

    1 = Balances with all paws on top of beam.
    2 = Puts paws on side of beam or wavers on beam.
    3 = One or two limbs slip off beam.
    4 = Three limbs slip off beam.
    5 = Rat attempts to balance on beam but falls off.
    6 = Rat drapes over beam and then falls off.
    7 = Rat falls off without attempting to balance on beam.

    Test rat before surgery until a baseline score of 1 is obtained.

3. Carry out MCAO by Et-1 infusion (see Basic Protocol 2).

4. Test Et-1-treated rat 1 and 3 days after surgery to demonstrate a deficit after MCAO. Test the rat 24 hr after surgery and at 2- to 3-day intervals for 30 days to show longer-term functional recovery with a test compound.

5. Analyze data is analyzed at each time point using analysis of variance (ANOVA) followed by post hoc test (Mann Whitney test).

SUPPORT PROTOCOL 3

USE OF THE STAIRCASE TEST TO MEASURE SKILLED PAW USE AFTER Et-1 MCAO

This test uses forelimb reaching and grasping to provide a quantitative measure of functional deficit after MCAO (Montoya et al., 1991).

Materials

Rat: untreated for training, and then Et-1-treated (see Basic Protocol 2) for testing
Staircase apparatus (Fig. 4.3.4; constructed as described in Montoya et al., 1991)
Food pellets (e.g., 45-mg Noyes sucrose reward pellets; Sandown Scientific)

1. One day prior to the first training session, feed untreated rat just enough food to maintain body weight at 95% of free-feeding body weight (generally 12 to 18 g of rat chow per day).

2. Place two pellets of food on each side of each of the six steps of the apparatus (Fig. 4.3.4). Place rat in the apparatus for 15-min trials and count the number of pellets missing (i.e., that the rat has eaten) and displaced (that rat has knocked off the steps and cannot reach (Fig. 4.3.5). Repeat twice daily with a minimum of 4 hr between trials until the rat learns to eat eight to ten pellets of food (usually about fourteen to sixteen trials).

3. Give rat free access to food for 2 days and then carry out MCAO by Et-1 infusion (see Basic Protocol 2). Ensure that rat recovers well from surgery (~3 to 5 days).

4. Repeat food deprivation (step 1) and testing (step 2). Test the animals twice a week for 30 days to determine if the compound improves the ischemia-induced deficits at later time points.

   The animal should exhibit a marked deficit in the ability to pick up pellets with the contralateral paw and therefore, the number of pellets eaten will decrease and the number of pellets displaced will increase. If a test compound has reduced damage, the animal should perform well (eat more pellets and displace less) at this stage.

5. Analyze data using analysis of variance (ANOVA) followed by a suitable post-hoc test (e.g., *t* test, Dunnett's test).

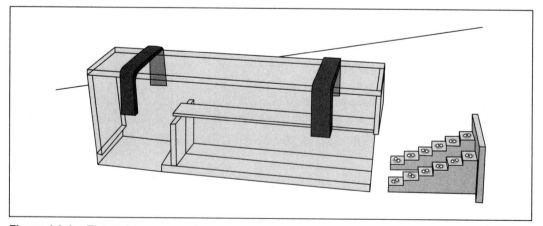

Figure 4.3.4 The staircase apparatus.

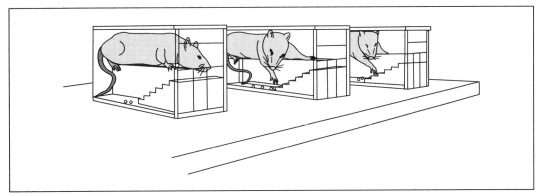

Figure 4.3.5 Three rats in the staircase apparatus. Two rats (middle and right) are clearly reaching for food pellets. In all three cases, some food pellets have been displaced to the bottom of the apparatus and cannot be reached.

BASIC PROTOCOL 3

THE TAMURA MODEL OF MIDDLE CEREBRAL ARTERY OCCLUSION (MCAO) TO TEST NEUROPROTECTIVE AGENTS

Materials

Male Sprague-Dawley rat, 275 to 300 g
Cefazolin (The Butler Company)
Isofluorane (Abbott Laboratories)
30% oxygen/70% nitrogen
Test compounds of interest
Artificial tears (The Butler Company)
Surgical scrub: Betadine and 70% (v/v) ethanol
EMLA cream (2.5% lidocaine and 2.5% prilocaine; Henry Schein)
Polysporin topical ointment (Henry Schein)

Anesthetic vaporizer and flow meter (e.g., Vetamac)
Rat anesthesia mask designed so that the rat can lay on its side with head held in place by an incisor bar
PE50 jugular cannula (polyethylene tubing; Becton Dickinson; optional)
Compact syringe pump (infusion pump; e.g., Harvard Apparatus model 975)
Flow-through swivel (e.g., Harvard Apparatus)
Homeothermic blanket system with rectal probe (Harvard Apparatus)
Elastikon porous tape
Surgical instruments (e.g., see Basic Protocol 1)
Portable hot bead sterilizer (Fine Science Tools)
Dissecting microscope (e.g., aus Jena 212 T OPM) or other suitable model
Hand-held drill with a 1.4-mm steel burr (Fine Science Tools)
Curved Fredman-Peason microrongeurs (Fine Science Tools)
440E bipolar coagulation unit (Radionics)
Extra-delicate mini-Vannas scissors (Fine Science Tools)
Surgical 3-0 nylon suture (Harvard Apparatus)

1. One day prior to surgery, administer a 40 mg/kg intraperitoneal injection of Cefazolin to a 275- to 300-g Sprague-Dawley rat (to prevent infection).

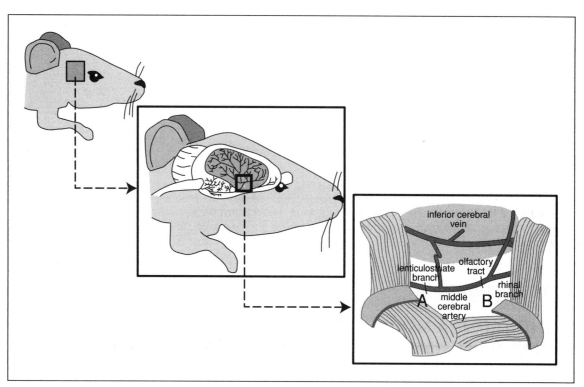

Figure 4.3.6 Position of craniectomy in rat, with proximal (**A**) and distal (**B**) sites for electrocoagulation of the MCA. For the Tamura model, it is important that the MCA be occluded at the proximal site in order to yield both cortical and subcortical damage.

2. On the day of surgery, anesthetize rat with 3% (v/v) isofluorane in 30% oxygen/70% nitrogen using an anesthetic vaporizer, flow meter, and rat anesthesia mask.

3. If test compounds are to be delivered preocclusion, place a PE50 jugular cannula in the jugular vein and exteriorize it at the nape of the neck. Deliver test compounds by means of an infusion pump connected to the jugular cannula by means of a feed-through swivel.

4. Place rat in the lateral position and hold the anterior portion of the face in the anesthesia mask by means of an incisor bar. Maintain anesthesia with 1% to 2% (v/v) isofluorane. Maintain body temperature at 37°C using a rectal probe and a small homeothermic blanket system.

5. Place artificial tears in the rat's eyes, close the right eyelid, and cover it with a small piece of Elastikon porous tape to reduce damage to the eye during surgery. Shave the area between the right lateral canthus of the eye and the base of the right ear using animal clippers. Cleanse the area with a surgical scrub mixture of Betadine and 70% ethanol.

6. Make a curved vertical 2-cm skin incision in the midpoint between the left orbit and the external auditory canal. Make an incision around the superior and posterior margins of the temporalis muscle, scrape the muscle from the lateral aspect of the skull, and reflect it forwards.

7. With the aid of a dissecting microscope, expose the proximal MCA through a subtemporal craniectomy using a hand-held drill with a 1.4-mm steel burr (Fig. 4.3.6). Use a curved Fredman-Peason microrongeurs to remove pieces of the skull without removing the zygomatic arch or orbital contents, and without transecting the facial nerve. Once the MCA is exposed, remove the dura that covers the MCA.

8. Occlude the MCA by micropolar coagulation from just proximal to the olfactory tract to the inferior cerebral vein (Fig. 4.3.6). Use a 440E bipolar coagulation unit according to manufacturer's instructions, at a setting of 15 to 20 (depending on the unit) to immediately effect coagulation.

   It is essential that the artery be fully occluded and that the occlusion be proximal to the lateral lenticulostriate branches if both cortical and subcortical damage are to be produced

9. Transect the MCA with extra-delicate mini-Vannus scissors. Close the incision site with surgical 3.0 nylon suture. Apply EMLA cream and Polysporin topical ointment to the incision site. Repeat injection of Cefazolin (step 1).

10. For delivery of test compounds postocclusion, place a PE50 jugular cannula in the jugular vein and exteriorize at the nape of the neck. Deliver test compounds by means of an infusion pump connected to the jugular cannula by means of a feed-through swivel.

11. After experimentation is complete, section brain tissue on a cryostat and stain with cresyl violet (see Basic Protocol 1, steps 12a to 14a) or perfusion fix for other histological procedures (see Basic Protocol 1, steps 12b to 16b).

BASIC PROTOCOL 4

THE SPONTANEOUSLY HYPERTENSIVE RAT MODEL OF MIDDLE CEREBRAL ARTERY OCCLUSION (MCAO) TO TEST NEUROPROTECTIVE AGENTS

Materials

Male spontaneously hypertensive rat, 300 to 325 g (Taconic or Harlan Bioproducts for Science)
Isofluorane (Abbott Laboratories)
30% oxygen/70% nitrogen
Artificial tears (The Butler Company)
Surgical scrub: Betadine and 70% (v/v) ethanol
0.9% (w/v) saline, sterile
EMLA cream (2.5% lidocaine and 2.5% prilocaine; Henry Schein)
Polysporin topical ointment (Henry Schein)
Test compounds of interest

Anesthetic vaporizer and flow meter (e.g., Vetmac)
Rat anesthesia mask designed so that the rat can lay on its side with head held in place by an incisor bar
Homeothermic blanket system with rectal probe (Harvard Apparatus)
Portable hot bead sterilizer (Fine Science Tools)
PE50 polyethylene tubing (Becton Dickinson)
Tygon tubing (0.010-mm i.d. × 0.30-mm o.d.)
Heart rate and blood pressure monitor (e.g., Gould model 11-2927-31 two-channel ink recorder)
Surgical instruments (e.g., see Basic Protocol 1)
Small hand-held retractor (Roboz Surgical Instruments)
Straight atraumatic aneurysm clamp, 0.75 × 4 mm (Roboz Surgical Instruments; optional)
3-0 silk suture (optional)
Dissecting microscope
Small hand-held drill with 1.4-mm diameter bit

Curved atraumatic aneurysm clamp, 1 × 6 mm (Roboz Surgical Instruments)
i-STAT portable clinical analyzer and test cartridges for blood gas, pH, and glucose analysis (SDI-sensor Device)
Light-duty electrocautery unit (e.g., Geiger-NY model 100 with style B cautery tips; George Tiemann)
Autoclip wound clips (Stoelting) with applier and remover (Fisher Scientific)
3-0 nylon suture
Compact syringe pump (infusion pump; e.g., Harvard Apparatus model 975)
Flow-through swivel (e.g., Harvard Apparatus)

1. Anesthetize a 300 to 325-g spontaneously hypertensive rat with 3% (v/v) isofluorane in 30% oxygen/70% nitrogen using an anesthetic vaporizer, flow meter, and rat anesthesia mask. Place rat in the lateral position and hold the anterior portion of the face in the anesthesia mask by means of an incisor bar. Maintain anesthesia with 1% to 2% (v/v) isofluorane. Maintain body temperature at 37°C using a rectal probe and a small homeothermic blanket system.

2. Cannulate the jugular vein with a length of PE50 polyethylene tubing and exteriorize the cannula at the nape of the neck for administration of test compounds.

3. Cannulate the femoral artery with a 20-cm length of Tygon tubing and connect the femoral artery cannula to a heart rate and blood pressure monitor. Continuously monitor heart rate and blood pressure throughout the experiment.

   The femoral cannula will be used to record physiological parameters such as blood pressure and heart rate, and to obtain arterial blood for blood gas analysis. Continuous monitoring is especially important when transient occlusion is used because the animal is anesthetized for the entire experiment. Blood gases, blood pH, and glucose levels are generally measured at the onset of occlusion and 1 and 2 hr after occlusion. Measurement of these physiological parameters is vital when test compounds are given during the period of occlusion. Variations in blood gases, blood pressure, and blood glucose can seriously affect the outcome of the study.

4. Place artificial tears in the rat's eyes during the surgical procedure to prevent drying of the cornea. Shave the surgical site (between the right-lateral canthus of the eye and the base of the right ear) with animal clippers and clean with a surgical scrub of Betadine and 70% ethanol. Prepare the ventral neck region with a surgical scrub to expose the carotid arteries.

5. Make a 10- to 20-mm longitudinal, ventral neck incision and separate muscle and fascia using a small hand-held retractor to expose the right carotid artery.

6a. *For transient MCAO:* Place a straight atraumatic aneurysm clamp on the right carotid artery to provide temporary occlusion of the vessel.

6b. *For permanent MCAO:* Ligate the right carotid artery by tying off the artery with two 3-0 silk sutures and cutting the artery between two silk ligatures.

7. Under direct visualization with a dissecting microscope, expose the right MCA by drilling an ∼2-mm-diameter burr hole 2 to 3 mm anterior to the fusion of the zygomatic arch with the squamosal bone using a hand-held drill and 1.4-mm bit. Bathe the area with saline to prevent heat injury of the underlying cerebral cortex. Do not remove any of the zygomatic arch or orbital contents, or transect the facial nerve.

8a. *For transient MCAO:* Remove dura from the area of the MCA and place a curved atraumatic aneurysm clamp on the ventral surface of the MCA just proximal to the formation of the MCA branches. Measure blood gases, blood pH, and glucose levels at the onset of occlusion. Keep clamp in place for the desired duration of occlusion and then remove.

The duration of occlusion depends on the objectives of the experiment. Occlusion for only 1 hr produces a very small infarct. Transient occlusion for a period of 2 hr produces an infarct with a volume of ~ 100 mm^3. Arterial occlusion for a period of 3 hr produces a maximal-size infarct. Most of the studies reported using transient occlusion to test the effects of neuroprotective agents have used a 2-hr period of occlusion.

8b. *For permanent MCAO:* Permanently occlude MCA by electrical cauterization with a light-duty electrocautery device. Measure blood gases, blood pH, and glucose levels at the onset of occlusion.

   It is essential that the MCA artery be fully occluded (permanent or transient) or clamped (transient model) if consistent lesions are to be produced. It is also essential that the ipsilateral carotid artery be ligated (permanent) or clamped (transient) properly to achieve consistent lesions.

9. Use a 3-0 nylon suture and wound clips to close the incision site. Apply EMLA cream and Polysporin topical ointment to the incision sites. Measure blood gases, blood pH, and glucose levels again at 1 and 2 hr after occlusion.

10. Administer test compounds intravenously by means a compact syringe pump hooked to the animal's jugular cannula (step 2) through a feed-through swivel.

11. After experimentation is complete, section brain tissue on a cryostat and stain with cresyl violet (see Basic Protocol 1, steps 12a to 14a) or perfusion fix for other histological procedures (see Basic Protocol 1, steps 12b to 16b).

SUPPORT PROTOCOL 4

CRESYL VIOLET STAINING OF BRAIN TISSUE

In order to visualize the damage following ischemia and the protection provided by test compounds, it is necessary to stain the brain sections. Cresyl violet is employed to make permanent Nissl stains of nervous tissue. The stain is not taken up by dead tissue, therefore only intact, viable cells are stained.

Materials (*see* APPENDIX 1 *for items with* ✓)

Slides with 30-μm cryostat sections from occluded brain (see Basic Protocol 1)
Industrial methylated spirits (IMS; Fisher Scientific)
Xylene (Fisher Scientific)
✓ Cresyl violet solution
✓ Acid formalin solution
DPX mountant (BDH)
Slide rack and staining dishes

1. Place slides with 30-μm cryostat sections in a rack and clean 5 min in IMS followed by 5 min in xylene.

2. Rehydrate sections by washing sequentially:

   once for 5 min in 100% IMS
   twice for 2 min in 100% IMS
   once for 2 min in 70% IMS
   once for 5 min in water.

3. Process sections as follows:

   a. Stain sections 5 min in cresyl violet solution.

   b. Wash 5 min in water.

   c. Wash 2 min in acid formalin solution.

Table 4.3.1 Troubleshooting Guide for Focal Cerebral Ischemia Studies

Problem	Possible cause	Solution
Monofilament MCA model:		
Upon waking, rat fails to show expected neurological signs	Suture not inserted far enough	Insert suture further
Highly variable infarct size	Diameter of suture too small	Increase diameter of suture
	Inconsistent placement of suture	More accurate measurement of distance that suture is inserted
	Suture tip has rough edges	Ensure that suture tip does not have rough edges
Animal dies from hemorrhage	Suture tip inserted too far	Decrease distance of insertion
Striatal infarct but no cortical infarct	Length of ischemic period insufficient	Increase period of ischemia
	Movement of suture from the origin of MCA due to animal movement upon awakening	Increase filament diameter
No infarct produced	Multiple reasons, including duration of ischemia, suture size, length of insertion	Try inserting different size filaments for different distances while using a laser Doppler monitor to measure blood flow; substantial decrease in blood flow indicates correct distance of insertion or appropriate filament diameter
Endothelin-1 MCA model:		
Animals seize or convulse	Dose of Et-1 too high	Reduce Et-1 dose
High mortality rate	Damage during surgery	Take greater care
	Dose of Et-1 too high	Reduce Et-1 dose
Mechanical damage to cortex	Drill hole too big or the drill is penetrating too deeply	Use small drill bit and take greater care
Bleeding or hemorrhaging after Et-1 infusion	Cannula nicking MCA	Check and alter stereotaxic coordinates
Writhing and irritation	Compound irritating the abdomen due to poor solubility or pH	Check that compound is completely dissolved and check pH
No damage/lesion	Rat strain and size an issue	Check strain and ensure body weight is between 280 and 320 g
	Et-1 denatured or concentration wrong	Check concentration and make fresh batch
	Cannula blocked	Check by running infusion pump while cannula is not in the brain; unblock or change cannula if necessary
	Incorrect infusion location: stereotaxic frame needs testing, coordinates wrong, or cannula bent	Check frame, coordinates, and cannula; if necessary, infuse blue dye to see if coordinates are correct

Problem	Possible Cause	Solution
Compound fails to protect	Compound not crossing blood-brain barrier or reaching sufficient levels in brain	Consult adsorption, distribution, metabolism and excretion (ADME) data/personnel to measure brain levels or administer centrally
	Compound given too late	Repeat experiment, giving compound at earlier times
	Compound does not protect against ischemic damage	Change pharmacological intervention
Tamura MCA model:		
Animal starts to bleed or dies from hemorrhage	Damage to MCA while exposing and coagulating artery	Take greater care during surgery; conduct coagulation slowly so as not to puncture MCA
Highly variable infarct size	Inconsistent coagulation of the MCA; some blood gets through or MCA not transected	Conduct more accurate surgery and occlusion of MCA proximal to lenticulostriate branches; carry out coagulation for a greater length of MCA; ensure MCA is transected after coagulation
Cortical but not striatal damage	Coagulation of MCA distal to lenticulostriate branches	Perform more accurate surgery and occlusion of the artery proximal to lenticulostriate branches
Animals have a poor recovery or die after surgery	Craniectomy too big and surgery too invasive; the zygomatic arch is damaged	Take greater care during surgery; avoid damage to zygomatic arch
Damage to cortical surface of brain	Cortical surface damage by drill or during removal of pieces of skull	Take greater care while drilling and removing skull pieces
SHR MCA model:		
Animal starts to bleed or dies from hemorrhage	Damage to MCA while exposing and coagulating or clamping artery	Take greater care during surgery; carry out coagulation or clamping slowly so as not to puncture MCA
Highly variable infarct size	Lesion too small due to strain of rat	Check supplier and use only SHR rats
	Position of MCAO wrong	Check and adjust if necessary
	If transient ischemia, ipsilateral carotid not fully clamped	Ensure CCA is fully clamped
No infarct produced	If permanent model, MCA not occluded properly	Ensure MCA is coagulated properly and transected after coagulation
	If transient model, MCA not clamped properly or time of occlusion is too short	Ensure MCA is fully clamped and, if necessary, increase time of occlusion
Animals have a poor recovery or die after surgery	Craniectomy too big and surgery too invasive	Take greater care during surgery
Damage to cortical surface of brain	Cortical surface damage by drill or during removal of pieces of skull	Take greater care while drilling and removing skull pieces
	Damage due to clamping or insertion of wire under MCA	Take greater care during surgery

d. Wash 5 min in water.

e. Dehydrate with one 3-min wash in 75% IMS followed by three 3-min washes in 100% IMS.

f. Clear sections with three 3-min washes in xylene.

g. Mount sections using DPX mountant and a coverslip.

References: Bederson et al., 1986; Belayev et al., 1996; Brint et al., 1988; Menzies et al., 1992; Robinson et al., 1990; Sharkey et al., 1993; Tamura et al., 1981; Zea Longa, 1989

Contributors: Michael J. O'Neill and James A. Clemens

UNIT 4.4

Inducing Photochemical Cortical Lesions in Rat Brain

BASIC PROTOCOL

See Table 4.4.1 at the end of this unit for troubleshooting procedures related to these methods.

CAUTION: There are two main potential hazards associated with argon lasers: the laser beam and the high-voltage power supply. Before operating a laser, appropriate safety measures should be implemented to prevent serious and irreversible injury to the eyes, skin, and other parts of the body. In addition, the surgery room should be approved for the use of a laser by the local nonionizing radiation safety officer and should conform to Occupational Safety and Health Administration (OSHA) regulations. Safety measures include but are not limited to understanding the concepts of nonionizing radiations, appropriate training on the particular laser to be used, using approved laser safety eyewear, using appropriate shielding of refractive surfaces and instruments during laser operation, and using an interlock safety system that shuts off the laser power if, for example, the cover is removed, the electrical components malfunction, the internal laser temperature is too high, or cooling water flow is too low.

NOTE: All surgical instruments and sutures should be sterilized before each surgery by autoclaving. Alternatively, soaking the instruments in glutaraldehyde or other noncorrosive antiseptic sterilizing solutions for ≥1 hr (before surgery) and rinsing them thoroughly in sterile water before use will achieve a similar degree of aseptic protection. The use of a portable hotbead sterilizer (e.g., Fine Science Tools, Harvard Apparatus, Stoelting) to sterilize instruments between animals is also useful.

Materials (*see APPENDIX 1 for items with* ✓)

275- to 300-g male or female rats (e.g., Charles River Laboratories, Harlan)
4% (v/v) isoflurane (Henry Schein) in 30% (v/v) oxygen/70% (v/v) nitrogen gas mixture
Lubricant gel (e.g., Henry Schein, Owens and Minor)
Lubricant ophthalmic ointment or artificial tears (e.g., The Butler Company, Henry Schein)
Antiseptic scrubbing solution (e.g., Betadine; Henry Schein, Owens and Minor)
70% (v/v) ethanol
Diluted styptic pencil solution, made by soaking tip of styptic pencil (generally available from pharmacies) in 10 ml sterile water for 2 to 4 min

Water, 37 to 40°C
Saline: 0.9% (w/v) NaCl, sterile
Mask with desired aperture for lesion cut with heavy-duty punch or knife from ≤0.25-mm-thick brass sheet
Superglue
Mineral oil, optional
✓ Photosensitizing dye solution
Topical antibiotic ointment (e.g., Panolog; Henry Schein)
Topical anesthetic cream, such as 2.5% lidocaine and 2.5% prilocaine (e.g., ELMA cream; Henry Schein)
Dry ice/isopentane bath, for infarct assessment without fixation

Thin brass sheet, small piece (optional) *or* black electrical tape
Laser safety alignment eyewear, with an optical density (OD) of 2 to 3 and wavelength range of 488 to 515 nm (e.g., Ritz Safety Equipment, Melles Griot, UVEX Safety, Kentek, Rockwell Laser Industries)
Argon ion laser (multiline or 514.5 nm), available new (e.g., Lexel, Coherent Laser, Melles Griot) or as refurbished equipment (e.g., Evergreen Laser)
Optical components for photochemical lesioning, including:
    Kinematic mount with mirror (e.g., Melles Griot)
    Plano-convex BK7 glass lens with 250-mm focal length and 30-mm diameter (e.g., Melles Griot)
    Lens holder with 30-mm nominal lens diameter (e.g., Melles Griot)
    Mounting post, 20 mm in diameter and 150 mm long (e.g., Melles Griot)
    Collet post holder (e.g., Melles Griot) for mounting post
    Optical rail carrier (e.g., Melles Griot) for 50-mm optical rail
    50 × 500–mm optical rail (e.g., Melles Griot)
Indoor refrigerated, recirculating cooling unit, high capacity (Electro Impulse)
Ruler
Surgical table or other suitable nonreflective surface
Transparent anesthesia-induction chamber (e.g., Stoelting, VetEquip, Harvard Apparatus)
Anesthetic vaporizer and flow meter for use with isoflurane (e.g., VetEquip, Vetamac, Stoelting, Harvard Apparatus)
Electric hair clippers (e.g., Jeffers, Harvard Apparatus, Stoelting, Roboz Surgical Instruments)
Thermocouple rectal temperature probe and feedback-controlled homeothermic heating blanket (e.g., Harvard Apparatus)
Stereotaxic frame (e.g., David Kopf Instruments) equipped with rat anesthesia nose cone (e.g., David Kopf Instruments, Stoelting, Harvard Apparatus) fitted over the incisor bar
Gauze, sterile (e.g., Henry Schein, Owens and Minor)
Dual fiber-optic illuminator (e.g., Stoelting, Harvard Apparatus)
Sterile surgical instruments (e.g., Roboz Surgical Instruments, Fine Science Tools, Miltex), including:
    Scalpel
    4.4-cm (1.75-in.) barraquer retractor (e.g., Roboz Surgical Instruments)
    Hemostats
Cotton swabs, sterile (e.g., Fisher, Henry Schein)
24-G, 1.9-cm (0.75-in.) intravenous catheter needle with needle guide (e.g., Caligor), sterile
1-ml syringes

High-optical density argon laser safety eyewear, with OD of 7 to 8 and wavelength range of 190 to 532 nm (e.g., Ritz Safety Equipment, Melles Griot, Uvex Safety, Kentek, Rockwell Laser Industries)

Syringe pump (e.g., Harvard Apparatus, Stoelting)

Braided-silk or nylon suture (gauge 4-0 or 3-0; length, 46 cm) with 1/4 or 3/8 curved, reverse cutting needle (e.g., Roboz Surgical Instruments, Henry Schein, Harvard Apparatus), sterile

Cage fitted with heating pad (e.g., Baxter, Harvard Apparatus), 37°C

1. *Optional:* Create a mask (for more accurate and reproducible alignment of the laser beam) by cutting a small piece of thin brass sheet to fit over a skull area slightly larger (≥ 1 mm all around) than the area to be irradiated (see Fig. 4.4.2B). Use a heavy-duty punch or a knife to make a hole of the desired size (smaller than the beam size) and shape that encompasses the stereotaxic coordinates of the cortical area to be lesioned. Use an autoclave or soak in sterilizing solution to sterilize the masks between experiments (recommended). Alternatively, use black electrical tape for masking.

2. Put on laser safety alignment eyewear and set up an argon ion laser and optical components as shown in Figure 4.4.1 on a surgical table or other suitable nonreflective surface. Connect the laser to a high-capacity indoor refrigerated, recirculating cooling unit. Align all components so that the laser beam hits the center of the lens and mirror to maximize the efficiency, precision, and reproducibility of irradiation. Use the optical rail (Fig. 4.4.1A) to move the plano-convex focusing lens within focal length distance of the mirror for maximum focusing power (Fig. 4.4.1B)

    CAUTION: *To allow safe visualization of the laser beam during all initial alignment procedures, the use of special argon alignment safety eyewear (2 to 3 OD; 488- to 515-nm wavelength) is highly recommended. Decreasing the laser power just enough to see the beam is an extra, added precaution. After alignment is complete, high-optical density argon laser safety eyewear should be worn at all times for all other laser operations. Never look or stare directly into a laser beam even with the safety eyewear on.*

3. While directing the 514.5-nm (green light) incident argon laser beam directly on a white piece of paper on the surgical table beneath the mirror, measure the approximate beam width directly with a ruler (information will be used later for adjustments). Using the kinematic mount with mirror, adjust the incidence angle of the laser beam to $\sim 7°$ from vertical so that the laser beam is perpendicular to the convex skull surface (Fig. 4.4.1 A).

    Prior to use each day, it is recommended to tune the argon laser to its maximal power with the internal tuning prism control buttons (vertical and horizontal adjustment). The appropriate high-optical density argon laser safety eyewear must be worn during tuning operations. Power should be monitored with an external power meter or the integrated power meter of the laser unit. This procedure should be done at the excitation transition closest to the absorption maximum of the photosensitizing dye to be used. Additionally, it is important to maintain the optical components free of dust and fingerprints. Use high-grade acetone to clean the mirror and lens.

4. Place a 275- to 300-g male or female rat in a transparent anesthesia-induction chamber (Plexiglas or plastic) and expose to 4% isoflurane in 30% oxygen/70% nitrogen for 3 min using an anesthetic vaporizer attached to a flow meter. Monitor the respiration rate and degree of anesthesia of the animal for slow, constant breathing and unresponsiveness to pinching of hindlimbs or tail with fingers, signs of an appropriately anesthetized animal.

5. Remove the anesthetized animal from the induction chamber, record body weight, and shave the top of the animal's head and the dorsal part of the base of the tail with electric hair clippers, taking care to avoid shaving the tip of the whiskers, especially if the animals are to be used in sensorimotor behavioral tests that require intact whiskers.

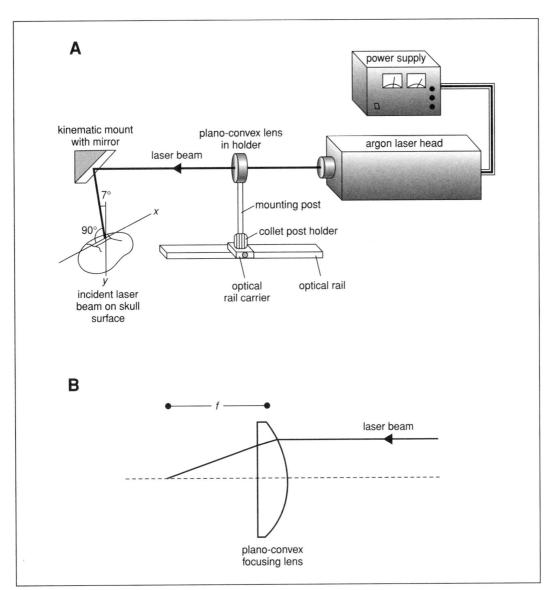

Figure 4.4.1 Schematic illustration of the laser and optical components necessary to create a photochemical cortical lesion. (**A**) Laser components. The argon laser head produces the laser beam. The plano-convex lens and mirror should be aligned so that the laser beam hits the center of each component. The collet post holder secures the mounting post, which holds the lens holder in place. The lens can be moved along the optical rail using the optical rail carrier. Using the kinematic mount with mirror, the incidence angle of the reflected laser beam should be adjusted at 90° from the anteroposterior horizontal plane over the skull (*x* axis) and about 7° from the vertical (*y* axis) so that the convex skull surface meets with the laser beam perpendicularly. (**B**) A plano-convex focusing lens. The distance between the plano-convex lens and the mirror should be within the focal length (*f*) of the lens for maximum power.

6. Insert a thermocouple rectal temperature probe ∼2 cm deep into the rectum of the animal using a lubricant gel and place the animal in a prone position on a feedback-controlled homeothermic heating blanket set to maintain the body temperature at 37°C. Maintain body temperature at 37°C throughout surgery to prevent variation in infarct size. (Hypothermia protects against ischemic brain injury, whereas hyperthermia exacerbates the degree of brain injury).

7. Position the head of the animal (prone position) in a stereotaxic frame equipped with a rat anesthesia nose cone fitted over the incisor bar. Insert the ear bars into the external canal and carefully tighten each bar symmetrically to prevent damage to the eardrums (signified by a popping sound and possible bleeding).

8. Decrease isoflurane to 2% and maintain this level during surgery (check for unresponsive to pinching of hindlimbs or tail with fingers). Apply a lubricant ophthalmic ointment or artificial tears in both eyes, and close eyelids to prevent dehydration during the surgical procedure.

9. Using sterile gauze soaked in antiseptic scrubbing solution, scrub the top of the shaved head by making a continuous spiral movement starting from the center of the surgical field to the outside to decrease the concentration of contaminants in the center of the surgical field. Rinse off excess antiseptic solution with sterile gauze soaked with 70% ethanol using the same inside-out circular motion.

10. Illuminate the surgical field with a dual fiber-optic illuminator (cold-light source or initial temperature monitoring using a thermocouple of the surgical field recommended) and make an ∼2-cm midline incision over the skull with a scalpel, starting between the eyes and continuing posteriorly between the ears. Use a small fan to cool the surgical area if necessary.

11. Expose the skull by gently separating the skin and underlying fascia with sterile cotton swabs. Apply a barraquer retractor to hold the skin back and identify the bone sutures delineating the interaural line, bregma, and the midline (Fig. 4.4.2). To stop superficial bleeding on the bone surface, apply a few drops of diluted styptic pencil solution over the skull and let stand for ∼2 to 5 min while cannulating the tail vein.

12. Scrub the shaved portion of the tail with antiseptic scrubbing solution and 70% ethanol as described in step 9. Use warm water to warm the base of the tail for ∼1 min to dilate the tail veins. To help visualize the veins, tie a rubber band around the base of the tail (area closest to the body) as a tourniquet.

13. To cannulate the lateral tail vein, slowly insert a sterile 24-G, 1.9-cm intravenous catheter needle at a 25° angle into the tail, ∼1 cm below the tourniquet. Tape the catheter in place and gently remove the tourniquet and the needle guide from the catheter. Let the catheter fill with blood and connect a 1-ml syringe filled with 0.2 ml sterile saline. To ensure correct catheter placement, slowly flush with 0.1 ml saline. There should be no resistance.

    Although it requires some initial practice, tail vein injection is preferred because it is less invasive and more rapid than femoral or jugular injections, which can interfere with subsequent locomotor behavioral testing. To prevent blood vessel damage and subsequent tail necrosis, it is important that the tail vein cannulation be performed at the very base of the tail (as close as possible to the body) where the vein diameter is larger.

14. Clean and dry the skull surface using sterile cotton swabs. Using the bone sutures as a stereotaxic guide (Fig. 4.4.2 A; also Paxinos and Watson, 1986), place a mask (made in step 1) with the desired aperture exactly over the brain area to lesion (Fig. 4.4.2B) to allow for more accurate and reproducible alignment of the laser beam. Use a small amount of superglue to temporarily fix the mask to the skull. Alternatively, use black electrical tape as a mask to protect the brain regions outside the area to be lesioned.

15. *Optional:* To optically smooth the surface of the dry skull, apply a small drop of mineral oil in the center of the shield's aperture.

16. Turn off the fiber-optic illuminator over the surgical field and reduce isoflurane to 1.5% (to prevent sudden death if the photosensitizing dye, e.g., Rose Bengal, is associated

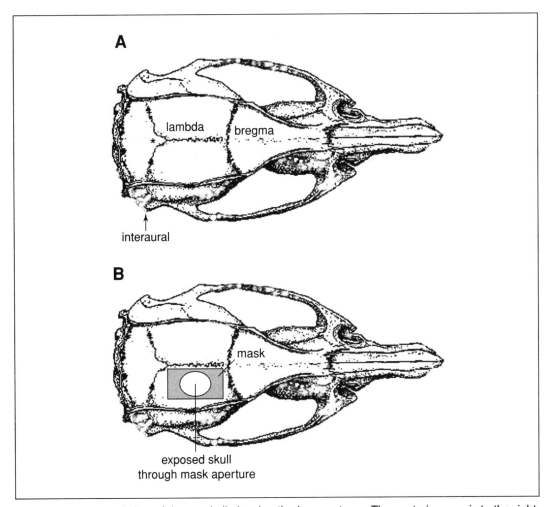

Figure 4.4.2 Dorsal view of the rat skull showing the bone sutures. The posterior area is to the right. (**A**) The stereotaxic positions of bregma, lambda (indicated by the asterisk), and the plane of the interaural line are illustrated. The distance between bregma and the interaural line is ~9 mm (adapted from Paxinos and Watson, 1986). (**B**) The stereotaxic coordinates are used to align the mask's aperture over a cortical brain region of interest. The laser beam can then be easily aligned with the mask's aperture for more precise lesioning of select cortical regions.

with hypotension). Put on the laser safety alignment eyewear. Open the laser aperture and reduce the laser power to a minimum (enough to visualize the laser beam).

17. Align the head of the animal so that the laser beam irradiates the center of the shield's aperture. Using the kinematic mount with mirror, adjust the incidence angle of the laser beam to ~7° from vertical so that the laser beam is perpendicular to the convex skull surface (Fig. 4.4.1A). Close the laser aperture.

18. Load a 1-ml syringe with 0.275 to 0.300 ml photosensitizing dye solution (minimize light exposure of the dye). Determine appropriate concentrations for the desired size of cortical injury in pilot studies. For example, for a dose of 20 mg/kg Rose Bengal, make a stock solution of 20 mg/ml and inject 0.1 ml/100 g animal weight. Remove air bubbles from the syringe and connect it to a syringe pump.

19. Flush tail vein catheter with 0.05 to 0.1 ml sterile saline to assure catheter patency. Remove the saline syringe and attach the dye syringe to the catheter. Do this carefully to prevent the introduction of air bubbles in the syringe or the catheter.

20. Remove laser safety alignment eyewear and put on high-optical density laser safety eyewear. Open the laser aperture and increase laser power to 200 mW/cm^2 of irradiated area.

    CAUTION: *High-optical density argon laser safety eyewear must be worn during laser operation and not the laser safety alignment eyewear, which has a much lower optical density and reduces light exposure less efficiently.*

21. Using the syringe pump (recommended for slow, steady injection), immediately begin the infusion of dye over a 1- to 2-min period. Continue the irradiation of the skull with the laser beam at 200 mW/cm^2 for 5 to 10 min, depending on the desired lesion size and depth. Confirm successful dye administration and intraluminal photochemical reaction by the decrease and eventual disappearance of bright yellow or orange fluorescence through the translucent bone over the region of the skull illuminated by the laser (observed with high-optical density laser safety eyewear). Close laser aperture.

    Slow dye administration over a 1.5- to 2-min period is recommended to prevent changes in blood pressure with Rose Bengal. Erythrosin B is not associated with fluctuations in blood pressure and can be injected over a 1-min period. The gas anesthetic should be maintained at ~1.5% during Rose Bengal infusion and then returned to ~2%.

22. Include the following control groups:

    untreated animals (no dye, no laser),
    animals receiving vehicle (sterile saline instead of dye) and subjected to laser irradiation of the exposed skull (to control for potential heat damage to the brain), and
    animals receiving a scalp incision and sensitizing dye injection without laser irradiation of the exposed skull (to control for potential interactions of neuroprotective drugs with the sensitizing dye).

23. Remove the shield with a pair of hemostats and the barraquer retractor. Clean the surface of the skull and the skin with sterile saline and cotton swabs. Use sterile 4-0 or 3-0 braided-silk or nylon suture to close the skin incision and then apply a topical antibiotic ointment around the wound to prevent infection. In addition, apply a topical anesthetic cream along the sutured skin to reduce postoperative pain.

24. Carefully remove the tail vein catheter and the rectal temperature probe. Discontinue anesthesia and remove the animal from the stereotaxic apparatus. Place the animal at the bottom of a cage fitted with a prewarmed heating pad for ≥30 min or until fully recovered from the anesthesia.

    Body temperature should be maintained close to 37°C during recovery from surgery to prevent variation in infarct size. Hypothermia protects against ischemic brain injury whereas hyperthermia exacerbates the degree of brain injury.

25. At an appropriate time point after surgery and any necessary testing (typically, 2 to 7 days after surgery), determine the infarct size. Deeply anesthetize animal with 4% to 5% isoflurane or with an intraperitoneal injection of 80 mg/kg ketamine and 8 mg/kg xylazine until the animal becomes unresponsive to pinching of the tail or hindlimb with fingers (~3 to 4 min). If carrying out assessment with fixation (step 26b) inject 200 U/mg heparin.

For assessment without fixation

26a. Remove rat brain (Gerfen, 2006) and immediately freeze by dropping the intact tissue in a dry ice/isopentane bath (−40°C for ~1 min). Store frozen brain at −70°C until analysis.

27a. Cut 30-μm coronal brain sections on a cryostat (Gerfen, 2006) across the whole brain to determine the volume of infarct. Mount sections on precoated glass microscope slides and dry sections for ≥1 hr.

Table 4.4.1 Troubleshooting Guide for the Photothrombotic Cortical Lesion Model in Rats

Problem	Possible cause	Solution
Animal responds to earbar insertion, skin incision, or tail vein cannulation	Anesthetic level too low	Increase anesthetic percentage
Animal stops breathing during anesthetization	Anesthetic level too high	Decrease anesthetic percentage
Animal stops breathing during Rose Bengal infusion	Anesthetic too high (Rose Bengal can produce hypotension)	Decrease anesthetic percentage during dye infusion to prevent apnea
	Infusion of Rose Bengal too quick	Infuse dye over a longer period (typically 1.5–2 min); use infusion pump
	Concentration of Rose Bengal too high	Reduce the dose of Rose Bengal (typically 10–20 mg/kg)
	Tail vein dye infusion failed	Ensure tail vein cannulation is patent
Lesion is very superficial and/or smaller than expected; or no brain damage; or skull does not show a yellow glow during laser irradiation	Dye concentration too low	Check dose, weight of animals, and volume injected; verify the concentration of dye solution; check that the dye is completely dissolved and only use filtered dye solution to prevent microvasculature clogging
	Photosensitizing dye degraded	Make fresh dye solution (stock dye solution is generally stable for several months if stored in the dark at ambient temperature)
	Laser power too low	Adjust irradiation laser power to 200 mW/cm^2; before experiments, tune laser beam to maximize laser power output
	Inappropriate laser wavelength	Use a multiline laser and adjust laser wavelength close to that of the absorption peak of the dye
	Laser beam is misaligned	Align the laser beam so that it hits the center of the lens and mirror; carefully align the center of the beam with the center of shield aperture on the skull; also ensure that the beam path is not obstructed
	Optical components dirty	Clean lens and mirror with high-grade acetone
Lesion is very white, irregular and/or larger than expected	Laser power too high	Adjust irradiation power to 200 mW/cm^2
	Brain tissue overheating	Decrease laser power and check brain tissue temperature with thermocouple probe inserted under skull, over the irradiated brain area; use a small fan over the skull surface
	Dye concentration too high	Decrease dye dose
	Laser irradiation too long	Decrease time of laser irradiation ($\leq$10 min)
Variable lesion location	Inconsistent placement of shield and/or laser beam over the skull	Use stereotaxic coordinate for more accurate placement of shield; make sure the center of laser beam (highest intensity) is aligned with the center of the shield aperture; make sure animals are same strain and age
	Size and shape of shields used are different	Reuse the same shield (brass is best) for all animals

28a. Stain sections with cresyl violet (*UNIT 4.3*) and determine the total volume of infarcted brain tissue (*UNIT 4.3*).

For assessment requiring tissue fixation

26b. Perfuse rat through the left ventricle, via the ascending aorta, with a flow rate of 20 to 50 ml/min (Gerfen, 2006) with 150 ml saline (or until perfusate becomes clear, without blood) followed by 250 ml cold 4% paraformaldehyde prepared fresh in 0.1 M phosphate buffer (Gerfen, 2006).

27b. Remove brain, postfix in cold 4% paraformaldehyde for ≤5 hr and prepare brain for histological evaluation (*UNIT 4.2*).

> *Overfixation of brain tissue may result in loss of antigenicity of several proteins and makes the tissue less permeable to antibodies or in situ probes (Liang et al., 2001).*

28b. Cut coronal brain sections at predetermined stereotaxic levels and determine infarct size as described (step 28a) or perform desired histological procedures such as immunohistochemistry or in situ hybridization (Liang et al., 2001).

References: Hecht, 1993; Laser Institute of America, 1993; Watson et al., 1985, 1995; Watson 1988;

Contributor: Marcelle Bergeron

UNIT 4.5

Traumatic Brain Injury in the Rat Using the Fluid-Percussion Model

BASIC PROTOCOL

Materials

Male Sprague-Dawley rats (250 to 300 g)
Halothane
Oxygen
Surgical skin cleanser (e.g., Betadine)
Artificial Tears (The Butler Co.)
Cresyl violet dye (optional)
Gel foam
Cyanoacrylate glue
Saline: sterile 0.9% (w/v) NaCl

Bone wax
Fluid-percussion device (Virginia Commonwealth University) consisting of:
    70-cm pendulum fitted with a 4.8-kg hammer head
    60-cm-long × 4.5-cm-diameter Plexiglas cylindrical fluid reservoir
    Cork-covered piston mounted on O rings bound to one end of the reservoir
    Transducer mounted to the other end of the reservoir
    5-cm-long × 5-mm-diameter metal tube connected proximally to the transducer, with a male Luer-Lok port at distal end
Oscilloscope with integrator (Tektronix or equivalent)
Strip chart recorder with pressure transducer (Gould Instrument Systems or equivalent)
4-way stopcock with Luer-Lok fittings

Pressure tubing: 4-foot (1.22-m) clear pressure-monitoring line with a male Luer-Lok port at one end and a female Luer-Lok port at the other end
Transducer calibration system (Spectramed or equivalent)
Carpenter's bubble level
Sharp knife
27-G, 0.5-in. (1.27-cm) sterile, disposable hypodermic needle
Fine file
Vaporizer
Vacuum trap for anesthetic vapors
Warming pad
Rectal temperature probe (Mallinckrodt Medical or equivalent)
Electric fur clippers
Scalpel and blade
Skin stapler and 9-mm staples
Stereotaxic head frame for rats (ASI or equivalent)
Skin marker
Electric hand drill (Dremel or equivalent) with a 5-mm trephine drill bit
Forceps, microdissecting

1. Anchor a fluid-percussion device (Fig. 4.5.1) to a flat immovable surface and connect an oscilloscope to the transducer with the attached strip chart recorder. Attach a 4-way stopcock to the metal tube located at the transducer end of the reservoir. Attach pressure tubing to the stopcock. Fill the device with sterile water. Be certain that all connections are tight and there are no air bubbles or leaks.

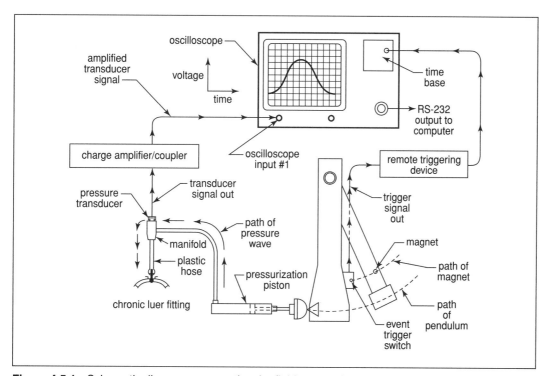

Figure 4.5.1 Schematic diagram representing the fluid-percussion model of traumatic brain injury. The levels of injury severity are confirmed using an oscilloscope connected to the percussion device. The pressurized piston and the plastic hose fitted to the animal's skull are filled with sterile water. Injury is induced when the pendulum strikes the piston, causing the water to move along the tube and onto the exposed dura mater. Figure courtesy of Dragonfly Research and Development, Inc., from their online Installation and Setup Manual.

2. Turn on oscilloscope and strip chart recorder. Attach a transducer calibration system to the strip chart recorder. Calibrate the recorder for a range of pressures from 0 to 2280 mmHg (3 atm or 303.87 kPa). Attach the distal end of the pressure tubing emanating from the fluid-percussion device to the pressure transducer connected to the strip chart recorder.

3. Using a carpenter's bubble level, verify that the pendulum arm is 90° perpendicular to the ground at the rest position. Also check that the pendulum head just abuts the piston. Check reservoir and all connections and tubing for air bubbles and remove any that are discovered by disconnecting the distal end of the tubing and forcing fluid through the reservoir and tubing until they have been expelled.

4. Place the pendulum at a desired starting height and then release the pendulum, allowing it to fall freely and strike the piston. Be certain to catch the pendulum on its return so as to avoid an inadvertent second strike. Identify the angles that will produce pressures of 0.5, 1, 1.5, 2, 2.5, and 3 atm (50.65, 101.29, 151.94, 202.58, 253.23, and 303.87 kPa). Simultaneously record the voltage generated for each of these forces using the oscilloscope. Set the pendulum at the appropriate angle calculated from the calibrated measurements.

   The equipment must be maintained so that the pendulum arm swings freely.

   It is critical that the device be calibrated so that the force resulting from a particular starting pendulum height will be precisely known. Calibration should be repeated before every experiment. Validating oscilloscope data must also be obtained during every experiment.

   There are three levels of injury: mild, moderate, and severe. When the animal is normothermic, mild injury occurs following a percussion of 0.1 to 1 atm (10.13 to 101.29 kPa), moderate after 1.5 to 2.0 atm, and severe after 2.5 to 3 atm. Each of these force levels yields increasing levels of systemic and neurologic dysfunction, histopathologic changes, and decreasing survival.

5. Using a sharp knife, cut off the hub of a 27-G sterile, disposable hypodermic needle 7 mm from the top of the female Luer-Lok port. Discard the remnant needle tip. Smooth the cut edge using a fine file.

6. Anesthetize a 250- to 300-g male Sprague-Dawley rat with a mixture of oxygen and 1% to 3% halothane using a vaporizer and a vacuum trap. Place rat prone on a warming pad and insert a rectal temperature probe. Adjust the warming pad temperature as needed to maintain normothermia.

   Each animal's body temperature must be accurately measured and maintained because it is well known that hypothermia can reduce the severity of injury and hyperthermia can exacerbate injury.

7. Constrain head in a stereotaxic head frame using blunt-tip ear bars. Apply Artificial Tears to the eyes to avoid dry-eye. Shave the scalp clean of fur and scrub the skin with a surgical skin cleanser.

8. Make a 1.5-cm sagittal incision from the midpoint between the ears towards the nose. Reflect the overlying skin and muscle to expose the cranium. Carefully detach the temporalis muscle from the skull to assure adequate exposure when gluing the hub to the skull.

9. Use a skin marker (or equivalent) to mark the skull at 2 mm lateral to the central suture and 3 mm posterior to the bregma suture (*UNIT 4.7*) and to position the hole parasagitally over the cerebral cortex. Locate the burr hole at the same site on each animal to ensure that the same part of the brain is injured.

10. Use an electric drill with a 5-mm diameter trephine drill bit to gently drill a shallow, circular track with the mark from step 9 as its center. Do not make more than a slight track

in the skull, as friction from the drill may cause heat injury to the brain. Use a manual trephine (5-mm outer diameter) to continue drilling through the skull to the depth of the skull's inner table (identified by its distinct pink appearance). Take care when applying downward pressure as the rat's skull is thin.

11. Using microdissecting forceps, carefully remove the round piece of bone. Arrest all bleeding. If desired, cut small pieces of gel foam, wet with saline, and place them on the bleeding area. Rewet the foam pieces before removing them. Be sure the dura is intact as determined visually, by the absence of pinkish or bloody cerebrospinal fluid seeping up through the craniotomy, or by gross pathological examination upon sacrifice.

12. Introduce the head cannula through the hole (fitting snugly) until it abuts the dural surface. Be sure that the cannula and skull surface are completely dry. Affix the cannula to the skull using several thin layers cyanoacrylate glue and allowing the glue to dry completely (∼5 min) between applications.

13. Fill the head cannula with sterile saline, making sure there are no air bubbles. Optionally, add a few drops of cresyl violet dye to the saline to facilitate observation of any leaks around the burr hole site during the injury.

14. At the resting position, position the pendulum at 90° perpendicular to the ground and verify that it just abuts the piston. Check the reservoir and all connections and tubing for air bubbles and remove any that are discovered. Remove bubbles by disconnecting the distal end of the tubing from the rat and forcing fluid through the reservoir and tubing until all air bubbles have been expelled. Reconnect the tubing to the rat.

15. Check that the tubing from the percussion device is completely filled with water and attach it to the head cannula. Hold the head cannula with forceps and twist the tubing on by hand while applying a slight degree of counter-twisting with the forceps to prevent the application of excessive torque, which would cause the head cannula to detach from the rat's skull. Alternatively, carefully align the tubing with one hand while twisting the female Luer-Lok end onto the cannula. Stop when resistance is felt.

    It is recommended that one person hold the cannula and tube in place while another person drops the pendulum.

16. Place the pendulum at the starting position (height based on the intended severity of the injury). Check to make sure there is an open path from the piston through to the cannula. Release the pendulum, allowing it to fall freely and strike the piston. Be sure to catch the pendulum on its return so as to avoid an inadvertent second strike. Check the oscilloscope to confirm that the proper force was applied. Inspect the rat's head for any evidence of leaking saline.

17. Remove the head cannula from the rat, fill the burr hole with bone wax, and staple the incision closed. Remove the animal from the stereotaxic head frame and turn off the anesthesia. Allow the animal to recover on a heating pad, but remove the rectal temperature probe before it fully regains consciousness.

18. Evaluate all animals according to the established testing parameters outlined in Figure 4.5.2 and 4.5.3.

References: Dixon et al., 1987; Lighthall et al., 1989; McIntosh et al., 1989; Toulmond et al., 1993

Contributors: Geoffrey S. F. Ling, Eleanor Y. Lee, and Audrey N. Kalehua

If unable, score 1. If able, score 0.

Inability to exit from a circle 50 cm in diameter when placed at the center. *Rat is considered to have exited the circle when the forelimbs are both outside. Rat is recovered in the center of the circle.*
    Within 30 min after injury   _____
    Within 60 min after injury   _____
    >60 min after injury   _____

Loss of righting reflex. *Rat is recovered lying on its left side and should right itself.*
    For 20 min after injury   _____
    For 40 min after injury   _____
    >60 min after injury   _____

Hemiplegia: inability of the rat to resist forced changes in position. *Rat is pushed back and forth laterally by the shoulders. It should resist equally in both directions. Variations in resistance are fairly easy to detect.*   _____

Flexion of hind limb when raised by the tail. *Rat should extend both hind limbs and reach upwards. Hind limbs should be straight, not flexed.*   _____

Inability to walk straight. *Rat can be enticed to walk with a hind limb pinch, food, or water.*   _____

Inability to move   _____

Loss of startle reflex. *Rat should flinch heavily in response to a loud noise about 20 cm above the head.*   _____

Loss of pinna reflex. *The external auditory meatus is touched with a Q-tip. The rat should shake its head.*   _____

Loss of seeking behavior. *A normal rat will explore the area and sniff unknown objects. A rat with a moderate disability or more will receive the point.*   _____

Prostration. *If prostrating, score 1. If not, score 0.*   _____

Loss of placing reflexes. *Rat is raised by the tail and placed back on the ground. Each limb should reach for the ground and should place on the floor with the palm down. The limb should not be tucked close to the body.*
    Right forelimb   _____
    Left forelimb   _____
    Right hind limb
    Left hind limb

Balance beam. *(1.5 cm wide) Rat is scored on how long it can balance.*
    < 20 sec.   _____
    < 40 sec.   _____
    < 60 sec.   _____

Beam walking. *Rat can be enticed to walk with food, water, or a hind limb pinch.*
    Failure on 2.5 cm wide beam   _____
    Failure on 5.0 cm wide beam   _____
    Failure on 8.0 cm wide beam   _____

NIH stroke severity score.
Based on observed deficits, a functional grade is assessed.   _____

Grade 0: Normal
Grade 1: Lethargy
Grade 2: Clear signs of paresis in at least one limb but able to walk
Grade 3: Severe paresis/paralysis, unable to walk
Grade 4: Dead

Figure 4.5.2 Neurological Severity Score (NSS).

> *In each category rat is scored on a scale of 0-4 with 0 being non-functional and 4 being normal.*
>
> a) **Forelimb Flexion upon suspension by tail.** *Rat is held by the tail. Forelimbs should extend towards the floor.*
>   - 4. normal: *Rat extends both forelimbs equally, reaching for the floor*
>   - 3. slightly impaired: *There is slight forelimb flexion*
>   - 2. moderately impaired: *There is moderate forelimb flexion*
>   - 1. severely impaired: *There is severe forelimb flexion*
>   - 0. non-functional: *Forelimb is tucked close to the body*
>
> b) **Decrease in resistance to lateral pulsion.** *Rat is pulled by each limb.*
>   - 4. normal: *Rat should resist equally when pulled to each side*
>   - 3. slightly impaired
>   - 2. moderately impaired
>   - 1. severely impaired
>   - 0. non-functional
>
> c) **Circling behavior upon spontaneous ambulation.** *Rat is placed on the ground and should be able to walk straight.*
>   - 4. normal: *Rat can walk straight even if it does partially circle*
>   - 3. slightly impaired: *Rat always walks to the paretic side*
>   - 2. moderately impaired: *Rat shows partial circle and always walks to the paretic side*
>   - 1. severely impaired: *Rat circles to the paretic side when attempting to walk*
>   - 0. non-functional: *Rat cannot walk, can only spin*
>
> d) **Ability to stand on an inclined plane.** *Rat is placed on the inclined plane and must stay on for 5 sec.*
>   - 4. normal: *ability to stand on 45° to 50° plane*
>   - 3. slightly impaired: *ability to stand on 40° to 45° plane*
>   - 2. moderately impaired: *ability to stand on 35° to 40° plane*
>   - 1. severely impaired: *ability to stand on 30° to 35° plane*
>   - 0. non-functional: *ability to stand on <30° plane or no ability*
>
> e) **Open Field Activity/Exploratory Behavior**
>   - 4. normal: *Rat explores the area, walking around and sniffing*
>   - 3. slightly impaired: *Rat sniffs and explores but not to a normal degree*
>   - 2. moderately impaired: *Rat either does not sniff or does not explore at all.*
>   - 1. severely impaired: *Rat neither sniffs nor explores*
>   - 0. non-functional: *Rat does not move*

Figure 4.5.3 Neurobehavioral Scale (NBS).

UNIT 4.6

Models of Neuropathic Pain in the Rat

NOTE: All protocols using live animals must be reviewed and approved by the Institutional Animal Care and Use Committee (IACUC). The ethical implications of pain research in general, and the creation of a neuropathic pain syndrome in particular, must be considered with the utmost care. In addition to the regulations of the National Institutes of Health and applicable U. S. federal law (or their foreign equivalents), adherence to the guidelines for pain research in animals adopted by the International Association for the Study of Pain (Zimmermann, 1983) is strongly recommended.

BASIC PROTOCOL 1

THE CHRONIC CONSTRICTION INJURY (CCI) MODEL OF NEUROPATHIC PAIN

Materials

    Saline, sterile
    Adult male rats (e.g., Sprague-Dawley)
    Anesthetic (e.g., sodium pentobarbital)
    Topical disinfectant (e.g., Betadine)
    Rat chow (e.g., phytoestrogen-free rat chow)

    4-0 chromic gut suture
    Animal clippers
    Animal heating pad
    ∼4-cm diameter rod (e.g., rolled-up paper toweling or gauze)
    Scalpel
    Rat-toothed forceps
    Blunt-tipped scissors
    Dissecting microscope
    No. 5 jeweler's forceps
    Microscissors
    Blunt-tipped curved forceps
    Sutures to close surgical incision
    Cages with solid floors and soft bedding (e.g., sawdust as opposed to corn cobs or coarse wood chips)

1. Prior to surgery, prepare the 4-0 chromic gut suture (required type) by cutting it into lengths of 6- to 9-cm and quickly immersing them in sterile saline.

2. Anesthetize the animal, e.g., with sodium pentobarbital i.p. (50 mg/kg). Keep the animal on a heating pad with feedback control from a rectal thermometer until it recovers from the anesthesia to avoid both hypothermia and hyperthermia during the procedure.

   Ketamine should be avoided because it is an N-methyl-D-aspartate (NMDA) receptor antagonist. These receptors are likely to play a significant role in the production of the neuropathic pain syndrome via an induction of the central sensitization phenomenon. Even a step as seemingly trivial as infiltrating the incision with lidocaine can affect the development of neuropathic pain symptoms (Dougherty et al., 1992).

3. Closely clip the fur on the posterior thigh and swab the skin with Betadine or a similar disinfectant. Fix the anesthetized rat on its side with the vertebral column straight and the hind leg positioned at a 90° angle relative to the backbone (i.e., the leg is neither pulled forward nor backward). Place a short rod, ∼4-cm diameter (e.g., rolled-up paper toweling or gauze), beneath the thigh to give support for the dissection.

4. Make a 3- to 4-cm incision down the center of the posterior thigh, starting at the sciatic notch (easily palpable by running a finger down the vertebral column) and ending just proximal to the knee. Make the incision parallel to the long axis of the femur and down the center of the biceps femoris muscle.

   Note that the rostral border of biceps femoris corresponds to the fascia overlying the femur, while the caudal border corresponds to the back edge of the thigh where it attaches with a sheet-like tendon that attaches along the proximal two-thirds of the tibia and actually covers the knee and most of the calf muscle.

5. Using rat-toothed forceps, pick up a fold of the biceps femoris muscle near the top of the incision and carefully pierce the full thickness of the muscle with blunt-tipped scissors (remember that the sciatic nerve is directly beneath the site being pierced). Using blunt dissection technique, spread the wound made through the muscle until the sciatic nerve can be visualized. Insert one blade of the blunt-tipped scissors into the muscle wound, angle the point slightly upwards, and then cut through the muscle in a line that is also parallel to the long axis of the femur.

6. Under a dissecting microscope, use no. 5 jeweler's forceps to pick up the fascia on one side of the nerve (not atop the nerve, to avoid damaging it) and cut down alongside the nerve with microscissors; repeat on the other side. Introduce curved, blunt-tipped forceps beneath the nerve. Do not isolate the nerve by poking forceps beneath it and spreading them; this stretches the nerve too much. Take care not to damage the branch of the femoral artery that supplies the nerve's circulation (usually present near the top of the exposure, but often difficult to find). Free as much length of nerve as possible proximal to its trifurcation. In a small number of animals, the tibial, peroneal, or sural branch will leave the common sciatic nerve much more proximally than normal—discard these animals.

7. Introduce the curved blunt-tipped forceps beneath the nerve. Without lifting the forceps (which would stretch the nerve), use the other hand to advance the end of a piece of 4-0 chromic gut suture until it is within the jaws of the forceps, then grasp it and pull it under.

8. Tie the ligature around the nerve with a single loop (not the non-slip double loop that a surgeon would generally use) that is initially very loose. Then, grasp the two ends very close to the loop and tighten until the loop is just barely snug, gently pulling the ligature up-and-down a short stretch of nerve while tightening the knot to the point where the ligature will not slide smoothly along the nerve.

9. Place a second single loop atop the first to complete the knot (the procedure is neater if care is taken to form a square knot). Use care when tying the top loop because the bottom loop will often become too loose or too tight in the process. If it does, loosen the knot slightly, grasp the "arms" of the knot between the bottom and top loops, and readjust the tightness of the bottom loop. The finished knot will often flop in one direction or another and create a slight kink in the nerve. Ignore this unless the kink looks severe, then re-tie.

10. Trim the two ends ("tails") that remain after tying the knot to ~0.5 mm. Place a total of four ligatures with ≤1 mm between them. Close the wound in layers, using resorbable suture for the muscle layer and suture or wound clips for the skin.

    It is important to note that tying the ligatures is not the immediate cause of the requisite nerve damage; the actual damage is created when the nerve swells and self-strangulates beneath the ligatures. If the ligatures are tied too loosely, little or no edema is evoked and the sensory abnormality is either absent or of small magnitude and short duration. If the ligatures are tied too tightly, the subsequent edema causes a total or near total transection of the nerve and the hind paw becomes insensate. Note also that because the ligatures are not the immediate cause of damage, the time of onset of the nerve damage is not known with exactness. As described in Bennett and Xie (1988), the nerve is definitely constricted by 24 hr post-surgery.

11. *Optional:* Create a contralateral control by duplicating the muscle dissection as closely as possible in the opposite leg without placing any ligatures (not desirable in all cases).

12. Allow the animal to recover in a warm, temperature-controlled setting in a cage with a solid floor and soft bedding (e.g., sawdust as opposed to corn cobs or coarse wood chips). Do not house them in cages with wire mesh floors.

    It is possible that attempts to create all three models of neuropathic pain may sometimes evoke autotomy, i.e., self-injurious biting of the toes and paw. In most experimental contexts, the response to autotomy should probably be to euthanize the animal promptly.

BASIC PROTOCOL 2

THE PARTIAL SCIATIC LIGATION (PSL) MODEL OF NEUROPATHIC PAIN

Additional Materials (*also see Basic Protocol 1*)

Ophthalmic needle with attached suture (3(8 curved, reversed-cutting mini-needle attached to 8-0 silicone-treated silk)

1. Anesthetize and clip the animal as described in Basic Protocol 1, steps 2 and 3. On a heating pad, fix the anesthetized rat on its ventral side with its hind limbs splayed.

2. Locate the sciatic notch by palpation along the vertebral column and using the scalpel, make a skin incision in the direction of the long axis of the femur, beginning just proximal to the notch and extending distally for ∼2 cm, such that the notch is in the center of exposure.

3. Grasp the muscle just proximal to the notch, pierce with blunt-tipped scissors (note that the sciatic nerve is directly below) and spread the blades of the scissors in the direction of the long axis of the femur, exposing the sciatic nerve as it emerges from beneath the sacroiliac bone and turns to run down into the leg; several large nerves to the thigh will be seen traveling caudally, roughly in parallel with the backbone.

4. Free the dorsal surface of the sciatic nerve of connective tissue at a site near the trochanter, just distal to the point where the posterior biceps semi-tendinosus branches from the sciatic.

5. Using no. 5 jeweler's forceps, grasp the epineurium to stabilize the nerve (without depressing or lifting it). Using an ophthalmic needle with attached suture, pierce the nerve from the side such that ∼50% of the nerve is trapped above the suture. Tightly ligate the suture, effectively lesioning all of the entrapped axons.

6. Close the wound in layers. See Basic Protocol 1, steps 11 and 12, for post-operative care considerations.

BASIC PROTOCOL 3

THE SPINAL NERVE LIGATION (SNL) MODEL OF NEUROPATHIC PAIN

The SNL model is distinct in that the injury is created at the level of the spinal nerve (L5 and L6), where the dorsal and ventral roots join distal to the dorsal root ganglia, but proximal to the lumbar plexus where the spinal nerves sort themselves into the various peripheral nerves.

Additional Materials (*also see Basic Protocol 1*)

Small rongeur (e.g., 2 mm jaw-width)
6-0 silk suture

1. Anesthetize and clip the animal as described in Basic Protocol 1, steps 1 and 2. On a heating pad, fix the anesthetized animal on its ventral surface with the hind limbs splayed.

2. Make a ∼3-cm longitudinal incision on the back ∼5 mm lateral to the midline and above the lower lumbar vertebrae and the rostral sacrum, exposing the paraspinal muscles on the left side (which is more convenient).

3. Using blunt-tipped scissors, isolate the paraspinal muscles and remove them from the L4 spinous process to the rostral part of the sacrum, exposing the space ventrolateral to

the articular processes, dorsal to the L6 transverse process, and medial to the ilium. Use a scraper to remove any remaining muscle and connective tissue to visualize the dorsal surface of the L6 transverse process and the rostral part of the dorsal sacrum.

The lumbosacral junction is the joint formed by the L6 and S1 vertebrae. The junction is easily identified because it can be seen to move when the backbone is flexed. The sacrum is formed from the fusion of the S1-S3 vertebrae and it is not jointed. Note, however, that in the young animal, the S1-S2 fusion is relatively weak and can be easily broken.

4. Using a small rongeur, remove the L6 transverse process as completely as possible, avoiding the L4 and L5 spinal nerves running just beneath it.

   It is easier to manipulate the spinal nerve after removal of the transverse process very close to the body of the vertebrae. Be very careful removing the transverse process because the. Once the transverse process is removed, one can usually see the ventral rami of the L4 and L5 spinal nerves, but in some animals, one must first remove a thin sheet of muscle.

   The L4 spinal nerve runs lateral and ventral to the L5 spinal nerve before they join together distally. There is a great deal of variability between rats in the location where these two nerves join together. Thus, in some animals, the L4 and L5 spinal nerves will have already merged at the location whose exposure is described above. Since L5, but not L4, is to be lesioned, such nerves must be carefully separated before proceeding. It is of the utmost importance to avoid damage to the L4 spinal nerve during this and all other procedures.

   Damage to the L4 spinal nerve can occur with a seemingly minor mechanical trauma (excessive touch, gentle stretch, or slight entrapment within a fragment of the epineurium). Even slight damage to the L4 spinal nerve essentially eliminates the subsequent expression of mechanical allodynia. If an extensive manipulation of the L4 spinal nerve is required to separate it from L5, it is advisable to simply discard the animal. Note also that the dorsal rami of the L4-L6 spinal nerves are located proximal to the ligation sites. Although they are not lesioned intentionally, it is probable that they are damaged during dissection of the paraspinal muscles.

5. Using a piece of 6-0 silk suture, tightly ligate the L5 spinal nerve (thereby lesioning all the axons within it). Also tightly ligate the L6 spinal nerve after using the rongeur to remove a bit of the rostral edge of the sacrum.

   Alternatively, leave the sacrum intact and insert a small glass hook beneath the sacrum and then gently pull the L6 spinal nerve outwards until it can be visualized, where it is then tightly ligated. If the ligature is tied too tightly and the nerve is actually cut, there are a number of points to consider. Tight ligation and transection may not be equivalent. It is conceivable that a tight ligation (1) helps the formation of neuroma, which may contribute to the development of neuropathic pain, and (2) prevents the injured axons from regenerating, which may contribute to a delay in recovery. However, there does not appear to be a big difference in behavior whether the spinal nerve is cut or ligated, at least during the initial postoperative period (1 to 2 weeks; J.M. Chung, unpub. observ.).

6. Close the wound in layers. See Basic Protocol 1, steps 11 and 12, for postoperative care considerations.

SUPPORT PROTOCOL 1

BEHAVIORAL ASSAYS: HEAT-HYPERALGESIA

Heat-hyperalgesia may be assessed with the paw-flick test using an apparatus like that described by Hargreaves et al. (1988). This is available commercially from a number of suppliers including the Department of Anesthesiology, University of California, San Diego; IITC Life Science Instruments (*http://www.iitcinc.com*); and Stoelting (*http://www.stoeltingco.com*). The Hargreaves apparatus is a modification of the tail-flick test (see UNIT 4.9) in which the rat is confined within a compartment made of clear plastic that is set atop a floor made of window glass and a radiant heat source is aimed at the plantar hind paw. The heat source and a timer are controlled by a photocell that receives light reflected from the paw. The nocifensive withdrawal reflex interrupts the reflection and stops the heat and timer. The heat source is adjusted at the beginning of the experiment to yield a paw flick in ∼10 sec. As a guide, if a finger tip is stimulated, one will have about the same latency to feeling a distinct pain sensation.

The rats are sometimes reluctant to place their weight on the affected hind paw and this creates two problems. First, when testing the nerve-injured side, the plantar surface of the hind paw must be in solid contact with the glass floor. If it is too high off the floor, it will receive too little heat; but, oddly, if it is just barely off the floor it will heat more quickly because the glass acts as a heat sink when the skin touches it (Bennett and Hargreaves, 1990; Hirata et al., 1990). Moreover, the everted posture of the affected hind paw turns part of the plantar surface away from contact with the glass. Note that rats are digitigrade and when standing, they will often be "up on their toes," i.e., with their weight distributed forward over the toe pads and metatarsal pads, and with the heel very slightly elevated. When at rest, the rat will be "back on its heels," i.e., with its weight shifted backwards and the whole plantar surface in contact with the floor. Therefore, it is critical to be certain that only that skin that is in good contact with the glass is being stimulated. Arrange the apparatus to afford a clear view of the plantar surface of the hind paw. Stimulate the fat part of the heel that lies distal to the medial and lateral pads (avoiding the long skinny calcaneus), because this part of the paw is least affected by the eversion. It is not useful to stimulate the pads because they are thickly cornified and relatively insensitive. Do not stimulate the mid-paw region that lies surrounded by the medial, lateral, and metatarsal pads, because the pads keep this skin from contacting the glass.

The second problem arises when testing the opposite side since the rat cannot flick its good paw while it is holding the other paw in the air. Make sure both paws are on the glass before testing. Do not attempt a paw-flick trial if the animal is about to step away, or grooming, or rearing; all four paws must be on the glass if it is to make an unimpeded withdrawal reflex. Rats never seem to tire of sniffing through the nose-high nickel-sized holes in the walls of their containers, and such holes will tend to keep the rat in the proper position for testing.

It is crucial that the temperature of the glass be ∼30°C (the approximate normal skin temperature of a rat's hind paw in a thermally neutral environment). If the glass is colder, it will markedly cool the paw and inflate the withdrawal latency.

It is known that for the tail-flick test, manipulations that change blood flow to the tail (e.g., vasoconstrictor drugs) yield spurious results, simply because a cold tail takes longer to heat to pain threshold and a hot one heats more quickly. Rats that have undergone a procedure to induce neuropathy may have normal, hot, or cold hind paws due to the nerve injury itself. Spurious hyperalgesia has not been detected in rats with hot paws (i.e., no correlation between skin temperature and degree of hyperalgesia), and there is no evidence that the hyperalgesia of rats with cold paws is seriously underestimated (G. J. Bennett, unpub. observ.). It is possible that with the heat source set to yield a baseline latency of ∼10 sec, the rate of heat increase is slow enough to bring the skin of all paws to a near constant temperature well before the pain threshold.

During baseline or post-injury test sessions, the first latency from each side is typically distinctly longer than all subsequent latencies in the session (the same first latency anomaly occurs in the tail flick test). A solution is to measure four to five latencies/side, but discard the first measurements when averaging the data.

The primary outcome measure of the paw-flick test is the mean withdrawal latency of each hind paw. Ancillary data are the percentage of withdrawals accompanied by paw licking, the percentage of rats yielding abnormally prolonged (>10 sec) withdrawals, and the mean duration of the withdrawal response.

As an alternative to the Hargreaves test, one can simply dip the hind paw into hot water of several temperatures that span the normal pain threshold (e.g., 42°, 44°, 46°C) and measure the latency or threshold for withdrawal (Attal et al., 1990). Cold allodynia (see Support Protocol 4) can also be assessed in this manner using ice water (4°C). Note, however, that it is important to attend closely to the position that the animal is held in for these tests. Holding the rat suspended in mid-air evokes a strong bilateral extensor reflex (the "I'm falling" reflex) that will interact strongly with the pain-evoked flexor reflex.

SUPPORT PROTOCOL 2

BEHAVIOR ASSAYS: MECHANO-HYPERALGESIA

Mechano-hyperalgesia may be assessed by two methods. The first, the pin-prick test, is described by Tal and Bennett (1994). The rat is confined within a clear plastic cage set upon an elevated wire mesh floor. The mesh should have holes ~1 cm in diameter; if the holes are larger, the nerve-injured paw will fall through the hole; if smaller, it will be difficult to see. The tip of an ordinary safety pin is pressed against the skin of the fat part of the plantar heel such that the skin is dimpled but not penetrated. Do not use a hypodermic needle because the point is too sharp. The normal response to pin-prick is a nocifensive withdrawal reflex of very small amplitude and short duration; following nerve injury, the response is greatly increased in amplitude and duration, and the animal will frequently lick the stimulated site. The symptomatic paw may sometimes have spots of insensitivity, and it may be necessary to stimulate in two or three places; if this is done, comparable stimulations should be done on the control side. The symptomatic paw should be stimulated only when it is resting or nearly resting on the floor because a strongly flexed limb allows little room for the detection of an additional flexor response.

The primary end point of the pin-prick test is the duration of the withdrawal as measured with a stop watch. An ancillary datum is the percentage of stimulus applications that result in hind paw licking. The normal response is so fast that it cannot be timed by hand; an arbitrary score of 0.5 sec can be assigned in these cases.

The second method uses the Randall-Sellito (1957) device, which consists of a motorized screw that slowly advances a conical stylus against the paw with increasing amounts of pressure. The stylus is placed against a standardized site (e.g., between the 3rd and 4th metatarsi) and the pressure needed to evoke a withdrawal response or a vocalization is designated as the pain threshold. Because the animal is held in restraint and may struggle independent of paw stimulation, the detection of a withdrawal response related to the paw pressure is often difficult. Using vocalization as an end point requires greater pressure, but the response is unequivocal. There is a potential problem with the Randall-Sellito method; if mechano-allodynia (see Support Protocol 3) is present, then the pain threshold of the rat may be exceeded as soon as the stylus contacts the skin. Randall-Sellito paw-pressure apparatuses can be obtained from IITC. Life Science Instruments; Stoelting, and a similar device can be obtained from Somedic Sales AB (*http://www.somedic.com*).

SUPPORT PROTOCOL 3

BEHAVIORAL ASSAYS: MECHANO-ALLODYNIA

Mechano-allodynia is assessed using von Frey hairs. The "hairs" are nylon monofilaments of different diameters that exert defined levels of force when they are pressed against the skin with sufficient force to cause the hair to bend. The set of monofilaments currently in use is known as the Semmes-Weinstein series (Semmes et al., 1960). There are several points to note about von

Frey hairs. First, the stimulus is sometimes reported in units of pressure (Newtons = force × area). This presumes that the area stimulated is defined, i.e., that the entire cross-sectional area of the hair is in equal contact with the skin, that there are no edge effects, and that the elasticity of the skin is negligible; these are all questionable assumptions.

Second, the nylon monofilaments are hydrophilic, and commonly encountered variations in humidity have a surprisingly large effect on their stiffness and the force that they generate. This probably matters little when performing a side-to-side comparison, or if all experimental groups are tested on the same day. But serious error may result if the experiment requires that comparable data be obtained over the course of several days or weeks. In the latter case, the testing room's humidity should be monitored carefully. For any given level of humidity, the force exerted by each hair can be calibrated by pressing it against a pan balance.

Third, the tips of the monofilaments are sometimes chipped such that a sharp point is applied instead of a relatively blunt tip. The tips should be checked with a dissecting microscope and cut flush if jagged.

Fourth, ordinal numbers 18 to 20 cannot be used because they are so stiff that they will lift the paw before they bend. Ordinal numbers 1 to 3 exert such little force that even the most allodynic animal rarely responds to them. Thus, testing uses ordinal numbers 4 to 17, inclusive.

von Frey hairs are applied perpendicular to the skin. Contact should be gentle (no jabbing) and pressure exerted slowly until the hair just starts to bend. A standardized site should be tested (e.g., the fat part of the plantar heel). Several testing protocols and methods of data analysis have been used successfully; for details see Tal and Bennett (1993), Kim and Chung (1992), and Chaplan et al. (1994). In all cases, the primary endpoint is an estimate of the threshold hair (i.e., the weakest hair) that evokes the hind paw withdrawal reflex.

Sets of von Frey hairs are available from: North Coast Medical (*http://www.ncmedical.com*); Somedic Sales AB; and Stoelting. An "electronic" von Frey device has been described recently that consists of a von Frey hair-like probe attached to a pressure gauge (Moller et al., 1998; commercially available from IITC Life Science Instruments, and Somedic Sales AB).

SUPPORT PROTOCOL 4

BEHAVIORAL ASSAYS: COLD ALLODYNIA

Two methods of measuring cold allodynia have been used. In the first method (Bennett and Xie, 1988), rats are placed for 20 min on a metal plate cooled to 4°C by water circulating beneath it. The number and duration of the nocifensive withdrawal reflexes that occur when the animal's symptomatic paw touches the floor are then measured and summed. These values can be compared to those obtained with the metal floor warmed to 30°C, which is approximately normal skin temperature (note that usual room temperature is significantly cooler than skin temperature). The values from the 30°C floor in Bennett and Xie (1988) were significantly different from zero, presumably due to the pain caused by contact with a hard surface, but the values significantly increased when the floor was at 4°C. The cold floor does not evoke any pain-related behaviors in normal rats. This method is time-consuming and exposes the animal to a cold environment, which may be a confounding factor because it will activate the sympathetic thermoregulatory mechanisms.

A better and quicker method for measuring cold allodynia has the rat standing on the elevated mesh floor used for mechanical testing (Choi et al., 1994). A drop (~5 µl) of acetone is expressed from a length of PE 240 tubing connected to a syringe. The acetone is applied via plastic tubing rather than a metal needle. A liquid-metal interface develops insufficient surface tension for the formation of a drop, while this is easily done with the liquid-plastic interface.

The drop of acetone (not the tip of the tubing) is touched to the skin of the fat part of the plantar heel. Surface tension causes the drop to spread along the skin and evaporation causes a sensation that is normally a gentle cooling (note that there is a delay of 1 to 2 sec before the cooling is felt). Normal rats either ignore the stimulus or respond with a very brief and small flick of the paw. On the nerve-injured side, the cooling evokes large, prolonged, and often repetitive withdrawal responses. The endpoint of the acetone test can be either the summed duration of evoked withdrawal responses as measured with a stopwatch, or the percentage of trials that evoke a withdrawal reflex.

References: Merskey and Bogduk, 1994; Zeltzer and Seltzer, 1994; Zimmerman, 1983

Contributors: Gary J. Bennett, Jin Mo Chung, Marie Honore, and Ze'ev Seltzer

UNIT 4.7

Dural Inflammation Model of Migraine Pain

For help in troubleshooting problems related to this protocol see Table 4.7.1 at the end of the unit.

NOTE: For experiments involving oral administration of test compounds, animals should be fasted overnight before experimentation to enhance the speed of oral absorption.

BASIC PROTOCOL

ASSESSMENT OF DURAL EXTRAVASATION IN RATS AS A MODEL OF MIGRAINE PAIN

Materials (see APPENDIX 1 for items with ✓)

    Male Wistar rats, 250 to 350 g, group housed
    50 mg/ml sodium pentobarbital solution (for anesthesia)
    2% (w/v) chlorhexidine surgical scrub antiseptic solution (Fort Dodge Animal Health)
✓ 50 mg/ml Evans Blue solution (for injection)
    Test compound in saline (0.9% NaCl) or other appropriate vehicle
    70% (v/v) glycerin in water

Electrodes (e.g., David Kopf model RNE 300×-50 mm)
Two electrode manipulators for stereotaxic apparatus (e.g., David Kopf model 1460)
Electrode holders (e.g., David Kopf model 1770 or 1771)
Clamping device (e.g., David Kopf model 1772)
Constant-current electrical tissue stimulator (Grass 888 or equivalent)
Electric hair clippers
Stereotaxic apparatus with rat adapter (e.g., David Kopf 1400 series)
Temperature controller with rectal temperature probe and small heating blanket (e.g., YSI model 73A controller with model 402 temperature probe, and a conventional electric heating pad)
Small hemostats
Cotton swabs
Fine-tipped felt pen
Hand-held electric drill (e.g., Dremel model 260)
Small ball-type dental bur (e.g., SS White Burs, model HP1)
Fluorescence microscope with spectrophotometer (e.g., Carl Zeiss)

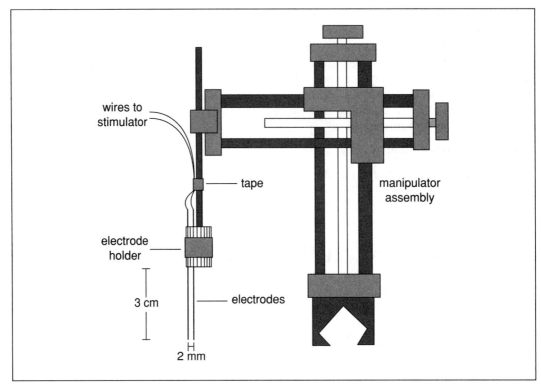

Figure 4.7.1 Arrangement of stimulating electrodes held in stereotaxic electrode manipulator.

1. Arrange the electrodes as shown in Figure 4.7.1. Make sure the electrodes are parallel and vertical with 2-mm separation. Position the tips the same vertical distance from the holder, extended far enough to reach the base of the skull when the rat is in the stereotaxic apparatus, and clamp the electrodes so they extend 3 cm beyond the base of the electrode holder. Connect the two leads from the stimulator to the electrode wires. Attach the insulated wires to the shaft of the electrode holder using adhesive tape (see Fig. 4.7.1), so that any tension, or pulling, on the wires to the stimulator will not be transferred to the electrodes.

2. Weigh the rat, and anesthetize it with an i.p. injection of 65 mg/kg sodium pentobarbital. After a few minutes, test by touching the surface of the eye with a cotton swab. (If the anesthetic depth is adequate, the animal will not blink.)

3. Shave the skin on the dorsal surface of the animal's skull, from between the eyes to the nape of the neck, using the electric clippers, taking care to avoid damaging the ears. Place the animal in the stereotaxic device, and insert and tighten the ear bars. Set the incisor bar 3.5 mm below the intra-aural line and hook the incisors over the bar. Gently lower the nose bar to hold the nose on the incisor bar, taking care not to allow the bar to push down on the nose with too much force and interfere with normal breathing.

4. Insert a rectal temperature probe 4 to 5 cm into the colon of the animal and attach this probe to a thermostatic controller that powers a heating pad upon which the animal rests. Set the controller to maintain the animal's core temperature at 37°C.

5. With the head of the anesthetized animal firmly held in the stereotaxic device, apply an antiseptic solution to the shaved portion of the skin, and make a 2- to 2.5-cm midline incision with a scalpel, extending the incision along the midline from the most caudal aspects of the eyes to an imaginary line drawn between the center of the ears.

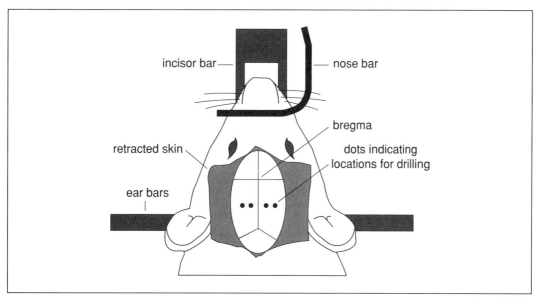

Figure 4.7.2 Rat head held in stereotaxic apparatus with the dorsal surface of skull exposed. The skull landmark bregma is indicated, as are the locations where holes will be drilled to allow electrodes to be lowered to the trigeminal ganglion.

6. Retract the skin. Use four small hemostats to weigh down the edges of skin and scrape the exposed skull with cotton swabs until the fascia is removed and the skull is dry.

   At this point, the lines (sutures) formed by various intersections of the bony plates that compose the dorsal surface of the skull should be visible. The right-angle intersection of the suture that runs in the mid-sagittal plane with a perpendicular suture located ∼1 cm caudal to the eyes is the skull landmark known as bregma. This is the point from which medial-lateral and rostral-caudal measurements are made. Often the bone suture lines that define bregma are irregular and do not meet at a perfect right angle in the mid-sagittal plane. In these cases, a "best guess" localization of bregma must be made by approximating the intersection that would occur if the sutures were more ideal.

7. Secure the uncapped fine-tipped felt pen in the clamping device held by another electrode manipulator. Using the manipulator and stereotaxic frame, carefully position the pen just over bregma and note the rostral-caudal and medial-lateral position as displayed on the stereotaxic rail and manipulator arm respectively.

8. Move the manipulator caudally 6 mm and secure it firmly to the rail. Using the pen and manipulator, make small marks on the dry skull 2 and 4 mm laterally in both directions. These four marks will be 6 mm caudal from bregma and 2 and 4 mm lateral from the mid-sagittal suture (see Fig. 4.7.2).

9. Remove the manipulator holding the pen from the stereotaxic frame, and cap the pen. Using a small hand-held drill equipped with a small ball-type dental bur, carefully drill through the skull at the location of each dot. Make sure the drill enters the skull vertically and the diameter of the hole drilled is only slightly larger than the electrode diameter, i.e., only ∼1 mm deep, just deep enough to penetrate the skull but not damage the underlying cortex.

10. Slide the electrode manipulator holding the two electrodes onto the stereotaxic frame and position the electrodes above the set of two holes over the control (unstimulated) hemisphere. Carefully lower the electrodes into the holes, making note of the vertical

reading on the electrode manipulator when the tips of the electrodes are at the outer surface of the skull. Slowly lower the electrodes into the brain to a depth of 10 mm below the surface of the skull (this will be ~9 mm below dura) to produce damage similar to that in the stimulated side.

11. Next, carefully remove the electrodes and, following the above procedure, lower them into the hemisphere containing the trigeminal ganglion to be stimulated, this time leaving them in place once they are at the proper depth. Immediately absorb any bleeding with gauze pads.

12. Test for proper electrode placement by delivering a single square wave pulse of stimulating current (1 mA, 4 msec duration) through the electrodes, and monitor an ipsilateral jaw twitch using an index finger placed along the ipsilateral side of the rat's head.

13. To inject test compound into the femoral vein shave the inside of the leg, lift the skin overlying the femoral vein with forceps, and make a small (~1 cm long) incision in the skin with scissors. Make the injection into the femoral vein, which is visible through the incision. Alternatively, make intravenous injections in the tail vein. Wait 8 min and slowly inject 50 mg/kg of the Evans Blue solution intravenously.

    Parenteral administration can occur at any interval before trigeminal stimulation. Compounds administered orally should be given when the animal is awake and at least 30 min before anesthesia, to allow time for the compound to move from the stomach to the intestines and for absorption to occur before the anesthesia inhibits GI motility. Test compounds with long half-lives can be administered many hours and even days before anesthesia and subsequent trigeminal stimulation.

14. Two minutes after the Evans Blue injection, electrically stimulate the trigeminal ganglion for 3 min using a constant current of 1 mA applied in a square wave pulse train with a pulse width of 4 msec and a frequency of 5 Hz.

    If animals are not surviving the stimulation period, using a lighter depth of anesthesia will often help.

15. Immediately after stimulation, raise the electrodes out of the brain and remove the animal from the stereotaxic apparatus. Confirm that the trigeminal ganglion was stimulated by noting the presence of blue dye in the ipsilateral ear, eye, eyelids, and ipsilateral portions of the nose and tongue (resulting from inflammation-induced leakage of plasma proteins).

16. Euthanize the animal by cutting the right atrium and injecting 40 ml normal saline at a rate of ~1 ml/sec into the left ventricle of the heart. Before injecting the saline, clamp the descending aorta with a hemostat at the level of the diaphragm to enhance rostral perfusion to clear the blood remaining in dural blood vessels.

    Variations in the interval between trigeminal stimulation and animal sacrifice can result in variations in drug potency. It is therefore important, if one wishes to compare the potencies of various compounds, that identical experimental conditions be used.

17. Using scissors, remove the top of the skull exposing the underlying dura. Carefully collect the supratentorial dural membrane (~0.75 cm in diameter sample) from each hemisphere separately.

    By removing the brain from the cranial vault, one can expose underlying dura, which will adhere to the skull. Using a pointed scalpel, a dural sample can be circumscribed and then gently scraped free from the skull.

18. Rinse the dural samples under a stream of deionized water. This can be easily done by leaning a microscope slide against a rubber stopper at a 45° angle. Hold the sample of dura

against the bottom end of the slide with forceps, and then gently spray it with a stream of water from a squirt bottle for a few seconds to wash off any blood remaining on the tissue surface.

19. Using the forceps, adjust the dural membrane so that it lies flat on the glass surface. Rotate the slide so that the first dural sample is now at the top of the slide. Take the sample from the other hemisphere and rinse it as described above, being careful not to disturb the first sample. Spread this sample flat so the slide is holding two dural samples, one at each end, lying flat on its surface Alternatively, put individual samples on separate slides.

20. Remove any excess water from the slide and allow it to dry (or use a slide warmer set at 37°C to speed the drying process). Once dry, apply a drop of a 70% glycerin solution to each dural sample and coverslip.

21. Measure the dural protein extravasation (the amount of Evans Blue dye in each dural sample) with a fluorescence microscope equipped with a grating monochromator and a spectrophotometer. Use an excitation wavelength of 535 nm and monitor the emission intensity at 600 nm. Avoid sampling fluorescence from dural tissue located near the electrode penetration sites (where the mechanical damage to the dura induces increased extravasation) and from areas containing large dural vessels.

    The authors use a microscope equipped with a computer-controlled motorized stage, which allows automated measurement of fluorescence intensity at multiple points (e.g., measure fluorescence at 25 points arranged in a 5 × 5 matrix with each point separated by 0.5 mm). Use the average fluorescence intensity for these 25 points to calculate the extravasation ratio.

22. Analyze the data and calculate a ratio of the fluorescence intensity of the stimulated side to the intensity of the control side; this number is called the extravasation ratio. In the case of a single-dose experiment, use a simple *t* test to compare groups of animals treated with drug or vehicle.

    Alternatively, multiple doses of a test agent could be given to different groups of animals and a dose-response curve could be generated. In this case, a one-way analysis of variance followed by a post hoc test of confidence intervals (such as the Dunnett's or Tukey-Kramer tests) could be used to determine those points that are significantly different from vehicle.

ALTERNATE PROTOCOL

ASSESSMENT OF DURAL EXTRAVASATION IN GUINEA PIGS

Follow the steps outlined for working with rats (see Basic Protocol), but using male Hartley guinea pigs weighing ~250 to 350 g, and making the following changes at the indicated steps:

2. Anesthetize the guinea pig with an i.p. injection of 45 to 55 mg/kg sodium pentobarbital.

    Keep in mind that guinea pigs are somewhat more sensitive to the anesthetic than rats, so extra care must be paid to maintaining an adequate, but not excessive, depth of anesthesia.

3. Set the incisor bar on the stereotaxic frame at 4.0 mm below the intra-aural line.

8. Make the marks on the skull 4 mm caudal to bregma and 3.2 and 5.2 mm lateral to the mid-sagittal line.

10. Lower the electrodes to a depth of 10.5 mm below dura, or 11.5 mm below the surface of the skull.

Contributors: Lee A. Phebus and Kirk W. Johnson

Table 4.7.1 Troubleshooting Guide for Dural Inflammation Assessment

Problem	Possible cause	Solution
Anesthesia not deep enough after first injection on some, but not all, animals	Absorption and/or metabolism of pentobarbital variable between individual animals	Supplement anesthetic slightly (an additional ~0.1 ml i.p.); try a new animal and slightly increase anesthetic dose; vary the injection site
No jaw twitch on single pulse test stimulation	Trigeminal ganglion not adequately stimulated	Check electrode connections; systematically move electrode placement using new animals for each location
Animals die during stimulation	Anesthesia too deep	Use less anesthetic
	Test compound compromises animal	Lower dose of test compound
Ipsilateral side of control animal's face does not turn blue	Evans Blue injection missed vein	Make sure entire dose of Evans Blue is injected intravenously
	Electrodes misplaced	Systematically adjust electrode placement
	Electrodes not delivering current	Check electrode connections; replace electrodes if necessary
Low extravasation ratio in control animals while ipsilateral side of face is blue	Electrodes slightly misplaced, not stimulating entire ganglion	Increase stimulation intensity; systematically move electrode placement
Stimulator occasionally unable to provide adequate current	Electrodes tips insulated by being jammed into base of skull	Raise electrodes slightly

UNIT 4.8

Animal Models of Painful Diabetic Neuropathy: The STZ Rat Model

BASIC PROTOCOL

INDUCTION OF DIABETES WITH STZ

STZ is injected into food-deprived rats, and the animals are then monitored for blood glucose levels.

NOTE: All protocols using live animals must first be reviewed and approved by an Institutional Animal Care and Use Committee (IACUC). Because of significant ethical implications involved with the induction of a chronic neuropathic pain state, in addition to the regulations of the National Institutes of Health and applicable federal laws, such research should adhere to the guidelines for pain research in animals adopted by the International Association for the Study of Pain (Zimmermann, 1992).

CAUTION: STZ is cytotoxic to both humans and animals. Preparation of this solution should be performed in a class II biosafety cabinet/hood that has been certified for preparation of cytotoxic materials. Prep and injection areas should be covered with disposable absorbent covers. Individuals handling these materials should follow standard precautions and wear a laboratory coat, mask, and gloves.

Materials *(see APPENDIX 1 for items with ✓)*

    Adult rats (e.g., ~225-g Sprague-Dawley; males preferred)
✓ STZ, premeasured
✓ Sodium citrate buffer, pH 5.5
    10% sucrose (Sigma) in tap water (store up to 3 days at 5° to 10°C)

Animal scale
Conical test tube or equivalent with screw top, sterile
Aluminum foil
1-ml TB or insulin syringes with 26- to 28-G needles
Cages with solid floors and soft bedding

1. Deprive adult rats of food (i.e., fast) 4 to 6 hr prior to STZ induction. Weigh all animals using an animal scale.

2. Just prior to use (i.e., within 15 to 20 min), prepare STZ solution as follows: Pour contents of one premeasured microcentrifuge tube of STZ into a sterile conical test tube covered with aluminum foil and containing sufficient sodium citrate buffer, pH 5.5, to yield a 50 mg/ml solution. Mix until STZ is completely dissolved. Store protected from light for no more than 15 to 20 min.

   STZ degrades rapidly in sodium citrate buffer and is sensitive to light.

3. Draw the 50 mg/ml STZ solution into a 1-ml insulin or TB syringe with 26- to 28-G needle so that the final dosage is 45 mg/kg rat. Immediately after filling the syringe (to minimize exposure of the syringe contents to light) inject i.p. Repeat for each animal in the experimental group.

4. House injected rats in cages with solid floors and soft bedding. House no more than two rats per cage and change the cage bedding daily. Supply rats with 10% sucrose water as

the sole water source for 48 hr after STZ injection to protect the rats from the sudden hypoglycemic period that occurs immediately after the lysis of the pancreatic islet cells by STZ.

If possible, diets containing primarily plant protein (e.g., soy, alfalfa) should be avoided. Such diets will contain phytoestrogens that may influence the appearance of neuropathic pain.

5. Measure body weights and blood glucose levels at 72 hr after STZ injection and subsequently at 1-week intervals to identify the onset and continued presence of diabetic hyperglycemia.

    Normal blood glucose levels in the rat range from 60 to 100 mg/dl. Diabetic hyperglycemia is defined as a nonfasting plasma glucose concentration >250 mg/dl. Blood glucose levels in STZ-injected animals will typically range from 300 to 550 mg/dl.

6. Assess painful neuropathy by testing for mechanical allodynia (see UNIT 4.6) and heat hyperalgesia (see UNIT 4.6), starting at ∼4 weeks after induction of diabetes. Always test the animals at the same time of day to minimize diurnal effects. Carefully control the ambient room temperature.

References: Fox et al., 1999; Sima and Shafrir, 2000

Contributor: Thomas J. Morrow

UNIT 4.9

Models of Nociception: Hot-Plate, Tail-Flick, and Formalin Tests in Rodents

A table at the end of this unit (Table 4.9.1) outlines troubleshooting procedures for the methods described in this protocol.

BASIC PROTOCOL 1

MEASUREMENT OF ACUTE PAIN USING THE HOT-PLATE TEST

Materials

Animal: 30- to 35-g male CD mice, group-housed (e.g., Charles River Labs) or 200- to 250-g male Sprague-Dawley rats
Test compounds in saline or other appropriate vehicle
Reference analgesic compound (e.g., morphine, to be given 10 mg/kg, i.p.)

1-ml syringes with 27-G needles
Hot-plate apparatus or metal plate in a controlled temperature water bath
Stopwatch or timer

1. Bring animals to test room and record their body weights. Allow animals to acclimate to test room for 15 to 30 min.

2. Set hot plate to 55°C for mice or 52.5°C for rats (or the temperature at which animals exhibit a near-maximal to maximal antinociceptive response in acute pain assays when dosed with 8 to 10 mg/kg morphine administered i.p.).

3. Prepare doses of test compounds, vehicle (control), and reference agent such as morphine. Inject animals with test compound, vehicle, or reference agent. Inject animals in a staggered fashion to allow a consistent time to pass between treatment and testing for each animal. Randomize the order in which different treatments are administered such that each condition is distributed throughout the duration of the entire experimental procedure.

4. When post-treatment time (e.g., 30 min) has elapsed, begin testing animals. Place a single animal on the hot plate and immediately start a stopwatch or timer. Observe the animal until it shows a nociceptive response (e.g., first licks one of its rear paws) or until the cutoff time is reached. Remove the animal from the hot plate. Record its latency to respond. For animals that do not respond prior to the cutoff time, record the cutoff time (e.g., 30 sec). Repeat for all animals in the order in which they were treated. Make similar measurements for vehicle-treated animals (controls).

5. Perform appropriate statistical analysis on the data. With multiple groups, use an ANOVA followed by post-hoc analysis (if appropriate) to compare the treatment groups to the controls (vehicle treatment).

Data from hot-plate tests are often expressed as percent maximum possible effect (%MPE). %MPE = (test − baseline)/(cutoff − baseline) × 100, where test is the latency to respond after treatment; baseline is the latency to respond prior to treatment; and cutoff is the preset time at which the test will be ended in the absence of a response. To determine a compound's %MPE, a baseline measure must be obtained for each animal prior to treatment with vehicle or test compound.

BASIC PROTOCOL 2

MEASUREMENT OF ACUTE PAIN USING THE TAIL-FLICK TEST

The tail-flick test is a test of acute nociception in which a high-intensity thermal stimulus is directed to the tail of a mouse or a rat. The time from onset of stimulation to a rapid flick/withdrawal of the tail from heat source is recorded.

Materials

Animals: 25- to 30-g male C57BL/6 or CD mice, *or* 200- to 250-g male Sprague-Dawley rats

Test compound and control solution of appropriate vehicle

Reference analgesic compound (e.g., morphine, to be administered subcutaneously at 10 mg/kg)

Tail-flick apparatus with automated timer or stopwatch for manual use (Fig. 4.9.1)
1-ml syringes with 30-G needles for injecting mice *or* 27-G needles for injecting rats
Towel for gentle restraint of the animals *or* Plexiglas tube for same purpose

1. Bring animals to the test room. Record the body weights of the animals and allow animals to acclimate to the test environment for 30 min.

2. Record baseline latencies of the animals. Test the animals' tail-flick response using a tail-flick apparatus, and adjust the intensity of the heat source to produce tail-flick latencies of 3 to 4 sec. For mice, focus the light beam ∼15 mm from the tip of the tail. For rats, stimulate an area ∼50 mm from the tip of the tail. In the absence of a withdrawal reflex, set the stimulus cutoff to 10 sec to avoid possible tissue damage. Make three separate determinations of tail-flick latency per animal and average the numbers to determine the baseline latency for the animal. Test the animals at 3- to 5-min intervals. To avoid tissue

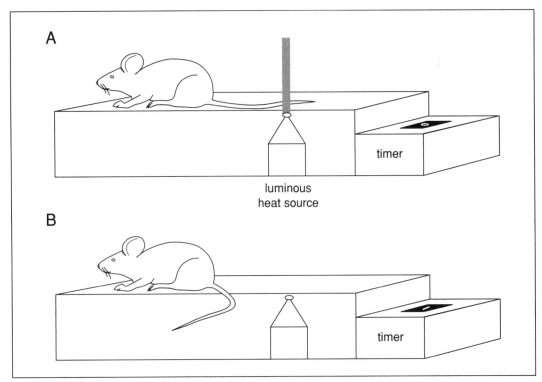

Figure 4.9.1 Apparatus used for the tail-flick test.

damage, vary the stimulated area 10 mm in both directions away from the original site for the second and third stimulation.

3. Prepare test agent, vehicle control, and reference compound such that the injection volume is 2 ml/kg for rats or 10 ml/kg for mice.

4. Inject the animals with the test agent, vehicle control, or reference compound using a 1-ml syringe equipped with a 30-G needle (for mice) or a 27-G needle (for rats). Stagger the injections to allow sufficient time to perform the tail-flick test at the same time interval after the injection of each animal. Randomize the treatment order.

   Another person may prepare and inject the agents in order to blind the person who performs the tail-flick testing of the treatment.

5. At 15, 30, and 60 min after injection of the agent, perform the tail-flick test. Record the time for the animal to show a tail-flick response, or assign a value of 10 sec (cutoff time) if no tail-flick is observed. Perform only one post-drug tail-flick test at each time point to minimize tissue damage to the tail. Test at the time of peak effect of the drug (if known; e.g., at 30 min after s.c. administration of morphine).

6. After sufficient data collection ($n = 8$ per group and dose), perform statistical analysis and calculate the means and standard errors for data presentation. For statistical analysis of multiple groups, use ANOVA followed by an appropriate post-hoc test.

   Data analyses may be performed on raw or transformed data. Data generated using the tail-flick test are often expressed as the percent maximal possible effect: %MPE = (post injection latency − baseline latency)/[cutoff (10 sec) − baseline latency]. To determine an ED_{50} value (dose that produces 50% effect), the data must be expressed as % effect.

BASIC PROTOCOL 3

MEASUREMENT OF PERSISTENT PAIN USING THE FORMALIN TEST

Materials (see APPENDIX 1 for items with ✓)

Animal: 25- to 30-g male C57BL/6 or CD mice, *or* 250-g male Sprague-Dawley rats, group-housed (e.g., Charles River Labs)
Test compounds in saline or appropriate vehicle
✓ 5% formalin

Individual clear containers to hold animals (e.g., shoebox cages with tops; see Fig. 4.9.2)
Mirrors
0.5-ml syringes with 28.5-G needles (for rats) *or* 50-μl Hamilton syringe with 30-G needles (for mice)
1-ml syringes with 27-G needles (for rats) or 26-G needles (for mice)
Hand-held counters (one per animal)
Clock or timer
Electronically controlled timing lights (optional)

1. Remove animal from home cage and record its body weight. Mark the rear paw that will be injected with formalin with a permanent black marker. Color both the top and bottom of the paw as completely as possible.

2. Place marked rats into the clear container that will be used during testing. Place mirrors behind and beside container to ensure that the animals' marked paws can be seen from all angles. Allow animals to acclimate for 15 to 30 min.

3. Prepare drug solutions to be tested. For each rat, load 0.5-ml syringes equipped with 28.5-G needles with 50 μl of 5% formalin; for each mouse, load 50-μl Hamilton syringes equipped with 30-G needles with 20 μl of 5% formalin (typically four animals per experiment).

4. Using 1-ml syringes with 27-G needles (rats) or 26-G needles (mice)—larger needles for suspensions, if necessary—inject animals with test compound (three doses, staggered at 0.5 log units, e.g., 0.1 mg/kg, 0.3 mg/kg, and 1.0 mg/kg) or appropriate vehicle. Allow 10 min (or another appropriate amount of time; or after formalin injection in some cases) for drug to take effect.

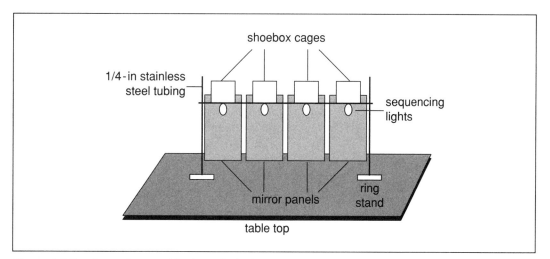

Figure 4.9.2 Apparatus used for formalin test.

5. Remove animal from its container and place it on the table. Grasp animal around its trunk with one hand and gently hold the tail with the other hand. Place animal in a "sitting" position, allowing the extension of the rear legs and paws. Gently extend the marked paw and with the free hand, pick up the formalin syringe. For rats, inject the solution into the dorsal surface of the paw by placing the needle above the toes and below the ankle and inserting it beneath the surface of the skin. For mice, inject the formalin solution under the skin in the middle of the paw on the plantar side.

6. Immediately return injected animal to its container, and start timer to mark the beginning of phase 1. Repeat formalin injections for remaining animals. Make every effort to complete injections as quickly as possible.

7a. *For rats:* Observe the first rat for 15 sec (time determined in pilot experiments) and count the number of times it flinches the injected paw (a single flinch is equal to an elevated foot, an episode of licking or biting, or several fast flinches). Repeat for remaining three rats. When the fourth rat has been observed, repeat the sequence starting with the first rat. Continue for 10 min.

   The phase 1 response (forceful flinching and licking or biting the injected paw) occurs during the first 10 min after formalin injection, and phase 2 (more subtle and almost reflexive flinching, and licking or biting the injected paw) occurs 30 to 50 min after formalin injection.

7b. *For mice:* Observe the mice for 1 min at 5-min intervals and record the time spent licking per second. Inject the next mouse 1 min after the first mouse and record the licking time. Inject a total of four mice at 1-min intervals and then observe them for 1 min at 5-min intervals for a total time of 40 min (after which most will have stopped licking).

   For mice, phase 1 occurs 0 to 9 min after formalin injection; however, phase 2 starts somewhat earlier. Therefore it is best to observe mice continuously from the start to 40 min after the formalin injection. The time spent licking the paw 10 to 40 min after the formalin injection is a measure of behaviors during phase 2.

   If the animals do not exhibit a sufficient number of quantifiable behaviors during the observation periods, the observation time may be extended (e.g., to 1 min instead of 15 sec for rats; to 2 min at 5-min intervals for mice). If the reference compound has little effect, it is suggested that the formalin concentration be decreased, for example to 2.5%.

8. Record the total number of flinches (rats) or the total time licking (mice) for each animal during phase 1. Reset counter to zero.

9a. *For rats:* Begin counting flinches for phase 2 at 30 min after the first formalin injection. Follow the observation procedure outlined in step 7a. Continue counting flinches for 20 min. Record the total number of flinches each rat exhibits during phase 2.

9b. *For mice:* Record phase 2 behavior as time spent licking from 10 to 40 min after the injection of formalin.

10. Repeat steps 1 to 9 until the desired number of animals (e.g., $n = 8$) for each treatment group has been studied. Include as many different treatments as possible in each run, i.e., do not examine all of the vehicle controls or any particular treatment group at the same time.

11. For simple dose-response studies (using flinches as dependent measure), perform an ANOVA followed by post-hoc analysis (if appropriate) to determine statistical significance. Analyze and present data either as total flinches per phase, or as flinches per unit time (e.g., flinches/min). If using a rating scale, employ nonparametric analysis to analyze results.

Contributors: Annika B. Malmberg and Anthony W. Bannon

Table 4.9.1 Troubleshooting Guide For the Hot-Plate, Tail-Flick, and Formalin Tests

Problem	Possible cause	Solution
Hot-plate test		
Response occurs too quickly to be measured reliably in untreated or control animals	Too high a stimulus intensity	Reduce stimulus intensity
No effect is observed for reference analgesic (e.g., morphine) used at "effective" dose recommended for the procedure	Too high a stimulus intensity	Reduce stimulus intensity
Response does not occur or occurs too close to cutoff time in untreated or control animals	Too low a stimulus intensity	Increase stimulus intensity
Effect of test compound or reference drug cannot be detected because control animals show a narrow window of response	Too low a stimulus intensity	Increase stimulus intensity
Response of control animals varies significantly from experiment to experiment	Handling effects	Keep all procedures prior to testing (e.g., housing, acclimation time, etc.) consistent from experiment to experiment
Results for test compounds are difficult to reproduce	Possibly handling effects	Keep all procedures prior to testing (e.g., housing, acclimation time, etc.) consistent from experiment to experiment
Tail-flick test		
No response is observed before cutoff in untreated animals	Too low a stimulus intensity and/or handling/restraint effects	Increase stimulus intensity Acclimate animals to test before experiment
Large variability in response is observed	Possibly too high or low a stimulus intensity and/or handling/restraint effects	Adjust stimulus intensity Acclimate animals Make several determinations of tail-flick latency and calculate mean latency
No effect is observed for reference agent	Too high a stimulus intensity	Reduce stimulus intensity
Formalin test		
Few responses are observed in control animals	Stressed animals	Increase acclimation period, be careful with injections, and handle animals prior to testing (e.g., the day before the test)
	Improper injection of formalin	Give consistent injections
	Too small (or too young) animals	Use heavier animals (typically at least 250 g)
	Possibly overly long observation period per animal	Increase sensitivity of assay by increasing the number of observations made for a given animal over a period of time; for example, observe animals for 15-sec intervals instead of 30-sec intervals
Difficulty detecting dose-response effects is observed	Insufficiently sensitive measure	Be sure to count licking and biting episodes as well as flinching to help increase sensitivity; consider using a rating scale instead of counts

UNIT 4.10

Place Preference Test in Rodents

The conditioned place preference procedure is widely used in rodents for assessing drug effects on motivational processes (Schechter and Calcagnetti, 1993).

BASIC PROTOCOL 1

PLACE PREFERENCE TEST IN RATS

Materials

    Male Wistar rats (180 to 200 g at the beginning of the experiment)
    Twelve nonexperimental rats (for priming boxes with rat odors)
    Test compound(s) dissolved in physiological saline *or* suspended in 0.2% hydroxypropylmethylcellulose in physiological saline (for parenteral administration)
    Vehicle

    Transparent plastic (e.g., macrolon) cages (41 × 25 × 25–cm)
    Video camera and videotape recorder
    Six two-compartment test apparatuses: experimental boxes (35 × 35 × 70–cm) with two compartments of equal size (35 × 35 × 35–cm) separated by a guillotine door (12 × 12–cm); each compartment is tactually and visually distinct (black-and-white striped walls with corrugated floors versus gray walls with smooth floors)
    60-W red lights
    Animal balance with accuracy of 1 g (e.g., Sartorius 1401.001.2)
    25-G × 5/8-in. (0.8 × 16–mm) syringes for intraperitoneal or subcutaneous injection (e.g., Terumo)
    23-G × 1-in. (0.6 × 25–mm) *or* 21-G × 1.5-in. (0.8 × 40–mm) needles for intraperitoneal or subcutaneous injection, respectively (e.g., Terumo)
    Luer gastric probes with olive extremity (70 mm long × 1.5 mm oval diameter) for oral administration

1. Have 24 experimental rats delivered to the laboratory at least 5 days prior to testing. House them in a colony room at six per cage in macrolon cages (home cages) on wood shavings and allow free access to standard rodent diet and tap water. Maintain environmental conditions at 21°C on a nonreversed light/dark cycle with illumination from 0700 to 1900. Mark the tail of each rat with a separate pattern, so that each rat can be uniquely identified.

2. In the test room, mount a single video camera above the group of six two-compartment test apparatuses and connect it to a videotape recorder. Install two 60-W red lights directly overhead and illuminating all six boxes, so that experiments can be conducted under dim lighting conditions.

3. Outline the experiment, taking into account eight conditioning sessions conducted over four days (Monday to Thursday, two sessions per day, one in the morning and one in the afternoon). Condition two successive series of twelve rats in one week (four home cages each containing six animals). Administer each of the two treatments (test compound or vehicle) to each rat each day. Evenly distribute the boxes, time of testing, and treatments between the different groups. However, use the same treatment schedule for all animals in the same home cage.

4. Before testing each morning, clean all parts of the test apparatus with water. Place two rats that are not included in the experiment into each box (one per compartment) for 30 min to prime the boxes with rat odors.

5. Move the home cages from the colony room and place them in a holding room adjacent to the test room. Allow free access to water, but remove food until after the second session has been completed in the afternoon.

6. Weigh each animal from two home cages in the holding room 30 min prior to the conditioning session. Immediately administer test compound to six animals (from one home cage) and vehicle to the other six (from a second home cage), each at 5 ml/kg.

7. With the guillotine door closed between the two compartments, place one rat in each compartment of all six test apparatuses. Always associate test compound administration with the gray compartment. Allow animals to explore for 45 min.

8. Remove the rats from the compartments in the same order as they were introduced and return them to their home cages in the holding room. Allow free access to water, but not food, between the daily sessions.

9. Wipe the floors of the boxes with absorbent paper to eliminate urine or feces left from the previous animal. Repeat steps 6 to 9 for the second series of twelve rats.

10. In the afternoon (5 hr later), repeat steps 6 to 9 for all 24 rats, except ensure that each rat is given the opposite treatment and is placed in the opposite compartment of the *same* apparatus as in the morning. Move all the animals to the colony room until the next day. Allow free access to food and water.

11. Repeat steps 4 to 10 each day for four days. Alternate the order of presentation of the two compartments from day to day (A-B, B-A, A-B, B-A), with half the animals starting the daily conditioning in the gray compartment (test compound) and half in the striped compartment (vehicle). Ensure that each animal is conditioned in the same two-compartment box at the same time each day.

12. On the morning immediately following the last conditioning day, bring the rats to the holding room and provide access to water but not to food. Weigh all 24 animals immediately before the test, but do not administer any injections.

13. Open the guillotine doors separating the two compartments in each box, allowing free access to both sides. Activate the video camera and place one rat in the gray compartment of each box, using the same two-compartment box used for conditioning. Allow rats to explore the boxes for 20 min and record the following measures for each rat:

    a. time spent in the drug-paired (gray) compartment

    b. number of crossings between the two compartments, calculated in blocks of 5 min and accumulated over the 30-min test period.

    Repeat for all 24 animals in a single morning session.

14. Compare data obtained from the treated group with that from the control group using a nonpaired Student's *t* test.

BASIC PROTOCOL 2

PLACE PREFERENCE TEST IN MICE

Materials

Male mice (Swiss albino, C57/BL6, hybrids 50% 129/sv, 50% C57/BL6, etc.) weighing 25 to 30 g at the beginning of the experiment (males preferred, but females acceptable)
Nonexperimental mice (for priming boxes with mouse odors)
Test compound
Vehicle

Experimental room with indirect white or red lights; sound proof with white noise (for behavioral experiments)

Video camera and videotape recorder or another detection system (e.g., Videotrack; ViewPoint)

Group housing cages (25 × 25 × 20–cm)

Two-compartment apparatus (Fig. 4.10.1). Connect each compartment (15 × 15 × 15–cm) to an intermediate triangular area by guillotine doors and cover with a transparent plastic cover to prevent mice from escaping the box. Boxes must be constructed of plastic with removable floors in order to adequately clean mouse feces and urine. Four distinctive sensory cues are necessary to differentiate each compartment: (1) wall and floor dressing (painted stripes and dots); (2) floor texture (plastic rough or smooth); (3) floor color (painted black or cream); (4) box orientation (60° angle). The combinations are as follows: (a) striped walls, rough and black floor; (b) dotted walls, smooth and cream color floor. The intermediate compartment is gray and has smooth walls and floor. At least two apparatuses can be used simultaneously in a given experiment; using four or six will further speed data collection. However, never use the same apparatus for male and female animals.

Animal balance for mice (with accuracy of 0.1 g)

Device for drug administration:

    For intraperitoneal or subcutaneous injection: 1-ml sterile syringes (e.g., Plastipak from Becton Dickinson)

    For intraperitoneal or subcutaneous injection: needles 26-G × 1/2-in. (0.45 × 12–mm) or 26-G × 3/8–in. (0.45 × 10–mm), e.g., Terumo or Microlance

    For oral administration: Luer gastric probes with olive extremity for mice (Panlab)

    For intracerebroventricular administration: a cannula must be implanted, preferably in the third ventricle, using stereotaxic surgery (Kopf stereotaxic apparatus; see APPENDIX 2A). In the case of the third ventricle, a cannula is positioned 0.3 mm caudal to bregma in the midline, and 3.0 mm deep. The cannula consists of a 30-G dental needle (Sofijet) bent at a right angle 4.5 mm from the tip, cut 6 to 7 mm from the bend, and connected to polyethylene tubing (Tygon) filled with artificial cerebrospinal fluid solution (see APPENDIX 1). A Hamilton glass microsyringe will be required in order to perform the central microinjection in a volume of 2 μl over 60 sec (see Hutcheson et al., 2001). The perfusion system is maintained an additional 60 sec to allow diffusion of the drug into the third ventricle.

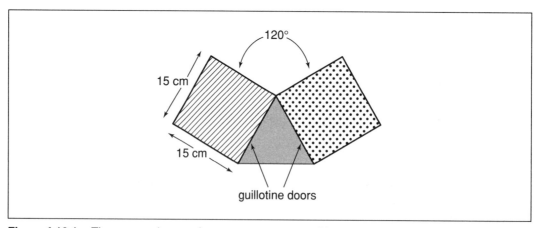

Figure 4.10.1 The mouse place preference test apparatus. Dimensions are expressed in centimeters.

1. Install room lighting (indirect white or red lights; installed on each wall and protected with a screen) in a small (4 to 5 m^2) experimental room lacking natural light to result in light intensity inside the experimental box of <50 lux. Mount a detection system (e.g., video camera installed in a fixed position on the ceiling of the room).

   The lighting must not be modified during the experiment since consistent lighting is crucial for the success of the procedure. It is also crucial to not change the placement of any objects within the room during the experiment or to introduce any additional tools or objects. The home cage of the mice must be in a separate chamber. The investigator should only enter the experimental room to introduce or remove mice from the place conditioning box.

2. Acclimate mice to the environmental conditions of the behavioral laboratory (21°C ± 1°C on a nonreversed light/dark cycle with illumination from 0700 to 1900 or 0800 to 2000) at least 5 days before starting the experiments. Handle the mice to habituate them to the researcher and experimental protocol (e.g., mark the tail for identification, weigh each animal, etc.). House mice in groups of four or five per cage (except when animals must be housed alone due to the experimental conditions, e.g., intracerebroventricular cannula implantation) with free access to food and water, except during the testing sessions. Maintain genetically modified male mice of the same litter together to avoid aggressiveness.

3. Clean the two-compartment apparatus with water (no detergent) before each animal exposure during the preconditioning, conditioning, and testing phases. Before starting the experiment, place two or three mice not belonging to any experimental group in the cleaned apparatus to establish a mouse odor.

4. Randomly assign one compartment to be paired to drug administration and the other to vehicle. Therefore, in each group of mice, 50% of the animals will be conditioned with the drug paired to one of the compartments, and the other 50% will be paired with the drug in the other one. Ensure that all treatments are counterbalanced as closely as possible between the compartments.

   An unbiased procedure is critical in this paradigm. Indeed, the sensory cues are designed such that animals do not exhibit an initial preference or aversion for any compartment in the preconditioning phase.

5. Conduct all experiments in the light phase between 0900 and 1900. Place mice in the holding room during the whole experiment, only transporting them into the experimental room during the phases of the test that require exposure to the apparatus.

6. Place mice in the middle of the neutral area and allow free access to each compartment. Record the location of the mouse for the next 18 min. Discard mice showing strong unconditioned aversion (<25% of the session time) or preference (>75% of the session time) for any compartment. Define entry into a given compartment as being when the head and the two forepaws are inside of this compartment.

7. After the session, randomly assign mice to be paired to drug or vehicle administration and to a particular compartment. Between each session, wipe the doors and the walls of the boxes with damp absorbent paper towels to eliminate urine or fecal matter from the previous mouse. At the end of the day, thoroughly clean the box with hot water.

8. Weigh animals 30 to 60 min before beginning the session on the first day of the conditioning phase (6 consecutive days of alternate drug or vehicle injection).

9. Inject an animal with drug or the appropriate vehicle just over the corresponding compartment of the box. Immediately after injection, place the mouse in the assigned compartment with the guillotine door closed. Confine the animal in the compartment for the designated period of time, depending on the pharmacokinetic characteristics of the drug (e.g., 20 min in the case of morphine or cocaine administration).

10. Between any two sessions, wipe the doors and the walls of the boxes with damp absorbent paper towels to eliminate urine or fecal matter from the previous mouse. At the end of the day, clean the box with hot water.

11. Administer test compound to mice on days 2, 4, and 6, and vehicle on days 3, 5, and 7. Administer vehicle every day to the control animals. During the conditioning session, inject each mouse and confine them in the respective compartment each day at approximately the same hour. Start the conditioning session every day at the same hour, and inject the animals in the same order and with a similar time schedule.

12. On the day of the testing phase, remove the guillotine doors to allow the mouse free access to each compartment for 18 min. Record the time spent in each compartment as well as the number of visits to each compartment.

13. Between two sessions, wipe the doors and the walls of the boxes with damp, absorbent paper towels to eliminate urine or fecal matter from the previous mouse. At the end of the day, clean the box with hot water.

14. Calculate the place preference score (PPS) using data obtained during the preconditioning and testing phases as follows: multiply the time spent in the conditioned compartment by the total time in sec (1080) and divide by the total time (1080) minus the time spent in the neutral area.

$$PPS = \frac{\text{Time in conditioned compartment (sec)} \times 1080}{1080 - \text{time in the neutral area (sec)}}$$

This score expresses the difference between the time spent in the compartment associated with the drug during the testing and the preconditioning phases. The time spent in the neutral area is proportionally shared and added to the time value of each conditioned compartment. Using this procedure, the addition of the time spent in the two conditioning compartments corresponds to 100% of the total time (1080 sec). Scores of the different groups can be compared using ANOVA and the subsequent post-hoc analysis.

Data can also be analyzed by comparing the time spent in the drug-associated compartment during the preconditioning and testing phases with a paired Student's t test. However, this procedure does not allow the comparison between the different experiments groups when a treatment is administered.

15. *Optional:* To evaluate the rewarding properties of a natural reward (e.g., food), food-restrict mice for 5 days before the experiments (Maldonado et al., 1997). Otherwise, use the same procedure as that used for drug conditioning, with the exception that mice have free access to food (normal mouse food plus sucrose) during the conditioning phase in the confined compartment on days 2, 4, and 6, and no access to food in the rewarding compartment on days 3, 5, and 7. Control animals have no access to food in the assigned compartments during the 6 conditioning days.

Reference: Schechter and Calcagnetti, 1993

Contributors: Sylvain Roux, Christelle Froger, Roger D. Porsolt, Olga Valverde, and Rafael Maldonado

UNIT 4.11

Intravenous Self-Administration of Ethanol in Mice

BASIC PROTOCOL

The apparatus described below was developed by Dr. Christopher Cunningham at Oregon Health Sciences University, and has been validated for use in this procedure (see Fig. 4.11.1). However, standard operant enclosures (e.g., ENV-307A mouse operant chamber, Med Associates) could presumably be substituted without undermining the basic behaviors measured here.

NOTE: Sterilize the subcutaneous saddle and surgical instruments by soaking in sterilization solution (e.g., Cidex) ≥ 10 hr before implanting.

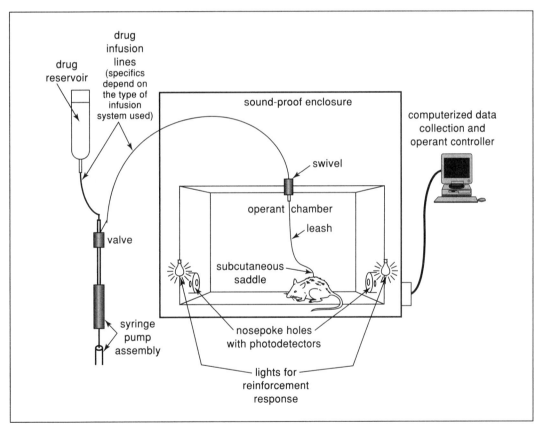

Figure 4.11.1 Operant intravenous self-administration apparatus used to deliver response-contingent drug infusions and collect data during i.v. self-administration sessions. Each nosepoke hole should have an aperture of 2.3 cm, and be mounted 1.5 cm above the wire mesh floor. A bulb (e.g., 5 VDC GE-47 bulb) should be mounted inside the hole to provide indirect illumination upon completion of a reinforced response. Responses consist of occlusion of an infrared photoemitter-receiver pair for at least 50 msec followed by nonocclusion; photoemitter/receivers are available at an electronics shop or from Med Associates.

Materials

Twelve to fifteen 15- to 35-g adult male or female mice, C57BL/6J or DBA/2J (The Jackson Laboratory) or other strains if desired

Mouse anesthetic cocktail: 17 mg/ml ketamine, 3 mg/ml acepromazine, 6 mg/ml xylazine

Low-viscosity cyanoacrylic cement (e.g., Nexaband, Closure Medical) in a 1-ml tuberculin syringe mated to an 18-G needle

Heparinized saline solution for injection

Standard rodent diet

Sterile saline

Brevital sodium for injection

95% to 100% ethanol diluted with 0.9% saline to produce a unit dose of 75 mg/kg body weight in 5 μl infusion volume (i.e., 28.3% to 75.6% v/v ethanol)

Isothermic heating pad

Scale accurate to 0.1 g

Rectal temperature probe, accurate to 0.1°C

Surgical scissors, scalpel, supplies (swabs, Betadine, hair clippers, 5-0 surgical silk)

Dissection microscope with fiber-optic light source

Fine-tip forceps (e.g., Dumont no. 55, Fine Science Tools)

Subcutaneous stainless steel mouse saddles constructed of 27-G hypodermic tubing and wire; 1 cm of tubing is exteriorized to provide an attachment point for drug-delivery tubing, and comes with removable cap; can be purchased from IITC (*http://www.iitcinc.com/*). Prior to surgery, mate the exterior portion of the saddle to a 1-ml tuberculin syringe filled with heparinized saline to permit delivery of anticoagulant during the surgery.

Caps for mouse saddles, constructed of PE-50 tubing cinched closed at one end (using heat and a pair of forceps) to create a seal. Be sure to test caps for a water-tight seal before use.

Cannulae composed of 50-mm Silastic tubing (Dow Corning; 0.012-in. i.d.; 0.025-in. o.d.) with a movable 3-mm cuff of PE-50 threaded over the Silastic, ~11 mm from the end of the catheter. The jugular end of the cannula should be beveled, and the catheter end secured to the mouse saddle with a 3-mm section of heat-shrink tubing.

6 × 4–in. piece of 1/2-in. thick Plexiglas, with a 0.25-in. slot running halfway down the long end

50-g weight

27-G, 1/2-in. hypodermic needle, bent 120°, 5 mm from the beveled tip using pliers

Polycarbonate caging (27.9 × 9.5 × 12.7–cm) with corn cob or Cell-sorb bedding

Drug delivery system (e.g., PHM-100 syringe pump from Med Associates)

Restraint tube (e.g., IITC restrainer model no. 84)

Two to three small hemostats (e.g., Halsted-Mosquito straight, from FST)

Four boxes with aluminum end walls and Plexiglas side walls (30 × 15 × 30–cm), with wire-mesh floors, and a hinged, drilled (for air holes) Plexiglas cover

Sound-attenuating cubicles (e.g., Med Associates, Env-021M) to house aluminum/Plexiglas boxes

Two nosepoke response devices located at either end of the boxes in the aluminum walls

Six to eight pairs of infrared photoemitter-receiver diodes mounted at regular intervals on the Plexiglas walls, 2.5 cm above the floo, optional for measuring locomotor activity (*UNIT 3.2*)

Leash for fluid delivery from swivel to mouse; this can be an aperture of Tygon tubing that will fit both the fluid-bearing swivel and the subcutaneous saddle

Fluid-bearing swivel (e.g., Instech Laboratories; stainless steel 25-G single channel)

Personal computer for event recording and response-contingent light and drug delivery

1. House mice on-site for at least 2 weeks prior to surgery. Begin procedure by injecting mice with mouse anesthetic cocktail (e.g., ~0.15 ml for a 25-g mouse). Maintain mice at 36°C throughout surgery by working on a heating pad and by checking temperature regularly with a rectal probe. For best results, perform all surgical work under a dissecting microscope, using aseptic technique.

2. Upon achieving surgical anesthesia, shave and disinfect the skin over the right external jugular vein, and ~1.0 cm^2 on the midline of the back halfway between neck and tail.

3. Make a 0.5-cm cut in the neck, isolate the right external jugular vein by blunt dissection using the forceps, and thread two lengths of 5-0 surgical silk under the vein, ~6 mm apart from each other. Tie the rostral suture to close off the jugular vein.

4. Make a 1.5-cm incision on the back, a little less than halfway down from the neck to the tail in a rostral-caudal orientation. Place the subcutaneous saddle with catheter attached over the thorax. Pull the tip of the cannula subcutaneously (using the straight forceps) until it exits through the ventral incision. Close the dorsal incision using cyanoacrylic cement; be sure to accurately mate the two sides of the incision before applying the glue, and make sure that the glue is completely dry before turning the mouse back over to work on the ventral incision (~30 sec).

5. Position the mouse so it is lying with the exteriorized portion of the saddle in the groove of the slotted Plexiglas and connected to the heparin delivery syringe. Place the slotted Plexiglas over the heating pad to keep it warm. Measure the distance from the end of the PE tubing portion of the catheter to the end of the catheter and slide the PE tubing so that it is 11 mm from the end of the catheter to result in the tip of the catheter lying just above the right atrium on the average sized mouse, once the catheter is inserted.

6. Retract the jugular vein using a 50-g weight attached to the tied rostral suture and make a cut in the vein using the bent 27-G needle. With the needle still in the vein, thread the beveled end of the catheter into the vein and remove the needle. Advance the catheter until the PE tubing portion of the catheter meets the entrance to the vein.

7. Quickly pull the caudal length of 5-0 silk tight, but not overly so, around the jugular vein with the catheter inside. Using the straight forceps, immediately insert a third length of 5-0 silk into the musculature just dorsal to the PE tubing and tie the silk tightly around the short length of PE tubing to secure the catheter.

8. Retract on the attached syringe to verify that blood is obtainable from the catheter. After testing, inject ~20 μl heparinized saline to prevent clotting inside the catheter. Close the ventral incision with the cyanoacrylic cement, and immediately cap the externalized portion of the catheter.

9. Ensure that the mouse is kept warm until completely recovered from the anesthetic (usually ~2 hr). Singly house mice after surgery and wait 72 hr before starting any experiments. House mice at 21°C with a standard light cycle (0700 to 1900 lights on) in polycarbonate caging (27.9 × 9.5 × 12.7–cm) with corn cob or Cell-sorb bedding, and free access to rodent chow and water, throughout the course of the experiment. If a mouse loses >20% body weight in the day after surgery, supplement food supply with a mash of lab chow and water placed in a petri dish on the floor of the cage.

10. Avoid dehydration by slowly introducing 1 ml sterile saline through the catheter. Flush the catheter daily with 20 μl heparinized saline. Following surgery, handle mice daily until they acclimate to the restraint tube, because excessive force applied to the external portion of the catheter can cause problems.

11. Forty-eight hours after surgery, verify catheter patency by infusing 20 μl Brevital injection solution and observing that the mouse immediately loses its righting reflex. Discard any

animal that exhibits a delayed or no response. Always replace the interior volume of the catheter with heparinized saline after infusing any other substance.

12. Preload the drug delivery system with the correct concentration of ethanol for the mouse. Ensure no bubbles are present in the system, and that the ethanol solution is primed to the end of the drug delivery leash.

13. During the light phase of the cycle, remove mouse from the home cage, and weigh it. Place the mouse into the restraint tube, making sure to allow the exteriorized portion of the catheter to protrude through the medial slot in the tube. Using a hemostat to brace the catheter, remove the catheter cap with forceps.

14. With the mouse still in the restraint tube, attach the mouse to the leash inside the operant chamber. Remove the mouse from the restraint tube, close the apparatus, and initialize the session. Immediately, manually activate the drug delivery system to administer a single 5-μl reinforcer, which is sufficient to prime the ethanol to the tip of the catheter.

15. Give each mouse a 2-hr session in which ethanol is available on a fixed ratio (FR)-3 schedule of reinforcement from one of the nosepoke holes (in other words, every third response is reinforced). Counterbalance which side is reinforced (i.e., when different dose groups or strains of mice are used, ensure that equal numbers of left- and right- reinforced mice exist within each group). Indicate alcohol delivery by a 2-sec illumination of the stimulus light during which ethanol is not available,

16. Record the number of responses made in each nosepoke hole during the 2-hr session. Record the number of infusions taken by mice during each 2-hr session, and if applicable, locomotor activity during the session. If desired, record when responses are made by cumulating responses made in 5-min blocks of the session.

17. At the end of the session, open the apparatus and place the mouse back into the restraint tube. Remove the leash and replace the cap, and return the mouse to the home cage. Replace fluid in delivery lines with water at the end of the day to prevent ethanol evaporation, always filling with fresh ethanol solution at the beginning of each day.

18. Repeat steps 12 to 17 for nine consecutive daily sessions.

19. Retest the catheter patency with a Brevital infusion following 9 days of acquisition, and discard mice when the catheter fails. Recheck catheter patency about every 7 days.

20. If stable behavior is acquired (this can be defined as <20% change in the number of reinforcers administered over 4 consecutive days, with >10 reinforcers administered per 2-hr session), vary the unit dose of ethanol to determine the effect on the rate of responding, or saline may be substituted for ethanol to examine behavior in extinction.

21. Using the number of responses made in each nosepoke hole, calculate the percentage made on the reinforced side each day. Graph collected or calculated data, either as a function of individual subject, or dose or strain groups of animals, depending on how much detail is required.

Reference: Grahame and Cunningham, 1997

Contributors: Nicholas J. Grahame and Christopher L. Cunningham

UNIT 4.12

Preclinical Models to Evaluate Potential Pharmacotherapeutic Agents in Treating Alcoholism and Studying the Neuropharmacological Bases of Ethanol-Seeking Behaviors in Rats

While a step-by-step procedural description appears in each of the four Basic Protocols, it is strongly recommended that less experienced investigators consult a seasoned researcher to assist with understanding the written procedures, data interpretation, and to provide some basic understanding of the theoretical foundation from which many of the principles inherent within the four Basic Protocols were derived.

BASIC PROTOCOL 1

TRAIN RATS TO INITIATE ETHANOL-MAINTAINED RESPONDING ON AN FR-4 SCHEDULE

In this protocol, the operant software is programmed to record number of responses (i.e., response rates) and earned reinforcers (i.e., stimuli) for each solution presented (saccharin, ethanol-saccharin cocktail, and ethanol) under each schedule of reinforcement, fixed ratio (FR)-1 or FR-4. While all data should be recorded and archived during the acquisition phase, the data obtained once the animals exhibit stable response rates are the primary concern of the investigator. Most operant software permits a direct transfer of these data to spreadsheets (e.g., Excel). Once the number of responses and reinforcers of the saccharin, ethanol-saccharin cocktail, and ethanol have been determined for the post-stabilization period, analysis of these data will allow evaluation of the reinforcing efficacy (i.e., strength) of the various reinforcers. The effects of drug treatments on these measures are also easily determined following the post-stabilization period. The amount of absolute ethanol intake in g/kg should also be evaluated.

Materials

Ten 2- to 3-month-old naive outbred rats or rats selectively bred for alcohol consumption
   ($\sim$200 to 300 g, female or male)
Standard rodent diet
0.10% (w/v) saccharin (Fisher Scientific) solution in distilled water
2% (v/v) ethanol/0.075% (w/v) saccharin reinforcer
5% and 10% (v/v) ethanol solutions

Wire-mesh stainless-steel cages or plastic tubs
Personal computer with standard operant software packages (e.g., from Coulbourn
   Instruments or Med Associates) to record responses and control reinforcements
Stopwatch
Ten standard operant chambers (Coulbourn Instruments) equipped with two removable
   levers and two dipper fluid delivery systems enclosed in sound-attenuated cubicles.
   Dipper presentations should provide 1.5-sec access to a 0.10-ml dipper solution,
   followed by a 3-sec time-out period; the amount of the earned reinforcer delivered

following the various response requirement in Basic Protocols 1 to 4 is 0.10 ml. Above each lever, a stimulus light (red, green, or yellow) is present and is illuminated upon presentation of the stimulus delivery/reinforcer.

Animal balance for weighing rats

1. One week prior to experiment, handle 10 rats daily and allow them to become acclimated to the vivarium and light cycle (either a normal or reversed 12-hr light/dark cycle is appropriate). House animals individually in wire-mesh stainless steel cages or plastic tubs. Maintain the vivarium at 21°C and provide ad libitum access to food and water, except during the first days of the training phase. Weigh animals once weekly. If animals are being trained after a stereotaxic surgical procedure, delay the beginning of the initiation phase for at least 1 week.

2. Provide food ad libitum to the rats; the only exception being that no food should be available during the 60-min operant session (which should take place between 10 a.m. and 3 p.m.). To begin the initiation phase, place only the right lever, dipper assembly, and fluid delivery system (i.e., trough) in the operant chamber. Fill the liquid reservoir with 15 ml of 0.10% saccharin solution.

3. Select and define the testing parameters for the operant session: for example, determine if the data will be gathered for 30 (generally adequate for outbred rats) or 60 min (more data, increases the power of the analysis with genetically selected high-response rates; see June et al., 1998). Then, determine if these data will be gathered in 2-, 5-, or 10-min blocks. Program this information in the computer using the commercial Coulbourn or Med Associates software.

4. Prior to any experimentation, use a stopwatch to determine an appropriate duration of stimulus delivery. During the first phase of initiation, use an FR-1 schedule, synonymous in this unit with the term continuous reinforcement (CRF) schedule, that provides the rat with one 0.1-ml dose of liquid reinforcer for a specified duration of stimulus delivery (e.g., 1.5 to 3.0 sec) each time the animal makes a single lever press. Use the stopwatch data to define the duration of stimulus delivery and response requirements and program this information in the operant software.

5. Deprive each animal of fluid for 22 hr prior to experimentation. Once the testing parameters, duration of stimulus delivery, and response requirements have been specified, remove the fluid-deprived rat from its home cage and place it in the operant chamber for the 1-hr session. Close the door to the operant chamber and then close the door to the sound-attenuated chamber (the animal will be visible through the peep-hole in the sound-attenuated chamber). During the 1-hr session, simply observe the rat. Complete daily observations during the beginning of this initial training phase to learn about the different response profiles that occur during the acquisition phase of the different types of response-contingent behaviors.

6. To facilitate consistent and robust lever pressing, deprive rats of fluid for 22 hr prior to the 1-hr operant session for at least 2 to 3 days; however, once the rats are shaped and responding rapidly on the CRF schedule (i.e., for two or three sessions), rats should receive no additional fluid restriction throughout the experiment.

7. After the last day of fluid deprivation, train the rats for 5 additional days under the 60-min daily access paradigm to respond for the 0.1% saccharin solution on an FR-1 schedule. On day 6, discontinue the FR-1 schedule and increase the response requirement to an FR-4 schedule, which provides the rat with a 0.1-ml dose of liquid reinforcer each time the animal makes four lever presses. Continue until stable responding is obtained (about day 14), defined as at least five consecutive sessions that show neither systematic increase nor

decrease in responding (±20% of the average responses for 5 consecutive days). Allow 2 or 3 additional sessions for animals that fail to exhibit stability before excluding them from the experiment.

8. Train rats to respond for a 2% ethanol/0.075% saccharin reinforcer (a "cocktail") on an FR-4 schedule (0.1 ml/four responses). After development of stable FR-4 responding (i.e., typically five sessions) for the ethanol/saccharin cocktail, expose rats to six successive FR-4 sessions in which the drinking solution is alternated daily from the ethanol/saccharin cocktail to 5% ethanol only (0.1 ml/four responses).

9. After the 6-day alternation procedure, raise the ethanol concentration in the ethanol-saccharin cocktail solution each day from 2% to 5%, 7%, 9%, and 10% and concurrently decrease the concentration of saccharin from 0.075%, to 0.05%, 0.025%, and 0.0125% until saccharin is eventually eliminated at the 10% ethanol concentration level. Obtain stable responding for the 10% ethanol solution for at least 2 weeks before beginning experimental drug treatments (the response rates for the last 5 days before the drug treatment should be no more than ±20% of the average daily response rate for the sample of 10 rats).

10. Divide the number of lever presses obtained under the FR-4 schedule by 4 to determine the number of earned reinforcers. Calculate the volume (ml) of solution consumed. Calculate the grams of ethanol consumed: one earned reinforcer using a 0.10-ml dipper cup volume = 0.10 ml volume of 10% ethanol. Divide grams of ethanol consumed by the weight of the rat to obtain a result in g/kg.

   For example, a rat presses the lever 350 times for a 10% ethanol solution in a 60-min session.

   350/4 = 87 earned reinforcers, and 87 × 0.10 ml = 8.7 ml total solution

   8.7 ml total solution × 10% ethanol × 0.793 g/ml = 0.69 g absolute ethanol consumed.

   0.69 g ethanol consumed/0.370 kg rat = 1.86 g/kg dose of absolute ethanol

11. To determine the pattern of ethanol intake prior to drug treatment (baseline phase), analyze data by a repeated measures analysis of variance (ANOVA) for the factors day (day 1, 2, and 3), and response interval (0 to 10, 11 to 20, 21 to 30, 31 to 40, 41 to 50, 51 to 60 min, and total: 10 to 60 min). Analyze both factors as repeating variables. Conduct separate analyses on each dependent measure (i.e., number of responses, reinforcers, and intake in g/kg of the ethanol and saccharin/sucrose solutions).

12. Analyze the data from the drug treatment phase by repeated measures ANOVA with drug treatment, response interval, and consumption day (day 1, 2, and 3) as the independent factors. Conduct separate analyses on each dependent measure, i.e., number of responses, reinforcers, and intake (g/kg) of the ethanol and saccharin/sucrose solutions. Determine the drug treatment × response interval interaction to obtain a detailed analysis of the profile of antagonism by a test agent across the full length of the 60-min session. Make post-hoc comparisons using the Newman-Keuls test (or similar measures) during the baseline and drug treatment phases.

   Illustrative example: Using 4 doses of "drug x" and one placebo treatment (e.g., saline), the 5 experimental manipulations would become the factor "drug treatment condition", comprising 5 levels. The pooled no-injection baseline (e.g., average of 5 days before the drug treatment) would be the 6th drug treatment level (i.e., no-injection control). Hence, the analysis for response rate would be a 6 × 7 repeated measures ANOVA. The latter factor with 7 levels would be response interval (0 to 10, 11 to 20, 21 to 30, 31 to 40, 41 to 50, 51 to 60 min, total: 0 to 60 min). If the experimenter is not interested in evaluating the time-course data, then the analysis would be reduced to a one-way repeated measures ANOVA for only the 60-min data.

SUPPORT PROTOCOL 1

BLOOD ALCOHOL CONTENT (BAC)

To ensure animals are consuming pharmacologically relevant amounts of ethanol during operant sessions, tail blood samples are collected on a subset or all animals ($n = 10$) under baseline conditions (i.e., when animals are not receiving a drug treatment). The Analox Analyzer (Analox Instruments) can be used to measure BAC and is highly recommended, although other methods may also be used. However, compared with many previously described HPLC methods (June et al., 1999), this new methodology is more efficient in measuring BAC content and real-time data can be obtained.

Materials

Rat (see Basic Protocols 1 to 4)
Heparin-coated microcentrifuge tubes
Clinical analyzer (GL-5 MicroStat, Analox Instruments)
Alcohol reagent buffer solutions, pH 7.4 (Analox Instruments)
Alcohol oxidase enzymes (Analox Instruments)
Clark-type amperometric oxygen electrode

1. After the first 30 min of an operant session, collect ~100 µl of whole blood from the tip of the rat's tail into a heparin-coated microcentrifuge tubes. Immediately microcentrifuge the blood 5 min at 1100 rpm, room temperature. Collect duplicate 5-µl plasma samples with a pipet and inject them directly into an analyzer (Clark-type amperometric oxygen electrode).

2. Use the mean of the duplicate samples as an index of the level of BAC content for each rat. Correlate data obtained from the BAC measurements (mg/dl) with ethanol response rates and absolute ethanol intake (g/kg).

BASIC PROTOCOL 2

TRAIN RATS TO INITIATE SACCHARIN-MAINTAINED RESPONDING ON AN FR-4 SCHEDULE

The goal of this protocol is to match the behavioral patterns and response rates obtained with the 10% (v/v) alcohol group in Basic Protocol 1.

For materials, see Basic Protocol 1

1. Follow Basic Protocol 1, steps 1 to 7 using 10 naive rats.

2. On day 14, after obtaining stable responding on the FR-4 schedule, maintain rats on the 0.10% saccharin reinforcer continuing to use only the right lever in the chamber for 1 to 2 additional weeks (day 21 to 28) to ensure optimal stabilization.

   Based on the author's experience with outbred and alcohol selectively-bred rats, this saccharin concentration should be in the range of 0.025% to 0.10% (see June et al., 1998). It is recommended to start with 0.05% saccharin (i.e., a moderate concentration) and then adjust the concentration accordingly.

3. Analyze the saccharin-maintained responding data in an identical way as the ethanol-maintained responding data as discussed in steps 11 and 12 of Basic Protocol 1.

BASIC PROTOCOL 3

TRAIN RATS TO LEVER PRESS CONCURRENTLY FOR ETHANOL AND SACCHARIN UNDER AN FR-4 SCHEDULE

In a concurrent procedure, a 10% ethanol reinforcer is presented in the operant chamber simultaneously with a saccharin reinforcer (e.g., 0.025% to 0.1%) that provides response rates and patterns similar to those obtained with ethanol (see Basic Protocol 2; Samson and Grant, 1985; Petry and Heyman, 1995; June et al., 1998). In general, high- to very high-responding rats that lever-press for ethanol or saccharin tend to perform substantially better in a concurrent protocol situation (see June et al., 1998, 1999, 2001). Thus, beginning this protocol with a sample size two times that of Basic Protocols 1 and 2 (i.e., $n = 20$) will provide the researcher with a greater opportunity to obtain the desired sample of rats capable of responding similarly for two concurrently presented reinforcers. It is important to note that while the experimental demands inherent within the concurrent protocol are much greater than Basic Protocols 1 and 2, the capacity to show that a pharmacological agent can selectively alter ethanol consumption in the absence of modifying another highly palatable oral reinforcer is regarded as an optimal test for preclinical screening of agents that might hold promise in reducing alcohol intake in humans (Carroll et al., 1991; Petry, 1997; June et al., 1999; Rodefer et al., 1999).

Materials (see Basic Protocol 1)

1. Perform Basic Protocol 1, steps 1 to 9 using 20 naive rats.

2. After obtaining stable responding on the FR-4 schedule for 10% ethanol (around day 35 to 40), maintain rats on the ethanol reinforcer for an additional 1 week, continuing to use only the right lever in the chamber.

3. After the final week of ethanol stabilization with the single right lever, insert the left lever and dipper assembly system into the operant chamber. Fill the left reservoir with ∼15 ml of saccharin solution. Use the saccharin concentration determined in Basic Protocol 2 that matched the patterns and response rates of the 10% ethanol.

4. With 10% ethanol in the right dipper assembly and saccharin in the left, train rats for 10 days in 60-min daily sessions under the FR-4 schedule. Alternate the position of the levers and associated dippers for each reinforcer for each session (or for alternating 2-day and 1-day sessions) to avoid the establishment of lever preference. Following a 10-day evaluation, determine if responding for the two reinforcers is similar or equal (generally rates of saccharin responding that are at least 70% that of ethanol and no greater than 25% higher than ethanol). Adjust the saccharin concentration up or down, if necessary, to match the ethanol response rates. Use animals that respond only within these limits.

5. Obtain stable responding for the concurrently presented 10% ethanol and saccharin reinforcers for at least 2 weeks before beginning experimental drug treatments. 6. Analyze data as described in Basic Protocol 1, steps 11 and 12. Collect BAC data (see Support Protocol 1) as necessary throughout the drug-treatment phase.

BASIC PROTOCOL 4

TRAIN RATS TO LEVER PRESS CONCURRENTLY FOR ALCOHOL AND AN ISOCALORIC ALTERNATIVE SOLUTION UNDER AN FR-4 SCHEDULE

Additional Materials (also see Basic Protocol 1)

10% (w/v) sucrose solution

1. Follow Basic Protocol 1, steps 1 to 6, using 10 naive rats. After the last day of fluid deprivation, train the rats for 5 additional days under the 60-min daily access paradigm to respond for the 10% sucrose solution on an FR-1 schedule.

2. Continue the FR-1 schedule on the 10% sucrose solution for 7 consecutive days. Around day 8, after obtaining stable responding on the FR-1 schedule for the sucrose (i.e., ±20% of the average responses for 5 consecutive days), gradually add increasing concentrations of ethanol (2.5%, 5%, 7.5%, 10%, 15%, 20%, and 25%) to the sucrose solution. Maintain animals on each of the sucrose plus ethanol cocktails for 2 days (for a total of 14 days of sucrose plus ethanol).

3. After completing the induction phase, insert the right lever and dipper assembly into the operant chamber. Present 10% ethanol + 10% sucrose on one lever and water on the second lever to begin the concurrent schedule presentation.

4. On the second day of the concurrent schedule presentation, begin gradually increasing the concentration of sucrose in the water reservoir over 5 days (i.e., 2%, 4%, 6%, 8%, and 10%) while continuing to present the 10% ethanol/10% sucrose cocktail on the second lever. Place both solutions on alternating sides (i.e., maintaining the left/right orientation) of the operant chamber to ensure sampling of both alternatives and prevention of establishing a lever preference. Use the following schedule to proceed with concurrent schedule presentation over the next 5 days after the initial concurrent 10% ethanol + 10% sucrose (lever 1) and water (lever 2) choice:

    Day 2: 10% ethanol + 10% sucrose (right lever); 2% sucrose (left lever)
    Day 3: 10% ethanol + 10% sucrose (left lever); 4% sucrose (right lever)
    Day 4: 10% ethanol + 10% sucrose (right lever); 6% sucrose (left lever)
    Day 5: 10% ethanol + 10% sucrose (left lever); 8% sucrose (right lever)
    Day 6: 10% ethanol + 10% sucrose (right lever); 10% sucrose (left lever).

5. On day 7, begin training on the concurrent schedule with the two isocaloric alternatives. Over the next 5 days, the rats should receive 10% ethanol + 10% sucrose + 0.25% saccharin (w/v) (lever 1), and 10% (w/v) sucrose (lever 2).

    The addition of the saccharin to the cocktail will increase the palatability of the ethanol and, in turn, increase the efficacy (i.e., reward strength) of the ethanol cocktail.

6. Gradually fade the sucrose by daily reducing the concentration (i.e., 7.5%, 5%, 0%) in the 10% ethanol + 10% sucrose + 0.25% saccharin cocktail (lever 1) while continuing to present the 10% sucrose (lever 2). Place both solutions on alternating sides of the operant chamber to ensure sampling of both alternatives and prevention of establishing a lever preference. The concurrent schedule presentation would proceed as follows over the next 6 days:

    Day 1: 10% ethanol + 10% sucrose + 0.25% saccharin (right lever); 10% sucrose (left lever)
    Day 2: 10% ethanol + 7.5% sucrose + 0.25% saccharin (left lever); 10% sucrose (right lever)
    Day 3: 10% ethanol + 5% sucrose + 0.25% saccharin (right lever); 10% sucrose (left lever)
    Day 4: 10% ethanol + 2.5% sucrose + 0.25% saccharin (left lever); 10% sucrose (right lever)
    Day 5: 10% ethanol + 0% sucrose + 0.25% saccharin (right lever); 10% sucrose (left lever)
    Day 6: 10% ethanol (left lever); 10% sucrose (right lever).

7. Continue the 10% ethanol (lever 1) and 10% sucrose (lever 2) regimen for 2 additional days.

8. On day 9, increase the reinforcement schedule to an FR-4 schedule and increase the 10% sucrose solution to a 14.2% solution. (The 10% ethanol and the 14.2% sucrose solutions are isocaloric.) Maintain the animals on the FR-4 schedule for 7 days with the two isocaloric fluids. Place solutions on alternating sides of the operant chamber to ensure sampling of both alternatives and prevention of establishing a lever preference.

    This period will stabilize the animals under the FR-4 schedule for the two isocaloric fluids. The stability criteria outlined in Basic Protocol 1 (i.e., neither systematic increase nor decrease in responding over at least five consecutive sessions using a reference point of ±20%) should also be applied to the two-lever situation. However, each solution must be viewed individually for stabilization purposes; for example, stabilization should occur across time in a "single" solution, not "between" the two solutions. After attaining the stabilization phase, experimental manipulations may begin.

9. Record the number of responses and reinforcers of the concurrently presented ethanol and sucrose solutions under the FR-1 and FR-4 schedules of reinforcement by the operant software. Collect BAC data (see Support Protocol 1) as necessary throughout the drug-treatment phase.

    The number of responses and reinforcers of the sucrose and ethanol solutions during the stabilization and post-stabilization periods are the principal data of interest. For investigators who are interested in conducting an interval-by-time analyses, the operant software can be programmed to collect data in 1-, 2-, 5-, or 10-min blocks/bins for within-session analyses. Even if the investigator is not sure that he will analyze the within-session data, the author suggests these data be recorded. The amount of absolute ethanol intake (g/kg) should also be evaluated and recorded as noted in Basic Protocols 1 and 3.

References: Samson, 1987; Weiss and Koob, 1991

Contributor: Harry L. June

UNIT 4.13

Experimental Autoimmune Encephalomyelitis (EAE)

STRATEGIC PLANNING

This section discusses the major subtypes of the EAE model, active EAE and passive or adoptively transferred EAE, and the strengths of the different models. In active EAE, susceptible strains of mice are immunized with an appropriate myelin antigen or peptide emulsified in complete Freund's adjuvant (CFA). Mice also are usually administered pertussis toxin (PT) on the day of immunization and 48 hr later. Depending on the strain of mouse and the antigen used, EAE will manifest 10 to 15 days after the initial immunization. With active immunization, 75% to 80% incidence of disease can be expected, and the course of disease is often monophasic. An important exception is the active immunization model induced in SJL or (SJL × SWR) F1 mice using the proteolipid protein (PLP) peptide 139–151 emulsified in CFA as the inducing antigen. In this instance, the mice develop a relapsing and remitting disease course that occurs without the use of PT.

In passive or adoptively transferred EAE, the disease is induced by injecting mice with activated, myelin-specific T cells, which are generated by immunizing mice with the inducing myelin protein or peptide emulsified in CFA. The draining lymph nodes are removed and the T cells activated again with the inducing antigen in vitro. The activated T cells are then transferred into naive recipients, which then develop EAE. The most common forms of adoptively transferred EAE are in the SJL or (PL × SJL) F1 mouse using myelin basic protein (MBP) as the inducing antigen (e.g., Racke et al., 1992). In these two forms of EAE, the mice develop a relapsing and remitting form of disease, which makes this model very conducive to studies examining interventions for established disease.

Another advantage to the adoptive transfer model is that EAE can be broken down into its various pathophysiologic components. These components include the initial priming of encephalitogenic T cells in donor mice that have been immunized with the myelin antigen, the subsequent activation and expansion of encephalitogenic T cells in vitro, and the subsequent capability of these T cells to enter the CNS and cause the clinical manifestations of EAE. If an experimental therapeutic agent prevented the priming of encephalitogenic T cells, it probably would not be effective if given to recipient mice that have received T cells that were already primed and activated.

With the advent of transgenic technology, mice have been generated that have the majority of their T cells specific for myelin antigens. The mice that have been most thoroughly characterized in this regard are those with T cells specific for the N-terminal epitope of MBP, Ac1-11 (Goverman et al., 1993). This strain is available from the Jackson Laboratory and allows study of the adoptive transfer model of EAE using a very well defined T cell for inducing EAE. It is important to note that when working with these mice, a very clean mouse facility is absolutely required, as these mice develop a high incidence of spontaneous EAE in conventional animal facilities. Several investigators have used the Vβ8 T cell receptor transgenic mouse for EAE studies, and this mouse is also available through the Jackson Laboratory (Ratts et al., 1999). This mouse does not develop spontaneous EAE, but can easily be induced to develop active EAE by immunization with MBP Ac1-11 peptide emulsified in CFA. In the author's experience, this protocol induces 100% incidence of disease and does not require PT in the induction protocol.

When testing genetically manipulated mice, the choice of EAE model is most likely going to be influenced by the genetic background of the transgenic or mutant mouse strain. The susceptibility of many mouse strains to MBP and PLP is shown in Table 4.13.1. In addition, many mutant strains are available on the 129 or B6 background, and in some cases, (B6 × 129) F1 mice. B6 mice and (B6 × 129) F1 mice are resistant to the induction of MBP-induced EAE, but the 129 mice are susceptible to this antigen for both active and passive disease. B6 mice are susceptible to EAE induction with myelin oligodendrocyte glycoprotein (MOG) or the MOG peptide 35-55 (see Table 4.13.2). Most reports using MOG in the B6 strain employed the active EAE model and the 35-55 peptide.

For testing the therapeutic potential of a reagent in the EAE model, the chronic, relapsing forms of disease are most advantageous because of the relevance of being able to test the therapeutic reagent in the setting of established or ongoing disease. The SJL mouse for relapsing disease has been most widely used for this purpose. Adoptively transferred EAE using MBP as the inducing antigen with either SJL or (PL × SJL) F1 mice produces a chronic, relapsing model that should be quite reproducible within a particular laboratory. For many laboratories, the active EAE model in the SJL mouse with PLP peptide 139-151 is of greatest utility because it produces a relapsing-remitting disease following a single immunization and is technically much simpler than the adoptive transfer model.

NOTE: Female mice are generally used because they can be group housed without difficulty. Male mice will often fight and need to be housed separately. In SJL mice, females are more

Table 4.13.1 Susceptibility of Inbred Strains of Mice to Actively Induced EAE[a]

Strain	MBP[b]	PLP[b]	MOG[b]
B6	No	Yes, but very mild	Yes
SJL/J	Yes	Yes	Yes
DBA/1	Yes, mild	Yes, but very mild	Unknown
PL/J	Yes	Yes	Unknown
C3H	Yes	Yes, but mild	Unknown
Balb/c	No	Yes (depends on substrain)	Unknown
B10	No	Unknown	Unknown
B10.PL	Yes	Yes	Unknown
B10.A	Yes	Unknown	Unknown
B10.RIII	Yes	Unknown	Unknown
B10.S	No	No	No
AKR	Yes	Yes, but mild	Unknown
129	Yes	Unknown	Yes

[a] Inbred mouse strains can be obtained from the Jackson Laboratory, Taconic Farms, or Harlan Sprague Dawley.
[b] Abbreviations: MBP, myelin basic protein; MOG, myelin oligodendrocyte glycoprotein; PLP, proteolipid protein.

Table 4.13.2 Myelin Peptides Used To Induce EAE

Peptide[a]	Predominant mouse strains when used	Encephalitogenic sequence
MBP Ac1–11	PL/J, B10.PL	ASQKRPSQRSK
MBP 89–101	SJL	FKNIVTPRTPPP
PLP 139–151	SJL	HSLGKWLGHPDKF
MOG 35–55	B6	MEVGWYRSPFSRVVHLYRNGK

[a] These peptides are usually custom-synthesized. The author has had success using peptides synthesized by CS Bio.

susceptible to the development of EAE, but there are mouse strains (e.g., B10.PL) where males are more susceptible.

BASIC PROTOCOL

ACTIVE INDUCTION OF EAE IN MICE

Materials (see APPENDIX 1 for items with ✓)

    Myelin antigen (MBP, PLP, MOG, or other peptide; see Table 4.13.2). Peptides can be custom synthesized or obtained from a commercial source such as CS Bio

✓ PBS
    Complete Freund's adjuvant (CFA; Difco)
    Female mice from EAE susceptible strain (see Table 4.13.1), 8 to 12 weeks of age, group housed
    Methoxyflurane (anesthesia)
    Pertussis toxin (PT; List Biological Laboratories)

    Omni Mixer (Omni International, Popper & Sons; can also use 2- or 5-ml Micro-mate interchangeable glass syringes connected by a 7/8-in. stainless steel, 18-G micro-emulsifying needle)

1-ml tuberculin syringes
25-G needles
Bell jar
Electric hair clippers
Animal balance (for weighing mice), accurate to 0.1 g

1. For whole-protein induction of EAE, dissolve either MBP, PLP, or MOG (see Table 4.13.2) in PBS at a concentration of 8 mg/ml. For antigenic peptides, dissolve them at a concentration of 4 mg/ml in PBS.

2a. *To use an Omni Mixer:* Place 1:1 (v/v) antigen and CFA in the small mixing cup and mix for ~3 min at high speed until a thick, white emulsion forms. Mix enough antigen in PBS and CFA for two times the number of mice to be immunized with 100 µl for a total antigenic dose of 400 µg/mouse for whole protein or 200 µg/mouse for peptide. For example, for 20 mice, mix 2 ml of the myelin antigen in PBS and 2 ml of CFA.

2b. *To use a micro-emulsifying needle:* Use 2- or 5-ml Micro-mate glass syringes and 18-G stainless steel micro-emulsifying needle. Attach the syringes to a double-ended micro-emulsifying needle and push the plungers back and forth to mix the antigen with the CFA until a thick, white emulsion forms. For example, for 20 mice, draw up 2 ml of the myelin antigen in a 5-ml syringe and 2 ml of CFA in another 5-ml syringe.

3. Transfer the emulsion into 1-ml tuberculin syringes and attach a 25-G needle. Prepare the emulsion just prior to immunization and keep on ice until used.

4. Anesthetize mice (up to five at one time) by placing them in bell jar with methoxyflurane-moistened cotton swabs or paper towels to the point where they are unresponsive to painful stimuli such as a tail pinch. Shave mice with electric hair clippers over their flanks in preparation for immunization. Inject mice subcutaneously in the four shaved areas with ~25 µl of emulsion per site (100 µl/mouse).

5. Inject mice (i.p. or i.v. via the tail vein, anesthetic not required) with 200 ng PT dissolved in 200 µl PBS and return mice to their home cage. Repeat 200-ng PT injection 48 hr later. Dilate the tail vein by warming the tail with a heating lamp or warm water to facilitate tail vein injection.

6. Weigh animals daily on an animal balance (10% loss of body weight expected in 4 to 7 days of an acute EAE episode). Monitor daily for signs of EAE beginning 1 week after immunization. Grade mice using the following score (see Figure 4.13.1 for explanation).

    0: no abnormality
    1: a limp tail
    2: mild hindlimb weakness
    3: severe hindlimb weakness
    4: complete hindlimb paralysis
    5: quadriplegia or premoribund state
    6: death

ALTERNATE PROTOCOL

ADOPTIVE TRANSFER OF EAE IN MICE

Additional Materials (*also see Basic Protocol; see* APPENDIX 1 *for items with* ✓)

✓ HBSS
✓ Complete EAE medium
75% ethanol

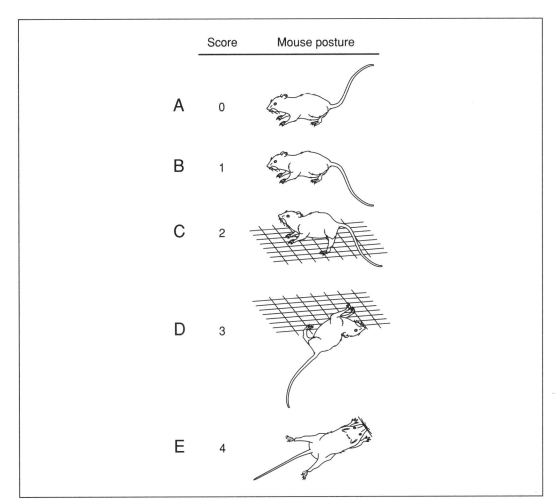

Figure 4.13.1 Clinical scoring of EAE. (**A**) Mice that are removed from a cage will normally respond by having their tail stand straight up. When picking up a mouse by the tail, one can feel that the tail has tone. Such a mouse is normal and represents a clinical score of 0. (**B**) A normal mouse when placed on top of the cage will not misstep between the bars of the cage top. When a mouse has reached a score of 1, the tail no longer stands up on end. When the mouse is picked up by the tail, there is a distinct lack of tone in the tail. However, the mouse walks normally, and when put on the underside of the cage top, can climb to the top without difficulty. (**C**) When a mouse has reached a clinical score of 2, the mouse has a limp tail and shows signs of hindlimb dysfunction. This is most easily detected by placing the mouse on the underside of the cage, where the mouse is able to hang on, but because it does not have the same dexterity of hindlimb movement, it has difficulty climbing to the top side of the cage. When placed on the top side of the cage, the mouse may misstep and the hindlimb foot may protrude between the bars of the cage. (**D**) When the mouse reaches a clinical score of 3, the mouse can no longer hold on to the cage with its hindlimbs when placed on the underside of the cage. However, when ambulating, the mouse still can move the hindlimbs. (**E**) When the mouse reaches a clinical score of 4, the hindlimbs drag behind and are not used by the mouse for movement. A moribund mouse is still alive, but really makes little spontaneous movement and receives a clinical score of 5. These animals are routinely euthanized.

50-ml conical tubes
Styrofoam dissecting board
Surgical scissors
Jeweler's curved forceps
4.5 × 4.5–cm stainless steel wire mesh screen

60 × 15–mm petri dish, sterile
3-ml plastic syringe
24-well tissue culture plates
Humidified 37°C, 5% CO_2 incubator

1. Perform Basic Protocol, steps 1 to 4 for active induction of EAE. Prepare HBSS and EAE media prior to lymph node harvest. Put ∼25 ml HBSS in a 50-ml conical tube on ice.

2. Sacrifice the mice with an overdose of anesthesia. Pin mouse supine on the dissecting board and spray with 75% ethanol prior to removal of lymph nodes.

3. Using surgical scissors, cut mouse along the midline and cut along all four extremities. Tease back skin and pull it back to expose the axillary and inguinal lymph nodes. Using the jeweler's curved forceps, dissect out the inguinal and axillary lymph nodes (6 nodes total). Place each set of nodes into the 50-ml conical tube with the cold HBSS while collecting lymph nodes from all the mice in the particular experiment.

4. Bend a 4.5 × 4.5–cm wire mesh screen at the corners to form a platform and place it in a sterile 60 × 15–mm petri dish. Place 10 to 20 lymph nodes on the screen and press them through the screen with a plunger from a 3-ml plastic syringe. Wash cells remaining on screen with 2 to 4 ml of HBSS and then transfer the media containing the cells to a new 50-ml conical tube. Repeat this until all the lymph nodes have been disrupted into a single cell suspension.

5. Centrifuge the cell suspension for 10 min at 200 × g (1000 rpm), 4°C using an Eppendorf A-4-62 swinging bucket rotor. Decant the supernatant and disrupt the pellet by tapping the 50-ml conical tube between the thumb and index finger. Resuspend the cell pellet in 20 ml of HBSS and centrifuge again (this completes the first wash). Repeat the wash, but resuspend the cells from the second wash in complete EAE medium at 4×10^6 cells/ml (see Phelan, 1998). Culture cells with the immunizing myelin antigen (25 to 50 μg/ml of MBP, PLP, or MOG) in 24-well tissue culture plates (2 ml/well) for 96 hr in a humidified 37°C, 5% CO_2 incubator.

6. Aspirate cells from the 24-well plates, place in 50-ml conical tubes, and centrifuge for 10 min at 200 × g (1000 rpm), 4°C. Wash pellets two times with complete HBSS (as in step 5) and count (see Phelan, 1998). In the pilot experiment, inject 3×10^7 cells/mouse i.v. in the tail vein or i.p.

7. Monitor animals daily for signs of EAE as described in the Basic Protocol, step 6 for the active induction of EAE (100% incidence of EAE in 6 to 7 days expected).

Because this is a biological system where numerous factors influence the severity of disease, the number of cells required to cause reproducible EAE may have to be adjusted.

References: Martin et al., 1992; Zamvil and Steinman, 1990

Contributor: Michael K. Racke

UNIT 4.14

Models of Amyotrophic Lateral Sclerosis

Amyotrophic lateral sclerosis (ALS) is an adult-onset chronic neuromuscular disease characterized pathologically by the relatively selective progressive degeneration of cortical motor neurons (upper motor neurons) and motor neurons in the brainstem and spinal cord (lower motor neurons). The two experimental models described in this unit, organotypic cultures of spinal cord (Basic Protocol 1) and transgenic mice expressing a mutated human *SOD1* gene (Basic Protocol 2), can be used to screen putative therapeutic agents against neurodegeneration. Please see Table 4.14.1 at the end of this unit for troubleshooting information.

NOTE: All reagents and equipment coming in contact with live cells must be sterile, and proper aseptic technique should be followed accordingly.

BASIC PROTOCOL 1

SPINAL CORD ORGANOTYPIC CULTURES

This model addresses whether a chronic loss in glutamate uptake could produce elevations in extracellular glutamate and potentially lead to a slow loss of motor neurons, as seen in ALS.

Materials *(see APPENDIX 1 for items with* ✓ *)*

✓ Incubation medium, 37°C
    GBSS (Invitrogen Life Technologies) supplemented with 6.4 mg/ml glucose and filter sterilized before each use
    70%, 95% (v/v), and absolute ethanol
    5% (v/v) Betadine
    8-day-old litter of Sprague Dawley rat pups (Charles River Laboratories)
✓ 0.1 M sodium phosphate buffer, pH 7.4
    100% (v/v) methanol, ice cold (for SMI-32 immunostaining only)
    0.1% (v/v) Triton X-100 in 0.1 M sodium phosphate buffer, pH 7.4 (for Islet-1 immunostaining only)
    TBS: 50 mM Tris·Cl/1.5% (w/v) NaCl, pH 7.4
    1% and 5% (v/v) horse serum (heat inactivated) in TBS
    Monoclonal SMI-32 antibody (Sternberger Monoclonals)
    Monoclonal Islet-1 antibody (Vector Laboratories)
    Biotinylated horse anti-mouse antibody (Vector Laboratories)
    ABC reagent (Vector Laboratories) diluted 1:100 in TBS
    0.5 mg/ml diaminobenzidine (DAB)/0.075% (v/v) hydrogen peroxidase
    Xylene
    Permount

6-well (35 mm) tissue culture plates
Sterilized dissection instruments, including:
        17.8-cm (7-in.) microdissecting forceps, serrated
        12-cm (4.75-in.) microdissecting tweezers, straight
        Two 10.5-cm (4.125-in.) microdissecting tweezers, tips at 45° angle
        16-cm (6.25-in.) microspatula, 1.3 × 0.3–cm (0.5 × 0.125–in.) spoon end
        14-cm (5.5-in.) operating scissors, straight, sharp-blunt
        Microdissecting tweezers, curved
        11.5-cm (4.5-in.) microdissecting scissors, straight
Millicell CM 30-mm inserts (Millipore)
Petri dishes or 10-cm tissue culture plates, sterile
McIlwain tissue chopper (Stoelting)
Double- and single-edged razor blades, sterile
Aclar plastic film (Honeywell Specialty Films), cut into 5.5 × 3.5–cm pieces, sterilized by autoclaving
100-ml beakers
Glass transfer pipets (∼2-mm i.d.; sterile): made by breaking a 2.3-cm (5.75-in.) Pasteur pipet where it narrows (∼1.7 cm from the tip) and flaming the broken end for smoothness

Dissecting bench setup including:
    Styrofoam dissection surface and pins
    250-ml beaker containing 5% (v/v) Betadine
    250-ml beaker containing 70% (v/v) ethanol
    Spray bottle with sterile saline
    Spray bottle with 70% (v/v) ethanol
    Cotton gauze, numerous 4 × 4–cm pieces, soaked in 70% (v/v) ethanol
Magnifying dissecting glasses
Incubator, humidified 37°C, 5% CO_2
Pasteur pipets, sterile
Fine brush
Netwell inserts (Fisher)
12-well tissue culture plates
Shaker, 4°C
Glass slides
Coverslips
Microscope with 40 and 100× magnification

1. Working in a laminar flow hood, pipet 1 ml of 37°C incubation medium into each well of a 6-well tissue culture plate. Set up five plates per litter (average of 10 pups). Using sterile 17.8-cm microdissecting forceps, place a Millicell CM 30-mm insert (Millipore membranes highly recommended for successful cultures) into each medium-filled well. Cover the prepared plates and set aside in the hood.

2. Fill two petri dishes or 10-cm tissue culture plates per rat pup with sterile supplemented GBSS, cover dishes, and set aside in the hood.

3. Wipe a McIlwain tissue chopper with 70% ethanol. Set slice thickness to 350 μm. Attach a sterile double-edged razor blade to the chopper and position the blade for even cuts with sufficient force to partially, but not completely, cut through a 5.5 × 3.5–cm piece of Aclar plastic film.

4. Prepare two 100-ml beakers with 25 to 50 ml of 70% ethanol. Place a pair of 12-cm straight microdissecting tweezers, two pairs of 10.5-cm angled microdissecting tweezers, and a 16-cm microspatula in one beaker and glass transfer pipets in the second beaker.

5. Place 17.8-cm forceps and a pair of 14-cm operating scissors in 5% Betadine and place curved microdissecting tweezers, 11.5-cm microdissecting scissors, and a single-edged razor blade in 70% ethanol. Wipe a Styrofoam dissection surface with 70% ethanol.

6. Wearing a gown, gloves, hood, and mask, quickly dip each 8-day-old rat pup in a 250-ml beaker containing 70% ethanol, then in a 250-ml beaker containing 5% Betadine, and then in ethanol again, and rapidly decapitate pup with operating scissors. Pin pups supine on ethanol-soaked gauze on the dissecting board.

    The initial animal dissection (of up to four pups at a time) should be performed outside the hood, and the procedures completed inside the hood (step 8)

7. Use the razor blade to make a single longitudinal midline cut extending from the suprasternal notch to the pubic symphysis in each animal. Expose the thoracic cavities by removing ventral ribs and overlying skin with forceps and scissors. Remove the contents of thoracic and abdominal cavities using forceps, exposing the full length of the vertebral column.

8. Use a spray bottle with sterile saline to rinse the cavities. Remove the carcasses from the board, dip them into 5% Betadine, and use a spray bottle with 70% ethanol to rinse the

cavities. Pin pups supine onto clean 70% ethanol-soaked cotton gauze on the dissecting board and place the board in the laminar flow hood.

9. Make a transverse cut with the 11.5-cm microdissecting scissors between the two most caudal full-length ribs and through the vertebral column and spinal cord. Carefully lift the ventral vertebral column up and away from the lumbosacral spinal cord by making small, distally directed cuts with the microdissecting scissors held parallel to the spinal cord.

10. Make a second transverse cut through the distal end of the spinal cord near the conus and carefully lift the cord from the vertebral column using the curved microdissector tweezers. Use the meninges attached to the cord to lift the cord. Place the cord from each pup into a separate GBSS-filled petri dish (from step 2).

11. While wearing a pair of magnifying dissecting glasses, enter the hood with head and shoulders, carefully dissect the cord free of surrounding meninges, and transect the remaining nerve roots adjacent to the cord.

12. Remove the magnifying glasses. Lift the cord onto a piece of Aclar film using the microspatula. Align the cord longitudinally along the length of the Aclar film with the dorsal spinal cord vessel on top and move any excess GBSS away from the cord with the microspatula.

13. Place the Aclar film directly on the plastic disk of the tissue chopper chopping stage and cut the cord transversely in 350-μm sections (generally twelve to fifteen organotypic slices per lumbar cord). Stop the chopper at the point at which the cord narrows distally.

14. Lift the Aclar film from the chopper and place the cord sections into a fresh GBSS-filled petri dish (from step 2) by turning the film upside down against the petri dish and "washing" the sectioned cord into the dish.

15. Using magnification inside the hood as in step 11, carefully separate each transverse section of the cord with curved tweezers. Transfer individual sections of cord (only intact, undamaged sections) to the Millicell inserts (from step 1) using glass transfer pipets. Create suction on the top of the pipet with the thumb, lift a single cord section into the pipet; let the cord section settle to the bottom of the GBSS in the pipet and lightly touch the surface of the membrane with the section, releasing a minimal amount of GBSS with the section. Remove excess GBSS on the membrane with bulb suction and a sterile Pasteur pipet. Place five cord sections around the circumference of each membrane in each well.

16. Incubate the culture plates in a humidified incubator at 37°C, 5% CO_2. Change the incubation medium, along with any added pharmacological agents (after a 7-day recovery period), twice weekly. Use sterile Pasteur pipets attached to a vacuum aspirator to remove the medium and carefully refill each well with 1 ml prewarmed medium. Keep the medium off the surface of the Millicell membrane.

17. Rinse cultures three times, for 10 min each, with 4 ml of 0.1 M sodium phosphate buffer. Permeabilize cultures by incubating 10 min in 4 ml ice-cold 100% methanol or 0.1% Triton X-100 in 0.1 M sodium phosphate buffer for SMI-32 and Islet-1 staining, respectively. Rinse three times with 4 ml TBS. Incubate cultures 60 min in 4 ml 5% horse serum in TBS to block nonspecific staining.

18. Remove cultures from the Millicell inserts and use a fine brush to transfer them to Netwell inserts in a 12-well tissue culture plate. Incubate cultures 60 min in monoclonal SMI-32 antibody (1:8000) or monoclonal Islet-1 antibody (1:100) diluted in 1% horse serum in TBS. Incubate some cultures (as negative controls) in blocking solution with not primary

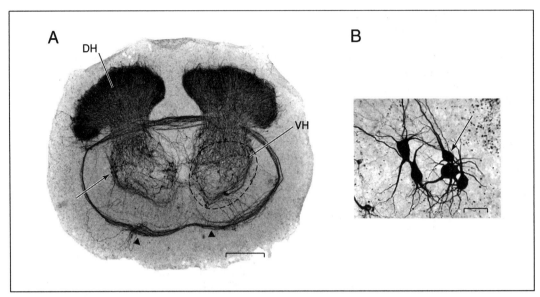

Figure 4.14.1 Identification of individual motor neurons in spinal cord organotypic cultures. (**A**) Pools of ventral-horn motor neurons can be readily visualized with SMI-32 immunolabeling. The large cells have a typical medial and lateral ventral-horn location. Lacking a target (muscle), their large-caliber axons encircle the slice cultures. (**B**) A higher-power view demonstrates the large soma characteristic of motor neurons. In this whole-mount stain, the individual motor neurons can be easily counted. Scale bars equal ∼1 mm and 50 μm, respectively. Abbreviations and symbols: DH, dorsal horn; VH, ventral horn; arrowheads, motor axons; arrows, motor neurons.

antibody added. Then place the plates on a shaker at 4°C and continue to incubate overnight. Rinse the plates three times, for 10 min each, in 4 ml TBS at room temperature with agitation.

Using the Netwell inserts allows smaller volumes (as little as 250 μl/well) to be used and conserves expensive antibodies. Up to 15 sections can be arranged in each Netwell insert, so sections from an entire 6-well plate (6 × 5 = 30) are probed in two wells of a 12-well plate.

19. Incubate cultures 60 min with biotinylated horse anti-mouse antibody diluted 1:2000 in 1% horse serum in TBS. Then incubate cultures 60 min in 4 ml ABC reagent diluted 1:100 in TBS.

20. Rinse cultures three times, for 10 min each, in 4 ml TBS. Develop color reaction in 2 ml 0.5 mg/ml DAB/0.075% hydrogen peroxide. Incubate until desired stain intensity develops (intense brown; usually 3 to 10 min). Rinse three times, for 10 min each, in 4 ml TBS. Mount sections on glass slides, for convenience in the same orientation and in a straight line. Allow them to air dry for a few hours.

21. Wash the mounted sections and dehydrate them by incubating for a few seconds in water and then in an alcohol series consisting of two incubations each in 70%, 95%, and absolute ethanol. Clear the sections by incubating for a few seconds in xylene.

22. Add a small drop of Permount to the glass slide. Carefully place a coverslip on the drop, avoiding air bubbles. Remove any excess mounting solution with a paper towel. Observe each section under 40× magnification to get oriented. Use 100× magnification for the actual counting and measurement of motor neurons, which are darkly stained large cells (with >25-μm somas) with processes that are located in the ventral horn (Figure 4.14.1A and B).

BASIC PROTOCOL 2

ASSESSMENT OF THE CLINICAL EFFICACY OF THERAPEUTIC AGENTS IN A MUTANT SOD1 TRANSGENIC MOUSE MODEL OF AMYOTROPHIC LATERAL SCLEROSIS

The effect of the familial amyotrophic lateral sclerosis (FALS)-linked superoxide dismutase (SOD) mutant proteins has been studied in transgenic mice expressing various *SOD1* mutations. These mice develop clinical symptoms similar to those seen in ALS patients (i.e., weakness of the limbs, muscle wasting, motor neuron degeneration, and death; Lee et al., 1996) and are a widely accepted model of the familial form of the disease.

CAUTION: Phenol is extremely caustic and chloroform is a suspected carcinogen. Wear appropriate protective equipment and work in a fume hood.

CAUTION: Sodium azide is poisonous. Wear gloves and handle cautiously.

Materials *(see APPENDIX 1 for items with ✓)*

Male SOD1 transgenic mice: B6SJL-TgN(SOD1-G93A)1Gur (GIH; survival: 132 ± 11 days) or B6SJL-TgN(SOD1-G93A)1Gur^{del} (GIL; survival: 251 ± 28 days) mice (The Jackson Laboratory)
Female B6SJL mice (The Jackson Laboratory)
Inhalational anesthetic (halothane or isoflurane)
✓ Tail buffer
10 mg/ml proteinase K, stored in aliquots at −20°C
✓ Buffered phenol, pH 8.0
✓ 25:24:1 (v/v/v) phenol/chloroform/isoamyl alcohol
24:1 (v/v) chloroform/isoamyl alcohol
70% and 100% (v/v) ethanol
✓ TE buffer, optional
5 U/μl *Taq* DNA polymerase supplied with 10× PCR amplification buffer with 15 mM MgCl$_2$ (Roche Diagnostics)
20 mM dNTP mix (Invitrogen Life Technologies)
676 μM mouse SOD1 primer (5′-GTTACATATAGGGGTTTACTTCATAATCTG-3′), custom synthesis
676 μM human SOD1 primer (5′-CCAAGATGCTTAACTCTTGTAATCAATGGC-3′), custom synthesis
676 μM mouse or human (m/h) SOD1 primer (5′-CAGCAGTCACATTGCCCAA/GGTCTCCAACATG-3′), custom synthesis
Mineral oil
5× loading buffer: 10% (w/v) Ficoll 400/0.05 M disodium EDTA, pH 8.0/0.5% (w/v) SDS/0.125% (w/v) bromphenol blue
Therapeutic agent
4% (w/v) chloral hydrate
Cryoprotective solution: 30% (w/v) sucrose/5% (v/v) glycerol/0.02% (w/v) sodium azide in 0.1 M PBS (see APPENDIX 1 for PBS)

55°C water bath
Microcentrifuge, room temperature and 4°C
Thermal cycler
Microscope with 250× magnification

1. Mate male SOD1 transgenic mice with female B6SJL (i.e., wild-type) mice. Remove male when pups are born. Wean pups at 3 to 4 weeks of age. Anesthetize pups with an inhalant

anesthetic (APPENDIX 2B) for 1 to 2 min. Sex and number the pups using the ear punch method (or toe clipping or electronic tags) and use an ethanol-flamed scalpel to cut a 0.5- to 1-cm section from the tip of each tail. Transfer each tail piece into a separate, labeled microcentrifuge tube. Use immediately or store up to 6 to 9 months at −70°C.

2. Cut tail into small, readily digestible pieces and add 700 µl tail buffer. Add 35 µl of 10 mg/ml proteinase K (must be added fresh) and incubate 6 to 24 hr in a 55°C water bath, until the tissue almost entirely disintegrates.

3. Add an equal volume of buffered phenol and mix gently. Microcentrifuge 5 min at 18,000 × g, room temperature. Repeat the phenol extraction until the sample looks relatively clean by visual inspection (no white precipitate at the boundary of the aqueous and organic phases).

4. Transfer the aqueous (top) layer to a fresh microcentrifuge tube and repeat extraction with an equal volume of 25:24:1 phenol/chloroform/isoamyl alcohol. Transfer the aqueous (top) layer to a fresh microcentrifuge tube and repeat extraction with an equal volume of 24:1 chloroform/isoamyl alcohol Transfer the aqueous (top) layer to a fresh tube and add 2 vol room temperature 100% ethanol. Microcentrifuge 10 min at 14,000 × g, 4°C, to recover DNA. Decant supernatant.

5. Gently rinse the pellet well with 70% ethanol. Decant ethanol and air dry the pellet. Resuspend pellet in 50 to 100 µl TE. Quantify DNA (Gallagher, 2006) and dilute to 200 to 500 ng/µl in water. Alternatively, store extracted DNA at −4° to −20°C for up to 1 year.

6. Prepare a reaction mixture containing the following (25 µl final volume), multiplying the volumes by the number of reactions plus extra for losses due to handling:

    2.5 µl of 10× PCR amplification buffer with 15 mM $MgCl_2$
    2.5 µl of 20 mM dNTP mix (2 mM final)
    0.37 µl mouse SOD1 primer (10 µM final)
    0.37 µl human SOD1 primer (10 µM final)
    0.37 µl m/h SOD1 primer (10 µM final)
    0.2 µl of 5 U/µl *Taq* polymerase (1 U final)
    17.69 µl H_2O.

    Dispense 24 µl of the master mix into each PCR tube and then add 1 µl of mouse genomic DNA (from step 5; 200 to 500 ng final). Always include a nontransgenic mouse DNA sample in the PCR as a negative control.

7. Cover reaction mixtures with a drop of mineral oil and carry out PCR in a thermal cycler using the following amplification cycles:

1 cycle:	3 min	94°C	(initial step)
	30 sec	94°C	(denature)
30 cycles:	30 sec	50°C	(anneal)
	30 sec	72°C	(extend)
	10 min	72°C	(final step)
	indefinitely	4°C	(hold).

    The annealing temperature can range from 50° to 55°C if problems occur during the amplification.

8. Prepare a 1.3% agarose gel in 1× TAE containing 0.5 µg/ml ethidium bromide (Voytas, 2000). Add 6 µl of 5× loading buffer to each reaction tube and load 25 to 30 µl reaction

product per well. Electrophorese 1 hr at 80 to 100 V or until the mouse PCR product (~800 bp) is resolved from the human SOD1 gene product (~600 bp).

Transgenic mice have both a human and mouse PCR gene product; nontransgenic mice have only a mouse PCR product.

9. Select sex-, age-, and weight-matched mice randomly across litters for the various treatment groups (i.e., transgenic and nontransgenic animals treated with either therapeutic agent or vehicle alone; 20 to 25 mice for each treatment group). House females in groups of one to six in macrolon cages (32 × 21 × 14–cm), but house males individually to prevent fighting between the males from different litters.

   Ideally, G1H/+ mice, those that express high levels of the mutant SOD, should be no older than 50 days when entered into the study. If G1L/+ (i.e., low expressor) mice are used, then treatment can be started later, between 1 and 4 months of age.

10. Choose the method of administration (intravenously, by subcutaneous or intraperitoneal injection, or oral) and prepare the therapeutic agent for administration. Dissolve water-soluble compounds in sterile water for oral, subcutaneous, or intraperitoneal injection, but for intravenous injection use sterile physiological saline to avoid hemolysis.

11. Weigh the mice daily (to compute the correct dosage) and administer the same dose of therapeutic agent (mg/kg body weight) at the same time each day.

    The average adult male mouse weighs ~25 g and the average female weighs ~20 g. The volume of therapeutic agent should be 0.05 to 0.08 ml. Thus for a dosage of 100 mg/kg, 2.5 mg (i.e., 100 mg/kg × 0.025 kg) or 2 mg (i.e., 100 mg/kg × 0.02 kg) of therapeutic agent would be given to male and female mice, respectively. If a solution containing 2.5 mg/0.08 ml is prepared, then 0.08 ml and 0.064 ml would be administered to males and females, respectively. The dosage is typically based on in vitro studies.

12. Test extension reflex daily when administering therapeutic agent by suspending the mouse in the air by its tail. Score the hindlimb extension as follows:

    Normal extension of both hindpaws is given a score of 2
    Extension of only one hindpaw is given a score of 1
    The absence of any hindlimb extension is given a score of 0.

13. Monitor rotarod activity (UNIT 3.3) at least once a week starting at 12 weeks of age by measuring the length of time and the speed at which the mouse remains on the rotating axle without falling. Increase rpm from 0 to 40 within 10 min. Test each mouse four times on the rotarod, with at least a 20-min rest between tests.

14. Sacrifice three or four mice from each treatment protocol at set time points (e.g., 60, 90, and 120 days of age). Anesthetize mice with an intraperitoneal injection (0.5 to 0.8 ml) of 4% chloral hydrate and perfuse transcardially with 4% paraformaldehyde (Gerfen, 2006). Remove brain and spinal cord and postfix with 4% paraformaldehyde for 24 to 48 hr.

15. Transfer tissue to 7 ml cryoprotective solution. Proceed immediately or store tissue indefinitely at 4°C in cryoprotective solution.

16. Embed the lumbar region (L3) of the spinal cord (Zeller, 1989) in paraffin and cut 7-μm sections on a sliding microtome (Gerfen, 2006). Alternatively cut on a cryostat into 50-μm-thick sections.

17. Stain sections with thionin (Gerfen, 2006) and count the number of motor neurons in the ventral horn region with distinct nuclei and nucleoli using a microscope at a magnification of 250×. Count the number of motor neurons in approximately ten sections per animal per treatment group and average the total.

Table 4.14.1 Troubleshooting Animal Models of Amyotrophic Lateral Sclerosis

Problem	Possible cause	Solution
Spinal cord organotypic cultures:		
Cord sticks to tissue chopper blade during cutting	Tissue chopper blade is dirty	Clean blade with 70% ethanol between cords
Sections slide towards center of Millicell insert or hook up to one another when transferred	Too much fluid carried with the sections; sections will not adhere to a wet membrane.	Carry as little GBSS as possible during section transfer
Medium yellow and cloudy	Bacterial contamination	Remove affected plates from the incubator to prevent cross-contamination; filter sterilize all solutions and additives, and maintain strict aseptic techniques
Sections poorly stained	Sections floating off the membrane during the staining	Totally submerge sections in the solutions throughout the staining; if a section comes off the membrane and starts floating, poke it back down or use a Pasteur pipet to apply solution on top of the floating section to make it sink back in
One part of the section not stained as intensely as the rest	During the different stages of staining, one section touched another and blocked exposure to the staining solutions	Observe sections throughout the staining procedure to make sure that each one is totally submerged in the solutions and that they do not touch one another; if sections are too crowded in the staining inserts, use more inserts and fewer sections per insert (never >15 sections in one insert during immunostaining)
Variable control cultures make it difficult to perform a meaningful statistical analysis	Large variation in motor neuron number between control cultures is normal	Increase the statistical sample size
SOD1 mouse model:		
DNA precipitate absent	Digestion of tissue not complete	Cut tail into smaller pieces and/or incubate longer at 55°C
	Too much tissue prevented separation of organic and aqueous phases	Dilute with more tail buffer and repeat extraction; use smaller tail sample.
	Tissue improperly handled before digestion	Freeze tail samples immediately or add tail buffer to protect DNA from nucleases
PCR product absent	PCR conditions not optimal	Lower the annealing temperature
	DNA template contaminated with cellular proteins, SDS, phenol, or salt	Repeat organic extractions
	Too much template used	Dilute template DNA
	Too little template was used	Use more template
Mouse PCR product absent	High transgene copy number competed for primers	Increase mouse and mouse/human primer concentrations Carry out human and mouse PCR in separate tubes
PCR product present in negative control	PCR master mix contaminated	Repeat PCR using new reagents
Therapeutic agent had no effect	Motor neuron damage too rapid	Begin treatment earlier; use higher dose or bi-daily administration; use different mode of delivery; use low-expressor (G1 L/+) mice
Therapeutic agent had negative effect on controls	Agent is toxic	Use lower dose

18. Sacrifice mice when they are unable to right themselves within 30 sec when placed on their sides on a flat surface or if severe infections develop in one or both eyes. When paresis becomes marked, place moistened food pellets on the bottom of the cage.

19. Using the data from the daily observation for survival, weekly weight measurements, and motor function tests (steps 12 and 13), calculate survival probability and onset probability using standard Kaplan-Meier analysis and commercial software packages, e.g., StatView (SAS Institute).

    A decrease in performance by ≥30% of prior baseline rotarod activity (total running distance) was defined by the authors as the onset of disease-related weakness.

20. Perform statistical analyses using either a two-tailed unpaired Student's t test or an analysis of variance (ANOVA) with $p < 0.05$ to test for significant differences between the different treatment groups for the number of surviving motor neurons (step 17).

References: Barneoud et al., 1997; Gurney et al., 1994, 1996

Contributors: Mandy Jackson, Raquelli Ganel, and Jeffrey D. Rothstein

UNIT 4.15

Measurement of Panic-Like Responses Following Intravenous Infusion of Sodium Lactate in Panic-Prone Rats

BASIC PROTOCOL 1

PRIMING OF ANXIETY IN RATS BY ADMINISTRATION OF UROCORTIN IN THE BASOLATERAL NUCLEUS OF THE AMYGDALA

Materials

Heparin
Saline: 0.9% (w/v) NaCl, sterile
Acetone
Cidex
Male Wistar rats (250 to 300 g; Harlan Sprague-Dawley), singly housed
99.9% isoflurane
Betadine
Alcohol pads
Lidocaine
Racellets: hemostatic cotton pellets with epinephrine
Cranioplastic cement (liquid and powder)
Buprenex
0.05 N sodium lactate solution
Urocortin in 1% (w/v) BSA
Bovine serum albumin (BSA), optional
India ink

0.01- and 0.02-in. Tygon tubing
26-G guide cannulae (10-mm length; Plastic Products)

Surgical tools:
    Scalpel
    Large blunt-tipped scissors
    Small pointed scissors
    Vannas scissors
    Small and large trocars
    Curved forceps
    Vessel cannulation forceps
    Mosquito hemostats
    Needle holder
Telemetric probe (PXT 50; Data Sciences)
Gas anesthetic machine:
    Anesthetic cone mask
    Gas anesthetic regulator
    Plexiglas anesthetic induction chamber
Animal balance (accurate to 1 g)
Electric shaver
Thermoblanket with electric source
Rectal probe
Surgical tape
Sterile surgical drape
1- and 5-ml syringes, sterile
23-G needle
Sterile suture 4.0
Sterile gauze
Plastic hemostats
Straight pins
AM/FM radio
Telemetric magnet
Telemetric probe (PXT 50; Data Sciences)
Sterile 3.0 suture with 1/2-in. tapered needle
Small animal stereotaxic instrument (dual arms)
Anesthetic mask for incisor bar
Cauterizer
Dremel drill with no. 105 tip
2.4-mm screws (Plastic Products)
Ceramic or glass wells
Dummy cannulae (Plastic Products)
Internal cannulae (33-G; Plastic Products)
Spandex jacket (Harvard Apparatus)
Plastic shoebox
Social-interaction box (36-in. length × 36-in. width × 12-in. height with open top)
Computer with telemetry software
Infusion pump (200 nl to 1 ml per minute capacity)
10-μl microsyringes
PE 20 tubing

1. Prepare 40 ml of heparinized saline (1 ml heparin/40 ml of 0.9% NaCl) in a sterile beaker. Make catheters by soaking a 3-cm section of a 30-cm piece of 0.02-in. Tygon tubing in acetone for ~1 min, then insert 0.5 cm of a 4-cm section of 0.01-in. Tygon tubing into the 0.02-in. Tygon tubing. Let dry overnight. Sterilize 26-G guide cannulae and surgical tools. If necessary, sterilize telemetric probe by soaking overnight in Cidex and then rinsing with sterile saline two times.

2. Fill the gas anesthetic chamber with 99.9% isoflurane. Turn on machine and set the pressure dial to 1 to 2 and adjust the flow meter between 2.5 and 3.5. Check the tubing between the induction chamber and the anesthetic mask to verify flow and to check for leakage. Next, weigh the animal on a scale accurate to 1 g and place in the induction chamber. Watch the animal closely and allow sufficient time for anesthetic to take effect (~5 min).

3. Remove the rat from the anesthetic chamber and immediately place the nose cone over the snout. Shave the head and neck region then carefully place the animal on its back on top of the thermal blanket to maintain body temperature (monitor by inserting a rectal probe). Next, shave the abdominal as well as the leg area. Clean the nape of the neck and the ventral area with a Betadine wash and then with alcohol.

4. Ensure that the animal is fully anesthetized by verifying a lack of response to a tail and foot pinch. Apply a 2- to 3-in. piece of tape to each ankle area to secure legs during the surgical procedure. Place a sterile surgical drape over the animal (covering the abdomen and the leg) and cut a 1 × 1–in. oval section out over the upper thigh region and another over the abdominal area.

5. Fill a 1-ml syringe equipped with a 23-G needle with heparinized saline. Insert the tip of the needle into the catheter (prepared in step 1) and fill with saline, making sure that the line is clear and that there are no leaks. Leave the needle in the catheter and make sure the syringe is still 3/4 filled.

6. Make a 1- to 1 1/2-in. incision above the region where the femoral artery exits the abdomen (where the leg is exposed through the drape). Carefully tease away the connective and fatty tissue with small pointed scissors until the femoral nerve, vein, and artery are exposed. Using the curved forceps, gently separate the femoral nerve from the artery and vein. Do not overly stretch the nerve or damage it in any way. After separating the nerve, continue using the forceps to separate the artery and vein.

7. Fold an 8-in. piece of sterile suture in half and, using a pair of mosquito forceps, carefully pull half the suture between the artery and vein and cut the suture at the top of the loop. Use one piece to gently pull the artery away. Secure both ends of the suture with a pair of mosquito forceps and place the forceps on the abdomen. Repeat the process with the other piece of suture; however, place the forceps toward the knee. Repeat this process to secure the vein with a second 8-in. piece of suture.

8. Fill a 1-ml syringe with lidocaine and expel the solution into the cavity on top of the artery and vein. Leave on for a couple of minutes to allow the vein to dilate, and then absorb the excess with sterile gauze.

9. Starting with the vein, carefully tie a knot in the suture closest to the investigator (furthest from the aorta). Secure the ends of the suture back with the forceps and pull very gently with both sets of forceps making the vein taut.

10. Using the vannas scissors, make a small incision in, but do not sever, the vein. Take great care to ensure that the incision is only large enough to insert the catheter. Carefully insert the tip of a pair of curved forceps into the incision. Next, while holding the first pair of forceps, use the vessel cannulation forceps and gently insert the free end of the catheter into the vein under the tip of the first pair of forceps.

11. Once the tip of the catheter is inserted, very gently feed the 0.01-in. Tygon tubing piece of the catheter into the vein until the 0.02-in. section of the tubing is even with the incision area. Release the suture from the forceps over the portion of the vein containing the catheter (the forceps situated on the abdomen) and tie a knot around the vein, making sure the catheter is secure, but not closed off. Flush the line with saline from the attached syringe. Tie the suture at the other end as well, again making sure the catheter is secured,

but not closed off. Repeat flushing of the line with saline from the attached syringe, but this time as the saline is going in, clamp the catheter ~1 in. from the tip of the needle with a pair of plastic hemostats. Remove the syringe and place a straight pin in the end of the catheter to seal it off.

12. Insert the large trocar under the leg and route it dorsally to the nape of the neck (the trocar should slide easily between the muscle and skin layer of the animal). Make a small incision in the nape of the neck and allow the trocar to come through. Insert the catheter into the hollow opening and route it through the trocar until it appears through the opening at the nape of the neck. Gently pull the catheter through the skin, while watching the venous area where it is attached, and completely straighten the line. Make sure it is not too taut; a little bit of slack is necessary for natural movement of the animal. Place one or two sutures in the nape of the neck to secure the catheter in place.

13. Make a 1- to 1 1/2-in. incision into the abdominal skin. Using the large blunt-tipped scissors, make a cut of the same size in the abdominal wall. Insert the sterile telemetric probe into the peritoneal cavity, leaving the catheter portion exposed for manipulation. Pierce the abdominal wall above the femoral artery with the small trocar and thread the catheter section of the probe through the trocar, and then remove the trocar leaving the catheter in place. Turn on the radio to the AM band and tune to the very last frequency. Swipe the magnet over the probe until a high-pitched tone is heard coming from the radio.

14. Repeat the same procedure for the arterial catheterization as was done to the vein. Take special care not to damage the tip of the catheter. Once the catheter is correctly inserted (~1 in.) into the artery, listen for the change of tone on the radio from a constant pitch to a pulsating pitch, indicating that the catheter is undamaged and transmitting signal. After successful insertion, secure with the suture.

15. Once the arterial catheter is secured, use sterile 3.0 suture with the attached 1/2-in. tapered needle to stitch the probe to the inner abdominal wall. Suture the entire abdominal incision closed, followed by the skin layer, and finally the leg. Swipe the magnet back over the abdomen until no sound is heard from the radio (this turns the probe off). Remove the tape from the legs and the anesthetic mask and immediately place the animal into the stereotaxic instrument with the incisor bar set at -3.3 mm.

16. Place animal in stereotaxic instrument, clean scalp area with Betadine and alcohol. Make a 1-cm incision in the scalp. Clean away connective tissue and expose the top of the skull so that bregma is clearly visible. If tissue bleeding occurs on the skull, dip 3 to 5 racellets in water and put them directly on the skull. Leave in place for 45 to 60 sec. If skull bleeding occurs, use the cauterizer to seal vessels.

17. Once the area is completely cleaned, use a pair of mosquito forceps (curved usually work best) to pull the scalp away from the top of the head. Do this on both sides. Using the Dremel drill, place one hole just to the left and anterior to bregma, another just to the left and posterior to lambda, and a final one just to the right and posterior to lambda. Place a 2.4-mm screw into each hole.

18. Insert a sterile guide cannula into each manipulator on the stereotaxic apparatus. Determine the area where the skull sutures meet on bregma and dot it with a nonsmearing pen. Move the right stereotaxic arm so that the tip of the guide cannula rests directly in the center of the dot on the skull. On a sheet of paper, record all three coordinates for the right side. Next, move the right arm -2.1 mm anterior and $+5.0$ mm lateral. Repeat this procedure for the left side.

19. Drill a hole in the skull where the cannulae will enter, then lower them into place, -8.5 mm ventral. In a ceramic or glass well, mix a small amount of the powder with the liquid portion

of the cranioplastic cement (add just enough liquid to create a thick consistency with the powder). Working quickly, place the cement on the skull, building it up around the guide cannulae. After the first layer is dry, repeat.

20. After the second layer of cement is thoroughly dry, carefully release the cannulae from the manipulators and insert dummy cannulae into the guides. Inject the animal with 0.03 mg/kg buprenex (s.c.) and remove from the stereotaxic instrument. Put the spandex jacket on the rat, secure the venous catheter to the jacket, and then place the animal in a plastic shoebox without bedding. Make sure the animal is placed on a thermal blanket and monitor every 5 min until it is fully recovered from the anesthetic.

21. After animals have fully recovered from surgery (∼72 hr), place each animal in the social-interaction box for 5 min to familiarize them with the behavioral test (see Support Protocol).

22. After 48 hr, measure a cardiovascular and behavioral response before treatment. Bring the rat into the testing room (keep the animal in its home cage unless undergoing behavioral assessment in the social-interaction test), and place on a receiving plate near the computer with the telemetry software. Unwrap the venous line from the jacket, remove the pin, and clamp the line. Insert a syringe containing heparinized saline, unclamp the line, flush with saline, and then re-clamp.

23. Fill a 5-ml syringe with a 0.05 N sodium lactate solution and attach it to a 24-in. piece of 0.02-in. Tygon tubing. Place the syringe in an infusion pump set to deliver 1 ml in a 5-min period. Attach the Tygon tubing, equipped with a connector at the free end, to the venous line and remove the clamp from the catheter.

24. Swipe the magnet over the animal to turn the probe on. Turn on the computer and record in real time. Leave the rat untouched and obtain its baseline heart rate, blood pressure, and respiration (this will take 15 to 30 min). Once a steady cardiovascular state is reached, turn on the infusion pump, and infuse a total of 10 ml/kg 0.05 N sodium lactate solution.

25. Immediately following the 0.05 N sodium lactate infusion, turn off the pump, flush the venous line with heparinized saline, replace the pin, and secure the catheter to the jacket. Turn off the probe and then assess the animal in the social-interaction test (see Support Protocol). After behavioral testing, return the animal to the animal facility.

26. After 48 hr post-challenge, bring the rat to the testing room. Fill two 10-μl microsyringes plus two 24-in. pieces of PE 20 (with attached injection cannulae) with urocortin (6 fmol/100 nl). Slide the tubing on the syringes and situate them on an infusion pump. Set the pump to deliver 100 nl over a 30-sec period. Turn the pump on and verify the delivery of urocortin. Remove the dummy cannulae from the guide cannulae and place the injection cannulae into the guides. Turn the pump on for 30 sec, then wait 1 min before removing the injectors. Remove the injection cannulae and turn the pump on again for verification of urocortin delivery. Place the dummy cannulae back into the guides. If sham-priming is desired, then replace the urocortin with vehicle (1% BSA).

27. After 30 min, assess the animal in the social-interaction test (see Support Protocol), then return it to the animal facility. After 24 hr, repeat step 26, but do not perform any behavioral testing. After an additional 24 hr, repeat step 26, but include the behavioral assessment. Continue the experiment with only those animals that show a significant decrease in social interaction (SI) time between day 1 and day 3 (primed animals).

28. After an additional 48 hr, repeat steps 22 to 25 using the primed rats. Upon completion of the experiment, euthanize the rats and inject 100 nl of India ink into the guide cannulae. Remove the brains of the animals and section into 40-μm slices (Gerfen, 2006) to verify placement of cannulae.

29. Obtain data (cardiovascular as well as behavioral) for baseline lactate infusion and experimental lactate infusion. Analyze the 2 days with a paired t test. Use data only from animals that show correct bilateral placement of cannulae.

   If successfully primed, rats will exhibit a significant increase in heart rate (HR), blood pressure (BP), and respiratory rate (RR), as well as reductions in SI following lactate infusions; a constellation of responses similar to those seen in humans with panic disorder.

ALTERNATE PROTOCOL

LACTATE CHALLENGE IN BICUCULLINE METHIODIDE-PRIMED RATS

Additional Materials *(also see Basic Protocol 1)*
   Bicuculline methiodide (BMI)

1. Repeat Basic Protocol 1, steps 1 to 25.

2. After 48 hr, bring the rat to the testing room and place the cage on a receiving plate. Fill two 10-μl microsyringes plus two 24-in. pieces of PE 20 attached to injection cannulae with BMI (6 pmol/100 nl). Slide the tubing on the syringes and situate them on an infusion pump. Set the pump to deliver 100 nl over a 30-sec period. Turn the pump on and verify the delivery of BMI. If sham priming is desired, replace the BMI with vehicle (sterile saline). Swipe the magnet over the animal to turn on the probe. Turn on the computer and record from the probe.

3. Remove the dummy cannulae from the guide cannulae and place the injection cannulae into the guides. Allow the animal to obtain a steady cardiovascular state. Turn on the pump for 30 sec then wait 1 min before removing the injectors. Remove the injection cannulae and turn the pump on again for verification of BMI delivery. Continue recording from the probe for 12 min.

4. After 12 min of recording, place the dummy cannulae back into the guides, swipe the magnet over the animal (turning off the probe), and assess in the social-interaction test (see Support Protocol). Inject BMI daily (6 pmol/100 nl) for 3 additional days.

5. Repeat steps 2 to 4 for the fifth and final injection. Compare the changes in HR, BP, and SI time from the first injection of BMI to the fifth injection. Use only animals that how a significant increase in HR and BP as well as a decrease in SI time on day 5 as compared to day 1 (primed animals) to continue the experiments.

6. After 48 hr, use only the rats determined as primed and repeat Basic Protocol 1, steps 28 and 29.

BASIC PROTOCOL 2

DEVELOPING PANIC-PRONE STATE IN RATS BY ADMINISTRATION OF L-ALLYLGLYCINE IN THE DORSOMEDIAL HYPOTHALAMUS

Additional Materials *(also see Basic Protocol 1)*
   Hydrogen peroxide, optional
   Bicuculline methiodide/saline solution
   L-allylglycine or D-allylglycine solution

Osmotic pump cannula (Plastics One)
Tygon tubing (0.020-in. i.d., 0.060-in. o.d.)
25-G needle
10-μl Hamilton syringe
Microsyringe pump (Harvard Apparatus)
Cannula holder (Plastics One)
Osmotic Alzet minipumps (14 day, model 2002)
Alzet brain infusion kit

1. Repeat Basic Protocol 1, steps 1 to 15. However, after removing tape and anesthetic mask, place rat back in cage and allow it to recover. Run a baseline lactate challenge followed by a social interaction test (i.e., pre-surgery); see Basic Protocol 1, steps 22 to 25.

2. Fill the gas anesthetic chamber with 99.9% isoflurane. Turn on machine and set the pressure dial to 1 to 2 and adjust the flow meter between 2.5 and 3.5. Check the tubing between the induction chamber and the anesthetic mask to verify flow and to check for leakage. Next, weigh the animal on a scale accurate to 1 g and place in the induction chamber. Watch the animal closely and allow sufficient time for anesthetic to take effect (∼5 min).

3. Remove the rat from the chamber and immediately position it in the ear bars of the stereotaxic apparatus, securing the nose cone in place, and setting the incisor bar at +5.0 mm. Clean the previously shaved scalp area with Betadine and alcohol. Ensure the animal is fully anesthetized by verifying a lack of response to a tail and foot pinch.

4. Make a 1- to 2-cm incision in the scalp (dissecting bregma). Clean away the connective tissue, exposing the top of the skull so that bregma is clearly visible. As tissue bleeding occurs, dip 3 to 5 racellets in water and put directly on the skull where necessary. Leave in place 1 to 2 min. If bleeding persists, use the cauterizer and/or apply hydrogen peroxide. Let sit for 60 sec, then wipe clear.

5. Once the area is cleaned, use a pair of mosquito forceps to pull the skin and the incision edges away from the center. Using the Dremel drill, make one hole anterior to bregma and one posterior to bregma, opposite the side where the pump cannula will be placed. Mount a screw into each hole.

6. Connect the osmotic pump cannula with a 6-cm piece of Tygon tubing (0.020-in. i.d., 0.060-in. o.d.) and then insert a 1-meter strand of PE-20 tubing into the loose end of the Tygon tubing. Using a 25-G needle, fill a 1-ml syringe with bicuculline methiodide/saline solution (prepared so as to deliver 50 pmol/100 nl). Fill the tubing/pump cannula assembly to the tip and attach to a pre-filled 10-μl Hamilton syringe, replacing the 1-ml syringe and needle. Put the Hamilton syringe in a microsyringe pump. Prime the assembly and check for leaks. Fit the pump cannula into a cannula holder and then mount holder onto the manipulator arm of the stereotaxic apparatus angled at 10×. Prime the assembly again.

7. Mark a dot on bregma with a nonsmearing pen to serve as the point of reference for the stereotaxic coordinates for pump cannula placement. Move the pump cannula tip to the center of the black dot just resting on the skull. On prepared surgery sheet, record all coordinates for the manipulator. Next, move the manipulator −1.2 mm anterior and +2.1 mm lateral. Lower the manipulator so that the cannula tip just touches the skull. Draw a circle around the cannula on the skull. Drill a hole in the skull at that circle after raising the cannula out of the way.

8. Turn on the telemetry probe as described in Basic Protocol 1 and obtain a 5- to 10-min baseline of heart rate, blood pressure, and respiration rate. Lower the pump cannula into the brain, −9.0 mm ventral. Stop any bleeding as in step 4. Microinject 100 nl of BMI solution (50 pmol) into the putative DMH site, over a period of 30 sec. Monitor HR, RR,

and BP (all usually rise within 5 to 10 min; ≥50 beats/min in HR considered the significant indicator of a valid DMH site).

9. Withdraw the pump cannula from the brain by raising the manipulator arm and flush with sterile saline. Cut the Tygon tubing to ∼4 to 6 cm and fill the pump cannula/tubing with L-allylglycine (active isomer) or D-allylglycine (inactive isomer; negative control) solution sufficient to deliver the glutamic acid decarboxylase (GAD) inhibitor at an infusion rate of 3.5 nmol/0.5 ml/hr.

10. Using the same allylglycine solution as was used to flush and fill the pump cannula, fill the osmotic minipump in the following manner:

    a. Weigh the empty pump and flow moderator.

    b. Fill the osmotic pump with allylglycine using the filling syringe needle provided in the Alzet brain infusion kit.

    c. Insert the flow moderator into the pump, blot any excess liquid, and weigh again.

    d. Subtract the empty pump weight from the filled pump weight to obtain the filling wt./vol.

    e. Divide the filling wt./vol by the mean filling wt./vol (given in the Alzet pump specification sheet) to calculate a percentage of how full the pump is. Any percentage >94% full will be sufficient for the pump to operate efficiently at the specifications needed.

    For example:
    filled pump + flow moderator = 1.4681
    empty pump + flow moderator = 1.2265
    filling pump wt./vol = 0.2416
    mean filling wt./vol = 0.2432
    % full = 0.2416/0.2432 = 99.3%.

11. After the osmotic pump has been filled and the pump cannula/Tygon tubing assembly has been sufficiently flushed, remove the protective plastic cap from the pump and join the pump to the tubing (while the pump cannula/tubing still rests in the manipulator holder), being careful not to allow any bubbles in the line or to move the manipulator arm, which would distort the measured coordinates.

12. Lower the pump cannula with osmotic pump attached into the previously determined reactive site within the DMH (−9.0 ventral) to deliver the allylglycine to the reactive site at the specified flow rate within 4 hr of implantation.

13. In a ceramic or glass well, mix a small amount of cranioplastic cement with cranioplastic liquid to create a thick consistency. Working quickly, apply the cement to the skull building it up around the pump cannula and embedded screws. After the first layer is dry, repeat. Build the cement up until the screws are covered and the cannula is adhered to the skull.

14. After the cement is thoroughly dry, carefully release the cannula from the holder. Make an incision in the nape of the neck adjacent to where the femoral catheter exits. Inject the rat s.c. with 0.03 mg/kg buprenex for postoperative pain. With a hemostat, clear away an area and a path right under the skin where the pump and tubing can be placed. Carefully place the pump in this area and suture the exposed insertion hole to cover the pump and most of the tubing.

15. Take the rat out of the anesthetic mask and stereotaxic apparatus and allow to recover in its home cage 3 to 4 days (required for reliable lactate sensitivity) placed atop a warming blanket. Monitor its respiration rate and tail pinch/eye blink reflexes until fully recovered.

16. Re-challenge the animal with i.v. sodium lactate (see Basic Protocol 1, steps 21 to 24) and monitor for significant increases in HR, BP, and RR, as well as decreases in SI following lactate, but not saline, infusions, all of which indicate successfully prepared rats.

SUPPORT PROTOCOL

BEHAVIORAL ASSESSMENT OF RATS IN THE SOCIAL INTERACTION TEST

This behavioral protocol is a modified version of the social interaction test (File, 1980), a validated test for assessing anxiety-like behaviors in rats. The test does not require any type of behavioral conditioning and can easily be administered following treatment. In addition, the testing period lasts for a total of only 5 min. This paradigm is useful when a nonconditioning test is necessary.

NOTE: When using the dorsomedial hypothalamus model (i.e., Sprague-Dawley rats), it is not necessary to run the paradigm in red light. Normal room and house lighting is fine.

Materials

Rat subject
Light source (e.g., red light)
Social interaction box
Camera
VCR
Monitor
Video tape

1. Place the social interaction box on top of an elevated stable surface. Run the cables from the camera (analog or digital) to the VCR and from the VCR to the monitor according the manufacturer's specifications. Turn on camera, VCR, and monitor. Mount the camera to a wall, ceiling, or rod so that the lens points directly down and is centered. Focus the image.

2. Prior to testing (48 hr), place the experimental rat in the box alone for 5 min. Remove the animal and thoroughly clean the box. Repeat for each animal undergoing experimentation.

3. On the day of the test turn on the red light and turn off any other light source. Turn on the camera, VCR, and monitor and insert a blank tape into the VCR. Begin recording and place the experimental animal in the box with a novel untreated partner rat (matched for weight, age, and sex) at the center of the box. Record for 5 min then remove the animals from the box. Clean box thoroughly.

4. Only score the experimental rat. Record social interaction (sniffing, grooming, and climbing upon the partner) as the total time the experimental animal initiates and maintains interaction with the partner rat during the 5-min test (an increase in SI time indicates a decrease in anxiety and vice versa).

References: Shekhar et al., 2001, 2002

Contributors: Tammy J. Sajdyk, Stanley R. Keim, Shelley R. Thielen, Stephanie D. Fitz, and Anantha Shekhar

UNIT 4.16

Chemoconvulsant Model of Chronic Spontaneous Seizures

BASIC PROTOCOL

KAINATE INDUCTION OF CHRONIC SPONTANEOUS SEIZURES IN RATS

Materials (see APPENDIX 1 for items with ✓)

Male Sprague-Dawley rats, 180 to 250 g (Harlan)
Tattoo dye (approved for animals) or ear-piercing tool, optional
✓ 2.5 mg/ml kainic acid in 0.9% (w/v) NaCl
0.9% (w/v) NaCl (sterile saline)
Lactated Ringer's solution, 35°C
Apple slices and/or mashed rat chow

Animal scale
35°C water bath or incubator
Standard rat cages (∼43-cm length × 22-cm width × 20-cm height), with plastic lids approved by the Association for Assessment and Accreditation of Laboratory Animal Care (AAALAC; e.g., Allentown Caging)
Disposable bed liners and/or paper towels
Syringes and needles suitable for intraperitoneal and subcutaneous injections

1. Weigh each rat and record the weight (should be 180 to 250 g). Use male rats unless examining the effects of hormones and epilepsy. If two or more animals are placed in one cage during the injections, mark each rat with an identifying feature (e.g., tattoo dye or ear piercing).

2. Calculate the amount of kainate each rat will receive for a single injection (5 mg/kg body weight in a volume of 0.2 ml per 100 g of rat at a concentration of 2.5 mg/ml, a maximum of ten injections per animal) and prepare sufficient kainate solution for the experiment. For control rats, administer saline in the same volume as the kainate solution (0.2 ml per 100 g of rat). Place the lactated Ringer's solution in the 35°C water bath or incubator before beginning injections.

3. Remove all loose wood-chip bedding from cages and replace with a disposable bed-liner or paper towels to prevent rats from inhaling and choking on the bedding during seizures. Use plastic lids (approved by AAALAC) to cover cages. If only metal lids are available, turn them upside down on the cages to prevent the rat from injuring itself on the water and food dispensers. Remove any object from the cage that may cause injury to the animal during convulsive SE.

4a. *For saline-treated control animals:* Administer 0.2 ml saline per 100 g body weight by intraperitoneal injection just lateral to the linea alba (midline), record the time of each injection, volume administered, and any potential seizure-like behavior for each rat (see Fig. 4.16.1). Give each saline-treated rat the same number and volume of injections as its matched kainate-treated rat.

4b. *For kainate-treated animals:* Administer 0.2 ml kainate solution per 100 g body weight by intraperitoneal injection just lateral to the linea alba (midline), record the time of each

Date:
Kainate lot number:

Rat Identification	kainate-1	kainate-2	kainate-3	kainate-4	kainate-5	kainate-6
Weight (kg)						
Injection amount						
Time (hr)						
1: dose amount						
Seizure type: Class III						
Class IV						
Class V						
Time (hr)						
2: dose amount						
Seizure type: Class III						
Class IV						
Class V						
Time (hr)						
3: dose amount						
Seizure type: Class III						
Class IV						
Class V						
Time (hr)						
4: dose amount						
Seizure type: Class III						
Class IV						
Class V						
Time (hr)						
5: dose amount						
Seizure type: Class III						
Class IV						
Class V						
Time (hr)						
6: dose amount						
Seizure type: Class III						
Class IV						
Class V						
Time (hr)						
7: dose amount						
Seizure type: Class III						
Class IV						
Class V						
Time (hr)						
8: dose amount						
Seizure type: Class III						
Class IV						
Class V						

Total seizures:
Amount of Ringer's:

Figure 4.16.1 Sample worksheet used during injections to tally seizure types and to determine doses. A similar table may be made to track saline-treated rats; however, the "Seizure type" cells may be replaced with a single cell for comments.

injection, dose administered, and all seizure-like behavior for each rat. Give all rats a full dose for the first injection; however, for the remaining injections, use the flowchart to determine the next course of action (see Fig. 4.16.2).

To minimize the possibility of inserting the needle into the bladder or other major organs, tip the rat to shift its internal organs toward the thoracic cavity. Use alternate sides for the site of the injection to decrease the risk of peritonitis.

5. Determine the severity of motor seizure activity according to a modified Racine scale (Racine, 1972; Ben Ari, 1985):

    a. Class I (facial automatisms) and class II (head nodding). Use only wet dog shakes and the number of class III/IV/V seizures as the criteria for further injections because class I and II activities can be difficult to identify unambiguously.

    b. "Wet dog shakes"—after the first few injections, rats will shake like a wet dog until motor seizures develop.

    c. Class III—rats display forelimb clonus with a lordotic posture.

    d. Class IV—forelimb clonus continues along with rearing.

    e. Class V—animals have a class III or IV stage seizure and fall over. Some rats will fall to one side first and then show evidence of forelimb clonus.

    f. Jumping—the rat spontaneously begins to jump in one place or throughout the cage.

6. For each seizure-like activity, make a tally mark in the appropriate box (Fig. 4.16.1). At the end of each hour, count the total number of class III/IV/V seizures. Use these numbers to determine the appropriate course of action for subsequent kainate injections (see below and Fig. 4.16.2).

7. Each and every hour (until a rat has experienced ≥ 3 consecutive hr of convulsive SE) assess the severity of the seizures and well-being of the rats to determine whether the next dose of kainate given should be (a) full (5 mg/kg), (b) half (2.5 mg/kg), or (c) omitted (no dose given; Fig. 4.16.2). Adjust the injections in accordance with the animal's behavior to prevent death.

    If rats begin to exhibit excessive inactivity (i.e., they appear lethargic and are barely moving) or hyperactivity (i.e., exaggerated jumping, pacing, or running in circles), or have had ≥ 10 class IV and/or class V seizures per hour, skip the next injection until the assessment in the following hour. These behaviors generally precede death; therefore, be vigilant about monitoring the behavior of the animal.

    If the total number of class IV and V seizures is between five and nine seizures per hour, then reduce the next injection to 2.5 mg/kg. Once the dose is reduced by half, determine the severity of seizures every 30 min, and not every hour, because SE may not continue for >3 hr if the dose is reduced by half. Depending on the behavior of the rat (e.g., the severity of seizures), omit the next dose or give another half dose in 30 min.

    If a rat shows (a) only "wet dog shakes," (b) only class III seizures, or (c) <5 class IV and/or class V seizures per hour, then administer a full injection every hour until the animal has experienced an episode of convulsive SE that consists of >3 hr of continuous seizures. In rare cases, should a rat never show ≥ 4 class IV and/or class V seizures in each hourly assessment, terminate treatment 3 hr after the first observed motor seizure.

    Generally, once seizures have occurred for ~1 hr, class IV and/or class V seizures will predominate. Wet dog shakes and class III seizures will become less likely to occur. Thus, class IV and V seizures are primarily used as markers to determine the next course of action.

    Using these criteria, an animal could conceivably receive only a single dose of kainate if it begins to have ≥ 10 class IV/V seizures per hour for >3 continuous hr.

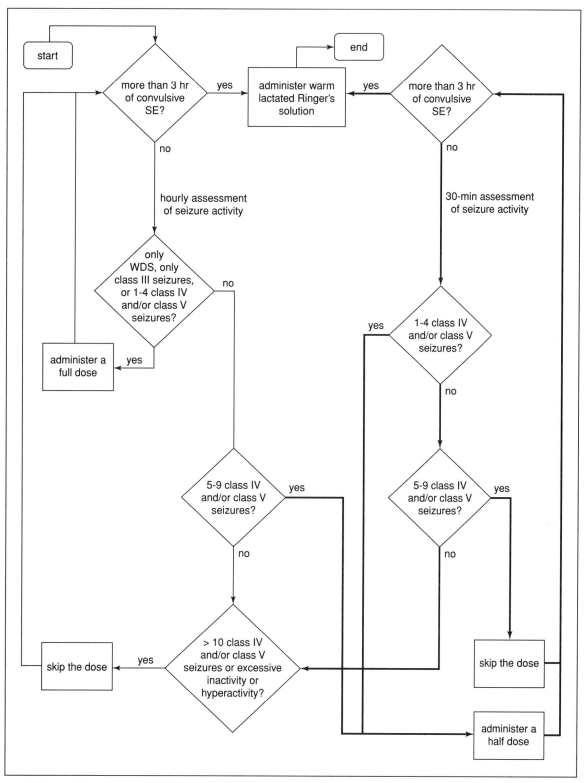

Figure 4.16.2 A use-case flow chart for determining subsequent dose administration for each kainate-treated rat. Following the first full-dose injection, seizure-like activity is monitored and recorded for 1 hr. The recorded seizure activity is then utilized with the seizure assessment flowchart to determine the next course of action. Two finite state loops—an hourly assessment loop on the left side of figure (thin lines) and a 30-min assessment loop on the right side of figure, (thick lines)—determine the process (rectangles) for each possible decision (diamonds). The start box represents the first hour after the initial kainate injection and all remaining hours until each rat experiences >3 hr of convulsive SE. SE = status epilepticus; WDS = wet dog shakes.

Table 4.16.1 Assessment Index for Aggression in Rats with Kainate-Induced Epilepsy

Index number	Animal behavior
0	Little to no resistance other than flight and mild vocalization
1	Excitable (i.e., jumpy) with mild to strong flight tendencies, mild resistance to restraint (i.e., turning around but not actively biting), and mild to moderate vocalization
2	Hyperexcitable (i.e., jumpy and running) with strong flight tendencies and occasional turning and facing, strong resistance to restraint (i.e., turning and biting), and moderate vocalization
3	Standing on hind limbs with tail shaking (prior to capture), hyperexcitable, actively attempting to bite, strong resistance to restraint, and fervent vocalization

8. Continue to observe rats for 3 hr after the onset of seizures to ensure that SE has been achieved. Continue treatment until class IV/V seizures are elicited for >3 hr, possibly yielding a total dose of 15 to 30 mg/kg.

9. Once an animal has experienced >3 consecutive hr of convulsive SE (defined as when treated rats have had at least one convulsive seizure during each consecutive 1-hr period or continuous clonic seizures), inject the rat with warm lactated Ringer's solution (1.5 to 2.5 ml per 100 g body weight). Inject one bolus of fluid subcutaneously between the shoulder blades and another bolus subcutaneously just above the hind limb. Provide apple slices to aid the rat's recovery from hypoglycemia and/or dehydration. For the first week after treatment, continue to administer warm lactated Ringer's solution and apple slices (or moistened rat chow) daily until the rat begins eating normally and gaining weight. Use disposable bed-liners and/or paper towels as bedding. Place kainate-treated rats in a Healthguard or a microinsulator rack to reduce the risk of injury from the food and water dispensers.

Approximately 8% of the rats have an additional one to two motor seizures during the next 72 hr following treatment, although most of the rats have no additional seizures until the onset of chronic epilepsy (see Fig. 3B from Hellier et al., 1999). Although a few rats have seizures within the first week after kainate treatment, most rats begin to have spontaneous motor seizures 1 to 2 months after kainate treatment. Once seizures begin, the seizure rate is low, <0.1 seizures/hr, but increases over time, ranging from 0.5 to 1.2 seizures/hr. Animal care is the same as in the above step 9 throughout the lifespan of rats with kainate-induced epilepsy (up to 20 to 22 months after kainate treatment).

Rats with kainate induced epilepsy become more aggressive as they become more epileptic. Table 4.16.1 is an assessment index for aggression that describes possible abnormal behaviors in these rats.

Contributors: Jennifer L. Hellier and F. Edward Dudek

APPENDIX 1
Reagents and Solutions

This appendix includes recipes for all reagents, media, and solutions used in *Short Protocols in Neuroscience: Systems and Behavioral Methods*. Recipes are listed alphabetically, with the unit(s) in which the solution is used listed parenthetically. No unit numbers are indicated for commonly used solutions. In some cases (e.g., PBS), there is a recipe for the commonly used version followed by other formulations that are unique to specific units. Recipes for some solutions include reagents for which separate recipes are included elsewhere in this appendix; these are indicated by a ✓ or (see recipe). It is critical that the correct recipe be used for each protocol and that the correct base or stock solutions be used in preparing the final solution.

When preparing solutions, use deionized distilled water and reagents of the highest available grade. Sterilization—by filtration through a 0.22-μm filter or by autoclaving—is recommended for most applications, especially for cell culture. Discard any reagent that shows evidence of contamination, precipitation, or discoloration or has exceeded the recommended shelf life. General information about the molarities and specific gravities of acids and bases can be found in Table A.1.1.

CAUTION: Follow standard laboratory safety guidelines and heed manufacturers' precautions when working with hazardous chemicals.

Acetylcholine/choline standards (UNIT 2.9)
  Stock solutions: Prepare diluent by adding 3.0 ml glacial acetic acid to 1 liter HPLC-grade water. Adjust pH to 5.25 with dilute NaOH. Store at 4°C. Prepare 2.0 mM acetylcholine stock solution by dissolving 36.3 mg acetylcholine chloride in 100 ml diluent and 2 mM choline stock solution by dissolving 27.9 g choline chloride in 100 ml diluent. Store at 4°C.
  Working solutions: On day of assay, prepare fresh 20 μM working standards and other standard concentrations as needed by diluting.

Acid, concentrated stock solutions
  See Table A.1.1.

Acid formalin solution (UNIT 4.3)
  20 ml 40% (w/v) formaldehyde
  1 ml glacial acetic acid
  Bring volume to 500 ml with H_2O
  Store up to 6 months at room temperature

Ammonium hydroxide, concentrated stock solution
  See Table A.1.1.

Table A.1.1 Molarities and Specific Gravities of Concentrated Acids and Bases[a]

Acid/base	Molecular weight	% by weight	Molarity (approx.)	M solution (ml/liter)	Specific gravity
Acetic acid (glacial)	60.05	99.6	17.4	57.5	1.05
Ammonium hydroxide	35.0	28	14.8	67.6	0.90
Formic acid	46.03	90	23.6	42.4	1.205
		98	25.9	38.5	1.22
Hydrochloric acid	36.46	36	11.6	85.9	1.18
Nitric acid	63.01	70	15.7	63.7	1.42
Perchloric acid	100.46	60	9.2	108.8	1.54
		72	12.2	82.1	1.70
Phosphoric acid	98.00	85	14.7	67.8	1.70
Sulfuric acid	98.07	98	18.3	54.5	1.835

[a] *CAUTION:* Handle strong acids and bases carefully.

Argon-purged Locke's solution (UNIT 2.5)
154 mM NaCl
5.6 mM KCl
1 mM $MgCl_2$
2.3 mM $CaCl_2$
9.4 mM HEPES
5.6 mM glucose
Store up to 3 months at 4°C
Before use, purge and degas

Artificial cerebrospinal fluid (aCSF)
Monkey
Dissolve in 900 ml H_2O:
268 mg $Na_2HPO_4 \cdot 7H_2O$ (1.0 mM final)
8.6 g NaCl (147 mM final)
0.22 g KCl (3 mM final)
0.19 g $CaCl_2 \cdot 2H_2O$ (1.3 mM final)
0.20 g $MgCl_2 \cdot 6H_2O$ (1.0 mM final)
35 mg ascorbic acid (weigh accurately; 0.20 mM final)
Add H_2O to 1 liter
Filter sterilize
Make fresh the day of the experiment

If this solution has been prepared correctly, the pH should be 7.4 ± 0.1. To achieve the desired pH, the ascorbic acid must be weighed accurately, as even small variations may affect pH drastically. If necessary, adjust pH to 7.4 with small volumes of 100 mM mono- or dibasic sodium phosphate. If pH is significantly different from 7.4, prepare new solution.

Rodent
Dissolve in 900 ml H_2O:
7.25 g NaCl (124 mM final)
0.37 g KCl (5 mM final)
0.015 g $CaCl_2 \cdot 2H_2O$ (0.1 mM final)
0.65 g $MgCl_2 \cdot 6H_2O$ (3.2 mM final)
2.18 g $NaHCO_3$ (26 mM final)
1.8 g glucose (10 mM final)
Adjust pH to 7.4 with 0.1 M NaOH or HCl
H_2O to 1 liter
Filter sterilize
Prepare fresh on day of experiment

aCSF is a balanced salt solution which can vary according to species and cell type. A balanced salt solution is an isotonic solution containing appropriate ions and buffers suitable for maintaining neurons in vitro. Consult a physiology reference text to formulate an appropriate solution for the animal being studied. A general Ringer's solution (e.g., Life Technologies) is often substituted for aCSF, but the requirement for serum should be considered.

Artificial cerebrospinal fluid (aCSF), 10× (UNIT 1.4)
1.24 M NaCl
30 mM KCl
260 mM $NaHCO_3$
12.5 mM NaH_2PO_4
10 mM $MgSO_4$
100 mM D-glucose
Store up to 1 week at 4°C

Prepare a 1× working solution on the day of the experiment and add $CaCl_2$ to 2 mM final. Bubble solution constantly with a mixture of 95% O_2/5% CO_2.

Artificial cerebrospinal fluid (aCSF, rat) (UNIT 2.7)
10× stock solution:
0.284 g Na_2HPO_4 (20 mM final)
0.952 g $MgCl_2$ (10 mM final)
0.133 g $CaCl_2$ (12 mM final)
0.201 g KCl (27 mM final)
8.470 g NaCl (1.45 M final)
H_2O to 100 ml
Filter sterilize, adjust to pH 7.4 if necessary with 1 N NaOH or phosphoric acid, and store up to 1 month at 4°C.

Working Solution:
Dilute the stock solution 1:10, filter sterilize, and adjust to pH 7.4 prior to the start of experiments.

A typical microdialysis experiment in rats requires 100 to 200 ml perfusate.

Assay buffer (UNIT 2.4)
✓ 50 mM Tris·Cl, pH 7.4
✓ 1 mM EDTA
1.25 mM $CaCl_2$ (as $CaCl_2 \cdot 2H_2O$)
Store components without DTT up to 1 month (or longer) at 4°C
Add 1 mM dithiothreitol (DTT) on day of use

Assay mix 1 (UNIT 2.4)
✓ 50 mM Tris·Cl, pH 7.4
100 µM L-arginine
4 µCi/ml L-[^{3}H]arginine or -[^{14}C]arginine (purified, see Support Protocol 5; add immediately before use)
4 mM nicotine adenine dinucleotide phosphate (reduced form; NADPH; add immediately before use)

Solutions containing NADPH should not be stored longer than one day. The L-arginine concentration in assay mix 1 will result in a final concentration of 25 µM in the reaction mixture; if other known sources of L-arginine are introduced (e.g., L-arginine is often present in purified NOS preparations to prevent enzyme inactivation), this should be considered in the calculation of conversion rate. L-[^{3}H]arginine (or L-[^{14}C] arginine) should be added to the buffer in the appropriate quantity (i.e., sufficient for the planned assay) immediately before use, since diluted radioisotope solutions are subject to more rapid decomposition and should not be stored. It is recommended that 100,000 to 250,000 dpm be included per assay tube.

Assay mix 2 (UNIT 2.4)
✓ 50 mM Tris·Cl, pH 7.4
400 nM calmodulin (Sigma)
40 µM tetrahydrobiopterin (BH_4; Calbiochem)
Keep on ice and dispose of any assay mix 2 that is not used on day of preparation

Calmodulin and BH_4 are required for assay of membrane-associated NOS preparations; soluble NOS preparations that are not subjected to extensive purification steps will exhibit full activity without these agents. For assays of iNOS activity, neither calmodulin nor calcium are needed and may be omitted from this solution. BH_4 is kept at $-80°C$ as a 10 mM solution in 10 mM HCl and must be freshly added to the buffer just prior to assay initiation.

Beads/antibody/[^{125}I]cAMP mixture (UNIT 2.1)

Prepare anti-cAMP antibody: Add 5 ml of 50 mM sodium acetate, pH 5.8, to a vial of lyophilized anti-rabbit cAMP antibody (Sigma) and dilute 1:25 with 50 mM sodium acetate. Store unused antibody at 0°C to 4°C for 30 days.

Prepare beads: Add 8 ml of 50 mM sodium acetate, pH 5.8, to a 500-mg bottle of SPA PVT antibody-binding beads, anti-rabbit reagent (Amersham Biosciences).

This is sufficient for ~4 plates and is stable at 4°C for 30 days.

Prepare [^{125}I]cAMP: Dilute adenosine 3′,5′-cyclic phosphoric acid 2′-O-succinyl, [^{125}I]-iodotyrosine methyl ester (Perkin Elmer Life and Analytical Sciences) in 50 mM acetate buffer to a concentration equivalent to ~20,000 cpm/well.

Prepare beads/antibody/[^{125}I]cAMP mixture: Combine equal amounts of antibody, bead, and ^{125}I-cAMP mixtures. Use 2 ml of each for a total of 6 ml per 96-well microtiter plate.

Biotinylated ^{125}I-labeled rhinovirus protease 3C substrate stock solution, 50 nM (UNIT 2.1)

Dissolve 100 μCi of lyophilized biotinylated T-L-F-Q-G-P-V[^{125}I] (Amersham Biosciences, custom synthesis) in a total volume of 1 ml of Milli-Q-purified water. Store 50-μl aliquots at −20°C. Avoid multiple freeze/thaw cycles. To calculate the concentration of the stock radioligand, use the technical information supplied with the reagent. For example, for a specific activity = 2000 Ci/mmol and a radioactive concentration following reconstitution = 100 μCi/ml:

$$\frac{100 \ \mu Ci}{ml} \times \frac{1 \ mmol}{2000 \ Ci} \times \frac{1 \ Ci}{10^6 \ \mu Ci}$$

$$= 5.0 \times 10^{-8} \ mmol/ml = 50 \ nM$$

Borate buffer, 50 mM, pH 10.0 (UNIT 2.5)

61.1 g potassium tetraborate (Sigma)
14.91 g potassium chloride
3.5 liters water
Adjust pH to 10.0
Bring volume to 4 liters with water
Store up to 3 months at 4°C

Table A.1.2 Ca^{2+} Standard Solutions for Dye Calibration

Desired [Ca^{2+}] (nM)	Vol Ca^{2+}/EGTA solution (ml)	Vol 100 mM EGTA (ml)
0	0.000	10.000
20	0.735	9.265
50	1.656	8.344
100	2.841	7.159
200	4.425	5.575
300	5.435	4.565
400	6.135	3.865
500	6.649	3.351
750	7.485	2.515
1000	7.988	2.012
1500	8.562	1.438
2000	8.881	1.119

Buffered formalin, 10% (v/v) (UNIT 4.2)

To make 10 liters: Dissolve 50.5 g of disodium hydrogen phosphate (Na_2HPO_4) in 8 liters of distilled water. When dissolved, add 30.5 g of potassium dihydrogen phosphate (KH_2PO_4) and stir to dissolve. Add 1000 ml of 40% formaldehyde (formalin). Bring to 10 liters with distilled water.

Buffered formalin, 10% (UNIT 4.3)

Dissolve 50.5 g of disodium hydrogen phosphate (Na_2HPO_4) in 8 liters of water. When dissolved, add 30.5 g of potassium dihydrogen phosphate (KH_2PO_4) and stir to dissolve. Add 1000 ml 40% formaldehyde (formalin). Bring volume to 10 liters with distilled water. Store up to 6 months at room temperature.

Ca^{2+} calibration solutions (UNIT 2.13)

Mix 1 vol of the appropriate Ca^{2+} standard solution (see recipe; Table A.1.2) with 1 vol of 10× intracellular buffer solution (see recipe). Add 8 vol H_2O (final 1× Ca^{2+} and buffer). Adjust pH of each solution to 7.1 with 1 M KOH. Store up to 12 months at 4°C.

Small volumes of the potassium salts of the appropriate dye are added to each calibration solution after the pH has been adjusted (see UNIT 2.13, Support Protocol 1).

Ca^{2+}/EGTA solution (UNIT 2.13)

Combine 3.804 g EGTA with 1.001 g $CaCO_3$. Heat in a minimal volume of water to dissolve. Add 20 ml of 1 M KOH and bring volume to 100 ml with deionized water. Store up to 12 months at 4°C.

Ca^{2+}-free HEPES buffer (UNIT 2.13)

Prepare as for 1× HEPES buffer (see recipe), but omit $CaCl_2$.

Ca^{2+} standard solutions *(UNIT 2.13)*

To achieve Ca^{2+} standard solutions of the desired concentration, dilute Ca^{2+}/EGTA solution (see recipe) in 100 mM EGTA (see recipe) using the volumes specified in Table A.1.2. Store up to 1 year in plastic containers at 4°C.

These solutions can be purchased ready-made from Molecular Probes. However, they are relatively simple to make and are stable for many months if stored as indicated.

The mixtures of solutions for the Ca^{2+} standards are generated using the equation:

$$V_{Ca} = \frac{K_s \times V_f \times [Ca^{2+}]}{1 + (K_s \times [Ca^{2+}])}$$

where K_s represents an affinity constant for Ca^{2+}/EGTA, V_{Ca} is the volume of Ca^{2+}/EGTA solution, and V_f is the final volume. K_s will vary at different temperatures and pH. The volumes in Table A.1.2 are derived using a value of $3.969 \times 10^6 M^{-1}$ (at 22°C, pH 7.1), and a final volume of 10 ml.

Calcium- and magnesium-free Hank's buffered salt solution (CMF-HBSS)

HBSS (see recipe below) prepared omitting all calcium and magnesium salts.

5-Carboxyamidotryptamine, 10 mM *(UNIT 2.1)*
- ✓ 313 μl of saturation assay buffer
- 1.0 mg of 5-carboxyamidotryptamine (5-CT; mol. wt. 319)
- Prepare fresh daily and store at 4°C

Casein buffer *(UNIT 2.2)*

Mix the following ingredients:
- 0.25% (w/v) casein from bovine milk (Sigma)
- ✓ 1× PBS
- 0.05% (v/v) Tween 20 (Boehringer-Mannheim)
- 0.01% (w/v) thimerosal (Sigma)
- Adjust pH to 8.6 with 5 M NaOH and stir overnight

The final pH, after the casein has dissolved, is usually close to 7.4, and should be adjusted to this pH with 5 M HCl or 5 M NaOH, if necessary.

After the overnight stirring, filter the casein buffer through a large Whatman no. 1 filter paper and then through a 0.45-μm sterile filter unit. Store (without EDTA or protease inhibitors) up to 6 months at 4°C. Just before use, add AEBSF to 1 μM, E-64 to 2 μM, EDTA to 1 mM, leupeptin to 1 μM, and pepstatin to 1 μM (all protease inhibitors available from Boehringer-Mannheim).

Alternatively, Casein Blocker can be purchased (Pierce) and diluted 1:4 in PBS containing 0.625% (v/v) Tween 20 and 0.01% (w/v) thimerosal.

Charcoal suspension *(UNIT 2.14)*
- 20 mg/ml carbon (Decolorizing Norit, Neutral; Fisher Scientific)
- 5 mg/ml bovine serum albumin (BSA)
- Store at 4°C
- Stir before using

Chemiluminescent (CL) cocktail *(UNIT 2.5)*
- 4 mg cytochrome c (horse heart, type VI; Sigma; 10 μg/ml final)
- ✓ 3.2 ml 250 μg/ml luminol stock solution (1 μg/ml final)
- ✓ 50 mM borate buffer, pH 10.0 to make 400 ml
- Add reagents while stirring buffer
- Store up to 2 days at 4°C in an amber bottle
- Degas under vacuum for 10 min before use

$^{36}Cl^-$ flux assay buffer *(UNIT 2.12)*

Prepare synaptoneurosome preparation buffer (see recipe), but adjust to pH 7.4 at 30°C for uptake, and at 25°C for efflux.

Complete EAE medium *(UNIT 4.13)*
- 500 ml RPMI without L-glutamine (Life Technologies)
- 0.5 ml 50 mM 2-mercaptoethanol
- 5.0 ml 100 mM sodium pyruvate
- 5.0 ml 10 mM non-essential amino acids
- 5.0 ml 200 mM L-glutamine
- 5.0 ml 10,000 U/ml penicillin/10,000 mg/ml streptomycin
- 6.25 ml 1 M HEPES solution
- 50.0 ml fetal bovine serum
- Store up to 1 month 4°C
- *Total volume 576.75 ml*

Cresyl violet solution *(UNIT 4.3)*
- 2.5 g cresyl violet
- 1.5 ml 19% acetic acid
- Bring volume to 500 ml with H_2O
- Store up to 6 months at room temperature

Denhardt's solution, 100×
- 10 g Ficoll 400
- 10 g polyvinylpyrrolidone
- 10 g bovine serum albumin (Pentax Fraction V; Miles Laboratories)
- H_2O to 500 ml
- Filter sterilize
- Store at −20°C in 25-ml aliquots

Diethylpyrocarbonate (DEPC) treatment of solutions

Add 0.2 ml DEPC to 100 ml of the solution to be treated (excluding Tris solutions). Shake vigorously to get the DEPC into solution. Autoclave the solution to inactivate the remaining DEPC.

Many investigators keep the solutions they use for RNA work separate to ensure that "dirty" pipets do not go into them.

CAUTION: *Wear gloves and use a fume hood when using DEPC, as it is a suspected carcinogen.*

Dilution buffer *(UNIT 2.1)*
- ✓ Prepare DMEM/F12 /FBS without FBS and containing 20 mM HEPES, pH 7.4
- Store up to 4 weeks at 4°C

Dissection buffer (UNIT 1.1)
    119 mM NaCl
    2.5 mM KCl
    1.3 mM $MgSO_4$ or $MgCl_2$
    2.5 mM $CaCl_2$
    1.0 mM NaH_2PO_4
    26.2 mM $NaHCO_3$
    11.0 mM glucose (dextrose)
    Store in a tightly capped container 2 to 3 days at room temperature *or* several weeks at 4°C

The calcium chloride must be in solution before the sodium phosphate or bicarbonate is added. In experiments where it is desirable to elevate the calcium concentration to ∼10 mM or more, it will be necessary to use magnesium chloride instead of magnesium sulfate, and to remove the sodium phosphate completely. This does not appear to have an adverse affect on the health of the slices.

If there is still trouble dissolving the calcium, try carbogenating the solution for 5 to 10 min before adding the calcium, or use a nonbicarbonate buffer. Calcium precipitates quickly from buffer that is left in an open container without carbonization.

Contamination of the buffer chemicals can result in the failure of tissue slices to survive. If suspected, make buffer with all new chemicals.

Dissection medium (for interface cultures) (UNIT 1.7)
    50 ml 2× minimal essential medium (MEM; e.g., Life Technologies; store up to 6 months at 4°C)
    1 ml 100× penicillin-streptomycin (Life Technologies)
    120 mg Tris base (10 mM)
    H_2O to 100 ml
    Store up to 1 month at 4°C

DMEM/F12/10% FBS medium (UNIT 2.1)
To a 3:1 mixture of Dulbecco's Modified Eagle Medium (DMEM, high-glucose formulation, without HEPES or phenol red; Life Technologies) and nutrient mixture F-12 (Life Technologies), add the following:
    2.2 g/liter sodium bicarbonate (26 mM final)
    0.1 μg/liter sodium selenite (578 pM final)
    4.9 mg/liter ethanolamine·HCl (80 mM final)
    20 ml/liter 1 M HEPES, pH 7.55 (omit for Basic Protocol 5)
    50 μg/ml gentamicin (e.g., Life Technologies) to inhibit bacterial growth
    250 ng/ml Fungizone (amphotericin B and sodium desoxycholate) to inhibit fungal growth if necessary
    10 ml/liter of 100× MEM nonessential amino acids (e.g., Life Technologies)
    10% (v/v) fetal bovine serum (FBS)
    Store up to 4 weeks at 4°C

DNase I, RNase-free (1 mg/ml)
Prepare a solution of 0.1 M iodoacetic acid plus 0.15 M sodium acetate and adjust pH to 5.3. Filter sterilize. Add sterile solution to lyophilized RNase-free DNase I (e.g., Worthington) to give a final concentration of 1 mg/ml. Heat 40 min at 55°C and then cool. Add 1 M $CaCl_2$ to a final concentration of 5 mM. Store at −80°C in small aliquots.

Drug solutions (UNIT 3.13)
Prepare drug solutions fresh prior to use in water sterilized by passing through a 0.2 μm-filter. Neutralize pH prior to injection by titrating with 1 M NaOH or 1 M HCl. If acidified water is used for mice, match the pH of the test solutions with control solutions. Stabilize drug solutions with 0.1% sodium metabisulfite. Inject intraperitoneally at 1 ml/kg with sterile syringes and 23- to 25-G needles, with solutions at room temperature, or deliver orally via gavage. As an example, preparation of lithium chloride for injection is presented here.

0.15 M lithium chloride: Dissolve 6.36 g LiCl (Sigma) per liter of sterile water. Use 76 mg/kg in rats (3.0 ml for a 250-g rat).

Dissolution of relatively insoluble drugs can be aided by adding modified cyclodextrins (Research Biochemicals), by dissolving in 10% (v/v) ethanol or in up to 60% propylene glycol, or by extended mixing and/or heating.

Dulbecco's modified Eagle medium (DMEM), supplemented
Dulbecco's modified Eagle medium, high-glucose formulation (e.g., Life Technologies), containing:
    5%, 10%, or 20% (v/v) FBS, heat-inactivated (optional; see recipe)
    1% (v/v) nonessential amino acids
    2 mM L-glutamine
    100 U/ml penicillin
    100 μg/ml streptomycin sulfate
    Filter sterilize and store up to 1 month at 4°C

DMEM containing this set of additives is sometimes called "complete DMEM." The percentage of serum used is indicated after the medium name—e.g., "DMEM/5% FBS." Absence of a number indicates no serum is used. DMEM is also known as Dulbecco's minimum essential medium.

Two common nutrient mixtures, Ham's F-12 nutrient mixture and N2 supplements (both available commercially, e.g., from Life Technologies), are often added to DMEM either individually or in combination; the resulting media are known as DMEM/F-12, DMEM/N2, and DMEM/F-12/N2.

Dye solution, 1 mM (UNIT 2.13)
Cation-sensitive dyes are available from many sources (e.g., Molecular Probes, Texas Fluorescence Labs). Molecular Probes offers the most comprehensive range, and also offers dyes packaged in small aliquots that are suitable for single-day use.

Acetoxymethyl (AM) esters: AM dyes are relatively insoluble in aqueous solution and do not crystallize readily. Instead, they are supplied as an oily residue. It is rarely practical to weigh these oils, so it is usually preferable to dissolve the entire contents of the vial at 1 mM, by adding an appropriate volume of anhydrous dimethyl sulfoxide (DMSO) and triturating vigorously until no trace of the oil remains. The stock solution can be aliquoted into suitable containers, which should then be stored at −20°C in the dark until use (up to several months).

AM esters are susceptible to hydrolysis, and DMSO is quite hygroscopic. The use of anhydrous DMSO prolongs the storage life of the dye aliquots. However, if low volume usage is anticipated, purchasing the small packages is recommended as the oil is more stable than the DMSO solution.

Dye salts: Store pentapotassium salts of dyes as 1 mM aqueous solutions. Store up to 6 months in the dark at −20°C.

Dye stocks should be kept dark at all times.

ECF buffer (UNIT 2.15)
122 mM NaCl
25 mM $NaHCO_3$
3 mM KCl
1.4 mM $CaCl_2$
1.2 mM $MgSO_4$
10 mM glucose
0.4 mM K_2HPO_4
10 mM HEPES
Adjust pH to 7.4 with concentrated HCl or NaOH
Store up to ∼1 month at 4°C

Just prior to use, bubble for ∼20 min with 95% (v/v) O_2/5% (v/v) CO_2 to attain pH 7.4 and filter through a 0.2-μm syringe filter unit. Add test and reference compounds to achieve the desired concentration (optional) and warm to 37°C.

EDTA (ethylenediaminetetraacetic acid), 0.5 M (pH 8.0)
Dissolve 186.1 g disodium EDTA dihydrate in 700 ml water. Adjust pH to 8.0 with 10 M NaOH (∼50 ml; add slowly). Add water to 1 liter and filter sterilize. Store up to 6 months at room temperature.

Begin titrating before the sample is completely dissolved. EDTA, even in the disodium salt form, is difficult to dissolve at this concentration unless the pH is increased to between 7 and 8.

EGTA, 100 mM (UNIT 2.13)
Add 20 ml of 1 M KOH to 3.804 g EGTA. Bring volume to 100 ml with deionized water. Store up to 12 months at 4°C.

Eosin, 2% (w/v) (UNIT 4.2)
Dissolve 2 g of eosin (Sigma) in 100 ml of tap water. Add 4 crystals of thymol dissolved in 5 ml of hot water.

Thymol stops growth of fungi and bacteria and therefore reduces artifacts without producing detrimental effects on the staining.

Et-1 solution (UNIT 4.3)
Dissolve a 0.5-mg frozen pellet of endothelin 1 (Et-1; Calbiochem-Novabiochem) in 1 to 2 ml saline containing 5% (v/v) acetic acid. Divide into 100-μl aliquots (10 to 20 microcentrifuge tubes) and store up to 1 month at −20°C. On day of surgery, thaw one or two microcentrifuge tubes and dilute solution as necessary using saline.

Depending on the experiment, Et-1 is infused at 10 to 300 pmol in a volume of 1 to 3 μl.

Ethidium bromide staining solution
Concentrated stock (10 mg/ml): Dissolve 0.2 g ethidium bromide in 20 ml H_2O. Mix well and store at 4°C in dark or in a foil-wrapped bottle. Do not sterilize.

Working solution: Dilute stock to 0.5 μg/ml or other desired concentration in electrophoresis buffer (e.g., 1× TBE or TAE) or in H_2O.

Ethidium bromide working solution is used to stain agarose gels to permit visualization of nucleic acids under UV light. Gels should be placed in a glass dish containing sufficient working solution to cover them and shaken gently or allowed to stand for 10 to 30 min. If necessary, gels can be destained by shaking in electrophoresis buffer or H_2O for an equal length of time to reduce background fluorescence and facilitate visualization of small quantities of DNA.

Alternatively, a gel can be run directly in ethidium bromide by using working solution (made with electrophoresis buffer) as the solvent and running buffer for the gel.

CAUTION: *Ethidium bromide is a mutagen and must be handled carefully.*

Evans Blue solution, 50 mg/ml (UNIT 4.7)
Dissolve 100 mg of Evans Blue powder (Sigma) in 2 ml sterile isotonic saline. Vortex to dissolve the powdered dye. Make fresh batches daily.

Extracellular artificial cerebrospinal fluid (aCSF) (UNIT 1.8)
125 mM NaCl
25 mM $NaHCO_3$
11 mM glucose
3 mM KCl
1.25 mM NaH_2PO_4
2 mM $CaCl_2$
1 mM $MgCl_2$
Adjust pH to 7.4 by bubbling with carbogen (95% O_2/5% CO_2)
Prepare fresh each day

The osmolarity of this solution is ∼305 mmol/kg.

Other versions of aCSF may also be suitable; the exact composition will depend on the experiment.

Extracellular solution (UNIT 1.11)
125 mM NaCl

2.5 mM KCl
1.25 mM NaH_2PO_4
25 mM $NaHCO_3$
2 mM $CaCl_2$
2 mM $MgCl_2$
10 mM dextrose
Store up to 1 week at 4°C
Bubble with carbogen (95% O_2/5% CO_2) just prior to use and throughout the experiment

This solution is also known as artificial cerebrospinal fluid (aCSF).

Ferrous ammonium sulfate (FAS) solution, 10 mM (UNIT 2.5)
    0.0039 g FAS (Sigma)
✓ 1 ml Argon-purged Locke's solution, degassed
    Prepare immediately before use

Fetal bovine serum (FBS)

Thaw purchased fetal bovine serum (shipped on dry ice and kept frozen until needed). Store 3 to 4 weeks at 4°C. If FBS is not to be used within this time, aseptically divide into smaller aliquots and refreeze until used. Store ≤1 year at −20°C. To inactivate FBS, heat serum 30 min to 1 hr in a 56°C water bath.

Repeated thawing and refreezing should be avoided as it may cause denaturation of the serum.

Inactivated FBS (FBS that has been treated with heat to inactivate complement protein and thus prevent an immunological reaction against cultured cells) is useful for a variety of purposes. It can be purchased commercially or made in the lab as described above.

[^{3}H]Flumazenil stock solution, 100 nM (UNIT 2.10)

Prepare a 100 nM working concentration of [^{3}H]flumazenil (NEN [^{3}H]Ro 15–1788; 74 Ci/mmol) in a total volume of 1.1 ml of 50 mM Tris citrate buffer (see recipe); determine the dilutions required using one of the following methods. Make fresh for each assay.

There are two ways to conceptualize this, both requiring knowledge of the specific activity and concentration of the radioligand. This information is usually listed on the source vial and the package insert. The following example is described in terms of a vial of [^{3}H]flumazenil with a specific activity of 74 Ci/mmol and a concentration of 1 μCi/μl.

Method 1: Since the specific activity of [^{3}H]flumazenil is 74 Ci/mmol, and the concentration is 1 μCi/μl (or 1 Ci/liter), then the molar concentration of [^{3}H]flumazenil in the source vial is:

$$\frac{1 \text{ Ci/liter}}{74 \text{ Ci/mmol}} = 0.0135 \text{ mmol/liter} = 13.5 \text{ μM}$$

Since 1.1 ml of a 100 nM solution of [^{3}H]flumazenil is needed, the dilution would be (mass × volume / mass × volume):

$$\frac{(100 \times 10^{-9} \text{ M}) \times (1.1 \times 10^{-3} \text{ liter})}{13.5 \times 10^{-6} \text{ M}} = 8.1 \times 10^{-6} \text{ liter}$$

Method 2: Alternatively, the volume of [^{3}H]flumazenil to be taken from the source vial may be calculated as follows:

(working concentration of radioligand)
× (specific activity of radioligand)
× (volume of working solution)
× (radioligand concentration in source vial)
= volume of source vial stock to yield working concentration

Thus,

$(100 \times 10^{-9} \text{ mol/liter}) \times (74 \times 10^{-6} \text{ Ci}/10^{-9} \text{ mol})$
$\times (1.1 \times 10^{-3} \text{ liter}) \times (1 \times 10^{-6} \text{ liter}/1 \times 10^{-6} \text{ Ci})$
$= 8.1 \times 10^{-6}$ liter

Flunitrazepam (unlabeled), 10 mM (UNIT 2.10)
31.5 mg flunitrazepam (MW 315.3; Research Biochemicals) in
10 ml of 100% ethanol.
Store indefinitely at −20°C

Dilute 1/100 (to 100 μM) just before use by adding 10 μl of 10 mM stock to 990 μl of 50 mM Tris citrate buffer (see recipe).

Flux buffer (UNIT 2.1)
Modified Earle's Balanced Salt Solution (Na^+-free; Life Technologies) with:
1 mM $CaCl_2$
5 mM K_2HPO_4
1.2 mM $MgSO_4$
0.3 mM H_3PO_4
5.55 mM D-glucose
25 mM 2(*N*-morpholino)ethanesulfonic acid (MES)
Titrate to pH 6.0 with KOH
Add choline chloride to bring to 300 mOsm/kg (∼125 mM)
Store up to 4 weeks at 4°C

Formalin, 5% (v/v) (UNIT 4.9)

Dilute 10% buffered formalin solution (Sigma) 1:1 in distilled water. Store in an amber vial up to several weeks at room temperature.

10% buffered formalin solutions contain ∼4% formaldehyde. Therefore, 5% formalin will contain ∼2% formaldehyde.

GABA (unlabeled), 10 mM (UNIT 2.10)
    2.062 mg GABA (γ-aminobutyric acid, MW 103.1; Research Biochemicals)
✓ 2 ml of 50 mM Tris citrate buffer
    Make fresh for each assay

Gel loading buffer, 6×
0.25% (w/v) bromphenol blue
0.25% (w/v) xylene cyanol FF
40% (w/v) sucrose *or* 15% (w/v) Ficoll 400 *or* 30% (v/v) glycerol
Store at 4°C (room temperature if Ficoll is used)

This buffer does not need to be sterilized. Sucrose, Ficoll 400, and glycerol are essentially interchangeable in this recipe.
Other concentrations (e.g., 10×) can be prepared if more convenient.

Griess reagents (UNIT 2.4)

Reagent A: 1% sulfanilamide in 5% phosphoric acid

Dissolve 10 g of sulfanilamide (Sigma) in 50 ml of concentrated *o*-phosphoric acid in a 1-liter reagent bottle. Bring volume to 1 liter with ultrapure deionized water (e.g., Milli-Q-purified water), add stir bar, and mix on a magnetic stir plate until completely dissolved.

Reagent B: 0.1% N-(1-napthyl)-ethylenediamine dihydrochloride

Dissolve 1 g *N*-(1-napthyl)-ethylenediamine dihydrochloride in 1 liter of ultrapure deionized water (e.g., Milli-Q-purified water).

Store Griess reagents A and B unmixed at 4°C. These solutions should be stable for at least 1 to 2 months, although solution A has a tendency to darken.

[γ-^{35}S]GTP in ligand solution (UNIT 2.1)

Purchase [γ-^{35}S]GTP in stable aqueous solution from PerkinElmer Life and Analytical Sciences. Divide the purchased sample into smaller aliquots convenient for use in a single assay and freeze at −20°C (do not subject to multiple freeze-thaw cycles). To calculate the amount of radioligand stock solution required for the assay, use the following parameters contained on the technical information sheet supplied with the product—e.g., specific activity (SA), 1250 Ci/mmol; radioactive concentration (RAC), 1 mCi/ml; decay rate constant $= e^{(-0.693/\text{half-life}) \times \text{time}}$; calibration date, March 9. Assume an assay date of March 15 for this assay.

Calculate as follows (for final concentration of 500 pM [γ-^{35}S]GTP in each 200-μl well):

Nominal radioactive concentration in stock (M) = RAC ÷ SA

= (1 mCi/ml ÷ 1250 Ci/mmol) × (1/1000) = 8 × 10^{-7} M

Use the basic equation, (concentration$_1$ × volume$_1$) = (concentration$_2$ × volume$_2$), to calculate the following:

(1) concentration per well = (final conc. × total vol.) ÷ vol. added

= (5 × 10^{-10} M × 200 μl) ÷ 50 μl = 2 × 10^{-9} M

(2) volume diluted stock = (conc./well × total vol. required) ÷ stock concentration

= (2 × 10^{-9} M × 5.4 ml) ÷ (8 × 10^{-7} M × 1000 μl/ml) = 13.5 μl

(3) Adjust the volume of [γ-^{35}S]GTP based upon the decay rate of the ligand. The fraction radioactivity remaining = $e^{-(\text{assay date} - \text{calibration date} \times 0.693 \div \text{half life})}$

= $e^{(-(15-9) \times (0.693) \div 87.4 \text{ days})}$ = 0.95

(4) Therefore the adjusted volume = volume of diluted stock ÷ decay factor

= 13.5 μl ÷ 0.95 = 14.16 μl

(5) Add 14.16 μl of radioactive stock to 5.4 ml of assay buffer

Guanidine buffer (UNIT 2.2)

Dissolve ∼600 g guanidine hydrochloride (Sigma) in ∼750 ml of 50 mM Tris base, with warming. Add 50 mM Tris base with stirring until refractive index drops to 1.4251 (52.4% for sucrose refractometer), the refractive index for 5.5 M guanidine. If a refractometer is not available, determine solid content by lyophilization of a small sample in a centrifugal vacuum concentrator unit. Allow the solution to cool to room temperature and then adjust pH to 8.0 with 5 M HCl (pH will rise to >8.4 upon cooling to 4°C). Add 1 g activated charcoal and gently stir for 1 hr. Filter through 0.22-μm sterile filter unit. Store at 4°C for several months.

For biochemical assays of large numbers of tissue homogenates, most commercial grades of guanidine can be used at a more reasonable cost than higher grades by following the above procedure. Removal of colored material by charcoal treatment can reduce background color or fluorescence in the detection phase. Assaying the strength of the guanidine hydrochloride solution ensures that the denaturant is at appropriate strength to solubilize all but the most insoluble cellular debris. Filtration removes not only the charcoal, but also particulate matter, which can result in detection inaccuracies.

HBSS (Hanks' balanced salt solution)

0.40 g KCl (5.4 mM final)
0.09 g Na$_2$HPO$_4$·7H$_2$O (0.3 mM final)
0.06 g KH$_2$PO$_4$ (0.4 mM final)
0.35 g NaHCO$_3$ (4.2 mM final)
0.14 g CaCl$_2$ (1.3 mM final)
0.10 g MgCl$_2$·6H$_2$O (0.5 mM final)
0.10 g MgSO$_4$·7H$_2$O (0.6 mM final)
8.0 g NaCl (137 mM final)
1.0 g D-glucose (5.6 mM final)
0.2 g phenol red (0.02%; optional)
Add H$_2$O to l liter and adjust pH to 7.4 with 1 M HCl or 1 M NaOH
Filter sterilize and store up to 1 month at 4°C

HBSS may be made or purchased without Ca^{2+} and Mg^{2+} (CMF-HBSS). These components are optional and usually have no effect on an experiment; in a few cases, however, their presence may be detrimental. Consult individual protocols to see if the presence or absence of these components is recommended.

HBSS (UNIT 4.13)

415.0 ml sterile H$_2$O
50.0 ml 10× HBSS without Ca^{2+} and Mg^{2+} (e.g., Life Technologies)
20.0 ml fetal bovine serum
5.0 ml 1 M HEPES solution

5.0 ml 2.8% sodium bicarbonate solution
5.0 ml 10,000 U/ml penicillin/10,000 mg/ml streptomycin
Store up to 1 month, 4°C
Total volume 500 ml

HEPES buffer (UNIT 2.13)

10× stock solution
Dissolve in 900 ml H_2O:
80 g NaCl (1.37 M)
3.7 g KCl (50 mM)
10 g glucose (56 mM)
47.7 g HEPES (200 mM)
0.8 g KH_2PO_4 (5.9 mM)
0.8 g Na_2HPO_4 (5.6 mM)
Adjust pH to 7.4 with NaOH
Bring volume to 1 liter with H_2O
Store up to 1 month at 4°C

1× working solution
To ∼500 ml H_2O add:
100 ml 10× stock solution
10 ml 140 mM $CaCl_2$ (1.4 mM final)
10 ml 90 mM $MgSO_4$ (0.9 mM final)
20 ml 500 mM of $NaHCO_3$ (10 mM final)
Adjust volume to 1 liter with H_2O

The $NaHCO_3$ stock should be made daily, while the other stocks are stable for months at 4°C.

High-Ba^{2+} solution (UNIT 1.6)
40 mM BaOH
50 mM NaOH
2 mM KOH
5 mM HEPES
Titrate with concentrated methanesulfonic acid to pH of 7.5
Store up to 1 month at room temperature

If any precipitation is detected, filter the solution through a 0.45-μm filter. This solution is used for studying voltage-dependent Ca^{2+} channels.

High-K^+ solution (UNIT 1.6)
2 mM NaCl
96 mM KCl
1 mM $MgCl_2$
1 mM $CaCl_2$
5 mM HEPES
Adjust pH to 7.5 with 5 M KOH
Store up to 1 week at room temperature without sterilization

This solution is used for the study of inwardly rectifying K^+ channels.

Solutions with intermediate concentrations of K^+ can be obtained by mixing the high-K^+ solution with ND96 (see recipe).

Homogenization buffer (UNIT 2.4)
✓ 50 mM Tris·Cl, pH 7.4
✓ 0.1 mM EDTA
100 μg/ml phenylmethylsulfonyl fluoride (PMSF)
10 μg/ml leupeptin
10 μg/ml soybean trypsin inhibitor
2 μg/ml aprotinin

Store components without DTT up to 1 month (or longer) at 4°C
Add 1 mM dithiotheritol (DTT) immediately before use

Various combinations of protease inhibitors are reported in the literature; however, none has been proven to be more efficacious than the one described here.

HPLC mobile phase (UNIT 2.11)
6.9 g/liter NaH_2PO_4
250 mg/liter 1-heptanesulfonic acid sodium salt
80 mg/liter EDTA
70 ml/liter HPLC-grade methanol (7% v/v)
Adjust pH to 3.7 with HCl
Filter through 0.2-μm filter (Rainin Nylon-66, 47-mm diameter) under negative pressure
Store up to 6 months at 4°C
Use only HPLC-grade reagents

[^{14}C]p-Hydroxy-loracarbef, 20 mM (UNIT 2.1)

Obtain reagent custom synthesized (this was performed for the authors at Lilly Research Laboratories by D. O'Bannon and W. Wheeler). Store as a solid, desiccated, at −20°C.

Prepare a 20 mM solution in flux buffer (see recipe) according to the following formula:

$$3.74 \text{ ml} \times \frac{20 \times 10^{-3} \text{ mmol}}{\text{ml}} \times \frac{365.77 \text{ mg}}{\text{mmol}} = 27.36 \text{ mg}$$

5-Hydroxytryptamine (unlabeled), 10 mM and 1 mM (UNIT 2.1)

10 mM stock solution
✓ 1 ml saturation assay buffer
3.87 mg of 5-hydroxytryptamine, creatine sulfate complex (mol. wt. 387.4; Sigma)
Prepare fresh daily and store at 4°C

1 mM solution
✓ 900 μl of saturation assay buffer
100 μl of 10mM stock solution
Prepare fresh daily and store at 4°C

[^3H]5-Hydroxytryptamine, 1 μM (UNIT 2.1)

Prepare a 1 μM stock solution of 5-hydroxy [^3H]tryptamine trifluoroacetate (Amersham Biosciences) in a total volume of 1 ml of 1× saturation assay buffer (see recipe). To calculate the concentration of the purchased stock radioligand, use the technical information supplied with the reagent. For example, for a specific activity = 123 Ci/mmol and a radioactive concentration = 1 mCi/ml:

$$\frac{1 \text{ mmol}}{123 \text{ Ci}} \times \frac{1 \text{ mCi}}{\text{ml}} \times \frac{1 \text{ Ci}}{1000 \text{ mCi}}$$

$$= 8.13 \times 10^6 \text{ mmol/ml} = 8.13 \text{ mM}$$

For 1 ml of 1 µM stock solution, volume$_1$ × concentration$_1$ = volume$_2$ × concentration$_2$:

$$\frac{1 \text{ ml} \times 1 \text{ mM}}{8.13 \text{ mM}} = \frac{123 \text{ ml stock added to } 877 \text{ ml}}{1 \times \text{saturation assay buffer}}$$

Prepare stock solution daily and store on ice during the working day.

Hypertonic solution 1 *(UNIT 1.6)*
96 mM NaCl
2 mM KCl
1 mM MgCl$_2$
5 mM HEPES
100 mM sucrose
Adjust pH to 7.5 with 5 M NaOH
Store up to 2 to 3 weeks at 4°C
Sterilization by 0.45-µm filtration is advisable.

Hypertonic solution 2 *(UNIT 1.6)*
200 mM potassium glutamate
20 mM NaCl
1 mM MgCl$_2$
10 mM HEPES
Adjust pH to 7.5 with 5 M NaOH
Store up to 2 to 3 weeks at 4°C
Sterilization by 0.45 µm filtration is advisable.

Incubation buffer *(UNIT 2.14)*
Per liter:
6.08 g NaCl (108 mM final)
0.35 g KCl (4.7 mM final)
0.30 g MgSO$_4$·7H$_2$O (1.2 mM final)
0.16 g KH$_2$PO$_4$ (1.2 mM final)
0.44 g CaCl$_2$·2H$_2$O (3.0 mM final)
4.77 g HEPES (20 mM final)
0.19 g EDTA·2H$_2$O (0.5 mM final)
1.80 g glucose (10 mM final)
Adjust to pH 7.4 with 10 N NaOH

Incubation medium *(UNIT 4.14)*
25% (v/v) horse serum, heat inactivated
✓ 25% (v/v) HBSS without calcium chloride, magnesium chloride, and magnesium sulfate and supplemented with an additional 25.6 mg/ml D-glucose
50% (v/v) modified essential medium (MEM) with Hanks' salts and 25 mM HEPES and without L-glutamine (Invitrogen Life Technologies)
2 mM L-glutamine
Filter sterilize through a 0.2-µm filter and store ≤1 week at 4°C

The final pH of the medium is 7.2, although initial studies using pH 7.4 or 7.8 revealed similar motor neuron survival.

Interface culture medium *(UNIT 1.7)*
25 ml 2× minimal essential medium (MEM; e.g., Life Technologies; store up to 6 months at 4°C)
25 ml horse serum, mycoplasma screened (heat-inactivate complement 30 min at 56°C; store up to 18 months at −20°C; thaw slowly at room temperature)
2.5 ml 10× Hanks' balanced salt solution (e.g., Life Technologies; store up to 6 months at room temperature)
1 ml 100× penicillin-streptomycin (e.g., Life Technologies)
60 mg Tris base (5 mM)
7.5% (w/v) NaHCO$_3$
H$_2$O to 100 ml
Store up to 1 month at 4°C

Internal pipet solution, cell-attached (for K$^+$ currents) *(UNIT 1.11)*
125 mM NaCl
10 mM HEPES
2.5 mM KCl
2 mM CaCl$_2$
1 mM MgCl$_2$
1 µM tetrodotoxin (TTX)
Adjust pH to 7.3 with 1 M NaOH
Store up to 1 month at 4°C

Internal pipet solution, whole-cell *(UNIT 1.11)*
120 mM KMeSO$_4$ or K-gluconate
20 mM KCl
10 mM HEPES
0.2 mM EGTA
2 mM MgCl$_2$
4 mM Na$_2$-ATP
0.3 mM Tris-GTP
Adjust pH to 7.3 with 1 M KOH
Store in 3-ml aliquots up to 6 months at −20°C

These storage conditions will preserve ATP/GTP activity.

KMeSO$_4$ produces less blockade of the K$^+$ currents mediating slow afterhyperpolarizing potentials than K-gluconate (Velumian et al., 1997).

Intracellular buffer *(UNIT 2.13)*
10× stock solution
Dissolve in 90 ml H$_2$O:
8.95 g KCl (1.2 M)
0.29 g NaCl (50 mM)
0.1 g KH$_2$PO$_4$ (7.4 mM)
0.42 g NaCO$_3$ (50 mM)
4.77 g HEPES (200 mM)
Do not adjust pH
Bring volume to 100 ml with H$_2$O
Store up to 1 month at 4°C

1× working solution: dilute 1:9 with H$_2$O before use.

This buffer is used for cell-free calibrations, and approximates the concentration of the predominant intracellular ions.

Intracellular patch-pipet solution *(UNIT 1.8)*
125 mM potassium gluconate
5 mM NaCl
10 mM HEPES
10 mM EGTA
2 mM MgCl$_2$
2 mM Na$_2$ATP
Adjust pH to 7.2 with KOH
Filter with 0.2-µm filter

Store in aliquots at −20°C; use thawed aliquots the same day.

The osmolarity of this solution is ∼280 mmol/kg.

Other versions of this solution may also be suitable; the exact composition will depend on the experiment.

[^{125}I]Iodocyanopindolol, 500 nM (UNIT 2.1)

(−)-3-[^{125}I]iodocyanopindolol is available from Amersham Biosciences as a solution in methanol containing 0.01% phenol. This stock is stored at −20°C, which provides reagent stability without freezing of the solution. The concentration of the purchased stock radioligand, on the reference date, can be calculated using the information supplied by the manufacturer:

Specific activity = 2000 Ci/mmol
Radioactive concentration = 1 mCi/ml
Thus:
(1 mmol/2000 Ci) × (1 mCi/ml) × (1 Ci/1000 mCi) = 5×10^{-7} mmol/ml = 500 nM

To determine concentrations on days other than the reference date, a correction using the fraction of radioactivity remaining can be performed:
fraction remaining = $e^{(-0.693/\text{Half-life}) \times \text{Days past reference date}}$

Original concentration × Fraction remaining = concentration on day of use

For example, if the radioligand is being used 10 days after the reference date, the concentration of the stock is (half life for ^{125}iodine = 60 days):
fraction remaining = $e^{(-0.693/60) \times 10}$ = 0.891
Concentration = 500 nM × 0.891 = 445.5 nM

K^+/low Na^+ buffer (UNIT 2.11)
✓ Krebs-HEPES buffer, modified
Increase KCl to 20 mM and reduce NaCl to 112 mM.

Kainic acid solution (UNIT 4.16)
2.5 mg/ml kainic acid (Sigma-Aldrich)
0.9% (w/v) NaCl
Store up to 2 days at 4°C
To calculate the amount of kainic acid required for each experiment, use the following formula:
(dose in mg/kg) × (no. of rats) × (average weight of rats in kg) × (no. of injections) = amount of kainic acid (mg)
Example: dose = 5 mg/kg; no. of rats = 5; average weight = 0.2 kg; no. of injections = 10.
(5 mg/kg) × (5 rats) × (0.2 kg) × (10 injections/rat) = 50 mg of kainic acid
Prepare a 2.5 mg/ml solution for the established dose volume (0.2 ml per 100 g body weight). Thus, for this example, dissolve 50 mg of kainate in 20 ml of 0.9% NaCl.
Store unused kainic acid desiccated and protected from light at 2° to 8°C.

Ketamine/xylazine cocktail, 10 ml (APPENDIX 2A)
0.65 ml of 100 mg/ml ketamine stock solution
0.22 ml of 20 mg/ml xylazine stock solution
9.13 ml sterile isotonic saline

This working solution is stored at room temperature and is stable until the expiration dates of its respective components. The recipe is adequate for anesthetizing ∼20 adult mice.

NOTE: The Drug Enforcement Administration (DEA) has placed ketamine into Schedule III of the Controlled Substances Act. Therefore, registration with the DEA and your state board of pharmacy may be required in order to purchase and use this drug.

Krebs-bicarbonate buffer (UNIT 2.11)
118 mM NaCl
4.7 mM KCl
1.3 mM $CaCl_2$
1.2 mM $MgCl_2$
1 mM NaH_2PO_4
25 mM $NaHCO_3$
11.1 mM glucose
✓ 4 μM disodium EDTA
300 u ascorbic acid
Adjust pH to 7.4
Prepare fresh daily

Krebs-bicarbonate buffer (UNIT 2.15)
128 mM NaCl
24 mM $NaHCO_3$
4.2 mM KCl
1.5 mM $CaCl_2$
0.9 mM $MgCl_2$
2.4 mM NaH_2PO_4
9 mM glucose
10 U/ml heparin
Store up to ∼1 month at 4°C

Just prior to use, bubble for ∼20 min with 95% (v/v) O_2/5% (v/v) CO_2 to attain pH 7.4 and filter through a 0.2-μm syringe filter unit. Add test compound to achieve the desired concentration (optional) and warm to 37°C (or cool to 4°C; see Support Protocol 7).

To study the effects of test compound-protein binding effects on BBB permeability, 4% (w/v) albumin may be added to the buffer.

Krebs-Henseleit buffer (UNIT 2.14)
Per liter:
6.92 g NaCl (118.5 mM final)
0.35 g KCl (4.7 mM final)
0.29 g $MgSO_4 \cdot 7H_2O$ (1.2 mM final)
0.37 g $CaCl_2 \cdot 2H_2O$ (2.5 mM final)
0.16 g KH_2PO_4 (1.2 mM final)
2.09 g $NaHCO_3$ (24.9 mM final)
3.42 g glucose (19 mM final)
Maintain pH at 7.4 by continuous bubbling with 95% O_2/5% CO_2

Krebs-HEPES buffer, modified (UNIT 2.11)
127 mM NaCl
5 mM KCl
1.3 mM NaH_2PO_4

2.5 mM CaCl$_2$
1.2 mM MgSO$_4$
15 mM HEPES
10 mM glucose
100 μM ascorbic acid
1 μM domperidone (Research Biochemicals)
10 μM pargyline (Research Biochemicals)
Adjust pH to 7.4 with NaOH
Prepare fresh daily

LB medium (Luria broth) and LB plates
Per liter add:
10 g tryptone
5 g yeast extract
5 g NaCl
1 ml 1 M NaOH
15 g agar or agarose (for plates only)

Autoclave solution. Add filter-sterilized additives (see below) after the solution has cooled to 55°C. For LB plates, pour agar-containing solution into sterile petri dishes in a tissue culture hood. Store autoclaved medium up to 1 month at room temperature. Store antibiotic-containing plates in the dark up to 2 weeks at 4°C.

Contamination is easily identified by the clouding of solutions or the presence of growth on plates.
Additives:
Antibiotics (if required):
Ampicillin to 50 μg/ml
Tetracycline to 12 μg/ml
Galactosides (if required):
5-bromo-4-chloro-3-indolyl-β-D-galactoside (Xgal) to 20 μg/ml
Isopropyl-1-thio-β-D-galactoside (IPTG) to 0.1 mM

LCS (lipid cleaning solution) (UNIT 2.14)
18.64 g KCl (500 mM final)
0.45 g myoinositol (5 mM final)

Dissolve in 250 ml H$_2$O. When the solution is clear, add 250 ml methanol [50% (v/v) final] and mix.

Low-Cl$^-$ buffer (UNIT 2.12)
137 mM D-glucuronic acid (27 g/liter) *or* 137 mM D-gluconic acid, sodium salt (30 g/liter)
5 mM KHSO$_4$ (681 mg/liter)
4.2 mM NaHCO$_3$ (353 mg/liter)
0.44 mM KH$_2$PO$_4$ (60 mg/liter)
1.4 mM MgSO$_4$·7H$_2$O (247 mg/liter)
20 mM 2-{[tris(hydroxymethyl)methyl]amino}-1-ethanesulfonic acid (TES; 4.6 g/liter)
Adjust pH to 7.4 with NaOH
Store up to several days at 4°C

The buffer is also stable at least 1 month at −18°C. Glucuronic acid has some effect on the pH, while gluconic acid may form complexes with Mg^{2+} and Ca^{2+}.
Immediately before use add:
1 mM CaCl$_2$·2H$_2$O (147 mg/liter)
10 mM glucose (1.8 g/liter)
0.1% (w/v) bovine serum albumin
1 mM furosemide
Check the pH of the buffer again just before use

Add furosemide from a 1 M stock solution prepared by adding grains of Tris to furosemide in water until all furosemide is dissolved (store stock up to several weeks at −20C).

This buffer is supplemented with varying concentrations of NaCl for calibration (Support Protocol 4). Isoosmotic replacement with sodium glucuronate or gluconate is necessary to compensate for the ionic change.

Low-Cl$^-$ Ringer's solution (UNIT 2.12)

Prepare as for Ringers solution (see recipe) but substitute 119 mM sodium gluconate (25.7 g/liter) for NaCl, and 2.5 mM potassium gluconate (0.58 g/liter) for KCl.

This buffer is supplemented with varying concentrations of NaCl for calibration (Support Protocol 4). Isoosmotic replacement with sodium gluconate is necessary to compensate for the ionic change.

Luminol stock solution, 250 μg/ml (UNIT 2.5)
10 mg 5-amino-2,3-dihydro-1,4-phthalazinedione (luminol; Sigma)
✓ 40 ml 50 mM borate buffer, pH 10.0
Stir ≥30 min
Store in an amber bottle for up to 2 months at 4°C

A stock solution of luminol is prepared to minimize the possible concentration fluctuation in the CL cocktail due to error introduced by weighing small amounts of luminol before each experiment.

Lysis buffer, pH 4.7 at 25°C (UNIT 2.3)
10 g sodium dodecyl sulfate (SDS)
25 ml N,N-dimethylformamide (DMF)
25 ml H$_2$O
1 ml glacial acetic acid
1.25 ml 1 N HCl

Dissolve SDS into combined water/DMF solution. Adjust pH to 4.7 with glacial acetic acid or 1 N HCl (Hansen et al., 1989). Store at room temperature for up to 1 month.

Methyl-5-hydroxytryptamine, 10 mM (UNIT 2.1)
1.0 mg of α-methyl-5-hydroxytryptamine or 2-methyl-5-hydroxytryptamine (mol. wt. 306.32; Research Biochemical)
✓ 327 μl of saturation assay buffer.
Prepare fresh daily and store at 4°C during the day.

Miniature SPA buffer (UNIT 2.1)
To prepare 500 ml, combine the following at room temperature:
✓ 25 ml 1 M Tris·Cl, pH 7.5 (50 mM final)
5 ml 1 M MgCl$_2$ (10 mM final)
470 ml Milli-Q-purified H$_2$O
Adjust pH after additions, if necessary, and adjust final volume to 500 ml
Filter using a 0.22-μm sterile filter unit

Store up to 3 to 4 weeks at room temperature; check pH periodically.
For larger volumes, scale up accordingly.

Mobile phase buffer I *(UNIT 2.9)*
Prepare in HPLC grade H_2O:
75 mM $NaH_2PO_4 \cdot H_2O$ (10.4 g/liter)
2.1 mM 1-octanesulfonic acid (235 mg/l)
50 μM disodium EDTA (19 mg/liter)
100 μl/liter triethylamine
11% (v/v) acetonitrile
Adjust pH to 3.25 using 85% phosphoric acid
Store in a sealed container up to 1 week at 4°C
Use HPLC-grade chemicals.

Mobile phase buffer II *(UNIT 2.9)*
Prepare in HPLC grade H_2O:
75 mM $NaH_2PO_4 \cdot H_2O$ (10.4 g/liter)
3.5 to 5.5 mM 1-octanesulfonic acid (0.822 to 1.29 g/liter)
50 μM disodium EDTA (19 mg/l)
100 μl/liter triethylamine
7% to 13% (v/v) acetonitrile
Adjust pH to 5.25 with sodium hydroxide
Store in a sealed container up to 1 week at 4°C
Use HPLC-grade chemicals.

Mobile phase buffer III *(UNIT 2.9)*
Prepare 0.2 M K_2HPO_4 in HPLC-grade water (34.8 g K_2HPO_4 per liter). To this buffer, add tetramethylammonium hydroxide to 0.001 M (181 mg/liter). Adjust pH to 8.0 using 0.2 M KH_2PO_4. Store in a sealed container up to 1 week at 4°C.
Use HPLC-grade chemicals.

Mobile phase buffer IV *(UNIT 2.9)*
Prepare 35 mM Na_2HPO_4 in HPLC-grade water (4.978 g/liter). Adjust pH to 8.5 with sodium hydroxide. Store in a sealed container up to 1 week at 4°C.
Use HPLC-grade chemicals.

Mobile phase buffer V *(UNIT 2.9)*
Prepare in HPLC-grade H_2O:
0.1 M Na_2HPO_4 (14.2 g/liter)
0.13 mM disodium EDTA (48 mg/liter)
33% (v/v) methanol
Store up to several months at 4°C in a tightly capped glass container

Use HPLC-grade chemicals. The sodium phosphate component may require a brief period of mixing to completely dissolve. Adjust the pH of the buffer to 5.88 (25°C) with H_3PO_4. Filter and degas the buffer before use (see Basic Protocol 3, step 2).

Monkey artificial cerebrospinal fluid (aCSF)
(UNIT 2.8)
Dissolve in 900 ml H_2O:
268 mg $Na_2HPO_4 \cdot 7H_2O$ (1.0 mM final)
8.6 g NaCl (147 mM final)
0.22 g KCl (3 mM final)
0.19 g $CaCl_2 \cdot 2H_2O$ (1.3 mM final)
0.20 g $MgCl_2 \cdot 6H_2O$ (1.0 mM final)

35 mg ascorbic acid (weigh accurately; 0.20 mM final)
Add H_2O to 1 liter
Filter sterilize
Make fresh the day of the experiment

Monoamine standards *(UNIT 2.9)*
Stock solutions (prepare in HPLC-grade H_2O):
1 mM dopamine (18.96 mg/100 ml)
1 mM serotonin (38.74 mg/100 ml)
1 mM 3,4-dihydroxyphenylacetic acid (DOPAC; 16.81 mg/100 ml)
1 mM hydroxyvalproic acid (HVA; 18.22 mg/100 ml)
1 mM 5-hydroxyindoleacetic acid (5-HIAA; 19.12 mg/100 ml)
Store up to 1 month at −20°C

Working solutions: Prepare 10 μM working standards on day of assay by diluting 1 ml of stock solutions in 100 ml of the perfusate buffer used to collect the microdialysate of interest. Prepare other concentrations as needed to bracket the estimated concentrations of the samples.

The monoamines can be obtained from a variety of sources including Sigma and Research Biochemicals.

Muscarinic assay buffer *(UNIT 2.1)*
To 1500 ml of water add:
9.54 g HEPES (20 mM final)
11.70 g NaCl (100 mM final)
2.04 g $MgCl_2$ (5 mM final)

Prepare the solution using reagents that are at room temperature. Adjust the final volume to 2000 ml. Adjust pH to 7.4 with NaOH. Store up to 4 weeks at 4°C.

Muscimol, 100 μM *(UNIT 2.10)*
1 mg muscimol (Sigma; MW 114.1)
✓ 876 μl 50 mM Tris citrate buffer

Dilute 10 mM stock 1/100 (e.g., 10 μl of 10 mM of stock/990 μl Tris citrate buffer) to yield 100 μM working solution at time of assay. Store 10 mM stock up to 1 month at −20°C.

Na^+-free HEPES buffer *(UNIT 2.13)*
5× stock solution
Dissolve in 900 ml H_2O:
136.6 g N-methyl-D-glucamine (700 mM)
5 g glucose (28 mM)
23.85 g HEPES (100 mM)
0.4 g KH_2PO_4 (3 mM)
0.44 g K_2HPO_4 (2.5 mM)
Adjust pH to 7.4 with 10 N HCl
Bring volume to 1 liter with H_2O

All Na^+ salts are replaced by K^+ equivalents except for NaCl, which is substituted with N-methyl-D-glucamine. Na^+-free solutions can also be made with Li^+, Tris, or choline.
1× working solution
To ∼500 ml H_2O add:
200 ml 5× stock solution
10 ml 140 mM $CaCl_2$ (1.4 mM final)

10 ml 90 mM $MgSO_4$ (0.9 mM final)
20 ml 500 mM $KHCO_3$ (10 mM final)
Adjust the volume to 1 liter with H_2O

ND96 standard extracellular solution (UNIT 1.6)
96 mM NaCl
2 mM KCl
1 mM $MgCl_2$
1 mM $CaCl_2$
5 mM HEPES
Adjust pH to 7.5 with 5 M NaOH
Store up to 1 week at room temperature without sterilization

Neurotransmitter standards (UNIT 2.8)

Dopamine (DA) stock: 10 mg/10 ml final concentration

Dilution: 1 µl diluted to 1000 µl in aCSF (10 µg or 10,000 ng per 1000 µl)

Glutamate stock: 30 mg/30 ml final concentration

Dilution: 50 µl diluted to 1000 µl in aCSF (50 µg or 50,000 ng per 1000 µl)

γ-*Aminobutyric acid (GABA) stock*: 30 mg/30 ml final concentration

Dilution: 50 µl diluted to 1000 µl in aCSF (50 µg or 50,000 ng per 1000 µl)

For testing probe recovery efficiency in vitro, add 10^{-7} M mix of DA, glutamate, and GABA (Sigma) to aCSF.

Nitrate reductase buffer (UNIT 2.4)
✓ 200 mM Tris·Cl, pH 7.6
0.1 mM flavin adenine dinucleotide (FAD)
1 mM nicotine adenine dinucleotide phosphate, reduced form (NADPH)
Store up to 2 weeks at $-20°C$

OPA stock solution (UNIT 2.9)

To prepare 10 ml of OPA stock solution, first dissolve 27 mg o-phthaldialdehyde (OPA; 20 mM final concentration) in a mixture of 1 ml methanol (10% v/v final concentration) and 12 µl 2-mercaptoethanol (0.12% v/v final concentration). Finally, add 9 ml of 0.1 M (38.1 g/liter) sodium tetraborate, pH 9.4 ($NaB_4O_7·10H_2O$). Store ≤1 week at 4°C in a foil-wrapped capped glass container.

Paraformaldehyde, 8% and 4% (UNIT 2.3)

8% stock solution

Heat ~800 ml water to 65°C, then add 80 g paraformaldehyde (Sigma) in a fume hood. Stir while slowly adding 2 to 4 ml of 5 N NaOH (add by drops until paraformaldehyde completely dissolves). After solution cools, filter through Whatman no. 1 filter paper to remove insoluble debris. Add water to 1 liter. Store up to 1 week at 4°C or aliquot and store up to 6 months at $-20°C$.

4% working solution

For 4% paraformaldehyde, dilute 8% paraformaldehyde stock 1:1 in PBS (see recipe) just before use.

CAUTION: *Formaldehyde is toxic. Appropriate care should be taken when handling formaldehyde. All steps should be done in a fume hood. Under no circumstances should any tools used during paraformaldehyde preparation (i.e., glassware, stirrers, spatulas, thermometers) be used for medium preparation.*

PBS (phosphate-buffered saline)
10× stock solution (per liter):
80 g NaCl
2 g KCl
11.5 g $Na_2HPO_4·7H_2O$
2 g KH_2PO_4

1× working solution:
137 mM NaCl final
2.7 mM KCl final
4.3 mM $Na_2HPO_4·7H_2O$ final
1.4 mM KH_2PO_4 final
If needed, adjust to desired pH (usually 7.2 to 7.4) with 1 M NaOH or 1 M HCl
Filter sterilize and store up to 1 month at 4°C

Without adjustment, the pH will generally be ~7.3.

PBS (UNIT 2.11)

Mix 20 mM Na_2HPO_4 and 20 mM NaH_2PO_4 (both prepared in 0.9% NaCl) until a pH of 7.4 is reached. Store up to 2 weeks at 4°C.

Perfusion solution (UNIT 1.11)
110 mM choline chloride
2.5 mM KCl
1.25 mM NaH_2PO_4
25 mM $NaHCO_3$
0.5 mM $CaCl_2$
7 mM $MgCl_2$
7 mM dextrose
1.3 mM ascorbic acid
3 mM pyruvic acid
Store up to 1 week at 4°C

Prior to use, chill the solution until a thin layer of ice has formed, then stir the ice into a thin slurry using a hand-held ice grinder (Braun). Bubble with carbogen before and during perfusion.

Sucrose may be substituted for choline chloride in equi-osmolar amounts. (Total osmolarity should be verified with an osmometer to be ~300 mOsm.) Ascorbate and pyruvate may improve tissue resistance to hypoxia during the slicing procedure (MacGregor et al., 1996; Lee et al., 2001).

Phenol, buffered

Add 0.5 g of 8-hydroxyquinoline to a 2-liter glass beaker containing a stir bar. Gently pour in 500 ml of liquefied phenol or crystals of redistilled phenol that have been melted in a water bath at 65°C (the phenol should turn yellow). Add 500 ml of 50 mM Tris base (unadjusted pH ~10.5). Cover the beaker with aluminum foil. Stir 10 min at low speed with magnetic stirrer at room temperature. Stop stirring and let phases separate at room temperature. Gently decant the top (aqueous) phase into a suitable waste receptacle. Using a pipet and suction bulb, remove residual aqueous phase that cannot be decanted. Add 500 ml of 50 mM Tris·Cl, pH 8.0 (see recipe below). Repeat two successive equilibrations with 500 ml of 50 mM Tris·Cl, pH 8.0 (i.e., the same procedure as with the Tris base). Check phenol phase with pH paper to verify that pH is 8.0; if not, perform additional equilibrations. Add 250 ml of 50 mM Tris·Cl, pH 8.0, or TE buffer, pH 8.0 (see recipe below), and store up to 2 months at 4°C in brown glass bottles or clear glass bottles wrapped in aluminum foil.

CAUTION: *Phenol can cause severe burns to skin and damage clothing. Gloves, safety glasses, and a lab coat should be worn whenever working with phenol, and all manipulations should be carried out in a fume hood. A glass receptacle should be available exclusively for disposing of used phenol and chloroform.*

In this recipe, the 8-hydroxyquinoline acts as an antioxidant.

Phenol/chloroform/isoamyl alcohol, 25:24:1 (v/v/v)
25 vol buffered phenol (bottom yellow phase of stored solution; see recipe above)
24 vol chloroform
1 vol isoamyl alcohol
Store up to 2 months at 4°C

Photosensitizing dye solution (UNIT 4.4)
Mix together 0.1 g or 0.2 g Rose Bengal (or Erythrosin B; Sigma) and 10 ml sterile saline (0.9% [w/v] NaCl). Filter through Whatman 3MM filter paper to remove insoluble crystals and impurities. Store in the dark up to 2 months at room temperature.

Rose Bengal and Erythrosin B are prepared and administered in the same fashion.

Different concentrations of sensitizing dye should be tested at first to determine the appropriate dose that will result in the desired extent of cortical injury. To deliver 10 mg/kg, make a 10 mg/ml concentration. To deliver 20 mg/kg, make a 20 mg/ml concentration.

o-Phthalaldehyde (OPA) solution, 5 mg/ml (UNIT 2.5)
Dissolve 50 mg reagent grade *o*-phthalaldehyde (Sigma) in 0.5 ml methanol, and dilute this to a final volume of 10 ml with 400 mM potassium borate, pH 9.9 (see recipe). Store up to several weeks at 4°C without loss of activity.

PIPES saline (UNIT 1.10)
6.7206 g NaCl (115 mM final)
0.3733 g KCl (5 mM final)
6.028 g PIPES free acid (20 mM final)
1 ml 1 M $CaCl_2$ stock solution (BDH Lab Suppliers; 1 mM final)
4 ml 1 M $MgCl_2$ (Sigma; 4 mM final)
4.504 g D-glucose (25 mM final)
H_2O to 900 ml
Adjust pH to 7.0 with NaOH
Add H_2O to 1 liter
Store ≤1 year at 4°C

Plate blocking buffer (UNIT 2.2)
73 mM sucrose
0.25% (w/v) bovine serum albumin (BSA)
50 mM Na_2HPO_4
0.01% (w/v) thimerosal (Sigma)
Adjust pH to 7.4 using 5 M NaOH
Store up to 6 months at 4°C

Plate coating buffer (UNIT 2.2)
100 mM Na_2HPO_4
0.05% (w/v) NaN_3
Adjust pH to 8.5 using 5 M NaOH
Store up to 6 months at 4°C

Antibody coating buffers are almost always prepared at a pH greater than 8.0. A buffer of 15 mM sodium carbonate, pH 9.6, is sometimes also used.

Polyethylenimine, 0.03% (UNIT 2.10)
Polyethylenimine is commercially available as a 50% (w/v) aqueous solution (Sigma). Make a 10% (w/v) stock solution by weighing 10 g of polyethylenimine and diluting in Milli-Q-purified water to 100 ml. At the time of assay, calculate that each 2 1/4 × 12 1/4-in. filter strip will require 20 ml of soaking solution. For example, 20 filter strips will require 400 ml of solution. Add 1.2 ml of 10% polyethylenimine solution to 400 ml of purified water (yielding 0.03% solution), and soak the filters in this mixture for ~15 min before the start of filtration. At the end of filtration, remove the filter disks from the strips before they dry. Glass fiber filters soaked in polyethylenimine (especially at higher concentrations, >3%) become stiff and tough when dry, making handling difficult.

Poly-L-lysine-coated glass coverslips (UNIT 2.12)
Prepare a stock solution by dissolving 25 mg/ml polylysine in 4.73 ml water (both poly-L-lysine and poly-D-lysine are used to coat tissue culture surfaces; check specific protocol for choice of isomer) and filter sterilize through a 0.22-μm filter. Store in 100-μl aliquots at –20°C. When ready to use, dilute one aliquot in 40 ml water to prepare 13 μg/ml working solution. Sterilize coverslips by autoclaving prior to coating. Dip coverslips in the working solution, then incubate 15 min to several hours in a humidified 37°C, 5% CO_2 incubator. Allow surface to dry.

Store coated coverslips up to 3 months at 4°C. Use diluted solutions only once, but unused diluted aliquots can be stored up to 3 months at 4°C.

Polylysine-coated tissue culture surfaces
To coat coverslips: Prepare a stock solution by dissolving 25 mg/ml polylysine in 4.73 ml water (both poly-L-lysine and poly-D-lysine are used to coat tissue culture surfaces; check specific protocol for choice of isomer) and filter sterilize through a 0.22-μm filter. Store in 100-μl aliquots at −20°C. When ready to use, dilute one aliquot in 40 ml water to prepare 13 μg/ml working solution. Sterilize coverslips by autoclaving prior to

Table A.1.3 Preparation of 0.1 M Sodium and Potassium Phosphate Buffers[a]

Desired pH	Solution A (ml)	Solution B (ml)	Desired pH	Solution A (ml)	Solution B (ml)
5.7	93.5	6.5	6.9	45.0	55.0
5.8	92.0	8.0	7.0	39.0	61.0
5.9	90.0	10.0	7.1	33.0	67.0
6.0	87.7	12.3	7.2	28.0	72.0
6.1	85.0	15.0	7.3	23.0	77.0
6.2	81.5	18.5	7.4	19.0	81.0
6.3	77.5	22.5	7.5	16.0	84.0
6.4	73.5	26.5	7.6	13.0	87.0
6.5	68.5	31.5	7.7	10.5	90.5
6.6	62.5	37.5	7.8	8.5	91.5
6.7	56.5	43.5	7.9	7.0	93.0
6.8	51.0	49.0	8.0	5.3	94.7

[a]Adapted by permission from CRC (1975).

coating. Dip coverslips in the working solution, then incubate 15 min to several hours in a humidified 37°C, 5% CO_2 incubator. Allow surface to dry.

To coat culture dishes or 8-well chamber slides: Prepare a stock solution by dissolving 100 mg polylysine in 100 ml water (both poly-L-lysine and poly-D-lysine are used to coat tissue culture surfaces; check specific protocol for choice of isomer) and filter sterilize through a 0.22-μm filter. Store in 5-ml aliquots at −20°C. When ready to use, dilute 1 part stock solution with 9 parts water to prepare 100 μg/ml working solution. Fill tissue culture dishes or slide wells with the working solution and incubate 1 hr in a humidified 37°C, 5% CO_2 incubator, then remove solution by vacuum aspiration and allow surface to dry.

Store coated tissue culture ware up to 3 months at 4°C. Use diluted solutions only once, but unused diluted aliquots can be stored up to 3 months at 4°C.

Potassium borate, 400 mM, pH 9.9 (UNIT 2.5)

Add 12.22 g $K_2B_4O_7·H_2O$ (Sigma) to 90 ml water and adjust pH to 9.9. Bring final volume to 100 ml with water. Store up to 3 months at 4°C.

Potassium phosphate buffer, 0.1 M

Solution A: 27.2 g KH_2PO_4 per liter (0.2 M final)
Solution B: 34.8 g K_2HPO_4 per liter (0.2 M final)

Referring to Table A.1.3 for desired pH, mix the indicated volumes of solutions A and B, then dilute with water to 200 ml. Filter sterilize if necessary. Store up to 3 months at room temperature.

This buffer may be made as a 5- or 10-fold concentrate simply by scaling up the amount of potassium phosphate in the same final volume. Phosphate buffers show concentration-dependent changes in pH, so check the pH of the concentrate by diluting an aliquot to the final concentration.

Protease assay buffer (UNIT 2.1)

2.5 ml 0.2 M HEPES, pH 7.0 (10 mM final)
0.5 ml 1 M DTT (1 mM final)
0.5 ml 10% (w/v) Nonidet P-40 (NP-40; 0.1% w/v final)

Adjust pH to 7.0, bring volume to 50 ml, and filter using a 0.22-μm sterile filter. Store solution at room temperature.

0.2 M HEPES: Dissolve 11.195 g of HEPES (mol. wt. 238.3) in 200 ml Milli-Q-purified water. Adjust pH to 7.0 with sodium hydroxide, bring volume to 250 ml, and sterile filter. Store at room temperature.

1 M DTT: Store in 1-ml aliquots at −20°C.

S(−)-Propranolol, 1.5 mM (UNIT 2.1)

Stock solution (15 mM): Dissolve 4.44 mg of S(−)-propranolol hydrochloride (mol. wt. 295.8; Sigma-Aldrich) in 1 ml of Milli-Q-purified water. Divide into aliquots and store frozen up to 6 months at −20°C. Avoid refreezing of aliquots.

Working solution (1.5 mM)
10 μl of 15 mM S(−)-propranolol stock
✓ 90 μl miniature SPA assay buffer
Prepare fresh daily and store at 4°C during the day

Protease assay buffer (UNIT 2.1)

2.5 ml 0.2 M HEPES, pH 7.0 (10 mM final)
0.5 ml 1 M DTT (1 mM final)
0.5 ml 10% (w/v) Nonidet P-40 (NP-40; 0.1% w/v final)

Adjust pH to 7.0, bring volume to 50 ml, and filter using a 0.22-μm sterile filter. Store solution at room temperature.

0.2 M HEPES: Dissolve 11.195 g of HEPES (mol. wt. 238.3) in 200 ml Milli-Q-purified water.

Adjust pH to 7.0 with sodium hydroxide, bring volume to 250 ml, and sterile filter. Store at room temperature.

1 M DTT: Store in 1-ml aliquots at −20°C.

Proteinase K solution *(UNIT 1.10)*

Prepare 100-μl aliquots of 5 mM $CaCl_2$ containing 3 mg proteinase K (Sigma) per aliquot. Store ≤3 weeks at 4°C.

Protein kinase A (PKA) *(UNIT 2.14)*

Dissolve the contents of one vial of PKA (3′,5′-cyclic AMP-dependent protein kinase from bovine heart; Sigma) to a concentration of ∼6 mg/ml in H_2O. Store in 50-μl aliquots at −70°C. Dilute 1:100 in TE cAMP buffer (see recipe) before using.

Rabbit anti-G$\alpha_{q/11}$ antibody, diluted *(UNIT 2.1)*

Perform 1:200 dilution of Rabbit G$_{\alpha q/11}$ antibody (Santa Cruz Biotechnology) as follows.

Well dilution = (amount added per well/volume in well) = 20/290 = 1/14.5

Dilution needed for assay solution = (final concentration/well dilution) = 200/14.5 = 13.8

Volume of concentrated stock needed = 3000/13.8 = 217 μl

Add 217 μl of concentrated antibody stock to (3000 − 217) = 2783 μl of buffer

If testing M_1, M_3, or M_5 receptors, use the $G_{\alpha q/11}$ antibody. If testing M_2 or M_4 receptors, use $G_{\alpha i\text{-}3}$ antibody (Santa Cruz Biotechnology) prepared in the same manner as the $G_{\alpha q/11}$ antibody.

Rat tail collagen type II, 4 mg/ml *(UNIT 2.1)*

Dissolve 100 mg lyophilized solid in 25 ml 0.5% acetic acid (4 mg/ml) as this is insoluble at neutral pH. Store at 4°C. Just before use, prepare a 1:10 dilution of the 4 mg/ml stock by adding 500 μl to 4.5 ml of 0.5% acetic acid. Use to coat plates as outlined in the procedure. Do not store the diluted material. Prepare smaller dilution volumes if necessary.

Ringers solution, oxygenated *(UNIT 2.12)*

119 mM NaCl (6.95 g/liter)
2.5 mM KCl (0.19 g/liter)
1.3 mM $MgSO_4$ (0.16 g/liter)
2.5 mM $CaCl_2 \cdot 2H_2O$ (0.37 g/liter)
26 mM $NaHCO_3$ (2.18 g/liter)
1.0 mM NaH_2PO_4 (1.14 g/liter)
11 mM glucose (1.98 g/liter)
Bubble continuously with 95% O_2/5% CO_2 to keep pH at 7.4

10× stock solutions can be made of the salts and stored for 2 weeks at 4°C. Glucose should be added the day of an experiment.

RNase A stock solution, DNase-free, 2 mg/ml

Dissolve RNase A (e.g., Sigma) in DEPC-treated H_2O (see recipe above) to 2 mg/ml. Boil 10 min in a 100°C water bath. Store up to 1 year at 4°C.

The activity of the enzyme varies from lot to lot; therefore, prepare several 10-ml aliquots of each dilution to facilitate standardization.

Roller-tube culture medium *(UNIT 1.7)*

50 ml Eagle's basal medium (based on Earle's or Hanks' balanced salt solution; without glutamine, which tends to precipitate; store up to 6 months at 4°C), containing 10 μg/ml phenol red

25 ml Earle's or Hanks' balanced salt solution (BSS; store up to 6 months at room temperature), containing 10 μg/ml phenol red

25 ml horse serum, mycoplasma screened (heat-inactivate complement 30 min at 56°C; store up to 18 months at −20°C; thaw slowly at room temperature)

27.7 mM glucose
1 mM L-glutamine
Store up to ∼1 month at 4°C

BSS with Earle's salts contains more bicarbonate than Hanks' and will offer more buffering capacity against the greater amounts of CO_2 and lactic acid produced by larger cultures or cocultures.

For most studies, a semisynthetic, serum-based culture medium provides the best results. A full description of serum-free media is given in Annis et al. (1990) and Wray et al. (1991).

Saturation assay buffer *(UNIT 2.1)*

✓ 100 ml 10× TME assay buffer
1 g L-ascorbic acid (0.1% w/v final)
1.0 ml 10 mM pargyline (see below)
Dilute to 1 liter with Milli-Q-purified water
Check to ensure that the pH is 7.8; adjust with 1 N NaOH if pH drops to 7.2 to 7.5

IMPORTANT NOTE: *Saturation assay buffer must be prepared fresh each day to avoid the oxidation of the ascorbic acid. It is stable throughout the day at room temperature.*

10 mM pargyline stock: Dissolve 195.7 mg pargyline (N-methyl-N-propargylbenzylamine, MW 195.7; Sigma) in 100 ml Milli-Q-purified water. Store in 1-ml aliquots at −20°C.

SDS sample buffer (for discontinuous systems)

Prepare the 2× or 4× mixture described in Table A.1.4. Mix well, divide into ten 1-ml aliquots, and store at −80°C.

To avoid reducing proteins to subunits (if desired), omit 2-ME (reducing agent) and add 10 mM iodoacetamide to prevent disulfide interchange.

Serum-free medium *(UNIT 1.7)*

Prepare roller tube culture medium (see recipe) as described, except substitute 25 ml of additional Eagle's basal medium for the horse serum.

This medium may be applied for up to 3 days to cultures that have already been maintained in vitro for >14 days.

Table A.1.4 Preparation of SDS Sample Buffer

Ingredient	2×	×	Final conc. in 1× buffer
0.5 M Tris·Cl, pH 6.8[a]	2.5 ml	5.0 ml	62.5 mM
SDS	0.4 g	0.8 g	2% (w/v)
Glycerol	2.0 ml	4.0 ml	10% (v/v)
Bromphenol blue	20 mg	40 mg	0.1% (w/v)
2-Mercaptoethanol	400 µl	800 µl	~300 mM
H$_2$O	to 10 ml	to 10 ml	

[a]See recipe below.

Sodium acetate, 3 M

Dissolve 408 g sodium acetate trihydrate (NaC$_2$H$_3$O$_2$·3H$_2$O) in 800 ml H$_2$O
Adjust pH to 4.8, 5.0, or 5.2 (as desired) with 3 M acetic acid (see Table A.1.1)
Add H$_2$O to 1 liter
Filter sterilize
Store up to 6 months at room temperature

Sodium acetate, 0.15 M, pH 7.0 (UNIT 2.5)

Dissolve 20.41 g sodium acetate in 900 ml water. Adjust pH to 7.0 and bring total volume to 1 liter with water. Store up to 3 months at 4°C.

Sodium citrate buffer, pH 5.5 (UNIT 4.8)

Dissolve 210.14 mg citric acid monohydrate (C$_6$H$_8$O$_7$·H$_2$O; mol. wt. 210.14; Sigma) in 100 ml distilled water to make a 10 mM citric acid solution. Dissolve 294.10 mg trisodium citrate dihydrate (C$_8$H$_5$O$_7$Na$_3$·2H$_2$O; mol. wt. 294.12; Sigma) in 100 ml distilled water to make a 10 mM sodium citrate solution. Add citric acid solution to the sodium citrate solution until the pH is 5.5 as monitored by a pH meter. Prepare fresh before experiment.

Sodium phosphate buffer, 0.1 M

Solution A: 27.6 g NaH$_2$PO$_4$·H$_2$O per liter (0.2 M final)

Solution B: 53.65 g Na$_2$HPO$_4$·7H$_2$O per liter (0.2 M final)

Referring to Table A.1.3 for desired pH, mix the indicated volumes of solutions A and B, then dilute with water to 200 ml. Filter sterilize if necessary. Store up to 3 months at room temperature.

This buffer may be made as a 5- or 10-fold concentrate simply by scaling up the amount of sodium phosphate in the same final volume. Phosphate buffers show concentration-dependent changes in pH, so check the pH of the concentrate by diluting an aliquot to the final concentration.

Solvent for dopamine standards (UNIT 2.11)

5 ml 2 N perchloric acid
✓ 400 µl 100 mM EDTA
100 µl 100 mM NaHSO$_3$
H$_2$O to 100 ml
Store up to 3 months at 4°C

SPQ/MQAE loading buffer (UNIT 2.12)

Immediately before use, add 1 mM CaCl$_2$·2H$_2$O (147 mg/liter) to SPQ/MQAE wash buffer (see recipe).

SPQ/MQAE wash buffer (UNIT 2.12)

137 mM NaCl (8 g/liter)
5 mM KCl (373 mg/liter)
4.2 mM NaHCO$_3$ (353 mg/liter)
0.44 mM KH$_2$PO$_4$ (60 mg/liter)
1 mM MgCl$_2$·6H$_2$O (203 mg/liter)
20 mM TES (4.6 g/liter) *or* 20 mM *N*-2-hydroxyethylpiperazine-*N'*-2-ethanesulfonic acid (HEPES; 4.8 g/liter)
Adjust pH to 7.4 with NaOH
Store up to several days at 4°C
The buffer is also stable at least 1 month at −18°C.
Immediately before use add:
10 mM glucose (1.8 g/liter)
0.1% (w/v) bovine serum albumin

SSC (sodium chloride/sodium citrate), 20×

Dissolve in 900 ml H$_2$O:
175 g NaCl (3 M final)
88 g trisodium citrate dihydrate (0.3 M final)
Adjust pH to 7.0 with 1 M HCl
Adjust volume to 1 liter
Filter sterilize
Store up to 6 months at room temperature

SSPE (sodium chloride/sodium phosphate/EDTA), 20×

175.2 g NaCl
27.6 g NaH$_2$PO$_4$·H$_2$O
7.4 g disodium EDTA
800 ml H$_2$O
Adjust pH to 7.4 with 6 M NaOH, then bring volume to 1 liter with H$_2$O
Filter sterilize
Store up to 6 months at room temperature

The final sodium concentration of 20× SSPE is 3.2 M.

Steroid hormone (UNIT 3.5)

Prepare the following three solutions as needed:
1 mg β-estradiol 3-benzoate in 10 ml sesame oil (100 μg/ml)
50 mg progesterone in 10 ml sesame oil (5 mg/ml)
100 mg testosterone propionate in 10 ml sesame oil (10 mg/ml)
Store indefinitely at room temperature

For animals with body weights very different from those of the average adult rat, hormone concentrations may need to be varied so as to yield an injection volume of ∼0.1 ml.

CAUTION: *Avoid direct contact with steroid hormones and treat them as toxic substances. Powdered steroids should be handled under a hood and as far away as possible from areas in which radioimmunoassay of steroid hormones is performed, because laboratory contamination by the steroids will interfere with assays.*

Streptavidin SPA beads, 20 mg/ml (UNIT 2.1)

Dissolve a 500-mg bottle of streptavidin SPA beads (Amersham Biosciences) in 25 ml stop solution (i.e., 10% orthophosphoric acid) to make a 20 mg/ml slurry stock. Store up to 3 weeks at 4°C without loss of activity.

STZ (UNIT 4.8)

Calculate the amount of streptozotocin (STZ; Sigma) needed for all injections in a single experiment, using a final dosage of 45 mg/kg/rat (based on weight measured 1 day prior to the experiment). Weigh out this amount of STZ into a microcentrifuge tube (do not add buffer). Cover the tube with aluminum foil (STZ is light sensitive) and store at −20°C in a desiccator until ready for use in the sodium citrate buffer (see recipe). The stock bottle of STZ powder should also be stored at −20°C in a desiccator.

CAUTION: *Streptozotocin is cytotoxic to both humans and animals. The weighing of STZ powder and preparation of the STZ solution should be performed in a class II biosafety cabinet/hood that has been certified for preparation of cytotoxic materials. Prep and injection areas should be covered with disposable absorbent covers. Individuals handling these materials should follow standard precautions and wear a laboratory coat, mask, and gloves.*

Sucrose, 320 mM (UNIT 2.10)

Take 109.5 g of >99% pure sucrose (MW = 342.3) and add to Milli-Q-purified water to yield a final volume of 1 liter. Store up to 2 months at 0° to 4°C. If contamination becomes a problem, make fresh batches more frequently.

Sucrose/PBS/EDTA solution (UNIT 2.5)

0.25 M sucrose
10 mM sodium phosphate
✓ 0.1 mM EDTA
Adjust pH to 8.1 using 0.1 N HCl
Filter and store up to 1 month and 4°C

Synaptoneurosome preparation buffer (UNIT 2.12)

118.5 mM NaCl (6.93 g/liter)
4.7 mM KCl (0.35 g/liter)
1.18 mM $MgSO_4$ (0.14 g/liter)
1.0 mM $CaCl_2$ (0.15 g/liter)
20 mM HEPES (4.77 g/liter)
9 mM Tris (1.09 g/liter)
10 mM glucose (1.8 g/liter)

10× stock solutions of the salts can be made ahead of time and stored at 4°C for 2 weeks. Concentrated stock solutions of HEPES and Tris can also be made. Glucose should be added the day of an experiment.

Tail buffer (UNIT 4.14)

✓ 50 mM Tris·Cl, pH 8.0
✓ 100 mM EDTA, pH 8.0
100 mM NaCl
1% (w/v) SDS
Filter sterilize through a 0.2-μm filter and store indefinitely at room temperature

TBA/acetate/DTPA solution (UNIT 2.5)

Dissolve 1 mM diethylenetriaminepentaacetic acid (DPTA) in 2 M sodium acetate buffer, pH 3.5. Then add 0.2% (w/v) thiobarbituric acid (TBA). Store up to 3 months at 4°C.

TBE (Tris/borate/EDTA) electrophoresis buffer, 10×

108 g Tris base (890 mM final)
55 g boric acid (890 mM final)
960 ml H_2O
40 ml 0.5 M EDTA, pH 8.0 (20 mM final; see recipe above)
Store up to 6 months at room temperature

TCA/TBA/HCl solution (UNIT 2.5)

15% (w/v) trichloroacetic acid
0.375% (w/v) thiobarbituric acid
0.25 N HCl
Store up to 3 months at 4°C

This solution may be mildly heated to assist in the dissolution of the thiobarbituric acid.

TE (Tris/EDTA) buffer

✓ 10 mM Tris·Cl, pH 7.4, 7.5, or 8.0 (or other pH)
✓ 1 mM EDTA, pH 8.0
Store up to 6 months at room temperature

TE cAMP buffer (UNIT 2.14)

For 1 liter:
✓ 50 mM Tris·Cl, pH 7.5
✓ 4 mM EDTA
Adjust pH to 7.5 with 6 N HCl

Thrombin solution, ∼150 U/ml (UNIT 1.7)

Dissolve 50 U/mg thrombin (Merck) in glucose-enriched Gey's BSS at a concentration of ∼150 U/ml and filter sterilize through bottle-top filters. Freeze in 0.5- or 1-ml aliquots at −20°C for up to several years.

TME assay buffer (UNIT 2.1)

10× buffer:
60.65 g Tris base (500 mM)

12.37 g MgSO$_4$ (anhydrous; 100 mM)
1.86 g disodium EDTA (5 mM)

Adjust pH to 7.8 with 1 M HCl, bring volume to 1 liter, and filter using a 0.22-μm sterile filter. Store 10× solution at room temperature.

For 1× buffer: Prepare a 1:10 mixture by diluting 50 ml of 10 × TME assay buffer with 450 ml Milli-Q-purified water. Check pH and adjust to pH 7.8 if necessary.

Tris-buffered saline (TBS)
✓ 100 mM Tris·Cl, pH 7.5
0.9% (w/v) NaCl
Store up to several months at 4°C

Tris citrate buffer, pH 7.4, 50 mM (UNIT 2.10)
30.285 g Tris base
5 liters H$_2$O
Adjust pH to 7.4 with citric acid
Store up to 2 weeks at 0° to 4°C

Tris·Cl, 1 M

Dissolve 121 g Tris base in 800 ml H$_2$O
Adjust to desired pH with concentrated HCl
Adjust volume to 1 liter with H$_2$O
Filter sterilize if necessary
Store up to 6 months at 4°C or room temperature

Approximately 70 ml HCl is needed to achieve a pH 7.4 solution, and ~42 ml for a solution that is pH 8.0.

IMPORTANT NOTE: *The pH of Tris buffers changes significantly with temperature, decreasing approximately 0.028 pH units per 1°C. Tris-buffered solutions should be adjusted to the desired pH at the temperature at which they will be used. Because the pK$_a$ of Tris is 8.08, Tris should not be used as a buffer below pH ~7.2 or above pH ~9.0.*

Tris·Cl (pH 8.0), 50 mM (UNIT 4.14)
Dissolve 60.5 g Tris base in 800 ml H$_2$O
Adjust to desired pH with concentrated HCl
Adjust volume to 1 liter with H$_2$O
Filter sterilize if necessary
Store up to 6 months at 4°C or room temperature

Reference: CRC Handbook, 1975

Approximately 70 ml HCl is needed to achieve a pH 7.4 solution, and ~42 ml for a solution that is pH 8.0.

IMPORTANT NOTE: *The pH of Tris buffers changes significantly with temperature, decreasing approximately 0.028 pH units per 1°C. Tris-buffered solutions should be adjusted to the desired pH at the temperature at which they will be used. Because the pK$_a$ of Tris is 8.08, Tris should not be used as a buffer below pH ~7.2 or above pH ~9.0.*

Trypsin solution (UNIT 1.10)

Prepare 500-μl aliquots of PIPES saline (see recipe), without glucose, containing 15 mg trypsin (Sigma, type XI) per aliquot. Store ≤3 weeks at 4°C.

Uptake buffer (UNIT 2.11)
25 mM HEPES
120 mM NaCl
5 mM KCl
2.5 mM CaCl$_2$
1.2 mM MgSO$_4$
Store up to 2 months at 4°C
On day of experiment, add:
1 μM pargyline (Research Biochemicals)
2 mg/ml glucose
0.2 mg/ml ascorbic acid

Wash buffer (UNIT 2.2)
✓ 1× PBS (or Life Technologies)
0.05% (w/v) Tween 20
Adjust pH to 7.4 using 5 M NaOH
Filter sterilize through a 0.22-μm filter unit
Store up to 6 months at 4°C

Wheat germ agglutinin (WGA) SPA beads, 20 mg/ml (UNIT 2.1)
500-mg WGA SPA beads (Amersham Biosciences; one bottle)
✓ 25 ml of 1× TME assay buffer
Store the 20 mg/ml slurry stock for up to 3 weeks at 4°C without loss of activity.

Use 1× TME buffer instead of saturation assay buffer so that the slurry can be used for longer than just the day of preparation.

APPENDIX 2
Animal Techniques

APPENDIX 2A
High Precision Stereotaxic Surgery in Mice

Stereotaxic surgery is a well-established technique for localizing specific structures within the brains of living animals. Among other applications, this technique can include the direct introduction of fluids or the chronic implantation of cannulae within these structures. These surgeries allow the assessment of the effects of blood-brain barrier-impermeable substances on the central nervous system or the structure-specific effects of a substance within the brain. The development of ever-increasing numbers of transgenic and knockout strains of mice has made them highly desirable as experimental subjects. While stereotaxic surgeries have been carried out in mice for some time (e.g., see UNIT 2.7), the relatively recent development of higher precision stereotaxic frames capable of 10- or even 1-μm precision has greatly improved the accuracy and usefulness of these techniques.

These protocols use bony features on the surface of the skull as reference points for determining the location of brain structures within. These features are bregma and lambda, the intersections of the sagittal suture with the coronal and lambdoidal sutures, respectively. A mouse stereotaxic atlas (e.g., Slotnick and Leonard, 1975) allows the determination of the anterior-posterior (AP), medial-lateral (ML), and dorsal-ventral (DV) coordinates of brain structures from measurements of these skull features.

BASIC PROTOCOL

SITE-SPECIFIC CENTRAL MICROINJECTION IN THE ANESTHETIZED MOUSE

For applications in which the mouse does not need to be awake and behaving (e.g., biochemical or molecular biological assays) or where the injected substance has a relatively long duration of action (e.g., antisense oligonucleotides or viral vectors), this protocol describes the delivery of substances to specific regions within the brains of mice.

Materials (see APPENDIX 1 for items with ✓)

✓ Ketamine/xylazine cocktail
Hair shaver or depilatory cream (optional)
Ophthalmic ointment or mineral oil
Betadine
Bone wax

Mouse stereotaxic frame (Cartesian Research) equipped with:
    Sighting scope
    Drill with #74 bit
    Monoinjector with 24-G Hamilton syringe (model #8800 for injections of up to 5 μl)
Heating pad

Forceps
Scalpel
Surgical scissors
Agricola-style retractors (Fine Science Tools)
Cotton swabs
Sutures

1. Anesthetize the mouse with an intraperitoneal injection of an appropriate dose of ketamine/xylazine cocktail (~20 µl/g body weight; 750 µl maximum). After the animal has reached full anesthesia (~5 min), remove the fur from the top of the skull with a shaver or depilatory cream or by plucking with fingertips or a pair of forceps.

   Mice of different strains, ages, and adiposities will have differing sensitivities to the anesthetic.

2. Swing the nose clamp and incisor bar down and out of the way, and then lock one of the ear bars at ~3.5 mm from center. Set a heating pad on low (~35°C) and place it on the stereotaxic frame to maintain body temperature, then raise the animal to the level of the ear bars. While supporting the animal's head from beneath, guide the tip of the locked ear bar into the external auditory meatus and hold firmly. Slide the other ear bar into the opposite ear and apply steadily increasing pressure until small popping sounds are heard; this popping indicates that the ear bars are properly inserted to the tympanic membranes.

3. Swing the incisor bar up and slide it just into the mouth while holding the jaw open with a pair of blunt forceps. Apply downward pressure to the bridge of the nose while sliding the incisor bar back into the mouth. Stop when the incisors drop into the hole in the bar. Swing the nose clamp into position on the bridge of the nose and tighten firmly. Optionally apply ophthalmic ointment or mineral oil to the eyes to keep them from drying out and protect them from accidental spills.

4. Disinfect the scalp with Betadine and make an incision along the midline with a scalpel. Using surgical scissors, extend this incision forward to the back of the eyes and backward to between the ears. Use caution to avoid damaging the muscles that insert on the external occipital crest on the back of the skull. Set Agricola-style retractors in the incision to expose the skull, and scrape away the transparent pericranial tissues with a cotton swab.

5. Fit the monoinjector to the stereotaxic frame and take DV measurements at bregma and lambda by lowering the tip of the injection needle until it is just touching these structures. Raise or lower the incisor bar/nose clamp to bring the bregma and lambda DV coordinates within 0.1 mm of each other. Zero the DV scale at bregma. Fit the sighting scope to the stereotaxic frame, align the crosshairs with bregma and zero the AP and ML scales.

6. Fit the drill to the stereotaxic frame, set it at the target AP and ML coordinates and drill a hole through the skull at a rate of ~25 to 50 µm/sec.

   More accurate injections can be achieved by using a correction factor for determining the target coordinates. To get this correction factor, divide the bregma-to-lambda distance of each animal by the bregma-to-lambda distance from a stereotaxic atlas. Multiply the coordinates given by the atlas by this number to determine the corrected coordinates.

7. Fit the monoinjector to the stereotaxic frame. Lower the needle into the hole to the target DV coordinate and inject brain tissue at <0.5 µl/min (or intracerbroventricularly at 1 µl/min). Leave the injection needle in place for at least 30 sec after completion of the injection and before removal of the needle.

8. Remove the monoinjector and seal the hole with bone wax. Suture the incision and remove the animal from the stereotaxic frame. Keep the animal warm during recovery by placing its cage on a heating pad set on low (~35°C). House animals individually until the incision has healed to prevent them from chewing each other's stitches.

ALTERNATE PROTOCOL

CANNULA IMPLANTATION AND CENTRAL MICROINJECTION IN THE BEHAVING MOUSE

For applications in which behavior is to be assessed during or soon after injection, this protocol describes the implantation of cannulae and subsequent central injection in awake, behaving mice.

Additional Materials *(also see Basic Protocol)*

Filter-sterilized food coloring (dark colors such as blue, black, or green work best)
Cyanoacrylate glue and accelerator (hobby supplies store)
Dental acrylic
Injectate

Cannula and wire plug (see Support Protocol)
Cannula holder
Injection needle (see Support Protocol)
Polyethylene tubing (0.58-mm i.d.)
Micro-syringe
Syringe pump

1. Anesthetize the mouse with an intraperitoneal injection of an appropriate dose of ketamine/xylazine cocktail (see Basic Protocol, step 1).

2. Prior to the surgery, slip a wire plug into a cannula and lightly clamp them into the cannula holder.

3. Prop the animal up on a heating pad set on low (~35°C). Lock ear bars firmly into ear sockets of the stereotaxic frame (see Basic Protocol, step 2). Slide incisor bar into place and tighten nose clamp firmly on the bridge of the nose (see Basic Protocol, step 3).

4. Disinfect the scalp with Betadine and make an incision along the midline with a scalpel (see Basic Protocol, step 4). Set retractors in the incision to expose the skull. Score the surface of the skull with a scalpel, use caution to avoid the skull sutures. Dab the skull with filter-sterilized food coloring to dye transparent pericranial tissues and scrape away the pericranial tissues with a cotton swab. Remove the dye with an ethanol-soaked swab.

5. Fit the cannula holder to the stereotaxic frame and, using the tip of the wire plug, take DV measurements at bregma and lambda. Level the head so these DV measurements are within 0.1 mm of each other. Zero the AP, ML, and DV scales at bregma (see Basic Protocol, step 5).

6. Put 2 or 3 drops of cyanoacrylate glue on the skull and spread to cover the exposed bone. Add 2 drops of glue accelerator and allow it to dry. Score the dried surface of the glue with a scalpel.

    While it generally is not necessary, the adhesion of the dental acrylic (step 10) can be improved through the use of anchoring screws. The mouse skull is relatively thin and fragile, but the cyanoacrylate glue reinforces it sufficiently that a pair of small jeweler's screws can be anchored approximately 5 to 6 mm apart. Small guide holes should be drilled for the screws to avoid splitting the skull, and care should be taken to only advance the screws to about 1 mm depth. This should firmly anchor the screws without depressing the meninges and the surface of the brain. The screws should be placed so as to avoid the weaker areas of the skull near its sutures and not interfere with the positioning of the cannulae.

7. Fit the drill to the stereotaxic frame, using a correction factor for precision set the frame at the target AP and ML coordinates, and drill a hole through the glue and skull. Fit

the cannula holder to the stereotaxic frame. Lower the wire plug/cannula assembly into the hole to the target DV coordinate. Clean the surface of the glue thoroughly with an ethanol-soaked swab and allow it to dry completely.

8. Mix the dental acrylic and apply in a small mound around the cannula, taking care to avoid cementing the cannula holder to the skull or creating sharp edges that may interfere with post-operative healing. Allow the cement to dry fully before releasing the cannula from the holder and removing the cannula holder from the stereotaxic frame. Fit a small length of plastic tubing over the plug cap and exposed portion of the cannula to act as a retaining sleeve.

9. Suture the incision behind the mound of cement, and remove the animal from the stereotaxic frame. Keep the animal warm while it recovers from the anesthetic by placing its cage on a heating pad set on low (~35°C). House the animals individually after surgery to prevent them from chewing each other's stitches or removing each other's wire plugs.

10. After a week, prepare for injection by attaching an injection needle (see Support Protocol) to a length of 0.58-mm i.d. polyethylene tubing. Fill the tubing with water, making sure to purge any air bubbles. Attach the free end of the tubing to a micro-syringe mounted on a syringe pump. Draw 1 µl of air into the tubing, then an appropriate amount of injectate.

11. Remove the wire plug with forceps while an assistant restrains the animal. Fit the injection needle into the cannula and inject while allowing the animal to roam freely. After the injection, restrain the animal, remove the injection needle and replace the wire plug.

SUPPORT PROTOCOL

CANNULA, WIRE PLUG, AND INJECTION NEEDLE MANUFACTURE

Protocols for inexpensive and simple construction of 8-mm cannulae and 9-mm wire plugs, and injection needles, which are appropriate for target structures in the dorsal part of the brain, are described. However, the lengths and diameters of the tubing from which they are constructed can be modified for specific applications. It is important that the wire plugs and injection needles fit within the interior diameter of the cannulae, that they extend beyond the tips of the cannulae to maintain patency, and that the cannulae are long enough to reach the target brain structure and long enough to protrude above the skull to be securely affixed with dental acrylic.

NOTE: To further reduce the damage to the brain, smaller gauge cannulae, wire plugs and injectors can be used. For example, 28-G thin-walled tubing and 33-G tubing can be substituted for the 24-G thin-walled tubing and 30-G tubing recommended here. The smaller tubing is more prone to bending and occlusion, however, and will increase the difficulty of parts manufacture and animal infusion.

Materials

Ethanol
Acid flux (welding supply store)
Dremel tool with abrasive wheel and cutting disk (Small Parts)
Small bench vise
24-G thin-walled stainless steel hypodermic tubing (Small Parts)
30-G stainless steel hypodermic tubing and/or solid core wire (Small Parts)
Hemostats
Soldering iron (40-W minimum)
Silver bearing solder (welding supply store)
Outside micrometer (Small Parts)

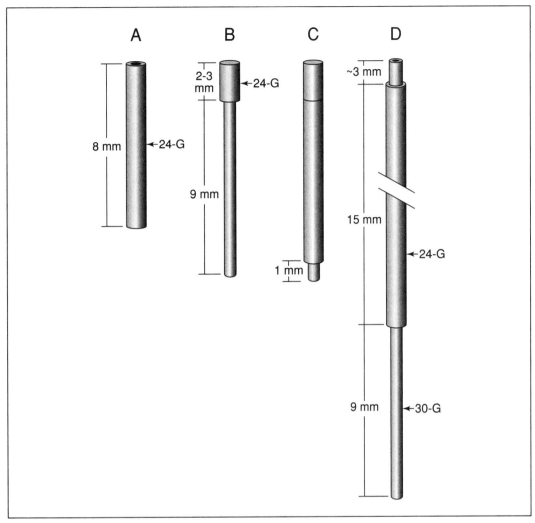

Figure A.2A.1 Diagram of the cannula, wire plug, and injector needle. (**A**) Cannula: stainless steel hypodermic tubing is cut to length and abraded with an abrasive wheel. (**B**) Wire plug: a short stainless steel cap is soldered onto a length of smaller diameter tubing or solid-core wire such that when it is seated within the cannula, (**C**) its tip protrudes. (**D**) Injection needle: a length of stainless steel tubing is soldered onto smaller diameter tubing. Like the wire plug, the tip of the injection needle extends beyond the end of the cannula. However, the smaller diameter tubing is left open so that it can be attached to a length of polyethylene tubing and fluids can be administered through it by means of a syringe pump.

Construct cannula (see Figure A.2A.1A)

1a. Clamp the Dremel tool in the bench vise. Using the cutting disk, cut 24-G stainless steel hypodermic tubing to 8-mm lengths as measured with the outside micrometer.

   Slide the 8-mm lengths onto a longer piece of 30-G stainless steel hypodermic tubing or a 30-G syringe needle. Using the abrasive wheel on the Dremel tool, rough up the outside of the 8-mm lengths as they spin freely.

2a. Place the finished cannulae in a small container of ethanol and vortex well to remove loose metal particles. Remove from ethanol and allow to air dry before use.

Construct wire plug (see Figure A.42.1B)

1b. Cut ~15-mm lengths of 30-G stainless steel hypodermic tubing or wire and an equal number of 2- to 3-mm lengths of 24-G stainless steel hypodermic tubing.

2b. Clamp one of the finger loops of the hemostat into the bench vise. Slide the 24-G piece of stainless steel hypodermic tubing onto the 30-G piece of stainless steel hypodermic tubing, flush one end, and clamp in hemostat. Add a drop of acid flux to the tubing and solder the 24-G piece of stainless steel hypodermic tubing into place with a soldering iron and silver bearing solder.

3b. Slide plug into a 9-mm piece of 24-G stainless steel hypodermic tubing and cut off the protruding excess. Place the finished wire plugs in a small container of ethanol and vortex well. Air dry before using.

Construct injection needle (see Figure A.2A.1D)

1c. Cut ~30-mm lengths of 30-G tubing and an equal number of 15-mm lengths of 24-G tubing. Slide 24-G piece onto 30-G piece, ~3 mm from one end, and clamp in hemostats. Add flux and solder 24-G piece, use caution not to block the open end of the 30-G piece, with a soldering iron and silver bearing solder.

2c. Slide plug into a 9-mm piece of 24-G tubing and cut off the protruding excess. Place the finished injection needles in a small container of ethanol and vortex well. Air dry before using.

References: Flecknell, 1996; Slotnick and Leonard, 1975

Contributors: Jaime Athos and Daniel R. Storm

APPENDIX 2B

Mouse and Rat Anesthesia and Analgesia

Management of rodent anesthesia and analgesia presents a formidable challenge due to the animals' small size and variability in sensitivity to anesthetics and analgesics. A common mistake is to focus on the surgical procedure, while giving little attention to selection of an appropriate anesthetic or the animal's physiological status during the procedure and postoperative period (see Table A.2B.1 for normal physiological parameters). Proper preparation prior to surgery, coupled with appropriate monitoring of the animal both during and after the procedure, will contribute to successful creation of the surgical model. See Table A.2B.4 at the end of this unit for troubleshooting information.

NOTE: An external heat source should be provided to all anesthetized animals to maintain body temperature. This can be accomplished by using recirculating warm water blankets, heat lamps, or pocket warmers.

BASIC PROTOCOL 1

INJECTABLE ANESTHESIA FOR MOUSE AND RAT

Because mice and rats have high metabolic rates, they require higher anesthetic dosages than larger animals with low metabolic rates to achieve an effective level of anesthesia. The duration of anesthesia is typically shorter, and mice and rats are less likely to survive respiratory arrest from overdosage. The high metabolic rate also increases the risk of hypothermia and dehydration due to exposed membranes. Usually a group of mice or rats are anesthetized

Table A.2B.1 Normal Physiological Parameters[a] of Mice and Rats

	Mouse	Rat
Temperature	36.5°–38.0°C	37.0°–38.0°C
Heart rate	300–650 beats/min	250–370 beats/min
Respiratory rate	150–220 breaths/min	70–115 breaths/min
Tidal volume[b]	0.09–0.23 ml	0.6–2.0 ml
Minute ventilation[b]	24 ml	200 ml
Body surface area	$10.5 \times (\text{wt in g})^{2/3}$	$10.5 \times (\text{wt in g})^{2/3}$
Blood volume	75 ml/kg	58 ml/kg

[a]Published parameters vary, which probably reflects differences in strain, age, and gender of the species.

[b]Tidal volume is the volume of gas drawn into the respiratory tract with each breath. Minute ventilation is the volume of gas breathed in 1 min (tidal volume × respiratory rate). Rodents maintain minute ventilation by high respiratory rate and low tidal volume. During anesthesia, both parameters fall, causing a decrease in minute ventilation.

simultaneously or serially to allow surgical procedures on a group of animals. Regardless of which injectable anesthetic is used, each animal must be weighed and dosed accordingly. Rodents do not require fasting prior to administration of anesthetic as they cannot vomit.

Materials

Laboratory mice or rats
Anesthetic of choice (see Table A.2B.2)
Petroleum-based artificial tear ointment (optional)
Lactated Ringer's solution, warmed to 37°C (optional) (e.g., A. J. Buck or J. A. Webster)

Warm water recirculating heating pad or heat lamp, 38° ± 2°C
Sterile drape (optional)
Cages for mice or rats
Laboratory scale or balance (capacity 800 g; readability 0.1 g)
1- or 3-ml syringe
20- or 22-G needle
21- or 23-G needle (optional)

1. Prepare the operating area so it is clean, uncluttered, and provides adequate room for instruments, equipment, heating pad or lamp, and, if required, a stereotaxic apparatus. Ensure that the area is well lit. Set up a heating pad or lamp and turn it on. If using a heating pad, cover it with a sterile drape.

2. Retrieve the animal from its cage, place it in a tared container, and determine its weight. Return the animal to its cage.

3. Calculate the appropriate dosage of the chosen anesthetic using Table A.2B.2. Fill either a 1- or 3-ml syringe with selected anesthetic. Manually restrain the animal and inject the selected dose intraperitoneally using a 22-G needle for mice or 20-G needle for rats.

4. Return the animal to a different cage and monitor for the depth of anesthesia (e.g., loss of righting reflex, loss of palpebral blink reflex, rate and depth of respiration). Do not put the animal in a cage containing other animals.

   Many variables influence anesthesia, such as strain, gender, body weight, age, and choice of injectable anesthetic. Wait at least 10 min before administering a second dose. Overdosage of injectable anesthetics is a frequent cause of death.

5. *Optional:* Apply a petroleum-based artificial tear ointment to the eyes to prevent corneal desiccation.

Table A.2B.2 Guidelines for Injectable Anesthetics in Rodents

Agent	Species	Dosage (mg/kg) and route of administration[a]	Duration of surgical anesthesia (min)[b]
Pentobarbital[c]	Mouse	40–70, i.p.	20–40
	Rat	30–50, i.p.	15–60
Ketamine/xylazine[d]	Mouse	60–100 ketamine + 5–7.5 xylazine, i.p.	20–25
	Rat	50–90 ketamine + 5–10 xylazine, i.p.	60–80
Ketamine/xylazine/acepromazine[e]	Mouse	25–30, i.p.	20–35
	Rat	30–40, i.p.	20–45
Ketamine/medetomidine[f]	Mouse	NA	NA
	Rat	75 ketamine + 0.5 medetomidine, i.p.	15–60
Tribromoethanol[g]	Mouse	125–160, i.p.	15–30
	Rat	300, i.p.	15–20
Atropine[h]	Mouse	0.02–0.05 mg/kg, s.c., i.p.	30–40
	Rat	0.04 mg/kg, s.c., i.p.	15–20

[a] Dosages are from a variety of sources. The dose may need to be adjusted for individual situations. Abbreviations: i.p., intraperitoneally; s.c., subcutaneously.

[b] Duration of surgical anesthesia (unconscious, nonresponsive to painful stimuli) is not the same as "sleep time" (quiet, not moving, but responsive to stimuli), which is generally much longer.

[c] Dilute 1:10 (v/v) with sterile saline. This is a controlled substance. Barbiturate classification (long, short, ultrashort) is misleading, as species differences in barbiturate pharmacokinetics are responsible for significant variation in duration of action. As barbiturates do not provide analgesia, they are often combined with sedatives or tranquilizers to provide deep anesthesia and smooth recovery.

[d] Ketamine and xylazine can be safely mixed and given as a single injection. After mixing, dilute 1:10 (v/v) with sterile saline. Ketamine is a controlled drug.

[e] Mix 1.5 ml (100 mg/ml) ketamine, 1.5 ml (20 mg/ml) xylazine, and 0.5 ml (2 mg/ml) acepromazine together; stable at room temperature (shelf life of the ingredients). Do not use in preweanling animals. Ketamine is a controlled drug.

[f] Ketamine/medetomidine provides marked differences in surgical anesthesia between male and female rats. Males have loss of reflexes for 25–60 min; females, 140–150 min. Sleep time is also marked; males 135–160 min, females 240–300 min. Heavy urination occurs, which may lead to dehydration, indicating the need for parenteral fluids. Wetting of fur may also lead to hypothermia. Ketamine is a controlled drug. NA, data not available.

[g] Make a 100% (w/v) stock solution by dissolving 5 g of 2,2,2-tribromoethanol in 5 ml 2-methyl-2-butanol (*tert*-amyl alcohol). Gentle heating (50°C) provides better solubility. The anesthetic solution should be freshly prepared from the stock solution by adding 1.25 ml of stock solution to 48.75 ml of sterile saline. Both stock and anesthetic solutions are stable 2 to 4 months at 4°C in a dark bottle. Anesthetic solutions should be made fresh weekly; filter the solution using a 0.2-μm filter. *CAUTION:* Stored solutions often deteriorate to irritant solutions that cause peritonitis and/or death following i.p. administration. Add 1 drop of Congo Red (0.1% w/v) to 5 ml of anesthetic solution. Purple color developing at pH < 5 indicates decomposition (Papaioannou and Fox, 1993).

[h] Atropine is an anticholinergic, not an anesthetic. Atropine blocks acetylcholine at muscarinic receptors. Desirable effects include reduction in bronchial secretions and protection of the heart from vagal stimulation, which may occur during surgical procedures. If used, it should be given 5 to 10 min before the anesthetic agent.

6. *Optional:* Immediately prior to making the surgical incision, inject 60 ml/kg prewarmed (37°C) lactated Ringer's solution under the skin (subcutaneously between the shoulder blades) to provide maintenance hydration. Use a 23-G needle for mice, or a 21-G needle for rats.

BASIC PROTOCOL 2

INHALANT ANESTHESIA USING ISOFLURANE FOR MOUSE AND RAT

The most consistent and reliable anesthetic protocols for rodents involve inhalant anesthesia. Delivery of volatile anesthetic agents to rodents is difficult, due to their small size and the need for special equipment. A high level of technical competence is needed, and the equipment for scavenging anesthetic gas to ensure operator safety may not be available. Isoflurane is a relatively safe anesthetic for both the surgeon and the animal and provides a high margin of safety for the animal.

CAUTION: All procedures with isoflurane require a scavenging device for expired anesthetic gases. Frequently, a chemical fume hood, down draft table, or safety cabinet is used for continuous exhaust of anesthetic gases away from personnel (see Critical Parameters).

Materials

Isoflurane
Laboratory mice or rats
Petroleum-based artificial tear ointment (optional)
Lactated Ringer's solution, warmed to 37°C (optional)

20-, 30-, or 60-cc syringe (optional)
Latex glove (optional)
Commercial face mask (optional; Kent Scientific)
Warm water re-circulating heating pad or heat lamp, 30° to 40° ± 1°C
Sterile drape (optional)
Precision vaporizer (Kent Scientific or SurgiVet)
Oxygen tank with flow meter (Kent Scientific or SurgiVet)
Induction chamber (20 cm length × 10 cm height × 10 cm width; 2 liters) with inlet and outlet ports

1. Prepare the operating area so it is clean, uncluttered, and provides adequate room for instruments, equipment, heating pad or lamp, and, if required, a stereotaxic apparatus. Ensure that the area is well lit.

2. (*Optional*) Remove the plunger from a 6-cc syringe (for mice) or a 20- or 30-cc syringe (for rats). Cut the syringe barrel off at approximately the 4- to 5-cc mark. Cut a finger from a latex glove and stretch it over the cut end of the syringe (Fig. A.2B.1A). With scissors, cut a small hole (1- to 3-mm diameter) in the tip of the finger (Fig. A.2B.1B).

3. Plug in and turn on a heating pad or lamp. If using a heating pad, cover it with a sterile drape.

4. Pour isoflurane into the precision vaporizer to the fill line. Connect the vaporizer and oxygen tank to an induction chamber. Turn the oxygen valve on to provide high flow (100% oxygen) to precharge the chamber.

5. Retrieve the animal from its cage and place it in the induction chamber. Expose the animal to 100% oxygen for 2 to 3 min, then adjust the precision vaporizer with the flow meter to provide 5% isoflurane. Monitor the animal until it becomes ataxic, recumbent, and immobile, which signifies the onset of anesthesia.

6. Remove the animal from the induction chamber and place the face mask over the muzzle of the animal to maintain anesthetic delivery. If necessary, use Stomahesive Paste (ConvaTec) to seal the face mask around the animal's muzzle for a tight fit.

7. Connect the tubing from the induction chamber to the face mask and adjust the flow meter on the vaporizer to deliver 1.0% to 1.5% isoflurane for maintenance and oxygen flow rate at least three times the minute ventilation rate (10 ml/kg or ∼0.2 ml/mouse) to lower CO_2.

8. *Optional:* Apply a petroleum-based artificial tear ointment to the eyes to prevent corneal desiccation.

9. *Optional:* Immediately prior to making the surgical incision, inject 60 ml/kg prewarmed (37°C) lactated Ringer's solution under the skin (subcutaneously between the shoulder

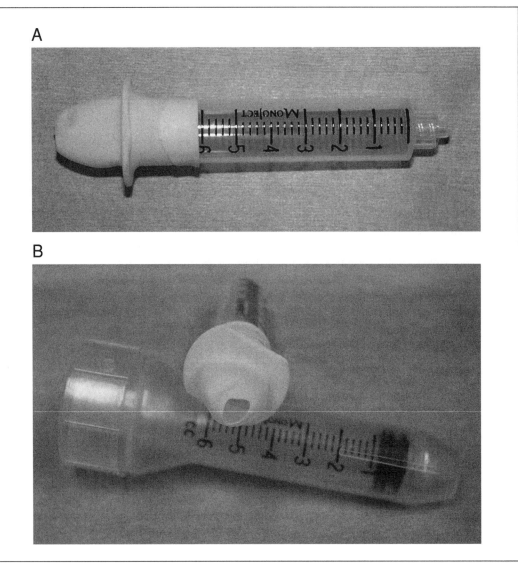

Figure A.2B.1 Assembly of a simple face mask for use in rodent anesthesia. (**A**) Syringe with attached rubber glove finger. (**B**) Hole cut in glove finger tip to fit over animal's muzzle.

blades) to provide maintenance hydration. Use a 23-G needle for mice, or a 21-G needle for rats.

10. At the end of the procedure, turn the vaporizer off and deliver 100% oxygen to the animal for min.

BASIC PROTOCOL 3

ANALGESIA FOR MICE AND RATS

During general anesthesia, the cerebral cortex, but not the rest of the nervous system, is depressed. A scalpel incision stimulates nociceptors, sending waves of impulses to the spinal cord. Such activity sensitizes the nociceptor system to further input from injured tissue, and such sensitization outlasts the duration of the original stimulus. Once the animal is in pain, there is an increase in sympathetic activity; epinephrine increases the sensitivity of injured neurons to stimulation. A reflex increase in muscle tension surrounding the area of damaged tissue also commonly occurs, which adds to the pain.

Analgesia, in the strictest sense, is the absence of pain. Practically, this means trying to reduce the intensity of the pain perceived. The goal is to diminish the pain as much as possible without undue depression of the animal. Intraoperative as well as postoperative analgesics should be provided during experimental manipulations. Many commonly used injectable anesthetics, such as pentobarbital or ketamine, have poor analgesic potency. Immobility does not mean that pain or distress is not present. Many injectable anesthetic combinations (e.g., ketamine/xylazine) include an analgesic drug (xylazine). A common mistake is to assume that the analgesic component remains after the procedure is completed and the animal has recovered. In fact, most analgesics have very short half-lives in rodents. An additional analgesic dose may be given at the end of the procedure.

Selection of an analgesic regimen depends upon the type of procedure performed, the nature of the pain (severity and duration), and the potential influence on research objectives. Many options are available (topical, regional, central, and peripherally acting), along with other means of protecting injured tissue (e.g., soft bedding, Elizabethan collars). It is difficult to make general recommendations concerning analgesia choices, because species and strain of animal, choice of anesthetic, and invasiveness of the procedure all warrant consideration. For example, if using isoflurane, buprenorphine should be given before surgery; if using an injectable anesthetic, the analgesic should be given at the end of surgery. If the procedure is simple, like a jugular catheterization, a nonsteroidal anti-inflammatory drug (NSAID) would be beneficial. When major surgery is completed, administration of buprenorphine (for 24 to 36 hr) followed by an NSAID alone for 24 hr is helpful. Good postoperative care is essential both for the animal's welfare and because it is good scientific practice.

Preoperative Medications

Selection of preoperative medications will help manage the animal by eliminating unnecessary anxiety and distress. Comments are provided here for preoperative medications, along with points to consider for analgesic selection. Table A.2B.3 lists guidelines for injectable analgesic agents in rodents.

Preoperative stress and pain can be relieved by administration of tranquilizers or analgesics, respectively. Benzodiazepines (diazepam, midazolam), classified as minor tranquilizers, are commonly used as co-induction agents in combination with injectable anesthetics. They do not provide analgesia, but have minimal effects on the cardiopulmonary, renal, and hepatic systems. However, benzodiazepines potentiate the respiratory depression induced by opioids, so attention should be given to the selected injectable anesthetic combinations. α-Adrenoceptor agonists (xylazine, detomidine, medetomidine) generally provide sedation and analgesia with some side effects, such as bradycardia, hypothermia, and respiratory depression. In mice and rats, sedation outlasts analgesia.

Because mice and rats frequently hold their breath and have stress-induced catecholamine release, they are particularly susceptible to respiratory acidosis and hypoxemia. Dopram is a CNS respiratory stimulant that can be incorporated into the preoperative regimen by placing one full-strength drop on the animal's tongue. Atropine (0.05 mg/kg intraperitoneally or subcutaneously) can be given $\sim$10 min before anesthetic induction to prevent bradycardia; it acts in concert with Dopram as a respiratory stimulant.

As a word of caution, the use of anticholinergics (atropine) in rodents is controversial. Atropine protects the heart from vagal inhibition or opioid-induced bradycardia, but also increases the viscosity of airway secretions. In rodents, increased viscosity of secretions may inadvertently

Table A.2B.3 Guidelines for Injectable Analgesics in Rodents

Agent	Species	Dosage (mg/kg) and route of administration[a]	Frequency of dosage
Buprenorphine[b]	Mouse	0.05–0.1 mg/kg, i.p., s.c.	Every 6–12 hr
	Rat	0.01–0.05 mg/kg, i.p., s.c.	Every 8 hr
Butorphanol[c]	Mouse	1.0–5.0 mg/kg, s.c.	Every 4 hr
	Rat	2.0–2.5 mg/kg, s.c.	Every 4 hr
Flunixin (Banamine)	Mouse	2.5 mg/kg, s.c.	Every 12 hr
	Rat	1.1 mg/kg, s.c.	Every 12 hr
Ketoprofen	Mouse[d]	5 mg/kg, s.c.	Every 12–24 hr
	Rat	5 mg/kg, s.c.	Every 12–24 hr
Aspirin	Mouse	400 mg/kg, s.c.	Every 24 hr
	Rat	400 mg/kg, s.c.	Every 24 hr
Acetaminophen	Mouse	300 mg/kg, p.o.	Every 4 hr
	Rat	100–300 mg/kg, p.o.	Every 4 hr

[a]Dosages are from a variety of sources. The dose may need to be adjusted for individual situations. Published dosages for oral administration are generally much larger and, in some cases, are not effective (Flecknell et al., 1999). For mice, use a 1-ml syringe with a 22-G needle; for rats, use a 6-ml syringe with a 22-G needle. Abbreviations: i.p., intraperitoneally; p.o., perorally; s.c., subcutaneously.

[b]Buprenorphine is a partial agonist that is difficult to reverse if ventilation becomes compromised. It may be prudent to begin with a low published dose and redose if necessary. This is a controlled substance.

[c]This is a controlled substance.

[d]Published data for clinical trials is not available for mice; however, experimental tests using analgesiometry tests indicate that ketoprofen is an effective analgesic in mice (Flecknell, 1998).

obstruct small airways or the trachea. The respiratory pattern should be monitored closely when an anticholinergic is used, and the animal should be ventilated vigorously when an inhalant is used. Similarly, the heart rate should be monitored if anticholinergics are not used.

Analgesic Drugs

Analgesics are best provided preemptively because sensitization of primary afferents increases the transmission of nociceptive impulses towards the CNS. If given before surgery, the analgesic diminishes the hyperexcitability of neurons that occurs with nociceptive stimulation, thus reducing the amount of postoperative analgesics needed. Some analgesics also potentiate the anesthetic effect, decreasing the amount (often by 40% to 60%) of inhalant anesthesia needed for surgical anesthesia. For example, many opioids have been shown to reduce the dosage of anesthetic required for surgical anesthesia.

Narcotic (opioid) analgesia: Opioids act as agonists of pre- and postsynaptic receptors in the CNS; the affinity of the opioid for the receptor correlates with analgesic efficacy. The most effective of these works at u and κ receptors. The primary advantage of opioids is profound analgesia, but they also induce CNS depression characterized by hypothermia, bradycardia, and respiratory depression. If clinically effective dosages are given, however, opioids have minimal effect on the cardiovascular system. Bradycardia can be managed by atropine, using 0.04 mg/kg for mice or 0.05 mg/kg for rats.

Opioid agonists/antagonists: Butorphanol has some analgesic activity at κ receptors with marked antagonist activity at the μ receptor. It has a relatively short (3 to 4 hr) duration of effect. Buprenorphine hydrochloride is a partial μ agonist with prolonged duration (8 to 12 hr) of activity. Use of buprenorphine has an "anesthetic-sparing" effect, in that less anesthesia is required for the animal. Using a narcotic agent prior to surgery will also reduce the analgesic dosage required in the postoperative stages to keep the animal comfortable. One side effect, however, is that the animal will be sedated and, therefore, less likely to eat and drink while receiving these types of analgesics. This is an important point if weight and activity are to be monitored in the postoperative period as a means to assess pain and/or discomfort. Several studies suggest that buprenorphine should be used in rats no longer than 48 hr postoperatively.

α_2 agonist analgesia: α_2 adrenergic agonists (e.g., xylazine, medetomidine) are generally regarded as sedative hypnotics, and some are also potent analgesics. α_2 adrenergic agonists have a wide range of effects, including sedation, visceral and somatic analgesia, mild to moderate muscle relaxation, peripheral vasoconstriction, bradycardia, and hypothermia. Due to the inhibitory effect on sympathetic outflow, they also depress gastrointestinal and endocrine functions. Xylazine, the most commonly used α_2 agonist in mice and rats, has wide variation in species sensitivity and response. Xylazine has a rapid (3 to 5 min) onset of action and a short duration. Duration of sedation and analgesia is dosage dependent. Mice and rats are sedated for 1 to 2 hr, but analgesia is brief (15 min). Xylazine should be supplemented with other analgesic agents if prolonged (>30 min) analgesia is needed. Medetomidine is the most potent and selective α_2 agonist. Medetomidine has fewer side effects, because it is more selective for α_2 receptors. Although α_2 agonist drugs are excellent additions to anesthetic injectable cocktails due to their prolonged sedative effects, their role as sole analgesic agents remains to be established in mice and rats. The detrimental side effects can be reversed with the α_2 adrenergic antagonist atipamezole (1.0 mg/kg subcutaneously, intraperitoneally).

Before using a reversal agent, it is important to carefully consider the advantages and disadvantages of doing so. Once an α_2 adrenergic antagonist is given, duration of anesthesia is markedly reduced and recovery of reflexes occurs within 5 min of administration. Advantages, other than faster recovery time, are reversal of bradycardia and respiratory depression, thus reducing the occurrence of complications associated with prolonged sedation. If given too soon, however, rapid loss of anesthesia with concomitant loss of muscle relaxation may occur prior to the end of the procedure. Another major disadvantage is the loss of analgesia for animals undergoing a painful procedure. In this situation, the analgesic can be augmented with another opioid or NSAID (see below) before the α_2 agonist is reversed.

Local anesthesia: Local anesthetics prevent depolarization of nervous tissue by blocking Na^+ channels in the cell membrane, preventing activation of peripheral nociceptors. Properties of the anesthetics, size of dose, and accuracy of injection placement influence duration of response. To reduce pain from local irritants, such as the ear bars from the stereotaxic apparatus, 2% lidocaine gel can be placed into the ear canal and on the ear bars. Topical anesthetics can also enhance analgesia during surgery. Infiltrating 0.1 to 0.2 ml of 2% mepivicaine or lidocaine into the skin along the edges of the incision will enhance analgesia and reduce the needed anesthetic dosage. If use of systemic analgesics is contraindicated by the experimental study, topical anesthesia provides pain relief at the incision site. Working with a veterinarian, pain medications can be formulated into topical gels and creams. Organogel, a lecithin-based matrix, is used to compound a 40% lidocaine cream (38% stronger than 2% commercial preparations) for studies confounded by narcotics and nonsteroidal drugs.

Nonsteroidal anti-inflammatory drugs (NSAIDs): NSAIDs act peripherally to suppress inflammation and decrease production of kinins and prostaglandins. Use of NSAIDs allows faster anesthetic recovery and less postrecovery sedation while providing analgesia. Banamine may be given twice daily, beginning with the initial anesthesia. Administering Banamine in warm lactated Ringer's solution is an excellent way to provide both fluids and analgesic simultaneously. Fluids provide needed hydration and protect against renal damage. NSAIDs are a good choice for craniotomies requiring a stereotaxic apparatus. Generally, postoperative pain is from the mechanical pressure (trauma) of the ear bars, not the craniotomy; NSAIDs are ideal for this type of pain. If NSAIDs (Banamine, Ketoprofen) are to be used, one should consider including an H_2-blocking agent, such as Zantac (0.5 mg/g intramuscularly) or Pepcid (2.5 mg/kg subcutaneously) to protect against gastrointestinal damage. Because animals are more sensitive to the acute renal effects of NSAIDs after surgery, administration of warmed fluids with an NSAID is advised to help protect the kidneys.

Oral analgesia: Concerns about causing additional stress through handling animals following surgery and giving them injections have led many to try oral preparations of analgesic agents. These preparations also offer convenience, especially if the agent is placed in the drinking water. Several factors may influence the decision to use the oral route. First, high dosages are generally needed, due to metabolism of the drug following oral administration; high dosages of buprenorphine have been associated with consumption of bedding and gastric distension. Second, palatability is a concern. Water consumption following addition of acetaminophen reduced overall intake of water by rats in one study or failure to completely eat an analgesic-treated Jell-O cube. Third, surgery may depress food and water intake. An additional consideration is the method of housing; if more than one animal is housed per cage, one cannot ensure that each animal consumes his or her required dosage of drug.

Postoperative nutrition

Mice and rats have high energy needs. Nutritional support is critical upon recovery to avoid hypoglycemia. Placing a drop of undiluted 50% dextrose on the tongue, or subcutaneously injecting 3 to 15 ml (depending on the size of the animal) of warmed 5% dextrose solution is beneficial. Volume deficits can be corrected by subcutaneous or intraperitoneal injection of warmed lactated Ringer's solution.

References: Bishop, 1998; Dorsch and Dorsch, 1999; Flecknell, 1996; Kohn et al., 1997

Contributor: Judith A. Davis

Table A.2B.4 Troubleshooting for Anesthesia and Analgesia

Problem	Possible cause	Solution
Bradycardia, cardiovascular failure (blue mucous membranes, animal's limbs are cool to touch)	Impending circulatory crisis, hypothermia, pain, pulmonary obstruction, blood loss; pulse oximetry can be used for monitoring; oxygen saturation is 95%–98%; a reduction of >5% indicates a potential problem; a reduction of >10% requires immediate action	Place a drop of atropine on the tongue, provide heat, make sure airway is clear (the mouth may need to be suctioned with the tongue pulled out), assist in ventilation by gentle rapid compression of thorax between thumb and forefinger. Give 0.1 mg epinephrine if no response to atropine.
Hypoventilation, respiratory arrest	Respiratory rate below 40% of the preanesthetic rate signals impending respiratory failure; check for pulmonary obstruction, depression of respiratory center by anesthetic overdosage	Make sure airway is clear, pull tongue out. Place animal in palm of hand and tip its head down and up. Movement of abdominal contents against the diaphragm will compress and expand the chest. Once the animal takes a breath, place a drop of Dopram on tongue and provide 100% oxygen.
Hyperventilation	May be due to lightening level of anesthesia	Stop. If using volatile agents, check the vaporizer. Is it full? Is there a full tank of oxygen? Is the animal completely connected to the system?
Hypoxia (blue membranes), <85% oxygen saturation	Oxygen tank is empty, animal is rebreathing expired gases, obstruction.	If using volatile agents, is oxygen being delivered? Turn vaporizer to zero and increase oxygen flow rate. If using injectable agents, consider an appropriate antagonist (reversal agent), such as yohimbine or atapamezole for α2 adrenergics or naloxone for opioids. Check if head and neck of animal are extended, open animal's mouth and pull the tongue forward. Check for copious salivary secretions; if present, suction secretions from oral cavity.
Hypothermia (can happen quickly), peripheral limbs cool to touch	Change in respiratory pattern	Wrap in plastic, place heat lamp over surgical area, use warmed fluids.
Cardiopulmonary failure	Manipulations of the viscera stimulate vagus nerve, resulting in bradycardia; accidental compression of the thorax (e.g., instruments, surgeon resting hand on chest).	Ensure an open airway. If the heart has stopped, hold the chest between thumb and finger, rapidly compress the area over the heart ~90 times/min. Once respiration and heartbeat have returned, consider a reversal agent if anesthetic overdosage is suspected.

APPENDIX 2C

Animal Health Assurance

A program to assure the health of laboratory animals and to provide a stable environment for their maintenance is necessary to reduce unwanted variables or complicating factors in experimentation.

To assure animal health throughout a research study, the animals involved should be procured in good health, monitored for continued health status, and protected from pathogenic organisms through testing of any biological materials. This unit addresses all three of these health management practices. Essential in support of these practices is a diagnostic laboratory capable of testing for rodent and rabbit pathogens.

Unexpected disease or death has obvious deleterious effects but subclinical diseases or conditions may also have profound immunological impact or result in complications that make data analysis difficult. Both situations should be prevented.

QUARANTINE AND STABILIZATION

To prevent introducing pathogenic organisms into an animal colony, newly arrived animals should undergo clinical evaluation and quarantine. This allows the animals time to recover from the stress of shipping and to acclimate to the new environment. Landi et al. (1982) found that mice have altered immune functions with elevated corticosterone levels for 48 hr following shipment by truck or air.

Quarantine procedures will vary depending on the source of the animals. Animals received from commercial sources with an established pathogen-free history may undergo a limited quarantine involving clinical examination upon receipt and observation for 24 to 48 hr. A continuing dialogue with the commercial vendor to keep the facility manager and veterinarian informed of changes in the health profile is essential.

Animals arriving from noncommercial sources or sources with unknown health profiles require more rigorous procedures. These animals should be isolated from the animal colony for a minimum of 4 weeks. Sentinel animals, free of known pathogens, are placed in the cages with the quarantined animals. At the end of 4 weeks, the sentinel animals are submitted to the diagnostic laboratory for complete necropsy, serologic testing, bacterial cultures, and endoparasitic and ectoparasitic examinations. The animals should be released from quarantine and placed in the animal colony only after receipt of an acceptable health report from the laboratory. Some animal facilities allow research studies to be carried out on animals during the quarantine period, with the understanding that the animals must be maintained in quarantine for the duration of the study, and that proper procedures be followed to prevent traffic between the quarantine area and the animal colony.

HEALTH MONITORING

Monitoring the health of research animals is a necessary prerequisite to sound science and cost-effective research. Programs traditionally include one element which relies on observational data where animals are monitored for overt signs of disease such as diarrhea, sneezing, lethargy, hair loss, and weight loss. A second element uses diagnostic tests to identify clinical conditions. The animal care and veterinary staff, working in conjunction with investigators and their technicians, normally observe, diagnose, and treat clinical conditions.

Unfortunately, these methods alone do not satisfy the needs of modern research conducted with genetically and microbiologically defined animals. Monitoring animals for evidence of subclinical infections is necessary to detect and eliminate disease to minimize the impact on research results. In the following paragraphs, surveillance methodologies used to monitor colonies of rodents for the presence of unwanted pathogens are discussed.

When developing a program for health monitoring of laboratory animals, the first decision to make is what pathogens should be prevented from being introduced into the colony or should be detected at an early stage of infection. Common pathogens of mice, rats, hamsters, and rabbits are presented in Table A.2C.1. The investigator and veterinarian should discuss the various pathogens, their prevalence, potential impact on research, and the feasibility of including select agents in the monitoring program. Facility, staff, and budgetary factors must be considered. A decision for including an agent in a monitoring program must be coupled with a predetermined plan of action to be implemented upon detection of the organism in the animal colony. Monitoring pathogens that are known to be present in the colony, or for which no action is planned, is an unnecessary waste of resources. Once critical pathogens have been identified, the appropriate test and frequency of sampling must be determined.

An animal facility that is maintained free of specifically identified pathogens is considered "specific pathogen free" (spf). In addition to the pathogen monitoring schemes described here, it is often helpful to employ facilities such as micro-isolation housing, laminar-flow rack housing, and individually ventilated isolator housing, to minimize the introduction of new pathogens.

While serologic monitoring of rodents for viral disease is common and widely available, a well-rounded program includes testing for internal and external parasites as well as bacterial pathogens. Access to pathology and histopathological services is necessary for difficult diagnoses. There are many commercial clinical laboratories with expertise in laboratory animal pathology and clinical pathology. Early consultation with the institutional laboratory animal veterinarian and a testing laboratory will ensure an appropriate program of health monitoring.

TESTING OF BIOLOGICAL MATERIAL

The final basic component of an adequate health assurance program is the testing of biological material to be introduced into live animals. Pathogens, particularly viruses, are known to be carried by a variety of biological materials that are commonly injected, implanted, or otherwise presented to laboratory animals. Tumors, tissues, serum, blood cells, or cultured cells are capable of transmitting viral pathogens to susceptible species. To preclude the introduction of a pathogen into an animal colony via these mechanisms, biological material should be tested and found free of these agents prior to their use in the animal facility.

The most common method for detecting viral contaminants in biological material is through the antibody production test, commonly referred to as the mouse antibody production (MAP) test. Rat antibody production (RAP) and hamster antibody production (HAP) tests are also available and should be employed when materials will be introduced into these species. Antibody production tests involve the inoculation of the test material into pathogen-free animals of the appropriate species, followed in 4 weeks by serologic testing of inoculated animals for evidence of infection. In addition to the initial testing, material that initially tested negative, but which may have been exposed to animals or their products from uncharacterized or contaminated sources, should be retested. Most commercial laboratories perform antibody production testing.

References: Baker et al., 1979; Foster et al., 1981–1983; LaRegina and Lonigo, 1988; Manning et al., 1994; Percy and Berthold, 2001

Contributors: John Donovan and Patricia Brown

Table A.2C.1 Important Pathogenic Microorganisms of Rodents[a,b]

Bacterial	Viral	Parasitic/other
MICE		
Citrobacter rodentium (IT)	Ectromelia (integument)	**Pinworms**
Bordetella bronchiseptica (RT)	GD VII (CNS)	*Aspicularis tetraptera*
Clostridium piliforme (DT)	Lactic dehydrogenase elevating virus (liver)	*Syphacia obvelata*
Corynebacterium kutscheri (RT)	Lymphocytic choriomeningitis virus (CNS)	**Skin and fur mites**
Helicobacter spp. (IT)	Mouse hepatitis virus (DT)	*Myobia musculi*
Mycoplasma spp. (RT)	Mouse parvovirus (LT; IT)	*Myocoptes musculinus*
Pasteurella spp. (RT)	Pneumonia virus of mice (RT)	*Radfordia affinis*
Pseudomonas spp. (IT; RT)	Reo-3 (IT)	**Endoparasite**
Salmonella spp. (IT)	Sendai (RT)	*Pneumocystis carinii*
Streptobacillus moniliformis (RT)	Epizootic diarrhea of infant mice (IT)	
Streptococcus pneumonia (RT)		
RATS		
As in mice except *Citrobacter rodentium* and with the addition of cilia-associated respiratory bacillus (RT)	H-1 (RDC), Kilham rat virus (RDC)	**Pinworms**
	Pneumonia virus of mice (RT)	*Aspicularis tetraptera*
	Reo-3 (IT)	*Syphacia muris*
	Sialodacrioadenitis virus (salivary glands)	*Syphacia obvelata*
		Skin and fur mites
		Notoedres muris (ear mite)
		Radfordia ensifera
		Endoparasite
		Pneumocystis carinii
HAMSTERS		
As in mice except *Mycoplasma* spp., *Corynebacterium*, and *Streptobacillus moniliformis* and with the addition of *Clostridium dificile* (IT) and *Lawsonia intracellularis* (IT)	Lymphocytic choriomeningitis virus (CNS; UT)	**Pinworms**
	Pneumonia virus of mice (RT)	*Syphacia obvelata*
	Reo-3 (IT)	**Skin mites**
	Sendai (RT)	*Demodex aurati*
	Hamster papovavirus (tumors)	*Demodex criceti*
		Tape worms
		Hymenolepis nana
		Hymenolepis diminuta
RABBITS		
Clostridium piliforme (DT)	Rotavirus (IT)	**Ear mites**
Bordetella bronchiseptica (RT), middle ear		*Psoroptes cuniculi*
Lawsonia intracellularis (IT)		**Protozoa**
Pasteurella multocida (RT), middle ear		*Eimeria* spp. (DT)
Treponema paralviscunicuki (UT)		*Encephalitozoon* (CNS; kidney)
Clostridia spp. (IT)		

[a] Adapted from Percy and Barthold (2001). The scientific name is indicated first, followed by the primary target site.

[b] Abbreviations: CNS, central nervous system; DT, digestive tract; IT, intestinal tract; LT, lymphoid tissue; RDC, rapidly dividing cells; RT, respiratory tract; UT, urogenital tract.

APPENDIX 2D

Handling and Restraint

NOTE: Wear disposable gloves when handling animals in the following protocols.

BASIC PROTOCOL 1

MOUSE HANDLING AND MANUAL RESTRAINT

1. Remove the mouse from the cage by gently grasping the tail (with preferred hand) at the base. Place the animal on a wire-bar cage lid to permit grasping (Fig. A.2D.1).

2. Approach the back of the neck from the rear with the free hand. Firmly grasp the skin behind the ears with the thumb and index finger.

3. Transfer the tail from preferred hand to beneath the little finger of the hand holding the scruff of the neck (Fig. A.2D.1).

4. Observe or inject the restrained mouse.

BASIC PROTOCOL 2

RAT HANDLING AND MANUAL RESTRAINT

1. Remove the rat from the cage by gently grasping around the thorax with the thumb and middle finger behind the two front legs.

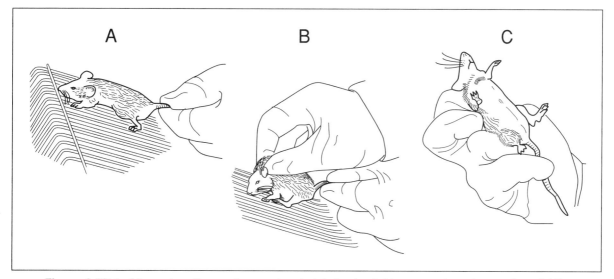

Figure A.2D.1 Mouse handling and manual restraint. Apply slight, rearward traction on the tail (**A**). Grasp skin behind ears with thumb and index finger (**B**). Transfer the tail from the preferred hand to beneath the little finger of the hand holding the scruff of the neck (**C**).

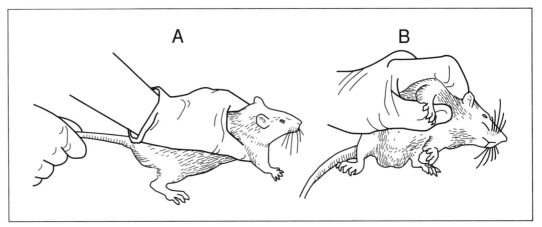

Figure A.2D.2 Rat handling and manual restraint. Grasp around the thorax with the thumb and middle finger behind the front two legs (**A**). Grasp the loose skin on the back of the neck with thumb and index finger (**B**).

2. Retain control of the animal with the middle and last two fingers around the thorax. Grasp the loose skin on the back of the neck with the thumb (still under animal's leg) and index finger (Fig. A.2D.2).

3. Observe or inject the restrained rat.

ALTERNATE PROTOCOL 1

RODENT RESTRAINERS

Plastic restraining devices of various sizes, shapes, and designs are available. An example of the most common is depicted in Figure A.2D.3. When placing an animal in a plastic restrainer, consideration should be given to the size of the restrainer relative to the size of the animal. A restrainer that is too large will permit the animal to turn around, while one that is too small may not allow adequate respiration. The length of time in the restrainer is also a concern. Most restrainers do not allow for adequate dissipation of body heat, causing the animal's core temperature to rise. Plastic restrainers are particularly useful for intravenous injection and sampling because both hands are free for tail immobilization and syringe manipulation. This method of restraint is most useful for bleeding the mouse and rat. It is also valuable for intravenous injection of the mouse.

1. Prepare a clean, open restrainer, with all parts readily accessible.

2. Manually restrain the mouse, rat, or hamster as described in the basic protocols.

3. Holding the tail, place the animal's head near the entry of the restrainer (Fig. A.2D.3).

4. As the animal's thorax is released, maintain tension with the tail.

5. Insert the securing block when the animal is fully in the device.

References: Fox et al., 2002; Poole, 1999; Suckow et al., 2001

Contributors: John Donovan and Patricia Brown

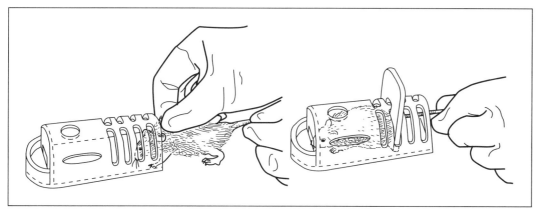

Figure A.2D.3 Rodent restrainers. With animal under control as described for handling and manual restraint, place the head at opening of the box while maintaining tension on the tail. Allow animal to crawl in and place the securing block appropriately.

APPENDIX 2E

Animal Identification

Basic animal information is usually maintained on cage cards which are utilized to identify single- or group-housed rodents where individual identification is not necessary. In some cases it is necessary to individually identify rodents. Specific procedures for marking or identifying individual rodents are described in the following protocols.

NOTE: Wear disposable gloves when handling animals in the following protocols.

BASIC PROTOCOL 1

EAR NOTCH OR PUNCH FOR MOUSE, RAT, AND HAMSTER

Materials

Ear punch—either hole (National Band and Tag) or notch (Harvard Apparatus)

1. Manually restrain the mouse, rat, or hamster (*APPENDIX 2D*).
2. Place the ear punch in the preselected location on the ear (Fig. A.2E.1).
3. Engage punch quickly and firmly to ensure a clean cut.
4. Return the animal to its cage.

BASIC PROTOCOL 2

EAR TAG FOR MOUSE, RAT, AND HAMSTER

Materials

Ear tags (National Band and Tag)

1. Manually restrain the animal (*APPENDIX 2D*), allowing access to the ears.
2. With the free hand, grasp a metal tag between the thumb and forefinger. Position the coupling device of the tag near the base of the caudal one-half of the ear.

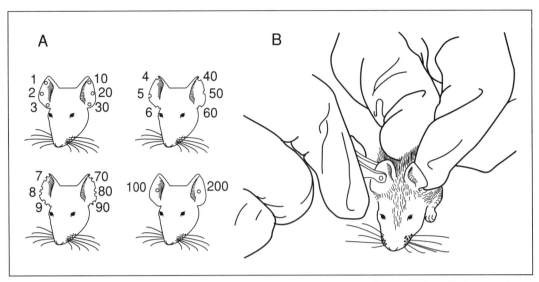

Figure A.2E.1 Ear punch identification of rodents. Determine the location of the identifying punch(es) according to the illustrated code (**A**). Restraining the animal manually, place punch in the desired position and firmly and quickly punch hole (**B**).

3. Couple the tag by quickly pressing the thumb and forefinger together.

4. Repeat in the other ear if necessary to ensure long-term identification.

BASIC PROTOCOL 3

TATTOO FOR MOUSE AND RAT

Materials

Micro-tattooing forceps (Ketchum Manufacturing)
Green tattoo paste (Ketchum Manufacturing)
25- to 30-G, 0.5-in. needles

1. Prepare the micro-tattooing forceps with the tattooing paste and needles according to the manufacturer's instructions.

2. Manually restrain the animal (*APPENDIX 2D*), allowing access to the paws or tail.

3. Tattoo the toe pad, foot pad, or tail with unique pattern.

BASIC PROTOCOL 4

SUBCUTANEOUS TRANSPONDER FOR MOUSE AND RAT

Materials

70% ethanol
Gauze sponge or swab
12-G implantation device loaded with sterilized transponders, and associated receiving unit (BioMedic Data Systems or AVID Identification Systems)

1. Restrain the animal for subcutaneous injection (*APPENDIX 2D*).

2. Swab the mid-dorsal region between the shoulder blades with 70% ethanol on a gauze sponge or swab.

3. Inject the implant using subcutaneous injection technique.

4. Scan the animal using the receiving unit to confirm transponder function and identification information.

5. Return the animal to the cage.

References: Ball et al., 1991; Fox et al., 2002

Contributors: John Donovan and Patricia Brown

APPENDIX 3
Selected Suppliers of Reagents and Equipment

Listed below are addresses and phone numbers of commercial suppliers who have been recommended for particular items used in our manuals because: (1) the particular brand has actually been found to be of superior quality, or (2) the item is difficult to find in the marketplace. Consequently, this compilation may not include some important vendors of biological supplies. For comprehensive listings, see *Linscott's Directory of Immunological and Biological Reagents* (Santa Rosa, CA), *The Biotechnology Directory* (Stockton Press, New York), the annual Buyers' Guide supplement to the journal *Bio/Technology*, as well as various sites on the Internet.

A.C. Daniels
72-80 Akeman Street
Tring, Hertfordshire, HP23 6AJ, UK
(44) 1442 826881
FAX: (44) 1442 826880

A.D. Instruments
5111 Nations Crossing Road #8
Suite 2
Charlotte, NC 28217
(704) 522-8415   FAX: (704) 527-5005
http://www.us.endress.com

A.J. Buck
11407 Cronhill Drive
Owings Mill, MD 21117
(800) 638-8673   FAX: (410) 581-1809
(410) 581-1800
http://www.ajbuck.com

A.M. Systems
131 Business Park Loop
P.O. Box 850
Carlsborg, WA 98324
(800) 426-1306   FAX: (360) 683-3525
(360) 683-8300
http://www.a-msystems.com

Aaron Medical Industries
7100 30th Avenue North
St. Petersburg, FL 33710
(727) 384-2323   FAX: (727) 347-9144
www.aaronmed.com

Abbott Laboratories
100 Abbott Park Road
Abbott Park, IL 60064
(800) 323-9100   FAX: (847) 938-7424
http://www.abbott.com

ABCO Dealers
55 Church Street Central Plaza
Lowell, MA 01852
(800) 462-3326   (978) 459-6101
http://www.lomedco.com/abco.htm

Aber Instruments
5 Science Park
Aberystwyth, Wales SY23 3AH, UK
(44) 1970 636300
FAX: (44) 1970 615455
http://www.aber-instruments.co.uk

ABI Biotechnologies
See Perkin-Elmer

ABI Biotechnology
See Apotex

Access Technologies
Subsidiary of Norfolk Medical
7350 N. Ridgeway
Skokie, IL 60076
(877) 674-7131   FAX: (847) 674-7066
(847) 674-7131
http://www.norfolkaccess.com

Accurate Chemical and Scientific
300 Shames Drive
Westbury, NY 11590
(800) 645-6264   FAX: (516) 997-4948
(516) 333-2221
http://www.accuratechemical.com

AccuScan Instruments
5090 Trabue Road
Columbus, OH 43228
(800) 822-1344   FAX: (614) 878-3560
(614) 878-6644
http://www.accuscan-usa.com

AccuStandard
125 Market Street
New Haven, CT 06513
(800) 442-5290   FAX: (877) 786-5287
http://www.accustandard.com

Ace Glass
1430 NW Boulevard
Vineland, NJ 08360
(800) 223-4524   FAX: (800) 543-6752
(609) 692-3333

ACO Pacific
2604 Read Avenue
Belmont, CA 94002
(650) 595-8588   FAX: (650) 591-2891
http://www.acopacific.com

Acros Organic
See Fisher Scientific

Action Scientific
P.O. Box 1369
Carolina Beach, NC 28428
(910) 458-0401   FAX: (910) 458-0407

AD Instruments
1949 Landings Drive
Mountain View, CA 94043
(888) 965-6040   FAX: (650) 965-9293
(650) 965-9292
http://www.adinstruments.com

Adaptive Biosystems
15 Ribocon Way
Progress Park
Luton, Bedsfordshire LU4 9UR, UK
(44)1 582-597676
FAX: (44)1 582-581495
http://www.adaptive.co.uk

Adobe Systems
1585 Charleston Road
P.O. Box 7900
Mountain View, CA 94039
(800) 833-6687   FAX: (415) 961-3769
(415) 961-4400
http://www.adobe.com

Advanced Bioscience Resources
1516 Oak Street, Suite 303
Alameda, CA 94501
(510) 865-5872   FAX: (510) 865-4090

Advanced Biotechnologies
9108 Guilford Road
Columbia, MD 21046
(800) 426-0764   FAX: (301) 497-9773
(301) 470-3220
http://www.abionline.com

Advanced ChemTech
5609 Fern Valley Road
Louisville, KY 40228
(502) 969-0000
http://www.peptide.com

Advance Machining and tooling
9850 Businesspark Avanue
San Diego, CA 92131
(858)530-0751   FAX:(858)530-611
http://www.amtmgf.com

Advanced Magnetics
See PerSeptive Biosystems

Advanced Process Supply
See Naz-Dar-KC Chicago

Advanced Separation Technologies
37 Leslie Court
P.O. Box 297
Whippany, NJ 07981
(973) 428-9080   FAX: (973) 428-0152
http://www.astecusa.com

Advanced Targeting Systems
11175-A Flintkote Avenue
San Diego, CA 92121
(877) 889-2288   FAX: (858) 642-1989
(858) 642-1988
http://www.ATSbio.com

Advent Research Materials
Eynsham, Oxford OX29 4JA, UK
(44) 1865-884440
FAX: (44) 1865-84460
www.advent-rm.com

Advet
Industrivagen 24
S-972 54 Lulea, Sweden
(46) 0920-211887
FAX: (46) 0920-13773

Aesculap
1000 Gateway Boulevard
South San Francisco, CA 94080
(800) 282-9000
http://www.aesculap.com

Affinity Chromatography
307 Huntingdon Road
Girton, Cambridge CB3 OJX, UK
(44) 1223 277192
FAX: (44) 1223 277502
http://www.affinity-chrom.com

Affinity Sensors
See Labsystems Affinity Sensors

Affymetrix
3380 Central Expressway
Santa Clara, CA 95051
(408) 731-5000   FAX: (408) 481-0422
(800) 362-2447
http://www.affymetrix.com

Agar Scientific
66a Cambridge Road
Stansted CM24 8DA, UK
(44) 1279-813-519
FAX: (44) 1279-815-106
http://www.agarscientific.com

A/G Technology
101 Hampton Avenue
Needham, MA 02494
(800) AGT-2535   FAX: (781) 449-5786
(781) 449-5774
http://www.agtech.com

Agen Biomedical Limited
11 Durbell Street
P.O. Box 391
Acacia Ridge 4110
Brisbane, Australia
61-7-3370-6300   FAX: 61-7-3370-6370
http://www.agen.com

Agilent Technologies
395 Page Mill Road
P.O. Box 10395
Palo Alto, CA 94306
(650) 752-5000
http://www.agilent.com/chem

Agouron Pharmaceuticals
10350 N. Torrey Pines Road
La Jolla, CA 92037
(858) 622-3000   FAX: (858) 622-3298
http://www.agouron.com

Agracetus
8520 University Green
Middleton, WI 53562
(608) 836-7300   FAX: (608) 836-9710
http://www.monsanto.com

AIDS Research and Reference
Reagent Program
U.S. Department of Health and Human Services
625 Lofstrand Lane
Rockville, MD 20850
(301) 340-0245   FAX: (301) 340-9245
http://www.aidsreagent.org

AIN Plastics
249 East Sanford Boulevard
P.O. Box 151
Mt. Vernon, NY 10550
(914) 668-6800   FAX: (914) 668-8820
http://www.tincna.com

Air Products and Chemicals
7201 Hamilton Boulevard
Allentown, PA 18195
(800) 345-3148   FAX: (610) 481-4381
(610) 481-6799
http://www.airproducts.com

ALA Scientific Instruments
1100 Shames Drive
Westbury, NY 11590
(516) 997-5780   FAX: (516) 997-0528
http://www.alascience.com

Aladin Enterprises
1255 23rd Avenue
San Francisco, CA 94122
(415) 468-0433   FAX: (415) 468-5607

Aladdin Systems
165 Westridge Drive
Watsonville, CA 95076
(831) 761-6200   FAX: (831) 761-6206
http://www.aladdinsys.com

Alcide
8561 154th Avenue NE
Redmond, WA 98052
(800) 543-2133   FAX: (425) 861-0173
(425) 882-2555
http://www.alcide.com

Aldrich Chemical
P.O. Box 2060
Milwaukee, WI 53201
(800) 558-9160   FAX: (800) 962-9591
(414) 273-3850   FAX: (414) 273-4979
http://www.aldrich.sial.com

Alexis Biochemicals
6181 Cornerstone Court East, Suite 103
San Diego, CA 92121
(800) 900-0065   FAX: (858) 658-9224
(858) 658-0065
http://www.alexis-corp.com

Alfa Laval
Avenue de Ble 5 - Bazellaan 5
BE-1140 Brussels, Belgium
32(2) 728 3811
FAX: 32(2) 728 3917 or
32(2) 728 3985
http://www.alfalaval.com

Alice King Chatham Medical Arts
11915-17 Inglewood Avenue
Hawthorne, CA 90250
(310) 970-1834   FAX: (310) 970-0121
(310) 970-1063

Allegiance Healthcare
800-964-5227
http://www.allegiance.net

Allelix Biopharmaceuticals
6850 Gorway Drive
Mississauga, Ontario
L4V 1V7 Canada
(905) 677-0831   FAX: (905) 677-9595
http://www.allelix.com

Allentown Caging Equipment
Route 526, P.O. Box 698
Allentown, NJ 08501
(800) 762-CAGE   FAX: (609) 259-0449
(609) 259-7951
http://www.acecaging.com

Alltech Associates
Applied Science Labs
2051 Waukegan Road
P.O. Box 23
Deerfield, IL 60015
(800) 255-8324   FAX: (847) 948-1078
(847) 948-8600
http://www.alltechweb.com

Alomone Labs
HaMarpeh 5
P.O. Box 4287
Jerusalem 91042, Israel
972-2-587-2202   FAX: 972-2-587-1101
US: (800) 791-3904
FAX: (800) 791-3912
http://www.alomone.com

Alpha Innotech
14743 Catalina Street
San Leandro, CA 94577
(800) 795-5556   FAX: (510) 483-3227
(510) 483-9620
http://www.alphainnotech.com

Altec Plastics
116 B Street
Boston, MA 02127
(800) 477-8196   FAX: (617) 269-8484
(617) 269-1400

Alza
1900 Charleston Road P.O. Box 7210
Mountain View, CA 94043
(800) 692-2990   FAX: (650) 564-7070
(650) 564-5000
http://www.alza.com

Alzet
c/o Durect Corporation
P.O. Box 530
10240 Bubb Road
Cupertino CA 95015
(800)692-2990 (408)367-4036
FAX: (408)865-1406
http://www.alzet.com

Amac
160B Larrabee Road
Westbrook, ME 04092
(800) 458-5060   FAX: (207) 854-0116
(207) 854-0426

Amaresco
30175 Solon Industrial Parkway
Solon, Ohio 44139
(800) 366-1313   FAX: (440) 349-1182
(440) 349-1313

Ambion
2130 Woodward Street, Suite 200
Austin, TX 78744
(800) 888-8804   FAX: (512) 651-0190
(512) 651-0200
http://www.ambion.com

American Association of Blood Banks College of American Pathologists
325 Waukegan Road
Northfield, IL 60093
(800) 323-4040   FAX: (847) 8166
(847) 832-7000
http://www.cap.org

American Bio-Technologies
See Intracel Corporation

American Bioanalytical
15 Erie Drive
Natick, MA 01760
(800) 443-0600   FAX: (508) 655-2754
(508) 655-4336
http://www.americanbio.com

American Cyanamid
P.O. Box 400
Princeton, NJ 08543
(609) 799-0400   FAX: (609) 275-3502
http://www.cyanamid.com

American HistoLabs
7605-F Airpark Road
Gaithersburg, MD 20879
(301) 330-1200   FAX: (301) 330-6059

American International Chemical
17 Strathmore Road
Natick, MA 01760
(800) 238-0001   (508) 655-5805
http://www.aicma.com

American Laboratory Supply
See American Bioanalytical

American Medical Systems
10700 Bren Road West
Minnetonka, MN 55343
(800) 328-3881   FAX: (612) 930-6654
(612) 933-4666
http://www.visitams.com

American Qualex
920-A Calle Negocio
San Clemente, CA 92673
(949) 492-8298   FAX: (949) 492-6790
http://www.americanqualex.com

American Radiolabeled Chemicals
11624 Bowling Green
St. Louis, MO 63146
(800) 331-6661   FAX: (800) 999-9925
(314) 991-4545   FAX: (314) 991-4692
http://www.arc-inc.com

American Scientific Products
See VWR Scientific Products

American Society for Histocompatibility and Immunogenetics
P.O. Box 15804
Lenexa, KS 66285 (913) 541-0009
FAX: (913) 541-0156
http://www.swmed.edu/home_pages/ASHI/ashi.htm

American Type Culture Collection (ATCC)
10801 University Boulevard
Manassas, VA 20110
(800) 638-6597   FAX: (703) 365-2750
(703) 365-2700
http://www.atcc.org

Amersham
See Amersham Pharmacia Biotech

Amersham International
Amersham Place
Little Chalfont, Buckinghamshire
HP7 9NA, UK
(44) 1494-544100
FAX: (44) 1494-544350
http://www.apbiotech.com

Amersham Medi-Physics
Also see Nycomed Amersham
3350 North Ridge Avenue
Arlington Heights, IL 60004
(800) 292-8514   FAX: (800) 807-2382
http://www.nycomed-amersham.com

Amersham Pharmacia Biotech
800 Centennial Avenue
P.O. Box 1327
Piscataway, NJ 08855
(800) 526-3593   FAX: (877) 295-8102
(732) 457-8000
http://www.apbiotech.com

Amgen
1 Amgen Center Drive
Thousand Oaks, CA 91320
(800) 926-4369   FAX: (805) 498-9377
(805) 447-5725
http://www.amgen.com

Amicon
Scientific Systems Division
72 Cherry Hill Drive
Beverly, MA 01915
(800) 426-4266   FAX: (978) 777-6204
(978) 777-3622
http://www.amicon.com

Amika
8980F Route 108
Oakland Center
Columbia, MD 21045
(800) 547-6766   FAX: (410) 997-7104
(410) 997-0100
http://www.amika.com

Amoco Performance Products
See BPAmoco

AMPI
See Pacer Scientific

Amrad
576 Swan Street
Richmond, Victoria 3121, Australia
613-9208-4000
FAX: 613-9208-4350
http://www.amrad.com.au

Amresco
30175 Solon Industrial Parkway
Solon, OH 44139
(800) 829-2805   FAX: (440) 349-1182
(440) 349-1199

Anachemia Chemicals
3 Lincoln Boulevard
Rouses Point, NY 12979
(800) 323-1414   FAX: (518) 462-1952
(518) 462-1066
http://www.anachemia.com

Ana-Gen Technologies
4015 Fabian Way
Palo Alto, CA 94303
(800) 654-4671   FAX: (650) 494-3893
(650) 494-3894
http://www.ana-gen.com

Analox Instruments USA
P.O. Box 208
Lunenburg MA 01462
(978) 582-9368   FAX: (978) 582-9588
http://www.analox.com

Analytical Biological Services
Cornell Business Park 701-4
Wilmington, DE 19801
(800) 391-2391   FAX: (302) 654-8046
(302) 654-4492
http://www.ABSbioreagents.com

Analytical Genetics Testing Center
7808 Cherry Creek S. Drive, Suite 201
Denver, CO 80231
(800) 204-4721   FAX: (303) 750-2171
(303) 750-2023
http://www.geneticid.com

AnaSpec
2149 O'Toole Avenue, Suite F
San Jose, CA 95131
(800) 452-5530   FAX: (408) 452-5059
(408) 452-5055
http://www.anaspec.com

Ancare
2647 Grand Avenue
P.O. Box 814
Bellmore, NY 11710
(800) 645-6379   FAX: (516) 781-4937
(516) 781-0755
http://www.ancare.com

Ancell
243 Third Street North
P.O. Box 87
Bayport, MN 55033
(800) 374-9523   FAX: (651) 439-1940
(651) 439-0835
http://www.ancell.com

Anderson Instruments
500 Technology Court
Smyrna, GA 30082
(800) 241-6898   FAX: (770) 319-5306
(770) 319-9999
http://www.graseby.com

Andreas Hettich
Gartenstrasse 100
Postfach 260
D-78732 Tuttlingen, Germany
(49) 7461 705 0
FAX: (49) 7461 705-122
http://www.hettich-centrifugen.de

Anesthetic Vaporizer Services
10185 Main Street
Clarence, NY 14031
(719) 759-8490
www.avapor.com

Animal Identification and Marking Systems (AIMS)
13 Winchester Avenue
Budd Lake, NJ 07828
(908) 684-9105   FAX: (908) 684-9106
http://www.animalid.com

Annovis
34 Mount Pleasant Drive
Aston, PA 19014
(800) EASY-DNA   FAX: (610) 361-8255
(610) 361-9224
http://www.annovis.com

Apotex
150 Signet Drive
Weston, Ontario M9L 1T9, Canada
(416) 749-9300   FAX: (416) 749-2646
http://www.apotex.com

Apple Scientific
11711 Chillicothe Road, Unit 2
P.O. Box 778
Chesterland, OH 44026
(440) 729-3056   FAX: (440) 729-0928
http://www.applesci.com

Applied Biosystems
See PE Biosystems

Applied Imaging
2380 Walsh Avenue, Bldg. B
Santa Clara, CA 95051
(800) 634-3622   FAX: (408) 562-0264
(408) 562-0250
http://www.aicorp.com

Applied Photophysics
203-205 Kingston Road
Leatherhead, Surrey, KT22 7PB
UK
(44) 1372-386537

Applied Precision
1040 12th Avenue Northwest
Issaquah, Washington 98027
(425) 557-1000
FAX: (425) 557-1055
http://www.api.com/index.html

Appligene Oncor
Parc d'Innovation
Rue Geiler de Kaysersberg, BP 72
67402 Illkirch Cedex, France
(33) 88 67 22 67
FAX: (33) 88 67 19 45
http://www.oncor.com/prod-app.htm

Applikon
1165 Chess Drive, Suite G
Foster City, CA 94404
(650) 578-1396   FAX: (650) 578-8836
http://www.applikon.com

Appropriate Technical Resources
9157 Whiskey Bottom Road
Laurel, MD 20723
(800) 827-5931   FAX: (410) 792-2837
http://www.atrbiotech.com

APV Gaulin
100 S. CP Avenue
Lake Mills, WI 53551
(888) 278-4321   FAX: (888) 278-5329
http://www.apv.com

Aqualon
See Hercules Aqualon

Aquebogue Machine and Repair Shop
Box 2055
Main Road
Aquebogue, NY 11931
(631) 722-3635   FAX: (631) 722-3106

Archer Daniels Midland
4666 Faries Parkway
Decatur, IL 62525
(217) 424-5200
http://www.admworld.com

Archimica Florida
P.O. Box 1466
Gainesville, FL 32602
(800) 331-6313   FAX: (352) 371-6246
(352) 376-8246
http://www.archimica.com

Arcor Electronics
1845 Oak Street #15
Northfield, IL 60093
(847) 501-4848

Arcturus Engineering
400 Logue Avenue
Mountain View, CA 94043
(888) 446 7911   FAX: (650) 962 3039
(650) 962 3020
http://www.arctur.com

Argonaut Technologies
887 Industrial Road, Suite G
San Carlos, CA 94070
(650) 998-1350   FAX: (650) 598-1359
http://www.argotech.com

Ariad Pharmaceuticals
26 Landsdowne Street
Cambridge, MA 02139
(617) 494-0400   FAX: (617) 494-8144
http://www.ariad.com

Armour Pharmaceuticals
See Rhone-Poulenc Rorer

Aronex Pharmaceuticals
8707 Technology Forest Place
The Woodlands, TX 77381
(281) 367-1666   FAX: (281) 367-1676
http://www.aronex.com

Artisan Industries
73 Pond Street
Waltham, MA 02254
(617) 893-6800
http://www.artisanind.com

ASI Instruments
12900 Ten Mile Road
Warren, MI 48089
(800) 531-1105   FAX: (810) 756-9737
(810) 756-1222
http://www.asi-instruments.com

Aspen Research Laboratories
1700 Buerkle Road
White Bear Lake, MN 55140
(651) 264-6000   FAX: (651) 264-6270
http://www.aspenresearch.com

Associates of Cape Cod
704 Main Street
Falmouth, MA 02540 (800) LAL-TEST
FAX: (508) 540-8680
(508) 540-3444
http://www.acciusa.com

Astra Pharmaceuticals
See AstraZeneca

AstraZeneca
1800 Concord Pike
Wilmington, DE 19850
(302) 886-3000   FAX: (302) 886-2972
http://www.astrazeneca.com

AT Biochem
30 Spring Mill Drive
Malvern, PA 19355
(610) 889-9300   FAX: (610) 889-9304

ATC Diagnostics
See Vysis

ATCC
See American Type Culture Collection

Athens Research and Technology
P.O. Box 5494
Athens, GA 30604
(706) 546-0207   FAX: (706) 546-7395

Atlanta Biologicals
1425-400 Oakbrook Drive
Norcross, GA 30093
(800) 780-7788 or (770) 446-1404
FAX: (800) 780-7374 or (770) 446-1404
http://www.atlantabio.com

Atomergic Chemical
71 Carolyn Boulevard
Farmingdale, NY 11735
(631) 694-9000   FAX: (631) 694-9177
http://www.atomergic.com

Atomic Energy of Canada
2251 Speakman Drive
Mississauga, Ontario
L5K 1B2 Canada
(905) 823-9040   FAX: (905) 823-1290
http://www.aecl.ca

ATR
P.O. Box 460
Laurel, MD 20725
(800) 827-5931   FAX: (410) 792-2837
(301) 470-2799
http://www.atrbiotech.com

Aurora Biosciences
11010 Torreyana Road
San Diego, CA 92121
(858) 404-6600   FAX: (858) 404-6714
http://www.aurorabio.com

Automatic Switch Company
A Division of Emerson Electric
50 Hanover Road
Florham Park, NJ 07932
(800) 937-2726   FAX: (973) 966-2628
(973) 966-2000
http://www.asco.com

Avanti Polar Lipids
700 Industrial Park Drive
Alabaster, AL 35007
(800) 227-0651   FAX: (800) 229-1004
(205) 663-2494   FAX: (205) 663-0756
http://www.avantilipids.com

Aventis
BP 67917
67917 Strasbourg Cedex 9, France
33 (0) 388 99 11 00
FAX: 33 (0) 388 99 11 01
http://www.aventis.com

Aventis Pasteur
1 Discovery Drive
Swiftwater, PA 18370
(800) 822-2463   FAX: (570) 839-0955
(570) 839-7187
http://www.aventispasteur.com/usa

Avery Dennison
150 North Orange Grove Boulevard
Pasadena, CA 91103
(800) 462-8379   FAX: (626) 792-7312
(626) 304-2000
http://www.averydennison.com

Avestin
2450 Don Reid Drive
Ottawa, Ontario K1H 1E1, Canada
(888) AVESTIN   FAX: (613) 736-8086
(613) 736-0019
http://www.avestin.com

AVIV Instruments
750 Vassar Avenue
Lakewood, NJ 08701
(732) 367-1663   FAX: (732) 370-0032
http://www.avivinst.com

Axon Instruments
1101 Chess Drive
Foster City, CA 94404
(650) 571-9400   FAX: (650) 571-9500
http://www.axon.com

Azon
720 Azon Road
Johnson City, NY 13790
(800) 847-9374   FAX: (800) 635-6042
(607) 797-2368
http://www.azon.com

BAbCO
1223 South 47th Street
Richmond, CA 94804
(800) 92-BABCO   FAX: (510) 412-8940
(510) 412-8930
http://www.babco.com

Bacharach
625 Alpha Drive
Pittsburgh, PA 15238
(800) 736-4666   FAX: (412) 963-2091
(412) 963-2000
http://www.bacharach-inc.com

Bachem Bioscience
3700 Horizon Drive
King of Prussia, PA 19406
(800) 634-3183   FAX: (610) 239-0800
(610) 239-0300
http://www.bachem.com

Bachem California
3132 Kashiwa Street
P.O. Box 3426
Torrance, CA 90510
(800) 422-2436   FAX: (310) 530-1571
(310) 539-4171
http://www.bachem.com

Baekon
18866 Allendale Avenue
Saratoga, CA 95070
(408) 972-8779   FAX: (408) 741-0944

Baker Chemical
See J.T. Baker

Bangs Laboratories
9025 Technology Drive
Fishers, IN 46038
(317) 570-7020   FAX: (317) 570-7034
www.bangslabs.com

Bard Parker
See Becton Dickinson

Barnstead/Thermolyne
P.O. Box 797
2555 Kerper Boulevard
Dubuque, IA 52004
(800) 446-6060   FAX: (319) 589-0516
http://www.barnstead.com

Barrskogen
4612 Laverock Place N
Washington, DC 20007
(800) 237-9192   FAX: (301) 464-7347

BAS
See Bioanalytical Systems

BASF
Specialty Products
3000 Continental Drive North
Mt. Olive, NJ 07828
(800) 669-2273   FAX: (973) 426-2610
http://www.basf.com

Baum, W.A.
620 Oak Street
Copiague, NY 11726
(631) 226-3940   FAX: (631) 226-3969
http://www.wabaum.com

Bausch & Lomb
One Bausch & Lomb Place
Rochester, NY 14604
(800) 344-8815   FAX: (716) 338-6007
(716) 338-6000
http://www.bausch.com

Baxter
Fenwal Division
1627 Lake Cook Road
Deerfield, IL 60015
(800) 766-1077   FAX: (800) 395-3291
(847) 940-6599   FAX: (847) 940-5766
http://www.powerfulmedicine.com

Baxter Healthcare
One Baxter Parkway
Deerfield, IL 60015
(800) 777-2298   FAX: (847) 948-3948
(847) 948-2000
http://www.baxter.com

Baxter Scientific Products
See VWR Scientific

Bayer
Agricultural Division
Animal Health Products
12707 Shawnee Mission Pkwy.
Shawnee Mission, KS 66201
(800) 255-6517   FAX: (913) 268-2803
(913) 268-2000
http://www.bayerus.com

Bayer
Diagnostics Division (Order Services)
P.O. Box 2009
Mishiwaka, IN 46546
(800) 248-2637   FAX: (800) 863-6882
(219) 256-3390
http://www.bayer.com

Bayer Diagnostics
511 Benedict Avenue
Tarrytown, NY 10591
(800) 255-3232   FAX: (914) 524-2132
(914) 631-8000
http://www.bayerdiag.com

Bayer Plc
Diagnostics Division
Bayer House, Strawberry Hill
Newbury, Berkshire RG14 1JA, UK
(44) 1635-563000
FAX: (44) 1635-563393
http://www.bayer.co.uk

BD Immunocytometry Systems
2350 Qume Drive
San Jose, CA 95131
(800) 223-8226   FAX: (408) 954-BDIS
http://www.bdfacs.com

BD Labware
Two Oak Park
Bedford, MA 01730
(800) 343-2035   FAX: (800) 743-6200
http://www.bd.com/labware

BD PharMingen
10975 Torreyana Road
San Diego, CA 92121
(800) 848-6227   FAX: (858) 812-8888
(858) 812-8800
http://www.pharmingen.com

BD Transduction Laboratories
133 Venture Court
Lexington, KY 40511
(800) 227-4063   FAX: (606) 259-1413
(606) 259-1550
http://www.translab.com

BDH Chemicals
Broom Road
Poole, Dorset BH12 4NN, UK
(44) 1202-745520
FAX: (44) 1202- 2413720

BDH Chemicals
See Hoefer Scientific Instruments

BDIS
See BD Immunocytometry Systems

Beckman Coulter
4300 North Harbor Boulevard
Fullerton, CA 92834
(800) 233-4685   FAX: (800) 643-4366
(714) 871-4848
http://www.beckman-coulter.com

Beckman Instruments
Spinco Division/Bioproducts Operation
1050 Page Mill Road
Palo Alto, CA 94304
(800) 742-2345   FAX: (415) 859-1550
(415) 857-1150
http://www.beckman-coulter.com

**Becton Dickinson Immunocytometry
& Cellular Imaging**
2350 Qume Drive
San Jose, CA 95131
(800) 223-8226   FAX: (408) 954-2007
(408) 432-9475
http://www.bdfacs.com

Becton Dickinson Labware
1 Becton Drive
Franklin Lakes, NJ 07417
(888) 237-2762   FAX: (800) 847-2220
(201) 847-4222
http://www.bdfacs.com

Becton Dickinson Labware
2 Bridgewater Lane
Lincoln Park, NJ 07035
(800) 235-5953   FAX: (800) 847-2220
(201) 847-4222
http://www.bdfacs.com

Becton Dickinson Primary
Care Diagnostics
7 Loveton Circle
Sparks, MD 21152
(800) 675-0908   FAX: (410) 316-4723
(410) 316-4000
http://www.bdfacs.com

Behringwerke Diagnostika
Hoechster Strasse 70
P-65835 Liederback, Germany
(49) 69-30511   FAX: (49) 69-303-834

Bellco Glass
340 Edrudo Road
Vineland, NJ 08360
(800) 257-7043   FAX: (856) 691-3247
(856) 691-1075
http://www.bellcoglass.com

Bender Biosystems
See Serva

Beral Enterprises
See Garren Scientific

Berkeley Antibody
See BAbCO

Bernsco Surgical Supply
25 Plant Avenue
Hauppague, NY 11788
(800) TIEMANN   FAX: (516) 273-6199
(516) 273-0005
http://www.bernsco.com

**Beta Medical and Scientific
(Datesand Ltd.)**
2 Ferndale Road
Sale, Manchester M33 3GP, UK
(44) 1612 317676
FAX: (44) 1612 313656

**Bethesda Research Laboratories
(BRL)**
See Life Technologies

Biacore
200 Centennial Avenue, Suite 100
Piscataway, NJ 08854
(800) 242-2599   FAX: (732) 885-5669
(732) 885-5618
http://www.biacore.com

Bilaney Consultants
St. Julian's
Sevenoaks, Kent TN15 0RX, UK
(44) 1732 450002
FAX: (44) 1732 450003
http://www.bilaney.com

Binding Site
5889 Oberlin Drive, Suite 101
San Diego, CA 92121
(800) 633-4484   FAX: (619) 453-9189
(619) 453-9177
http://www.bindingsite.co.uk

BIO 101
See Qbiogene

Bio Image
See Genomic Solutions

Bioanalytical Systems
2701 Kent Avenue
West Lafayette, IN 47906
(800) 845-4246   FAX: (765) 497-1102
(765) 463-4527
http://www.bioanalytical.com

Biocell
2001 University Drive
Rancho Dominguez, CA 90220
(800) 222-8382   FAX: (310) 637-3927
(310) 537-3300
http://www.biocell.com

Biocoat
See BD Labware

BioComp Instruments
650 Churchill Road
Fredericton, New Brunswick
E3B 1P6 Canada
(800) 561-4221   FAX: (506) 453-3583
(506) 453-4812
http://131.202.97.21

BioDesign
P.O. Box 1050
Carmel, NY 10512
(914) 454-6610   FAX: (914) 454-6077
http://www.biodesignofny.com

BioDiscovery
4640 Admiralty Way, Suite 710
Marina Del Rey, CA 90292
(310) 306-9310   FAX: (310) 306-9109
http://www.biodiscovery.com

Bioengineering AG
Sagenrainstrasse 7
CH8636 Wald, Switzerland
(41) 55-256-8-111
FAX: (41) 55-256-8-256

Biofluids
Division of Biosource International
1114 Taft Street
Rockville, MD 20850
(800) 972-5200   FAX: (301) 424-3619
(301) 424-4140
http://www.biosource.com

BioFX Laboratories
9633 Liberty Road, Suite S
Randallstown, MD 21133
(800) 445-6447   FAX: (410) 498-6008
(410) 496-6006
http://www.biofx.com

BioGenex Laboratories
4600 Norris Canyon Road
San Ramon, CA 94583
(800) 421-4149   FAX: (925) 275-0580
(925) 275-0550
http://www.biogenex.com

Bioline
2470 Wrondel Way
Reno, NV 89502
(888) 257-5155   FAX: (775) 828-7676
(775) 828-0202
http://www.bioline.com

Bio-Logic Research & Development
1, rue de l-Europe
A.Z. de Font-Ratel
38640 CLAIX, France
(33) 76-98-68-31
FAX: (33) 76-98-69-09

Biological Detection Systems
See Cellomics or Amersham

Biomeda
1166 Triton Drive, Suite E
P.O. Box 8045
Foster City, CA 94404
(800) 341-8787   FAX: (650) 341-2299
(650) 341-8787
http://www.biomeda.com

BioMedic Data Systems
1 Silas Road
Seaford, DE 19973
(800) 526-2637   FAX: (302) 628-4110
(302) 628-4100
http://www.bmds.com

Biomedical Engineering
P.O. Box 980694
Virginia Commonwealth University
Richmond, VA 23298
(804) 828-9829   FAX: (804) 828-1008

Biomedical Research Instruments
12264 Wilkins Avenue
Rockville, MD 20852
(800) 327-9498
(301) 881-7911
http://www.biomedinstr.com

Bio/medical Specialties
P.O. Box 1687
Santa Monica, CA 90406
(800) 269-1158   FAX: (800) 269-1158
(323) 938-7515

BioMerieux
100 Rodolphe Street
Durham, North Carolina 27712
(919) 620-2000
http://www.biomerieux.com

BioMetallics
P.O. Box 2251
Princeton, NJ 08543
(800) 999-1961   FAX: (609) 275-9485
(609) 275-0133
http://www.microplate.com

Biomol Research Laboratories
5100 Campus Drive
Plymouth Meeting, PA 19462
(800) 942-0430   FAX: (610) 941-9252
(610) 941-0430
http://www.biomol.com

Bionique Testing Labs
Fay Brook Drive
RR 1, Box 196
Saranac Lake, NY 12983
(518) 891-2356   FAX: (518) 891-5753
http://www.bionique.com

Biopac Systems
42 Aero Camino
Santa Barbara, CA 93117
(805) 685-0066   FAX: (805) 685-0067
http://www.biopac.com

Bioproducts for Science
See Harlan Bioproducts for Science

Bioptechs
3560 Beck Road Butler, PA 16002
(877) 548-3235   FAX: (724) 282-0745
(724) 282-7145
http://www.bioptechs.com

BIOQUANT-R&M Biometrics
5611 Ohio Avenue
Nashville, TN 37209
(800) 221-0549 (615) 350-7866
FAX: (615) 350-7282
http://www.bioquant.com

Bio-Rad Laboratories
2000 Alfred Nobel Drive
Hercules, CA 94547
(800) 424-6723   FAX: (800) 879-2289
(510) 741-1000   FAX: (510) 741-5800
http://www.bio-rad.com

Bio-Rad Laboratories
Maylands Avenue
Hemel Hempstead, Herts HP2 7TD, UK
http://www.bio-rad.com

BioRobotics
3-4 Bennell Court
Comberton, Cambridge CB3 7DS, UK
(44) 1223-264345
FAX: (44) 1223-263933
http://www.biorobotics.co.uk

BIOS Laboratories
See Genaissance Pharmaceuticals

Biosearch Technologies
81 Digital Drive
Novato, CA 94949
(800) GENOME1   FAX: (415) 883-8488
(415) 883-8400
http://www.biosearchtech.com

BioSepra
111 Locke Drive
Marlborough, MA 01752
(800) 752-5277   FAX: (508) 357-7595
(508) 357-7500
http://www.biosepra.com

Bio-Serv
1 8th Street, Suite 1
Frenchtown, NJ 08825
(908) 996-2155   FAX: (908) 996-4123
http://www.bio-serv.com

BioSignal
1744 William Street, Suite 600
Montreal, Quebec H3J 1R4, Canada
(800) 293-4501   FAX: (514) 937-0777
(514) 937-1010
http://www.biosignal.com

Biosoft
P.O. Box 10938
Ferguson, MO 63135
(314) 524-8029   FAX: (314) 524-8129
http://www.biosoft.com

Biosource International
820 Flynn Road
Camarillo, CA 93012
(800) 242-0607   FAX: (805) 987-3385
(805) 987-0086
http://www.biosource.com

BioSpec Products
P.O. Box 788
Bartlesville, OK 74005
(800) 617-3363   FAX: (918) 336-3363
(918) 336-3363
http://www.biospec.com

Biosure
See Riese Enterprises

Biosym Technologies
See Molecular Simulations

Biosys
21 quai du Clos des Roses
602000 Compiegne, France
(33) 03 4486 2275
FAX: (33) 03 4484 2297

Bio-Tech Research Laboratories
NIAID Repository
Rockville, MD 20850
http://www.niaid.nih.gov/ncn/repos.htm

Biotech Instruments
Biotech House
75A High Street
Kimpton, Hertfordshire SG4 8PU, UK
(44) 1438 832555
FAX: (44) 1438 833040
http://www.biotinst.demon.co.uk

Biotech International
11 Durbell Street
Acacia Ridge, Queensland 4110
Australia
61-7-3370-6396
FAX: 61-7-3370-6370
http://www.avianbiotech.com

Biotech Source
Inland Farm Drive
South Windham, ME 04062
(207) 892-3266   FAX: (207) 892-6774

Bio-Tek Instruments
Highland Industrial Park
P.O. Box 998
Winooski, VT 05404
(800) 451-5172   FAX: (802) 655-7941
(802) 655-4040
http://www.biotek.com

Biotecx Laboratories
6023 South Loop East
Houston, TX 77033
(800) 535-6286   FAX: (713) 643-3143
(713) 643-0606
http://www.biotecx.com

BioTherm
3260 Wilson Boulevard
Arlington, VA 22201
(703) 522-1705   FAX: (703) 522-2606

Bioventures
P.O. Box 2561
848 Scott Street
Murfreesboro, TN 37133
(800) 235-8938   FAX: (615) 896-4837
http://www.bioventures.com

BioWhittaker
8830 Biggs Ford Road
P.O. Box 127
Walkersville, MD 21793
(800) 638-8174   FAX: (301) 845-8338
(301) 898-7025
http://www.biowhittaker.com

Biozyme Laboratories
9939 Hibert Street, Suite 101
San Diego, CA 92131
(800) 423-8199   FAX: (858) 549-0138
(858) 549-4484
http://www.biozyme.com

Bird Products
1100 Bird Center Drive
Palm Springs, CA 92262
(800) 328-4139   FAX: (760) 778-7274
(760) 778-7200
http://www.birdprod.com/bird

B & K Universal
2403 Yale Way
Fremont, CA 94538
(800) USA-MICE   FAX: (510) 490-3036

BLS Ltd.
Zselyi Aladar u. 31
1165 Budapest, Hungary
(36) 1-407-2602   FAX: (36) 1-407-2896
http://www.bls-ltd.com

Blue Sky Research
3047 Orchard Parkway
San Jose, CA 95134
(408) 474-0988   FAX: (408) 474-0989
http://www.blueskyresearch.com

Blumenthal Industries
7 West 36th Street, 13th floor
New York, NY 10018
(212) 719-1251   FAX: (212) 594-8828

BOC Edwards
One Edwards Park
301 Ballardvale Street
Wilmington, MA 01887
(800) 848-9800   FAX: (978) 658-7969
(978) 658-5410
http://www.bocedwards.com

Boehringer Ingelheim
900 Ridgebury Road
P.O. Box 368
Ridgefield, CT 06877
(800) 243-0127   FAX: (203) 798-6234
(203) 798-9988
http://www.boehringer-ingelheim.com

Boehringer Mannheim
Biochemicals Division
See Roche Diagnostics

Boekel Scientific
855 Pennsylvania Boulevard
Feasterville, PA 19053
(800) 336-6929   FAX: (215) 396-8264
(215) 396-8200
http://www.boekelsci.com

Bohdan Automation
1500 McCormack Boulevard
Mundelein, IL 60060
(708) 680-3939   FAX: (708) 680-1199

BPAmoco
4500 McGinnis Ferry Road
Alpharetta, GA 30005
(800) 328-4537   FAX: (770) 772-8213
(770) 772-8200
http://www.bpamoco.com

Brain Research Laboratories
Waban P.O. Box 88
Newton, MA 02468
(888) BRL-5544   FAX: (617) 965-6220
(617) 965-5544
http://www.brainresearchlab.com

Braintree Scientific
P.O. Box 850929
Braintree, MA 02185
(781) 843-1644   FAX: (781) 982-3160
http://www.braintreesci.com

Brandel
8561 Atlas Drive
Gaithersburg, MD 20877
(800) 948-6506   FAX: (301) 869-5570
(301) 948-6506
http://www.brandel.com

Branson Ultrasonics
41 Eagle Road
Danbury, CT 06813
(203) 796-0400   FAX: (203) 796-9838
http://www.plasticsnet.com/branson

B. Braun Biotech
999 Postal Road
Allentown, PA 18103
(800) 258-9000   FAX: (610) 266-9319
(610) 266-6262
http://www.bbraunbiotech.com

B. Braun Biotech International
Schwarzenberg Weg 73-79
P.O. Box 1120
D-34209 Melsungen, Germany
(49) 5661-71-3400
FAX: (49) 5661-71-3702
http://www.bbraunbiotech.com

B. Braun-McGaw
2525 McGaw Avenue
Irvine, CA 92614
(800) BBRAUN-2   (800) 624-2963
http://www.bbraunusa.com

B. Braun Medical
Thorncliffe Park
Sheffield S35 2PW, UK
(44) 114-225-9000
FAX: (44) 114-225-9111
http://www.bbmuk.demon.co.uk

Brenntag
P.O. Box 13788
Reading, PA 19612-3788
(610) 926-4151   FAX: (610) 926-4160
http://www.brenntagnortheast.com

Bresatec
See GeneWorks

Bright/Hacker Instruments
17 Sherwood Lane
Fairfield, NJ 07004
(973) 226-8450   FAX: (973) 808-8281
http://www.hackerinstruments.com

Brinkmann Instruments
Subsidiary of Sybron
1 Cantiague Road
P.O. Box 1019
Westbury, NY 11590
(800) 645-3050   FAX: (516) 334-7521
(516) 334-7500
http://www.brinkmann.com

Bristol-Meyers Squibb
P.O. Box 4500
Princeton, NJ 08543
(800) 631-5244   FAX: (800) 523-2965
http://www.bms.com

Broadley James
19 Thomas
Irvine, CA 92618
(800) 288-2833   FAX: (949) 829-5560
(949) 829-5555
http://www.broadleyjames.com

Brookhaven Instruments
750 Blue Point Road
Holtsville, NY 11742
(631) 758-3200   FAX: (631) 758-3255
http://www.bic.com

Brownlee Labs
See Applied Biosystems
Distributed by Pacer Scientific

Bruel & Kjaer
Division of Spectris Technologies
2815 Colonnades Court
Norcross, GA 30071
(800) 332-2040   FAX: (770) 847-8440
(770) 209-6907
http://www.bkhome.com

Bruker Analytical X-Ray Systems
5465 East Cheryl Parkway
Madison, WI 53711
(800) 234-XRAY   FAX: (608) 276-3006
(608) 276-3000
http://www.bruker-axs.com

Bruker Instruments
19 Fortune Drive
Billerica, MA 01821
(978) 667-9580   FAX: (978) 667-0985
http://www.bruker.com

BTX
Division of Genetronics
11199 Sorrento Valley Road
San Diego, CA 92121
(800) 289-2465   FAX: (858) 597-9594
(858) 597-6006
http://www.genetronics.com/btx

Buchler Instruments
See Baxter Scientific Products

Buckshire
2025 Ridge Road
Perkasie, PA 18944
(215) 257-0116

Burdick and Jackson
Division of Baxter Scientific Products
1953 S. Harvey Street
Muskegon, MI 49442
(800) 368-0050   FAX: (231) 728-8226
(231) 726-3171
http://www.bandj.com/mainframe.htm

Burleigh Instruments
P.O. Box E
Fishers, NY 14453
(716) 924-9355   FAX: (716) 924-9072
http://www.burleigh.com

Burns Veterinary Supply
1900 Diplomat Drive
Farmer's Branch, TX 75234
(800) 92-BURNS   FAX: (972) 243-6841
http://www.burnsvet.com

Burroughs Wellcome
See Glaxo Wellcome

The Butler Company
5600 Blazer Parkway
Dublin, OH 43017
(800) 551-3861   FAX: (614) 761-9096
(614) 761-9095
http://www.wabutler.com

Butterworth Laboratories
54-56 Waldegrave Road
Teddington, Middlesex
TW11 8LG, UK
(44)(0)20-8977-0750
FAX: (44)(0)28-8943-2624
http://www.butterworth-labs.co.uk

Buxco Electronics
95 West Wood Road #2
Sharon, CT 06069
(860) 364-5558   FAX: (860) 364-5116
http://www.buxco.com

C/D/N Isotopes
88 Leacock Street Pointe-Claire, Quebec
Canada H9R 1H1
(800) 697-6254   FAX: (514) 697-6148

C.M.A./Microdialysis AB
73 Princeton Street
North Chelmsford, MA 01863
(800) 440-4980 FAX: (978) 251-1950
(978) 251-1940
http://www.microdialysis.com

Calbiochem-Novabiochem
P.O. Box 12087-2087
La Jolla, CA 92039
(800) 854-3417 FAX: (800) 776-0999
(858) 450-9600
http://www.calbiochem.com

California Fine Wire
338 South Fourth Street
Grover Beach, CA 93433
(805) 489-5144 FAX: (805) 489-5352
http://www.california.com

Calorimetry Sciences
155 West 2050 North
Spanish Fork, UT 84660
(801) 794-2600 FAX: (801) 794-2700
http://www.calscorp.com

Caltag Laboratories
1849 Bayshore Highway, Suite 200
Burlingame, CA 94010
(800) 874-4007 FAX: (650) 652-9030
(650) 652-0468
http://www.caltag.com

Cambridge Electronic Design
Science Park, Milton Road
Cambridge CB4 0FE, UK
44 (0) 1223-420-186
FAX: 44 (0) 1223-420-488
http://www.ced.co.uk

Cambridge Isotope Laboratories
50 Frontage Road
Andover, MA 01810
(800) 322-1174 FAX: (978) 749-2768
(978) 749-8000
http://www.isotope.com

Cambridge Research Biochemicals
See Zeneca/CRB

Cambridge Technology
109 Smith Place
Cambridge, MA 02138
(6l7) 441-0600 FAX: (617) 497-8800
http://www.camtech.com

Camlab
Nuffield Road
Cambridge CB4 1TH, UK
(44) 122-3424222
FAX: (44) 122-3420856
http://www.camlab.co.uk/home.htm

Campden Instruments
Park Road
Sileby Loughborough
Leicestershire LE12 7TU, UK
(44) 1509-814790
FAX: (44) 1509-816097
http://www.campden-inst.com/home.htm

Cappel Laboratories
See Organon Teknika Cappel

Carl Roth GmgH & Company
Schoemperlenstrasse 1-5
76185 Karlsrube
Germany
(49) 72-156-06164
FAX: (49) 72-156-06264
http://www.carl-roth.de

Carl Zeiss
One Zeiss Drive
Thornwood, NY 10594
(800) 233-2343 FAX: (914) 681-7446
(914) 747-1800
http://www.zeiss.com

Carlo Erba Reagenti
Via Winckelmann 1
20148 Milano
Lombardia, Italy
(39) 0-29-5231
FAX: (39) 0-29-5235-904
http://www.carloerbareagenti.com

Carolina Biological Supply
2700 York Road
Burlington, NC 27215
(800) 334-5551 FAX: (336) 584-76869
(336) 584-0381
http://www.carolina.com

Carolina Fluid Components
9309 Stockport Place
Charlotte, NC 28273
(704) 588-6101 FAX: (704) 588-6115
http://www.cfcsite.com

Cartesian Technologies
17851 Skypark Circle, Suite C
Irvine, CA 92614
(800) 935-8007
http://cartesiantech.com

Cayman Chemical
1180 East Ellsworth Road
Ann Arbor, MI 48108
(800) 364-9897 FAX: (734) 971-3640
(734) 971-3335
http://www.caymanchem.com

CB Sciences
One Washington Street, Suite 404
Dover, NH 03820
(800) 234-1757 FAX: (603) 742-2455
http://www.cbsci.com

CBS Scientific
P.O. Box 856
Del Mar, CA 92014
(800) 243-4959 FAX: (858) 755-0733
(858) 755-4959
http://www.cbssci.com

CCR (Coriell Cell Repository)
See Coriell Institute for Medical Research

Cedarlane Laboratories
5516 8th Line, R.R. #2
Hornby, Ontario L0P 1E0, Canada
(905) 878-8891 FAX: (905) 878-7800
http://www.cedarlanelabs.com

CE Instruments
Grand Avenue Parkway
Austin, TX 78728
(800) 876-6711 FAX: (512) 251-1597
http://www.ceinstruments.com

CEL Associates
P.O. Box 721854
Houston, TX 77272
(800) 537-9339 FAX: (281) 933-0922
(281) 933-9339
http://www.cel-1.com

Cel-Line Associates
See Erie Scientific

Celite World Minerals
130 Castilian Drive
Santa Barbara, CA 93117
(805) 562-0200 FAX: (805) 562-0299
http://www.worldminerals.com/celite

Cell Genesys
342 Lakeside Drive
Foster City, CA 94404
(650) 425-4400 FAX: (650) 425-4457
http://www.cellgenesys.com

Cell Systems
12815 NE 124th Street, Suite A
Kirkland, WA 98034
(800) 697-1211 FAX: (425) 820-6762
(425) 823-1010

Cellmark Diagnostics
20271 Goldenrod Lane
Germantown, MD 20876
(800) 872-5227 FAX: (301) 428-4877
(301) 428-4980
http://www.cellmark-labs.com

Cellomics
635 William Pitt Way
Pittsburgh, PA 15238
(888) 826-3857 FAX: (412) 826-3850
(412) 826-3600
http://www.cellomics.com

Celltech
216 Bath Road
Slough, Berkshire SL1 4EN, UK
(44) 1753 534655
FAX: (44) 1753 536632
http://www.celltech.co.uk

Cellular Products
872 Main Street
Buffalo, NY 14202
(800) CPI-KITS   FAX: (716) 882-0959
(716) 882-0920
http://www.zeptometrix.com

CEM
P.O. Box 200
Matthews, NC 28106
(800) 726-3331

Centers for Disease Control
1600 Clifton Road NE
Atlanta, GA 30333
(800) 311-3435   FAX: (888) 232-3228
(404) 639-3311
http://www.cdc.gov

CERJ
Centre d'Elevage Roger Janvier
53940 Le Genest Saint Isle
France

Cetus
See Chiron

Chance Propper
Warly, West Midlands B66 1NZ, UK
(44)(0)121-553-5551
FAX: (44)(0)121-525-0139

Charles River Laboratories
251 Ballardvale Street
Wilmington, MA 01887
(800) 522-7287   FAX: (978) 658-7132
(978) 658-6000
http://www.criver.com

Charm Sciences
36 Franklin Street
Malden, MA 02148
(800) 343-2170   FAX: (781) 322-3141
(781) 322-1523
http://www.charm.com

Chase-Walton Elastomers
29 Apsley Street
Hudson, MA 01749
(800) 448-6289   FAX: (978) 562-5178
(978) 568-0202
http://www.chase-walton.com

ChemGenes
Ashland Technology Center
200 Homer Avenue
Ashland, MA 01721
(800) 762-9323   FAX: (508) 881-3443
(508) 881-5200
http://www.chemgenes.com

Chemglass
3861 North Mill Road
Vineland, NJ 08360
(800) 843-1794   FAX: (856) 696-9102
(800) 696-0014
http://www.chemglass.com

Chemicon International
28835 Single Oak Drive
Temecula, CA 92590
(800) 437-7500   FAX: (909) 676-9209
(909) 676-8080
http://www.chemicon.com

Chem-Impex International
935 Dillon Drive
Wood Dale, IL 60191
(800) 869-9290   FAX: (630) 766-2218
(630) 766-2112
http://www.chemimpex.com

Chem Service
P.O. Box 599
West Chester, PA 19381-0599
(610) 692-3026   FAX: (610) 692-8729
http://www.chemservice.com

Chemsyn Laboratories
13605 West 96th Terrace
Lenexa, Kansas 66215
(913) 541-0525   FAX: (913) 888-3582
http://www.tech.epcorp.com/ChemSyn/
chemsyn.htm

Chemunex USA
1 Deer Park Drive, Suite H-2
Monmouth Junction, NJ 08852
(800) 411-6734
http://www.chemunex.com

Cherwell Scientific Publishing
The Magdalen Centre
Oxford Science Park
Oxford OX44GA, UK
(44)(1) 865-784-800
FAX: (44)(1) 865-784-801
http://www.cherwell.com

ChiRex Cauldron
383 Phoenixville Pike
Malvern, PA 19355
(610) 727-2215   FAX: (610) 727-5762
http://www.chirex.com

Chiron Diagnostics
See Bayer Diagnostics

Chiron Mimotopes Peptide Systems
See Multiple Peptide Systems

Chiron
4560 Horton Street
Emeryville, CA 94608
(800) 244-7668   FAX: (510) 655-9910
(510) 655-8730
http://www.chiron.com

Chrom Tech
P.O. Box 24248
Apple Valley, MN 55124
(800) 822-5242   FAX: (952) 431-6345
http://www.chromtech.com

Chroma Technology
72 Cotton Mill Hill, Unit A-9
Brattleboro, VT 05301
(800) 824-7662   FAX: (802) 257-9400
(802) 257-1800
http://www.chroma.com

Chromatographie
ZAC de Moulin No. 2
91160 Saulx les Chartreux
France
(33) 01-64-54-8969
FAX: (33) 01-69-0988091
http://www.chromatographie.com

Chromogenix
Taljegardsgatan 3
431-53 Mlndal, Sweden
(46) 31-706-20-70
FAX: (46) 31-706-20-80
http://www.chromogenix.com

Chrompack USA
c/o Varian USA
2700 Mitchell Drive
Walnut Creek, CA 94598
(800) 526-3687   FAX: (925) 945-2102
(925) 939-2400
http://www.chrompack.com

Chugai Biopharmaceuticals
6275 Nancy Ridge Drive
San Diego, CA 92121
(858) 535-5900   FAX: (858) 546-5973
http://www.chugaibio.com

Ciba-Corning Diagnostics
See Bayer Diagnostics

Ciba-Geigy
See Ciba Specialty Chemicals or
Novartis Biotechnology

Ciba Specialty Chemicals
540 White Plains Road
Tarrytown, NY 10591
(800) 431-1900   FAX: (914) 785-2183
(914) 785-2000
http://www.cibasc.com

Ciba Vision
Division of Novartis AG
11460 Johns Creek Parkway
Duluth, GA 30097
(770) 476-3937
http://www.cvworld.com

Cidex
Advanced Sterilization Products
33 Technology Drive
Irvine, CA 92618
(800) 595-0200   (949) 581-5799
http://www.cidex.com

Cinna Scientific
Subsidiary of Molecular Research Center
5645 Montgomery Road
Cincinnati, OH 45212
(800) 462-9868   FAX: (513) 841-0080
(513) 841-0900
http://www.mrcgene.com

Cistron Biotechnology
10 Bloomfield Avenue
Pine Brook, NJ 07058
(800) 642-0167   FAX: (973) 575-4854
(973) 575-1700
http://www.cistronbio.com

Clark Electromedical Instruments
See Harvard Apparatus

Clay Adam
See Becton Dickinson Primary Care Diagnostics

CLB (Central Laboratory of the Netherlands)
Blood Transfusion Service
P.O. Box 9190
1006 AD Amsterdam, The Netherlands
(31) 20-512-9222
FAX: (31) 20-512-3332

Cleveland Scientific
P.O. Box 300
Bath, OH 44210
(800) 952-7315   FAX: (330) 666-2240
http://www.clevelandscientific.com

Clonetics
Division of BioWhittaker
http://www.clonetics.com
Also see BioWhittaker

Clontech Laboratories
1020 East Meadow Circle
Palo Alto, CA 94303
(800) 662-2566   FAX: (800) 424-1350
(650) 424-8222   FAX: (650) 424-1088
http://www.clontech.com

Closure Medical Corporation
5250 Greens Dairy Road
Raleigh, NC 27616
(919) 876-7800   FAX: (919) 790-1041
http://www.closuremed.com

CMA Microdialysis AB
73 Princeton Street
North Chelmsford, MA 01863
(800) 440-4980   FAX: (978) 251-1950
(978) 251 1940
http://www.microdialysis.com

Cocalico Biologicals
449 Stevens Road
P.O. Box 265
Reamstown, PA 17567
(717) 336-1990   FAX: (717) 336-1993

Coherent Laser
5100 Patrick Henry Drive
Santa Clara, CA 95056
(800) 227-1955   FAX: (408) 764-4800
(408) 764-4000
http://www.cohr.com

Cohu
P.O. Box 85623
San Diego, CA 92186
(858) 277-6700   FAX: (858) 277-0221
http://www.COHU.com/cctv

Cole-Parmer Instrument
625 East Bunker Court
Vernon Hills, IL 60061
(800) 323-4340   FAX: (847) 247-2929
(847) 549-7600
http://www.coleparmer.com

Collaborative Biomedical Products and Collaborative Research
See Becton Dickinson Labware

Collagen Aesthetics
1850 Embarcadero Road
Palo Alto, CA 94303
(650) 856-0200   FAX: (650) 856-0533
http://www.collagen.com

Collagen Corporation
See Collagen Aesthetics

College of American Pathologists
325 Waukegan Road
Northfield, IL 60093
(800) 323-4040   FAX: (847) 832-8000
(847) 446-8800
http://www.cap.org/index.cfm

Colonial Medical Supply
504 Wells Road
Franconia, NH 03580
(603) 823-9911   FAX: (603) 823-8799
http://www.colmedsupply.com

Colorado Serum
4950 York Street
Denver, CO 80216
(800) 525-2065   FAX: (303) 295-1923
http://www.colorado-serum.com

Columbia Diagnostics
8001 Research Way
Springfield, VA 22153
(800) 336-3081   FAX: (703) 569-2353
(703) 569-7511
http://www.columbiadiagnostics.com

Columbus Instruments
950 North Hague Avenue
Columbus, OH 43204
(800) 669-5011   FAX: (614) 276-0529
(614) 276-0861
http://www.columbusinstruments.com

Computer Associates International
One Computer Associates Plaza
Islandia, NY 11749
(631) 342-6000   FAX: (631) 342-6800
http://www.cai.com

Connaught Laboratories
See Aventis Pasteur

Connectix
2955 Campus Drive, Suite 100
San Mateo, CA 94403
(800) 950-5880   FAX: (650) 571-0850
(650) 571-5100
http://www.connectix.com

Contech
99 Hartford Avenue
Providence, RI 02909
(401) 351-4890   FAX: (401) 421-5072
http://www.iol.ie/~burke/contech.html

Continental Laboratory Products
5648 Copley Drive
San Diego, CA 92111
(800) 456-7741   FAX: (858) 279-5465
(858) 279-5000
http://www.conlab.com

ConvaTec
Professional Services
P.O. Box 5254
Princeton, NJ 08543
(800) 422-8811
http://www.convatec.com

Cooper Instruments & Systems
P.O. Box 3048
Warrenton, VA 20188
(800) 344-3921   FAX: (540) 347-4755
(540) 349-4746
http://www.cooperinstruments.com

Cora Styles Needles 'N Blocks
56 Milton Street
Arlington, MA 02474
(781) 648-6289   FAX: (781) 641-7917

Coriell Cell Repository (CCR)
See Coriell Institute for Medical Research

Coriell Institute for Medical Research
Human Genetic Mutant Repository
401 Haddon Avenue
Camden, NJ 08103
(856) 966-7377   FAX: (856) 964-0254
http://arginine.umdnj.edu

Corion
8 East Forge Parkway
Franklin, MA 02038
(508) 528-4411   FAX: (508) 520-7583
(800) 598-6783
http://www.corion.com

Corning and Corning Science Products
P.O. Box 5000
Corning, NY 14831
(800) 222-7740   FAX: (607) 974-0345
(607) 974-9000
http://www.corning.com

Costar
See Corning

Coulbourn Instruments
7462 Penn Drive
Allentown, PA 18106
(800) 424-3771   FAX: (610) 391-1333
(610) 395-3771
http://www.coulbourninst.com

Coulter Cytometry
See Beckman Coulter

Covance Research Products
465 Swampbridge Road
Denver, PA 17517
(800) 345-4114   FAX: (717) 336-5344
(717) 336-4921
http://www.covance.com

Coy Laboratory Products
14500 Coy Drive
Grass Lake, MI 49240
(734) 475-2200   FAX: (734) 475-1846
http://www.coylab.com

CPG
3 Borinski Road
Lincoln Park, NJ 07035
(800) 362-2740   FAX: (973) 305-0884
(973) 305-8181

CPL Scientific
43 Kingfisher Court
Hambridge Road
Newbury RG14 5SJ, UK
(44) 1635-574902
FAX: (44) 1635-529322
http://www.cplscientific.co.uk

CraMar Technologies
8670 Wolff Court, #160
Westminster, CO 80030
(800) 4-TOMTEC
http://www.cramar.com

Crescent Chemical
1324 Motor Parkway
Hauppauge, NY 11788
(800) 877-3225   FAX: (631) 348-0913
(631) 348-0333
http://www.creschem.com

Crist Instrument
P.O. Box 128
10200 Moxley Road
Damascus, MD 20872
(301) 253-2184   FAX: (301) 253-0069
http://www.cristinstrument.com

Cruachem
See Annovis
http://www.cruachem.com

CS Bio
1300 Industrial Road
San Carlos, CA 94070
(800) 627-2461   FAX: (415) 802-0944
(415) 802-0880
http://www.csbio.com

CS-Chromatographie Service
Am Parir 27
D-52379 Langerwehe, Germany
(49) 2423-40493-0
FAX: (49) 2423-40493-49
http://www.cs-chromatographie.de

Cuno
400 Research Parkway
Meriden, CT 06450
(800) 231-2259   FAX: (203) 238-8716
(203) 237-5541
http://www.cuno.com

Curtin Matheson Scientific
9999 Veterans Memorial Drive
Houston, TX 77038
(800) 392-3353   FAX: (713) 878-3598
(713) 878-3500

CWE
124 Sibley Avenue
Ardmore, PA 19003
(610) 642-7719   FAX: (610) 642-1532
http://www.cwe-inc.com

Cybex Computer Products
4991 Corporate Drive
Huntsville, AL 35805
(800) 932-9239   FAX: (800) 462-9239
http://www.cybex.com

Cygnus
400 Penobscot Drive
Redwood City, CA
(650) 369-4300   FAX: (650) 599-2503
http://www.cygn.com/homepage.html

Cygnus Technology
P.O. Box 219
Delaware Water Gap, PA 18327
(570) 424-5701   FAX: (570) 424-5630
http://www.cygnustech.com

Cymbus Biotechnology
Eagle Class, Chandler's Ford
Hampshire SO53 4NF, UK
(44) 1-703-267-676
FAX: (44) 1-703-267-677
http://www.biotech.cymbus.com

Cytogen
600 College Road East
Princeton, NJ 08540
(609) 987-8200   FAX: (609) 987-6450
http://www.cytogen.com

Cytogen Research and Development
89 Bellevue Hill Road
Boston, MA 02132
(617) 325-7774   FAX: (617) 327-2405

CytRx
154 Technology Parkway
Norcross, GA 30092
(800) 345-2987   FAX: (770) 368-0622
(770) 368-9500
http://www.cytrx.com

Dade Behring
Corporate Headquarters
1717 Deerfield Road
Deerfield, IL 60015
(847) 267-5300   FAX: (847) 267-1066
http://www.dadebehring.com

Dagan
2855 Park Avenue
Minneapolis, MN 55407
(612) 827-5959   FAX: (612) 827-6535
http://www.dagan.com

Dako
6392 Via Real
Carpinteria, CA 93013
(800) 235-5763   FAX: (805) 566-6688
(805) 566-6655
http://www.dakousa.com

Dako A/S
42 Produktionsvej
P.O. Box 1359
DK-2600 Glostrup, Denmark
(45) 4492-0044   FAX: (45) 4284-1822

Dakopatts
See Dako A/S

Damon, IEC
See Thermoquest

Dan Kar Scientific
150 West Street
Wilmington, MA 01887
(800) 942-5542   FAX: (978) 658-0380
(978) 988-9696
http://www.dan-kar.com

DataCell
Falcon Business Park
40 Ivanhoe Road
Finchampstead, Berkshire
RG40 4QQ, UK
(44) 1189 324324
FAX: (44) 1189 324325
http://www.datacell.co.uk
In the US:
(408) 446-3575   FAX: (408) 446-3589
http://www.datacell.com

DataWave Technologies
380 Main Street, Suite 209
Longmont, CO 80501
(800) 736-9283   FAX: (303) 776-8531
(303) 776-8214

Datex-Ohmeda
3030 Ohmeda Drive
Madison, WI 53718
(800) 345-2700   FAX: (608) 222-9147
(608) 221-1551
http://www.us.datex-ohmeda.com

DATU
82 State Street
Geneva, NY 14456
(315) 787-2240   FAX: (315) 787-2397
http://www.nysaes.cornell.edu/datu

David Kopf Instruments
7324 Elmo Street
P.O. Box 636
Tujunga, CA 91043
(818) 352-3274   FAX: (818) 352-3139

Decagon Devices
P.O. Box 835
950 NE Nelson Court
Pullman, WA 99163
(800) 755-2751   FAX: (509) 332-5158
(509) 332-2756
http://www.decagon.com

Decon Labs
890 Country Line Road
Bryn Mawr, PA 19010
(800) 332-6647   FAX: (610) 964-0650
(610) 520-0610
http://www.deconlabs.com

Decon Laboratories
Conway Street
Hove, Sussex BN3 3LY, UK
(44) 1273 739241
FAX: (44) 1273 722088

Degussa
Precious Metals Division
3900 South Clinton Avenue
South Plainfield, NJ 07080
(800) DEGUSSA   FAX: (908) 756-7176
(908) 561-1100
http://www.degussa-huls.com

Deneba Software
1150 NW 72nd Avenue
Miami, FL 33126
(305) 596-5644   FAX: (305) 273-9069
http://www.deneba.com

Deseret Medical
524 West 3615 South
Salt Lake City, UT 84115
(801) 270-8440   FAX: (801) 293-9000

Devcon Plexus
30 Endicott Street
Danvers, MA 01923
(800) 626-7226   FAX: (978) 774-0516
(978) 777-1100
http://www.devcon.com

Developmental Studies Hybridoma Bank
University of Iowa
436 Biology Building
Iowa City, IA 52242
(319) 335-3826   FAX: (319) 335-2077
http://www.uiowa.edu/~dshbwww

DeVilbiss
Division of Sunrise Medical Respiratory
100 DeVilbiss Drive
P.O. Box 635 Somerset, PA 15501
(800) 338-1988   FAX: (814) 443-7572
(814) 443-4881
http://www.sunrisemedical.com

Dharmacon Research
3200 Valmont Road, #5
Boulder, CO 80301
(800) 235-9880   FAX: (303) 415-9879
(303) 415-9880
http://www.dharmacon.com

DiaCheM
Triangle Biomedical
Gardiners Place
West Gillibrands, Lancashire
WN8 9SP, UK
(44) 1695-555581
FAX: (44) 1695-555518
http://www.diachem.co.uk

Diagen
Max-Volmer Strasse 4
D-40724 Hilden, Germany
(49) 2103-892-230
FAX: (49) 2103-892-222

Diagnostic Concepts
6104 Madison Court
Morton Grove, IL 60053
(847) 604-0957

Diagnostic Developments
See DiaCheM

Diagnostic Instruments
6540 Burroughs
Sterling Heights, MI 48314
(810) 731-6000   FAX: (810) 731-6469
http://www.diaginc.com

Diamedix
2140 North Miami Avenue
Miami, FL 33127
(800) 327-4565   FAX: (305) 324-2395
(305) 324-2300

DiaSorin
1990 Industrial Boulevard
Stillwater, MN 55082
(800) 328-1482   FAX: (651) 779-7847
(651) 439-9719
http://www.diasorin.com

Diatome US
321 Morris Road
Fort Washington, PA 19034
(800) 523-5874   FAX: (215) 646-8931
(215) 646-1478
http://www.emsdiasum.com

Difco Laboratories
See Becton Dickinson

Digene
1201 Clopper Road
Gaithersburg, MD 20878
(301) 944-7000   (800) 344-3631
FAX: (301) 944-7121
www.digene.com

Digi-Key
701 Brooks Avenue South
Thief River Falls, MN 56701
(800) 344-4539   FAX: (218) 681-3380
(218) 681-6674
http://www.digi-key.com

Digitimer
37 Hydeway
Welwyn Garden City, Hertfordshire
AL7 3BE, UK
(44) 1707-328347
FAX: (44) 1707-373153
http://www.digitimer.com

Dimco-Gray
8200 South Suburban Road
Dayton, OH 45458
(800) 876-8353   FAX: (937) 433-0520
(937) 433-7600
http://www.dimco-gray.com

Dionex
1228 Titan Way
P.O. Box 3603
Sunnyvale, CA 94088
(408) 737-0700   FAX: (408) 730-9403
http://dionex2.promptu.com

Display Systems Biotech
1260 Liberty Way, Suite B
Vista, CA 92083
(800) 697-1111   FAX: (760) 599-9930
(760) 599-0598
http://www.displaysystems.com

Diversified Biotech
1208 VFW Parkway
Boston, MA 02132
(617) 965-8557   FAX: (617) 323-5641
(800) 796-9199
http://www.divbio.com

DNA ProScan
P.O. Box 121585
Nashville, TN 37212
(800) 841-4362   FAX: (615) 292-1436
(615) 298-3524
http://www.dnapro.com

DNAStar
1228 South Park Street
Madison, WI 53715
(608) 258-7420   FAX: (608) 258-7439
http://www.dnastar.com

DNAVIEW
Attn: Charles Brenner
http://www.wco.com
~cbrenner/dnaview.htm

Doall NYC
36-06 48th Avenue
Long Island City, NY 11101
(718) 392-4595   FAX: (718) 392-6115
http://www.doall.com

Dojindo Molecular Technologies
211 Perry Street Parkway, Suite 5
Gaitherbusburg, MD 20877
(877) 987-2667
http://www.dojindo.com

Dolla Eastern
See Doall NYC

Dolan Jenner Industries
678 Andover Street
Lawrence, MA 08143
(978) 681-8000   (978) 682-2500
http://www.dolan-jenner.com

Dow Chemical
Customer Service Center
2040 Willard H. Dow Center
Midland, MI 48674
(800) 232-2436   FAX: (517) 832-1190
(409) 238-9321
http://www.dow.com

Dow Corning
Northern Europe
Meriden Business Park
Copse Drive
Allesley, Coventry CV5 9RG, UK
(44) 1676 528 000
FAX: (44) 1676 528 001

Dow Corning
P.O. Box 994
Midland, MI 48686
(517) 496-4000
http://www.dowcorning.com

Dow Corning (Lubricants)
2200 West Salzburg Road
Auburn, MI 48611
(800) 248-2481   FAX: (517) 496-6974
(517) 496-6000

Dremel
4915 21st Street
Racine, WI 53406
(414) 554-1390
http://www.dremel.com

Drummond Scientific
500 Parkway
P.O. Box 700
Broomall, PA 19008
(800) 523-7480   FAX: (610) 353-6204
(610) 353-0200
http://www.drummondsci.com

Duchefa Biochemie BV
P.O. Box 2281
2002 CG Haarlem, The Netherlands
31-0-23-5319093
FAX: 31-0-23-5318027
http://www.duchefa.com

Duke Scientific
2463 Faber Place
Palo Alto, CA 94303
(800) 334-3883   FAX: (650) 424-1158
(650) 424-1177
http://www.dukescientific.com

DuPont Biotechnology Systems
See NEN Life Science Products

DuPont Medical Products
See NEN Life Science Products

DuPont Merck Pharmaceuticals
331 Treble Cove Road
Billerica, MA 01862
(800) 225-1572   FAX: (508) 436-7501
http://www.dupontmerck.com

DuPont NEN Products
See NEN Life Science Products

Dyets
2508 Easton Avenue
P.O. Box 3485
Bethlehem, PA 18017
(800) 275-3938   FAX: (800) 329-3938

Dynal
5 Delaware Drive
Lake Success, NY 11042
(800) 638-9416   FAX: (516) 326-3298
(516) 326-3270
http://www.dynal.net

Dynal AS
Ullernchausen 52,
0379 Oslo, Norway
47-22-06-10-00   FAX: 47-22-50-70-15
http://www.dynal.no

Dynalab
P.O. Box 112
Rochester, NY 14692
(800) 828-6595   FAX: (716) 334-9496
(716) 334-2060
http://www.dynalab.com

Dynarex
1 International Boulevard
Brewster, NY 10509
(888) DYNAREX   FAX: (914) 279-9601
(914) 279-9600
http://www.dynarex.com

Dynatech
See Dynex Technologies

Dynex Technologies
14340 Sullyfield Circle
Chantilly, VA 22021
(800) 336-4543   FAX: (703) 631-7816
(703) 631-7800
http://www.dynextechnologies.com

Dyno Mill
See Willy A. Bachofen

E.S.A.
22 Alpha Road
Chelmsford, MA 01824
(508) 250-7000   FAX: (508) 250-7090

E.W. Wright
760 Durham Road
Guilford, CT 06437
(203) 453-6410   FAX: (203) 458-6901
http://www.ewwright.com

E-Y Laboratories
107 N. Amphlett Boulevard
San Mateo, CA 94401
(800) 821-0044   FAX: (650) 342-2648
(650) 342-3296
http://www.eylabs.com

Eastman Kodak
1001 Lee Road
Rochester, NY 14650
(800) 225-5352   FAX: (800) 879-4979
(716) 722-5780   FAX: (716) 477-8040
http://www.kodak.com

ECACC
See European Collection of Animal Cell Cultures

EC Apparatus
See Savant/EC Apparatus

Ecogen, SRL
Gensura Laboratories
Ptge. Dos de Maig
9(08041) Barcelona (34) 3-450-2601
FAX: (34) 3-456-0607
http://www.ecogen.com

Ecolab
370 North Wabasha Street
St. Paul, MN 55102
(800) 35-CLEAN   FAX: (651) 225-3098
(651) 352-5326
http://www.ecolab.com

ECO PHYSICS
3915 Research Park Drive, Suite A-3
Ann Arbor, MI 48108
(734) 998-1600   FAX: (734) 998-1180
http://www.ecophysics.com

Edge Biosystems
19208 Orbit Drive
Gaithersburg, MD 20879-4149
(800) 326-2685   FAX: (301) 990-0881
(301) 990-2685
http://www.edgebio.com

Edmund Scientific
101 E. Gloucester Pike
Barrington, NJ 08007
(800) 728-6999   FAX: (856) 573-6263
(856) 573-6250
http://www.edsci.com

EG&G
See Perkin-Elmer

Ekagen
969 C Industry Road
San Carlos, CA 94070
(650) 592-4500   FAX: (650) 592-4500

Elcatech
P.O. Box 10935
Winston-Salem, NC 27108
(336) 544-8613   FAX: (336) 777-3623
(910) 777-3624
http://www.elcatech.com

Electron Microscopy Sciences
321 Morris Road
Fort Washington, PA 19034
(800) 523-5874   FAX: (215) 646-8931
(215) 646-1566
http://www.emsdiasum.com

Electron Tubes
100 Forge Way, Unit F
Rockaway, NJ 07866
(800) 521-8382   FAX: (973) 586-9771
(973) 586-9594
http://www.electrontubes.com

Elicay Laboratory Products, (UK) Ltd.
4 Manborough Mews
Crockford Lane
Basingstoke, Hampshire
RG 248NA, England
(256) 811-118   FAX: (256) 811-116
http://www.elkay-uk.co.uk

Eli Lilly
Lilly Corporate Center
Indianapolis, IN 46285
(800) 545-5979   FAX: (317) 276-2095
(317) 276-2000
http://www.lilly.com

ELISA Technologies
See Neogen

Elkins-Sinn
See Wyeth-Ayerst

EMBI
See European Bioinformatics Institute

EM Science
480 Democrat Road
Gibbstown, NJ 08027
(800) 222-0342   FAX: (856) 423-4389
(856) 423-6300
http://www.emscience.com

EM Separations Technology
See R & S Technology

Endogen
30 Commerce Way
Woburn, MA 01801
(800) 487-4885   FAX: (617) 439-0355
(781) 937-0890
http://www.endogen.com

ENGEL-Loter
HSGM Heatcutting Equipment
& Machines
1865 E. Main Street, No. 5
Duncan, SC 29334
(888) 854-HSGM   FAX: (864) 486-8383
(864) 486-8300
http://www.engelgmbh.com

Enzo Diagnostics
60 Executive Boulevard
Farmingdale, NY 11735
(800) 221-7705   FAX: (516) 694-7501
(516) 694-7070
http://www.enzo.com

Enzogenetics
4197 NW Douglas Avenue
Corvallis, OR 97330
(541) 757-0288

The Enzyme Center
See Charm Sciences

Enzyme Systems Products
486 Lindbergh Avenue
Livermore, CA 94550
(888) 449-2664   FAX: (925) 449-1866
(925) 449-2664
http://www.enzymesys.com

Epicentre Technologies
1402 Emil Street
Madison, WI 53713
(800) 284-8474   FAX: (608) 258-3088
(608) 258-3080
http://www.epicentre.com

Erie Scientific
20 Post Road
Portsmouth, NH 03801
(888) ERIE-SCI   FAX: (603) 431-8996
(603) 431-8410
http://www.eriesci.com

ES Industries
701 South Route 73
West Berlin, NJ 08091
(800) 356-6140   FAX: (856) 753-8484
(856) 753-8400
http://www.esind.com

ESA
22 Alpha Road
Chelmsford, MA 01824
(800) 959-5095   FAX: (978) 250-7090
(978) 250-7000
http://www.esainc.com

Ethicon
Route 22, P.O. Box 151
Somerville, NJ 08876
(908) 218-0707
http://www.ethiconinc.com

Ethicon Endo-Surgery
4545 Creek Road
Cincinnati, OH 45242
(800) 766-9534   FAX: (513) 786-7080

Eurogentec
Parc Scientifique du Sart Tilman
4102 Seraing, Belgium
32-4-240-76-76   FAX: 32-4-264-07-88
http://www.eurogentec.com

European Bioinformatics Institute
Wellcome Trust Genomes Campus
Hinxton, Cambridge CB10 1SD, UK
(44) 1223-49444
FAX: (44) 1223-494468

European Collection of Animal Cell Cultures (ECACC)
Centre for Applied Microbiology &
Research
Salisbury, Wiltshire SP4 0JG, UK
(44) 1980-612 512
FAX: (44) 1980-611 315
http://www.camr.org.uk

Evergreen Scientific
2254 E. 49th Street
P.O. Box 58248
Los Angeles, CA 90058
(800) 421-6261   FAX: (323) 581-2503
(323) 583-1331
http://www.evergreensci.com

Exalpha Biologicals
20 Hampden Street
Boston, MA 02205
(800) 395-1137   FAX: (617) 969-3872
(617) 558-3625
http://www.exalpha.com

Exciton
P.O. Box 31126
Dayton, OH 45437
(937) 252-2989   FAX: (937) 258-3937
http://www.exciton.com

Extrasynthese
ZI Lyon Nord
SA-BP62
69730 Genay, France
(33) 78-98-20-34
FAX: (33) 78-98-19-45

Factor II
1972 Forest Avenue
P.O. Box 1339
Lakeside, AZ 85929
(800) 332-8688   FAX: (520) 537-8066
(520) 537-8387
http://www.factor2.com

Falcon
See Becton Dickinson Labware

Fenwal
See Baxter Healthcare

Filemaker
5201 Patrick Henry Drive
Santa Clara, CA 95054
(408) 987-7000   (800) 325-2747

Fine Science Tools
202-277 Mountain Highway
North Vancouver, British Columbia
V7J 3P2 Canada
(800) 665-5355   FAX: (800) 665 4544
(604) 980-2481   FAX: (604) 987-3299

Fine Science Tools
373-G Vintage Park Drive
Foster City, CA 94404
(800) 521-2109   FAX: (800) 523-2109
(650) 349-1636   FAX: (630) 349-3729

Fine Science Tools
Fahrtgasse 7-13
D-69117 Heidelberg, Germany
(49) 6221 905050
FAX: (49) 6221 600001
http://www.finescience.com

Finn Aqua
AMSCO Finn Aqua Oy
Teollisuustiez, FIN-04300
Tuusula, Finland
358 025851   FAX: 358 0276019

Finnigan
355 River Oaks Parkway
San Jose, CA 95134
(408) 433-4800   FAX: (408) 433-4821
http://www.finnigan.com

Fisher Chemical Company
Fisher Scientific Limited
112 Colonnade Road Nepean
Ontario K2E 7L6, Canada
(800) 234-7437   FAX: (800) 463-2996
http://www.fisherscientific.com

W.F. Fisher & Son
220 Evans Way, Suite #1
Somerville, NJ 08876
(908) 707-4050   FAX: (908) 707-4099

Fisher Scientific
2000 Park Lane
Pittsburgh, PA 15275
(800) 766-7000   FAX: (800) 926-1166
(412) 562-8300
http://www3.fishersci.com

Fitzco
5600 Pioneer Creek Drive
Maple Plain, MN 55359
(800) 367-8760   FAX: (612) 479-2880
(612) 479-3489
http://www.fitzco.com

5 Prime → 3 Prime
See 2000 Eppendorf-5 Prime
http://www.5prime.com

Fleisch (Rusch)
2450 Meadowbrook Parkway
Duluth, GA 30096
(770) 623-0816   FAX: (770) 623-1829
http://ruschinc.com

Flow Cytometry Standards
P.O. Box 194344
San Juan, PR 00919
(800) 227-8143   FAX: (787) 758-3267
(787) 753-9341
http://www.fcstd.com

Flow Labs
See ICN Biomedicals

Flow-Tech Supply
P.O. Box 1388
Orange, TX 77631
(409) 882-0306   FAX: (409) 882-0254
http://www.flow-tech.com

Fluid Marketing
See Fluid Metering

Fluid Metering
5 Aerial Way, Suite 500
Sayosett, NY 11791
(516) 922-6050   FAX: (516) 624-8261
http://www.fmipump.com

Fluorochrome
1801 Williams, Suite 300
Denver, CO 80264
(303) 394-1000   FAX: (303) 321-1119

Fluka Chemical
See Sigma-Aldrich

FMC BioPolymer
1735 Market Street
Philadelphia, PA 19103
(215) 299-6000   FAX: (215) 299-5809
http://www.fmc.com

FMC BioProducts
191 Thomaston Street
Rockland, ME 04841
(800) 521-0390   FAX: (800) 362-1133
(207) 594-3400   FAX: (207) 594-3426
http://www.bioproducts.com

Forma Scientific
Milcreek Road
P.O. Box 649
Marietta, OH 45750
(800) 848-3080   FAX: (740) 372-6770
(740) 373-4765
http://www.forma.com

Fort Dodge Animal Health
800 5th Street NW
Fort Dodge, IA 50501
(800) 685-5656   FAX: (515) 955-9193
(515) 955-4600
http://www.ahp.com

Fotodyne
950 Walnut Ridge Drive
Hartland, WI 53029
(800) 362-3686   FAX: (800) 362-3642
(262) 369-7000   FAX: (262) 369-7013
http://www.fotodyne.com

Fresenius HemoCare
6675 185th Avenue NE, Suite 100
Redwood, WA 98052
(800) 909-3872
(425) 497-1197
http://www.freseniusht.com

Fresenius Hemotechnology
See Fresenius HemoCare

Fuji Medical Systems
419 West Avenue
P.O. Box 120035
Stamford, CT 06902
(800) 431-1850   FAX: (203) 353-0926
(203) 324-2000
http://www.fujimed.com

Fujisawa USA
Parkway Center North
Deerfield, IL 60015-2548
(847) 317-1088   FAX: (847) 317-7298

Ernest F. Fullam
900 Albany Shaker Road
Latham, NY 12110
(800) 833-4024   FAX: (518) 785-8647
(518) 785-5533
http://www.fullam.com

Gallard-Schlesinger Industries
777 Zechendorf Boulevard Garden City,
NY 11530
(516) 229-4000   FAX: (516) 229-4015
http://www.gallard-schlessinger.com

Gambro
Box 7373
SE 103 91 Stockholm, Sweden
(46) 8 613 65 00
FAX: (46) 8 611 37 31
In the US: COBE Laboratories
225 Union Boulevard
Lakewood, CO 80215
(303) 232-6800   FAX: (303) 231-4915
http://www.gambro.com

Garner Glass
177 Indian Hill Boulevard
Claremont, CA 91711
(909) 624-5071   FAX: (909) 625-0173
http://www.garnerglass.com

Garon Plastics
16 Byre Avenue
Somerton Park, South Australia 5044
(08) 8294-5126   FAX: (08) 8376-1487
http://www.apache.airnet.com.au/~garon

Garren Scientific
9400 Lurline Avenue, Unit E
Chatsworth, CA 91311
(800) 342-3725   FAX: (818) 882-3229
(818) 882-6544
http://www.garren-scientific.com

GATC Biotech AG
Jakob-Stadler-Platz 7
D-78467 Constance, Germany
(49) 07531-8160-0
FAX: (49) 07531-8160-81
http://www.gatc-biotech.com

Gaussian
Carnegie Office Park
Building 6, Suite 230
Carnegie, PA 15106
(412) 279-6700   FAX: (412) 279-2118
http://www.gaussian.com

G.C. Electronics/A.R.C. Electronics
431 Second Street
Henderson, KY 42420
(270) 827-8981   FAX: (270) 827-8256
http://www.arcelectronics.com

GDB (Genome Data Base, Curation)
2024 East Monument Street, Suite 1200
Baltimore, MD 21205
(410) 955-9705   FAX: (410) 614-0434
http://www.gdb.org

GDB (Genome Data Base, Home)
Hospital for Sick Children
555 University Avenue
Toronto, Ontario
M5G 1X8 Canada
(416) 813-8744   FAX: (416) 813-8755
http://www.gdb.org

Gelman Sciences
See Pall-Gelman

Gemini BioProducts
5115-M Douglas Fir Road
Calabasas, CA 90403
(818) 591-3530   FAX: (818) 591-7084

Gen Trak
5100 Campus Drive
Plymouth Meeting, PA 19462
(800) 221-7407   FAX: (215) 941-9498
(215) 825-5115
http://www.informagen.com

Genaissance Pharmaceuticals
5 Science Park
New Haven, CT 06511
(800) 678-9487   FAX: (203) 562-9377
(203) 773-1450
http://www.genaissance.com

GENAXIS Biotechnology
Parc Technologique
10 Avenue Ampère
Montigny le Bretoneux
78180 France
(33) 01-30-14-00-20
FAX: (33) 01-30-14-00-15
http://www.genaxis.com

GenBank
National Center for Biotechnology Information
National Library of Medicine/NIH
Building 38A, Room 8N805
8600 Rockville Pike
Bethesda, MD 20894
(301) 496-2475   FAX: (301) 480-9241
http://www.ncbi.nlm.nih.gov

Gene Codes
640 Avis Drive
Ann Arbor, MI 48108
(800) 497-4939   FAX: (734) 930-0145
(734) 769-7249
http://www.genecodes.com

Genemachines
935 Washington Street
San Carlos, CA 94070
(650) 508-1634   FAX: (650) 508-1644
(877) 855-4363
http://www.genemachines.com

Genentech
1 DNA Way
South San Francisco, CA 94080
(800) 551-2231   FAX: (650) 225-1600
(650) 225-1000
http://www.gene.com

General Scanning/GSI Luminomics
500 Arsenal Street
Watertown, MA 02172
(617) 924-1010   FAX: (617) 924-7327
http://www.genescan.com

General Valve
Division of Parker Hannifin Pneutronics
19 Gloria Lane
Fairfield, NJ 07004
(800) GVC-VALV
FAX: (800) GVC-1-FAX
http://www.pneutronics.com

Genespan
19310 North Creek Parkway, Suite 100
Bothell, WA 98011
(800) 231-2215   FAX: (425) 482-3005
(425) 482-3003
http://www.genespan.com

Généthon Human Genome Research Center
1 bis rue de l'Internationale
91000 Evry, France
(33) 169-472828
FAX: (33) 607-78698
http://www.genethon.fr

Genetic Microsystems
34 Commerce Way
Wobum, MA 01801
(781) 932-9333   FAX: (781) 932-9433
http://www.genticmicro.com

Genetic Mutant Repository
See Coriell Institute for Medical Research

Genetic Research Instrumentation
Gene House
Queenborough Lane
Rayne, Braintree, Essex CM7 8TF, UK
(44) 1376 332900
FAX: (44) 1376 344724
http://www.gri.co.uk

Genetics Computer Group
575 Science Drive
Madison, WI 53711
(608) 231-5200   FAX: (608) 231-5202
http://www.gcg.com

Genetics Institute/American Home Products
87 Cambridge Park Drive
Cambridge, MA 02140
(617) 876-1170   FAX: (617) 876-0388
http://www.genetics.com

Genetix
63-69 Somerford Road
Christchurch, Dorset BH23 3QA, UK
(44) (0) 1202 483900
FAX: (44)(0) 1202 480289
In the US: (877) 436 3849
US FAX: (888) 522 7499
http://www.genetix.co.uk

Gene Tools
One Summerton Way
Philomath, OR 97370
(541) 9292-7840   FAX: (541) 9292-7841
http://www.gene-tools.com

GeneWorks
P.O. Box 11, Rundle Mall
Adelaide, South Australia 5000, Australia
1800 882 555   FAX: (08) 8234 2699
(08) 8234 2644
http://www.geneworks.com

Genome Systems (INCYTE)
4633 World Parkway Circle
St. Louis, MO 63134
(800) 430-0030   FAX: (314) 427-3324
(314) 427-3222
http://www.genomesystems.com

Genomic Solutions
4355 Varsity Drive, Suite E
Ann Arbor, MI 48108
(877) GENOMIC   FAX: (734) 975-4808
(734) 975-4800
http://www.genomicsolutions.com

Genomyx
See Beckman Coulter

Genosys Biotechnologies
1442 Lake Front Circle, Suite 185
The Woodlands, TX 77380
(281) 363-3693   FAX: (281) 363-2212
http://www.genosys.com

Genotech
92 Weldon Parkway
St. Louis, MO 63043
(800) 628-7730   FAX: (314) 991-1504
(314) 991-6034

GENSET
876 Prospect Street, Suite 206
La Jolla, CA 92037
(800) 551-5291   FAX: (619) 551-2041
(619) 515-3061
http://www.genset.fr

Gensia Laboratories Ltd.
19 Hughes
Irvine, CA 92718
(714) 455-4700   FAX: (714) 855-8210

Genta
99 Hayden Avenue, Suite 200
Lexington, MA 02421
(781) 860-5150   FAX: (781) 860-5137
http://www.genta.com

GENTEST
6 Henshaw Street
Woburn, MA 01801
(800) 334-5229   FAX: (888) 242-2226
(781) 935-5115   FAX: (781) 932-6855
http://www.gentest.com

Gentra Systems
15200 25th Avenue N., Suite 104
Minneapolis, MN 55447
(800) 866-3039   FAX: (612) 476-5850
(612) 476-5858
http://www.gentra.com

Genzyme
1 Kendall Square
Cambridge, MA 02139
(617) 252-7500   FAX: (617) 252-7600
http://www.genzyme.com
See also R&D Systems

Genzyme Genetics
One Mountain Road
Framingham, MA 01701
(800) 255-7357   FAX: (508) 872-9080
(508) 872-8400
http://www.genzyme.com

George Tiemann & Co.
25 Plant Avenue
Hauppauge, NY 11788
(516) 273-0005   FAX: (516) 273-6199

GIBCO/BRL
A Division of Life Technologies
1 Kendall Square
Grand Island, NY 14072
(800) 874-4226   FAX: (800) 352-1968
(716) 774-6700
http://www.lifetech.com

Gilmont Instruments
A Division of Barnant Company
28N092 Commercial Avenue
Barrington, IL 60010
(800) 637-3739   FAX: (708) 381-7053
http://barnant.com

Gilson
3000 West Beltline Highway
P.O. Box 620027
Middletown, WI 53562
(800) 445-7661
(608) 836-1551
http://www.gilson.com

Glas-Col Apparatus
P.O. Box 2128
Terre Haute, IN 47802
(800) Glas-Col   FAX: (812) 234-6975
(812) 235-6167
http://www.glascol.com

Glaxo Wellcome
Five Moore Drive
Research Triangle Park, NC 27709
(800) SGL-AXO5   FAX: (919) 248-2386
(919) 248-2100
http://www.glaxowellcome.com

Glen Mills
395 Allwood Road
Clifton, NJ 07012
(973) 777-0777   FAX: (973) 777-0070
http://www.glenmills.com

Glen Research
22825 Davis Drive
Sterling, VA 20166
(800) 327-4536   FAX: (800) 934-2490
(703) 437-6191   FAX: (703) 435-9774
http://www.glenresearch.com

Glo Germ
P.O. Box 189
Moab, UT 84532
(800) 842-6622   FAX: (435) 259-5930
http://www.glogerm.com

Glyco
11 Pimentel Court
Novato, CA 94949
(800) 722-2597   FAX: (415) 382-3511
(415) 884-6799
http://www.glyco.com

Gould Instrument Systems
8333 Rockside Road
Valley View, OH 44125
(216) 328-7000   FAX: (216) 328-7400
http://www.gould13.com

Gralab Instruments
See Dimco-Gray

GraphPad Software
5755 Oberlin Drive #110
San Diego, CA 92121
(800) 388-4723   FAX: (558) 457-8141
(558) 457-3909
http://www.graphpad.com

Graseby Anderson
See Andersen Instruments
http://www.graseby.com

Grass Instrument
A Division of Astro-Med
600 East Greenwich Avenue
W. Warwick, RI 02893
(800) 225-5167   FAX: (877) 472-7749
http://www.grassinstruments.com

Greenacre and Misac Instruments
Misac Systems
27 Port Wood Road
Ware, Hertfordshire SF12 9NJ, UK
(44) 1920 463017
FAX: (44) 1920 465136

Greer Labs
639 Nuway Circle
Lenois, NC 28645
(704) 754-5237
http://greerlabs.com

Greiner
Maybachestrasse 2
Postfach 1162
D-7443 Frickenhausen, Germany
(49) 0 91 31/80 79 0
FAX: (49) 0 91 31/80 79 30
http://www.erlangen.com/greiner

GSI Lumonics
130 Lombard Street Oxnard, CA 93030
(805) 485-5559   FAX: (805) 485-3310
http://www.gsilumonics.com

GTE Internetworking
150 Cambridge Park Drive
Cambridge, MA 02140
(800) 472-4565   FAX: (508) 694-4861
http://www.bbn.com

GW Instruments
35 Medford Street
Somerville, MA 02143
(617) 625-4096   FAX: (617) 625-1322
http://www.gwinst.com

H & H Woodworking
1002 Garfield Street
Denver, CO 80206
(303) 394-3764

Hacker Instruments
17 Sherwood Lane
P.O. Box 10033
Fairfield, NJ 07004
800-442-2537   FAX: (973) 808-8281
(973) 226-8450
http://www.hackerinstruments.com

Haemenetics
400 Wood Road
Braintree, MA 02184
(800) 225-5297   FAX: (781) 848-7921
(781) 848-7100
http://www.haemenetics.com

Halocarbon Products
P.O. Box 661
River Edge, NJ 07661
(201) 242-8899   FAX: (201) 262-0019
http://halocarbon.com

Hamamatsu Photonic Systems
A Division of Hamamatsu
360 Foothill Road
P.O. Box 6910
Bridgewater, NJ 08807
(908) 231-1116   FAX: (908) 231-0852
http://www.photonicsonline.com

Hamilton Company
4970 Energy Way
P.O. Box 10030
Reno, NV 89520
(800) 648-5950   FAX: (775) 856-7259
(775) 858-3000
http://www.hamiltoncompany.com

Hamilton Thorne Biosciences
100 Cummings Center, Suite 102C
Beverly, MA 01915
http://www.hamiltonthorne.com

Hampton Research
27631 El Lazo Road
Laguna Niguel, CA 92677
(800) 452-3899   FAX: (949) 425-1611
(949) 425-6321
http://www.hamptonresearch.com

Harlan Bioproducts for Science
P.O. Box 29176
Indianapolis, IN 46229
(317) 894-7521   FAX: (317) 894-1840
http://www.hbps.com

Harlan Sera-Lab
Hillcrest, Dodgeford Lane
Belton, Loughborough
Leicester LE12 9TE, UK
(44) 1530 222123
FAX: (44) 1530 224970
http://www.harlan.com

Harlan Teklad
P.O. Box 44220
Madison, WI 53744
(608) 277-2070   FAX: (608) 277-2066
http://www.harlan.com

Harrick Scientific Corporation
88 Broadway
Ossining, NY 10562
(914) 762-0020   FAX: (914) 762-0914
http://www.harricksci.com

Harrison Research
840 Moana Court
Palo Alto, CA 94306
(650) 949-1565   FAX: (650) 948-0493

Harvard Apparatus
84 October Hill Road
Holliston, MA 01746
(800) 272-2775   FAX: (508) 429-5732
(508) 893-8999
http://harvardapparatus.com

Harvard Bioscience
See Harvard Apparatus

Haselton Biologics
See JRH Biosciences

Hazelton Research Products
See Covance Research Products

Health Products
See Pierce Chemical

Heat Systems-Ultrasonics
1938 New Highway
Farmingdale, NY 11735
(800) 645-9846   FAX: (516) 694-9412
(516) 694-9555

Heidenhain Corp
333 East State Parkway
Schaumberg, IL 60173
(847) 490-1191   FAX: (847) 490-3931
http://www.heidenhain.com

Hellma Cells
11831 Queens Boulevard
Forest Hills, NY 11375
(718) 544-9166   FAX: (718) 263-6910
http://www.helmaUSA.com

Hellma
Postfach 1163
D-79371 Müllheim/Baden, Germany
(49) 7631-1820
FAX: (49) 7631-13546
http://www.hellma-worldwide.de

Henry Schein
135 Duryea Road, Mail Room 150
Melville, NY 11747
(800) 472-4346   FAX: (516) 843-5652
http://www.henryschein.com

Heraeus Kulzer
4315 South Lafayette Boulevard
South Bend, IN 46614
(800) 343-5336
(219) 291-0661
http://www.kulzer.com

Heraeus Sepatech
See Kendro Laboratory Products

Hercules Aqualon
Aqualon Division
Hercules Research Center, Bldg. 8145
500 Hercules Road
Wilmington, DE 19899
(800) 345-0447   FAX: (302) 995-4787
http://www.herc.com/aqualon/pharma

Heto-Holten A/S
Gydevang 17-19
DK-3450 Allerod, Denmark
(45) 48-16-62-00
FAX: (45) 48-16-62-97
Distributed by ATR

Hettich-Zentrifugen
See Andreas Hettich

Hewlett-Packard
3000 Hanover Street
Mailstop 20B3
Palo Alto, CA 94304
(650) 857-1501   FAX: (650) 857-5518
http://www.hp.com

HGS Hinimoto Plastics
1-10-24 Meguro-Honcho
Megurouko
Tokyo 152, Japan
3-3714-7226   FAX: 3-3714-4657

Hitachi Scientific Instruments
Nissei Sangyo America
8100 N. First Street
San Elsa, CA 95314
(800) 548-9001   FAX: (408) 432-0704
(408) 432-0520
http://www.hii.hitachi.com

Hi-Tech Scientific
Brunel Road
Salisbury, Wiltshire, SP2 7PU
UK
(44) 1722-432320
(800) 344-0724 (US only)
http://www.hi-techsci.co.uk

Hoechst AG
See Aventis Pharmaceutical

Hoefer Scientific Instruments
Division of Amersham-Pharmacia
Biotech
800 Centennial Avenue
Piscataway, NJ 08855 (800) 227-4750
FAX: (877) 295-8102
http://www.apbiotech.com

Hoffman-LaRoche
340 Kingsland Street
Nutley, NJ 07110
(800) 526-0189   FAX: (973) 235-9605
(973) 235-5000
http://www.rocheUSA.com

Holborn Surgical and Medical Instruments
Westwood Industrial Estate
Ramsgate Road Margate, Kent CT9 4JZ
UK
(44) 1843 296666
FAX: (44) 1843 295446

Honeywell
101 Columbia Road
Morristown, NJ 07962
(973) 455-2000   FAX: (973) 455-4807
http://www.honeywell.com

Honeywell Specialty Films
P.O. Box 1039
101 Columbia Road
Morristown, NJ 07962
(800) 934-5679   FAX: (973) 455-6045
http://www.honeywell-specialtyfilms.com

Hood Thermo-Pad Canada
Comp. 20, Site 61A, RR2
Summerland, British Columbia
V0H 1Z0 Canada
(800) 665-9555   FAX: (250) 494-5003
(250) 494-5002
http://www.thermopad.com

Horiba Instruments
17671 Armstrong Avenue
Irvine, CA 92714
(949) 250-4811   FAX: (949) 250-0924
http://www.horiba.com

Hoskins Manufacturing
10776 Hall Road
P.O. Box 218
Hamburg, MI 48139
(810) 231-1900   FAX: (810) 231-4311
http://www.hoskinsmfgco.com

Hosokawa Micron Powder Systems
10 Chatham Road
Summit, NJ 07901
(800) 526-4491   FAX: (908) 273-7432
(908) 273-6360
http://www.hosokawamicron.com

HT Biotechnology
Unit 4
61 Ditton Walk
Cambridge CB5 8QD, UK
(44) 1223-412583

Hugo Sachs Electronik
Postfach 138
7806 March-Hugstetten, Germany
D-79229(49) 7665-92000
FAX: (49) 7665-920090

Human Biologics International
7150 East Camelback Road, Suite 245
Scottsdale, AZ 85251
(480) 990-2005   FAX: (480)-990-2155
http://www.humanbiological.com

Human Genetic Mutant Cell Repository
See Coriell Institute for Medical Research

HVS Image
P.O. Box 100
Hampton, Middlesex TW12 2YD, UK
FAX: (44) 208 783 1223
In the US: (800) 225-9261
FAX: (888) 483-8033
http://www.hvsimage.com

Hybaid
111-113 Waldegrave Road
Teddington, Middlesex TW11 8LL, UK
(44) 0 1784 42500
FAX: (44) 0 1784 248085
http://www.hybaid.co.uk

Hybaid Instruments
8 East Forge Parkway
Franklin, MA 02028
(888)4-HYBAID   FAX: (508) 541-3041
(508) 541-6918
http://www.hybaid.com

Hybridon
155 Fortune Boulevard
Milford, MA 01757
(508) 482-7500   FAX: (508) 482-7510
http://www.hybridon.com

HyClone Laboratories
1725 South HyClone Road
Logan, UT 84321
(800) HYCLONE   FAX: (800) 533-9450
(801) 753-4584   FAX: (801) 750-0809
http://www.hyclone.com

Hyseq
670 Almanor Avenue
Sunnyvale, CA 94086
(408) 524-8100   FAX: (408) 524-8141
http://www.hyseq.com

IBF Biotechnics
See Sepracor

IBI (International Biotechnologies)
See Eastman Kodak
For technical service (800) 243-2555
(203) 786-5600

ICN Biochemicals
See ICN Biomedicals

ICN Biomedicals
3300 Hyland Avenue
Costa Mesa, CA 92626
(800) 854-0530   FAX: (800) 334-6999
(714) 545-0100   FAX: (714) 641-7275
http://www.icnbiomed.com

ICN Flow and Pharmaceuticals
See ICN Biomedicals

ICN Immunobiochemicals
See ICN Biomedicals

ICN Radiochemicals
See ICN Biomedicals

ICONIX
100 King Street West, Suite 3825
Toronto, Ontario M5X 1E3 Canada
(416) 410-2411   FAX: (416) 368-3089
http://www.iconix.com

ICRT (Imperial Cancer Research Technology)
Sardinia House
Sardinia Street
London WC2A 3NL, UK
(44) 1712-421136
FAX: (44) 1718-314991

Idea Scientific Company
P.O. Box 13210
Minneapolis, MN 55414
(800) 433-2535   FAX: (612) 331-4217
http://www.ideascientific.com

IEC
See International Equipment Co.

IITC
Life Sciences
23924 Victory Boulevard
Woodland Hills, CA 91367
(888) 414-4482 (818) 710-1556
FAX: (818) 992-5185
http://www.iitcinc.com

IKA Works
2635 N. Chase Parkway, SE
Wilmington, NC 28405
(910) 452-7059   FAX: (910) 452-7693
http://www.ika.net

Ikegami Electronics
37 Brook Avenue
Maywood, NJ 07607
(201) 368-9171   FAX: (201) 569-1626

Ikemoto Scientific Technology
25-11 Hongo
3-chome, Bunkyo-ku
Tokyo 101-0025, Japan
(81) 3-3811-4181
FAX: (81) 3-3811-1960

Imagenetics
See ATC Diagnostics

Imaging Research
c/o Brock University
500 Glenridge Avenue
St. Catharines, Ontario
L2S 3A1 Canada
(905) 688-2040   FAX: (905) 685-5861
http://www.imaging.brocku.ca

Imclone Systems
180 Varick Street
New York, NY 10014
(212) 645-1405   FAX: (212) 645-2054
http://www.imclone.com

IMCO Corporation LTD., AB
P.O. Box 21195
SE-100 31
Stockholm, Sweden
46-8-33-53-09   FAX: 46-8-728-47-76
http://www.imcocorp.se

IMICO
Calle Vivero, No. 5-4a Planta
E-28040, Madrid, Spain
(34) 1-535-3960   FAX: (34) 1-535-2780

Immunex
51 University Street
Seattle, WA 98101
(206) 587-0430   FAX: (206) 587-0606
http://www.immunex.com

Immunotech
130, av. Delattre de Tassigny
B.P. 177
13276 Marseilles Cedex 9
France
(33) 491-17-27-00
FAX: (33) 491-41-43-58
http://www.immunotech.fr

Imperial Chemical Industries
Imperial Chemical House
Millbank, London SW1P 3JF, UK
(44) 171-834-4444
FAX: (44)171-834-2042
http://www.ici.com

Inceltech
See New Brunswick Scientific

Incstar
See DiaSorin

Incyte
6519 Dumbarton Circle
Fremont, CA 94555
(510) 739-2100   FAX: (510) 739-2200
http://www.incyte.com

Incyte Pharmaceuticals
3160 Porter Drive
Palo Alto, CA 94304
(877) 746-2983   FAX: (650) 855-0572
(650) 855-0555
http://www.incyte.com

Individual Monitoring Systems
6310 Harford Road
Baltimore, MD 21214

Indo Fine Chemical
P.O. Box 473
Somerville, NJ 08876
(888) 463-6346   FAX: (908) 359-1179
(908) 359-6778
http://www.indofinechemical.com

Industrial Acoustics
1160 Commerce Avenue
Bronx, NY 10462
(718) 931-8000   FAX: (718) 863-1138
http://www.industrialacoustics.com

Inex Pharmaceuticals
100-8900 Glenlyon Parkway
Glenlyon Business Park
Burnaby, British Columbia
V5J 5J8 Canada
(604) 419-3200   FAX: (604) 419-3201
http://www.inexpharm.com

Ingold, Mettler, Toledo
261 Ballardvale Street
Wilmington, MA 01887
(800) 352-8763   FAX: (978) 658-0020
(978) 658-7615
http://www.mt.com

Innogenetics N.V.
Technologie Park 6
B-9052 Zwijnaarde
Belgium
(32) 9-329-1329   FAX: (32) 9-245-7623
http://www.innogenetics.com

Innovative Medical Services
1725 Gillespie Way
El Cajon, CA 92020
(619) 596-8600   FAX: (619) 596-8700
http://www.imspure.com

Innovative Research
3025 Harbor Lane N, Suite 300
Plymouth, MN 55447
(612) 519-0105   FAX: (612) 519-0239
http://www.inres.com

Innovative Research of America
2 N. Tamiami Trail, Suite 404
Sarasota, FL 34236
(800) 421-8171   FAX: (800) 643-4345
(941) 365-1406   FAX: (941) 365-1703
http://www.innovrsrch.com

Inotech Biosystems
15713 Crabbs Branch Way, #110
Rockville, MD 20855 (800) 635-4070
FAX: (301) 670-2859
(301) 670-2850
http://www.inotechintl.com

INOVISION
22699 Old Canal Road
Yorba Linda, CA 92887
(714) 998-9600   FAX: (714) 998-9666
http://www.inovision.com

Instech Laboratories
5209 Militia Hill Road
Plymouth Meeting, PA 19462
(800) 443-4227   FAX: (610) 941-0134
(610) 941-0132
http://www.instechlabs.com

Instron
100 Royall Street
Canton, MA 02021
(800) 564-8378   FAX: (781) 575-5725
(781) 575-5000
http://www.instron.com

Instrumentarium
P.O. Box 300
00031 Instrumentarium
Helsinki, Finland
(10) 394-5566
http://www.instrumentarium.fi

Instruments SA
Division Jobin Yvon
16-18 Rue du Canal
91165 Longjumeau, Cedex, France
(33)1 6454-1300
FAX: (33)1 6909-9319
http://www.isainc.com

Instrutech
20 Vanderventer Avenue, Suite 101E
Port Washington, NY 11050
(516) 883-1300   FAX: (516) 883-1558
http://www.instrutech.com

Integrated DNA Technologies
1710 Commercial Park
Coralville, Iowa 52241
(800) 328-2661   FAX: (319) 626-8444
http://www.idtdna.com

Integrated Genetics
See Genzyme Genetics

Integrated Scientific Imaging Systems
3463 State Street, Suite 431
Santa Barbara, CA 93105
(805) 692-2390   FAX: (805) 692-2391
http://www.imagingsystems.com

Integrated Separation Systems (ISS)
See OWL Separation Systems

IntelliGenetics
See Oxford Molecular Group

Interactiva BioTechnologie
Sedanstrasse 10
D-89077 Ulm, Germany
(49) 731-93579-290
FAX: (49) 731-93579-291
http://www.interactiva.de

Interchim
213 J.F. Kennedy Avenue
B.P. 1140
Montlucon
03103 France
(33) 04-70-03-83-55
FAX: (33) 04-70-03-93-60

Interfocus
14/15 Spring Rise
Falcover Road
Haverhill, Suffolk CB9 7XU, UK
(44) 1440 703460
FAX: (44) 1440 704397
http://www.interfocus.ltd.uk

Intergen
2 Manhattanville Road
Purchase, NY 10577
(800) 431-4505   FAX: (800) 468-7436
(914) 694-1700   FAX: (914) 694-1429
http://www.intergenco.com

Intermountain Scientific
420 N. Keys Drive
Kaysville, UT 84037
(800) 999-2901   FAX: (800) 574-7892
(801) 547-5047   FAX: (801) 547-5051
http://www.bioexpress.com

International Biotechnologies (IBI)
See Eastman Kodak

International Equipment Co. (IEC)
See Thermoquest

International Institute for the Advancement of Medicine
1232 Mid-Valley Drive
Jessup, PA 18434
(800) 486-IIAM   FAX: (570) 343-6993
(570) 496-3400
http://www.iiam.org

International Light
17 Graf Road
Newburyport, MA 01950
(978) 465-5923   FAX: (978) 462-0759

International Market Supply (I.M.S.)
Dane Mill
Broadhurst Lane
Congleton, Cheshire CW12 1LA, UK
(44) 1260 275469
FAX: (44) 1260 276007

International Marketing Services
See International Marketing Ventures

International Marketing Ventures
6301 Ivy Lane, Suite 408
Greenbelt, MD 20770
(800) 373-0096   FAX: (301) 345-0631
(301) 345-2866
http://www.imvlimited.com

International Products
201 Connecticut Drive
Burlington, NJ 08016
(609) 386-8770   FAX: (609) 386-8438
http://www.mkt.ipcol.com

Intracel Corporation
Bartels Division
2005 Sammamish Road, Suite 107
Issaquah, WA 98027
(800) 542-2281   FAX: (425) 557-1894
(425) 392-2992
http://www.intracel.com

Invitrogen
1600 Faraday Avenue
Carlsbad, CA 92008
(800) 955-6288   FAX: (760) 603-7201
(760) 603-7200
http://www.invitrogen.com

In Vivo Metric
P.O. Box 249
Healdsburg, CA 95448
(707) 433-4819   FAX: (707) 433-2407

IRORI
9640 Towne Center Drive
San Diego, CA 92121
(858) 546-1300   FAX: (858) 546-3083
http://www.irori.com

Irvine Scientific
2511 Daimler Street
Santa Ana, CA 92705
(800) 577-6097   FAX: (949) 261-6522
(949) 261-7800
http://www.irvinesci.com

ISC BioExpress
420 North Kays Drive
Kaysville, UT 84037
(800) 999-2901   FAX: (800) 574-7892
(801) 547-5047
http://www.bioexpress.com

ISCO
P.O. Box 5347
4700 Superior
Lincoln, NE 68505
(800) 228-4373   FAX: (402) 464-0318
(402) 464-0231
http://www.isco.com

Isis Pharmaceuticals
Carlsbad Research Center
2292 Faraday Avenue
Carlsbad, CA 92008
(760) 931-9200
http://www.isip.com

Isolabs
See Wallac

ISS
See Integrated Separation Systems

J & W Scientific
See Agilent Technologies

J.A. Webster
86 Leominster Road
Sterling, MA 01564
(800) 225-7911   FAX: (978) 422-8959
http://www.jawebster.com

J.T. Baker
See Mallinckrodt Baker
222 Red School Lane
Phillipsburg, NJ 08865
(800) JTBAKER   FAX: (908) 859-6974
http://www.jtbaker.com

Jackson ImmunoResearch Laboratories
P.O. Box 9
872 W. Baltimore Pike
West Grove, PA 19390
(800) 367-5296   FAX: (610) 869-0171
(610) 869-4024
http://www.jacksonimmuno.com

The Jackson Laboratory
600 Maine Street
Bar Harbor, ME 04059
(800) 422-6423   FAX: (207) 288-5079
(207) 288-6000
http://www.jax.org

Jaece Industries
908 Niagara Falls Boulevard
North Tonawanda, NY 14120
(716) 694-2811   FAX: (716) 694-2811
http://www.jaece.com

Jandel Scientific
See SPSS

Janke & Kunkel
See Ika Works

Janssen Life Sciences Products
See Amersham

Janssen Pharmaceutica
1125 Trenton-Harbourton Road
Titusville, NJ 09560
(609) 730-2577   FAX: (609) 730-2116
http://us.janssen.com

Jasco
8649 Commerce Drive
Easton, MD 21601
(800) 333-5272   FAX: (410) 822-7526
(410) 822-1220
http://www.jascoinc.com

Jena Bioscience
Loebstedter Str. 78
07749 Jena, Germany
(49) 3641-464920
FAX: (49) 3641-464991
http://www.jenabioscience.com

Jencons Scientific
800 Bursca Drive, Suite 801
Bridgeville, PA 15017
(800) 846-9959   FAX: (412) 257-8809
(412) 257-8861
http://www.jencons.co.uk

JEOL Instruments
11 Dearborn Road
Peabody, MA 01960
(978) 535-5900   FAX: (978) 536-2205
http://www.jeol.com/index.html

Jewett
750 Grant Street
Buffalo, NY 14213
(800) 879-7767   FAX: (716) 881-6092
(716) 881-0030
http://www.JewettInc.com

John's Scientific
See VWR Scientific

John Weiss and Sons
95 Alston Drive
Bradwell Abbey
Milton Keynes, Buckinghamshire
MK1 4HF UK
(44) 1908-318017
FAX: (44) 1908-318708

Johnson & Johnson Medical
2500 Arbrook Boulevard East
Arlington, TX 76004
(800) 423-4018
http://www.jnjmedical.com

Johnston Matthey Chemicals
Orchard Road
Royston, Hertfordshire SG8 5HE, UK
(44) 1763-253000
FAX: (44) 1763-253466
http://www.chemicals.matthey.com

Jolley Consulting and Research
683 E. Center Street, Unit H
Grayslake, IL 60030
(847) 548-2330   FAX: (847) 548-2984
http://www.jolley.com

Jordan Scientific
See Shelton Scientific

Jorgensen Laboratories
1450 N. Van Buren Avenue
Loveland, CO 80538
(800) 525-5614   FAX: (970) 663-5042
(970) 669-2500
http://www.jorvet.com

JRH Biosciences and JR Scientific
13804 W. 107th Street
Lenexa, KS 66215
(800) 231-3735   FAX: (913) 469-5584
(913) 469-5580

Jule Bio Technologies
25 Science Park, #14, Suite 695
New Haven, CT 06511
(800) 648-1772   FAX: (203) 786-5489
(203) 786-5490
http://hometown.aol.com/precastgel/index.htm

K.R. Anderson
2800 Bowers Avenue
Santa Clara, CA 95051
(800) 538-8712   FAX: (408) 727-2959
(408) 727-2800
http://www.kranderson.com

Kabi Pharmacia Diagnostics
See Pharmacia Diagnostics

Kanthal H.P. Reid
1 Commerce Boulevard
P.O. Box 352440
Palm Coast, FL 32135
(904) 445-200   FAX: (904) 446-2244
http://www.kanthal.com

Kapak
5305 Parkdale Drive
St. Louis Park, MN 55416
(800) KAPAK-57   FAX: (612) 541-0735
(612) 541-0730
http://www.kapak.com

Karl Hecht
Stettener Str. 22-24
D-97647 Sondheim
Rhön, Germany
(49) 9779-8080   FAX: (49) 9779-80888

Karl Storz
Köningin-Elisabeth Str. 60
D-14059 Berlin, Germany
(49) 30-30 69 09-0
FAX: (49) 30-30 19 452
http://www.karlstorz.de

KaVo EWL
P.O. Box 1320
D-88293 Leutkirch im Allgäu, Germany
(49) 7561-86-0   FAX: (49) 7561-86-371
http://www.kavo.com/english/startseite.htm

Keithley Instruments
28775 Aurora Road
Cleveland, OH 44139
(800) 552-1115   FAX: (440) 248-6168
(440) 248-0400
http://www.keithley.com

Kemin
2100 Maury Street, Box 70
Des Moines, IA 50301
(515) 266-2111   FAX: (515) 266-8354
http://www.kemin.com

Kemo
3 Brook Court, Blakeney Road
Beckenham, Kent BR3 1HG, UK
(44) 0181 658 3838
FAX: (44) 0181 658 4084
http://www.kemo.com

Kendall
15 Hampshire Street
Mansfield, MA 02048
(800) 962-9888   FAX: (800) 724-1324
http://www.kendallhq.com

Kendro Laboratory Products
31 Pecks Lane
Newtown, CT 06470
(800) 522-SPIN   FAX: (203) 270-2166
(203) 270-2080
http://www.kendro.com

Kendro Laboratory Products
P.O. Box 1220
Am Kalkberg
D-3360 Osterod, Germany
(55) 22-316-213
FAX: (55) 22-316-202
http://www.heraeus-instruments.de

Kent Laboratories
23404 NE 8th Street
Redmond, WA 98053
(425) 868-6200   FAX: (425) 868-6335
http://www.kentlabs.com

Kent Scientific
457 Bantam Road, #16
Litchfield, CT 06759
(888) 572-8887   FAX: (860) 567-4201
(860) 567-5496
http://www.kentscientific.com

Keuffel & Esser
See Azon

Keystone Scientific
Penn Eagle Industrial Park 320 Rolling Ridge Drive
Bellefonte, PA 16823
(800) 437-2999   FAX: (814) 353-2305
(814) 353-2300 Ext 1
http://www.keystonescientific.com

Kimble/Kontes Biotechnology
1022 Spruce Street
P.O. Box 729
Vineland, NJ 08360
(888) 546-2531   FAX: (856) 794-9762
(856) 692-3600
http://www.kimble-kontes.com

Kinematica AG
Luzernerstrasse 147a
CH-6014 Littau-Luzern, Switzerland
(41) 41 2501257   FAX: (41) 41 2501460
http://www.kinematica.ch

Kin-Tek
504 Laurel Street
LaMarque, TX 77568
(800) 326-3627
FAX: (409) 938-3710
http://www.kin-tek.com

Kipp & Zonen
125 Wilbur Place
Bohemia, NY 11716
(800) 645-2065   FAX: (516) 589-2068
(516) 589-2885
http://www.kippzonen.thomasregister.com/olc/kippzonen

Kirkegaard & Perry Laboratories
2 Cessna Court
Gaithersburg, MD 20879
(800) 638-3167   FAX: (301) 948-0169
(301) 948-7755
http://www.kpl.com

Kodak
See Eastman Kodak

Kontes Glass
See Kimble/Kontes Biotechnology

Kontron Instruments AG
Postfach CH-8010
Zurich, Switzerland
41-1-733-5733   FAX: 41-1-733-5734

David Kopf Instruments
P.O. Box 636
Tujunga, CA 91043
(818) 352-3274   FAX: (818) 352-3139

Kraft Apparatus
See Glas-Col Apparatus

Kramer Scientific Corporation
711 Executive Boulevard
Valley Cottage, NY 10989
(845) 267-5050   FAX: (845) 267-5550

Kulite Semiconductor Products
1 Willow Tree Road
Leonia, NJ 07605
(201) 461-0900   FAX: (201) 461-0990
http://www.kulite.com

Lab-Line Instruments
15th & Bloomingdale Avenues
Melrose Park, IL 60160
(800) LAB-LINE   FAX: (708) 450-5830
FAX: (800) 450-4LAB
http://www.labline.com

Lab Products
742 Sussex Avenue
P.O. Box 639
Seaford, DE 19973
(800) 526-0469   FAX: (302) 628-4309
(302) 628-4300
http://www.labproductsinc.com

LabRepco
101 Witmer Road, Suite 700
Horsham, PA 19044
(800) 521-0754   FAX: (215) 442-9202
http://www.labrepco.com

Lab Safety Supply
P.O. Box 1368
Janesville, WI 53547
(800) 356-0783   FAX: (800) 543-9910
(608) 754-7160   FAX: (608) 754-1806
http://www.labsafety.com

Lab-Tek Products
See Nalge Nunc International

Labconco
8811 Prospect Avenue
Kansas City, MO 64132
(800) 821-5525   FAX: (816) 363-0130
(816) 333-8811
http://www.labconco.com

Labindustries
See Barnstead/Thermolyne

Labnet International
P.O. Box 841
Woodbridge, NJ 07095
(888) LAB-NET1   FAX: (732) 417-1750
(732) 417-0700
http://www.nationallabnet.com

LABO-MODERNE
37 rue Dombasle
Paris
75015 France
(33) 01-45-32-62-54
FAX: (33) 01-45-32-01-09
http://www.labomoderne.com/fr

Laboratory of Immunoregulation
National Institute of Allergy and
Infectious Diseases/NIH
9000 Rockville Pike
Building 10, Room 11B13
Bethesda, MD 20892
(301) 496-1124

Laboratory Supplies
29 Jefry Lane
Hicksville, NY 11801
(516) 681-7711

Labscan Limited
Stillorgan Industrial Park
Stillorgan
Dublin, Ireland
(353) 1-295-2684
FAX: (353) 1-295-2685
http://www.labscan.ie

Labsystems
See Thermo Labsystems

Labsystems Affinity Sensors
Saxon Way, Bar Hill
Cambridge CB3 8SL, UK
44 (0) 1954 789976
FAX: 44 (0) 1954 789417
http://www.affinity-sensors.com

Labtronics
546 Governors Road
Guelph, Ontario N1K 1E3, Canada
(519) 763-4930   FAX: (519) 836-4431
http://www.labtronics.com

Labtronix Manufacturing
3200 Investment Boulevard
Hayward, CA 94545
(510) 786-3200   FAX: (510) 786-3268
http://www.labtronix.com

Lafayette Instrument
3700 Sagamore Parkway North
P.O. Box 5729
Lafayette, IN 47903
(800) 428-7545   FAX: (765) 423-4111
(765) 423-1505
http://www.lafayetteinstrument.com

Lambert Instruments
Turfweg 4
9313 TH Leutingewolde
The Netherlands
(31) 50-5018461   FAX: (31)
50-5010034
http://www.lambert-instruments.com

Lancaster Synthesis
P.O. Box 1000
Windham, NH 03087
(800) 238-2324   FAX: (603) 889-3326
(603) 889-3306
http://www.lancastersynthesis-us.com

Lancer
140 State Road 419
Winter Springs, FL 32708
(800) 332-1855   FAX: (407) 327-1229
(407) 327-8488
http://www.lancer.com

LaVision GmbH
Gerhard-Gerdes-Str. 3
D-37079
Goettingen, Germany
(49) 551-50549-0
FAX: (49) 551-50549-11
http://www.lavision.de

Lawshe
See Advanced Process Supply

LC Laboratories
165 New Boston Street
Woburn, MA 01801
(781) 937-0777   FAX: (781) 938-5420
http://www.lclaboratories.com

LC Packings
80 Carolina Street
San Francisco, CA 94103
(415) 552-1855   FAX: (415) 552-1859
http://www.lcpackings.com

LC Services
See LC Laboratories

LECO
3000 Lakeview Avenue
St. Joseph, MI 49085
(800) 292-6141   FAX: (616) 982-8977
(616) 985-5496
http://www.leco.com

Lederle Laboratories
See Wyeth-Ayerst

Lee Biomolecular Research Laboratories
11211 Sorrento Valley Road, Suite M
San Diego, CA 92121
(858) 452-7700

The Lee Company
2 Pettipaug Road
P.O. Box 424
Westbrook, CT 06498
(800) LEE-PLUG   FAX: (860) 399-7058
(860) 399-6281
http://www.theleeco.com

Lee Laboratories
1475 Athens Highway
Grayson, GA 30017
(800) 732-9150   FAX: (770) 979-9570
(770) 972-4450
http://www.leelabs.com

Leica
111 Deer Lake Road
Deerfield, IL 60015
(800) 248-0123   FAX: (847) 405-0147
(847) 405-0123
http://www.leica.com

Leica Microsystems
Imneuenheimer Feld 518
D-69120
Heidelberg, Germany
(49) 6221-41480
FAX: (49) 6221-414833
http://www.leica-microsystems.com

Leinco Technologies
359 Consort Drive
St. Louis, MO 63011
(314) 230-9477   FAX: (314) 527-5545
http://www.leinco.com

Leitz U.S.A.
See Leica

LenderKing Metal Products
8370 Jumpers Hole Road
Millersville, MD 21108
(410) 544-8795   FAX: (410) 544-5069
http://www.lenderking.com

Letica Scientific Instruments
Panlab s.i., c/Loreto 50
08029 Barcelona, Spain
(34) 93-419-0709
FAX: (34) 93-419-7145
www.panlab-sl.com

Leybold-Heraeus Trivac DZA
5700 Mellon Road
Export, PA 15632
(412) 327-5700

LI-COR
Biotechnology Division
4308 Progressive Avenue
Lincoln, NE 68504
(800) 645-4267   FAX: (402) 467-0819
(402) 467-0700
http://www.licor.com

Life Science Laboratories
See Adaptive Biosystems

Life Science Resources
Two Corporate Center Drive
Melville, NY 11747
(800) 747-9530   FAX: (516) 844-5114
(516) 844-5085
http://www.astrocam.com

Life Sciences
2900 72nd Street North
St. Petersburg, FL 33710
(800) 237-4323   FAX: (727) 347-2957
(727) 345-9371
http://www.lifesci.com

Life Technologies
9800 Medical Center Drive
P.O. Box 6482
Rockville, MD 20849
(800) 828-6686   FAX: (800) 331-2286
http://www.lifetech.com

Lifecodes
550 West Avenue
Stamford, CT 06902
(800) 543-3263   FAX: (203) 328-9599
(203) 328-9500
http://www.lifecodes.com

Lightnin
135 Mt. Read Boulevard
Rochester, NY 14611
(888) MIX-BEST   FAX: (716) 527-1742
(716) 436-5550
http://www.lightnin-mixers.com

Linear Drives
Luckyn Lane, Pipps Hill
Basildon, Essex SS14 3BW, UK
(44) 1268-287070
FAX: (44) 1268-293344
http://www.lineardrives.com

Linscott's USA
6 Grove Street
Mill Valley, CA 94941
(415) 389-9674   FAX: (415) 389-6025
http://www.linscottsdirectory.com

Linton Instrumentation
Unit 11, Forge Business Center
Upper Rose Lane
Palgrave, Diss, Norfolk IP22 1AP, UK
(44) 1-379-651-344
FAX: (44) 1-379-650-970
http://www.lintoninst.co.uk

List Biological Laboratories
501-B Vandell Way
Campbell, CA 95008
(800) 726-3213   FAX: (408) 866-6364
(408) 866-6363
http://www.listlabs.com

LKB Instruments
See Amersham Pharmacia Biotech

Lloyd Laboratories
604 West Thomas Avenue
Shenandoah, IA 51601
(800) 831-0004   FAX: (712) 246-5245
(712) 246-4000
http://www.lloydinc.com

Loctite
1001 Trout Brook Crossing
Rocky Hill, CT 06067
(860) 571-5100   FAX: (860)571-5465
http://www.loctite.com

Lofstrand Labs
7961 Cessna Avenue
Gaithersburg, MD 20879
(800) 541-0362   FAX: (301) 948-9214
(301) 330-0111
http://www.lofstrand.com

Lomir Biochemical
99 East Main Street
Malone, NY 12953
(877) 425-3604   FAX: (518) 483-8195
(518) 483-7697
http://www.lomir.com

LSL Biolafitte
10 rue de Temara
7810C St.-Germain-en-Laye, France
(33) 1-3061-5260   FAX: (33) 1-3061-5234

Ludl Electronic Products
171 Brady Avenue
Hawthorne, NY 10532
(888) 769-6111   FAX: (914) 769-4759
(914) 769-6111
http://www.ludl.com

Lumigen
24485 W. Ten Mile Road
Southfield, MI 48034
(248) 351-5600   FAX: (248) 351-0518
http://www.lumigen.com

Luminex
12212 Technology Boulevard
Austin, TX 78727
(888) 219-8020   FAX: (512) 258-4173
(512) 219-8020
http://www.luminexcorp.com

LYNX Therapeutics
25861 Industrial Boulevard
Hayward, CA 94545 (510) 670-9300
FAX: (510) 670-9302
http://www.lynxgen.com

Lyphomed
3 Parkway North
Deerfield, IL 60015
(847) 317-8100   FAX: (847) 317-8600

M.E.D. Associates
See Med Associates

Macherey-Nagel
6 South Third Street, #402
Easton, PA 18042
(610) 559-9848   FAX: (610) 559-9878
http://www.macherey-nagel.com

Macherey-Nagel
Valencienner Strasse 11
P.O. Box 101352
D-52313 Dueren, Germany
(49) 2421-969141
FAX: (49) 2421-969199
http://www.macherey-nagel.ch

Mac-Mod Analytical
127 Commons Court
Chadds Ford, PA 19317
800-441-7508   FAX: (610) 358-5993
(610) 358-9696
http://www.mac-mod.com

Mallinckrodt Baker
222 Red School Lane
Phillipsburg, NJ 08865
(800) 582-2537   FAX: (908) 859-6974
(908) 859-2151
http://www.mallbaker.com

Mallinckrodt Chemicals
16305 Swingley Ridge Drive
Chesterfield, MD 63017
(314) 530-2172   FAX: (314) 530-2563
http://www.mallchem.com

Malven Instruments
Enigma Business Park
Grovewood Road
Malven, Worchestershire
WR 141 XZ, United Kingdom

Marinus
1500 Pier C Street
Long Beach, CA 90813
(562) 435-6522   FAX: (562) 495-3120

Markson Science
c/o Whatman Labs Sales
P.O. Box 1359
Hillsboro, OR 97123
(800) 942-8626   FAX: (503) 640-9716
(503) 648-0762

Marsh Biomedical Products
565 Blossom Road
Rochester, NY 14610
(800) 445-2812   FAX: (716) 654-4810
(716) 654-4800
http://www.biomar.com

Marshall Farms USA
5800 Lake Bluff Road
North Rose, NY 14516
(315) 587-2295
e-mail: info@marfarms.com

Martek
6480 Dobbin Road
Columbia, MD 21045
(410) 740-0081   FAX: (410) 740-2985
http://www.martekbio.com

Martin Supply
Distributor of Gerber Scientific
2740 Loch Raven Road
Baltimore, MD 21218
(800) 282-5440   FAX: (410) 366-0134
(410) 366-1696

Mast Immunosystems
630 Clyde Court
Mountain View, CA 94043
(800) 233-MAST   FAX: (650) 969-2745
(650) 961-5501
http://www.mastallergy.com

Matheson Gas Products
P.O. Box 624
959 Route 46 East
Parsippany, NJ 07054
(800) 416-2505   FAX: (973) 257-9393
(973) 257-1100
http://www.mathesongas.com

Mathsoft
1700 Westlake Avenue N., Suite 500
Seattle, WA 98109
(800) 569-0123   FAX: (206) 283-8691
(206) 283-8802
http://www.mathsoft.com

Matreya
500 Tressler Street
Pleasant Gap, PA 16823
(814) 359-5060   FAX: (814) 359-5062
http://www.matreya.com

Matrigel
See Becton Dickinson Labware

Matrix Technologies
22 Friars Drive
Hudson, NH 03051
(800) 345-0206   FAX: (603) 595-0106
(603) 595-0505
http://www.matrixtechcorp.com

MatTek Corp.
200 Homer Avenue
Ashland, Massachusetts 01721
(508) 881-6771   FAX: (508) 879-1532
http://www.mattek.com

Maxim Medical
89 Oxford Road
Oxford OX2 9PD
United Kingdom
44 (0)1865-865943
FAX: 44 (0)1865-865291
http://www.maximmed.com

Mayo Clinic
Section on Engineering
Project #ALA-1, 1982
200 1st Street SW
Rochester, MN 55905
(507) 284-2511   FAX: (507) 284-5988

McGaw
See B. Braun-McGaw

McMaster-Carr
600 County Line Road
Elmhurst, IL 60126
(630) 833-0300   FAX: (630) 834-9427
http://www.mcmaster.com

McNeil Pharmaceutical
See Ortho McNeil Pharmaceutical

MCNC
3021 Cornwallis Road
P.O. Box 12889
Research Triangle Park, NC 27709
(919) 248-1800   FAX: (919) 248-1455
http://www.mcnc.org

MD Industries
5 Revere Drive, Suite 415
Northbrook, IL 60062
(800) 421-8370   FAX: (847) 498-2627
(708) 339-6000
http://www.mdindustries.com

MDS Nordion
447 March Road
P.O. Box 13500
Kanata, Ontario K2K 1X8, Canada (800)
465-3666   FAX: (613) 592-6937
(613) 592-2790
http://www.mds.nordion.com

MDS Sciex
71 Four Valley Drive
Concord, Ontario
Canada L4K 4V8
(905) 660-9005   FAX: (905) 660-2600
http://www.sciex.com

Mead Johnson
See Bristol-Meyers Squibb

Med Associates
P.O. Box 319
St. Albans, VT 05478
(802) 527-2343   FAX: (802) 527-5095
http://www.med-associates.com

Medecell
239 Liverpool Road
London N1 1LX, UK
(44) 20-7607-2295
FAX: (44) 20-7700-4156
http://www.medicell.co.uk

Media Cybernetics
8484 Georgia Avenue, Suite 200
Silver Spring, MD 20910
(301) 495-3305   FAX: (301) 495-5964
http://www.mediacy.com

Mediatech
13884 Park Center Road
Herndon, VA 20171
(800) cellgro
(703) 471-5955
http://www.cellgro.com

Medical Systems
See Harvard Apparatus

Medifor
647 Washington Street
Port Townsend, WA 98368
(800) 366-3710   FAX: (360) 385-4402
(360) 385-0722
http://www.medifor.com

MedImmune
35 W. Watkins Mill Road
Gaithersburg, MD 20878
(301) 417-0770   FAX: (301) 527-4207
http://www.medimmune.com

Medoc Advanced Medical Systems
1502 West Highway 54, Suite 404
Durham, NC 27707
(919) 402-9600   FAX: (919) 402-9607
http://www.medoc-web.com

MedProbe AS
P.O. Box 2640
St. Hanshaugen
N-0131 Oslo, Norway
(47) 222 00137   FAX: (47) 222 00189
http://www.medprobe.com

Megazyme
Bray Business Park
Bray, County Wicklow
Ireland
(353) 1-286-1220
FAX: (353) 1-286-1264
http://www.megazyme.com

Melles Griot
4601 Nautilus Court South
Boulder, CO 80301
(800) 326-4363   FAX: (303) 581-0960
(303) 581-0337
http://www.mellesgriot.com

Menzel-Glaser
Postfach 3157
D-38021 Braunschweig, Germany
(49) 531 590080
FAX: (49) 531 509799

E. Merck
Frankfurterstrasse 250
D-64293 Darmstadt 1, Germany
(49) 6151-720

Merck
See EM Science

Merck & Company
Merck National Service Center
P.O. Box 4
West Point, PA 19486
(800) NSC-MERCK
(215) 652-5000
http://www.merck.com

Merck Research Laboratories
See Merck & Company

Merck Sharpe Human Health Division
300 Franklin Square Drive
Somerset, NJ 08873
(800) 637-2579   FAX: (732) 805-3960
(732) 805-0300

Merial Limited
115 Transtech Drive
Athens, GA 30601
(800) MERIAL-1   FAX: (706) 548-0608
(706) 548-9292
http://www.merial.com

Meridian Instruments
P.O. Box 1204
Kent, WA 98035
(253) 854-9914   FAX: (253) 854-9902
http://www.minstrument.com

Meta Systems Group
32 Hammond Road
Belmont, MA 02178
(617) 489-9950   FAX: (617) 489-9952

Metachem Technologies
3547 Voyager Street, Bldg. 102
Torrance, CA 90503
(310) 793-2300   FAX: (310) 793-2304
http://www.metachem.com

Metallhantering
Box 47172
100-74 Stockholm, Sweden
(46) 8-726-9696

MethylGene
7220 Frederick-Banting, Suite 200
Montreal, Quebec H4S 2A1, Canada
http://www.methylgene.com

Metro Scientific
475 Main Street, Suite 2A
Farmingdale, NY 11735
(800) 788-6247   FAX: (516) 293-8549
(516) 293-9656

Metrowerks
980 Metric Boulevard
Austin, TX 78758
(800) 377-5416
(512) 997-4700
http://www.metrowerks.com

Mettler Instruments
Mettler-Toledo
1900 Polaris Parkway
Columbus, OH 43240
(800) METTLER   FAX: (614) 438-4900
http://www.mt.com

Miami Serpentarium Labs
34879 Washington Loop Road
Punta Gorda, FL 33982
(800) 248-5050   FAX: (813) 639-1811
(813) 639-8888
http://www.miamiserpentarium.com

Michrom BioResources
1945 Industrial Drive
Auburn, CA 95603
(530) 888-6498   FAX: (530) 888-8295
http://www.michrom.com

Mickle Laboratory Engineering
Gomshall, Surrey, UK
(44) 1483-202178

Micra Scientific
A division of Eichrom Industries
8205 S. Cass Ave, Suite 111
Darien, IL 60561
(800) 283-4752   FAX: (630) 963-1928
(630) 963-0320
http://www.micrasci.com

MicroBrightField
74 Hegman Avenue
Colchester, VT 05446
(802) 655-9360   FAX: (802) 655-5245
http://www.microbrightfield.com

Micro Essential Laboratory
4224 Avenue H
Brooklyn, NY 11210
(718) 338-3618   FAX: (718) 692-4491

Micro Filtration Systems
7-3-Chome, Honcho
Nihonbashi, Tokyo, Japan
(81) 3-270-3141

Micro-Metrics
P.O. Box 13804
Atlanta, GA 30324
(770) 986-6015   FAX: (770) 986-9510
http://www.micro-metrics.com

Micro-Tech Scientific
140 South Wolfe Road
Sunnyvale, CA 94086
(408) 730-8324   FAX: (408) 730-3566
http://www.microlc.com

MicroCal
22 Industrial Drive East
Northampton, MA 01060
(800) 633-3115   FAX: (413) 586-0149
(413) 586-7720
www.microcalorimetry.com

Microfluidics
30 Ossipee Road
P.O. Box 9101
Newton, MA 02164
(800) 370-5452   FAX: (617) 965-1213
(617) 969-5452
http://www.microfluidicscorp.com

Microgon
See Spectrum Laboratories

Microlase Optical Systems
West of Scotland Science Park
Kelvin Campus, Maryhill Road
Glasgow G20 0SP, UK
(44) 141-948-1000
FAX: (44) 141-946-6311
http://www.microlase.co.uk

Micron Instruments
4509 Runway Street
Simi Valley, CA 93063
(800) 638-3770   FAX: (805) 522-4982
(805) 552-4676
http://www.microninstruments.com

Micron Separations
See MSI

Micro Photonics
4949 Liberty Lane, Suite 170
P.O. Box 3129
Allentown, PA 18106
(610) 366-7103   FAX: (610) 366-7105
http://www.microphotonics.com

MicroTech
1420 Conchester Highway
Boothwyn, PA 19061
(610) 459-3514

Midland Certified Reagent Company
3112-A West Cuthbert Avenue
Midland, TX 79701
(800) 247-8766   FAX: (800) 359-5789
(915) 694-7950   FAX: (915) 694-2387
http://www.mcrc.com

Midwest Scientific
280 Vance Road
Valley Park, MO 63088
(800) 227-9997   FAX: (636) 225-9998
(636) 225-9997
http://www.midsci.com

Miles
See Bayer

Miles Laboratories
See Serological

Miles Scientific
See Nunc

Millar Instruments
P.O. Box 230227
6001-A Gulf Freeway
Houston, TX 77023
(713) 923-9171   FAX: (713) 923-7757
http://www.millarinstruments.com

MilliGen/Biosearch
See Millipore

Millipore
80 Ashbury Road
P.O. Box 9125
Bedford, MA 01730
(800) 645-5476   FAX: (781) 533-3110
(781) 533-6000
http://www.millipore.com

Miltenyi Biotec
251 Auburn Ravine Road, Suite 208
Auburn, CA 95603
(800) 367-6227   FAX: (530) 888-8925
(530) 888-8871
http://www.miltenyibiotec.com

Miltex
6 Ohio Drive
Lake Success, NY 11042
(800) 645-8000   FAX: (516) 775-7185
(516) 349-0001

Milton Roy
See Spectronic Instruments

Mini-Instruments
15 Burnham Business Park
Springfield Road
Burnham-on-Crouch, Essex CM0 8TE, UK
(44) 1621-783282
FAX: (44) 1621-783132
http://www.mini-instruments.co.uk

Mini Mitter
P.O. Box 3386
Sunriver, OR 97707
(800) 685-2999   FAX: (541) 593-5604
(541) 593-8639
http://www.minimitter.com

Misonix
1938 New Highway
Farmingdale, NY 11735
(800) 645-9846   FAX: (516) 694-9412
http://www.misonix.com

Mitutoyo (MTI)
See Dolla Eastern

MJ Research
Waltham, MA 02451
(800) PELTIER   FAX: (617) 923-8080
(617) 923-8000
http://www.mjr.com

Modular Instruments
228 West Gay Street
Westchester, PA 19380
(610) 738-1420   FAX: (610) 738-1421
http://www.mi2.com

Molecular Biology Insights
8685 US Highway 24
Cascade, CO 80809-1333
(800) 747-4362   FAX: (719) 684-7989
(719) 684-7988
http://www.oligo.net

Molecular Biosystems
10030 Barnes Canyon Road
San Diego, CA 92121
(858) 452-0681   FAX: (858) 452-6187
http://www.mobi.com

Molecular Devices
1312 Crossman Avenue
Sunnyvale, CA 94089
(800) 635-5577   FAX: (408) 747-3602
(408) 747-1700
http://www.moldev.com

Molecular Designs
1400 Catalina Street
San Leandro, CA 94577
(510) 895-1313   FAX: (510) 614-3608

Molecular Dynamics
928 East Arques Avenue
Sunnyvale, CA 94086
(800) 333-5703   FAX: (408) 773-1493
(408) 773-1222
http://www.apbiotech.com

Molecular Probes
4849 Pitchford Avenue
Eugene, OR 97402
(800) 438-2209   FAX: (800) 438-0228
(541) 465-8300   FAX: (541) 344-6504
http://www.probes.com

Molecular Research Center
5645 Montgomery Road
Cincinnati, OH 45212
(800) 462-9868   FAX: (513) 841-0080
(513) 841-0900
http://www.mrcgene.com

Molecular Simulations
9685 Scranton Road
San Diego, CA 92121
(800) 756-4674   FAX: (858) 458-0136
(858) 458-9990
http://www.msi.com

Monoject Disposable Syringes & Needles/Syrvet
16200 Walnut Street
Waukee, IA 50263
(800) 727-5203   FAX: (515) 987-5553
(515) 987-5554
http://www.syrvet.com

Monsanto Chemical
800 North Lindbergh Boulevard
St. Louis, MO 63167
(314) 694-1000   FAX: (314) 694-7625
http://www.monsanto.com

Moravek Biochemicals
577 Mercury Lane
Brea, CA 92821
(800) 447-0100   FAX: (714) 990-1824
(714) 990-2018
http://www.moravek.com

Moss
P.O. Box 189
Pasadena, MD 21122
(800) 932-6677   FAX: (410) 768-3971
(410) 768-3442
http://www.mosssubstrates.com

Motion Analysis
3617 Westwind Boulevard
Santa Rosa, CA 95403
(707) 579-6500   FAX: (707) 526-0629
http://www.motionanalysis.com

Mott
Farmington Industrial Park
84 Spring Lane
Farmington, CT 06032
(860) 747-6333   FAX: (860) 747-6739
http://www.mottcorp.com

MSI (Micron Separations)
See Osmonics

Multi Channel Systems
Markwiesenstrasse 55
72770 Reutlingen, Germany
(49) 7121-503010
FAX: (49) 7121-503011
http://www.multichannelsystems.com

Multiple Peptide Systems
3550 General Atomics Court
San Diego, CA 92121
(800) 338-4965   FAX: (800) 654-5592
(858) 455-3710   FAX: (858) 455-3713
http://www.mps-sd.com

Murex Diagnostics
3075 Northwoods Circle
Norcross, GA 30071
(707) 662-0660   FAX: (770) 447-4989

MWG-Biotech
Anzinger Str. 7
D-85560 Ebersberg, Germany
(49) 8092-82890   FAX: (49) 8092-21084
http://www.mwg_biotech.com

Myriad Industries
3454 E Street
San Diego, CA 92102
(800) 999-6777   FAX: (619) 232-4819
(619) 232-6700
http://www.myriadindustries.com

Nacalai Tesque
Nijo Karasuma, Nakagyo-ku
Kyoto 604, Japan
81-75-251-1723
FAX: 81-75-251-1762
http://www.nacalai.co.jp

Nalge Nunc International
Subsidiary of Sybron International
75 Panorama Creek Drive
P.O. Box 20365
Rochester, NY 14602
(800) 625-4327   FAX: (716) 586-8987
(716) 264-9346
http://www.nalgenunc.com

Nanogen
10398 Pacific Center Court
San Diego, CA 92121
(858) 410-4600   FAX: (858) 410-4848
http://www.nanogen.com

Nanoprobes
95 Horse Block Road
Yaphank, NY 11980
(877) 447-6266   FAX: (631) 205-9493
(631) 205-9490
http://www.nanoprobes.com

Narishige USA
1710 Hempstead Turnpike
East Meadow, NY 11554
(800) 445-7914   FAX: (516) 794-0066
(516) 794-8000
http://www.narishige.co.jp

National Bag Company
2233 Old Mill Road
Hudson, OH 44236
(800) 247-6000   FAX: (330) 425-9800
(330) 425-2600
http://www.nationalbag.com

National Band and Tag
Department X 35, Box 72430
Newport, KY 41032
(606) 261-2035   FAX: (800) 261-8247
https://www.nationalband.com

National Biosciences
See Molecular Biology Insights

National Diagnostics
305 Patton Drive
Atlanta, GA 30336
(800) 526-3867   FAX: (404) 699-2077
(404) 699-2121
http://www.nationaldiagnostics.com

National Institute of Standards and Technology
100 Bureau Drive
Gaithersburg, MD 20899
(301) 975-NIST   FAX: (301) 926-1630
http://www.nist.gov

National Instruments
11500 North Mopac Expressway
Austin, TX 78759
(512) 794-0100   FAX: (512) 683-8411
www.ni.com

National Labnet
See Labnet International

National Scientific Instruments
975 Progress Circle
Lawrenceville, GA 300243
(800) 332-3331   FAX: (404) 339-7173
http://www.nationalscientific.com

National Scientific Supply
1111 Francisco Boulvard East
San Rafael, CA 94901
(800) 525-1779   FAX: (415) 459-2954
(415) 459-6070
http://www.nat-sci.com

Naz-Dar-KC Chicago
Nazdar
1087 N. North Branch Street
Chicago, IL 60622
(800) 736-7636   FAX: (312) 943-8215
(312) 943-8338
http://www.nazdar.com

NB Labs
1918 Avenue A
Denison, TX 75021
(903) 465-2694   FAX: (903) 463-5905
http://www.nblabslarry.com

NEB
See New England Biolabs

NEN Life Science Products
549 Albany Street
Boston, MA 02118
(800) 551-2121   FAX: (617) 451-8185
(617) 350-9075
http://www.nen.com

NEN Research Products, Dupont (UK)
Diagnostics and Biotechnology Systems
Wedgewood Way
Stevenage, Hertfordshire SG1 4QN, UK
44-1438-734831
44-1438-734000
FAX: 44-1438-734836
http://www.dupont.com

Neogen
628 Winchester Road
Lexington, KY 40505 (800) 477-8201
FAX: (606) 255-5532
(606) 254-1221
http://www.neogen.com

Neosystems
380, 11012 Macleod Trail South
Calgary, Alberta T2J 6A5 Canada
(403) 225-9022   FAX: (403) 225-9025
http://www.neosystems.com

Neuralynx
2434 North Pantano Road
Tucson, AZ 85715
(520) 722-8144   FAX: (520) 722-8163
http://www.neuralynx.com

Neuro Probe
16008 Industrial Drive
Gaithersburg, MD 20877
(301) 417-0014   FAX: (301) 977-5711
http://www.neuroprobe.com

Neurocrine Biosciences
10555 Science Center Drive
San Diego, CA 92121
(619) 658-7600   FAX: (619) 658-7602
http://www.neurocrine.com

Nevtek
HCR03, Box 99
Burnsville, VA 24487
(540) 925-2322   FAX: (540) 925-2323
http://www.nevtek.com

New Brunswick Scientific
44 Talmadge Road
Edison, NJ 08818
(800) 631-5417   FAX: (732) 287-4222
(732) 287-1200
http://www.nbsc.com

New England Biolabs (NEB)
32 Tozer Road
Beverly, MA 01915 (800) 632-5227
FAX: (800) 632-7440
http://www.neb.com

New England Nuclear (NEN)
See NEN Life Science Products

New MBR
Gubelstrasse 48
CH8050 Zurich, Switzerland
(41) 1-313-0703

Newark Electronics
4801 N. Ravenswood Avenue
Chicago, IL 60640
(800) 4-NEWARK   FAX: (773) 907-5339
(773) 784-5100
http://www.newark.com

Newell Rubbermaid
29 E. Stephenson Street
Freeport, IL 61032
(815) 235-4171   FAX: (815) 233-8060
http://www.newellco.com

Newport Biosystems
1860 Trainor Street
Red Bluff, CA 96080
(530) 529-2448   FAX: (530) 529-2648

Newport
1791 Deere Avenue
Irvine, CA 92606
(800) 222-6440   FAX: (949) 253-1800
(949) 253-1462
http://www.newport.com

Nexin Research B.V.
P.O. Box 16
4740 AA Hoeven, The Netherlands
(31) 165-503172
FAX: (31) 165-502291

NIAID
See Bio-Tech Research Laboratories

Nichols Institute Diagnostics
33051 Calle Aviador
San Juan Capistrano, CA 92675
(800) 286-4NID   FAX: (949) 240-5273
(949) 728-4610
http://www.nicholsdiag.com

Nichols Scientific Instruments
3334 Brown Station Road
Columbia, MO 65202
(573) 474-5522   FAX: (603) 215-7274
http://home.beseen.com
technology/nsi_technology

Nicolet Biomedical Instruments
5225 Verona Road, Building 2
Madison, WI 53711
(800) 356-0007   FAX: (608) 441-2002
(608) 273-5000
http://nicoletbiomedical.com

N.I.G.M.S. (National Institute of General Medical Sciences)
See Coriell Institute for Medical Research

Nikon
Science and Technologies Group
1300 Walt Whitman Road
Melville, NY 11747
(516) 547-8500   FAX: (516) 547-4045
http://www.nikonusa.com

Nippon Gene
1-29, Ton-ya-machi
Toyama 930, Japan
(81) 764-51-6548
FAX: (81) 764-51-6547

Noldus Information Technology
751 Miller Drive
Suite E-5
Leesburg, VA 20175
(800) 355-9541   FAX: (703) 771-0441
(703) 771-0440
http://www.noldus.com

Nordion International
See MDS Nordion

North American Biologicals (NABI)
16500 NW 15th Avenue
Miami, FL 33169
(800) 327-7106   (305) 625-5305
http://www.nabi.com

North American Reiss
See Reiss

Northwestern Bottle
24 Walpole Park South
Walpole, MA 02081
(508) 668-8600   FAX: (508) 668-7790

NOVA Biomedical
Nova Biomedical 200
Prospect Street Waltham, MA 02454
(800) 822-0911   FAX: (781) 894-5915
http://www.novabiomedical.com

Novagen
601 Science Drive
Madison, WI 53711
(800) 526-7319   FAX: (608) 238-1388
(608) 238-6110
http://www.novagen.com

Novartis
59 Route 10
East Hanover, NJ 07936
(800)526-0175   FAX: (973) 781-6356
http://www.novartis.com

Novartis Biotechnology
3054 Cornwallis Road
Research Triangle Park, NC 27709
(888) 462-7288   FAX: (919) 541-8585
http://www.novartis.com

Nova Sina AG
Subsidiary of Airflow Lufttechnik GmbH
Kleine Heeg 21
52259 Rheinbach, Germany
(49) 02226 920-0
FAX: (49) 02226 9205-11

Novex/Invitrogen
1600 Faraday
Carlsbad, CA 92008
(800) 955-6288   FAX: (760) 603-7201
http://www.novex.com

Novo Nordisk Biochem
77 Perry Chapel Church Road
Franklington, NC 27525
(800) 879-6686   FAX: (919) 494-3450
(919) 494-3000
http://www.novo.dk

Novo Nordisk BioLabs
See Novo Nordisk Biochem

Novocastra Labs
Balliol Business Park West
Benton Lane
Newcastle-upon-Tyne
Tyne and Wear NE12 8EW, UK
(44) 191-215-0567
FAX: (44) 191-215-1152
http://www.novocastra.co.uk

Novus Biologicals
P.O. Box 802
Littleton, CO 80160
(888) 506-6887   FAX: (303) 730-1966
http://www.novus-biologicals.com/main.html

NPI Electronic
Hauptstrasse 96
D-71732 Tamm, Germany
(49) 7141-601534
FAX: (49) 7141-601266
http://www.npielectronic.com

NSG Precision Cells
195G Central Avenue
Farmingdale, NY 11735
(516) 249-7474 FAX: (516) 249-8575
http://www.nsgpci.com

Nu Chek Prep
109 West Main
P.O. Box 295
Elysian, MN 56028
(800) 521-7728 FAX: (507) 267-4790
(507) 267-4689

Nuclepore
See Costar

Numonics
101 Commerce Drive
Montgomeryville, PA 18936
(800) 523-6716 FAX: (215) 361-0167
(215) 362-2766
http://www.interactivewhiteboards.com

NYCOMED AS Pharma
c/o Accurate Chemical & Scientific
300 Shames Drive
Westbury, NY 11590
(800) 645-6524 FAX: (516) 997-4948
(516) 333-2221
http://www.accuratechemical.com

Nycomed Amersham
Health Care Division
101 Carnegie Center
Princeton, NJ 08540
(800) 832-4633 FAX: (800) 807-2382
(609) 514-6000
http://www.nycomed-amersham.com

Nyegaard
Herserudsvagen 5254
S-122 06 Lidingo, Sweden
(46) 8-765-2930

Ohmeda Catheter Products
See Datex-Ohmeda

Ohwa Tsusbo
Hiby Dai Building
1-2-2 Uchi Saiwai-cho
Chiyoda-ku
Tokyo 100, Japan
03-3591-7348 FAX: 03-3501-9001

Oligos Etc.
29970 SW Town Centre
Loop West, Suite B419
Wilsonville, OR 97070
(800) 888-2358 FAX: (800) 869-0813

Olis Instruments
130 Conway Drive
Bogart, GA 30622
(706) 353-6547 (800) 852-3504
http://www.olisweb.com

Olympus America
2 Corporate Center Drive
Melville, NY 11747
(800) 645-8160 FAX: (516) 844-5959
(516) 844-5000
http://www.olympusamerica.com

Omega Engineering
One Omega Drive
P.O. Box 4047
Stamford, CT 06907
(800) 848-4286 FAX: (203) 359-7700
(203) 359-1660
http://www.omega.com

Omega Optical
3 Grove Street
P.O. Box 573
Brattleboro, VT 05302
(802) 254-2690 FAX: (802) 254-3937
http://www.omegafilters.com

Omnetics Connector Corporation
7260 Commerce Circle
East Minneapolis, MN 55432
(800) 343-0025 (763) 572-0656
FAX: (763) 572-3925
http://www.omnetics.com

Omni International
6530 Commerce Court
Warrenton, VA 20187
(800) 776-4431 FAX: (540) 347-5352
(540) 347-5331
http://www.omni-inc.com

Omnion
2010 Energy Drive
P.O. Box 879
East Troy, WI 53120
(262) 642-7200 FAX: (262) 642-7760
http://www.omnion.com

Omnitech Electronics
See AccuScan Instruments

Oncogene Research Products
P.O. Box Box 12087
La Jolla, CA 92039-2087
(800) 662-2616 FAX: (800) 766-0999
http://www.apoptosis.com

Oncogene Science
See OSI Pharmaceuticals

Oncor
See Intergen

Operon Technologies
1000 Atlantic Avenue
Alameda, CA 94501
(800) 688-2248 FAX: (510) 865-5225
(510) 865-8644
http://www.operon.com

Optiscan
P.O. Box 1066
Mount Waverly MDC, Victoria
Australia 3149
61-3-9538 3333 FAX: 61-3-9562 7742
http://www.optiscan.com.au

Optomax
9 Ash Street
P.O. Box 840
Hollis, NH 03049
(603) 465-3385 FAX: (603) 465-2291

Opto-Line Associates
265 Ballardvale Street
Wilmington, MA 01887
(978) 658-7255 FAX: (978) 658-7299
http://www.optoline.com

Orbigen
6827 Nancy Ridge Drive
San Diego, CA 92121
(866) 672-4436 (858) 362-2030
(858) 362-2026
http://www.orbigen.com

Oread BioSaftey
1501 Wakarusa Drive
Lawrence, KS 66047
(800) 447-6501 FAX: (785) 749-1882
(785) 749-0034
http://www.oread.com

Organomation Associates
266 River Road West
Berlin, MA 01503
(888) 978-7300 FAX: (978)838-2786
(978) 838-7300
http://www.organomation.com

Organon
375 Mount Pleasant Avenue
West Orange, NJ 07052
(800) 241-8812 FAX: (973) 325-4589
(973) 325-4500
http://www.organon.com

Organon Teknika (Canada)
30 North Wind Place
Scarborough, Ontario
M1S 3R5 Canada
(416) 754-4344 FAX: (416) 754-4488
http://www.organonteknika.com

Organon Teknika Cappel
100 Akzo Avenue
Durham, NC 27712
(800) 682-2666 FAX: (800) 432-9682
(919) 620-2000 FAX: (919) 620-2107
http://www.organonteknika.com

Oriel Corporation of America
150 Long Beach Boulevard
Stratford, CT 06615
(203) 377-8282 FAX: (203) 378-2457
http://www.oriel.com

OriGene Technologies
6 Taft Court, Suite 300
Rockville, MD 20850
(888) 267-4436   FAX: (301) 340-9254
(301) 340-3188
http://www.origene.com

OriginLab
One Roundhouse Plaza
Northhampton, MA 01060
(800) 969-7720   FAX: (413) 585-0126
http://www.originlab.com

Orion Research
500 Cummings Center
Beverly, MA 01915
(800) 225-1480   FAX: (978) 232-6015
(978) 232-6000
http://www.orionres.com

Ortho Diagnostic Systems
Subsidiary of Johnson & Johnson
1001 U.S. Highway 202
P.O. Box 350
Raritan, NJ 08869
(800) 322-6374   FAX: (908) 218-8582
(908) 218-1300

Ortho McNeil Pharmaceutical
Welsh & McKean Road
Spring House, PA 19477
(800) 682-6532
(215) 628-5000
http://www.orthomcneil.com

Oryza
200 Turnpike Road, Unit 5
Chelmsford, MA 01824
(978) 256-8183   FAX: (978) 256-7434
http://www.oryzalabs.com

OSI Pharmaceuticals
106 Charles Lindbergh Boulevard
Uniondale, NY 11553
(800) 662-2616   FAX: (516) 222-0114
(516) 222-0023
http://www.osip.com

Osmonics
135 Flanders Road
P.O. Box 1046
Westborough, MA 01581
(800) 444-8212   FAX: (508) 366-5840
(508) 366-8212
http://www.osmolabstore.com

Oster Professional Products
150 Cadillac Lane
McMinnville, TN 37110
(931) 668-4121   FAX: (931) 668-4125
http://www.sunbeam.com

Out Patient Services
1260 Holm Road
Petaluma, CA 94954
(800) 648-1666   FAX: (707) 762-7198
(707) 763-1581

OWL Scientific Plastics
See OWL Separation Systems

OWL Separation Systems
55 Heritage Avenue
Portsmouth, NH 03801
(800) 242-5560   FAX: (603) 559-9258
(603) 559-9297
http://www.owlsci.com

Oxford Biochemical Research
P.O. Box 522
Oxford, MI 48371
(800) 692-4633   FAX: (248) 852-4466
http://www.oxfordbiomed.com

Oxford GlycoSystems
See Glyco

Oxford Instruments
Old Station Way
Eynsham
Witney, Oxfordshire OX8 1TL, UK
(44) 1865-881437
FAX: (44) 1865-881944
http://www.oxinst.com

Oxford Labware
See Kendall

Oxford Molecular Group
Oxford Science Park
The Medawar Centre
Oxford OX4 4GA, UK
(44) 1865-784600
FAX: (44) 1865-784601
http://www.oxmol.co.uk

Oxford Molecular Group
2105 South Bascom Avenue, Suite 200
Campbell, CA 95008
(800) 876-9994   FAX: (408) 879-6302
(408) 879-6300
http://www.oxmol.com

OXIS International
6040 North Cutter Circle
Suite 317
Portland, OR 97217
(800) 547-3686   FAX: (503) 283-4058
(503) 283-3911
http://www.oxis.com

Oxoid
800 Proctor Avenue
Ogdensburg, NY 13669
(800) 567-8378   FAX: (613) 226-3728
http://www.oxoid.ca

Oxoid
Wade Road
Basingstoke, Hampshire RG24 8PW, UK
(44) 1256-841144
FAX: (4) 1256-814626
http://www.oxoid.ca

Oxyrase
P.O. Box 1345
Mansfield, OH 44901
(419) 589-8800   FAX: (419) 589-9919
http://www.oxyrase.com

Ozyme
10 Avenue Ampère
Montigny de Bretoneux
78180 France
(33) 13-46-02-424
FAX: (33) 13-46-09-212
http://www.ozyme.fr

PAA Laboratories
2570 Route 724
P.O. Box 435
Parker Ford, PA 19457
(610) 495-9400   FAX: (610) 495-9410
http://www.paa-labs.com

Pacer Scientific
5649 Valley Oak Drive
Los Angeles, CA 90068
(323) 462-0636   FAX: (323) 462-1430
http://www.pacersci.com

Pacific Bio-Marine Labs
P.O. Box 1348
Venice, CA 90294
(310) 677-1056   FAX: (310) 677-1207

Packard Instrument
800 Research Parkway
Meriden, CT 06450
(800) 323-1891   FAX: (203) 639-2172
(203) 238-2351
http://www.packardinst.com

Padgett Instrument
1730 Walnut Street
Kansas City, MO 64108
(816) 842-1029

Pall Filtron
50 Bearfoot Road
Northborough, MA 01532
(800) FILTRON   FAX: (508) 393-1874
(508) 393-1800

Pall-Gelman
25 Harbor Park Drive
Port Washington, NY 11050
(800) 289-6255   FAX: (516) 484-2651
(516) 484-3600
http://www.pall.com

PanVera
545 Science Drive
Madison, WI 53711
(800) 791-1400   FAX: (608) 233-3007
(608) 233-9450
http://www.panvera.com

Parke-Davis
See Warner-Lambert

Parr Instrument
211 53rd Street
Moline, IL 61265
(800) 872-7720   FAX: (309) 762-9453
(309) 762-7716
http://www.parrinst.com

Partec
Otto Hahn Strasse 32
D-48161 Munster, Germany
(49) 2534-8008-0
FAX: (49) 2535-8008-90

PCR
See Archimica Florida

PE Biosystems
850 Lincoln Centre Drive
Foster City, CA 94404
(800) 345-5224   FAX: (650) 638-5884
(650) 638-5800
http://www.pebio.com

Pel-Freez Biologicals
219 N. Arkansas
P.O. Box 68
Rogers, AR 72757
(800) 643-3426   FAX: (501) 636-3562
(501) 636-4361
http://www.pelfreez-bio.com

Pel-Freez Clinical Systems
Subsidiary of Pel-Freez Biologicals
9099 N. Deerbrook Trail
Brown Deer, WI 53223
(800) 558-4511   FAX: (414) 357-4518
(414) 357-4500
http://www.pelfreez-bio.com

Peninsula Laboratories
601 Taylor Way
San Carlos, CA 94070
(800) 650-4442   FAX: (650) 595-4071
(650) 592-5392
http://www.penlabs.com

Pentex
24562 Mando Drive
Laguna Niguel, CA 92677
(800) 382-4667   FAX: (714) 643-2363
http://www.pentex.com

PeproTech
5 Crescent Avenue
P.O. Box 275
Rocky Hill, NJ 08553
(800) 436-9910   FAX: (609) 497-0321
(609) 497-0253
http://www.peprotech.com

Peptide Institute
4-1-2 Ina, Minoh-shi
Osaka 562-8686, Japan
81-727-29-4121   FAX: 81-727-29-4124
http://www.peptide.co.jp

Peptide Laboratory
4175 Lakeside Drive
Richmond, CA 94806
(800) 858-7322   FAX: (510) 262-9127
(510) 262-0800
http://www.peptidelab.com

Peptides International
11621 Electron Drive
Louisville, KY 40299
(800) 777-4779   FAX: (502) 267-1329
(502) 266-8787
http://www.pepnet.com

Perceptive Science Instruments
2525 South Shore Boulevard, Suite 100
League City, TX 77573
(281) 334-3027   FAX: (281) 538-2222
http://www.persci.com

Perimed
4873 Princeton Drive
North Royalton, OH 44133
(440) 877-0537   FAX: (440) 877-0534
http://www.perimed.se

Perkin-Elmer
761 Main Avenue
Norwalk, CT 06859
(800) 762-4002   FAX: (203) 762-6000
(203) 762-1000
http://www.perkin-elmer.com
See also PE Biosystems

PerSeptive Bioresearch Products
See PerSeptive BioSystems

PerSeptive BioSystems
500 Old Connecticut Path
Framingham, MA 01701
(800) 899-5858   FAX: (508) 383-7885
(508) 383-7700
http://www.pbio.com

PerSeptive Diagnostic
See PE Biosystems
(800) 343-1346

Pettersson Elektronik AB
Tallbacksvagen 51
S-756 45 Uppsala, Sweden
(46) 1830-3880   FAX: (46) 1830-3840
http://www.bahnhof.se/~pettersson

PGC Scientifics
7311 Governors Way
Frederick, MD 21704
(800) 424-3300   FAX: (800) 662-1112
(301) 620-7777   FAX: (301) 620-7497
http://www.pgcscientifics.com

Pharmacia Biotech
See Amersham Pharmacia Biotech

Pharmacia Diagnostics
See Wallac

Pharmacia LKB Biotech
See Amersham Pharmacia Biotech

Pharmacia LKB Biotechnology
See Amersham Pharmacia Biotech

Pharmacia LKB Nuclear
See Wallac

Pharmaderm Veterinary Products
60 Baylis Road
Melville, NY 11747
(800) 432-6673
http://www.pharmaderm.com

Pharmed (Norton)
Norton Performance Plastics
See Saint-Gobain Performance Plastics

PharMingen
See BD PharMingen

Phenomex
2320 W. 205th Street
Torrance, CA 90501
(310) 212-0555   FAX: (310) 328-7768
http://www.phenomex.com

PHLS Centre for Applied Microbiology and Research
See European Collection of Animal Cell Cultures (ECACC)

Phoenix Flow Systems
11575 Sorrento Valley Road, Suite 208
San Diego, CA 92121
(800) 886-3569   FAX: (619) 259-5268
(619) 453-5095
http://www.phnxflow.com

Phoenix Pharmaceutical
4261 Easton Road, P.O. Box 6457
St. Joseph, MO 64506
(800) 759-3644   FAX: (816) 364-4969
(816) 364-5777
http://www.phoenixpharmaceutical.com

Photometrics
See Roper Scientific

Photon Technology International
1 Deerpark Drive, Suite F
Monmouth Junction, NJ 08852
(732) 329-0910   FAX: (732) 329-9069
http://www.pti-nj.com

Physik Instrumente
Polytec PI
23 Midstate Drive, Suite 212
Auburn, MA 01501
(508) 832-3456   FAX: (508) 832-0506
http://www.polytecpi.com

Physitemp Instruments
154 Huron Avenue
Clifton, NJ 07013
(800) 452-8510   FAX: (973) 779-5954
(973) 779-5577
http://www.physitemp.com

Pico Technology
The Mill House, Cambridge Street
St. Neots, Cambridgeshire
PE19 1QB, UK
(44) 1480-396-395
FAX: (44) 1480-396-296
www.picotech.com

Pierce Chemical
P.O. Box 117
3747 Meridian Road
Rockford, IL 61105
(800) 874-3723   FAX: (800) 842-5007
FAX: (815) 968-7316
http://www.piercenet.com

Pierce & Warriner
44, Upper Northgate Street
Chester, Cheshire CH1 4EF, UK
(44) 1244 382 525
FAX: (44) 1244 373 212
http://www.piercenet.com

Pilling Weck Surgical
420 Delaware Drive
Fort Washington, PA 19034
(800) 523-2579   FAX: (800) 332-2308
www.pilling-weck.com

PixelVision
A division of Cybex Computer Products
14964 NW Greenbrier Parkway
Beaverton, OR 97006
(503) 629-3210   FAX: (503) 629-3211
http://www.pixelvision.com

P.J. Noyes
P.O. Box 381
89 Bridge Street
Lancaster, NH 03584
(800) 522-2469   FAX: (603) 788-3873
(603) 788-4952
http://www.pjnoyes.com

Plas-Labs
917 E. Chilson Street
Lansing, MI 48906
(800) 866-7527   FAX: (517) 372-2857
(517) 372-7177
http://www.plas-labs.com

Plastics One
6591 Merriman Road, Southwest
P.O. Box 12004
Roanoke, VA 24018
(540) 772-7950   FAX: (540) 989-7519
http://www.plastics1.com

Platt Electric Supply
2757 6th Avenue South
Seattle, WA 98134
(206) 624-4083   FAX: (206) 343-6342
http://www.platt.com

Plexon
6500 Greenville Avenue
Suite 730
Dallas, TX 75206
(214) 369-4957   FAX: (214) 369-1775
http://www.plexoninc.com

Polaroid
784 Memorial Drive
Cambridge, MA 01239
(800) 225-1618   FAX: (800) 832-9003
(781) 386-2000
http://www.polaroid.com

Polyfiltronics
136 Weymouth St.
Rockland, MA 02370
(800) 434-7659   FAX: (781) 878-0822
(781) 878-1133
http://www.polyfiltronics.com

Polylabo Paul Block
Parc Tertiare de la Meinau
10, rue de la Durance
B.P. 36
67023 Strasbourg Cedex 1
Strasbourg, France
33-3-8865-8020
FAX: 33-3-8865-8039

PolyLC
9151 Rumsey Road, Suite 180
Columbia, MD 21045
(410) 992-5400   FAX: (410) 730-8340

Polymer Laboratories
Amherst Research Park
160 Old Farm Road
Amherst, MA 01002
(800) 767-3963   FAX: (413) 253-2476
http://www.polymerlabs.com

Polymicro Technologies
18019 North 25th Avenue
Phoenix, AZ 85023
(602) 375-4100   FAX: (602) 375-4110
http://www.polymicro.com

Polyphenols AS
Hanabryggene Technology Centre
Hanaveien 4-6
4327 Sandnes, Norway
(47) 51-62-0990
FAX: (47) 51-62-51-82
http://www.polyphenols.com

Polysciences
400 Valley Road
Warrington, PA 18976
(800) 523-2575   FAX: (800) 343-3291
http://www.polysciences.com

Polyscientific
70 Cleveland Avenue
Bayshore, NY 11706
(516) 586-0400   FAX: (516) 254-0618

Polytech Products
285 Washington Street
Somerville, MA 02143
(617) 666-5064   FAX: (617) 625-0975

Polytron
8585 Grovemont Circle
Gaithersburg, MD 20877
(301) 208-6597   FAX: (301) 208-8691
http://www.polytron.com

Popper and Sons
300 Denton Avenue
P.O. Box 128
New Hyde Park, NY 11040
(888) 717-7677   FAX: (800) 557-6773
(516) 248-0300   FAX: (516) 747-1188
http://www.popperandsons.com

Porphyrin Products
P.O. Box 31
Logan, UT 84323
(435) 753-1901   FAX: (435) 753-6731
http://www.porphyrin.com

Portex
See SIMS Portex Limited

Powderject Vaccines
585 Science Drive
Madison, WI 53711
(608) 231-3150   FAX: (608) 231-6990
http://www.powderject.com

Praxair
810 Jorie Boulevard
Oak Brook, IL 60521
(800) 621-7100
http://www.praxair.com

Precision Dynamics
13880 Del Sur Street
San Fernando, CA 91340
(800) 847-0670   FAX: (818) 899-4-45
http://www.pdcorp.com

Precision Scientific Laboratory Equipment
Division of Jouan
170 Marcel Drive
Winchester, VA 22602
(800) 621-8820   FAX: (540) 869-0130
(540) 869-9892
http://www.precisionsci.com

Primary Care Diagnostics
See Becton Dickinson Primary Care Diagnostics

Primate Products
1755 East Bayshore Road, Suite 28A
Redwood City, CA 94063
(650) 368-0663   FAX: (650) 368-0665
http://www.primateproducts.com

5 Prime → 3 Prime
See 2000 Eppendorf-5 Prime
http://www.5prime.com

Princeton Applied Research
PerkinElmer Instr.: Electrochemistry
801 S. Illinois
Oak Ridge, TN 37830
(800) 366-2741   FAX: (423) 425-1334
(423) 481-2442
http://www.eggpar.com

Princeton Instruments
A division of Roper Scientific
3660 Quakerbridge Road
Trenton, NJ 08619
(609) 587-9797   FAX: (609) 587-1970
http://www.prinst.com

Princeton Separations
P.O. Box 300
Aldephia, NJ 07710
(800) 223-0902   FAX: (732) 431-3768
(732) 431-3338

Prior Scientific
80 Reservoir Park Drive
Rockland, MA 02370
(781) 878-8442   FAX: (781) 878-8736
http://www.prior.com

PRO Scientific
P.O. Box 448
Monroe, CT 06468
(203) 452-9431   FAX: (203) 452-9753
http://www.proscientific.com

Professional Compounding Centers of America
9901 South Wilcrest Drive
Houston, TX 77099
(800) 331-2498   FAX: (281) 933-6227
(281) 933-6948
http://www.pccarx.com

Progen Biotechnik
Maass-Str. 30
69123 Heidelberg, Germany
(49) 6221-8278-0
FAX: (49) 6221-8278-23
http://www.progen.de

Prolabo
A division of Merck Eurolab
54 rue Roger Salengro
94126 Fontenay Sous Bois Cedex
France
33-1-4514-8500
FAX: 33-1-4514-8616
http://www.prolabo.fr

Proligo
2995 Wilderness Place Boulder, CO 80301
(888) 80-OLIGO   FAX: (303) 801-1134
http://www.proligo.com

Promega
2800 Woods Hollow Road
Madison, WI 53711
(800) 356-9526   FAX: (800) 356-1970
(608) 274-4330   FAX: (608) 277-2516
http://www.promega.com

Protein Databases (PDI)
405 Oakwood Road
Huntington Station, NY 11746
(800) 777-6834   FAX: (516) 673-4502
(516) 673-3939

Protein Polymer Technologies
10655 Sorrento Valley Road
San Diego, CA 92121
(619) 558-6064   FAX: (619) 558-6477
http://www.ppti.com

Protein Solutions
391 G Chipeta Way
Salt Lake City, UT 84108
(801) 583-9301   FAX: (801) 583-4463
http://www.proteinsolutions.com

Prozyme
1933 Davis Street, Suite 207
San Leandro, CA 94577
(800) 457-9444   FAX: (510) 638-6919
(510) 638-6900
http://www.prozyme.com

PSI
See Perceptive Science Instruments

Pulmetrics Group
82 Beacon Street
Chestnut Hill, MA 02167
(617) 353-3833   FAX: (617) 353-6766

Purdue Frederick
100 Connecticut Avenue
Norwalk, CT 06850
(800) 633-4741   FAX: (203) 838-1576
(203) 853-0123
http://www.pharma.com

Purina Mills
LabDiet
P. O. Box 66812
St. Louis, MO 63166
(800) 227-8941   FAX: (314) 768-4894
http://www.purina-mills.com

Qbiogene
2251 Rutherford Road
Carlsbad, CA 92008
(800) 424-6101   FAX: (760) 918-9313
http://www.qbiogene.com

Qiagen
28159 Avenue Stanford
Valencia, CA 91355
(800) 426-8157   FAX: (800) 718-2056
http://www.qiagen.com

Quality Biological
7581 Lindbergh Drive
Gaithersburg, MD 20879
(800) 443-9331   FAX: (301) 840-5450
(301) 840-9331
http://www.qualitybiological.com

Quantum Appligene
Parc d'Innovation
Rue Geller de Kaysberg
67402 Illkirch, Cedex, France
(33) 3-8867-5425
FAX: (33) 3-8867-1945
http://www.quantum-appligene.com

Quantum Biotechnologies
See Qbiogene

Quantum Soft
Postfach 6613
CH-8023
Zürich, Switzerland
FAX: 41-1-481-69-51
profit@quansoft.com

Questcor Pharmaceuticals
26118 Research Road
Hayward, CA 94545
(510) 732-5551   FAX: (510) 732-7741
http://www.questcor.com

Quidel
10165 McKellar Court
San Diego, CA 92121
(800) 874-1517   FAX: (858) 546-8955
(858) 552-1100
http://www.quidel.com

R-Biopharm
7950 Old US 27 South
Marshall, MI 49068
(616) 789-3033   FAX: (616) 789-3070
http://www.r-biopharm.com

R. C. Electronics
6464 Hollister Avenue
Santa Barbara, CA 93117
(805) 685-7770   FAX: (805) 685-5853
http://www.rcelectronics.com

R & D Systems
614 McKinley Place NE
Minneapolis, MN 55413
(800) 343-7475   FAX: (612) 379-6580
(612) 379-2956
http://www.rndsystems.com

R & S Technology
350 Columbia Street
Peacedale, RI 02880
(401) 789-5660   FAX: (401) 792-3890
http://www.septech.com

RACAL Health and Safety
See 3M
7305 Executive Way
Frederick, MD 21704
(800) 692-9500   FAX: (301) 695-8200

Radiometer America
811 Sharon Drive
Westlake, OH 44145
(800) 736-0600 FAX: (440) 871-2633
(440) 871-8900
http://www.rameusa.com

Radiometer A/S
The Chemical Reference Laboratory
kandevej 21
DK-2700 Brnshj, Denmark
45-3827-3827 FAX: 45-3827-2727

Radionics
22 Terry Avenue
Burlington, MA 01803
(781) 272-1233 FAX: (781) 272-2428
http://www.radionics.com

Radnoti Glass Technology
227 W. Maple Avenue
Monrovia, CA 91016
(800) 428-l4l6 FAX: (626) 303-2998
(626) 357-8827
http://www.radnoti.com

Rainin Instrument
Rainin Road
P.O. Box 4026
Woburn, MA 01888
(800)-4-RAININ FAX: (781) 938-1152
(781) 935-3050
http://www.rainin.com

Rank Brothers
56 High Street
Bottisham, Cambridge
CB5 9DA UK
(44) 1223 811369
FAX: (44) 1223 811441
http://www.rankbrothers.com

Rapp Polymere
Ernst-Simon Strasse 9
D 72072 Tübingen, Germany
(49) 7071-763157
FAX: (49) 7071-763158
http://www.rapp-polymere.com

Raven Biological Laboratories
8607 Park Drive
P.O. Box 27261
Omaha, NE 68127
(800) 728-5702 FAX: (402) 593-0995
(402) 593-0781
http://www.ravenlabs.com

Razel Scientific Instruments
100 Research Drive
Stamford, CT 06906
(203) 324-9914 FAX: (203) 324-5568

Reagents International
See Biotech Source

Receptor Biology
10000 Virginia Manor Road, Suite 360
Beltsville, MD 20705
(888) 707-4200 FAX: (301) 210-6266
(301) 210-4700
http://www.receptorbiology.com

Regis Technologies
8210 N. Austin Avenue
Morton Grove, IL 60053
(800) 323-8144 FAX: (847) 967-1214
(847) 967-6000
http://www.registech.com

Reichert Ophthalmic Instruments
P.O. Box 123
Buffalo, NY 14240
(716) 686-4500 FAX: (716) 686-4545
http://www.reichert.com

Reiss
1 Polymer Place
P.O. Box 60 Blackstone, VA 23824
(800) 356-2829 FAX: (804) 292-1757
(804) 292-1600
http://www.reissmfg.com

Remel
12076 Santa Fe Trail Drive
P.O. Box 14428
Shawnee Mission, KS 66215
(800) 255-6730 FAX: (800) 621-8251
(913) 888-0939 FAX: (913) 888-5884
http://www.remelinc.com

Reming Bioinstruments
6680 County Route 17
Redfield, NY 13437
(315) 387-3414 FAX: (315) 387-3415

RepliGen
117 Fourth Avenue
Needham, MA 02494
(800) 622-2259 FAX: (781) 453-0048
(781) 449-9560
http://www.repligen.com

Research Biochemicals
1 Strathmore Road
Natick, MA 01760
(800) 736-3690 FAX: (800) 736-2480
(508) 651-8151 FAX: (508) 655-1359
http://www.resbio.com

Research Corporation Technologies
101 N. Wilmot Road, Suite 600
Tucson, AZ 85711
(520) 748-4400 FAX: (520) 748-0025
http://www.rctech.com

Research Diagnostics
Pleasant Hill Road
Flanders, NJ 07836
(800) 631-9384 FAX: (973) 584-0210
(973) 584-7093
http://www.researchd.com

Research Diets
121 Jersey Avenue
New Brunswick, NJ 08901
(877) 486-2486 FAX: (732) 247-2340
(732) 247-2390
http://www.researchdiets.com

Research Genetics
2130 South Memorial Parkway
Huntsville, AL 35801
(800) 533-4363 FAX: (256) 536-9016
(256) 533-4363
http://www.resgen.com

Research Instruments
Kernick Road Pernryn
Cornwall TR10 9DQ, UK
(44) 1326-372-753
FAX: (44) 1326-378-783
http://www.research-instruments.com

Research Organics
4353 E. 49th Street
Cleveland, OH 44125
(800) 321-0570 FAX: (216) 883-1576
(216) 883-8025
http://www.resorg.com

Research Plus
P.O. Box 324
Bayonne, NJ 07002
(800) 341-2296 FAX: (201) 823-9590
(201) 823-3592
http://www.researchplus.com

Research Products International
410 N. Business Center Drive
Mount Prospect, IL 60056
(800) 323-9814 FAX: (847) 635-1177
(847) 635-7330
http://www.rpicorp.com

Research Triangle Institute
P.O. Box 12194
Research Triangle Park, NC 27709
(919) 541-6000 FAX: (919) 541-6515
http://www.rti.org

Restek
110 Benner Circle
Bellefonte, PA 16823
(800) 356-1688 FAX: (814) 353-1309
(814) 353-1300
http://www.restekcorp.com

Rheodyne
P.O. Box 1909
Rohnert Park, CA 94927
(707) 588-2000 FAX: (707) 588-2020
http://www.rheodyne.com

Rhone Merieux
See Merial Limited

Rhone-Poulenc
2 T W Alexander Drive
P.O. Box 12014
Research Triangle Park, NC 08512
(919) 549-2000   FAX: (919) 549-2839
http://www.Rhone-Poulenc.com
Also see Aventis

Rhone-Poulenc Rorer
500 Arcola Road
Collegeville, PA 19426
(800) 727-6737   FAX: (610) 454-8940
(610) 454-8975
http://www.rp-rorer.com

Rhone-Poulenc Rorer
Centre de Recherche de Vitry-Alfortville
13 Quai Jules Guesde, BP14 94403
Vitry Sur Seine, Cedex, France
(33) 145-73-85-11
FAX: (33) 145-73-81-29
http://www.rp-rorer.com

Ribi ImmunoChem Research
563 Old Corvallis Road
Hamilton, MT 59840
(800) 548-7424   FAX: (406) 363-6129
(406) 363-3131
http://www.ribi.com

RiboGene
See Questcor Pharmaceuticals

Ricca Chemical
448 West Fork Drive
Arlington, TX 76012
(888) GO-RICCA   FAX: (800) RICCA-93
(817) 461-5601
http://www.riccachemical.com

Richard-Allan Scientific
225 Parsons Street
Kalamazoo, MI 49007
(800) 522-7270   FAX: (616) 345-3577
(616) 344-2400
http://www.rallansci.com

Richelieu Biotechnologies
11 177 Hamon
Montral, Quebec
H3M 3E4 Canada
(802) 863-2567   FAX: (802) 862-2909
http://www.richelieubio.com

Richter Enterprises
20 Lake Shore Drive
Wayland, MA 01778
(508) 655-7632   FAX: (508) 652-7264
http://www.richter-enterprises.com

Riese Enterprises
BioSure Division
12301 G Loma Rica Drive
Grass Valley, CA 95945
(800) 345-2267   FAX: (916) 273-5097
(916) 273-5095
http://www.biosure.com

Robbins Scientific
1250 Elko Drive
Sunnyvale, CA 94086
(800) 752-8585   FAX: (408) 734-0300
(408) 734-8500
http://www.robsci.com

Roboz Surgical Instruments
9210 Corporate Boulevard, Suite 220
Rockville, MD 20850
(800) 424-2984   FAX: (301) 590-1290
(301) 590-0055

Roche Diagnostics
9115 Hague Road
P.O. Box 50457
Indianapolis, IN 46256
(800) 262-1640   FAX: (317) 845-7120
(317) 845-2000
http://www.roche.com

Roche Molecular Systems
See Roche Diagnostics

Rocklabs
P.O. Box 18-142
Auckland 6, New Zealand
(64) 9-634-7696
FAX: (64) 9-634-7696
http://www.rocklabs.com

Rockland
P.O. Box 316
Gilbertsville, PA 19525
(800) 656-ROCK   FAX: (610) 367-7825
(610) 369-1008
http://www.rockland-inc.com

Rohm
Chemische Fabrik
Kirschenallee
D-64293 Darmstadt, Germany
(49) 6151-1801   FAX: (49) 6151-1802
http://www.roehm.com

Roper Scientific
3440 East Brittania Drive, Suite 100
Tucson, AZ 85706
(520) 889-9933   FAX: (520) 573-1944
http://www.roperscientific.com

Rosetta Inpharmatics
12040 115th Avenue NE
Kirkland, WA 98034
(425) 820-8900   FAX: (425) 820-5757
http://www.rii.com

ROTH-SOCHIEL
3 rue de la Chapelle
Lauterbourg
67630 France
(33) 03-88-94-82-42
FAX: (33) 03-88-54-63-93

Rotronic Instrument
160 E. Main Street
Huntington, NY 11743
(631) 427-3898   FAX: (631) 427-3902
http://www.rotronic-usa.com

Roundy's
23000 Roundy Drive
Pewaukee, WI 53072
(262) 953-7999   FAX: (262) 953-7989
http://www.roundys.com

RS Components
Birchington Road
Weldon Industrial Estate
Corby, Northants NN17 9RS, UK
(44) 1536 201234
FAX: (44) 1536 405678
http://www.rs-components.com

Rubbermaid
See Newell Rubbermaid

SA Instrumentation
1437 Tzena Way
Encinitas, CA 92024
(858) 453-1776   FAX: (800)-266-1776
http://www.sainst.com

Safe Cells
See Bionique Testing Labs

Sage Instruments
240 Airport Boulevard
Freedom, CA 95076
831-761-1000   FAX: 831-761-1008
http://www.sageinst.com

Sage Laboratories
11 Huron Drive
Natick, MA 01760
(508) 653-0844   FAX: 508-653-5671
http://www.sagelabs.com

Saint-Gobain Performance Plastics
P.O. Box 3660
Akron, OH 44309
(330) 798-9240   FAX: (330) 798-6968
http://www.nortonplastics.com

San Diego Instruments
7758 Arjons Drive
San Diego, CA 92126
(858) 530-2600   FAX: (858) 530-2646
http://www.sd-inst.com

Sandown Scientific
Beards Lodge
25 Oldfield Road
Hampden, Middlesex TW12 2AJ, UK
(44) 2089 793300
FAX: (44) 2089 793311
http://www.sandownsci.com

Sandoz Pharmaceuticals
See Novartis

Sanofi Recherche
Centre de Montpellier
371 Rue du Professor Blayac
34184 Montpellier, Cedex 04
France
(33) 67-10-67-10
FAX: (33) 67-10-67-67

Sanofi Winthrop Pharmaceuticals
90 Park Avenue
New York, NY 10016
(800) 223-5511   FAX: (800) 933-3243
(212) 551-4000
http://www.sanofi-synthelabo.com/us

Santa Cruz Biotechnology
2161 Delaware Avenue
Santa Cruz, CA 95060
(800) 457-3801   FAX: (831) 457-3801
(831) 457-3800
http://www.scbt.com

Sarasep
(800) 605-0267   FAX: (408) 432-3231
(408) 432-3230
http://www.transgenomic.com

Sarstedt
P.O. Box 468
Newton, NC 28658
(800) 257-5101   FAX: (828) 465-4003
(828) 465-4000
http://www.sarstedt.com

Sartorius
131 Heartsland Boulevard
Edgewood, NY 11717
(800) 368-7178   FAX: (516) 254-4253
http://www.sartorius.com

SAS Institute
Pacific Telesis Center
One Montgomery Street
San Francisco, CA 94104
(415) 421-2227   FAX: (415) 421-1213
http://www.sas.com

Savant/EC Apparatus
A ThermoQuest company
100 Colin Drive
Holbrook, NY 11741
(800) 634-8886   FAX: (516) 244-0606
(516) 244-2929
http://www.savec.com

Savillex
6133 Baker Road
Minnetonka, MN 55345
(612) 935-5427

Scanalytics
Division of CSP
8550 Lee Highway, Suite 400
Fairfax, VA 22031
(800) 325-3110   FAX: (703) 208-1960
(703) 208-2230
http://www.scanalytics.com

Schering Laboratories
See Schering-Plough

Schering-Plough
1 Giralda Farms
Madison, NJ 07940
(800) 222-7579   FAX: (973) 822-7048
(973) 822-7000
http://www.schering-plough.com

Schleicher & Schuell
10 Optical Avenue
Keene, NH 03431
(800) 245-4024   FAX: (603) 357-3627
(603) 352-3810
http://www.s-und-s.de/english-index.html

Science Technology Centre
1250 Herzberg Laboratories
Carleton University
1125 Colonel Bay Drive
Ottawa, Ontario, Canada K1S 5B6
(613) 520-4442   FAX: (613) 520-4445
http://www.carleton.ca/universities/stc

Scientific Instruments
200 Saw Mill River Road
Hawthorne, NY 10532
(800) 431-1956   FAX: (914) 769-5473
(914) 769-5700
http://www.scientificinstruments.com

Scientific Solutions
9323 Hamilton
Mentor, OH 44060
(440) 357-1400   FAX: (440) 357-1416
www.labmaster.com

Scion
82 Worman's Mill Court, Suite H
Frederick, MD 21701
(301) 695-7870   FAX: (301) 695-0035
www.scioncorp.com

Scott Specialty Gases
6141 Easton Road
P.O. Box 310
Plumsteadville, PA 18949
(800) 21-SCOTT   FAX: (215) 766-2476
(215) 766-8861
http://www.scottgas.com

Scripps Clinic and Research Foundation
Instrumentation and Design Lab
10666 N. Torrey Pines Road
La Jolla, CA 92037
(800) 992-9962   FAX: (858) 554-8986
(858) 455-9100
http://www.scrippsclinic.com

SDI Sensor Devices
407 Pilot Court, 400A
Waukesha, WI 53188
(414) 524-1000   FAX: (414) 524-1009

Sefar America
111 Calumet Street
Depew, NY 14043
(716) 683-4050   FAX: (716) 683-4053
http://www.sefaramerica.com

Seikagaku America
Division of Associates of Cape Cod
704 Main Street
Falmouth, MA 02540
(800) 237-4512   FAX: (508) 540-8680
(508) 540-3444
http://www.seikagaku.com

Sellas Medizinische Gerate
Hagener Str. 393
Gevelsberg-Vogelsang, 58285
Germany
(49) 23-326-1225

Sensor Medics
22705 Savi Ranch Parkway
Yorba Linda, CA 92887
(800) 231-2466   FAX: (714) 283-8439
(714) 283-2228
http://www.sensormedics.com

Sensor Systems LLC
2800 Anvil Street, North
Saint Petersburg, FL 33710
(800) 688-2181   FAX: (727) 347-3881
(727) 347-2181
http://www.vsensors.com

SenSym/Foxboro ICT
1804 McCarthy Boulevard
Milpitas, CA 95035
(800) 392-9934   FAX: (408) 954-9458
(408) 954-6700
http://www.sensym.com

Separations Group
See Vydac

Sepracor
111 Locke Drive
Marlboro, MA 01752
(877)-SEPRACOR   (508) 357-7300
http://www.sepracor.com

Sera-Lab
See Harlan Sera-Lab

Sermeter
925 Seton Court, #7
Wheeling, IL 60090
(847) 537-4747

Serological
195 W. Birch Street
Kankakee, IL 60901
(800) 227-9412   FAX: (815) 937-8285
(815) 937-8270

Seromed Biochrom
Leonorenstrasse 2-6
D-12247 Berlin, Germany
(49) 030-779-9060

Serotec
22 Bankside
Station Approach
Kidlington, Oxford OX5 1JE, UK
(44) 1865-852722
FAX: (44) 1865-373899
In the US: (800) 265-7376
http://www.serotec.co.uk

Serva Biochemicals
Distributed by Crescent Chemical

S.F. Medical Pharmlast
See Chase-Walton Elastomers

SGE
2007 Kramer Lane
Austin, TX 78758
(800) 945-6154   FAX: (512) 836-9159
(512) 837-7190
http://www.sge.com

Shandon/Lipshaw
171 Industry Drive
Pittsburgh, PA 15275
(800) 245-6212   FAX: (412) 788-1138
(412) 788-1133
http://www.shandon.com

Sharpoint
P.O. Box 2212
Taichung, Taiwan
Republic of China
(886) 4-3206320
FAX: (886) 4-3289879
http://www.sharpoint.com.tw

Shelton Scientific
230 Longhill Crossroads
Shelton, CT 06484
(800) 222-2092   FAX: (203) 929-2175
(203) 929-8999
http://www.sheltonscientific.com

Sherwood-Davis & Geck
See Kendall

Sherwood Medical
See Kendall

Shimadzu Scientific Instruments
7102 Riverwood Drive
Columbia, MD 21046
(800) 477-1227   FAX: (410) 381-1222
(410) 381-1227
http://www.ssi.shimadzu.com

Sialomed
See Amika

Siemens Analytical X-Ray Systems
See Bruker Analytical X-Ray Systems

Sievers Instruments
Subsidiary of Ionics
6060 Spine Road
Boulder, CO 80301 (800) 255-6964
FAX: (303) 444-6272
(303) 444-2009
http://www.sieversinst.com

SIFCO
970 East 46th Street
Cleveland, OH 44103
(216) 881-8600   FAX: (216) 432-6281
http://www.sifco.com

Sigma-Aldrich
3050 Spruce Street
St. Louis, MO 63103
(800) 358-5287   FAX: (800) 962-9591
(800) 325-3101   FAX: (800) 325-5052
http://www.sigma-aldrich.com

Silenus/Amrad
34 Wadhurst Drive
Boronia, Victoria 3155 Australia
(613)9887-3909   FAX: (613)9887-3912
http://www.amrad.com.au

Silicon Genetics
2601 Spring Street
Redwood City, CA 94063
(866) SIG SOFT   FAX: (650) 365 1735
(650) 367 9600
http://www.sigenetics.com

SIMS Deltec
1265 Grey Fox Road
St. Paul, Minnesota 55112
(800) 426-2448   FAX: (615) 628-7459
http://www.deltec.com

SIMS Portex
10 Bowman Drive
Keene, NH 03431
(800) 258-5361   FAX: (603) 352-3703
(603) 352-3812
http://www.simsmed.com

SIMS Portex Limited
Hythe, Kent CT21 6JL, UK
(44)1303-260551
FAX: (44)1303-266761
http://www.portex.com

Siris Laboratories
See Biosearch Technologies

Skatron Instruments
See Molecular Devices

SLM Instruments
See Spectronic Instruments

SLM-AMINCO Instruments
See Spectronic Instruments

Small Parts
13980 NW 58th Court
P.O. Box 4650
Miami Lakes, FL 33014
(800) 220-4242   FAX: (800) 423-9009
(305) 558-1038   FAX: (305) 558-0509
http://www.smallparts.com

Smith & Nephew
11775 Starkey Road
P.O. Box 1970
Largo, FL 33779
(800) 876-1261
http://www.smith-nephew.com

SmithKline Beecham
1 Franklin Plaza, #1800
Philadelphia, PA 19102
(215) 751-4000   FAX: (215) 751-4992
http://www.sb.com

Solid Phase Sciences
See Biosearch Technologies

SOMA Scientific Instruments
5319 University Drive, PMB #366
Irvine, CA 92612
(949) 854-0220   FAX: (949) 854-0223
http://somascientific.com

Somatix Therapy
See Cell Genesys

SOMEDIC Sales AB
Box 194
242 22 Hörby, Sweden
(46) 415-165-50
FAX: (46) 415-165-60
http://www.somedic.com

Sonics & Materials
53 Church Hill Road
Newtown, CT 06470
(800) 745-1105   FAX: (203) 270-4610
(203) 270-4600
http://www.sonicsandmaterials.com

Sonosep Biotech
See Triton Environmental Consultants

Sorvall
See Kendro Laboratory Products

Southern Biotechnology Associates
P.O. Box 26221
Birmingham, AL 35260
(800) 722-2255   FAX: (205) 945-8768
(205) 945-1774
http://SouthernBiotech.com

SPAFAS
190 Route 165
Preston, CT 06365
(800) SPAFAS-1   FAX: (860) 889-1991
(860) 889-1389
http://www.spafas.com

Specialty Media
Division of Cell & Molecular Technologies
580 Marshall Street
Phillipsburg, NJ 08865
(800) 543-6029   FAX: (908) 387-1670
(908) 454-7774
http://www.specialtymedia.com

Spectra Physics
See Thermo Separation Products

Spectramed
See BOC Edwards

SpectraSource Instruments
31324 Via Colinas, Suite 114
Westlake Village, CA 91362
(818) 707-2655   FAX: (818) 707-9035
http://www.spectrasource.com

Spectronic Instruments
820 Linden Avenue
Rochester, NY 14625
(800) 654-9955   FAX: (716) 248-4014
(716) 248-4000
http://www.spectronic.com

Spectrum Medical Industries
See Spectrum Laboratories

Spectrum Laboratories
18617 Broadwick Street
Rancho Dominguez, CA 90220
(800) 634-3300   FAX: (800) 445-7330
(310) 885-4601   FAX: (310) 885-4666
http://www.spectrumlabs.com

Spherotech
1840 Industrial Drive, Suite 270
Libertyville, IL 60048
(800) 368-0822   FAX: (847) 680-8927
(847) 680-8922
http://www.spherotech.com

SPSS
233 S. Wacker Drive, 11th floor
Chicago, IL 60606
(800) 521-1337   FAX: (800) 841-0064
http://www.spss.com

SS White Burs
1145 Towbin Avenue
Lakewood, NJ 08701
(732) 905-1100   FAX: (732) 905-0987
http://www.sswhiteburs.com

Stag Instruments
16 Monument Industrial Park
Chalgrove, Oxon OX44 7RW, UK
(44) 1865-891116
FAX: (44) 1865-890562

Standard Reference Materials Program
National Institute of Standards and Technology
Building 202, Room 204
Gaithersburg, MD 20899
(301) 975-6776   FAX: (301) 948-3730

Starplex Scientific
50 Steinway
Etobieoke, Ontario
M9W 6Y3 Canada
(800) 665-0954   FAX: (416) 674-6067
(416) 674-7474
http://www.starplexscientific.com

State Laboratory Institute of Massachusetts
305 South Street
Jamaica Plain, MA 02130
(617) 522-3700   FAX: (617) 522-8735
http://www.state.ma.us/dph

Stedim Labs
1910 Mark Court, Suite 110
Concord, CA 94520
(800) 914-6644   FAX: (925) 689-6988
(925) 689-6650
http://www.stedim.com

Steinel America
9051 Lyndale Avenue
Bloomington, MN 55420
(800) 852-4343   FAX: (952) 888-5132
http://www.steinelamerica.com

Stem Cell Technologies
777 West Broadway, Suite 808
Vancouver, British Columbia
V5Z 4J7 Canada
(800) 667-0322   FAX: (800) 567-2899
(604) 877-0713   FAX: (604) 877-0704
http://www.stemcell.com

Stephens Scientific
107 Riverdale Road
Riverdale, NJ 07457
(800) 831-8099   FAX: (201) 831-8009
(201) 831-9800

Steraloids
P.O. Box 689
Newport, RI 02840
(401) 848-5422   FAX: (401) 848-5638
http://www.steraloids.com

Sterling Medical
2091 Springdale Road, Ste. 2
Cherry Hill, NJ 08003
(800) 229-0900   FAX: (800) 229-7854
http://www.sterlingmedical.com

Sterling Winthrop
90 Park Avenue
New York, NY 10016
(212) 907-2000   FAX: (212) 907-3626

Sternberger Monoclonals
10 Burwood Court
Lutherville, MD 21093
(410) 821-8505   FAX: (410) 821-8506
http://www.sternbergermonoclonals.com

Stoelting
502 Highway 67
Kiel, WI 53042
(920) 894-2293   FAX: (920) 894-7029
http://www.stoelting.com

Stovall Lifescience
206-G South Westgate Drive
Greensboro, NC 27407
(800) 852-0102   FAX: (336) 852-3507
http://www.slscience.com

Stratagene
11011 N. Torrey Pines Road
La Jolla, CA 92037
(800) 424-5444   FAX: (888) 267-4010
(858) 535-5400
http://www.stratagene.com

Strategic Applications
530A N. Milwaukee Avenue
Libertyville, IL 60048
(847) 680-9385   FAX: (847) 680-9837

Strem Chemicals
7 Mulliken Way
Newburyport, MA 01950
(800) 647-8736   FAX: (800) 517-8736
(978) 462-3191   FAX: (978) 465-3104
http://www.strem.com

StressGen Biotechnologies
Biochemicals Division
120-4243 Glanford Avenue
Victoria, British Columbia
V8Z 4B9 Canada
(800) 661-4978   FAX: (250) 744-2877
(250) 744-2811
http://www.stressgen.com

Structure Probe/SPI Supplies
(Epon-Araldite)
P.O. Box 656
West Chester, PA 19381
(800) 242-4774   FAX: (610) 436-5755
http://www.2spi.com

Süd-Chemie Performance Packaging
101 Christine Drive
Belen, NM 87002
(800) 989-3374   FAX: (505) 864-9296
http://www.uniteddesiccants.com

Sumitomo Chemical
Sumitomo Building
5-33, Kitahama 4-chome
Chuo-ku, Osaka 541-8550, Japan
(81) 6-6220-3891
FAX: (81)-6-6220-3345
http://www.sumitomo-chem.co.jp

Sun Box
19217 Orbit Drive
Gaithersburg, MD 20879
(800) 548-3968   FAX: (301) 977-2281
(301) 869-5980
http://www.sunboxco.com

Sunbrokers
See Sun International

Sun International
3700 Highway 421 North
Wilmington, NC 28401
(800) LAB-VIAL   FAX: (800) 231-7861
http://www.autosamplervial.com

Sunox
1111 Franklin Boulevard, Unit 6
Cambridge, ON N1R 8B5, Canada
(519) 624-4413   FAX: (519) 624-8378
http://www.sunox.ca

Supelco
See Sigma-Aldrich

SuperArray
P.O. Box 34494
Bethesda, MD 20827
(888) 503-3187   FAX: (301) 765-9859
(301) 765-9888
http://www.superarray.com

Surface Measurement Systems
3 Warple Mews, Warple Way
London W3 ORF, UK
(44) 20-8749-4900
FAX: (44) 20-8749-6749
http://www.smsuk.co.uk/index.htm

SurgiVet
N7 W22025 Johnson Road, Suite A
Waukesha, WI 53186
(262) 513-8500   (888) 745-6562
FAX: (262) 513-9069
www.surgivet.com

Sutter Instruments
51 Digital Drive
Novato, CA 94949
(415) 883-0128   FAX: (415) 883-0572
http://www.sutter.com

Swiss Precision Instruments
1555 Mittel Boulevard, Suite F
Wooddale, IL 60191
(800) 221-0198   FAX: (800) 842-5164

Synaptosoft
3098 Anderson Place
Decatur, GA 30033
(770) 939-4366   FAX: 770-939-9478
http://www.synaptosoft.com

SynChrom
See Micra Scientific

Synergy Software
2457 Perkiomen Avenue
Reading, PA 19606
(800) 876-8376   FAX: (610) 370-0548
(610) 779-0522
http://www.synergy.com

Synteni
See Incyte

Synthetics Industry
Lumite Division
2100A Atlantic Highway
Gainesville, GA 30501
(404) 532-9756   FAX: (404) 531-1347

Systat
See SPSS

Systems Planning and Analysis (SPA)
2000 N. Beauregard Street
Suite 400
Alexandria, VA 22311
(703) 931-3500
http://www.spa-inc.net

3M Bioapplications
3M Center
Building 270-15-01
St. Paul, MN 55144
(800) 257-7459   FAX: (651) 737-5645
(651) 736-4946

T Cell Diagnostics and T Cell Sciences
38 Sidney Street
Cambridge, MA 02139
(617) 621-1400

TAAB Laboratory Equipment
3 Minerva House
Calleva Park
Aldermaston, Berkshire RG7 8NA, UK
(44) 118 9817775
FAX: (44) 118 9817881

Taconic
273 Hover Avenue
Germantown, NY 12526
(800) TAC-ONIC   FAX: (518) 537-7287
(518) 537-6208
http://www.taconic.com

Tago
See Biosource International

TaKaRa Biochemical
719 Alliston Way
Berkeley, CA 94710
(800) 544-9899   FAX: (510) 649-8933
(510) 649-9895
http://www.takara.co.jp/english

Takara Shuzo
Biomedical Group Division
Seta 3-4-1
Otsu Shiga 520-21, Japan
(81) 75-241-5100
FAX: (81) 77-543-9254
http://www.Takara.co.jp/english

Takeda Chemical Products
101 Takeda Drive
Wilmington, NC 28401
(800) 825-3328   FAX: (800) 825-0333
(910) 762-8666   FAX: (910) 762-6846
http://takeda-usa.com

TAO Biomedical
73 Manassas Court
Laurel Springs, NJ 08021
(609) 782-8622   FAX: (609) 782-8622

Tecan US
P.O. Box 13953
Research Triangle Park, NC 27709
(800) 33-TECAN   FAX: (919) 361-5201
(919) 361-5208
http://www.tecan-us.com

Techne
University Park Plaza
743 Alexander Road
Princeton, NJ 08540
(800) 225-9243   FAX: (609) 987-8177
(609) 452-9275
http://www.techneusa.com

Technical Manufacturing
15 Centennial Drive
Peabody, MA 01960
(978) 532-6330   FAX: (978) 531-8682
http://www.techmfg.com

Technical Products International
5918 Evergreen
St. Louis, MO 63134
(800) 729-4451   FAX: (314) 522-6360
(314) 522-8671
http://www.vibratome.com

Technicon
See Organon Teknika Cappel

Techno-Aide
P.O. Box 90763
Nashville, TN 37209
(800) 251-2629   FAX: (800) 554-6275
(615) 350-7030
http://www.techno-aid.com

Ted Pella
4595 Mountain Lakes Boulevard
P.O. Box 492477
Redding, CA 96049
(800) 237-3526   FAX: (530) 243-3761
(530) 243-2200
http://www.tedpella.com

Tekmar-Dohrmann
P.O. Box 429576 Cincinnati, OH 45242
(800) 543-4461   FAX: (800) 841-5262
(513) 247-7000   FAX: (513) 247-7050

Tektronix
142000 S.W. Karl Braun Drive
Beaverton, OR 97077
(800) 621-1966   FAX: (503) 627-7995
(503) 627-7999
http://www.tek.com

Tel-Test
P.O. Box 1421
Friendswood, TX 77546
(800) 631-0600   FAX: (281)482-1070
(281)482-2672
http://www.isotex-diag.com

TeleChem International
524 East Weddell Drive, Suite 3
Sunnyvale, CA 94089
(408) 744-1331   FAX: (408) 744-1711
http://www.gst.net/~telechem

Terrachem
Mallaustrasse 57
D-68219 Mannheim, Germany
0621-876797-0   FAX: 0621-876797-19
http://www.terrachem.de

Terumo Medical
2101 Cottontail Lane
Somerset, NJ 08873
(800) 283-7866   FAX: (732) 302-3083
(732) 302-4900
http://www.terumomedical.com

Tetko
333 South Highland Manor
Briarcliff, NY 10510
(800) 289-8385   FAX: (914) 941-1017
(914) 941-7767
http://www.tetko.com

TetraLink
4240 Ridge Lea Road
Suite 29
Amherst, NY 14226
(800) 747-5170 FAX: (800) 747-5171
http://www.tetra-link.com

TEVA Pharmaceuticals USA
1090 Horsham Road
P.O. Box 1090
North Wales, PA 19454
(215) 591-3000 FAX: (215) 721-9669
http://www.tevapharmusa.com

Texas Fluorescence Labs
9503 Capitol View Drive
Austin, TX 78747
(512) 280-5223 FAX: (512) 280-4997
http://www.teflabs.com

The Nest Group
45 Valley Road
Southborough, MA 01772
(800) 347-6378 FAX: (508) 485-5736
(508) 481-6223
http://world.std.com/~nestgrp

ThermoCare
P.O. Box 6069
Incline Village, NV 89450
(800) 262-4020
(775) 831-1201

Thermo Labsystems
8 East Forge Parkway
Franklin, MA 02038
(800) 522-7763 FAX: (508) 520-2229
(508) 520-0009
http://www.finnpipette.com

Thermometric
Spjutvagen 5A
S-175 61 Jarfalla, Sweden
(46) 8-564-72-200

Thermoquest
IEC Division
300 Second Avenue
Needham Heights, MA 02194
(800) 843-1113 FAX: (781) 444-6743
(781) 449-0800
http://www.thermoquest.com

Thermo Separation Products
Thermoquest
355 River Oaks Parkway
San Jose, CA 95134
(800) 538-7067 FAX: (408) 526-9810
(408) 526-1100
http://www.thermoquest.com

Thermo Shandon
171 Industry Drive
Pittsburgh, PA 15275
(800) 547-7429 FAX: (412) 899-4045
http://www.thermoshandon.com

Thomas Scientific
99 High Hill Road at I-295
Swedesboro, NJ 08085
(800) 345-2100 FAX: (800) 345-5232
(856) 467-2000 FAX: (856) 467-3087
http://www.wheatonsci.com/html/nt/
Thomas.html

Thomson Instrument
354 Tyler Road
Clearbrook, VA 22624
(800) 842-4752 FAX: (540) 667-6878
(800) 541-4792 FAX: (760) 757-9367
http://www.hplc.com

Thorn EMI
See Electron Tubes

Thorlabs
435 Route 206
Newton, NJ 07860
(973) 579-7227 FAX: (973) 383-8406
http://www.thorlabs.com

Tiemann
See Bernsco Surgical Supply

Timberline Instruments
1880 South Flatiron Court, H-2
P.O. Box 20356
Boulder, CO 80308
(800) 777-5996 FAX: (303) 440-8786
(303) 440-8779
http://www.timberlineinstruments.com

Tissue-Tek
A Division of Sakura Finetek USA
1750 West 214th Street
Torrance, CA 90501
(800) 725-8723 FAX: (310) 972-7888
(310) 972-7800
http://www.sakuraus.com

Tocris Cookson
114 Holloway Road, Suite 200
Ballwin, MO 63011
(800) 421-3701 FAX: (800) 483-1993
(636) 207-7651 FAX: (636) 207-7683
http://www.tocris.com

Tocris Cookson
Northpoint, Fourth Way
Avonmouth, Bristol BS11 8TA, UK
(44) 117-982-6551
FAX: (44) 117-982-6552
http://www.tocris.com

Tomtec
See CraMar Technologies

TopoGen
P.O. Box 20607
Columbus, OH 43220
(800) TOPOGEN
FAX: (800) ADD-TOPO
(614) 451-5810 FAX: (614) 451-5811
http://www.topogen.com

Toray Industries, Japan
Toray Building 2-1
Nihonbash-Muromach
2-Chome, Chuo-Ku
Tokyo, Japan 103-8666
(03) 3245-5115 FAX: (03) 3245-5555
http://www.toray.co.jp

Toray Industries, U.S.A.
600 Third Avenue
New York, NY 10016
(212) 697-8150 FAX: (212) 972-4279
http://www.toray.com

Toronto Research Chemicals
2 Brisbane Road
North York, Ontario M3J 2J8, Canada
(416) 665-9696 FAX: (416) 665-4439
http://www.trc-canada.com

TosoHaas
156 Keystone Drive
Montgomeryville, PA 18036
(800) 366-4875 FAX: (215) 283-5035
(215) 283-5000
http://www.tosohaas.com

Towhill
647 Summer Street
Boston, MA 02210
(617) 542-6636 FAX: (617) 464-0804

Toxin Technology
7165 Curtiss Avenue
Sarasota, FL 34231
(941) 925-2032 FAX: (9413) 925-2130
http://www.toxintechnology.com

Toyo Soda
See TosoHaas

Trace Analytical
3517-A Edison Way
Menlo Park, CA 94025
(650) 364-6895 FAX: (650) 364-6897
http://www.traceanalytical.com

Transduction Laboratories
See BD Transduction Laboratories

Transgenomic
2032 Concourse Drive
San Jose, CA 95131
(408) 432-3230 FAX: (408) 432-3231
http://www.transgenomic.com

Transonic Systems
34 Dutch Mill Road
Ithaca, NY 14850
(800) 353-3569 FAX: (607) 257-7256
http://www.transonic.com

Travenol Lab
See Baxter Healthcare

Tree Star Software
20 Winding Way
San Carlos, CA 94070
800-366-6045
http://www.treestar.com

Trevigen
8405 Helgerman Court
Gaithersburg, MD 20877
(800) TREVIGEN   FAX: (301) 216-2801
(301) 216-2800
http://www.trevigen.com

Trilink Biotechnologies
6310 Nancy Ridge Drive
San Diego, CA 92121
(800) 863-6801   FAX: (858) 546-0020
http://www.trilink.biotech.com

Tripos Associates
1699 South Hanley Road, Suite 303
St. Louis, MO 63144
(800) 323-2960   FAX: (314) 647-9241
(314) 647-1099
http://www.tripos.com

Triton Environmental Consultants
120-13511 Commerce Parkway
Richmond, British Columbia
V6V 2L1 Canada
(604) 279-2093   FAX: (604) 279-2047
http://www.triton-env.com

Tropix
47 Wiggins Avenue
Bedford, MA 01730
(800) 542-2369   FAX: (617) 275-8581
(617) 271-0045
http://www.tropix.com

TSI Center for Diagnostic Products
See Intergen

2000 Eppendorf-5 Prime
5603 Arapahoe Avenue
Boulder, CO 80303
(800) 533-5703   FAX: (303) 440-0835
(303) 440-3705

Tyler Research
10328 73rd Avenue
Edmonton, Alberta
T6E 6N5 Canada
(403) 448-1249   FAX: (403) 433-0479

UBI
See Upstate Biotechnology

Ugo-Basile
Via G. Borghi 43
21025 Comerio, Varese, Italy
(39) 332 744 574
FAX: (39) 332 745 488
http://www.ugobasile.com

UltraPIX
See Life Science Resources

Ultrasonic Power
239 East Stephenson Street
Freeport, IL 61032
(815) 235-6020   FAX: (815) 232-2150
http://www.upcorp.com

Ultrasound Advice
23 Aberdeen Road
London N52UG, UK
(44) 020-7359-1718
FAX: (44) 020-7359-3650
http://www.ultrasoundadvice.co.uk

UNELKO
14641 N. 74th Street
Scottsdale, AZ 85260
(480) 991-7272   FAX: (480)483-7674
http://www.unelko.com

Unifab Corp.
5260 Lovers Lane
Kalamazoo, MI 49002 (800) 648-9569
FAX: (616) 382-2825
(616) 382-2803

Union Carbide
10235 West Little York Road, Suite 300
Houston, TX 77040
(800) 568-4000   FAX: (713) 849-7021
(713) 849-7000
http://www.unioncarbide.com

United Desiccants
See Süd-Chemie Performance Packaging

United States Biochemical
See USB

United States Biological (US Biological)
P.O. Box 261
Swampscott, MA 01907
(800) 520-3011   FAX: (781) 639-1768
www.usbio.net

Universal Imaging
502 Brandywine Parkway
West Chester, PA 19380
(610) 344-9410   FAX: (610) 344-6515
http://www.image1.com

Upchurch Scientific
619 West Oak Street
P.O. Box 1529
Oak Harbor, WA 98277
(800) 426-0191   FAX: (800) 359-3460
(360) 679-2528   FAX: (360) 679-3830
http://www.upchurch.com

Upjohn
Pharmacia & Upjohn
http://www.pnu.com

Upstate Biotechnology (UBI)
1100 Winter Street, Suite 2300
Waltham, MA 02451
(800) 233-3991   FAX: (781) 890-7738
(781) 890-8845
http://www.upstatebiotech.com

USA/Scientific
346 SW 57th Avenue
P.O. Box 3565
Ocala, FL 34478
(800) LAB-TIPS   FAX: (352) 351-2057
(3524) 237-6288
http://www.usascientific.com

USB
26111 Miles Road
P.O. Box 22400
Cleveland, OH 44122
(800) 321-9322   FAX: (800) 535-0898
FAX: (216) 464-5075
http://www.usbweb.com

USCI Bard
Bard Interventional Products
129 Concord Road
Billerica, MA 01821
(800) 225-1332   FAX: (978) 262-4805
http://www.bardinterventional.com

UVP (Ultraviolet Products)
2066 W. 11th Street
Upland, CA 91786
(800) 452-6788   FAX: (909) 946-3597
(909) 946-3197
http://www.uvp.com

V & P Scientific
9823 Pacific Heights Boulevard, Suite T
San Diego, CA 92121
(800) 455-0644   FAX: (858) 455-0703
(858) 455-0643
http://www.vp-scientific.com

Valco Instruments
P.O. Box 55603
Houston, TX 77255
(800) FOR-VICI   FAX: (713) 688-8106
(713) 688-9345
http://www.vici.com

Valpey Fisher
75 South Street
Hopkin, MA 01748
(508) 435-6831   FAX: (508) 435-5289
http://www.valpeyfisher.com

Value Plastics
3325 Timberline Road
Fort Collins, CO 80525
(800) 404-LUER   FAX: (970) 223-0953
(970) 223-8306
http://www.valueplastics.com

Vangard International
P.O. Box 308
3535 Rt. 66, Bldg. #4
Neptune, NJ 07754
(800) 922-0784   FAX: (732) 922-0557
(732) 922-4900
http://www.vangard1.com

Varian Analytical Instruments
2700 Mitchell Drive
Walnut Creek, CA 94598
(800) 926-3000   FAX: (925) 945-2102
(925) 939-2400
http://www.varianinc.com

Varian Associates
3050 Hansen Way
Palo Alto, CA 94304
(800) 544-4636   FAX: (650) 424-5358
(650) 493-4000
http://www.varian.com

Vector Core Laboratory/National Gene Vector Labs
University of Michigan
3560 E MSRB II
1150 West Medical Center Drive
Ann Arbor, MI 48109
(734) 936-5843   FAX: (734) 764-3596

Vector Laboratories
30 Ingold Road
Burlingame, CA 94010
(800) 227-6666   FAX: (650) 697-0339
(650) 697-3600
http://www.vectorlabs.com

Vedco
2121 S.E. Bush Road
St. Joseph, MO 64504
(888) 708-3326   FAX: (816) 238-1837
(816) 238-8840
http://database.vedco.com

Ventana Medical Systems
3865 North Business Center Drive
Tucson, AZ 85705
(800) 227-2155   FAX: (520) 887-2558
(520) 887-2155
http://www.ventanamed.com

Verity Software House
P.O. Box 247
45A Augusta Road
Topsham, ME 04086
(207) 729-6767   FAX: (207) 729-5443
http://www.vsh.com

Vernitron
See Sensor Systems LLC

Vertex Pharmaceuticals
130 Waverly Street
Cambridge, MA 02139
(617) 577-6000   FAX: (617) 577-6680
http://www.vpharm.com

Vetamac
Route 7, Box 208
Frankfort, IN 46041
(317) 379-3621

Vet Drug
Unit 8
Lakeside Industrial Estate
Colnbrook, Slough SL3 0ED, UK

Vetus Animal Health
See Burns Veterinary Supply

Viamed
15 Station Road
Cross Hills, Keighley
W. Yorkshire BD20 7DT, UK
(44) 1-535-634-542
FAX: (44) 1-535-635-582
http://www.viamed.co.uk

Vical
9373 Town Center Drive, Suite 100
San Diego, CA 92121
(858) 646-1100   FAX: (858) 646-1150
http://www.vical.com

Victor Medical
2349 North Watney Way, Suite D
Fairfield, CA 94533
(800) 888-8908   FAX: (707) 425-6459
(707) 425-0294

Virion Systems
9610 Medical Center Drive, Suite 100
Rockville, MD 20850
(301) 309-1844   FAX: (301) 309-0471
http://www.radix.net/~virion

VirTis Company
815 Route 208
Gardiner, NY 12525
(800) 765-6198   FAX: (914) 255-5338
(914) 255-5000
http://www.virtis.com

Visible Genetics
700 Bay Street, Suite 1000
Toronto, Ontario M5G 1Z6, Canada
(888) 463-6844   (416) 813-3272
http://www.visgen.com

Vitrocom
8 Morris Avenue
Mountain Lakes, NJ 07046
(973) 402-1443   FAX: (973) 402-1445

VTI
7650 W. 26th Avenue
Hialeah, FL 33106
(305) 828-4700   FAX: (305) 828-0299
http://www.vticorp.com

VWR Scientific Products
200 Center Square Road
Bridgeport, NJ 08014
(800) 932-5000   FAX: (609) 467-5499
(609) 467-2600
http://www.vwrsp.com

Vydac
17434 Mojave Street
P.O. Box 867 Hesperia, CA 92345
(800) 247-0924   FAX: (760) 244-1984
(760) 244-6107
http://www.vydac.com

Vysis
3100 Woodcreek Drive
Downers Grove, IL 60515
(800) 553-7042   FAX: (630) 271-7138
(630) 271-7000
http://www.vysis.com

W&H Dentalwerk Bürmoos
P.O. Box 1
A-5111 Bürmoos, Austria
(43) 6274-6236-0
FAX: (43) 6274-6236-55
http://www.wnhdent.com

Wako BioProducts
See Wako Chemicals USA

Wako Chemicals USA
1600 Bellwood Road
Richmond, VA 23237
(800) 992-9256   FAX: (804) 271-7791
(804) 271-7677
http://www.wakousa.com

Wako Pure Chemicals
1-2, Doshomachi 3-chome
Chuo-ku, Osaka 540-8605, Japan
81-6-6203-3741   FAX: 81-6-6222-1203
http://www.wako-chem.co.jp/egaiyo/index.htm

Wallac
See Perkin-Elmer

Wallac
A Division of Perkin-Elmer
3985 Eastern Road
Norton, OH 44203
(800) 321-9632   FAX: (330) 825-8520
(330) 825-4525
http://www.wallac.com

Waring Products
283 Main Street
New Hartford, CT 06057
(800) 348-7195   FAX: (860) 738-9203
(860) 379-0731
http://www.waringproducts.com

Warner Instrument
1141 Dixwell Avenue
Hamden, CT 06514
(800) 599-4203   FAX: (203) 776-1278
(203) 776-0664
http://www.warnerinstrument.com

Warner-Lambert
Parke-Davis
201 Tabor Road
Morris Plains, NJ 07950
(973) 540-2000   FAX: (973) 540-3761
http://www.warner-lambert.com

Washington University Machine Shop
615 South Taylor
St. Louis, MO 63310
(314) 362-6186   FAX: (314) 362-6184

Waters Chromatography
34 Maple Street
Milford, MA 01757
(800) 252-HPLC   FAX: (508) 478-1990
(508) 478-2000
http://www.waters.com

Watlow
12001 Lackland Road
St. Louis, MO 63146 (314) 426-7431
FAX: (314) 447-8770
http://www.watlow.com

Watson-Marlow
220 Ballardvale Street
Wilmington, MA 01887
(978) 658-6168   FAX: (978) 988 0828
http://www.watson-marlow.co.uk

Waukesha Fluid Handling
611 Sugar Creek Road
Delavan, WI 53115
(800) 252-5200   FAX: (800) 252-5012
(414) 728-1900   FAX: (414) 728-4608
http://www.waukesha-cb.com

WaveMetrics
P.O. Box 2088
Lake Oswego, OR 97035
(503) 620-3001   FAX: (503) 620-6754
http://www.wavemetrics.com

Weather Measure
P.O. Box 41257
Sacramento, CA 95641
(916) 481-7565

Weber Scientific
2732 Kuser Road
Hamilton, NJ 08691
(800) FAT-TEST   FAX: (609) 584-8388
(609) 584-7677
http://www.weberscientific.com

Weck, Edward & Company
1 Weck Drive
Research Triangle Park, NC 27709
(919) 544-8000

Wellcome Diagnostics
See Burroughs Wellcome

Wellington Laboratories
398 Laird Road, Guelph
Ontario, Canada, N1G 3X7
(800) 578-6985   FAX: (519) 822-2849
http://www.well-labs.com

Wesbart Engineering
Daux Road
Billingshurst, West Sussex
RH14 9EZ, UK
(44) 1-403-782738
FAX: (44) 1-403-784180
http://www.wesbart.co.uk

Whatman
9 Bridewell Place
Clifton, NJ 07014
(800) 631-7290   FAX: (973) 773-3991
(973) 773-5800
http://www.whatman.com

Wheaton Science Products
1501 North 10th Street
Millville, NJ 08332
(800) 225-1437   FAX: (800) 368-3108
(856) 825-1100   FAX: (856) 825-1368
http://www.algroupwheaton.com

Whittaker Bioproducts
See BioWhittaker

Wild Heerbrugg
Juerg Dedual Gaebrisstrasse 8 CH
9056 Gais, Switzerland
(41) 71-793-2723
FAX: (41) 71-726-5957
http://www.homepage.swissonline.net/dedual/wild_heerbrugg

Willy A. Bachofen AG Maschinenfabrik
Utengasse 15/17
CH4005 Basel, Switzerland
(41) 61-681-5151
FAX: (41) 61-681-5058
http://www.wab.ch

Winthrop
See Sterling Winthrop

Wolfram Research
100 Trade Center Drive
Champaign, IL 61820
(800) 965-3726   FAX: (217) 398-0747
(217) 398-0700
http://www.wolfram.com

World Health Organization
Microbiology and Immunology Support
20 Avenue Appia
1211 Geneva 27, Switzerland
(41-22) 791-2602
FAX: (41-22) 791-0746
http://www.who.org

World Precision Instruments
175 Sarasota Center Boulevard
International Trade Center
Sarasota, FL 34240
(941) 371-1003   FAX: (941) 377-5428
http://www.wpiinc.com

Worthington Biochemical
Halls Mill Road
Freehold, NJ 07728
(800) 445-9603   FAX: (800) 368-3108
(732) 462-3838   FAX: (732) 308-4453
http://www.worthington-biochem.com

WPI
See World Precision Instruments

Wyeth-Ayerst
2 Esterbrook Lane
Cherry Hill, NJ 08003
(800) 568-9938   FAX: (858) 424-8747
(858) 424-3700

Wyeth-Ayerst Laboratories
P.O. Box 1773
Paoli, PA 19301
(800) 666-7248   FAX: (610) 889-9669
(610) 644-8000
http://www.ahp.com

Xenotech
3800 Cambridge Street
Kansas City, KS 66103
(913) 588-7930   FAX: (913) 588-7572
http://www.xenotechllc.com

Xillix Technologies
300-13775 Commerce Parkway
Richmond, British Columbia
V6V 2V4 Canada
(800) 665-2236   FAX: (604) 278-3356
(604) 278-5000
http://www.xillix.com

Xomed Surgical Products
6743 Southpoint Drive N
Jacksonville, FL 32216
(800) 874-5797   FAX: (800) 678-3995
(904) 296-9600   FAX: (904) 296-9666
http://www.xomed.com

Yakult Honsha
1-19, Higashi-Shinbashi 1-chome
Minato-ku Tokyo 105-8660, Japan
81-3-3574-8960

Yamasa Shoyu
23-8 Nihonbashi Kakigaracho
1-chome, Chuoku
Tokyo, 103 Japan
(81) 3-479 22 0095
FAX: (81) 3-479 22 3435

Yeast Genetic Stock Center
See ATCC

Yellow Spring Instruments
See YSI

YMC
YMC Karasuma-Gojo Building
284 Daigo-Cho, Karasuma Nisihiirr
Gojo-dori Shimogyo-ku
Kyoto, 600-8106, Japan
(81) 75-342-4567
FAX: (81) 75-342-4568
http://www.ymc.co.jp

YSI
1725-1700 Brannum Lane
Yellow Springs, OH 45387
(800) 765-9744   FAX: (937) 767-9353
(937) 767-7241
http://www.ysi.com

Zeneca/CRB
See AstraZeneca
(800) 327-0125   FAX: (800) 321-4745

Zivic-Miller Laboratories
178 Toll Gate Road
Zelienople, PA 16063
(800) 422-LABS   FAX: (724) 452-4506
(800) MBM-RATS   FAX: (724) 452-5200
http://zivicmiller.com

Zymark
Zymark Center
Hopkinton, MA 01748
(508) 435-9500   FAX: (508) 435-3439
http://www.zymark.com

Zymed Laboratories
458 Carlton Court
South San Francisco, CA 94080
(800) 874-4494   FAX: (650) 871-4499
(650) 871-4494
http://www.zymed.com

Zymo Research
625 W. Katella Avenue, Suite 30
Orange, CA 92867
(888) 882-9682   FAX: (714) 288-9643
(714) 288-9682
http://www.zymor.com

Zynaxis Cell Science
See ChiRex Cauldron

References

Adkison, K.D. and Shen, D.D. 1996. Uptake of valproic acid into rat brain is mediated by a medium-chain fatty acid transporter. *J. Pharmacol. Exp. Ther.* 276:1189–1200.

Alaouilsmali, M.H., Gervais, C., Brunette, S., Gouin, G., Hamil, M., Rando, R.F., and Bedard, J. 2000. A novel high throughput screening assay for HCVNS3 helicase activity. *Antiviral Res.* 46:181–193.

Alderton, W.K., Boyhan, A., and Lowe, P.N. 1998. Nitroarginine and tetrahydrobiopterin binding to the haem domain of neuronal nitric oxide synthase using a scintillation proximity assay. *Biochem. J.* 332:195–201.

Alexander, G.M., Grothusen, J.R., and Schwartzman, R.J. 1988. Flow dependent changes in the effective surface area of microdialysis probes. *Life Sci.* 42:595–601.

Alger, B.E., Dhanjal, S.S., Dingledine, R., Garthwaite, J., Henderson G., King, G.L., Lipton, P., North, A., Schwartzkroin, P.A., Sears, T.A., Segal, M., Whittingham, T.S., and Williams J. 1984. Appendix: Brain slice methods. *In* Brain Slices (R. Dingledine, ed.) pp. 381–437. Plenum, New York.

Allan, G.F., Hutchins, A., and Clancy, J. 1999. An ultrahigh-throughput screening assay for estrogen receptor ligands. *Anal. Biochem.* 275:243–247.

Alouani, S. 2000. Scintillation proximity binding assay. *Methods Mol. Biol.* 138:135–141

Altman, J. and Sudarshan, K. 1975. Postnatal development of locomotion in the laboratory rat. *Anim. Behav.* 23:896–920.

Amberg, G. and Lindefors, N. 1989. Intracerebral microdialysis: II. Mathematical studies of diffusion kinetics. *J. Pharmacol. Methods* 22:157–183.

Anagnostaras, S.G., Josselyn, S.A., Frankland, P.W., and Silva, A.J. 2000. Computer-assisted behavioral assessment of Pavlovian fear conditioning in mice. *Learn. Mem.* 7:58–72.

Anderson, A.C. and Patrick, J.R. 1934. Some early behavior patterns in the white rat. *Psych. Rev.* 41:480–496.

Anisman, H., deCatanzaro, D., and Remington, G. 1978. Escape performance following exposure to inescapable shock: Deficits in motor response maintenance. *J. Exp. Psychol. Anim. Behav. Process.* 4:197–218.

Anisman, H., Shanks, N., Zalcman, S., and Zacharko, R.M. 1992. Multisystem regulation of performance deficits induced by stressors: An animal model of depression. *In* Animal Models in Psychiatry, Vol. II (M. T. Martin-Iverson, ed.) pp. 1–55. Humana Press, Clifton, N.J.

Annett, L.E., Rogers, D.C., Hernandez, T.D., and Dunnett, T.B. 1992. Behavioral analysis of unilateral monamine depletions in the marmoset. *Brain* 115:825–856.

Attal, N., Jazat, F., Kayser, V., and Guilbaud, G. 1990. Further evidence for "pain-related" behaviours in a model of unilateral peripheral mononeuropathy. *Pain* 41:235–251.

Baker, H.J., Lindsey, J.R., and Manning, P.J. (eds.) 1979. The Laboratory Rat, Vol. I: Biology and Diseases. ACLAM Series. Academic Press, San Diego.

Baker, H.J., Lindsey, J.R., and Manning, P.J. (eds.) 1980. Vol. II: Research Applications. ACLAM Series. Academic Press, San Diego.

Ball, D.J., Argentieri, G., Krause, R., Lipinski, M., Robinson, R.L., Stoll, R.E., and Visscher, G.E. 1991. Evaluation of a microchip implant system used for identification in rats. *Lab. Anim. Sci.* 41:185–186.

Banks, M., Graber, P., Proudfoot, A.E.I., Arod, C.Y., Allet, B., Bernard, A.R., Sebille, E., McKinnon, M., Wells, T.N.C., and Solari, R. 1995. Soluble interluekin-5 receptor α-chain binding assays: Use for screening and analysis of interleukin-5 mutants. *Anal. Biochem.* 230:321–328.

Bard, A.G. and Faulkner, L.R. 1980. Electrical Methods, p. 153. John Wiley & Sons, New York.

Barker, L.M., Best, M.R. and Domjan, M. (eds.) 1977. Learning Mechanisms in Food Selection. Baylor University Press, Waco, Tex.

Barneoud, P., Lolivier, J., Sanger, D.J., Scatton, B., and Moser, P. 1997. Quantitative motor assessment in FALS mice: A longitudinal study. *Neuroreport* 8:2861–2865.

Bashir, Z.I., Alford, S., Davies, S.N., Randall, A.D., and Collingridge, G.L. 1991. Long-term potentiation of NMDA receptor-mediated synaptic transmission in the hippocampus. *Nature* 349:156–158.

Baudet, L., Roby, P., Boissonneault, M., Popoff, V., Johnson, R., Xie, H., and Hurt, S. 2003. FlashBlue GPCR scintillating beads. *In* 9th Annual Conference of Society for Biomolecular Screening. Society for Biomolecular Screening, Danbury, Ct.

Baum, E.Z., Johnston, S.H., Bebernitz, G.A., and Gluzman, Y. 1996. Development of a scintillation proximity assay for human cytomegalovirus protease using 33-phosphorus. *Anal. Biochem.* 237:129–134.

Bear, M.F., Connors, B.W., and Paradiso, M.A. 1996. Synaptic mechanisms of memory. *In* Neuroscience: Exploring the Brain. 1st ed. (T.S. Satterfield, ed.) pp. 546–575. Williams & Wilkins, Baltimore.

Becker, J.B., Breedlove, M.S., and Crews, D. (eds.) 1993. Behavioral Endocrinology. MIT Press, Cambridge, Mass.

Bederson, J.B., Pitts, L.H., Tsuji, M., Nishimura, M.C., Davis, R.L., and Bartkowski, H. 1986. Rat middle cerebral artery occlusion: Evaluation of the model and development of neurological examination. *Stroke* 17:472–476.

Bednar, R.A., Gaul, S.L., Hamill, T.G., Egbertson, M.S., Shafer, J.A., Hartman, G.D., Gould, R.J., and Bednar, B. 1998. Identification of low molecular weight GP IIb/IIIa antagonists that bind preferentially to activated platelets. *J. Pharmacol. Exp. Ther.* 285:1317–1326.

Bekkers, J.M. and Stevens, C.F. 1991. Excitatory and inhibitory autaptic currents in isolated hippocampal neurons maintained in cell culture. *Proc. Natl. Acad. Sci. U.S.A.* 88:7834–7838.

Belayev, L., Alonso, O.F., Busto, R., Zhao, W., and Ginsberg, M.D. 1996. Middle cerebral artery occlusion in the rat by intraluminal suture. Neurological and pathological evaluation of an improved model. *Stroke* 27:1616–1622.

Ben Ari, Y. 1985. Limbic seizure and brain damage produced by kainic acid: Mechanisms and relevance to human temporal lobe epilepsy. *Neuroscience* 14:375–403.

Bennett, G.J. and Hargreaves, K.M. 1990. Reply to Dr. Hirata and his colleagues. *Pain* 42:255.

Bennett, G.J. and Xie, Y.-K. 1988. A peripheral mononeuropathy in rat that produces disorders of pain sensation like those seen in man. *Pain* 33:87–107.

Bennett, J.P. and Yamamura, H.I. 1985. Neurotransmitter, hormone or drug receptor binding methods. *In* Neurotransmitter Receptor Binding (H.I. Yamamura, S.J. Enna, and M.J. Kuhar, eds.) pp. 61–89. Raven Press, New York.

Benveniste, H. 1989. Brain microdialysis. *J. Neurochem.* 52:1667–1679.

Benveniste, H. and Huttemeier, P.C. 1990. Microdialysis: Theory and application. *Prog. Neurobiol.* 35:195–215.

Beveridge, M., Park, Y.W., Hermes, J., Marenghi, A., Brophy, G., and Santos, A. 2000. Detection of p56(lck) kinase activity using scintillation proximity assay in 384-well format and imaging proximity assay in 384- and 1536-well format. *J. Biomol. Screen.* 5:205–211.

Bezanilla, F. 1985. A high capacity data recording device based on a digital audio processor and a video cassette recorder. *Biophys. J.* 47:437–441.

Birch, G.M., Black, T., Malcolm, S.K., Lai, M.T., Zimmerman, R.E., and Jaskunas, S.R. 1995. Purification of recombinant human rhinovirus 14 3C protease expressed in *Escherichia coli. Prot. Express. Purif.* 6:609–618.

Birzin, E.T. and Rohrer, S.P. 2002. High-throughput receptor-binding methods for somatostatin receptor 2. *Anal. Biochem.* 307:159–166.

Bishop, Y. (ed). 1998. The Veterinary Formulary, 4th ed. Pharmaceutical Press, London.

Blanton, M.G., Loturco, J.J., and Kriegstein, A.R. 1989. Whole cell recording from neurons in slices of reptilian mammalian cerebral cortex. *J. Neurosci. Methods* 30:203–210.

Blumstein, L.K. and Crawley, J.N. 1983. Further characterization of a simple automated exploratory model for the anxiolytic effects of benzodiazepines. *Pharmacol. Biochem. Behav.* 18:37–40.

Boje, K.M.K. 1995. Cerebrovascular permeability changes during experimental meningitis in the rat. *J. Pharmacol. Exp. Ther.* 274:1199–1203.

Bolivar, V.J., Pooler, O., and Flaherty, L. 2001. Inbred strain variation in contextual and cued fear conditioning behavior. *Mamm. Genome* 12:651–656.

Bonge, H., Hallen, S., Fryklund, J., and Sjostrom, J.E. 2000. Cytostar-T scintillating microplate assay for measurement of sodium-dependent bile acid uptake in transfected HEK-293 cells. *Anal. Biochem.* 282:94–101.

Bosse, R., Garlick, R., Brown, B., and Menard, L. 1998. Development of nonseparation binding and functional assays for G protein-coupled receptors for high throughput screening: Pharmacological characterization of the immobilized CCR5 receptor on FlashPlate. *J. Biomol. Screen.* 3:285–292.

Bradbury, M.W.B. 1979. The Concept of a Blood-Brain Barrier. John Wiley & Sons, New York.

Brandeis, R., Brandys, Y., and Yehuda, S. 1989. The use of the Morris water maze in the study of memory and learning. *Int. J. Neurosci.* 48:29–69.

Brandeis, R., Sapir, M., Kapon, Y., and Borelli, G. 1991. Improvement of cognitive function by MAO-B inhibitor L-deprenyl in aged rats. *Pharmacol. Biochem. Behav.* 39:297–304.

Brandish, P.E., Hill, L.A., Zheng, W., and Scolnick, E.M. 2003. Scintillation proximity assay of inositol phosphates in cell extracts: High-throughput measurement of G-protein-coupled receptor activation. *Anal. Biochem.* 313:311–318.

Braunwalder, A.F., Yarwood, D.R., Hall, T., Missbach, Lipson, K.E., and Sills, M.A. 1996a. A solid-phase assay for determination of protein tyrosine kinase activity of c-src using scintillating microtitration plates. *Anal. Biochem.* 234:23–26.

Braunwalder, A.F., Wennogle, L., Gay, B., Lipson, K.E., and Sills, M.A. 1996b. Application of scintillating microtiter plates to measure phosphopeptide interactions with the GRB2-SH2 binding domain. *J. Biomol. Screen.* 1:23–26.

Braveman, N.S. and Bronstein, P. 1985 (eds.) Experimental Assessments and Clinical Applications of Conditioned Taste Aversions. The New York Academy of Sciences, New York.

Bridges, R.S. 1984. A quantitative analysis of the roles of dosage, sequence, and duration of estradiol and progesterone exposure in the regulation of maternal behavior in the rat. *Endocrinology* 114:930–940.

Bridges, R.S., Zarrow, M.X., Gandelman, R., and Denenberg, V.H. 1972. Differences in maternal responsiveness between lactating and sensitized rats. *Dev. Psychobiol.* 5:123–127.

Brint, S., Jacewicz, M., Kiessling, M., Tanabe, J., and Pulsinelli, W. 1988. Focal brain ischaemia in the rat: Methods for reproducible neocortical infarction using tandem occlusion of the distal middle cerebral and ipsilateral common carotid arteries. *J. Cereb. Blood Flow Metab.* 8:474–485.

Brown, B. 1996. FlashPlate technology. *Pharm. Manufact. Int.* 23–25.

Brown, B.A., Cain, M., Broadbent, J., Tompkins, S., Henrich, G., Joseph, R., Casto, S., Harney, H., Greene, R., Delmondo, R., and Delmondo, S.N. 1997. FlashPlate technology. In High Throughput Screening: The Discovery of Bioactive Substances. (J.P. Devlin, ed.) pp. 317–328. Marcel Dekker, New York.

Bungay, P.M., Morrison, P.F., and Dedrick, R.L. 1990. Steady-state theory for quantitative microdialysis of solutes and water *in vivo* and *in vitro*. *Life Sci.* 46:105–119.

Bures, J., Bermudez-Rattoni, F., and Yamamoto, T. 1998. Conditioned Taste Aversion: Memory of a Special Kind. Oxford University Press, New York.

Burns, R.S., Chiueh, C.C., Markey, S.P., Ebert, M.H., Jacobowitz, D.M., and Kopin, I.J. 1983. A primate model of parkinsonism: Selective destruction of dopaminergic neurons in the pars compacta of the substantia nigra by *N*-methyl-4-phenyl-1,2,3,6-tetrahydropyridine. *Proc. Natl. Acad. Sci. U.S.A.* 890:4546–4550.

Caine, S.G., Lintz, R., Koob, G.F. 1993. Intravenous drug self-administration techniques in animals. *In* Behavioral Neuroscience (A. Sahgal, ed.) pp. 117–143. IRL Press, Oxford.

Carlsson, B. Haggblad, J. 1995. Quantitative determination of DNA-binding parameters for the human estrogen receptor in a solid-phase, nonseparation assay. *Anal. Biochem.* 232:172–179.

Carlsson, B., Ahola, H., and Haggblad, J. 1997. Application of a novel method for the comparison of DNA binding parameters of the two human thyroid hormone receptor subtypes hTRa1 and hTRb1. *J. Recept. Signal Transduct. Res.* 17:355–371.

Carpenter, J.W., Laethem, C., Hubbard, F.R., Eckols, T.K., Baez, M., D. McClure, Nelson, D.L., and Johnston, P.A., 2002. Configuring radioligand receptor binding assays for HTS using scintillation proximity assay technology. *Methods Mol. Biol.* 190:31–49.

Carroll, M.E., Carmona, G., and May, S.A. 1991. Modifying drug-reinforced behavior by altering the economic conditions of the drug and a non-drug reinforcer. *J. Exp. Anal. Behav.* 18:361–376.

Carter, R.J., Lione, L.A., Humby, T., Mangiarini, L., Mahal, A., Bates, G.P., Morton, A.J., and Dunnett, S.B. 1999. Characterization of progressive motor deficits in mice transgenic for the human Huntington's disease mutation. *J. Neurosci.* 19:3248–3257.

Chaplan, S.R., Bach, F.W., Pogrel, J.W., Chung, S.M., and Yaksh, T.L. 1994. Quantitative assessment of tactile allodynia in the rat paw. *J. Neurosci. Meth.* 53:55–63.

Chemical Rubber Company 1975. CRC Handbook of Biochemistry and Molecular Biology, Physical and Chemical Data, 3rd. ed. CRC Press, Boca Raton, Fla.

Cheng, Y.C. and Prusoff, W.H. 1973. Relationship between the inhibition constant (Ki) and the concentration of inhibitor which causes 50 percent inhibition (IC_{50}) of an enzymatic reaction. *Biochem. Pharmacol.* 22:3099–3108.

Cherry, J.A. 1993. Measurement of sexual behavior: Controls for variables. *Methods Neurosci.* 15:3–15.

Choi, Y., Yoon, Y.W., Na, H.S., Kim, S.H., and Chung, J.M. 1994. Behavioral signs of ongoing pain and cold allodynia in a rat model of neuropathic pain. *Pain* 59:369–376.

Church, W.H. and Justice, J.B. 1989. Online small-bore chromatography for neurochemical analysis in the brain. *In* Advances in Chromatography (J.C. Giddings, J.C. Grushka, and R.P. Brown, eds.). Marcel Dekker, New York.

Cini, M., Fariello, R.G., Bianchetti, A., and Moretti, A. 1994. Studies on lipid peroxidation in rat brain. *Neurochem. Res.* 19:283–288.

Clarke, K.A. and Still, J. 1999. Gait analysis in the mouse. *Physiol. Behav.* 66:723–729.

Claro, F., Del Abril, A., Segovia, S., and Guillamon, A. 1990. SBR: A computer program to record and analyze sexual behavior in rodents. *Physiol. Behav.* 58:589–593.

Clemens, J.A., Saunders, R.D., Ho, P.P., Phebus, L.A., and Panetta, J.A. 1993. The antioxidant LY231617 reduces global ischemic neuronal injury in rats. *Stroke* 24:716–723.

Cook, N.D. 1996. Scintillation proximity assay: A versatile high-throughput screening technology. *Drug Disc. Today* 1:287–294.

Corbit, J.D. and Stellar, E. 1964. Palatability, food intake and obesity in normal and hyperphagic rats. *J. Comp. Physiol. Psychol.* 58:63–67.

Cosford, R.J., Vinson, A.P., Kukoyi, S., and Justice, J.B. Jr. 1996. Quantitative microdialysis of serotonin and norepinephrine: Pharmacological influences on in vivo extraction fraction. *J. Neurosci. Methods* 68:39–47.

Costall, B., Jones, B.J., Kelly, M.E., Naylor, R.J., and Tomkins, D.M. 1989. Exploration of mice in a black and white test box: Validation as a model of anxiety. *Pharmacol. Biochem. Behav.* 32:777–785.

Crawley, J.N. 1981. Neuropharmacological specificity of a simple animal model for the behavioral actions of benzodiazepines. *Pharmacol. Biochem. Behav.* 15:695–699.

Crawley, J.N. 2000. Sensory abilities: Olfaction, vision, hearing, taste, touch, nociception. *In* What's Wrong with My Mouse?: Behavioral Phenotyping of Transgenic and Knockout Mice. pp. 65–82. John Wiley & Sons, New York.

Crowther, J.R. 1995. ELISA theory and practice. *In* Methods in Molecular Biology, Vol. 42. (J.M. Walker, series ed.) Humana Press, Totowa, N.J.

Cushing, A., PriceJones, M.J., Graves, R., Harris, A.J., Hughes, K.T., Bleakman, D., and Lodge, D. 1999. Measurement of calcium flux through ionotropic glutamate receptors using Cytostar-T scintillating microplates. *J. Neurosci. Methods* 90:33–36.

Cussac, D., Newman-Tancredi, A., Duqueyroix, D., Pasteau, V., and Millan, M.J., 2002. Differential activation of Gq/1 and Gi(3) proteins at 5-hydroxytryptamine(2C) receptors revealed by antibody capture assays: Influence of receptor reserve and relationship to agonist-directed trafficking. *Mol. Pharm.* 62:578–589.

Dairaghi, D.J., Oldham, E.R., Bacon, K.B., and Schall, T.J. 1997. Chemokine receptor CCR3 function is highly dependent on local pH and ionic strength. *J. Biol. Chem.* 272:28206–28209.

Davis, G.C., Williams, A.C., Markey, S.P., Ebert, M.N., Caine, E.D., Reichert, C.M., and Kopin, I.J. 1979. Chronic parkinsonism secondary to intravenous injection of meperidine analogs. *Psychiatry Res.* 1:249–254.

Davis, M. 1984. The mammalian startle response. *In* Neural Mechanisms of the Startle Behavior (R.C. Eaton, ed.). Plenum Press, New York.

DeLapp, N.W., McKinzie, J.H., Sawyer, B.D., Vandergriff, A., Falcone, J., McClure, D., and Felder, C.C. 1999. Determination of [S-35]guanosine-5'-O-(3-thio)triphosphate binding mediated by cholinergic muscarinic receptors in membranes from Chinese hamster ovary cells and rat striatum using an anti-G protein scintillation proximity assay. *J. Pharm. Exp. Ther.* 289:946–955.

Dillon, K.J., Smith, G.C., and Martin, N.M. 2003. A FlashPlate assay for the identification of PARP-1 inhibitors. *J. Biomol. Screen.* 8:347–352.

Dixon, C.E., Lyeth, B.G., Povlishock, J.T., Findling, R.L., Hamm, R.J., Marmarou, A., Young, H.F., and Hayes, R.L. 1987. A fluid percussion model of experimental brain injury in the rat. *J. Neurosurg.* 67:110–119.

Donnan, G.A., Kaczmarczyk, S.J., Mckenzie, J.S., Rowe, P.J., Kalnins, R.M., and Mendelsohn, F.A. 1987. Regional and temporal effects of 1-methyl-4-phenyl-1,2,3,6-tetrahydropyridine on dopamine uptake sites in mouse brain. *J. Neurol. Sci.* 81:261–271.

Donovan, J. and Brown, P. 1995. Parenteral injections. *In* Current Protocols in Immunology (J.E. Coligan, A.M. Kruisbeek, D.H. Margulies, E.M. Shevach, and W. Strober, eds.) pp. 1.6.1-1.6.10. John Wiley & Sons, New York.

Dorsch, J.A. and Dorsch, S.E. 1999. Understanding Anesthesia Equipment, 4th ed. Williams & Wilkins, Baltimore.

Dougherty, P.M., Garrison, C.J., and Carlton, S.M. 1992. Differential influence of local anesthetic upon two models of experimentally-induced peripheral mononeuropathy in the rat. *Brain Res.* 570:109–115.

Dykstra, K.H., Hsiao, J.K., Morrison, P.F., Bungay, P.M., Mefford, I.N., Scully, M.M., and Dedrick, R.L. 1992. Quantitative examination of tissue concentration profiles associated with microdialysis. *J. Neurochem.* 58:931–940.

Earnshaw, D.L. and Pope, A.J. 2001. FlashPlate scintillation proximity assays for characterization and screening of DNA polymerase, primase, and helicase activities. *J. Biomol. Screen.* 6:39–46.

Engblom, A.C. and Kerman, K.E.O. 1993. Determination of the intracellular free Cl⁻ concentration in rat brain synaptoneurosomes using a Cl⁻ sensitive fluorescent indicator. *Biochim. Biophys. Acta* 1153:262–266.

Eshleman, A.J., Calligaro, D.O., and Eldefrawi, M.E., 1993. Allosteric regulation by sodium of the binding of [^{3}H]cocaine and [^{3}H]GBR 12935 to rat and bovine striata. *Membr. Biochem.* 10:129–144.

Eshleman A.J., Neve, R.L., Janowsky, A., and Neve, K.A. 1995. Characterization of a recombinant human dopamine transporter in multiple cell lines. *J. Pharmacol. Exp. Ther.* 274:276–283.

Fanselow, M.S. 1990. Factors governing one-trial contextual conditioning. *Animal Learn. Behav.* 18:264–270.

Favaron, M., Manev, H., Alho, H., Bertolino, M., Ferret, B., Guidotti, A., and Costa, E. 1988. Gangliosides prevent glutamate and kainate neurotoxicity in primary neuronal cultures of neonatal rat cerebellum and cortex. *Proc. Natl. Acad. Sci. U.S.A.* 85:7351–7355.

Fazeli, M.S. and Collingridge, G.L. 1996. Cortical Plasticity LTP and LTD. Molecular and Cellular Neurobiology Series. 1st ed. BIOS Scientific Publishers, Oxford.

Feeney, D.M., Gonzalez, A.M., and Law, W.A. 1982. Amphetamine, haloperidol, and experience interact to affect rate of recovery after motor cortex injury. *Science* 217:855–857.

Felder, C.C., Kanterman, R.Y., Ma, A.L., and Axelrod, J. 1989. A transfected m1 muscarinic acetylcholine receptor stimulates adenylate cyclase via phosphatidylinositol hydrolysis. *J. Biol. Chem.* 264:20356–20362.

Fenwick, S., Jenner, W.N., Linacre, P., Rooney, R.M., and Wring, S.A. 1994. Application of the scintillation proximity assay technique to the determination of drugs. *Anal. Proc. Incl. Anal. Commun.* 31:103–106.

File, S.E. 1980. The use of social interaction as a method for detecting anxiolytic activity of chlordiazepoxide-like drugs. *J. Neurosci. Methods* 2:219–238.

File, S.E. 1993. The social interaction test of anxiety. *Neurosci. Protocols* 93-010-01-01-07.

File, S.E. and Hyde, J.R.G. 1978. Can social interaction be used to measure anxiety? *Br. J. Pharmacol.* 62:19–24.

File, S.E. and Seth, P. 2003. A review of 25 years of the social interaction test. *Eur. J. Pharmacol.* 463:35–53.

Fitch, T., Adams, B., Chaney, S., Gerlai, R. 2002. Force transducer-based movement detection in fear conditioning in mice: A comparative analysis. *Hippocampus* 12:4–17.

Flecknell, P.A. 1996. Laboratory Animal Anaesthesia: A Practical Introduction for Research Workers and Technicians. Academic Press, San Diego.

Flecknell, P.A. 1998. Analgesia in small mammals. *Semin. Avian Exotic Pet Med.* 7:41–47.

Flecknell, P.A., Roughan, J.V., and Stewart, R. 1999. Use of oral buprenorphine ("buprenorphine jello") for postoperative analgesia in rats—a clinical trial. *Lab. Anim.* 33:169–174.

Fleming, A.S. and Corter, C. 1995. Psychobiology of maternal behavior in non-human mammals: Role of sensory, experiential and neural factors. *In* Handbook of Parenting (M. Bornstein, ed.) pp. 59–86. L. Erlbaum Associates, N.J.

Fleming, A.S., Cheung, U., Myhal, N., and Kessler, Z. 1989. Effects of maternal hormones on "timidity" and attraction to pup-related odors in female rats. *Physiol. Behav.* 46:449–453.

Fleming, A.S., Korsmit, M., and Deller, M. 1994. Rat pups are potent reinforcers to the maternal animal: Effects of experience, parity, hormones and dopamine function. *Psychobiology* 22:44–53.

Forscher, P. and Oxford, G.S. 1985. Modulation of calcium channels by norepinephrine in internally perfused dialyzed avian sensory neurons. *J. Gen. Physiol.* 85:743–763.

Foster, H.L., Small, J.D., and Fox, J.G. (eds.) 1981–1983. The Mouse in Biomedical Research, Vols. I to IV. ACLAM Series. Academic Press, San Diego.

Fowler, A., Price-Jones, M., Hughes, K., Anson, J., Lingham, R., and Schulman, M. 2000. Development of a high throughput scintillation proximity assay for hepatitis C virus NS3 protease that reduces the proportion of competitive inhibitors identified. *J. Biomol. Screen.* 5:153–158.

Fox, A., Eastwood, C., Gentry, C., Manning, D., and Urban, L. 1999. Critical evaluation of the streptozotocin model of painful diabetic neuropathy in the rat. *Pain* 81:307–316.

Fox, J.G., Anderson, L.C., Lowe, F.M., and Quimby, F.W. (eds.) 2002. Laboratory Animal Medicine, 2nd ed., pp. 179–180. Academic Press, San Diego.

Frankland, P.W., Cestari, V., Filipkowski, R.K., McDonald, R.J., Silva, A.J. 1998. The dorsal hippocampus is essential for context discrimination but not for contextual conditioning. *Behav. Neurosci.* 112:863–874.

Freeman, M.E. 1994. The neuroendocrine control of the ovarian cycle of the rat. *In* The Physiology of Reproduction, Vol. 2 (E. Knobil and J.D. Neill, eds.) pp. 613–658. Raven Press, New York.

French, R.J. and Wonderlin, W.F. 1992. Software for acquisition and analysis of ion channel data: Choices, tasks, and strategies. *Methods Enzymol.* 207:711–728.

Frolik, C.A., Black, E.C., Chandrasekhar, S., and Adrian, M.D. 1998. Development of a scintillation proximity assay for high-throughput measurement of intact parathyroid hormone. *Anal. Biochem.* 265:216–224.

Gahwiler, B.H., Capogna, M., Debanne, D., McKinney, R.A., and Thompson, S.M. 1997. Organotypic slice cultures: A technique has come of age. *Trends Neurosci.* 20:471–477.

Gahwiler, B.H., Thompson, S.M., McKinney, R.A., Debanne, D., and Robertson, R.T. 1998. Organotypic slice cultures of neural tissue. *In* Culturing Nerve Cells (G. Banker K. Goslin, eds.), Chapter 17. MIT Press, Cambridge, Mass.

Gallagher, S.R. 2006. Quantitation of DNA and RNA with absorption and fluorescence spectroscopy. *In* Short Protocols In Neuroscience: Cellular and Molecular Methods. (C.R. Gerfen, M.A. Rogawski, D.R. Sibley, P. Skolnick, and S, Wray, eds.) pp. A2-24 to A2-27. John Wiley & Sons, Hoboken, N.J.

Garcia, J. and Ervin, F.R. 1968. Gustatory-visceral and telereceptor-cutaneous conditioning: Adaptation in internal and external milieus. *Commun. Behav. Biol.* 1:389–415.

Gerfen, C.R. 2006. Basic neuroanatomical methods. *In* Short Protocols In Neuroscience: Cellular and Molecular Methods. (C.R. Gerfen, M.A. Rogawski, D.R. Sibley, P. Skolnick, and S, Wray, eds.) pp. 1–2 to 1–7. John Wiley & Sons, Hoboken, N.J.

Gervay, J. and McReynolds, K.D. 1999. Utilization of ELISA technology to measure biological activities of carbohydrates relevant in disease states. *Curr. Med. Chem.* 6:129–153.

Gevi, M. and Domenici, E. 2002. A scintillation proximity assay amenable for screening and characterization of DNA gyrase B subunit inhibitors. *Anal. Biochem.* 300:34–39.

Gill, R., Foster, and A.C., Woodruff, G.N. 1987. Systemic administration of MK-801 protects against ischemia-induced hippocampal neurodegeneration in the gerbil. *J. Neurosci.* 7:3343–3349.

Goverman, J., Woods, A., Larson, L., Weiner, L.P., Hood, L., and Zaller, D.M. 1993. Transgenic mice that express a myelin basic protein-specific T cell receptor develop spontaneous autoimmunity. *Cell* 72:551–560.

Graham, F. 1975. The more or less startling effects of weak prestimuli. *Psychophysiology* 12:238–248.

Grahame, N.J. and Cunningham, C.L. 1997. Intravenous ethanol self-administration in C57BL/6J and DBA/2J mice. *Alcohol. Clin. Exp. Res.* 21:56–62.

Graves, R., Davies, R., Brophy, G., O'Beirne, G., and Cook, N. 1997a. Noninvasive, real-time method for the examination of thymidine uptake events: Application of the method to V-79 cell synchrony studies. *Anal. Biochem.* 248:251–257.

Graves, R., Davies, R., Owen, P., Clynes, M., Cleary, I., O'Beirne, G. 1997b. An homogeneous assay for measuring the uptake and efflux of radiolabelled drugs in adherent cells. *J. Biochem. Biophys. Methods* 34:177–187.

Gray, E.C. and Whittaker, V.J. 1962. The isolation of nerve endings from brain: An electron-microscope study of cell fragments derived by homogenization and centrifugation. *J. Anat.* 96:79–87.

Graziani, F., Aldegheri, L., and Terstappen, G.C. 1999. High throughput scintillation proximity assay for the identification of FKBP-12 ligands. *J. Biomol. Screen.* 4:3–7.

Grill, H.J. and Kaplan, J.M. 1990. Caudal brainstem participates in the distributed neural control of feeding. *In* Handbook of Behavioral Neurobiology: Neurobiology of Food and Fluid Intake (E.M. Stricker, ed.) pp. 125–150. Plenum, New York.

Grynkiewicz, G., Poenie, M., and Tsien, R.Y. 1985. A new generation of Ca^{2+} indicators with greatly improved fluorescence properties. *J. Biol. Chem.* 260:3440–3450.

Gurney, M.E., Pu, H., Chiu, A.Y., Dal Canto, M.C., Polchow, C.Y., Alexander, D.D., Caliendo, J., Hentati, A., Kwon, Y.W., Deng, H.-X., Chen, W., Zhai, P., Sufit, R.L., and Siddique, T. 1994. Motor neuron degeneration in mice that express a human Cu,Zn superoxide dismutase mutation. *Science* 264:1772–1775.

Gurney, M.E., Cuttings, F.B., Zhai, P., Doble, A., Taylor, C.P., Andrus, P.K., and Hall, E.D. 1996. Benefit of vitamin E, riluzole, and gabapentin in a transgenic model of familial amyotrophic lateral sclerosis. *Ann. Neurol.* 39:147–157.

Gyenes, M., Farrant, M., and Farb, D.H. 1988. "Run-down" of γ-aminobutyric $acid_A$ receptor function during whole-cell recording: A possible role for phosphorylation. *Mol Pharmacol.* 34:719–723.

Haggblad, J., Carlsson, B., Kivela, P., and Siitari, H. 1995. Scintillating microtitration plates as platform for determination of [^{3}H]estradiol binding constants for hER-HBD. *BioTechniques* 18:146–151.

Halliwell, B. and Gutteridge, J.M.C. 1991. Free Radicals in Biology and Medicine, 2nd ed. Clarendon, Oxford.

Hallman, H., Lange, J., Olson, L., Stromberg, I., Johnsson, G. 1985. Neurochemical and histochemical characterization of neurotoxic effects of 1-methyl-4-phenyl-1,2,3,6-tetrahydropyridine on brain catecholamine neurons in the mouse. *J. Neurochem.* 44:117–127.

Hamberger, A., Berthold, C.-H., Karlsson, B., Lehmann, A., and Nystrom, B. 1983. Extracellular GABA, glutamate, and glutamine in vivo perfusion dialysis of the rabbit hippocampus. *In* Neurology and Neurobiology, Vol. 7, Glutamine, Glutamate and GABA in the Central Nervous System (L. Herz, E. Krammer, E.G. McGee, and A. Schousboe, eds.), pp. 473–492. Wiley-Liss, New York.

Hamill, O.P., Marty, A., Neher, E., Sakmann, B., and Sigworth, F.J. 1981. Improved patch-clamp techniques for high-resolution current recording from cells and cell-free membrane patches. *Pfluegers Arch. Eur.J. Physiol.* 391:85–100.

Hancock, A.A., Vodenlich, A.D., Maldonado, C., and Janis, R. 1995. α_2-Adrenergic agonist-induced inhibition of cyclic AMP formation in transfected cell lines using a microtiter-based scintillation proximity assay. *J. Recept. Signal Transduct. Res.* 15:557–579.

Hanley, M. 1985. Peptide binding assays. *In* Neurotransmitter Receptor Binding (H.I. Yamamura, S.J. Enna, and M.J. Kuhar, eds.) pp. 91–101. Raven Press, New York.

Hanselman, J.C., Schwab, D.A., Rea, T.J., Bisgaier, C.L., and Pape, M.E. 1997. A cDNA-dependent scintillation proximity assay for quantifying apolipoprotein A-1. *J. Lipid Res.* 38:2365–2373.

Hansen, M.B, Nielson, S.E, and Berg, K. 1989. Re-examination and further development of a precise and rapid dye method for measuring cell growth/cell kill. *J. Immunol. Methods* 119:203–210.

Hargreaves, K., Dubner, R., Brown, F., Flores, C., and Joris, J. 1988. A new and sensitive method for measuring thermal nociception in cutaneous hyperalgesia. *Pain* 32:77–88.

Harootunian, A.T., Kao, J.P.Y., Eckert, B.K., and Tsien, R.Y. 1989. Fluorescence ratio imaging of cytosolic free Na^+ in individual fibroblasts and lymphocytes. *J. Biol. Chem.* 264:19458–19467.

Harris, D.W., Kenrick, M.K., Pither, R.J., Anson, J.G., Jones, D.A. 1996. Development of a high-volume in situ mRNA hybridization assay for the quantification of gene expression utilizing scintillating microplates. *Anal. Biochem.* 243:249–256.

Haugland, R.P. 1993. Intracellular ion indicators. *In* Fluorescent and Luminescent Probes for Biological Activity (W.T. Mason, ed.) pp. 34–43. Academic Press, London.

Hecht, J. 1993. Understanding Lasers. An Entry-Level Guide. 2nd ed. The Institute of Electrical and Electronics Engineers Press, New York.

Heikkila, R.E., Hess, A., and Duvoisin, R.C. 1985. Dopaminergic neurotoxicity of 1-methyl-4-phenyl-1,2,3,6-tetra-hydropyridine (MPTP) in the mouse: Relationships between monoamine oxidase, MPTP metabolism and neurotoxicity. *Life Sci.* 36:231–236.

Heinemann, S.H. 1995. Guide to data acquisition and analysis. *In* Single-Channel Recording (B. Sakmann and E. Neher, eds.) pp. 53–91. Plenum, New York.

Hellier, J.L., Patrylo, P.R., Dou, P., Nett, M., Rose, G.M., and Dudek, F.E. 1999. Assessment of inhibition and epileptiform activity in the septal dentate gyrus of freely behaving rats during the first week after kainate treatment. *J. Neurosci.* 19:10053-10064.

Higgins, G.A, Bradbury, A.J., Jones, B.J., and Oakley, N.R. 1988. Behavioural and biochemical consequences following activation of 5-HT_1-like and GABA receptors in the dorsal raphe nucleus of the rat. *Neuropharmacology* 27:993-1001.

Hille, B. 1992. Ion Channels of Excitable Membranes. Sinauer Associates, Sunderland, Mass.

Hirata, H., Pataky, A., Kajander, K., LaMotte, R.H., and Collins, J.G. 1990. A model of peripheral mononeuropathy in the rat. *Pain* 42:253-254.

Hodges, H., Green, S., Glenn, B. 1987. Evidence that the amygdala is involved in benzodiazepine and serotonergic effects on punished responding but not on discrimination. *Psychopharmacology* 92:491-504.

Hodgkin, A.L. and Huxley, A.F. 1952. A quantitative description of membrane current and its application to conduction and excitation in nerve. *J. Physiol.* 117:500-544.

Hoffman, H.S. and Searle, J.L. 1968. Acoustic and temporal factors in the evocation of startle. *J. Acoust. Soc. Am.* 43:269-282.

Holland, J.D., Singh, P., Brennand, J.C., Garman, A.J. 1994. A nonseparation microplate receptor binding assay. *Anal. Biochem.* 222:516-518.

Hollingsworth, E.B., McNeal, E.T., Burton, J.L., Williams, R.J., Daly, J.W., and Creveling, C.R. 1985. Biochemical characterization of a filtered synaptoneurosome preparation from guinea pig cerebral cortex: Cyclic adenosine $3',5'$-monophosphate-generating systems, receptors and enzymes. *J. Neurosci.* 5:2240-2253.

Holmes, G.M., Holmes, D.G., Sachs, B.D. 1988. An IBM-PC based data collection system for recording rodent sexual behavior and for general event recording. *Physiol. Behav.* 55:825-828.

Hood, C.M., Kelly, V.A., Bird, M.I., Britten, C.J. 1998. Measurement of a(1-3)fucosyltransferase activity using scintillation proximity. *Anal. Biochem.* 255:8-12.

Horn, R. Patlak, J. 1980. Single channel currents from excised patches of muscle membrane. *Proc. Natl. Acad. Sci. U.S.A.* 77:6930-6934.

Hosoya, K., Sugawara, M., Asaba, H., Terasaki, T. 1999. Blood-brain barrier produces significant efflux of L-aspartic acid but not D-aspartic acid: In vivo evidence using the brain efflux index method. *J. Neurochem.* 73:1206-1211.

Howard, J.L. and Pollard, G.T. 1991. Effects of drugs on punished behavior: Preclinical test for anxiolytics. In Psychopharmacology of Anxiolytics and Antidepressants (S.E. File, ed.) pp. 131-153. Pergamon Press, New York.

Hsiao, J.K., Ball, B.A., Morrison, P.F., Mefford, I.N., and Bungay, P. 1990. Effects of different semipermeable membranes on *in vitro* and *in vivo* performance of microdialysis probes. *J. Neurochem.* 54:1449-1452.

Hutcheson, D.M., Matthes, H.W.D., Valjent, E., Sanchez-Blazquez, P., Rodriguez-Diaz, M., Garzon, J., Kieffer, B.L., and Maldonado, R. 2001. Lack of dependence and rewarding effects of deltorphin II in mu-opioid receptor-deficient mice. *Eur. J. Pharmacol.* 13:153-161.

Iatridou, H., Foukaraki, E., Kuhn, M.A., Marcus, E.M., Haugland, R.P., and Katerinopoulos, H.E. 1994. The development of a new family of intracellular calcium probes. *Cell Calcium* 15:190-198.

Imperato, A. and Di Chiara, G. 1984. Trans-striatal dialysis coupled to reverse-phase high performance liquid chromatography with electrochemical detection: A new method for study of the in vivo release of endogenous dopamine and metabolites. *J. Pharmacol. Exp. Ther.* 4:966-977.

Irwin, S. 1968. Comprehensive observational assessment. Ia.A systematic, quantitative procedure for assessing the behavioral and physiological state of the mouse. *Psychopharmacol.* 20:222-257.

Isaac, J.T.R., Luthi, A., Palmer, M.J., Anderson, W.W., Benke, T.A., Collingridge, G.L. 1998. An investigation of the expression mechanism of LTP of AMPA receptor-mediated synaptic transmission at hippocampal CA1 synapses using failures analysis and dendritic recordings. *Neuropharmacology* 37:1399-2410.

Ison, J.R. Hoffman, H.S. 1983. Reflex modifications in the domain of startle: II. The anomalous history of a robust and ubiquitous phenomenon. *Psychol. Bull.* 94:3-17.

Jackson, M.B. 1993. Thermodynamics of Membrane Receptors and Channels. CRC Press, Boca Raton, Fla.

Jacobson, I., Sandberg, M., Hamberger, A. 1985. Mass transfer in brain dialysis devices—A new method for the estimation of extracellular amino acids concentration. *J. Neurosci. Methods* 15:263-268.

Janero, D.R. 1990. Malondialdehyde and thiobarbituric acid-reactivity as diagnostic indices of lipid peroxidation and peroxidative tissue injury. *Free Radical Biol. Med.* 9:515-540.

Janowsky, A., Berger, P., Vocci, F., Labarca, R., Skolnick, P., Paul, S.M. 1986. Characterization of sodium-dependent [^{3}H]GBR-12935 binding in brain: A radioligand for selective labelling of the dopamine transport complex. *J. Neurochem.* 46:1272-1276.

Javitch, J.A., Blaustein, R.O., Snyder, S.H. 1984. [3H]Mazindol binding associated with neuronal dopamine and norepinephrine uptake sites. *Mol. Pharmacol.* 26:35-44.

Jeffery, J.A., Sharom, J.R., Fazekas, M., Rudd, P., Welchner, E., Thauvette, L., White, P.W. 2002. An ATPase assay using scintillation proximity beads for high-throughput screening or kinetic analysis. *Anal. Biochem.* 304:55-62.

Jenh, C.-H., Zhang, M., Wiekowski, M., Tan, J.C., Fan, X.-D., Hegde, V., Patel, M., Bryant, R., Narula, S.K., Zavodny, P.J., Chou, C.C. 1998. Development of a CD28 receptor binding-based screen and identification of a biologically active inhibitor. *Anal. Biochem.* 256:47-55.

Jiang, Z.-Y., Woollard, A.C.S., and Wolff, S.P. 1991. Lipid hydroperoxide measurement by oxidation of Fe^{2+} in the presence of xylenol orange. Comparison with the TBA assay and an iodometric method. *Lipids* 26:853-856.

Johnson, R.A., Eshleman, A.J., Meyers, T. Neve, K.A., and Janowsky, A. 1998. [^{3}H]substrate- and cell-specific effects of uptake inhibitors on human dopamine and serotonin transporter-mediated efflux. *Synapse* 1:97-106.

Johnson, R.D. Justice J.B. 1983. Model studies for brain dialysis. *Brain Res. Bull.* 10:567-571.

June, H.L., Grey, C., Warren-Reese, C., Lawrence, A., Thomas, A., Cummings, R., Williams, L., McCane,

S.L., Durr, L.F., and Mason, D. 1998. The opioid receptor antagonist nalmefene reduces alcohol motivated behaviors: Preclinical studies in alcohol preferring (P) and outbred Wistar rats. *Alcohol. Clin. Exp. Res.* 22:2174–2185.

June, H.L., McCane, S., Zink, R.W., Portoghese, P., Li, T.K., Froehlich, J.C. 1999. The delta 2-opioid receptor antagonist naltriben reduces motivated responding for ethanol. *Psychopharmacology* 147:81–89.

June, H.L., Harvey, S.C., Foster, K.L., McKay, P.F., Cummings, R.C., Garcia, M., Mason, D., Grey, C., McCane, S., Williams, L., Johnson, T.B., He, X., Rock, S., Cook, J.M. 2001. $GABA_A$-receptors containing α_5 subunits in the CA1 and CA3 hippocampal fields regulate ethanol-motivated behaviors: An extended ethanol reward circuitry. *J. Neurosci.* 21:2166–2177.

Justice, J.B. Jr. 1993. Quantitative microdialysis of neurotransmitters. *J. Neurosci. Methods* 48:263–276.

Kahl, S.D., Hubbard, F.R., Sittampalam, G.S., and Zock, J.M. 1997. Validation of a high throughput scintillation proximity assay for 5-hydroxytryptamine1E receptor binding activity. *J. Biomol. Screen.* 2:33–39.

Kahl, S.D., Liu, X.J., Ling, N., De Souza, E.B., Gehlert, D.R. 1998. Characterization of [125I-Tyr0]-corticotropin releasing factor (CRF) binding to the CRF binding protein using a scintillation proximity assay. *J. Neurosci. Methods* 83:103–11.

Kakee, A., Terasaki, T., Sugiyama, Y. 1996. Brain efflux index as a novel method of analyzing efflux transport at the blood-brain barrier. *J. Pharmacol. Exp. Ther.* 277:1550–1559.

Kakee, A., Terasaki, T., Sugiyama, Y. 1997. Selective brain to blood efflux transport of para-aminohippuric acid across the blood-brain barrier: In vivo evidence by use of the brain efflux method. *J. Pharmacol. Exp. Ther.* 283:1018–1025.

Kalivas, P.W., Duffy, P., DuMars, L.A., and Skinner, C. 1988. Behavioral and neurochemical effects of acute and daily cocaine administration in rats. *J. Pharmacol. Exp. Ther.* 245:485–492.

Kariv, I., Stevens, M.E., Behrens, D.L., and Oldenburg, K.R. 1999. High throughput quantitation of cAMP production mediated by activation of seven transmembrane domain receptors. *J. Biomol. Screen.* 4:27–32.

Kay, A.R. Wong, R.K.S. 1986. Isolation of neurons suitable for patch-clamping from the adult mammalian central nervous systems. *J. Neurosci. Methods* 16:227–238.

Kendrick, K.M. 1989. Use of microdialysis in neuroendocrinology. *Methods Enzymol.* 168:182–205.

Kendrick, K.M. 1990. Microdialysis measurement of *in vivo* neuropeptide release. *J. Neurosci. Methods* 34:35–46.

Kennedy, L.T. and Hanbauer, I. 1983. Sodium-sensitive cocaine binding to rat striatal membrane: Possible relationship to dopamine uptake sites. *J. Neurochem.* 41:172–178.

Kim, J.J., Rison, R.A., Fanselow, M.S. 1993. Effects of amygdala, hippocampus and periaqueductal gray lesions on short- and long-term contextual fear. *Behav. Neurosci.* 107:1093–1098.

Kim, S.H. and Chung, J.M. 1992. An experimental model for peripheral neuropathy produced by segmental spinal nerve ligation in the rat. *Pain* 50:355–363.

Kirino, T. 1982. Delayed neuronal death in the gerbil hippocampus following ischemia. *Brain Res.* 239:57–69.

Kirino, T., Yoshihiko, T., Tamura, A. 1991. Induced tolerance to ischemia in gerbil hippocampal neurons. *J. Cereb. Blood Flow Metab.* 11:299–307.

Klosterhalfen, S. Klosterhalfen, W. 1985. Conditioned taste aversion and traditional learning. *Psych. Res.* 47:71–94.

Koh, J.Y. Choi, D.W. 1987. Quantitative determination of glutamate mediated cortical neuronal injury in cell culture by lactate dehydrogenase efflux assay. *J. Neurosci. Methods* 20:83–90.

Kohn, D.F., Wixson, S.K., White, W.J., Benson, G.J. (eds). 1997. Anesthesia and Analgesia in Laboratory Animals. American College of Laboratory Animal Medicine Series, Academic Press, New York.

Kolachana, B.S., Saunders, R.C., and Weinberger, D.R. 1994. An improved methodology for routine in vivo microdialysis in nonhuman primates. *J. Neurosci. Methods* 55:1–6.

Kolachana, B.S., Saunders, R.C., and Weinberger, D.R. 1996. In vivo characterization of extracellular GABA release in the caudate nucleus and prefrontal cortex of the rhesus monkey. *Synapse* 25:1–8.

Komesli, S., Vivien, D., Dutartre, P. 1998. Chimeric extracellular domain of type II transforming growth factor (TGF)-β receptor fused to the Fc region of human immunoglobulin as a TGF-β antagonist. *Eur.J. Biochem.* 254:505–513.

Konig, J.F. and Klippel, R.A. 1963. The Rat Brain: A Stereotaxic Atlas of the Forebrain and Lower Parts of the Brain Stem. Williams and Wilkins, Baltimore.

Kowski, T.J. and Wu, J.J. 2000. Fluorescence polarization is a useful technology for reagent reduction in assay miniaturization. *Comb. Chem. High Throughput Screen.* 3:437–444.

Kusuhara, H., Suzuki, H., Terasaki, T., Atsuyuki, K., Lemaire, M., Sugiyama, Y. 1997. P-glycoprotein mediates the efflux of quinidine across the blood-brain barrier. *J. Pharmacol. Exp. Ther.* 283:574–580.

Kwon, T.-W. Watts, B.M. 1963. Determination of malonaldehyde by ultraviolet spectrophotometry. *J. Food Sci.* 28:627–630.

Lagrost, L., Loreau, N., Gambert, P., Lallemant, C. 1995. Immunospecific scintillation proximity assay of cholesteryl ester transfer protein activity. *Clin. Chem.* 41:914–919.

Lambert, P.D., Wilding, J.P., Turton, M.D., Ghatel, M.A., and Bloom, S.R. 1994. Effects of food deprivation and streptozotocin-induced diabetes on hypothalamic neuropeptide Y release as measured by a radio-immunoassay-linked microdialysis procedure. *Brain Res.* 656:135–140.

Landi, M.S., Kreider, J.M., Lang, C.M., Bullock, L.P. 1982. Effects of shipping on the immune function of mice. *Am.J. Vet. Res.* 43:1654–1657.

Langmuir, I. 1918. The adsorption of gases on plane surfaces of glass, mica and platinum. *J. Am. Chem. Soc.* 40:1361–1403.

LaRegina, M.C. and Lonigro, J. 1988. Serologic screening for murine pathogens: Basic concepts and guidelines. *Lab. Anim.* 17:40–47.

Larsson, C.I. 1991. The use of an "internal standard" for control of the recovery in microdialysis. *Life Sci.* 49:PL73–78.

Laser Institute of America (LIA) Laser Safety Committee. 1993. Laser Safety Guide. 9th ed. (D.H. Sliney, ed.) LIA, Orlando, Fla.

Lee, A., Clancy, S., Fleming, A.S. 2000. Mother rats bar-press for pups: Effects of lesions of the mpoa and limbic sites on maternal behavior and operant responding for pup-reinforcement. *Behav. Brain Res.* 108:215–231.

Lee, C.S., Sauer, H., Bjorklund, A. 1996. Dopaminergic neuronal degeneration and motor impairments following axon terminal lesion by intrastriatal 6-hydroxydopamine in the rat. *Neuroscience* 72:641–653.

Lee, M.K., Borchelt, D.R., Wong, P.C., Sisodia, S.S., Price, D.L. 1996. Transgenic models of neurodegenerative diseases. *Curr. Opin. Neurobiol.* 6:651–660.

Le Quellec, A., Dupin, S., Genissel, P., Saivin, S., Marchand, B., and Houin, G. 1995. Microdialysis probes calibration: Gradient and tissue dependent changes in no net flux and reverse dialysis methods. *J. Pharmacol. Exp. Ther.* 33:11–16.

Leroux-Nicollet, I. Costentin, J. 1986. Acute locomotor effects of MPTP in mice and relationships with dopaminergic systems. *In* MPTP: A Neurotoxin Producing a Parkinsonian Syndrome (S.P. Markey, N. Castagnoli, A.J. Trevor, I.J. Kopin, eds.) pp. 419–424. Academic Press.

Levis, R.A. and Rae, J.L. 1995. Technology of patch-clamp electrodes. *In* Neuromethods, Vol. 26: Patch-Clamp Applications and Protocols (A. Boulton, G. Baker and W. Walz, eds.) pp. 1–36. Humana Press, Totowa, N.J.

Liang, J.D., Liu, J., McClelland, P., Bergeron, M. 2001. Cellular localization of BM88 mRNA in paraffin-embedded rat brain sections by combined immunohistochemistry and non-radioactive in situ hybridization. *Brain Res. Brain Res. Protoc.* 7:121–130.

Lighthall, J.W., Dixon, C.E., Anderson, T.E. 1989. Experimental models of brain injury. *J. Neurotrauma* 6:83–97.

Lindefors, N., Amberg, G., Ungerstedt, U. 1989. Intracerebral microdialysis: I. Experimental studies of diffusion kinetics. *J. Pharmacol. Methods* 22:141–156.Lonnroth, P. and Strindberg, L. 1995. Validation of the "internal reference technique" for calibrating microdialysis catheters *in situ. Acta Physiol. Scand.* 153:375–380.

Lister, R.G. 1987. The use of a plus-maze to measure anxiety in the mouse. *Psychopharmacology* 92:180–185.

Little, K.Y., Kirkman, J.A., Carroll, F.I., Breese, G.R., Duncan, G.E. 1993. [125I]RTI-55 binding to cocaine-sensitive dopaminergic and serotonergic uptake sites in the human brain. *J. Neurochem.* 61:1996–2006.

Liu, J.J., Hartman, D.S., and Bostwick, J.R. 2003. An immobilized metal ion affinity adsorption and scintillation proximity assay for receptor-stimulated phosphoinositide hydrolysis. *Anal. Biochem.* 318:91–99.

Logue, A.W. 1979. Taste aversion and the generality of the laws of learning. *Psych. Bull.* 86:276–296.

Lonnroth, P. and Strindberg, L. 1995. Validation of the "internal reference technique" for calibrating microdialysis catheters *in situ. Acta Physiol. Scand.* 153:375–380.

Lonnroth, P., Jansson, P.A., and Smith, U. 1987. A microdialysis method allowing characterization of intercellular water space in humans. *Am. J. Physiol.* 253:E228-E231.

Lonnroth, P., Jansson, P.A., Fredholm, B.B., Smith, U. 1989. Microdialysis of intercellular adenosine concentration in subcutaneous tissue in humans. *Am J. Physiol.* 256:E250-E255.

Lonnroth, P., Carlsten, J., Johnson, L., Smith, U. 1991. Measurements by microdialysis of free tissue concentrations of propranolol. *J. Chromatogr.* 568:419–425.

Lonstein, J.S., Wagner, C.K., De Vries, G.J. 1999. Comparison of the "nursing" and other parental behaviors of nulliparous and lactating female rats. *Horm. Behav.* 36:242–251.

Loskota, W.J., Lomax, P., and Verity, M.A. 1974. A Stereotaxic Atlas of the Mongolian Gerbil Brain. Ann Arbor Science Publishers, Ann Arbor, Mich.

Lovrien, R. Matulis, D. 1995. Assays for total protein. *In* Current Protocols in Protein Science (J.E. Coligan, B.M. Dunn, H.L. Ploegh, D.W. Speicher, P.T. Wingfield, eds.) pp. 3.4.1–3.4.24. John Wiley & Sons, New York.

Lubow, R.E. 1989. Latent Inhibition and Conditioned Attention Theory. Cambridge University Press, Cambridge.

Lubow, R.E. Gewirtz, J.C. 1995. Latent inhibition in humans: Data, theory, and implications for schizophrenia. *Psychol. Bull.* 117:87–103.

Lunte, S.M. Lunte, C.E. 1996. Microdialysis sampling from pharmacological studies: HPLC and CE analysis. *Adv. Chromatogr.* 36:383–432.

MacDonald, J.F., Mody, I., and Salter, M.W. 1989. Regulation of N-methyl-D-aspartate receptors revealed by intracellular dialysis of murine neurons in culture. *J. Physiol.* 414:17–34.

Madison, D.V. 1992. Whole-cell voltage-clamp techniques applied to the study of synaptic function in hippocampal slices. *In* Cellular Neurobiology: A Practical Approach (J. Chad and H. Wheal, eds.) pp. 137–149. IRL Press, Oxford.

Madras, B.K., Fahey, M.A., Bergman, J., Canfield, D.R., Spealman, R.D. 1989a. Effects of cocaine and related drugs in nonhuman primates. 1. [^{3}H]cocaine binding sites in caudate-putamen. *J. Pharmacol. Exp. Ther.* 251:131–141.

Madras, B.K., Spealman, R.D., Fahey, M.A., Neumeyer, J.L., Saha, J.K., Milius, R.A. 1989b. Cocaine receptors labeled by [^{3}H]2β-carbomethoxy-3β-(4-fluorophenyl)tropane. *Mol. Pharmacol.* 36:518–524.

Maidment, N.T., Brumbaugh, D.R., Rudolph, V.D., Erdelyi, E., Evans, C.J. 1989. Microdialysis of extracellular endogenous opioid peptides from rat brain in vivo. *Neuroscience* 33:549–557.

Maier, S.F. and Seligman, M.E.P. 1976. Learned helplessness: Theory and evidence. *J. Exp. Psychol. Gen.* 105:3–46.

Major, J.S. 1995. Challenges of high throughput screening against cell surface receptors. *J. Recept. Signal Transduct. Res.* 15:595–607.

Maldonado, R., Saiardi, A., Valverde, O., Samad, T.A., Roques, B.P., Borrelli, E. 1997. Absence of opiate rewarding effects in mice lacking dopamine D2 receptors. *Nature* 388:586–589.

Mallari, R., Swearingen, E., Liu, W., Ow, A., Young, S.W., S.G. Huang, S.G. 2003. A generic high-throughput screening assay for kinases: Protein kinase A as an example. *J. Biomol. Screen.* 8:198–204.

Mallick, H.N., Tilakaratna, P., Manchanda, S.K., and Kumar, V.M. 1993. A computer program for recording male sex behaviour in rats. *Indian J. Physiol. Pharmacol.* 37:151–156.

Mandine, E., Gofflo, D., V. Jean-Baptiste, Sarubbi, E., Touyer, G., Deprez, P., Lesuisse, D. 2001. Src homology-2 domain binding assays by scintillation proximity and surface plasmon resonance. *J.Mol. Recognit.* 14:254–260.

Manning, P.J., Ringler, D.H., and Newcomer, C.E. (eds.) 1994. The Biology of the Laboratory Rabbit, 2nd ed. ACLAM Series. Academic Press, San Diego.

Marchand, A.R., Luck, D., DiScala, G. 2003. Evaluation of an improved automated analysis of freezing behaviour in rats and its use in trace fear conditioning. *J. Neurosci. Meth.* 126:145–153.

Marsh, D., Dickenson, A., Hatch, D., Fitzgerald, M. 1999. Epidural opioid analgesia in infant rats. I. Mechanical and heat responses. *Pain* 82:23–32.

Martin, R., McFarland, H.F., and McFarlin, D.E. 1992. Immunological aspects of demyelinating diseases. *Annu. Rev. Immunol.* 10:153–187.

Martinez, Z.A., Halim, N.D., Oostwegel, J.L., Geyer, M.A., and Swerdlow, N.R. 2000. Ontogeny of phencyclidine and apomorphine-induced startle gating deficits in rats. *Pharmacol. Biochem. Behav.* 65:449–457.

Mason, P.A. and Romano, W.F. 1995. Recovery characteristics of a rigid, nonmetallic microdialysis probe for use in an electromagnetic field. *Bioelectromagnetics* 16:113–118.

Mason, W.T. 1993. Fluorescent and Luminescent Probes for Biological Activity. Academic Press, London.

Matochik, J.A., Sipos, M.L., Nyby, J.G., Barfield, R.J. 1994. Intracranial androgenic activation of male-typical behaviors in house mice: Motivation vs. performance. *Behav. Brain Res.* 60:141–149.

McDonald, O.B., Chen, W.J. Ellis, B. Hoffman, C. Overton, L. Rink, M. Smith, A., Marshall, C.J., and Wood, E.R. 1999. A scintillation proximity-assay for the Raf/MEK/ERK Kinase cascade: High-throughput screening and identification of selective enzyme inhibitors. *Anal. Biochem.* 268:318–329.

McIntosh, T.K., Vink, R., Noble, L., Yamakami, I., Fernyak, S., Soares, H., Faden, A.L. 1989. Traumatic brain injury in the rat: Characterization of a lateral fluid-percussion model. *Neuroscience* 28:233–244.

McMurtrey, A.E., Graves, R.J., Hooley, J., Brophy, G., Phillips, G.D.L. 1999. A novel 96-well scintillation proximity assay for the measurement of apoptosis. *Cytotechnology* 31:271–282.

Meisel, R.L. and Sachs, B.D. 1994. The physiology of male sexual behavior. *In* The Physiology of Reproduction, Vol. 2 (E. Knobil J.D. Neill, eds.) pp. 3–105. Raven Press, New York.

Menacherry, S., Hubert, W., Justice, J.B. Jr. 1992. In vivo calibration of microdialysis probes of exogenous compounds. *Anal. Chem.* 64:577–583.

Menzies, S.A., Hoff, J.T., Betz, A.L. 1992. Middle cerebral artery occlusion in rat: A neurological and pathological evaluation of a reproducible model. *Neurosurgery* 31:100–107.

Merskey, H. Bogduk, N. 1994. Classification of Chronic Pain: Descriptions of Chronic Pain Syndromes and Definitions of Pain Terms. pp. 209–213. IASP Press, Seattle.

Miklyaeva, E.I., Martens, D.J., and Whishaw, I.Q. 1995. Impairments and compensatory adjustments in spontaneous movement after unilateral dopamine depletion in rats. *Brain Res.* 681:23–40.

Milgram, N.W., Krames, L., Alloway, T.M. (eds.). 1977. Food Aversion Learning. Plenum, New York.

Miller, S., Yasuda, M., Coats, J.K., Jones, Y., Martone, M.E., Mayford, M. 2002. Disruption of dendritic translation of CaMKIIalpha impairs stabilization of synaptic plasticity and memory consolidation. *Neuron* 36:507–519.

Minta, A., Kao, J.P.Y., and Tsien, R.Y. 1989. Fluorescent indicators for cytosolic calcium based on rhodamine and fluorescein chromophores. *J. Biol. Chem.* 264:18171–18178.

Miyazawa, T., Yasuda, K., Fujimoto, K. 1987. Chemiluminescence-high-performance liquid chromatography of phosphatidylcholine hydroperoxide. *Anal. Lett.* 20:915–925.

Moghaddam, B. Bunney, B.S. 1989. Ionic composition of microdialysis perfusing solution alters the pharmacological responsiveness and basal outflow of striatal dopamine. *J. Neurochem.* 53:652–654.

Moita, M.A., Rosis, S., Zhou, Y., LeDoux, J.E., and Blair, H.T. 2003. Hippocampal place cells acquire location-specific responses to the conditioned stimulus during auditory fear conditioning. *Neuron* 37:485–497.

Moller, K.A., Johansson, B., Berge, O.-G. 1998. Assessing mechanical allodynia in the rat paw with a new electronic algometer. *J. Neurosci. Meth.* 84:41–47.

Montgomery, J., Ste-Marie, L., Boismenu, D., Vachon, L. 1995. Hydroxylation of aromatic compounds as indices of hydroxyl radical production: A cautionary note revisited. *Free Radical Biol. Med.* 19:927–933.

Montoya, C.P., Campbell-Hope, L.J., Pemberton, K.D., Dunnett, S.B. 1991. The "staircase test": A measure of independent forelimb reaching and grasping abilities in rats. *J. Neurosci. Methods* 36:219–228.

Morrison, P.F., Bungay P.M., Hsiao, J.K., Mefford, I.N., Dykstra, K.H., Dedrick, R.L. 1991. Quantitative microdialysis. *In* Microdialysis in the Neurosciences (T.E. Robinson and J.B. Justice, Jr., eds.) pp. 47–80. Elsevier Science Publishing, New York.

Moyer, J.R. Jr. Brown, T.H. 1998. Methods for whole-cell recording from visually preselected neurons of perirhinal cortex in brain slices from young and aging rats. *J. Neurosci. Methods* 86:35–54.

Nakayama, G. R., Nova, M.P., Parandoosh, Z. 1998. A scintillating microplate assay for the assessment of protein kinase activity. *J. Biomol. Screen.* 3:43–48.

Nare, B., Allocco, J.J., Kuningas, R., Galuska, S., Myers, R.W., Bednarek, M.A., and Schmatz, D.M. 1999. Development of a scintillation proximity assay for histone deacetylase using a biotinylated peptide derived from histone-H4. *Anal. Biochem.* 267:390–396.

Nathan, C. Xie, Q.W. 1994. Regulation of biosynthesis of nitric oxide. *J. Biol. Chem.* 269:13725–13728.

Neher, E. 1992. Correcting for liquid junction potentials in patch clamp experiments. *Methods Enzymol.* 207:123–131.

Nichols, J.S., Parks, D.J., Consler, T.G., Blanchard, S.G. 1998. Development of a scintillation proximity assay for peroxisome proliferator-activated receptor g ligand binding domain. *Anal. Biochem.* 257:112–119.

Nicholson, C. Rice, M.E. 1986. The migration of substances in the neuronal microenvironment. *Ann.N.Y. Acad. Sci.* 481:55–71.

Numan, M. 1994. Maternal behavior. *In* The Physiology of Reproduction, 2nd ed. (E. Knobil and J.D. Neill, eds.) pp. 221–302. Raven Press, New York.

Obrenovitch, T.P., Zikha, E., Urenjak, J. 1995. Intracerebral microdialysis: Electrophysiological evidence of a critical pitfall. *J. Neurochem.* 64:1884–1887.

O'Brien, R.A. 1986. Receptor Binding in Drug Research. Marcel Dekker, New York.

Ohmi, N., Wingfield, J.M., Yazawa, H., Inagaki, O. 2000. Development of a homogeneous time-resolved fluorescence assay for high throughput screening to identify Lck inhibitors: Comparison with scintillation

Ohno, K., Pettigrew, K.D., Rapoport, S.I. 1978. Lower limits of cerebrovascular permeability to nonelectrolytes in the conscious rat. *Am.J. Physiol.* 235:H299-H307.

Olson, R.J. and Justice, J.B., Jr. 1993. Quantitative microdialysis under transient conditions. *Anal. Chem.* 65:1017-1025.

Olton, D.S. 1985. The radial arm maze as a tool in behavioral pharmacology. *Physiol. & Behav.* 40:793-797.

Osborne, P.G., O'Connor, W.T., and Ungerstedt, U. 1991. Effect of varying the ionic concentration of a microdialysis perfusate on basal striatal dopamine levels in awake rats. *J. Neurochem.* 56:452-456.

Ossenkopp, K.P., Kavaliers, M., Sanberg, P.R. (eds.) 1996. Measuring Movement and Locomotion: From Invertebrates to Humans. R.G. Landes, Austin, Tex.

Owen, E.H., Christensen, S.C., Paylor, R., Wehner, J.M. 1997a. Identification of quantitative trait loci involved in contextual and auditory-cued fear conditioning in BXD recombinant inbred strains. *Behav. Neurosci.* 111:292-300.

Owen, E.H., Logue, S.F., Rasmussen, D.L., and Wehner, J.M. 1997b. Assessment of learning by the Morris water task and fear conditioning in inbred mouse strains and F1 hybrids: Implications of genetic background for single gene mutations and quantitative trait loci analyses. *Neuroscience* 80:1087-1099.

Pachter, J.A., Zhang, R., Mayer-Ezell, R. 1995. Scintillation proximity assay to measure binding of soluble fibronectin to antibody-captured a5b1 integrin. *Anal. Biochem.* 230:101-107.

Park, Y.W., Cummings, R.T., Wu, L., Zheng, S., Cameron, P.M., Woods, A., Zaller, D.M., Marcy, A.I., Hermes, J.D. 1999. Homogeneous proximity tyrosine kinase assays: Scintillation proximity assay versus homogeneous time-resolved fluorescence. *Anal. Biochem.* 269;94–104.

Parsons, L.H. and Justice, J.B. Jr. 1992. Extracellular concentration and in vivo recovery of dopamine in the nucleus accumbens using microdialysis. *J. Neurochem.* 58:212-218.

Parsons, L.H. and Justice, J.B. Jr. 1994. Quantitative approaches to in vivo brain microdialysis. *Crit. Rev. Neurobiol.* 8:189-220.

Patel, S., Morris, S.A., Adkins, C.E., O'Beirne, G., Taylor, C.W. 1997. Ca^{2+}-independent inhibition of inositol trisphosphate receptors by calmodulin: Redistribution of calmodulin as a possible means of regulating Ca^{2+} mobilization. *Proc. Natl. Acad. Sci. U.S.A.* 94:11627–11632.

Paxinos, G. Watson, C. 1986. The Rat Brain in Stereotaxic Coordinates. Academic Press, San Diego.

Paylor, R., Tracy, R., Wehner, J., Rudy, J.W. 1994. C57BL/6 and DBA/2 mice differ on contextual but not auditory fear conditioning. *Behav. Neurosci.* 108:810–817.

Pellegrino, L. Cushman, A.J. 1979. A Stereotaxic Atlas of the Rat Brain. Plenum, New York.

Pellis, S.M. Pellis, V.C. 1994. Development of righting when falling from a bipedal standing posture: Evidence for the dissociation of dynamic and static righting reflexes in rats. *Physiol. Behav.* 56:659–663.

Pellow, S., Chopin, P., File, S.E., Briley, M. 1985. Validation of open:closed arm entries in an elevated plus-maze as a measure of anxiety in the rat. *J. Neurosci. Methods* 14:149–167.

Percy, D.H. and Barthold, S.W. 2001. Pathology of Laboratory Rodents and Rabbits, 2nd ed. Iowa State University Press, Ames.

Perese, D.A., Ulman, J., Viola, J., Ewing, S.E., Bankiewicz, K.S. 1989. A 6-hydroxydopamine-induced selective parkinsonian rat model. *Brain Res.* 494:285–293.

Peters, R.H., Wellman, P.J., Gunion, M.W., Luttmers, L.L. 1979. Acids and quinine as dietary adulterants. *Physiol. Behav.* 22:1055–1059.

Peterson, M.F., Martin, W.H. Spencer, R.W., Tate, B.F. 1999. The use of beta, gamma-methyleneadenosine 5′-triphosphate to determine ATP competition in a scintillation proximity kinase assay. *Anal. Biochem.* 271:131–136.

Petry, N.M. 1997. Benzodiazepine-GABA modulation of concurrent ethanol and sucrose reinforcement in the rat. *Exp. Clin. Psychopharmacol.* 5:183–194.

Petry, N.M. Heyman, G.M. 1995. Behavioral economic analysis of concurrent ethanol/sucrose and sucrose reinforcement in the rat: Effects of altering variable-ratio requirements. *J. Exp. Anal. Behav.* 64:331–359.

Pettit, H.O. and Justice, J.B. 1991. Procedures for microdialysis with smallbore HPLC. *In* Microdialysis in the Neurosciences (T.E. Robinson and J.B. Justice, Jr., eds.) pp. 117–154. Elsevier Science Publishing, New York.

Pfaff, D.W., Schwartz-Giblin, S., McCarthy, M.M., Kow, L.-M. 1995. Cellular and molecular mechanisms of female reproductive behaviors. *In* The Physiology of Reproduction, Vol. 2 (E. Knobil and J.D. Neill, eds.) pp. 107–220. Raven Press, New York.

Phelan, M.C. 1998. Techniques for mammalian cell tissue culture. *In* Current Protocols in Neuroscience (J.N. Crawley, C.R. Gerfen, M.A. Rogawski, D.R. Sibley, P. Skolnick, and S. Wray, eds.) pp. A3B.1-A3B.13. John Wiley & Sons, Hoboken, N.J,

Picardo, M. Hughes, K.T. 1997. Scintillation proximity assays. *In* High Throughput Screening: The Discovery of Bioactive Substances. (J.P. Devlin, ed.), pp. 307–316. Marcel Dekker, New York.

Pitcher, G.M., Ritchie, J., Henry, J.L. 1999. Paw withdrawal threshold in the von Frey hair test is influenced by the surface on which the rat stands. *J. Neuro. Methods.* 87:185–193.

Pletnicov, M.V., Storozheva, Z.I., and Sherstnev, V.V. 1995. Developmental analysis of habituation of the acoustic startle response in the preweanling and adult rats. *Behav. Proc.* 34:269–277.

Porsolt, R.D., Le Pichon, M., Jalfre, M. 1977. Depression: A new animal model sensitive to antidepressant treatment. *Nature* 266:730–732.

Porsolt, R.D., Lenegre, A., McArthur, R.A. 1991. Pharmacological models of depression. *In* Animal Models in Psychopharmacology (B. Olivier, J. Mos, and J.L. Slangen, eds.) pp. 137–160. Birkhauser Verlag, Basel.

Poole, T.B. (eds.) 1999. The UFAW Handbook on the Care and Management of Laboratory Animals, 7th ed., . 325–326. Blackwell Science Ltd., Oxford.

Price, E.O. 1993. Practical considerations in the measurement of sexual behavior. *Methods Neurosci.* 15:16-31.

Pristupa, Z.B., Wilson, J.M., Hoffman, B.J., Kish, S.J., and Niznik, H.B. 1994. Pharmacological heterogeneity of the cloned and native human dopamine transporter: Dissociation of [^{3}H]WIN 35428 and [^{3}H]GBR 12935 binding. *Mol. Pharmacol.* 45:125–135.

Pulsinelli, W.A. and Brierley, J.B. 1979. A new model of bilateral hemispheric ischemia in the unanesthetized rat. *Stroke* 10:267–272.

Pusch, M. Neher, E. 1988. Rates of diffusional exchange between small cells and a measuring patch pipette. *Pfluegers Arch Eur. J. Physiol.* 411;204–211.

Racine, R.J. 1972. Modification of seizure activity by electrical stimulation. II. Motor seizure. *Electroencephalogr. Clin. Neurophysiol.* 32:281–294.

Racke, M.K., Dhib-Jalbut, S., Cannella, B., Albert, P.S., Raine, C.S., and McFarlin, D.E. 1992. Prevention and treatment of chronic relapsing experimental allergic encephalomyelitis by transforming growth factor-beta1. *J. Immunol.* 154:2959–2968.

Rakerd, B., Brigham, D.A., and Clemens, L.G. 1985. A microcomputer-based system for recording and analyzing behavioral data regarding the sexual activity of male rodents. *Physiol. Behav.* 35:999–1001.

Ramm, P. 1999. Imaging systems in assay screening. *Drug Disc. Today* 4:401–410.

Randall, L.O. and Sellito, J.J. 1957. A method for measurement of analgesic activity on inflamed tissue. *Arch. Int. Pharmacodyn.* 4:409–419.

Rapoport, S.I., Fredericks, W.R., Ohno, K., Pettigrew, K.D. 1980. Quantitative aspects of reversible osmotic opening of the blood-brain barrier. *Am.J. Physiol.* 238:R421-R431.

Ratts, R.B., Arredondo, L.R., Bittner, P., Perrin, P.J., Lovett-Racke, A.E., Racke, M.K. 1999. The role of CTLA-4 in tolerance induction and T cell differentiation in experimental autoimmune encephalomyelitis: i.p. antigen administration. *Int. Immunol.* 11:1881–1888.

Revusky, S.H. and Garcia, J. 1970. Learned associations over long delays. *In* Psychology of Learning and Motivation: Advances in Research and Theory, vol. 4 (G. Bower and J. Spence, eds.) pp. 1–84. Academic Press, New York.

Ricaurte, G.A., Irwin, I., Forno, L.S., DeLanney, L.E., Langston, E., Langston, J.W. 1987. Aging and 1-methyl-4-phenyl-1,2,3,6-tetrahydropyridine-induced degeneration of dopaminergic neurons in the substantia nigra. *Brain Res.* 403:43–51.

Rice, M.E., Gerhardt, G.A., Hierl, P.M., Nagy, G., Adams, R.N. 1985. Diffusion coefficients of neurotransmitters and their metabolites in brain extracellular fluid space. *Neuroscience* 15:891–902.

Robinson, M.J., MaCrae, I.M., Todd, M., Read, J.L., and McCulloch, J. 1990. Reduction of local cerebral blood flow to pathological levels by endothelin-1 applied to the middle cerebral artery in the rat. *Neurosci. Lett.* 118;269–272.

Robinson, T.E. Whishaw, I.Q. 1988. Normalization of extracellular dopamine in striatum following recovery from a partial unilateral 6-OHDA lesion of the substantia nigra: A microdialysis study in a freely moving rat. *Brain Res.* 450: 209–224.

Rodefer, J.S., Campbell, U.C., Cosgrove, K.P., Carroll, M.E. 1999. Naltrexone pretreatment decreases the reinforcing efficacy of ethanol and saccharin but not PCP or food under concurrent progressive-ratio schedules in rhesus monkeys. *Psychopharmacology* 147:81–89.

Rodgers, R.J., Cole, J.C., Aboualfa, K., Stephenson, L.H. 1995. Ethopharmacological analysis of the effects of putative "anxiogenic" agents in the mouse elevated plus-maze. *Pharmacol. Biochem. Behav.* 52:805– 813.

Rohlf, F.J. and Sokal, R.R. 1995. Statistical Tables. Freeman, New York.

Role, L.W. and Fischbach, G.D. 1987. Changes in the number of chick ciliary ganglia neurons processes with time in cell culture. *J. Cell. Biol.* 104:363–370.

Rosenthal, H.E. 1967. A graphic method for the determination and presentation of binding parameters in complex systems. *Anal. Biochem.* 20:525–532.

Rozin, P. Kalat, J.W. 1971. Specific hungers and poison avoidance as adaptive specializations of learning. *Psych. Rev.* 78:459–486.

Rudnick, G. 1997. Mechanisms of biogenic amine neurotransmitter transporters. *In* Neurotransmitter Transporters: Structure, Function and Regulation. (M.E.A. Reith, ed.) pp. 73–100. Humana Press, Totowa, N.J.

Rudy, J.W. and Morledge, P. 1994. Ontogeny of contextual fear conditioning in rats: Implications for consolidation, infantile amnesia, and hippocampal system function. *Behav. Neurosci.* 108:227–234.

Sam, P. Justice, J.B. Jr. 1996. Effect of general microdialysis-induced depletion of extracellular dopamine. *Anal. Chem.* 68:724–728.

Samson, H.H. 1987. Initiation of ethanol-maintained behavior: A comparison of animal models and their implication to human drinking. *In* Neurobehavioral Pharmacology: Advances in Behavioral Pharmacology, Vol. 6 (T. Thompson, P.B. Dews, J. Barret, eds.) pp. 221–248. Lawrence Erlbaum Associates, New Jersey.

Samson, H.H. and Grant, K.A. 1985. Chlordiazepoxide effects on ethanol self-administration: Dependence on concurrent conditions. *J. Exp. Anal. Behav.* 43:353–364.

Sandberg, M. Lindstrom, S. 1983. Amino acids in the dorsal lateral geniculate nucleus of the cat: Collection in vivo. *J. Neurosci. Methods* 9:65–74.Scheller, D. and Kolb, J. 1991. The internal reference technique in microdialysis: A practical approach to monitoring dialysis efficiency and to calculating tissue concentration from dialysate samples. *J. Neurosci. Methods* 40:31–37.

Sattler, R., Charlton, M.P., Hafner, M., Tymianski. M. 1997. Determination of the time course and extent of neurotoxicity at defined temperatures in cultured neurons using a modified multiwell plate fluorescence scanner. *J. Cereb. Blood Flow Metab.* 17:455–463.

Sauer, H.W. and Oertel, H. 1994. Progressive degeneration of nigrostriatal dopamine neurons following intrastriatal terminal lesions with 6-hydroxydopamine—A combined retrograde tracing and immunocytochemical study in the rat. *Neuroscience* 59:401–415.

Saunders, R.C., Aigner, T.G., and Frank, J.A. 1990. Magnetic resonance imaging of the rhesus monkey brain: Use for stereotactic neurosurgery. *Exp. Brain Res.* 81:443–446.

Saunders, R.C., Kolachana, B.S., and Weinberger, D.R. 1993. Local pharmacological manipulation of dopamine in prefrontal cortex and caudate in the rhesus monkey: An in vivo microdialysis study. *Exp. Brain Res.* 98:44–52.

Scatchard, G. 1949. The attractions of proteins for small molecules and ions. *Ann.N.Y. Acad. Sci.* 51:660–672.

Schechter, M.D. Calcagnetti, D.J. 1993. Trends in place preference conditioning with a cross-indexed bibliography 1957–1991. *Neurosci. Behav. Rev.* 17:21–41.

Scheller, D. Kolb, J. 1991. The internal reference technique in microdialysis: A practical approach to monitoring dialysis efficiency and to calculating tissue concentration from dialysate samples. *J. Neurosci. Methods* 40:31–37.

Schoemaker, H., Pimoule, C., Arbilla, S., Scatton, B., Javoy-Agid, F., Langer, S.Z. 1985. Sodium dependent [^{3}H]cocaine binding associated with dopamine uptake sites in the rat striatum and human putamen decrease after dopaminergic denervation and in Parkinsons disease. *Naunyn-Schmiedeberg's Arch Pharmakol.* 329:227–235.

Schreibmayer, W., Lester, H.A., and Dascal, N. 1994. Voltage clamp of *Xenopus laevis* oocytes utilizing agarose cushion electrodes. *Pflugers Arch.* 426:453–458.

Schulz, D.W. and Mailman, R.B. 1984. An improved, automated adenylate cyclase assay utilizing preparative HPLC: Effects of phosphodiesterase inhibitors. *J. Neurochem.* 42:764–774.

Schultz, W. 1982. Depletion of dopamine in the striatum as an experimental model of parkinsonism: Direct effects and adaptive mechanisms. *Prog. Microbiol.* 18:121–166.

Schweri, M.M., Skolnick, P., Rafferty, M.F., Rice, K.C., Janowski, A.J., and Paul, S.M. 1985. Threo-(+/-)-methylphenidate binding to 3,4-dihydroxyphenylethylamine uptake sites in corpus striatum: Correlation with the stimulant properties acid esters. *J. Neurochem.* 45:1062–1070.

Selye, H. 1976. The Stress of Life. McGraw-Hill, New York.

Semmes, J., Weinstein, S., Ghent, L., Teuber, H.-L. 1960. Somatosensory Changes After Penetrating Brain Wounds in Man. Harvard Univ. Press, Cambridge.

Sen, S., Jaakola, V.P., Heimo, H., Kivela, P., Scheinin, M., Lundstrom, K., Goldman, A. 2002. Development of a scintiplate assay for recombinant human alpha(2B)-adrenergic receptor. *Anal. Biochem.* 307:280–286.

Seubert, P., Vigo-Pelfrey, C., Esch, F., Lee, M., Dovey, H., Davis, D., Sinha, S., Schlossmacher, M.G., Whaley, J., Swindlehurst, C., McCormack, R., Wolfert, R., Selkoe, D.J., Lieberburg, I., Schenk, D. 1992. Isolation and quantitation of soluble Alzheimer's β-peptide from biological fluids. *Nature* 359:325–327.

Sharkey, J., Ritchie, I.M., Kelly, P.A.T. 1993. Perivascular microapplication of endothelin-1: A new model of focal cerebral ischaemia in the rat. *J. Cereb. Blood Flow Metab.* 13:865–871.

Sharpe, T.L. and Koperwas, J. 2000. Software assist for education and social science settings: Behavior evaluation strategies and taxonomies (BEST) and accompanying qualitative applications. Sage-Scolari Publishing. Thousand Oaks, Ca.

Shekhar, A., McCann, U.D., Meany, M., Blanchard, C.D., Davis, M., Frey, K.A., Liberzon, I., Overall, K., Shear, K., Tecott, L.H., and Winsky, L. 2001. Developing animal models of anxiety disorders: A consensus paper from the National Institute of Mental Health. *Psychopharmacol.* 157:327–339.

Shekhar, A., Ball, S.G., Sajdyk, T.J., Goddrad, A.W. 2002. Neurobiology of panic disorder. *Tren. Econ. Neurosci.* 4:36–41.

Shimizu, I. Prasad, C. 1991. Relationship between [^{3}H]mazindol binding to dopamine uptake sites and [^{3}H]dopamine uptake in rat striatum during aging. *J. Neurochem.* 56:575–579.

Shnerson, A., Willott, J.F. 1980. Ontogeny of the acoustic startle response in C57BL/6J mouse pups. *J. Comp. Physiol. Psych.* 94:34–40.

Sigworth, F.J. and Zhou, J. 1992. Analysis of nonstationary single-channel currents. *Methods Enzymol.* 207:746–762.

Sills, M.A., Weiss, D., Pham, Q., Schweitzer, R., Wu, X., Wu, J.Z.J. 2002. Comparison of assay technologies for a tyrosine kinase assay generates different results in high throughput screening. *J. Biomol. Screen.* 7:191–214.

Sima, A.A. and Shafrir, E. (eds.) 2000. Animal Models in Diabetes: A Primer. Taylor & Francis Group, London.

Skinner, R.H., Picardo, M., Gane, N.M., Cook, N.D., Morgan, L., Rowedder, J., Lowe, P.N. 1994. Direct measurement of the binding of RAS to neurofibromin using a scintillation proximity assay. *Anal. Biochem.* 223:259–265.

Slater, T.F., Sawyer, B., Strauli, U. 1963. Studies on succinate-tetrazolium reductase systems III. Points of coupling of four different tetrazolium salts. *Biochim. Biophys. Acta* 77:383–393.

Slotnick, B.M. Leonard, C.M. 1975. A Stereotaxic Atlas of the Albino Mouse Forebrain. DHEW Publication (ADM) 75–100, U.S. Government Printing Office, Rockville, Md.

Smith, A.D., Olson, R.J., and Justice, J.B. Jr. 1992. Quantitative microdialysis of dopamine in the striatum: Effect of circadian variation. *J. Neurosci. Methods* 44:33–41.

Smith, E.R., Damassa, D.A., and Davidson, J.M. 1977. Hormone administration: Peripheral and intracranial implants. *Methods Psychobiol.* 3:259–279.

Smith, L., PriceJones, M., Hughes, K., Egebjerg, J., Poulsen, F., Wiberg, F.C., and Shank, R.P. 2000. Effects of topiramate on kainate- and domoate-activated [C-14]guanidinium ion flux through GluR6 channels in transfected BHK cells using cytostar-T scintillating microplates. *Epilepsia* 41:S48-S51.

Smith, Q.R. 1989. Quantitation of blood-brain barrier permeability. *In* Implications of the Blood-Brain Barrier and Its Manipulation (E.A. Neuwelt, ed.) pp. 85–118. Plenum, New York.

Smith, Q.R., Momma, S., Aoyagi, M., Rapoport, S. 1987. Kinetics of neutral amino acid transport across the blood-brain barrier. *J. Neurochem.* 49:1651–1658.

Smith, Q.R., Ziylan, Y.Z., Robinson, P.J., and Rapoport, S.I. 1988. Kinetics and distribution volumes for tracers of different sizes in the brain plasma space. *Brain Res.* 462:1–9.

Smith, Q.R., Nagura, H., Takada, Y., Duncan, M.W. 1992. Facilitated transport of the neurotoxin, β-N-methylamino-L-alanine, across the blood-brain barrier. *J. Neurochem.* 58:1330–1337.

Snider, R.S. and Lee, J.C. 1961. A Stereotaxic Atlas of the Monkey Brain (*Macaca mulatta*). University of Chicago Press, Toronto.

Sokal, R.R. Rohlf, F.J. 1994. Biometry. Freeman, New York.

Sorg, G., Schubert, H.D., Buttner, F.H., and Heilker, R. 2002. Automated high throughput screening for serine kinase inhibitors using a LEADSeekertrade mark scintillation proximity assay in the 1536-well format. *J. Biomol. Screen.* 7:11–19.

Spanagel, R., Herz, A., Shippenberg, T.S. 1990. The effects of opioid peptides on dopamine release in the nucleus accumbens: An in vivo microdialysis study. *J. Neurochem.* 55:1734–1740.

Spencer-Fry, J.E., Brophy, G., O'Beirne, G., Cook, N.D. 1997. Kinetic characterization of p34cdc2/Cyclin B kinase-mediated phosphorylation of peptides derived from histone H1 using phosphocellulose filter binding and scintillation proximity assays. *J. Biomol. Screen.* 2:25–32.

Spiker, V.A. 1977. Taste aversion: A procedural analysis and an alternative paradigmatic classification. *Psych. Rec.* 27:753–769.

Stahle, L., Sebersvard, S., Ungerstedt, U. 1991. A comparison between three methods for estimating the extracellular concentration of endogenous and exogenous compounds by microdialysis. *J. Pharmacol. Methods* 25:42–52.

Stern, J.M. 1996. Somatosensation and maternal care in Norway rats. In Parental Care: Evolution, Mechanisms, and Adaptive Significance. Advances in the Study of Behavior, vol. 25, . 243–294. (J.S. Rosenblatt C.T. Snowden, eds.) Academic Press, New York.

Stoll, J., Wadhwani, K.C., and Smith, Q.R. 1993. Identification of the cationic amino acid transporter (system y+) of the rat blood-brain barrier. *J. Neurochem.* 60:1956–1959.

Stuart, G.J., Dodt, H.-U., Sakmann, B. 1993. Patch-clamp recordings from the soma and dendrites of neurones in brain slices using infrared video microscopy. *Pfluegers Arch. Eur.J. Physiol.* 423:511–518.

Stuhmer, W. 1998. Electrophysiologic recordings from *Xenopus* oocytes. *Methods Enzymol.* 293:280–300.

Su, J.-L., Stimpson, S., Edwards, C., Van Arnold, J., Burgess, S., Lin, P. 1997. Neutralizing IGF-1 monoclonal antibody with cross-species reactivity. *Hybridoma* 16:513–518.

Suckow, M.A., Danneman, P., Brayton, C. 2001. The Laboratory Mouse, p. 113. CRC Press, Boca Raton, Fla.

Sundstrom, E., Fredriksson, A., Archer, T. 1990. Chronic neurochemical and behavioral changes in MPTP-lesioned C56BL/6 mice: A model for Parkinson's disease. *Brain Res.* 528:181–188.

Sutherland, R.J., Whishaw, I.Q., and Regehr, J.C. 1982. Cholinergic receptor blockade impairs spatial localization by use of distal cues in the rat. *J. Comp. Physiol. Psychol.* 96:563–573.

Swandulla, D. Chow, R.H. 1992. Recording solutions for isolating specific ionic channel currents. In Practical Electrophysiological Methods (H. Kettenmann and R. Grantyn, eds.) pp. 164–168. Wiley-Liss, New York.

Swerdlow, N.R. Geyer, M.A. 1996. Using an animal model of deficient sensorimotor gating to study the pathophysiology and new treatments of schizophrenia. *Schizophr. Bull.* 20:91–103.

Tadepalli, S.M. Quinn, R.P. 1996. Scintillation proximity radioimmunoassay for the measurement of acyclovir. *J. Pharm. Biomed. Anal.* 15:157–163.

Takada, Y., Greig, N.H., Vistica, D.T., Rapoport, S.I., Smith, Q.R. 1991. Affinity of antineoplastic amino acid drugs for the large neutral amino acid transporter of the blood-brain barrier. *Cancer Chemother. Pharmacol.* 29:89–94.

Takasato, Y., Rapoport, S.I., Smith, Q.R. 1984. An in situ brain perfusion technique to study cerebrovascular transport in the rat. *Am.J. Physiol.* 247:H484-H493.

Tal, M. Bennett, G.J. 1993. Dextrorphan relieves neuropathic heat-evoked hyperalgesia. *Neurosci. Lett.* 151:107–110.

Tal, M. Bennett, G.J. 1994. Extra-territorial pain in rats with a peripheral mononeuropathy: Mechano-hyperalgesia and mechano-allodynia in the territory of an uninjured nerve. *Pain* 57:375–382.

Tamura, A., Graham, D.I., McCulloch, J., Teasdale G.M. 1981. Focal cerebral ischaemia in the rat: 1. Description of technique and early neuropathological consequences following middle cerebral artery occlusion. *J. Cereb. Blood Flow Metab.* 1:53–60.

Thiebot, M.-H., Dangoumau, L., Richard, G., Puech, A.J. 1991. Safety signal withdrawal: A behavioural paradigm sensitive to both "anxiolytic" and "anxiogenic" drugs under identical experimental conditions. *Psychopharmacology* 103:415–424.

Thiessen, D.D. Lindzey, G. 1967. Negative geotaxis in mice: effect of balancing practice on incline behaviour in C57BL/6J male mice. *Anim. Behav.* 15:113–116.

Thomas, J., Wang, J., Takubo, H., Sheng, J.G., Dejesus, S., Bankiewicz, K.S. 1994. A 6-hydroxydopamine-induced selective parkinsonian rat model: Further biochemical and behavioral characterization. *Exp. Neurol.* 126:159–167.

Thompson, A.C., Justice, J.B. Jr., and McDonald, J.K. 1995. Quantitative microdialysis of neuropeptide Y. *J. Neurosci. Methods* 60:189–198.

Toates, F.M. Rowland, N.E. (eds.) 1987. Feeding and Drinking, Vol. 1: Techniques in the Behavioral and Neural Sciences. Elsevier, Amsterdam.

Tossman, U., Jonsson, G., Ungerstedt, U. 1986. Regional distribution and extracellular levels of amino acids in rat central nervous system. *Acta Physiol. Scand.* 127:533–545.

Toulmond, S., Duval, D., Serrano, A., Scatton, B., Benavides, J. 1993. Biochemical and histological alterations induced by fluid percussion brain injury in the rat. *Brain Res.* 620:24–31.

Treit, D. 1985. Animal models for the study of anti-anxiety agents: A review. *Neurosci. Biobehav. Rev.* 9:203–222.

Treit, D. 1991. Defensive burying: A pharmacological animal model for specific fears In Animal Models of Psychiatric Disorders, Vol. 3 (P Soubrie, P. Simon, D. Widlocher, eds.) pp. 1–19. S. Karger, Basel.

Triguero, D., Buciak, J., Pardridge, W. 1990. Capillary depletion method for quantification of blood-brain barrier transport of circulating peptides and plasma proteins. *J. Neurochem.* 54:1882–1888.

Trost, L.C. and Lemasters, J.J. 1994. A cytotoxicity assay for tumor necrosis factor employing a multiwell fluorescence scanner. *Anal. Biochem.* 220:149–153.

Trussell, L.O. and Jackson, M.B. 1987. Dependence of an adenosine-activated potassium current on a GTP-binding protein in mammalian central neurons. *J. Neurosci.* 7:3306–3316.

Turlais, F., Hardcastle, A., Rowlands, M., Newbatt, Y., Bannister, A., Kouzarides, T., Workman, P., Aherne, G.W. 2001. High-throughput screening for identification of small molecule inhibitors of histone acetyltransferases using scintillating microplates (flashplate). *Anal. Biochem.* 298:62–68.

Ungerstedt, U. 1971a. Striatal dopamine release after amphetamine or nerve degeneration revealed by rotational behavior. *Acta Physiol. Scand. Suppl.* 367:57–68.

Ungerstedt, U. 1971b. Postsynaptic supersensitivity after 6-hydroxydopamine induced degeneration of the nigrostriatal dopamine system. *Acta Physiol. Scand. Suppl.* 367:69–93.

Ungerstedt, U. 1984. Measurement of neurotransmitter release by intracranial dialysis. *In* Measurement of Neurotransmitter Release In Vivo (C.A. Marsden, ed.) pp. 81–105. John Wiley & Sons, New York.

Ungerstedt, U. Arbuthnott, G.W. 1970. Quantitative recording of rotational behavior in rats after 6-hydroxydopamine lesions of the nigrostriatal dopamine system. *Brain Res.* 24:485–493.

Ungerstedt, U., Herrera-Marchintz, M., Jungnelius, U., Stahle, L., Tossman, U., Zetterstrom, T. 1982. Dopamine synaptic mechanisms reflected in studies combining behavioral recordings and brain dialysis. *In* Advances in Dopamine Research (M. Kotisaka, ed.) pp. 219–231. Pergamon Press, Elmsford, N.Y.

Van Wylen, D.G.L., Park, T.S., Rubio, R., and Berne, R.M. 1986. Increases in cerebral interstitial fluid adenosine concentration during hypoxia, local potassium infusion, and ischemia. *J. Cereb. Blood Flow Metab.* 6:522–528.

Vinson, P.N. and Justice, J.B. Jr. 1997. Effect of neostigmine on concentration and extraction fraction of acetylcholine using quantitative microdialysis. *J. Neurosci. Methods* 73:61–67.

Vogel, J.R., Beer, B., Clody, D.E. 1971. A simple and reliable conflict procedure for testing anti-anxiety agents. *Psychopharmacologia (Berl.)* 21:1–7.

Vorhees, C.V., Acuff-Smith, K.D., Moran, M.S., and Minck, D.R. 1994. A new method for evaluating air-righting reflex ontogeny in rats using prenatal exposure to phenytoin to demonstrate delayed development. *Neurotoxicol. Teratol.* 16:563–573.

Voytas, D. 2000. Agarose gel electrophoresis. *In* Current Protocols in Molecular Biology (F.M.Ausubel, R.Brent, R.E.Kingston, D.D.Moore, J.G.Seidman, J.A.Smith, K. Struhl, eds.) pp. 2.5A.1–2.5A.9. John Wiley & Sons, Hoboken, N.J.

Wages, S.A., Church, W.H., and Justice, J.B. 1986. Sampling considerations for on-line microbore liquid chromatography of brain dialysis. *Anal. Chem.* 58:1649–1656.

Walker, J.L., Resig, P., Guarnieri, S., Sisken, B.F., and Evans, J.M. 1994. Improved footprint analysis using video recording to assess functional recovery following injury to the rat sciatic nerve. *Rest. Neurol. Neurosci.* 6:189–193.

Wang, J., Skirboll, S., Aigner, T.G., Saunders, R.C., Hsiao, J., Bankiewicz, K.S. 1990. Methodology of microdialysis of neostriatum in hemiparkinsonian nonhuman primates. *Exp. Neurol.* 110:181–186.

Watson, B.D. 1998. Animal models of photochemically induced brain ischemia and stroke. *In* Cerebrovascular Disease, Part I, Introduction, Models, and Neuropathology. (M.D. Ginsberg and J. Bogousslavsky, eds.) pp. 52–73. Blackwell Science, Malden, Mass.

Watson, B.D., Dietrich, W.D., Busto, R., Wachtel, M.S., and Ginsberg, M.D. 1985. Induction of reproducible brain infarction by photochemically initiated thrombosis. *Ann. Neurol.* 17:497–504.

Watson, B.D., Dietrich, W.D., Prado, R., Nakayama, H., Kanemitsu, H., Futrell, N.N., Yao, H., Markgraf, C.G., Wester, P. 1995. Concepts and techniques of experimental stroke induced by cerebrovasculature photothrombosis. *In* Central Nervous System Trauma: Research Techniques (S.T. Ohnishi and T. Ohnishi, eds.) pp. 169–194. CRC Press, Boca Raton, Fla.

Watson, J., Selkirk, J.V., and Brown, A.M. 1998. Development of FlashPlate technology to measure [^{35}S]GTPgS binding to Chinese hamster ovary cell membranes expressing the cloned human 5-HT1B Receptor. *J. Biomol. Screen.* 3:101–105.

Weed, J.L. and Boone, J.L. 1992. A Macintosh computer system for collecting and analyzing rodent sexual behavior. *Physiol. Behav.* 52:183–185.

Wehner, J.M., Radcliffe, R.A., Rosmann, S.T., Christensen, S.C., Rasmussen, D.L., Fulker, D.W., and Wiles, M. 1997. Quantitative trait locus analysis of contextual fear conditioning in mice. *Nature Genet.* 17:331–334.

Weiner, I. 1990. Neural substrates of latent inhibition: The switching model. *Psychol. Bull.* 108:442–461.

Weiner, I. 2000. The latent inhibition model of schizophrenia. *In* Contemporary Issues in Modeling Psychopathology (M. Myslobodsky and I. Weiner, eds.). Kluwer Academic Publishers, Boston.

Weiner, I. Feldon, J. 1997. The switching model of latent inhibition: An update of neural substrates. *Behav. Brain Res.* 88:11–25.

Weiss, F. Koob, G.F. 1991. The neuropharmacology of ethanol self-administration. *In* Neuropharmacology of Ethanol. (R.F. Meyer, G.F. Koob, M.J. Lewis, and S. Paul, eds.) pp. 125–162. Birkhauser, Boston.

Weiss, J.M., Goodman, P.A., Losito, B.G., Corrigan, S., Charry, J.M., Bailey, W.H. 1981. Behavioral depression produced by an uncontrollable stressor: Relationship to norepinephrine, dopamine and serotonin levels in various regions of rat brain. *Brain Res. Brain Res. Rev.* 3:161–191.

Wellman, P.J. 1994. Laboratory Exercises in Physiological Psychology. Allyn and Bacon, Boston.

Westerink, B.H.C., Hofsteede, H.M., Damsma, G., de Vries, J.B. 1988. The significance of extracellular calcium for the release of dopamine, acetylcholine and amino acids in conscious rats, evaluated by brain microdialysis. *Naunyn-Schmiedeberg's Arch. Pharmacol.* 337:373–378.

Williams, J.B., Mallorga, P.J., Lemaire, W., Williams, D.L., Na, S., Patel, S., Conn, J.P., Pettibone, D.J., Austin, C., Sur, C. 2003. Development of a scintillation proximity assay for analysis of Na+/Cl--dependent neurotransmitter transporter activity. *Anal. Biochem.* 321:31–37.

Winer, B.J., Brown, D.R., Michels, K.M. 1991. Statistical Principles in Experimental Design. McGraw-Hill, New York.

Wood, J.H. and Wood, J.H. (eds.) 1980. Neurobiology of cerebrospinal fluid, Vol. 1. Plenum, New York.

Wray, S., Kusano, K., Gainer, H. 1991. Maintenance of LHRH and oxytocin neurons in slice explants cultured in serum-free media: Effects of tetrodotoxin on gene expression. *Neuroendocrinology* 54:327–39.

Wroblewski, F. LaDue, J.S. 1955. Lactic dehydrogenase activity in blood. *Proc. Soc. Exp. Biol. Med.* 90:210–213.

Wu, J.J. 2002. Comparison of SPA, FRET, and FP for Kinase Assays. *In* High Throughput Screening: Methods and Protocols, (W.P. Janzen, ed.) pp. 65–86. Humana Press Inc., Totowa, N.J.

Yamamoto, C. McIlwain, H. 1966a. Electrical activities in thin sections from the mammalian brain maintained in chemically-defined media in vitro. *J. Neurochem.* 13:1333–1343.

Yamamoto, C. McIlwain, H. 1966b. Potentials evoked in vitro in preparations from the mammalian brain. *Nature* 210;1055–1056.

Yang, S.C., Markey, S.P., Bankiewicz, K.S., London, W.T., and Lunn, G. 1988. Recommended safe practices for using the neurotoxin MPTP in animal experiments. *Lab. Anim. Sci.* 38:563–567.

Young, E.A., Owen, E.H., Meiri, K.F., and Wehner, J.M. 2000. Alterations in hippocampal GAP-43 phosphorylation and protein level following contextual fear conditioning. *Brain Res.* 860:95–103.

Zamvil, S.S., and Steinman, L. 1990. The lymphocyte in experimental allergic encephalomyelitis. *Annu. Rev. Immunol.* 8:579–621.

Zea Longa, E., Weinstein, P.R., Carlson, S., Cummins, R. 1989. Reversible middle cerebral artery occlusion without craniectomy in rats. *Stroke* 20;84–91.

Zeller, R. 1989. Fixation, embedding, and sectioning of tissues, embryos, and single cells. *In* Current Protocols in Molecular Biology (F.M.Ausubel, R.Brent, R.E.Kingston, D.D.Moore, J.G.Seidman, J.A.Smith, K. Struhl, eds.) pp. 14.1–14.8

Zeltzer, R. Seltzer, Z. 1994. A practical guide for the use of animal models for neuropathic pain. *In* Touch, Temperature, and Pain. (J. Boive, P. Hansson, U. Lindblom, eds.) pp. 337–379. IASP Press, Seattle.

Zetterstrom, T., Sharp, T., Marsden, C.A., and Ungerstedt, U. 1984. Effects of neuroleptic drugs on striatal dopamine release and metabolism in the awake rat: Studies by intracerebral dialysis. *Eur.J. Pharmacol.* 106:27–37.

Zhang, J.-R., Andrus, P.K., and Hall, E.D. 1994a. Age-related phospholipid hydroperoxides measured by HPLC-chemiluminescence and their relation to hydroxyl radical stress. *Brain Res.* 639:275–282.

Zhang, J.-R., Cazers, A.R., Hall, E.D. 1994b. HPLC-chemiluminescence and thermospray LC/MS study of hydroperoxides generated from phosphatidylcholine. *Free Radical Biol. Med.* 18:1–10.

Zhao, M., Hollingworth, S., and Baylor, S.M. 1996. Properties of tri- and tetracarboxylate Ca^{2+} indicators in frog skeletal muscle fibers. *Biophys. J.* 70:896–916.

Zheng, W., Carroll, S.S., Inglese, J., Graves, R., Howells, L., Strulovici, B. 2001. Miniaturization of a hepatitis C virus RNA polymerase assay using a -102 degrees C cooled CCD camera-based imaging system. *Anal. Biochem.* 290:214–220.

Zigmond, M.J. and Strickler, E.M. 1989. Animal models of parkinsonism using selective neurotoxins: Clinical and basic implications. *Int. Rev. Neurosci.* 31:1–79.

Zimmermann, M. 1983. Ethical guidelines for investigations of experimental pain in conscious animals. *Pain* 16:109–110.

Zimmermann, M. 1992. Ethical constraints in pain research. *In* Animal Pain: Ethical and Scientific Perspectives (T.R. Kuchel, M. Rose, and J. Burrell, eds.) pp. 13–18. Australian Council on the Care of Animals in Research and Teaching, Glen Osmond, SA, Australia.

Zlokovic, B.V. 1995. Cerebrovascular permeability to peptides: Manipulations of transport systems at the blood-brain barrier. *Pharm. Res.* 12:1395–1406.

INDEX

Page numbers in this book are hyphenated: the number before the hyphen refers to the chapter and the number after the hyphen refers to the page within the chapter (e.g., 4-3 is page 3 of Chapter 4). A range of pages is indicated by an arrow connecting the page numbers (e.g., 4-3→4-5 refers to pages 3 through 5 of Chapter 4).

A

Acetylcholine
  in dialysates, HPLC/EC assays, 2-97→2-99
  IMER preparation for assays, 2-99→2-100
Acetylcholine/choline assay kit, HPLC, 2-99
Acoustic startle response assays, 3-84→3-89
ACTH determination in stressed rodent, 3-53→54
Activity monitoring
  in cerebral ischemia models, 4-35→4-36
  Parkinson's model, primate, 4-12
  see also Behavior; Locomotor behavior; Motor coordination
Acute brain slice, *see* Brain slice
Acute pain, *see* Pain models
Adoptive transfer of encephalomyelitis (EAE), 4-96→4-97
Agarose cushion microelectrodes, fabrication, 1-56→1-58
Agarose/KCl bridges, fabrication, 1-56→1-58
Age factors for startle assays, 3-85
Alcoholism models
  blood alcohol content (BAC), 4-90
  mouse, intravenous self-administration, 4-83→4-86
  rats, lever press assays
    concurrent for alcohol and isocaloric alternative, 4-91→4-93
    concurrent for ethanol and saccharin, 4-91
    for ethanol, 4-87→4-89
    for saccharin, 4-90
Allodynia, behavior assays
  cold allodynia, 4-63
  mechano-allodynia, 4-63
ALS, *see* Amyotrophic lateral sclerosis
Amino acid detection in microdialysates, HPLC with OPA labeling, 2-100→2-101
Amplifiers for synaptic recording
  single-channel, 1-68→1-69
  whole-cell voltage clamp, 1-19
Amygdala, anxiety assays
  bicuculline methiodide administration, 4-107→4-112
  urocortin administration, 4-107→4-112
Amyloid peptides from brain homogenates, ELISA, 2-26→2-28
Amyotrophic lateral sclerosis (ALS)
  cultures, spinal cord organotypic, 4-99→4-103
  mouse models for therapeutic agent assays, SOD1 transgenic, 4-103→4-107
Analgesia in mice and rats
  characteristics and definitions, A2-10→A2-11
  drugs and guidelines, A2-11→A2-14
  medications, preoperative, A2-11→A2-12

nutrition, postoperative, A2-14
Analyte properties, and relative recovery in microdialysis, 2-65→2-66
Anatomical verification of microdialysis probes, 2-94
Anesthesia in mice and rats
  inhalant, using isoflurane, A2-8→A2-10
  injectable, A2-6→A2-8
Animal techniques
  behavior studies, CHAPTER 3
  rodent care, handling, and preparation, APPENDIX 2
  see also Behavior; Rodent techniques
Antibody capture assay for [γ-^{35}S]GTP binding, 2-16→2-20
Anti-G protein antibody, [γ-^{35}S]GTP binding assay, receptor-mediated, 2-16→2-20
Anxiety tests in animals
  defensive burying, 3-67→3-69
  elevated plus-maze test, 3-64→3-67
  Geller-Seifter conflict, 3-62→3-63
  light/dark exploration, 3-65
  social interaction, 3-63→3-64
  sodium lactate challenge in rats
    using bicuculline methiodide, 4-112
    using L-allylglycine, in hypothalamus, 4-112→4-114
    social interaction test, 4-115
    using urocortin, in amygdala, 4-107→4-112
  thirsty rat conflict, 3-67→3-69
  see also Behavior; Fear conditioning
L-Arginine, purification of radiolabeled, 2-44→2-45
Arterial occlusion, *see* Focal cerebral ischemia; Global cerebral ischemia
Autoimmune disease, EAE induction in mice, 4-93→4-97

B

Balance, in rodents, 3-13→3-19
  see also Motor coordination
Basolateral nucleus of amygdala
  bicuculline methiodide administration, 4-107→4-112
  urocortin administration, 4-107→4-112
Bathing solutions for synaptic recording single-channel
  experimental solution, 1-70→1-71
  physiological solution, 1-70
  whole-cell voltage clamp, 1-21
BCAO, *see* Bilateral carotid artery occlusion
Beam walking assay for rodents, 3-14→3-15, 3-17→3-18
Behavior assays
  alcoholism models, rodent, 4-83→4-93
  anxiety assays
    basic tests, 3-62→3-70
    fear conditioning, 3-76→3-83
    panic-like behavior in rodents, 4-107→4-116
  depression/despair tests, 3-54→3-61
  development, 3-2→3-8
  fear conditioning, cued and contextual, 3-76→3-83

flavor aversions, conditioned, 3-83→3-84
food intake assays, 3-19→3-22
latent inhibition (LI), 3-91→3-101
locomotor and sensorimotor
  basic studies, 3-8→3-19
  in focal cerebral ischemia, 4-35→4-36
  in global cerebral ischemia, 4-25
  in Parkinson's models, 4-8, 4-11→4-13, 4-16
pain studies
  cold allodynia, 4-63
  heat-hyperalgesia, 4-62→4-63
  mechano-allodynia, 4-63
  mechano-hyperalgesia, 4-63
parental, 3-32→3-46
place preference tests in rodents, 4-78→4-82
scales for, NBS and NSS, 4-56→4-57
sexual and reproductive behaviors, 3-22→3-32
social interaction
  anxiety assay, 3-63→3-64
  in panic-prone rats, 4-115
spatial memory tests, 3-70→3-75
startle response measurement, 3-84→3-89
stressor application, experimental, 3-46→3-54
see also Developmental assays; *specific behaviors*
Behavioral neuroscience defined/characterized, 3-1→3-2
Benzodiazepine site, saturation assays of radioligand binding, 2-103→2-106
Bilateral carotid artery occlusion (BCAO) models of global cerebral ischemia, 4-17→4-22
  see also Global cerebral ischemia
Binding assays
  [^{3}H]5-HT to 5-HT$_{1E}$ receptor, 2-2→2-5
  see also Radiolabeled; Radioligand
Biological material testing in rodents, A2-17→A2-18
Blind patch-pipet recording, 1-64→1-65
Blocking procedure for brain slices, 1-81→1-82
Blood alcohol content, 4-90
Blood-brain barrier (BBB), permeability assays
  in situ methods, 2-149→2-154
  in vivo methods, 2-143→2-149, 2-154→2-164
  see also Efflux; Influx; Perfusion; Permeability
Brain
  amygdala
    bicuculline methiodide administration, 4-112
    urocortin administration, 4-107→4-112
  BBB permeability assays, 2-143→2-164
  cerebral ischemia
    focal, 4-28→4-44
    global, 4-16→4-27

cortical lesions, photochemical-induced, 4-44→4-52
efflux index method, 2-161→2-163
homogenates, amyloid peptides, ELISA analysis, 2-26→2-28
hypothalamus, L-allylglycine administration, 4-112→4-114
slices, acute MEQ-loaded
 Cl⁻ uptake assays, 2-126→2-128
 fluorescent dye calibration, 2-130→2-131
 preparation, 2-128→2-129
synaptic plasticity in hippocampal slices, 1-33→1-60
traumatic injury, fluid percussion model, 4-52→4-57
see also Central nervous system; Hippocampal

Brain slice preparation
 blocking procedure, 1-81→1-82
 for dendrite assays, 1-80→1-83
 hippocampal, 1-29→1-33
 see also Hippocampal

Brain tissue, cerebral ischemia models
 preparation steps, 4-20
 staining, 4-24→4-25, 4-44

Burying, defensive, anxiety assay, 3-67→3-69

C

Ca^{2+} dyes, calibration for cation assays, 2-134→2-136
Ca^{2+} standard solutions for dye calibration, 2-135 (table)
Calibration of dyes for cation assays
 nonratiometric dyes, 2-137→2-138
 ratiometric dyes, 2-134→2-137
cAMP, see Cyclic adenosine monophosphate
Cannula construction
 microdialysis in rodents, 2-68→2-80
 for mouse techniques, A2-4→A2-5
 perfusion in rodents, in situ, 2-151
Capillary volume
 depletion for BBB influx assays, 2-160→2-161
 intravascular measurement, BBB assay, 2-152→2-153
Carbachol dilution series for SPA, 2-11
Carotid artery occlusion
 bilateral (BCAO), 4-17→4-22
 see also Global cerebral ischemia
Carrier-mediated transport assay using in situ brain perfusion, 2-143→2-149
Castration, rodent, 3-26→3-28
Catheter construction, for rat jugular vein, 2-160
Cation movement in primary cultures
 dye calibration for, 2-134→2-138
 measurement assay, 2-132→2-133
 strategies: wavelength, filters, detectors, 2-131→2-132
Cation-exchange columns, preparation and regeneration, 2-43→2-44
CCI, see Chronic constriction injury
Cell cultures
 cation movement assays, 2-131→2-138
 Cl⁻ efflux and uptake assays, 2-125→2-126
 CNS organotypic, 1-60→1-64
 fluorescent dye calibration for, 2-129→2-130
 neuronal, viability assays, 2-28→2-35
 spinal cord organotypic, 4-99→4-103

see also Cultured cells
Cell preservation, for PI viability assays, 2-31
Cell viability, in neuronal cultures
 colloidal dye exclusion, 2-32
 comparison and guidelines, 2-28→2-30, 2-34→2-35
 LDH assay, 2-30→2-31
 LDH assay, simplified, 2-31
 mitochondrial activity, MTT reduction assay, 2-30
 PI assays, 2-31
Cell-attached patches, bathing solutions, 1-70
Cell-free calibration of Mg^{2+} dyes, 2-137
Central nervous system (CNS)
 BBB permeability assays, 2-143→2-164
 brain homogenates, for ELISA for amyloid peptides, 2-26→2-28
 cell viability in neuronal cultures, 2-28→2-35
 cerebral ischemia
  focal, 4-28→4-44
  global, 4-16→4-27
 chloride movement assays, 2-119→2-131
 cortical lesions, photochemical-induced, 4-44→4-52
 neuron isolation, mammalian adult, 1-74→1-76
 neurotransmitters
  detection in dialysates, 2-95→2-103
  uptake and release, 2-109→2-119
 organotypic slice cultures, 1-60→1-64
 oxygen radicals and lipid peroxidation assays, neural tissues, 2-46→2-59
 startle response assays, 3-84→3-89
 synaptic current recording, 1-51→1-60, 1-68→1-74, 1-76→1-79
 traumatic injury, fluid percussion model, 4-52→4-57
see also Brain; Hippocampal; Synaptic
CER, see Conditioned emotional response
Cerebral ischemia
 focal, 4-17, 4-28→4-44
 global, 4-16→4-27
 see also Focal cerebral and Global cerebral ischemia
Cerebrospinal fluid (CSF) collection
 in BBB efflux assay, 2-161→2-163
 unit construction, 2-164
Chemical stimulation of neurotransmitter release, time sampling, 2-114→2-115
Chemiluminescence
 lipid hydroperoxide assay, HPLC, 2-53→2-55
 NO detection, 2-36→2-38
Chinese hamster ovary (CHO) cells, SPA methods
 cAMP accumulation, fast measurement, 2-10→2-12
 uptake assay of [¹⁴C]p-hydroxy-loracarbef, 2-14→2-16
Chloride movement, radioisotopic Cl⁻ assays
 in brain slice, uptake assays
  confocal imaging, 2-127→2-128
  epifluorescence imaging, 2-126→2-127
 in primary cultures
  efflux assay, photometry, 2-125
  uptake assay, 2-122
 in synaptoneurosomes
  efflux assay, photometry, 2-123→2-124

uptake/efflux assay, 2-119→2-121
see also Efflux assays; Uptake assays
CHO cells, see Chinese hamster ovary
Chronic constriction injury (CCI) pain model, 4-58→4-59
Chronic pain, see Pain models
Chronic spontaneous seizures, kainate-induced, in rats, 4-116→4-120
L-Citrulline assay of NOS activity, 2-41→2-43
Cl⁻ assays, see Chloride movement
CNS, see Central nervous system
Cold allodynia, behavior assays, 4-63
Colloidal dye exclusion, cell viability assay, 2-32
Columns, cation-exchange, 2-43→2-44
Commercial suppliers, APPENDIX 3
Concentric microdialysis probes
 construction, 2-70→2-71
 implantation and tethering, 2-74→2-77
Conditioned emotional response (CER), latent inhibition assay, 3-91→3-96
Conditioning
 fear conditioning, 3-76→3-83
 flavor aversions, 3-83→3-84
 latent inhibition, 3-91→3-96
 see also Behavior; Induction
Confocal imaging, Cl⁻ uptake in brain slices, 2-127→2-128
Constriction injury pain model, 4-58→4-59
Contextual fear conditioning, 3-76→3-83
Cortical lesions, photochemical-induced in rat brain, 4-44→4-52
Corticosterone determination in stressed rodent, 3-53→54
Cresyl violet staining of brain tissue, 4-44
CSF, see Cerebrospinal fluid
Cued and contextual fear conditioning, 3-76→3-82
Cultured cells
 cAMP assays, 2-138→2-140
 cation movement measured in
  dye calibration for, 2-134→2-138
  of [ions⁺]ᵢ using ratiometric dyes, 2-132→2-133
 Cl⁻ movement in
  efflux assay, photometry, 2-125→2-126
  fluorescent dye calibration for, 2-129→2-130
  uptake assay, 2-122
 neuronal, cell viability assays, 2-28→2-35
 see also Cell culture
Cyclic adenosine monophosphate (cAMP) assays
 in CHO cells, rapid measurement, 2-10→2-12
 in cultured cells, 2-138→2-140
 in synaptoneurosomes, [³H]cAMP accumulation, 2-140→2-141
Cynomolgus macaques, see Primates
Cytostar-T plates
 plate setup, 2-16
 in uptake assay of [¹⁴C]p-hydroxy-loracarbef, 2-14→2-16

D

Defensive burying, anxiety assay, 3-67→3-69
Dendrites, patch-clamp recording, 1-76→1-79
Depression/despair tests in rodents

behavioral despair test in rats, 3-54→3-55
forced swimming test in mice, 3-55→3-56
learned helplessness in mice
induction of, 3-58→3-60
shock administration, 3-60
shuttle boxes for escape testing, 3-61
tail suspension test in mice, 3-56→3-58
uncontrollable shock, 3-60
see also Long-term depression
Deproteination of NO samples, 2-39→2-40
Despair, see Depression/despair
Developmental assays in rodents
locomotor behavior, 3-7→3-8
physical landmarks, 3-2→3-4
reflexes, 3-5→3-7
Diabetic neuropathy, STZ rat model, 4-71→4-72
Dialysates, neurotransmitter detection in, 2-95→2-103
Dialysis
microdialysis techniques, 2-62→2-95
see also Dialysates; Microdialysis
DIC, see Differential interference contrast
Dienes, lipid-conjugated, spectrophotometry, 2-48→2-49
Difference assay, relative recovery in microdialysis, 2-62→2-63
Differential interference contrast (DIC) microscopy, in patch-pipet recording, 1-65→1-67
Disease models
alcoholism, 4-83→4-93
amyotrophic lateral sclerosis (ALS), 4-98→4-107
cerebral ischemia
focal, 4-28→4-44
global, 4-16→4-27
diabetic neuropathy, STZ rat, 4-71→4-72
encephalomyelitis, autoimmune (EAE), 4-93→4-98
Parkinson's disease, 4-3→4-16
seizures, chemoconvulsant model, 4-116→4-120
Disruption, latent inhibition, behavior, 3-96→3-98
Dissociation apparatus for neuron isolation, 1-75
Dopamine
detection, HPLC/EC assay, 2-96→2-97
transporter radioligands
binding assay, 2-116→2-119
incubation conditions, 2-118
uptake and release assays, 2-109→2-119
see also Neurotransmitters; Release assays; Uptake assays
Dorsomedial hypothalamus, L-allylglycine administration, 4-112→4-114
Drug effects
food intake suppression, 3-83→3-84
in latent inhibition assays, 3-96→3-98
neuroprotective agents for cerebral ischemia models, 4-17→4-24, 4-28→4-32, 4-37→4-41
in startle assays, 3-85→3-86
therapeutic agents for ALS, 4-103→4-107
see also Behavior
Dural extravasation assays
assay guidelines, 4-70
in guinea pigs, 4-69
in rats, 4-65→4-69

Dural inflammation models of migraine pain, 4-65→4-70
Dyes
calibration for cation assays, 2-134→2-138
exclusion, cell viability assay, 2-32
see also Staining

E

EAE (Experimental autoimmune encephalomyelitis), 4-93→4-97
Ear notch or punch, rodent identification, A2-21→A2-23
Efflux assays
of the BBB
brain efflux index method, 2-161→2-163
CSF collection units, construction, 2-164
of Cl⁻ movement
in primary cultures, photometry, 2-125
in synaptoneurosomes, 2-119→2-121
in synaptoneurosomes, photometry, 2-123→2-124
Electric shock in rodents
footshock as stressor, 3-48
learned helplessness, 3-60
Electrical stimulation of neurotransmitter release, 2-115→2-116
Electrochemical HPLC (HPLC/EC), neurotransmitter assays
detection in dialysates
acetylcholine assays, 2-97→2-99
dopamine and serotonin assays, 2-96→2-97
uptake and release studies, 2-113→2-114
Electrodes, in synaptic recording
agarose cusion microelectrodes, fabrication, 1-56→1-58
pulling, patch pipet fabrication, 1-3→1-4
single-channel configuration, 1-71→1-73
Elevated plus-maze anxiety assay, 3-64→3-67
Encephalomyelitis (EAE) induction in mice, 4-93→4-97
Endothelin 1, in MCAO model of focal cerebral ischemia, 4-33→4-35
Enzymatic time-course analysis of rhinovirus 3C protease, SPA, 2-13→2-14
Enzyme-linked immunosorbent assay (ELISA), of amyloid peptides from brain homogenates, 2-26→2-28
Eosin staining of brain tissue, 4-24→4-25
Epifluorescence microscopy
cell viability assay, 2-31
uptake of Cl⁻ movement in brain slices, 2-126→2-127
Estradiol treatment, in rodent, 3-31→3-32
Ethanol, in rodent alcoholism models, 4-87→4-91
Excitatory postsynaptic potential (EPSP), 1-34
see also Long-term potentiation; Synaptic plasticity
Experimental autoimmune encephalomyelitis (EAE)
assay guidelines, 4-93→4-95
induction in mice
active induction, 4-95→4-96
adoptive transfer, 4-96→4-97

Extracellular concentration for microdialysis, 2-83→2-84
Extracellular recording, LTP assay, 1-42→1-45
Extravasation, dural, migraine studies, 4-65→4-69

F

Fear conditioning in mice
contextual without auditory cue, 3-82→3-83
cued and contextual
discrimination protocol, 3-82
one-trial assay, 3-76→3-80
pre-exposure experiments, 3-80→3-81
see also Anxiety; Panic-like responses
Feeding, see Food intake
Female behavior, rodent
parental, 3-32→3-40
sexual and reproductive, 3-25→3-26, 3-28→3-32
Field potential recording
LTD assays, 1-41→1-42
LTP assays, 1-37→1-39
see also Long-term depression; Long-term potentiation; Synaptic plasticity
Filter binding, dopamine transporter radioligand, 2-119
Fire polishing, patch pipets
apparatus, construction
current source, 1-14→1-15
heater wire, 1-15→1-17
polishing method, 1-13→1-16
see also Patch pipets
Fixed-ratio schedules, for alcoholism models, 4-87→4-93
FlashPlates, see Plate setup
Flavor aversions, conditioned, 3-83→3-84
Flow rate, in microdialysis
effect on relative recovery, 2-64→2-65
for microdialysate concentration, 2-63
Fluid-percussion model, traumatic brain injury, 4-52→4-57
Fluorescence detection, with HPLC assay of TBA-malondialdehyde adduct, 2-58→2-59
Fluorescent dyes
calibrating for synaptoneurosomes and cultured cells, 2-129→2-130
in Cl⁻ movement assays, 2-123→2-128
Focal cerebral ischemia in rodents
assay guidelines, 4-42→4-43
brain tissue, staining with cresyl violet, 4-41
intraluminal monofilament occluders, fabrication, 4-32→4-33
MCAO rat, Et-1 treated, sensorimotor tests
horizontal and inclined balance beam, 4-35
staircase, for skilled paw use, 4-36
MCAO rat, for neuroprotective agent tests
intraluminal suture model, 4-28→4-32
spontaneously hypertensive model, 4-39→4-41
with stereotaxic infusion of endothelin 1, 4-33→4-35
Tamura model, 4-37→4-39
Food intake
measurement, 3-19→3-22
palatability and caloric density manipulations, 3-20

rat cage and feeding setup, 3-21
suppression, drug-induced, 3-83→3-84
Footprint analysis for rodents, 3-15→3-16, 3-18, 3-19
Formalin test, nociception models, 4-75→4-76
Four-vessel occlusion (4-VO) model of global cerebral ischemia, 4-22→4-24
FR-1 and FR-4, see Fixed-ratio schedule

G

GABA receptors
   and release of, 1-33→1-36
   saturation assays of radioligand binding, $GABA_A$ receptor, 2-106→2-109
   see also Synaptic plasticity
Geller-Seifter conflict, anxiety assay, 3-62→3-63
Gender factors for startle assays, 3-85
Gerbils
   global cerebral ischemia
      BCAO models, 4-17→4-22
      locomotor activity in, 4-25
   see also Rodents
Global cerebral ischemia in rodents
   assay guidelines, 4-27
   brain tissue
      preparation steps, 4-20
      staining with eosin and hematoxylin, 4-24→4-25
   clasp fabrication, atraumatic, for rat 4-VO, 4-25→4-27
   gerbil BCAO models, for neuroprotective agents
      not penetrating BBB, 4-21→4-22
      systemically active, 4-17→4-20
      tolerance-inducing, 4-22
   locomotor activity assay in ischemic gerbils, 4-25
   rat 4-VO models, for neuroprotective agents and neuronal degeneration, 4-22→4-24
Glutathione
   HPLC assay, 2-51→2-52
   standard solutions, 2-52
Gonadectomy, rodent, 3-26→3-28
G-protein activation, [γ-^{35}S]GTP binding assay, 2-16→2-20
Griess assays of nitrite and nitrite/nitrate, 2-40→2-41
Guinea pigs
   dural extravasation assay, 4-69
   see also Rodents

H

Hamsters
   identification method, A2-21→A2-23
   see also Rodents
Handling and restraint of rodents, A2-19→A2-20
Health monitoring of rodents, A2-16
Heat-hyperalgesia, behavior assays, 4-62→4-63
Helplessness, learned, 3-58→3-61
Hematoxylin staining of brain tissue, 4-24→4-25
Heterosynaptic vs. homosynaptic plasticity, 1-36
High-performance liquid chromatography (HPLC)
   glutathione assay, 2-51→2-52
   hydroxyl radical assay, 2-46→2-48

lipid hydroperoxide assay, chemiluminescent, 2-53→2-55
malondialdehyde assays
   with fluorescence detection of TBA-malondialdehyde adduct, 2-58→2-59
   with UV detection, 2-56→2-58
neurotransmitter detection in dialysates
   electrochemical assays, 2-96→2-99
   using fluorometric labeling with OPA, 2-100→2-101
   method guidelines, 2-101→2-103
neurotransmitter uptake and release, electrochemical assays, 2-113→2-114
vitamin E assay, 2-49→2-51
Hippocampal slices
   preparation
      acute mammalian, 1-29→1-33
      for dendritic recording, 1-80→1-83
      tissue cultures, CNS organotypic, 1-60→1-64
   synaptic plasticity studies, 1-33→1-60
Homosynaptic plasticity
   vs. heterosynaptic, 1-36
   NMDA receptor-dependent induction
      of LTD, 1-45→1-46
      of LTP, 1-45→1-46
Horizontal microdialysis probes
   construction, 2-74
   implantation, 2-77→2-78
Hormone treatments, rodent
   estradiol and progesterone
      for parental behavior, 3-30→3-31
      for sexual behavior, 3-31→3-32
   testosterone, for sexual behavior, 3-30→3-31
Hot-plate test, nociception models, 4-72→4-73
Housing factors for startle assays, 3-85
Hovering behavior, parental, 3-33
HPLC, see High-performance liquid chromatography
5-HT, see Hydroxytryptamine
Humoral indices, rodent stressor, 3-46
Hydroperoxide
   chemiluminescence HPLC assay of lipid, 2-53→2-55
   content in standards, xylenol orange assay, 2-55→2-56
6-Hydroxydopamine (6-OHDA), Parkinson's disease model, rat, 4-13→4-15
Hydroxyl radical, salicylate trapping and HPLC assay, 2-46→2-48
[^{14}C]p-Hydroxy-loracarbef
   dilution series, 2-16 (table)
   uptake in CHO cells, SPA assay, 2-14→2-16
Hydroxytryptamine
   [^{3}H]5-HT, mix dilutions for SPA, 2-3
   5-HT$_{1E}$ receptor
      binding to [^{3}H]5-HT, saturation analysis, 2-2→2-5
      pharmacological profile, 2-7→2-10
Hyperalgesia, behavior assays
   heat-hyperalgesia, 4-62→4-63
   mechano-hyperalgesia, 4-63
Hypertensive MCAO model of focal cerebral ischemia, 4-39→4-41
Hypodermic needle construct, cut 22-G, 2-151
Hypothalamus, L-allylglycine administration, 4-112→4-114

I

Identification techniques for rodents, A2-21→A2-23
IMER, see Immobilized enzyme reactor
Immediate shock control, fear conditioning, 3-82→3-83
Immobilized enzyme reactor (IMER) preparation, acetylcholine detection, 2-99→2-100
Immune studies
   EAE induction in mice, 4-93→4-97
   see also Autoimmune; Disease models; Drug effects
Immunoassays
   of analytes in tissue homogenates
      ELISA technique, 2-26→2-28
      tissue preparation for, 2-25→2-26
   of 8-isoprostaglandin $F_{2\alpha}$, 2-59
Implant, testosterone in rodent, 3-30→3-31
In situ methods
   BBB permeability assays, 2-143→2-164
   dye calibration
      Ca^{2+} dyes, 2-135→2-136
      Na^+ dyes, 2-136→2-137
Induction
   depression/despair in mice, 3-58→3-60
   of maternal behavior, 3-37→3-38
   suppressing food intake, 3-82→3-83
   of synaptic potentiation, 1-39→1-41, 1-45→1-46
   see also Behavior
Influx assays of the BBB
   catheter construction, 2-160
   intravenous, 2-154→2-160
   in situ perfusion
      in mice, 2-149→2-151
      in rats, simplified, 2-149
   see also Blood-brain barrier; Perfusion; Permeability; Uptake
Injection
   microinjection in mice, A2-1→A2-4
   needle manufacture for mouse study, A2-4→A2-5
   see also Anesthesia; Intravenous; Microinjection; Subcutaneous
Inositol phosphate assays in synaptoneurosomes, 2-141→2-143
Inside-out patches, bathing solutions, 1-71
Intact cell assays
   dopamine transporter radioligand binding, 2-117→2-119
   neurotransmitter release and uptake, 2-109→2-110
Interface culture from CNS tissue, 1-62→1-64
Internal standard assay, relative recovery in microdialysis, 2-63→2-64
Intracellular recording
   LTD and LTP in hippocampal slices, 1-45→1-49
   synaptic current in brain slices, 1-51→1-52
   see also Synaptic plasticity; Voltage clamp
Intraluminal monofilament occluders, fabrication, 4-32→4-33
Intraluminal suture model, MCAO, for focal cerebral ischemia, 4-28→4-32
Intravascular capillary volume, measurement, 2-152→2-153
Intravenous methods
   influx assay of the BBB in rats, 2-154→2-160

self-administration, mouse model of alcoholism, 4-83→4-86
Intruder/resident rodent stressor, 3-50
Inulin, radiolabeled, for intravascular capillary volume, 2-152→2-153
[Ion$^+$]$_i$
   cation movement measured in primary culture, 2-132→2-133
   properties of common indicators, 2-132→2-133
Ionic transport, chloride movement assays, 2-119→2-131
Ischemia, see Focal cerebral ischemia; Global cerebral ischemia
Isolation of neurons from mammalian CNS, 1-74→1-76
8-Isoprostaglandin $F_{2\alpha}$, immunoassay of, 2-59

J

Jugular vein catheter construction, for rats, 2-160

K

Kainate, inducing chronic spontaneous seizures in rats, 4-116→4-120

L

Lactate dehydrogenase (LDH) assays of cell viability
   LDH release into bathing medium, 2-30→2-31
   simplified for 96-well plates, 2-31
L-Allylglycine administration in hypothalamus, 4-112→4-114
Latent inhibition (LI) assays in rodents
   conditioned emotional response, 3-91→3-96
   disruption and potentiation, 3-96→3-98
   two-way avoidances, 3-98→3-101
LDH, see Lactate dehydrogenase
L-DOPA, motor behavior Parkinson's models, 4-8, 4-11→4-12
Learned helplessness, 3-58→3-61
Lesions
   cortical, photochemical-induced, 4-44→4-52
   Parkinson's disease models, 4-2→4-16
LI, see Latent inhibition
Licking of pups, rodents, 3-32
Light/dark exploration, anxiety assay, 3-65
Lipid-conjugated dienes, spectrophotometry, 2-48→2-49
Lipids, oxygen radicals in
   hydroperoxide, chemiluminescence HPLC, 2-53→2-55
   peroxidation, in neural tissues, 2-46→2-59
Locomotor behavior
   in global cerebral ischemia, 4-25
   in Parkinson's models, 4-8, 4-11→4-13, 4-16
   in rodents
     data analysis, 3-8, 3-12
     developmental assays, 3-7→3-8
     direct observation, interval and ordinal scales, 3-9→3-11
     quantitation, photocell-based systems, 3-11→3-12
   see also Motor coordination; Sensorimotor
Long-term depression (LTD), in hippocampal slices
   characterized, 1-35→1-36
   field potential recording, 1-41→1-42
   metaplasticity discussed, 1-42→1-43
   University of Bristol free program, 1-50
   see also Field potential; Synaptic plasticity
Long-term potentiation (LTP), in hippocampal slices
   characterized, 1-34→1-35
   data management
     software, commercial and other, 1-49→1-50
     University of Bristol, free program, 1-50
   expression mechanism assays
     minimal stimulation and failures analysis, 1-46→1-48
     peak-scaled nonstationary fluctuation, 1-48→1-49
   extracellular recording, 1-42→1-45
   field potential recording
     baseline stability, 1-37
     configuration, 1-37
     input-output curves, 1-38→1-39
     optimizing responses, 1-37
     stimulus intensity, 1-38
     two-input assays, 1-38
   induction methods, 1-39→1-41
   induction methods, homosynaptic, 1-45→1-46
   intracellular recording, 1-45→1-49
   metaplasticity discussed, 1-42→1-43
   in transgenic mice, considerations, 1-49
   see also Field potential; Synaptic plasticity; Voltage clamp
Lordosis in female rodent, 3-25→3-26
LTD, see Long-term depression
LTP, see Long-term potentiation

M

Macaques, see Primates
Magnetic resonance imaging (MRI), skull coordinates for microdialysis probe attachment, 2-90→2-91
Male reproductive behavior, rodent, 3-23→3-28, 3-30→3-31
Malondialdehyde
   HPLC assay with UV detection, 2-56→2-58
   thiobarbituric acid assay, 2-56
Marmosets, see Primates
Maternal behavior, see Parental
Maternal separation, rodent pups, 3-50→3-51
MCAO, see Middle cerebral artery occlusion
Mechano-allodynia, behavior assays, 4-63
Mechano-hyperalgesia, behavior assays, 4-63
Membrane properties, and relative recovery in microdialysis, 2-65
Membrane protein concentration, for SPA, 2-5
Memory tests, spatial, 3-70→3-75
   see also Spatial memory; Working memory
MEQ-loaded acute brain slice, Cl$^-$ uptake assays, 2-126→2-131
Metaplasticity, synaptic, 1-42→1-43
   see also Synaptic plasticity
1-Methyl-4-phenyl-1,2,3,6-tetrahydropyridine (MPTP), Parkinson's disease
   mouse models, 4-15→4-16
   primate models, 4-3→4-13
Mg^{2+} dyes, cell-free calibration, 2-137
Mice
   alcoholism models, 4-83→4-86
   ALS models, SOD1 transgenic, 4-103→4-107
   anesthesia and analgesia, A2-6→A2-15
   BBB perfusion assays, 2-149→2-151
   behavior
     depression/despair testing, 3-54→3-61
     nest construction, 3-36
   care and health, A2-16→A2-18
   encephalomyelitis (EAE) induction, 4-93→4-97
   handling and restraint, A2-19→A2-20
   identification methods, A2-21→A2-23
   microdialysis probe implantation, 2-79→2-80
   Parkinson's disease models, MPTP, 4-15→4-16
   physiological parameters, A2-7
   place preference test, 4-79→4-82
   stereotaxic surgery, high precision, A2-1→A2-4
   see also Animal models; Animal techniques; Rodents
Microdialysates, amino acid detection, 2-100→2-101
Microdialysis
   experimental design, 2-68→2-70
   perfusate composition, 2-67→2-68
   in primates
     guide assembly techniques, 2-88→2-90
     probes, 2-84→2-88, 2-94
     in vivo, 2-91→2-94
     in vivo, modified for awake monkey, 2-94→2-95
   in rodents
     extracellular concentrations, 2-83→2-84
     probes and cannula, 2-68→2-80
     in vitro, 2-80→2-81
     in vivo, 2-81→2-82
   in vivo relative recovery
     factors affecting, 2-64→2-67
     methods, empirical, 2-62→2-64
Microdialysis probes
   attachment position
     MRI for coordinates, 2-90
     MRI verification, postoperative, 2-91
   attachment to primate skull, 2-88→2-90
   basic schematic, 2-61
   for primates
     anatomical verification, 2-94
     preparation, 2-84→2-88
   relative recovery
     factors affecting, 2-64→2-67
     in vitro determination, 2-64
     in vivo determination, 2-62→2-64
   for rodents
     construction, 2-70→2-74
     implantation, 2-74→2-80
     surgical implantation in mice, 2-79→2-80
Microelectrodes, agarose cushion, fabrication, 1-56→1-58
Microinjection, site-specific/central
   in anesthetized mouse, A2-1→A2-2
   with cannula implant, in behaving mouse, A2-3→A2-4
Microscopy
   confocal, Cl$^-$ uptake, 2-127→2-128
   DIC in patch-pipet recording, 1-65→1-67
   epifluorescence
     cell viability assay, 2-31

Cl⁻ uptake in brain slices, 2-126→2-127
fluorescence, with HPLC of TBA-malondialdehyde adduct, 2-58→2-59
Middle cerebral artery occlusion (MCAO)
 focal cerebral ischemia models, 4-28→4-36
 see also Focal cerebral ischemia
Migraine pain, dural inflammation models, 4-65→4-70
Miniaturization of receptor-radioligand interactions using 384-well FlashPlates, 2-20→2-25
Mitochondrial activity, cell viability assays, 2-30
Monkeys
 microdialysis, 2-84→2-95
 see also Primates
Monofilament occluders, fabrication, 4-32→4-33
Morris water maze task, for spatial memory, 3-73→3-75
Motor coordination and balance
 beam walking, 3-14→3-15, 3-17→3-18
 data analysis, 3-17→3-19
 footprint analysis, 3-15→3-16, 3-18, 3-19
 L-DOPA in Parkinson's models, 4-11
 rotarod method, 3-13→3-14, 3-17
 see also Locomotor; Sensorimotor
Motor reflexes, developmental assays, 3-5→3-7
MPTP, see 1-Methyl-4-phenyl-1,2,3,6-tetrahydropyridine
MTT reduction, cell viability assays, 2-30
Multiple time point methods
 influx assay of the BBB in rats, 2-154→2-160
 sampling for vesicular release of neurotransmitters, 2-114→2-116
Muscarinic M1 receptor
 [γ-³⁵S]GTP binding assay of G protein activation, 2-16→2-20
 see also Scintillation proximity assay
[³H]Muscimol
 in saturation assays of radioligand binding to GABA_A receptor, 2-106→2-109
 working concentrations, 2-108

N

Na⁺ dyes, in situ calibration, 2-136→2-137
NBS (Neurobehavior scale), 4-57
Needle, construction
 cut 22-G hypodermic, 2-151
 for mouse injection, A2-4→A2-5
Nest construction
 characterized, 3-32
 mouse assay, 3-36
 rat assay, 3-35→3-36
Neural tissues
 lipid peroxidation measured in, 2-46→2-59
 oxygen radicals measured in, 2-46→2-59
Neurobehavior scale (NBS) for brain injury, 4-57
Neurological severity scale (NSS) for brain injury, 4-56
Neuronal cultures, primary, cell viability assays, 2-28→2-35
Neuronal preparations, Cl⁻ efflux and uptake, 2-119→2-131
Neurons

dendrites, patch-clamp recording, 1-76→1-79
isolation from mammalian CNS, 1-74→1-76
Neuropathic pain models, rat
 constriction injury (CCI) model, 4-58→4-59
 sciatic (PSL) model, 4-60
 spinal nerve ligation (SNL) model, 4-60→4-61
Neuroprotective agents, cerebral ischemia models
 focal, 4-28→4-32, 4-37→4-41
 global, 4-17→4-24
Neuroscience, behavioral, defined, 3-1→3-2
Neurotransmitters
 detection and quantification in dialysates, 2-95→2-103
 release assays
  using HPLC/EC, 2-113→2-114
  from intact attached cells, 2-109→2-110
  superfusion apparatus for time sampling, 2-114→2-116
  from synaptosomes, 2-112→2-113
 uptake assays
  in detached cells, 2-111
  using HPLC/EC, 2-113→2-114
  in intact attached cells, 2-109→2-110
  in synaptosomes, 2-111→2-112
Nigrostriatal lesions, in Parkinson's primates
 induction process, 4-9→4-11
 rotational behavior, 4-8, 4-9→4-11, 4-13
Nitric oxide (NO)
 chemiluminescent detection, 2-36→2-38
 L-citrulline assay, for NOS activity, 2-41→2-43
 deproteination of samples, 2-39→2-40
 Griess assays, nitrite and nitrite/nitrate, 2-40→2-41
 reducing agents for assays
  for converting nitrite to NO, 2-38→2-39
  nitrate-reducing, 2-39
Nitric oxide synthase (NOS)
 activity assay, L-citrulline, 2-41→2-43
 preparation, 2-45→2-44
NMDA receptor, inducing LTP, 1-39→1-40
NO detection, see Nitric oxide
Nociception
 acute pain in rodents
  assay guidelines, 4-77
  formalin test, 4-75→4-76
  hot-plate test, 4-72→4-73
  tail-flick test, 4-73→4-74
 see also Pain
Non-human primates, microdialysis, 2-84→2-95
Nonratiometric [Ca²⁺]_i dyes, calibration for cation assays, 2-137→2-138
NOS, see Nitric oxide synthase
NSS (Neurological severity scale), 4-56

O

Occlusion, in cerebral ischemia
 BCAO models, 4-17→4-22
 MCAO model, 4-28→4-36
 4-VO model, 4-22→4-24
6-OHDA, see 6-Hydroxydopamine
Operative response tests, parental, 3-42→3-45
Organotypic cultures, CNS

interface cultures, 1-62→1-64
roller-tube cultures, 1-61→1-62
spinal cord, 4-99→4-103
Outside-out patches, bathing solutions, 1-70→1-71
Ovariectomy in rodent, 3-28→3-30
Oxotremorine
 dilution series, 2-18 (table)
 in [γ-³⁵S]GTP binding assay, 2-16→2-20
Oxygen radicals, measurement in neural tissues, 2-46→2-59

P

Pain models
 migraine, dural inflammation model, 4-65→4-70
 neuropathic, diabetic, STZ rat model, 4-71→4-72
 neuropathic, rat models
  constriction injury (CCI) model, 4-58→4-59
  sciatic (PSL) model, 4-60
  spinal nerve ligation (SNL) model, 4-60→4-61
 nociception models, 4-72→4-77
 see also Behavior assays
Paired-pulse
 depression (PPD), 1-33→1-34
 facilitation (PPF), 1-33→1-34
 plasticity (PPP), 1-33→1-34
 see also Synaptic plasticity; Long-term depression
Panic-like responses, induced
 using bicuculline methiodide, 4-112
 using L-allylglycine, in hypothalamus, 4-112→4-114
 social interaction test, 4-115
 sodium lactate challenge in rats, 4-107→115
 urocortin administration in amygdala, 4-107→4-112
Parental behavior in rodents
 behavior induction
  hormone method for virgin rats, 3-37→3-38
  sensitizing nonlactating rodents, 3-37
 characteristics of, 3-32→3-33
 direct observation
  continuous, 3-34→3-35
  periodic spot-check, 3-33→3-34
 hovering over pups
  active, 3-33
  quiescent, 3-33
 licking of pups, 3-32
 maternal preferences
  basic motivations, 3-38
  conditioned, place preference, 3-41→3-42
  unconditioned, for pups or related cues, 3-38→3-40
 nest construction, 3-35→3-36
 operant response tests, 3-42→3-45
 pup retrieval, 3-36
 T-maze assay of motivation, 3-45→3-46
Parkinson's disease (PD)
 choosing species, lesions, and toxins, 4-2→4-3
 primate models, preclinical
  behavior monitoring, 4-8 (table), 4-11→4-13
  bilateral asymmetric, MPTP, 4-4→4-7
  rating scale for, 4-8 (table)
  systemic MPTP lesion, 4-9
  unilateral 6-OHDA lesion, 4-9→4-11

rodent models, preclinical
  behavior monitoring in mice, 4-16
  MPTP lesions in mice, induction, 4-15
  6-OHDA lesions in rats, 4-13→4-15
  species selection, 4-3
  toxins: MPTP vs. OHDA, 4-3→4-4
Partial sciatic ligation (PSL) pain model, 4-60
Patch-clamp recording
  assay guidelines, 1-67→1-68
  blind method, 1-64→1-65
  DIC method, 1-65→1-67
  from neuronal dendrites
    dendritic, 1-76→1-79
    dendritic and somatic simultaneous, 1-79
  from Xenopus oocytes, 1-58→1-60
  see also Patch pipet; Synaptic current; Synaptic plasticity; Voltage clamp
Patch formation for single-channel recording, 1-71→1-73
Patch pipet fabrication
  automated pulling of electrodes, 1-3→1-4
  calibrating puller filament, 1-5→1-7
  capacity transients, 1-12
  coating, elastomer, 1-10→1-12
  filling, 1-17→1-18
  fire polishing apparatus, 1-14→1-17
  fire polishing method, 1-13→1-16
  glass, cleaning, 1-7
  noise issues, 1-7→1-9, 1-12
  optimizing tip geometry, 1-4→1-6
  setup, mounting, and testing, 1-18
Patch-pipet recording, see Patch-clamp
PD, see Parkinson's disease
Peptides, amyloid, see Amyloid peptides
Perfusion
  BBB assays, in situ
    cannula construction, 2-152
    cerebral perfusion, 2-152
    for mice, 2-149→2-151
    for rats, 2-143→2-149
    for rats, simplified, 2-149
    transport assays using, 2-154
  brain slice preparation, 1-80→1-83
Permeability assays of the BBB
  brain efflux index method, 2-161→2-163
  brain perfusion, in situ, 2-143→2-151; 2-154
  capillary volume for
    depletion method, 2-160→2-161
    intravascular measurement, 2-152→2-153
  cerebral perfusion, in situ, 2-152
  influx assays, 2-143→2-160
PE50-stainless steel connectors, construction, 2-151
Pharmacological profile of 5-HT$_{1E}$ receptor, 2-7→2-10
Phosphoinositide turnover measured in synaptoneurosomes, 2-141→2-143
Photocell-based assay of locomotor behavior, 3-11→3-12
Photochemical cortical lesions, induced in rat, 4-44→4-52
Photometric efflux assays of Cl$^-$ movement
  in primary cultures, 2-125
  in synaptoneurosomes, 2-123→2-124
Physical landmarks of rodent development, 3-2→3-4
Physiological parameters of mice and rats, A2-7

see also Care; Handling; Rodent techniques
PI, see Propidium iodide
Pipets, see Patch pipets; Single-channel; Whole-cell; Voltage clamp
Place preference test
  in mice, 4-79→4-82
  in rats, 4-78→4-79
Plasma ACTH/corticosterone determination, 3-53→54
Plasticity, synaptic, see Synaptic plasticity
Plate setup
  for binding inhibition of [^{125}I]iodocyanopindolol to β$_2$ adrenergic receptor membranes, 2-22
  for dipeptide transport in Cytostar-T microplate, 2-16
  for 5-HT$_{1E}$ receptor binding analysis, 2-9
  for LDH assay, 96-well plates, 2-31
  for SPA methods
    Cytostar-T plates, 2-24
    Flash plates, 2-23→2-24
    ScintiStrips/ScintiPlates, 2-24
Postsynaptic current (PSC)
  intracellular recording, 1-50→1-52
  voltage clamp recording, 1-52→1-60
  see also Synaptic currents
Post-tetanic potentiation (PTP), 1-34
Potential recording in hippocampal slices, 1-50→1-60
Potentiation
  latent inhibition, behavior, 3-96→3-98
  see also Long-term potentiation; Potential recording; Synaptic plasticity
PPD, see Paired-pulse depression
PPF, see Paired-pulse facilitation
Preclinical disease models, see Disease
Primary cultures
  cation movement measured in
    dye calibration for, 2-134→2-138
    of [ions$^+$]$_i$ using ratiometric dyes, 2-132→2-133
  Cl$^-$ movement in
    efflux assay, photometry, 2-125→2-126
    fluorescent dye calibration for, 2-129→2-130
    uptake assay, 2-122
  see also Cell cultures; Cultured cells
Primates, non-human
  microdialysis, 2-84→2-95
  Parkinson's disease models, 4-3→4-13
Probe, microdialysis, see Microdialysis probe
Progesterone treatment, in rodent, 3-31→3-32
Propidium iodide (PI), cell viability assays, 2-31
Propranolol hydrochloride, see S(−)-propranolol hydrochloride
Proteins
  concentration in dopamine transporter radioligand binding, 2-119
  deproteination of NO samples, 2-39→2-40
PSC, see Postsynaptic current
PSL, see Partial sciatic ligation
Psychiatric disorders, see Behavior; Brain; Disease; Fear; Pain
PTP, see Post-tetanic potentiation
Pups, rodent
  retrieval, parental behavior, 3-36
  separation from parent, anxiety, 3-50→3-51

see also Development; Parental behavior

Q

Quarantine of rodents, A2-16

R

Radial arm maze task, for spatial memory, 3-70→3-73
Radiolabeled inulin, for intravascular capillary volume, 2-152→2-153
Radioligand binding analysis
  miniaturization of receptor-radioligand interactions, 2-20→2-25
  saturation assays
    to benzodiazepine site, 2-103→2-106
    to GABA$_A$ receptor, 2-106→2-109
Rapid trypan blue staining, cell viability assay, 2-32
Ratiometric dyes, calibration for cation assays
  of Ca^{2+} dyes, 2-134→2-135
  of Ca^{2+} dyes, in situ, 2-135→2-136
  of Mg^{2+} dyes, cell free, 2-136→2-137
  of Na$^+$ dyes, 2-136→2-137
Rats
  anesthesia and analgesia, A2-6→A2-15
  BBB perfusion assays, 2-143→2-149, 2-151→2-164
  behavior
    anxiety tests, 3-62→3-70
    cage and feeding setup, 3-21
    handling, for stressor assays, 3-52→3-53
    latent inhibition (LI) assays, 3-91→3-101
    nest construction, 3-35→3-36
    thirsty rat conflict, 3-67→3-69
  brain studies
    cortical lesions, photochemical-induced, 4-44→4-52
    focal cerebral ischemia models, 1-28→1-41, 1-35→1-36
    traumatic injury, fluid-percussion model, 4-52→4-57
  care and health, A2-16→A2-18
  cultures, spinal cord organotypic, 4-99→4-103
  diabetic neuropathy, STZ rat model, 4-71→4-72
  handling and restraint, A2-19→A2-20
  identification methods, A2-21→A2-23
  migraine, dural extravasation assay, 4-65→4-69
  neuropathic pain models, 4-57→4-64, 4-71→4-72
  nociception models, 4-72→4-77
  Parkinson's disease models, 6-OHDA, 4-13→4-15
  physiological parameters, A2-7
  place preference test, 4-78→4-79
  see also Animal techniques; Rodents
Reagents and Solutions, APPENDIX 1
Receptor binding
  [^{3}H]5-HT to 5-HT$_{1E}$ receptor, 2-2→2-5
  saturation assays of radioligand binding
    to benzodiazepine site, 2-103→2-106
    to GABA$_A$ receptor, 2-106→2-109
Receptor-radioligand interactions, miniaturization, using 384-well FlashPlates, 2-20→2-25
Receptors, in synaptic plasticity
  GABA release, 1-33→1-36

NMDA-dependent LTP induction, 1-39→1-40
Recording of synapse, *see* Extracellular recording; Field recording; Intracellular recording; Synaptic current; Synaptic plasticity; Voltage clamp
Reflexes, *see* Motor reflexes
Release assays of neurotransmitters
   using HPLC/EC, 2-113→2-114
   from intact attached cells, 2-109→2-110
   superfusion apparatus for time sampling, 2-114→2-116
   from synaptosomes, 2-112→2-113
Reproductive behavior, *see* Sexual and reproductive
Resident/intruder rodent stressor, 3-50
Restrainers for rodents, A2-20
Restraints, rodent stressor, 3-47
Rhesus macaques, *see* Primates
Rodents
   BBB permeability assays, 2-143→2-164
   behavior assays
      alcoholism models, 4-83→4-93
      anxiety tests, 3-62→3-70
      depression/despair tests, 3-54→3-61
      developmental, 3-1→3-8
      fear conditioning, cued and contextual, 3-76→3-83
      fear measurement, sodium lactate-induced, 4-107→4-116
      flavor aversions, conditioned, 3-83→3-84
      food intake, 3-19→3-22
      latent inhibition (LI), 3-91→3-101
      locomotor and motor coordination, 3-8→3-19
      parental, 3-32→3-46
      place preference tests, 4-78→4-82
      sexual and reproductive, 3-1→3-8
      spatial memory tests, 3-70→3-75
      startle response measurement, 3-84→3-89
      stressor application, experimental, 3-46→3-54
   brain
      cerebral ischemia, focal, 4-28→4-44
      cerebral ischemia, global, 4-16→4-27
      cortical lesions, photochemical-induced, 4-44→4-52
      traumatic injury, fluid percussion model, 4-52→4-57
   colony maintenance, 3-52
   cultures, spinal cord organotypic, 4-99→4-103
   disease models, preclinical
      ALS, 4-98→4-107
      autoimmune (EAE), 4-93→4-98
      Parkinson's disease, 4-13→4-16
   microdialysis techniques and preparation, 2-62→2-84
   pain models
      acute, nociception models, 4-72→4-77
      diabetic neuropathy, STZ rat model, 4-71→4-72
      migraine, dural inflammation model, 4-65→4-70
      neuropathic, 4-57→4-65, 4-71→4-72
      seizures, chemoconvulsant model, 4-116→4-120
   techniques for use
      analgesia in mice and rats, A2-10→A2-15
      anesthesia in mice and rats, A2-6→A2-10
      care and health of mice and rats, A2-16→A2-18
      handling and restraint of mice and rats, A2-19→A2-20
      identification of mice, rats, and hamsters, A2-21→A2-23
      microinjection, site-specific/central, mice, A2-1→A2-4
      stereotaxic surgery in mice, high precision, A2-1→A2-4
   see also Animal models; Animal techniques; *specific rodents*
Roller-tube culture from CNS tissue, 1-61→1-63
Rotarod assay, motor coordination, 3-13→3-14, 3-17
Rotational behavior in Parkinson's primates, 4-8, 4-13

S

Saccharin, in rodent alcoholism models, 4-90→4-91
Salicylate trapping, hydroxyl radical assay, 2-46→2-48
Saturable transport assay using in situ brain perfusion, 2-154
Saturation analysis
   of [^{3}H]5-HT binding, SPA, 2-2→2-5
   radioligand binding assays
      to benzodiazepine site, 2-103→2-106
      to GABA$_A$ receptor, 2-106→2-109
Sciatic pain model, 4-60
Scintillation proximity assay (SPA)
   cAMP accumulation in CHO cells with muscarinic M$_1$ receptor, rapid measurement, 2-10→2-12
   muscarinic M$_1$ receptor-mediated [γ-^{35}S]GTP binding assay, 2-16→2-20
   pharmacological profile for 5-HT$_{1E}$ receptor, 2-7→2-10
   references and applications, 2-25
   saturation binding analysis of [^{3}H]5-HT to 5-HT$_{1E}$ receptor, 2-2→2-7
   SPA beads, 2-23
   time-course analysis of rhinovirus 3C protease, 2-13→2-14
   uptake of [^{14}C]p-hydroxy-loracarbef, 2-14→2-16
Second messengers measured in signal transduction, 2-138→2-143
Seizures, kainate-induced chronic spontaneous, in rats, 4-116→4-120
Sensorimotor tests
   for MCAO ischemic rats, Et-1 treated, 4-35→4-36
   see also Locomotor behavior
Seratonergic compounds, for 5-HT$_{1E}$ receptor profile, 2-8
Serotonin
   detection by HPLC/EC, 2-96→2-97
   see also Release assays; Uptake assays; Neurotransmitters
Sexual and reproductive behavior in rodents
   developmental landmarks, 3-2→3-4
   in evolution, 3-22
   female assessment
      estradiol and progesterone treatment, 3-31→3-32
      lordosis, 3-25→3-26
      ovariectomy, 3-28→3-30
   male assessment
      castration procedure, 3-26→3-28
      mounts, intromissions, and ejaculations, 3-23→3-25
      testosterone implant, 3-30→3-31
      testosterone injection, 3-31
Shock, *see* Electric shock
Side-by-side microdialysis probes, construction, 2-71→2-73
Signal transduction, second messenger assays, 2-138→2-143
Single-channel patch pipets
   fabrication, 1-3→1-5, 1-7→1-8
   noise issues, 1-7→1-8
   optimizing tip geometry, 1-4→1-5
   see also Patch pipets; Whole-cell; Voltage clamp
Single-channel recording
   data acquisition
      basic, 1-71
      continuous, 1-73
      pulsed, 1-73→1-74
   patch formation for, 1-71→1-73
   setup and strategy, 1-68→1-71
   see also Patch-clamp; Synaptic, 1-68→1-71
Sleep deprivation assay, 3-51→3-52
SNL, *see* Spinal nerve ligation
Social interaction
   anxiety assay, 3-63→3-64
   in panic-prone rats, 4-115
Social isolation assay, 3-49
SOD1 transgenic mouse models of ALS, 4-103→4-107
Sodium lactate, anxiety tests in rats
   using bicuculline methiodide, 4-112
   using L-allylglycine, in hypothalamus, 4-112→4-114
   social interaction test, 4-115
   using urocortin, in amygdala, 4-107→4-112
Somatic recording of synapse, simultaneous with dentritic, 1-79
SPA, *see* Scintillation proximity assay
SPA beads, wheat germ agglutinin, 2-6→2-7
Spatial memory tests
   Morris water maze task
      spatial memory, 3-73→3-74
      spatial probe trial, 3-74→3-75
      working memory, 3-75
   radial arm maze task
      working memory, 3-70→3-72
      working *vs.* reference memory, 3-72→3-73
Spinal cord organotypic cultures, 4-99→4-103
Spinal nerve ligation (SNL) pain model, 4-60→4-61
Spontaneously hypertensive model, for focal cerebral ischemia, 4-39→4-41
S(−)-propranolol hydrochloride dilution series, 2-21
Stabilization of mice and rats, A2-16
Stained cells, preservation for PI viability assays, 2-31
Staining
   brain tissue, cerebral ischemia models
      with cresyl violet, 4-44
      with eosin and hematoxylin, 4-24→4-25
   for cell viability
      colloidal dye exclusion assay, 2-32
      PI and rapid PI assays, 2-31
   with fluorescent dyes

calibration for synaptoneurosomes and cultured cells, 2-129→2-130
in Cl⁻ movement assays, 2-123→2-128

Startle response assays
  rat handling guidelines, 3-89→3-90
  startle habituation, between-subjects tests, 3-88
  startle inhibition, prepulse test, 3-88→3-89
  startle reactivity
    basic method, 3-86→3-87
    between-subjects tests, 3-87

Stereotaxic infusion of endothelin 1, focal cerebral ischemia model, 4-33→4-35
Stereotaxic surgery in mice, high precision, A2-1→A2-4
Streptozotocin (STZ) rat model of diabetic neuropathy, 4-71→4-72
Stressors, rodent assays
  ACTH/corticosterone determination, 3-53→3-54
  alternate methods, 3-47
  dependent vs. independent measure, 3-46→3-47
  electric footshock, 3-48
  humoral indices, 3-46
  maternal separation, in pups, 3-50→3-51
  resident/intruder assay, 3-50
  restraint assay, 3-47
  sleep deprivation, 3-51→3-52
  social isolation, 3-49
  swim assay, 3-48→3-49
  see also Anxiety; Behavior; Conditioned fear; Panic-like behavior

STZ, see Streptozotocin
Subcutaneous methods
  implant, testosterone, 3-30→3-31
  injection, rodent
    estradiol and progesterone, 3-31→3-32
    testosterone, 3-30→3-31

Superfusion apparatus
  for Cl⁻ uptake assay in brain slice, 2-126
  for neurotransmitter time sampling, 2-114→2-116

Suppliers, APPENDIX 3
Surgery in mice, high precision stereotaxic, A2-1→A2-4
Swimming, in rodents
  forced, depression assay, 3-54→3-58
  as stressor for behavior, 3-54→3-58
Synapses
  field potential recording, 1-36→1-42
  LTD and LTP assays, 1-34→1-46
  response optimized for LTD/LTP assays, 1-37
Synaptic current recording
  intracellular, in brain slices, 1-51→1-52
  patch clamp, in Xenopus oocytes, 1-58→1-59
  single-channel, 1-68→1-74
  voltage clamp, in Xenopus oocytes, 1-53→1-60
Synaptic plasticity in hippocampal slices
  depression (LTD), 1-35→1-36
  field potential recording, 1-36→1-42
  homosynaptic vs. heterosynaptic, 1-36
  long-term potentiation, 1-34→1-46
  paired-pulse, 1-33→1-34
  PTP, 1-34
  types of, 1-33→1-36
  see also Field potential; Long-term depression; Long-term potentiation

Synaptoneurosomes
  Cl⁻ movement in
    cell preparation, 2-124→2-125
    efflux assay, photometry, 2-123→2-124
    fluorescent dye calibration, 2-129→2-130
    uptake/efflux assay, 2-119→2-121
  signal transduction, second messenger assays
    [³H]cAMP accumulation, 2-140→2-141
    inositol phosphates, 2-141→2-143
  see also Synaptosomes
Synaptosomes
  neurotransmitters in
    release assay, 2-112→2-113
    uptake assay, 2-111→2-112
  see also Synaptoneurosomes

T

Tail suspension, depression/despair tests, 3-54→3-58
Tail-flick test, nociception models, 4-73→4-74
Tamura model, MCAO, for focal cerebral ischemia, 4-37→4-39
Temperature effects
  dopamine transporter radioligand binding, 2-119
  relative recovery in microdialysis, 2-66
Testosterone treatment, in rodent
  subcutaneous implant, 3-30→3-31
  subcutaneous injection, 3-30→3-31
TEVC, see Two-electrode voltage clamp
Therapeutic agents for ALS, transgenic mouse assays, 4-103→4-107
Thiobarbituric acid assay of malondialdehyde, 2-56
Thirsty rat conflict, anxiety assay, 3-67→3-69
Time sampling method for vesicular release of neurotransmitters, 2-114→2-116
Tissue cultures, CNS organotypic, 1-60→1-64
Tissue factors, for relative recovery in microdialysis, 2-66→2-67
T-maze assay of parental motivation, 3-45→3-46
Toxins, MPTP and 6-OHDA in Parkinson's models, 4-2→4-16
Transgenic mouse models of ALS, 4-103→4-107
Transport assays using in situ brain perfusion, 2-154
Traumatic brain injury, fluid-percussion model, 4-52→4-57
Trypan blue exclusion, cell viability assays, 2-32
Two-electrode voltage clamp (TEVC) recording from Xenopus oocytes
  assay guidelines, 1-59→1-60
  microelectrodes and bridges, fabrication, 1-56→1-58
  recording method, 1-53→1-56
Two-input field potential recording, LTP assay, 1-38
Two-way avoidance, latent inhibition tests
  active avoidance, 3-98→3-99
  with context shift, 3-99→3-100
  disruption and potentiation, 3-100→3-101

U

Uptake assays
  of Cl⁻ movement
    in brain slices, 2-126→2-128
    in primary cultures, 2-122
    in synaptoneurosomes, 2-119→2-121
  [¹⁴C]p-hydroxy-loracarbef in CHO cells, SPA, 2-14→2-16
  of neurotransmitters
    in detached cells, 2-111
    using HPLC/EC, 2-113→2-114
    in intact attached cells, 2-109→2-110
    in synaptosomes, 2-111→2-112
Urocortin administration in amygdala, 4-107→4-112
UV detection for HPLC assay of malondialdehyde, 2-56→2-58

V

Vesicular release
  of neurotransmitters, 2-114→2-116
  see also Release assays
Viability, see Cell viability
Vitamin E, HPLC assay, 2-49→2-51
Voltage clamp recording
  computer, interface, and software, 1-20
  data
    leak subtraction, 1-27
    patch clamp, cable analysis, 1-28→1-29
    patch clamp, standard analysis, 1-28
    pulse sequences, 1-25→1-27
  patch-clamp setup, 1-22→1-25
  whole-cell recording/preparation, 1-19→1-25
  from Xenopus oocytes
    patch clamp recording, 1-58→1-60
    TEVC (two-electrode) recording, 1-53→1-60
  see also Patch clamp; Patch pipet; Single-channel; Whole-cell

W

Water maze task, for spatial memory, 3-73→3-75
WGA, see Wheat germ agglutinin
Wheat germ agglutinin (WGA) SPA beads, 2-6→2-7
Whole-cell patch pipets
  fabrication, 1-3→1-4, 1-9
  noise issues, 1-9
  optimizing tip geometry, 1-4
  see also Patch pipets; Single-channel; Voltage clamp
Whole-cell voltage clamp recording
  amplifier, 1-19
  bathing solution, 1-21
  computer, interface, and software, 1-20
  data analysis, 1-25→1-29
  electrode holder, 1-20
  electrophysiology setup, 1-19
  experimental solutions, 1-21→1-22
  patch electrodes, 1-20→1-21
  patch-clamp setup, 1-22→1-25
  sample preparation, 1-21
Wire plug, manufacture for mouse study, A2-4→A2-5
Working memory, spatial assays, 3-70→3-73, 3-75

X

Xenopus oocytes, synaptic recording from, 1-53→1-60

Xylenol orange assay of hydroperoxide content, 2-55→2-56